T0343258

INTEGRATED PEST MANAGEMENT

INTEGRATED PEST MANAGEMENT

Current Concepts and Ecological Perspective

Edited by

DHARAM P. ABROL

AMSTERDAM • BOSTON • HEIDELBERG • LONDON
NEW YORK • OXFORD • PARIS • SAN DIEGO
SAN FRANCISCO • SINGAPORE • SYDNEY • TOKYO
Academic Press is an imprint of Elsevier

Academic Press is an imprint of Elsevier
525 B Street, Suite 1900, San Diego, CA 92101-4495, USA
32 Jamestown Road, London NW1 7BY, UK
225 Wyman Street, Waltham, MA 02451, USA

British Library Cataloguing-in-Publication Data
A catalogue record for this book is available from the British Library

Library of Congress Cataloging-in-Publication Data
A catalog record for this book is available from the Library of Congress

ISBN: 978-0-12-398529-3

For information on all Academic Press publications
visit our website at elsevierdirect.com

Contents

18 Role of Genetically Modified Insect-Resistant Crops in IPM: Agricultural, Ecological and Evolutionary Implications

S. MOHANKUMAR AND T. RAMASUBRAMANIAN

19 Breeding for Disease and Insect-Pest Resistance

S.K. GUPTA AND M.K. PANDEY

20 Integrated Management of Rodent Pests

R.S. TRIPATHI

21 Eco-Friendly Management of Phytophagous Mites

RACHNA GULATI

22 IPM Extension: A Global Overview

RAJINDER PESHIN, K.S.U. JAYARATNE AND RAKESH SHARMA

About the Editor

Dr D.P. Abrol is a very dedicated scientist. He received his PhD from Haryana Agricultural University Hisar in 1986. After serving at this university for some time, he joined the Division of Entomology at the Sher-e-Kashmir University of Agricultural Sciences and Technology, Srinagar, India where he served in various capacities as Assistant and then as Associate Professor. Dr Abrol has vast experience and expertise on honeybee management, pollination biology, bee ecology, toxicology, economic entomology and pest management. His research findings have been widely appreciated. His research work has been referred to/cited in national and international journals and books. He has chaired several national and international symposia/conferences and delivered lead/guest lectures. He is a member of several scientific societies in India and abroad. He has authored 11 books and published over 200 original research papers, 10 reviews, 10 book chapters and over 64 popular articles on honeybee diseases pollination biology, toxicology and integrated pest management. He is referee of various national and international journals and expert in various selection/screening and evaluation committees of scientific bodies/institutions. He has completed several externally-funded research projects and has collaborated in research projects with international organizations in Poland and Switzerland. He has visited South Korea, Malaysia and several other countries as special invitee to these countries.

As well as receiving letters of appreciation from different organizations, he was the recipient of the *Young Scientist Award* in 1992, conferred by the Jammu and Kashmir State Council for Science and Technology, a prestigious State Award for his outstanding contributions in the field of Agricultural Sciences. He was also a recipient of the *Pran Vohra Award* in 1993, a prestigious Young Scientist Award conferred by Indian Science Congress Association Calcutta for his outstanding and innovative research in the field of agricultural sciences. He was also conferred in the Prof. T. N. Ananthakrishnan Award for 1997–1998, a prestigious National Award for his outstanding contributions in the field of Entomology by the T N Ananthakrisnan Foundation, GS Gill Research Institute in Chennai. Dr. Abrol also won the Dr. Rajinder Prasad Puruskar award for 1999–2000, a national award from Indian Council of Agricultural Research New Delhi for his Hindi book on beekeeping *Madhmakhi Palan- Sidhant Evam Vidhian'* and an award at the 11th Asian Apicultural Association in 2010 for outstanding contributions in apiculture. Dr. Abrol is presently working in Division of Entomology at the Sher-e-Kashmir University of Agricultural Sciences and Technology, Jammu as Professor and Head, engaging in teaching and research in entomology for over 27 years.

Preface

Integrated Pest Management – Current Concepts and Ecological Perspective attempts to cover integrated pest management from multidisciplinary, multi-country and multi-faceted components. The book contains chapters on diverse aspects of integrated pest management (IPM) from management of rodent pests, spiders, and mites to weather-based pest forecasting, application of remote sensing, the impact of climate change on pest problems, the role of semiochemicals, natural products, biological control of crop pests, weeds, the role of pheromones, biotechnological approaches and breeding crops for disease resistance.

I hope that the book will prove useful to all those in interested in promoting the cause of IPM in formal and informal applications in both developed and developing countries so that the sustainability in agricultural system and environmental protection for future generation is achieved. All the contributors deserve special appreciation for writing chapters in their respective fields in great depth with dedication. Last, but not least, thanks are also due to Kristi Gomez and Patricia Gonzalez of Academic Press for taking great pains in publishing this book in a very impressive manner.

Dharam P. Abrol

List of Contributors

G.I. Aradottir Biological Chemistry and Crop Protection Department, Rothamsted Research, Harpenden, Hertfordshire, AL5 2JQ, UK

Karen L. Bailey Saskatoon Research Centre, Saskatchewan, Canada

N. Balakrishnan Tamil Nadu Agricultural University, Coimbatore, India

T.J.A. Bruce Biological Chemistry and Crop Protection Department, Rothamsted Research, Harpenden, Hertfordshire, AL5 2JQ, UK

S.L. DeFauw Pennsylvania State University, PA, USA

P.J. English Mississippi State University, MS, USA

Kristina Falke Central Institution for Decision Support Systems in Crop Protection, Bad-Kreuznach, Germany

R.T. Gahukar Arag Biotech Pvt. Ltd., Nagpur, India

Rachna Gulati Chaudhary Charan Singh Haryana Agricultural University, Haryana, India

S.K. Gupta Sher-e-Kashmir University of Agricul-Sciences and Technology of Jammu, Jammu and Kashmir, India

V. Gupta Sher-e-Kashmir University of Agricultural Sciences and Technology of Jammu, Ludhiana, Jammu and Kashmir, India

V.K. Gupta Punjab Agricultural University, Punjab, India

Gerrit Hoogenboom Washington State University, WA, USA

M.S. Hunjan Punjab Agricultural University, Ludhiana, India

S.S. Hussaini National Bureau of Agriculturally Important Insects, Bangalore, India

K.S.U. Jayaratne North Carolina State University, NC, USA

J.N. Jenkins Genetics and Precision Agriculture Research Unit, MS, USA

Vikas Jindal Punjab Agricultural University, Punjab, India

C.S. Kalha Sher-e-Kashmir University of Agricultural Sciences and Technology of Jammu, Jammu and Kashmir, India

S.S. Kang Punjab Agricultural University, Ludhiana, India

Benno Kleinhenz Central Institution for Decision Support Systems in Crop Protection, Bad-Kreuznach, Germany

Sriyanka Lahiri North Carolina State University, NC, USA

Tracy C. Leskey USDA-ARS Appalachian Fruit Research Station, WV, USA

S. Mohankumar Tamil Nadu Agricultural University, Coimbatore, India

Anne L. Nielsen Rutgers University, NJ, USA

Rabiu Olatinwo Louisiana State University, LA, USA

David Orr North Carolina State University, NC, USA

M.K. Pandey Sher-e-Kashmir University of Agricultural Sciences and Technology of Jammu, Jammu, India

Bruce L. Parker The University of Vermont, VT, USA

Rajinder Peshin Sher-e-Kashmir University of Agricultural Sciences and Technology of Jammu, Jammu and Kashmir, India

Jaime C. Piñero Lincoln University, MO, USA

David Pimentel Cornell University, NY, USA

Chandra S. Prabhakar International Crops Research Institute for the Semi-Arid Tropics (ICRISAT), Andhra Pradesh, India

Paolo Racca Central Institution for Decision Support Systems in Crop Protection, Bad-Kreuznach, Germany

M. Raghuraman Banaras Hindu University, Uttar Pradesh, India

V.V. Ramamurthy Indian Agricultural Research Institute, New Delhi, India

T. Ramasubramanian Indian Council of Agricultural Research, Coimbatore, India

Cesar R. Rodriguez-Saona PE Marucci Center for Blueberry & Cranberry Research & Extension, NJ, USA; Rutgers University, NJ, USA

Dietmar Rossberg Julius Kühn Institute – Federal Research Centre for Cultivated Plants JKI, Kleinmachnow, Germany

K. Samiayyan Tamil Nadu Agricultural University, Coimbatore, India

R. Samiyappan Tamil Nadu Agricultural University, Coimbatore, India

Uma Shankar Sher-e-Kashmir University of Agricultural Sciences and Technology of Jammu, Jammu and Kashmir, India

Devinder Sharma Sher-e-Kashmir University of Agricultural Sciences and Technology of Jammu, Jammu and Kashmir, India

Hari C. Sharma International Crops Research Institute for the Semi-Arid Tropics (ICRISAT), Andhra Pradesh, India

R. Sharma Sher-e-Kashmir University of Agricultural Sciences and Technology of Jammu, Jammu and Kashmir, India

Rakesh Sharma Sher-e-Kashmir University of Agricultural Sciences and Technology of Jammu, Jammu and Kashmir, India

P.P. Singh Punjab Agricultural University, Ludhiana, India

Margaret Skinner The University of Vermont, VT, USA

L.E. Smart Biological Chemistry and Crop Protection Department, Rothamsted Research, Harpenden, Hertfordshire, AL5 2JQ, UK

Michael J Stout Louisiana State University, LA, USA

Jae Su Kim Chonbuk National University, Jeollabuk-do, Republic of Korea

Sunil Tewari PE Marucci Center for Blueberry & Cranberry Research & Extension, NJ, USA

R.S. Tripathi Central Arid Zone Research Institute, Jodhpur, India

Beate Tschöpe Central Institution for Decision Support Systems in Crop Protection, Bad-Kreuznach, Germany

J.L. Willers Genetics and Precision Agriculture Research Unit, MS, USA

Mark G. Wright University of Hawaii at Manoa, HI, USA

CHAPTER

1

Host-Plant Resistance in Pest Management

Michael J. Stout

Louisiana State University, LA, USA

1.1 INTRODUCTION – WHAT IS PLANT RESISTANCE?

The interactions of herbivorous arthropods with their plant hosts are complex and multifaceted (Rasmann and Agrawal, 2009), even when they take place in the simplified ecosystems characteristic of agriculture. The overall process by which a herbivore makes use of a plant usually involves a number of phases: a searching phase in which the herbivore moves, often in response to visual and odour cues, from a location lacking a host-plant to a potential host; a contact evaluation phase mediated by an expanded set of visual, physical, and chemical cues from the plant; and a host utilization phase in which the performance of the herbivore is influenced by interacting suites of nutrients, toxins, digestibility reducers, and other factors in the plant (Duffey and Stout, 1996; Schoonhoven et al., 1998). At each step in this process, the herbivore interacts not only with the potential host plant but also directly or indirectly with other organisms at the same trophic level, such as competing herbivores, with organisms at different trophic levels, such as predators and parasitoids, and with microorganisms. Plant resistance results from the expression by the plant of resistance-related plant traits that affect one or more aspects of the herbivore's interaction with the host plant and with other plant-associated organisms. Plant resistance may be defined as the 'sum of the genetically inherited qualities' that determine the ultimate degree of damage (yield loss) done to the plant by the herbivore (Painter 1951; Smith and Clement, 2012).

There is a sense in which plant resistance is always an element of a pest management programme. After all, it is the interaction between the pest herbivore and the crop host, in all its complexity, that the pest manager seeks to manipulate in order to minimize the impact of the pest on crop yield and quality. By influencing the expression of resistance-related traits, the genotype of the crop is the strongest influence on this crop–pest interaction. Crop genotype is thus the foundation on which management strategies are built (Wiseman, 1994). In the context of pest management, however,

D. P. Abrol (Ed): *Integrated Pest Management*.
DOI: http://dx.doi.org/10.1016/B978-0-12-398529-3.00002-6

'plant resistance' typically references the integrated management tactic in which a resistant plant genotype is intentionally employed, alone or in combination with other tactics, to reduce the impact of herbivorous arthropods on crop yield or quality.

Plant resistance as a tactic has several advantages over other pest management tactics (Adkisson and Dyck, 1980). The effects of plant resistance on the target pest are often constant and cumulative, and plant resistance is usually simple and inexpensive for farmers to implement once the resistant variety has been developed. In addition, plant resistance is usually compatible with other tactics, such as insecticide applications and biological control (Wilde, 2002; Wiseman, 1994). Perhaps most importantly, plant resistance does not have the negative environmental effects associated with the use of insecticides, and in fact, the use of resistant plant varieties can reduce insecticide use (Wilde, 2002). Despite these advantages, host-plant resistance has not been used to full advantage in many crops (Smith and Clement, 2012; Wilde, 2002).

This chapter provides an overview of the traditional approach to using plant resistance in crop protection and briefly summarizes the current use and importance of plant resistance, including case studies from rice (*Oryza sativa*). It will also highlight recent advances in our understanding of inducible plant resistance and discuss how these recent advances may engender novel approaches to using plant resistance in pest management. It mostly excludes here the current commercial use of resistant transgenic plants, because, at present, these primarily involve the insertion of foreign genes into crop plants and thus provide limited insight into naturally-occurring plant resistance. Several other monographs and reviews of various aspects of plant resistance have been published in the past decade (Smith, 2005; Smith and Clement, 2012; Stout and Davis, 2009; Wilde, 2002).

1.2 THE TRADITIONAL APPROACH TO PLANT RESISTANCE

The traditional approach to the use of host-plant resistance in integrated pest management involves four steps: screening (evaluation of genotypes for resistance), categorization (assignment of resistance phenomena to one or more categories of resistance), breeding (introgression of genes responsible for resistance into agronomically acceptable backgrounds), and implementation (integration of resistant varieties into management programmes).

1.2.1 Screening

Once the basic outlines of the pest–crop interaction are understood, the initial step in using host-plant resistance in a management programme is the evaluation (screening) of crop germplasm to identify genotypes (lines, accessions, cultivars, etc.) that express resistance to the insect or that express a phenotype putatively related to resistance (Quisenberry and Clement, 1999). The precise methods used in screening depend, of course, on the biology of the crop–pest interaction and on the type of resistance or resistance trait of interest. Generally, however, screening involves exposing all or parts of the plants to be evaluated to uniform populations of the pest, then evaluating injury, damage, infestation levels, insect performance, or other appropriate endpoints correlated with resistance. Screening for tolerance (see below) is more involved, as it entails measuring yields in environments with and without herbivores. Screening may be conducted in a greenhouse or laboratory, but most often field screening of genotypes is necessary at some point in the evaluation process. Natural or artificial infestations of pest insects may be used. Smith et al. (1994) provide a comprehensive overview of methodologies used in resistance screening.

The importance of an appropriate screening method cannot be overemphasized. The standard and modified 'seedbox' methods for screening rice genotypes for resistance to planthoppers, for example, involve planting seeds of the rice genotypes to be tested, including susceptible and resistant checks, in rows in a seedbox (60×40×20cm), infesting with an appropriate number of second instar nymphs 7 days after planting (standard seedbox method) or 20 days after planting (modified seedbox method), and rating for damage on a 0–9 point scale when susceptible checks have been killed by planthoppers. These methods, while they give only an incomplete picture of rice resistance to planthoppers (Horgan, 2009), have allowed high-throughput evaluation of thousands of rice lines for planthopper resistance. This has resulted in the identification of numerous lines with high levels of resistance to planthoppers, and these lines have been used in the development of resistant varieties that have been used to great benefit over the past 40 years in Asia (Cuong et al., 1997; Khush, 1989; see below).

Sources of resistance – in the form of resistant cultivars, landraces, accessions, or wild relatives of crop species – have been identified for virtually all major pests of all major crops when screenings of sufficient scale have been conducted (Clement and Quisenberry, 1999; Kennedy and Barbour, 1992), although levels of resistance are not always high. A large screening programme for insect pests of potato, for example, has identified resistance against all major insect pests of potato in the US, including green peach aphid, *Myzus persicae*, potato aphid, *Macrosiphum euphorbiae*, Colorado potato beetle, *Leptinotarsa decemlineata*, potato flea beetle, *Epitrix cucumeris*, and potato leafhopper, *Empoasca fabae* (Flanders et al., 1992). Similarly, for US maize, sources of resistance to the corn leaf aphid, *Rhopalosiphum maidis*, corn earworm, *Helicoverpa zea*, European corn borer, *Ostrinia nubilalis*, southwestern corn borer, *Diatraea grandiosella*, fall armyworm, *Spodoptera frugiperda*, chinch bug, *Blissus leucopterus leucopterus*, corn rootworms (*Diabrotica* sp.), and several other pests have been identified (Barry et al., 1999; Ortega et al., 1980).

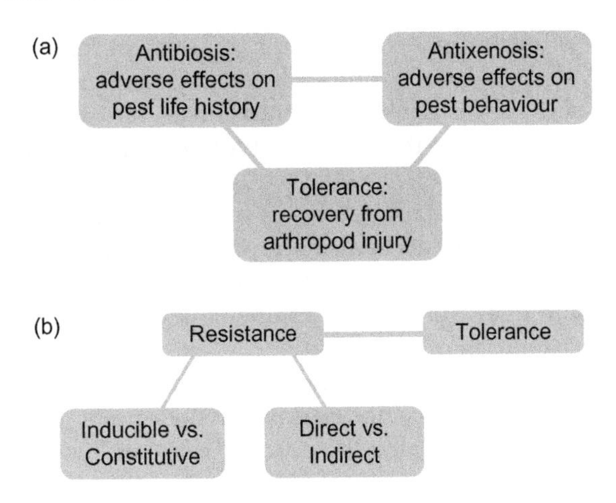

FIGURE 1.1 Painter's trichotomous scheme used for categorizing types of plant resistance in the applied literature (a) and the dichotomous scheme used for categorizing types of plant resistance in the fundamental literature (b).

1.2.2 Categorization of Resistance

Identification of resistant genotypes is often accompanied by research designed to allow the assignment of resistant genotypes to one or more of three categories of resistance originally defined by Painter (1951) (see also Horber, 1980; Kogan and Ortman, 1978; van Emden, 2002) (Figure 1.1a). The first category, 'antibiosis', is used to describe adverse effects of resistant plants on herbivore physiology and life history such as reduced growth, survival, and fecundity. The second category, 'antixenosis' (originally 'non-preference' in Painter, but later renamed by Kogan and Ortman in 1978), denotes plant traits affecting herbivore behaviour in ways that reduce the preference for, or acceptance of, a plant as a host by a herbivore. Finally, 'tolerance' refers to the ability of a plant to withstand herbivore injury such that agronomic yields or quality are reduced to a lesser extent than in a less tolerant plant subjected to equivalent injury. Although Painter originally described these as 'mechanisms', antibiosis, antixenosis, and tolerance are better thought of as types or 'functional categories' of effects (Horber, 1980; Smith, 2005); despite this fact, these terms are still frequently described as mechanisms in the literature (e.g. van Emden, 2002). Smith et al. (1994) describe techniques for distinguishing these types of resistance in plants.

Painter developed this trichotomous framework for classifying resistance when much less was known about the causal bases of resistance, and there is some question about the continuing utility of these categories (Stout, 2013). These categories are vaguely delineated, particularly antibiosis and antixenosis (Horber, 1980). This makes it difficult if not impossible to unambiguously assign some resistance phenomena to these categories; in fact, resistance quite frequently involves a combination of antibiosis, antixenosis, and/or tolerance (Horber, 1980; Smith, 2005; Smith and Clement, 2012). There are also some mechanisms of plant resistance that cannot be unequivocally assigned to any of these three categories, such as herbivore-induced emission of volatiles that attract parasitoids (see below; Stout, 2013). Finally, these categories in some cases may not be precise enough to represent their potential role in a management programme. For example, plant resistance that results in high levels of mortality in the early stages of insect development and plant resistance that merely slows the growth of an insect can both be described as 'antibiotic', but the two types might differ substantially in their value to a management programme and might be used quite differently. Nonetheless, these terms are still used widely in the applied plant resistance literature.

1.2.3 Breeding

Once a source of resistance has been identified, the gene or genes responsible for resistance must be introgressed (incorporated) into a genetic background suitable for growing in large-scale commercial agriculture. This step is, of course, not necessary or is greatly simplified when resistance is found in a variety or genotype with good agronomic traits. In such cases, the resistant variety can be recommended to growers in areas where the pest is expected to be a problem and used directly in a management programme. When the donor source of resistance is an unimproved line or accession, a wild species, or a variety with inferior characteristics, however, additional breeding steps are needed to develop a genotype that is both resistant and agronomically acceptable. Overviews of strategies for breeding for insect resistance in rice, corn, and wheat can be found in Khush (1989), Barry et al. (1999), and Berzonsky et al. (2003), respectively.

The difficulty of introducing resistance into an acceptable agronomic background is strongly influenced by the genetic basis of resistance, with resistance governed by a single gene in the plant (monogenic resistance) being much easier to transfer than polygenic resistance (Stout and Davis, 2009). Unfortunately, monogenic resistance to arthropods is fairly rare, and usually involves sucking insects or other small insects that develop extended and sometimes intimate associations with their host plants (e.g. aphids, gall midges). The level of resistance conferred by resistance genes in monogenic resistance is typically very high. Prominent examples include Hessian fly resistance and Russian wheat aphid resistance in wheat and brown planthopper resistance in rice (Berzonsky et al., 2003; Khush, 1989); another recent example is the resistance of lettuce, *Lactuca sativa*, to the lettuce aphid, *Nasonovia ribisnigri* (McCreight and Liu, 2012). Single-gene resistance to chewing insects has also been reported (e.g. Wearing et al., 2003), but is evidently much rarer.

One of the most important questions with respect to monogenic resistance in plant–arthropod interactions is the extent to which these examples of resistance conform to the gene-for-gene model so common in plant–pathogen interactions. In pathogen gene-for-gene interactions, resistance is governed by a specific resistance gene in the plant that directly or indirectly recognizes the product or products of a single corresponding avirulence gene in the pathogen (Walters and Heil, 2007). Perception of the avirulence gene product by the plant triggers a plant response that is often highly effective at killing the attacker or otherwise preventing the exploitation of the plant. The response triggered in the plant by the recognition of the attacker can involve changes in the expression of hundreds or thousands of genes and in comprehensive changes in plant metabolism (Kaloshian, 2004) and thus the mechanism by which the plant protects itself (the phenotype associated with resistance) can be very complex, even when expression of the resistance phenotype is regulated by a single gene. To date, three arthropod resistance genes have been cloned and sequenced, *Mi-1.2* from tomato, *Vat* from melon, and *Bph-14* from rice (Du et al., 2009; Smith and Clement, 2012). All three genes are members of the CC-NB-LRR disease resistance gene family, and the responses mediated by these genes share similarities with responses to pathogens, including activation of a salicylic acid-dependent signalling pathway. Furthermore, characterization of putative Hessian fly avirulence genes has revealed similarities with pathogen effector genes (Stuart et al., 2012). Thus, some arthropod–plant interactions do conform to the gene-for-gene model, although it is not yet clear whether all single-gene arthropod resistance in plants involves corresponding pairs of plant resistance genes and arthropod avirulence genes. Furthermore, elicitors of induced plant resistance have been characterized from chewing insects, but they do not elicit the extremely strong and specific types of responses elicited by avirulence gene products (Howe and Jander, 2008).

There are also plant–arthropod interactions that do not conform to the gene-for-gene model but in which a small number of plant genes have major effects on resistance. Thus, in cucurbits, resistance to many arthropods is strongly influenced by the presence or absence of cucurbitacins, the expression of which are controlled by one or a few genes (Balkema-Boomstra et al., 2003; Kennedy and Barbour, 1992), and resistance to stem sawflies in wheat is strongly influenced by the solid stem trait, which is controlled by three or fewer genes (Berzonsky et al., 2003). In addition, single-gene resistance to arthropods in some plants may be buttressed by the contributions of minor genes for resistance (Cohen et al., 1997).

Most commonly, plant resistance to arthropods is polygenic – that is, resistance is attributable to the contribution of many genes or quantitative trait loci (QTLs). The polygenic nature of most plant resistance is consistent with what is known about the biochemical and morphological basis of resistance, which typically involves the interaction of the arthropod with multiple plant traits (Rasmann and Agarwal, 2009; Duffey and Stout, 1996). The polygenic nature of most plant resistance has a number of important consequences (Stout and Davis, 2009). As with other polygenic traits, variation in polygenic plant resistance is continuous rather than discrete, and levels of resistance sufficiently high to be usable are often rare in germplasm collections. The low frequencies of usable levels of resistance in germplasm collections necessitate the evaluation of large numbers of genotypes. This difficulty is compounded by the environmental contingency of polygenic resistance and the inherent variability of insect populations (see Smith et al., 1994 for further discussion of the difficulties encountered in screening for polygenic resistance). After a source of resistance is identified, incorporation of polygenic resistance into an acceptable background by traditional breeding practices is also difficult because not all genes involved in resistance may be transferred during crosses, resulting in dilution of the desired resistance. As a result, it may take many years, diverse expertise and considerable infrastructure and resources to develop resistant varieties when the resistance is polygenic.

Over the past two decades, considerable progress has been made in using molecular tools to identify and map genes or genetic loci associated with (polygenic) resistance to arthropods in various plants. Smith and Clement (2012) list genes or QTLs associated with resistance to over 50 arthropod species in 22 crop species. The identification and mapping of resistance-associated genes and QTLs has facilitated the development of resistant varieties of some crops through the use of marker-assisted selection and related techniques (Jairin et al., 2009; Willcox et al., 2002).

Yields of resistant cultivars are sometimes lower than those of their susceptible counterparts. This may be the case because resistance entails substantial allocation costs; indeed, a number of studies have demonstrated costs associated with the expression of constitutive or inducible resistance (Strauss et al., 2002). Alternatively, low yields and other undesirable agronomic traits may occur in resistant plants because of genetic linkage between the resistance gene or genes and genes responsible for the undesirable traits (linkage drag) (Stout and Davis, 2009). These linkages may be difficult to eliminate during the breeding process.

1.2.4 Implementation

Once an agronomically suitable resistant variety has been developed or identified, it must then be deployed. Plant resistance has, generally speaking, been used in one of two ways: as a primary, stand-alone tactic and as an adjunct to other control methods. The former strategy is feasible (but perhaps not advisable) when resistance in a cultivar is sufficiently strong to suppress populations below economic thresholds. Resistance this strong is rare, and, when present, usually involves monogenic resistance. There are historically important examples of host-plant resistance used as a primary tactic. One of the most famous is the use of grape rootstocks resistant to grape phylloxera. This leaf- and root-galling pest, a native of North America, nearly destroyed the French wine industry in the late 19th century, and was brought under control using grapevine rootstocks derived from resistant North American *Vitis* species (Benheim et al., 2012). After over 100 years, use of resistant rootstocks remains the primary tactic used against phylloxera.

The risk associated with using strong, single-gene resistance as a primary tactic is, of course, that the target pest will evolve to overcome the resistance. Populations of an insect pest that have evolved the ability to overcome plant resistance are called biotypes. Biotypes are often identified using differential sets of plant cultivars possessing different resistance genes (Smith, 2005; Smith et al., 1994); cultivars susceptible to a given biotype are killed or severely damaged upon infestation (compatible interaction), while cultivars resistant to a given biotype are often left virtually unscathed (incompatible interaction). Biotypes have been documented in a number of crop–pest interactions in which resistance is conferred by a single major gene, such as lettuce-lettuce aphid, grape-grape phylloxera, wheat-Hessian fly, wheat-Russian wheat aphid, and rice-brown planthopper. The use of additional management tactics in conjunction with plant resistance may help delay the development of biotypes. For Hessian fly in wheat, resistant cultivars are often combined with the cultural practices of delaying planting until the fly-free date has passed and removing volunteer wheat from fields (Berzonsky et al., 2003).

Sometimes, significant economic benefits can be obtained simply by using moderately resistant varieties over large areas or, conversely, by avoiding the use of highly susceptible varieties over large areas, and this might be considered a special case of a primary use of plant resistance. The potential importance of an area-wide perspective was demonstrated recently by Hutchison et al. (2010). These authors showed that use throughout the US corn belt of transgenic maize varieties expressing an insecticidal protein from *Bacillus thuringiensis* has resulted in significant economic benefits for farmers who do not grow Bt varieties because of the area-wide suppression of the target pest, *Ostrinia nubilalis*, by the transgenic varieties. While most plant resistance does not result in the levels of control provided by Bt varieties, nonetheless, the principle may still apply. In Louisiana sugarcane, widespread use (ca. 90% of the acreage in Louisiana) of a variety highly susceptible to the sugarcane borer (*Diatreae saccharalis*) has forced producers to rely heavily on chemical insecticides for borer control, which in turn has probably contributed to the development of insecticide resistance in *D. saccharalis* populations (Akbar et al., 2008). Adoption of varieties with low levels of resistance would likely lead to area-wide reductions in sugarcane borer populations and reduced reliance on insecticides (T.E. Reagan, personal communication).

More commonly, the level of resistance found in a cultivar is not sufficiently high to be used as a primary strategy, and plant resistance must be combined or integrated with other management tactics, such as insecticide applications, cultural practices, and biological control. In such cases, the interactions of plant resistance with other management tactics are important to the net effectiveness of the management programme. Host-plant resistance may act independently of these other tactics to reduce pest populations, or antagonistic or synergistic interactions may occur. Numerous examples of these interactions are documented in Quisenberry and Schotzko (1994) and Wiseman (1994), and only a brief overview is presented here.

Interactions among plant resistance and biological control have received the most attention, and a number of studies have demonstrated antagonism between the two tactics. Plant resistance-related traits may interfere directly with the activities of natural enemies of herbivores, such as when tomato trichomes and their exudates entrap or interfere with the searching activities of predators or reduce the survival and parasitism rates of parasitoids (Simmons and Gurr, 2005). Resistance-related traits may also have indirect effects on natural enemies. This may occur, for example, when secondary chemicals accumulate or are actively sequestered in herbivore tissues, thereby toxifying parasitoids developing internally in the herbivore (Campbell and Duffey, 1979). Another type of indirect antagonism may result when plant resistance alters the movement or distribution of herbivores on plants in ways that reduce the searching efficiency of predators or parasitoids (Quisenberry and Schotzko, 1994). These types of antagonistic interactions can have substantial impacts on the net effectiveness of a management programme. Bartlett (2008) examined the combined effects of constitutive soybean resistance and predators (spined soldier bugs, *Podisus maculiventris*) on Mexican bean beetles (*Epilachna varivestis*) and soybean fitness in field cages. Soldier bugs were more likely to feed on Mexican bean beetles on susceptible plants than on resistant plants. Furthermore, the presence of predators on resistant plants did not result in increased seed production, whereas susceptible plants with predators produced significantly more seeds than susceptible plants without predators.

Although antagonistic interactions among plant resistance and biological control may be important in some crop–pest interactions, more recent research has tended to emphasize the compatibility of, and even synergism among, biological control and plant traits. Again, numerous mechanisms are involved in these positive interactions, but by far the most attention has been directed toward the herbivore-induced release of volatiles by plants. In many plants, feeding and/or oviposition by herbivores induces increased local and systemic emission of volatile compounds, usually complex blends of terpenes, phenolics, and green-leaf volatiles. These volatiles are potentially used as cues by natural enemies (parasitoids and predators) to locate herbivore-injured plants (Howe and Jander, 2008; Wu and Baldwin, 2010). This phenomenon, termed induced indirect defence, appears to be an adaptation to allow the active manipulation of the third trophic level by plants.

Plant resistance can also be integrated with tactics other than biological control. Management of the midge *Stenodiplosis sorghicola* in Australian sorghum involves widespread use of midge-resistant (antixenotic and antibiotic) sorghum, reduced insecticide use and consequent conservation of natural enemies, early planting to avoid high midge populations, and elimination of alternative midge hosts (Franzmann et al., 2008). Adkisson and Dyck (1980) describe how cultural practices such as stalk destruction and early crop defoliation were combined with the use of short-season, nectariless cotton varieties to dramatically reduce the need for insecticide applications in Texas cotton in the 1970s. On the other hand, it has occasionally been shown that feeding on plants with certain secondary chemicals can induce the activities of detoxicative enzymes in herbivore guts, thereby imparting greater tolerance of insecticides (Smith, 2005; Wiseman, 1994). Given the current and anticipated widespread use of transgenic crops containing Bt toxins, it might be particularly useful to investigate how expression of Bt toxins might be combined with natural plant resistance to create varieties with more effective, broader-based resistance (Meszaros et al., 2011).

Most investigations of interactions among plant resistance and other management tactics are conducted in the laboratory or greenhouse or on a small-plot scale. Thus, despite the considerable efforts expended in small-scale investigations of interactions among management tactics, resistant cultivars are, in most cases, deployed commercially with little consideration of methods for optimizing the benefits of the resistant variety in the context of a commercial-scale, multi-tactic management programme. Further large-scale studies of the integration of resistant varieties and other management tactics are needed (Stout and Davis, 2009).

1.3 CURRENT AND PAST USES OF PLANT RESISTANCE

The approach outlined in the preceding paragraphs for developing and using arthropod-resistant varieties has proven very successful. Wilde (2002) lists over 25 major crops for which varieties resistant to one or more insect pests have been developed, and Smith and Clement (2012) note that over 500 arthropod-resistant crop varieties had been developed by the mid-1970s. There are some crops for which the development of resistant cultivars has been essential to the economic viability of the crop. According to Wiseman (1999), production of sweet corn was not possible, even with heavy use of insecticides, before the introduction of corn earworm (*Helicoverpa zea*)-resistant corn in the early 1950s. Similarly, wheat production in areas where high populations of Hessian fly are present and grape production in areas where phylloxera is serious would probably not be profitable without resistant cultivars. For many crops, insect-resistant donor lines, or lines with resistant donors in their pedigrees, have been used widely for many years in breeding programmes, such that resistance has been diffused widely throughout the cultivars in current use and cultivars resistant to one or more insect pests comprise the majority of hectarage planted. For example, Barry et al. (1999)

estimated that 65% of commercial maize hybrids in use during the late 1990s possessed some resistance to the corn leaf aphid, while greater than 90% of maize hybrids possessed resistance to first-generation European corn borers. Pubescent soybean lines with resistance to the potato leafhopper, *Empoasca fabae*, were first incorporated into soybean breeding programmes in the 1930s, and use of pubescent lines has resulted in stable suppression of leafhopper populations over the past 70 years and has rendered this insect a virtual non-pest over that period of time (Boethel, 1999).

There have been some attempts to quantify the economic benefits of host-plant resistance, although recent attempts to do this are scarce. Wiseman (1999) cites several studies showing returns of $20 to $300 for every dollar invested in plant resistance research, and the economic benefits to farmers of Hessian fly resistant cultivars may exceed $100 per acre (Buntin and Raymer, 1989).

1.4 THE EVOLVING ROLE OF MECHANISTIC RESEARCH IN HOST-PLANT RESISTANCE

The mechanisms of plant resistance are the processes by which the aggregate expression of resistance-related traits by plants brings about reductions in herbivore damage, either by reducing the injury done to plants or by reducing the impact of herbivore injury on plant fitness (yield). The causal bases of plant resistance are almost always complex; even when genetic control of resistance is simple, the responses triggered during a resistance response involve changes in expression of dozens, even hundreds of genes. Historically, host-plant resistance research has been the province of empirically oriented scientists who have not emphasized the importance of understanding the mechanisms of plant resistance. Thus, Painter, in 1951, wrote: 'Hence, one must frequently deal with a number of causes or mechanisms which result in resistance rather than with a simple factor, and in attempting to breed resistant varieties a knowledge of these mechanisms may sometimes be of little use' and '...so far, experimenters have been able to utilize insect resistance in crop improvement and insect control without complete knowledge of the reasons why the plants are resistant' (Painter, 1951, pp. 24–25). Following Painter's lead, most plant resistance research published in the applied literature has retained this heavily empirical and practical orientation. This does not discount the fact that elegant work on resistance mechanisms has sometimes emerged from the applied entomologists – the discovery of DIMBOA as an insect resistance factor in maize is a striking and historically important example (Kogan, 1986) – but nonetheless, it is true that mechanistic research historically was not viewed as essential to the practical enterprise of developing resistant cultivars.

Over the past half century, roughly in parallel with the development of the applied literature on host-plant resistance described in the preceding paragraph, a second body of research and theory concerned with the interactions of arthropods and plants has also developed (Kogan, 1986). This second body of research and theory comprises fundamental research on plant–insect interactions directed primarily toward understanding the ecology and evolution of these interactions. This literature emphasizes the importance of secondary plant metabolites as mediators of reciprocal evolutionary relationships among plants and plant-feeding arthropods (Berenbaum and Zangerl, 2008). The framework for conceptualizing resistance phenomena in this literature is a dichotomous framework: 'resistance' is used broadly to denote those plant traits that reduce the extent of injury done to a plant by a herbivore, whereas 'tolerance' encompasses those plant traits or physiological processes that lessen the amount of fitness (yield) loss per unit injury (Figure 1.1b; Stout, 2013). Furthermore, resistance can be divided into 'constitutive' or 'inducible' and 'direct' or 'indirect' sub-categories. Constitutive plant resistance is resistance that is expressed regardless of the prior history of the plant, whereas inducible resistance is resistance only expressed, or expressed to a greater extent, after prior injury (i.e. expression of inducible defences is contingent on prior attack, whereas constitutive defences are not). Direct plant resistance refers to those plant traits that have direct (unmediated) effects on herbivore behaviour or biology. Indirect plant resistance, in contrast, depends for its effect on the actions of natural enemies as described above. In contrast to the applied literature, studies of the mechanisms of plant resistance have always held a central place in this literature.

The past few decades have seen significant improvements in the analytical tools needed to isolate, identify, and quantify resistance-related traits in plants. Likewise, advances in techniques for genetic manipulation (e.g. transformation, virus-induced gene silencing) have made altering plant phenotypes a much more rapid and efficient process. Use of these sophisticated tools, initially by researchers interested in basic aspects of plant resistance but now, increasingly, by applied scientists, has yielded unprecedented insights into the mechanisms of resistance and into the effects of resistance-related traits on individual herbivores and on communities of organisms centred on plants (Zheng and Dicke, 2008). Two examples will suffice to illustrate the possibilities. In wheat, virus-induced gene silencing of a WRKY transcription factor and an inducible gene for phenylalanine ammonia lyase resulted in greater susceptibility of silenced plants to the aphid *Diuraphis noxia* and greater fecundity of aphids on silenced plants (van Eck et al., 2010). In rice, silencing of genes responsible for emission of the volatile compounds caryophyllene and linalool profoundly affected interactions with two herbivores, the brown planthopper (*Nilaparvata lugens*) and rice leaf folder (*Cnaphalocrocis medinalis*), in the field. Population densities of brown planthoppers were twice as high on lines with reduced linalool emission than on wild types, whereas densities were lower on lines with suppressed caryophyllene emission. Population densities of the leaf folder were lower on lines with reduced linalool emission than on wild-type plants (Xiao et al., 2012). Populations of natural enemies were also affected: silencing volatile emission reduced parasitism of planthopper eggs by *Anagrus nilaparvatae* and also reduced populations of predatory spiders. Adoption of a mechanistic approach to studying plant resistance, including the increased use of these analytical and genetic tools, is now clearly the most efficient path to development of resistant varieties for crop protection.

1.5 INDUCED RESISTANCE AS A MANAGEMENT TOOL

Despite the fact that host-plant resistance possesses many advantages over other pest management tactics, holds considerable potential for reducing pesticide use, and has been used to great benefit in many crops, plant resistance remains an underutilized management tactic. Wiseman (1994) suggested several reasons for this situation. Principal among these was the failure of entomologists and breeders to complete the collaborative task of developing a resistant variety after resistant sources had been identified, a failure that perhaps stems from the time and money involved or the difficulty of transferring polygenic traits to improved varieties. Also, the general cost-effectiveness and ease of use of insecticides may discourage investments of time and money to develop resistant cultivars; a decreased interest in conventional plant resistance certainly appears to be a collateral effect of the widespread use and success of Bt crops. In addition, Wiseman cited failures in extending information about the value of proper use of resistant varieties to growers. These remain important problems that deserve attention. However, potentially the most fruitful path for increasing the use of plant resistance in crop protection is a greater exploitation of phenotypic plasticity in plant resistance.

In the early decades of research on plant resistance, the focus of research in both the basic and applied literature was on constitutive resistance. The influence of biotic and abiotic factors on the expression of resistance-related traits, though acknowledged – see, for example, the extended discussion of the influence of abiotic factors such as temperature and soil fertility on expression of plant resistance by Painter (1951) – was not emphasized. Over the past two decades, however, there has been a major shift in the emphasis of plant resistance research, and the extent and importance of phenotypic plasticity in plant resistance has increasingly been recognized. In particular, it is now widely recognized that many, if not most plants, respond to initial attack by arthropods in ways that increase the resistance of the plant to subsequent herbivores (induced direct and indirect resistance). In fact, it is now common for reviews of plant–insect interactions to focus on induced resistance to the near exclusion of constitutive resistance (e.g. Howe and Jander, 2008; Wu and Baldwin, 2010).

The ability of plants to respond to attack by insects and pathogens by rapidly changing their resistance phenotype requires the existence of hormonally mediated systems of recognition and response (Howe and Jander, 2008; Kim et al., 2011; Wu and Baldwin, 2010). Most of the research in this area has focused on pathways of response mediated by the plant hormones jasmonic acid (JA), ethylene, and salicylic acid (SA). JA has generally been viewed as a mediator of plant responses to insects, whereas SA has been viewed as a mediator of plant responses to pathogens. These two pathways are mutually inhibitory (Howe and Jander, 2008). Ethylene generally acts in concert with JA and serves to synergize or otherwise 'fine-tune' JA-induced responses (Wu and Baldwin, 2010). Levels of JA and SA increase locally (i.e. at the site of injury) and systemically in plants attacked by insects and pathogens and these increases in endogenous hormone levels are followed by increases in the expression of resistance-related genes and plant traits (Wu and Baldwin, 2010). Consistent with the putative role of SA and JA as mediators of plant responses and induced resistance, treatment of plants with exogenous JA or SA induces expression of resistance-related traits and resistance to insects or pathogens, and mutant or transgenic plants compromised in their ability to synthesize or perceive these hormones are generally unable to mount effective resistance responses (Wu and Baldwin, 2010; Zhou et al., 2009).

There is now abundant evidence, however, that regulation of the responses of plants to herbivores and pathogens is not as simple as the bifurcated scheme outlined above might at first suggest (Stout et al., 2006). For example, the distinction between insect- and pathogen-induced responses is not absolute. Many piercing-sucking insects activate the SA-mediated pathway and induce responses in plants similar to those induced by biotrophic pathogens, although the responses induced by these insects are not always effective against the inducing insect. Conversely, many necrotrophic pathogens induce the JA pathway and responses typically associated with chewing insects (Stout et al., 2006), and JA-related responses are effective against many pathogens (Nahar et al., 2011). Some arthropod herbivores induce both JA- and SA-associated responses (e.g. mites in tomato; Sarmento et al., 2011). The nature and extent of plant responses to attackers are influenced by factors present in the oral secretions of insects, which sometimes activate and sometimes suppress SA-, ethylene-, and JA-related responses (Howe and Jander, 2008; Kim et al., 2011).

Furthermore, data from a number of recent experiments indicate that plant hormones in addition to JA, SA, and ethylene are involved in mediating plant responses and induced resistance. For example, Thaler and Bostock (2004) found complex interactions, some negative and some positive, among JA, SA, and the stress-related hormone abscisic acid in tomato plants subjected to water and salt stress and insect and pathogen attack. Abscisic acid treatment enhanced resistance of rice plants to the fungal pathogen *Cochliobolus miyabeanus* but suppressed some ethylene-related responses (de Vleesschauwer et al., 2010). In poplars, treatment with cytokinin (another plant hormone) increased the wound-inducible accumulation of JA, thereby increasing poplar resistance to gypsy moth larvae (Dervinis et al., 2010). Similarly, GA synergized the positive effects of JA on the expression of trichomes (a resistance-related trait) in *Arabidopsis* (Traw and Bergelson, 2003). Thus, the picture of the regulation of induced responses emerging in the current literature is that of a complex regulatory network involving multiple signals originating from the attacking organism and multiple hormone signals interacting in positive or negative fashion in the attacked plant, with SA and JA playing central but not exclusive roles. Elucidation of these networks may reveal additional methods for activating or stimulating them.

There is clearly the potential for induced resistance to be used in crop protection, but there are challenges that must be overcome to achieve this potential. Identifying genotypes with greater responsiveness to herbivores or elicitors will require different methods of screening than are used to screen for constitutive resistance. In addition, one of the important questions that must be addressed before induced responses can be used effectively in agriculture is the extent to which induced resistance contributes to overall resistance (Smith and Clement, 2012); in other words, are preformed (constitutive) traits more important in protecting plants from herbivores, or are traits expressed after herbivory more important? Very little research has been conducted to explicitly address this question, although important insights have been gained from studies using plants incapable of responding to herbivores. Mutant rice and *Arabidopsis* plants deficient in JA signalling were vulnerable to attack by detritivorous isopods that normally do not feed on live plants (Farmer and Dubugnon, 2009). Silencing several genes in wild tobacco (*Nicotiana attenuata*) involved in the oxylipin (JA) pathway resulted in greater vulnerability to herbivores adapted to wild tobacco and, furthermore, resulted in infestation by novel herbivores that do not feed on wild-type tobacco (Kessler et al., 2004). The maize inbred line Mp708, which possesses high levels of resistance to a number of lepidopteran pests, showed constitutively elevated levels of JA and a resistance-related cysteine protease, and also responded more strongly to insect attack (Shivaji et al., 2010). Together, these studies point to the importance of a functioning inducible resistance pathway as a critical barrier to herbivory, and perhaps suggest a conflation of inducible and constitutive resistance in some plants. It is conceivable that, once the relative contributions of constitutive and induced resistance to overall plant resistance is better understood, novel strategies for using induced resistance in crop protection will suggest themselves.

The centrality of hormones in induced plant resistance also suggests the possibility of applying elicitors of plant hormonal systems to increase the resistance of plants in commercial agriculture. The ability to manipulate plant resistance selectively and at strategic time points during crop development is a tool with many potential applications (Stout et al., 2002; Walters and Fountaine, 2009). However, progress in harnessing this potential tool has been limited, particularly with reference to insect resistance. Most attempts to stimulate crop plant resistance to insects have consisted of applying JA under laboratory or greenhouse settings (e.g. Omer et al., 2001). Only more rarely has stimulation of crop resistance to insects been attempted under field settings (Black et al., 2003; El Wakeil et al., 2010; Hamm et al., 2010; Thaler et al., 2001); in these cases, elicitor-induced increases in resistance have been moderate and/or transient, but potentially useful as a component of a management programme. Currently, however, no elicitors of insect resistance are available commercially. On the other hand, efforts to develop commercial elicitors of pathogen resistance have been somewhat more successful, and several products are commercially available (Walters and Fountaine, 2009). Probably the most successful of these commercial elicitors are probenazole, used to manage rice blast disease (*Pyricularia oryzae*), and Actigard®, which elicits high levels of disease control in tobacco against several fungal pathogens (Walters and Foutaine, 2009). Recently, treating tomato seeds with JA was shown to increase plant resistance to spider mites, caterpillars, aphids, and a necrotrophic pathogen (Worrall et al., 2012). The authors suggested that the JA treatment primed (conditioned) plants to respond more strongly to subsequent attack. The impact of seed treatment on tomato resistance was long-lasting (at least 8 weeks), and no effects of the seed treatment on yields were noted. If the effectiveness of JA seed treatments can be confirmed in other crops under field conditions, seed treatments may be one method for increasing the use of inducible resistance in crop protection. Much work remains, however, to characterize those crop–pest interactions and cropping conditions for which the use of resistance elicitors might be effective and affordable.

Another potentially useful strategy for using induced resistance in pest management is the development of cultivars that exhibit high levels of indirect induced defence. The attraction of predators and parasitoids by herbivore-induced volatiles has been particularly well-studied in maize, and work with this species suggests avenues for using indirect defence in crop protection. Both feeding and oviposition by leaf-feeding and stem-boring caterpillars induce the systemic emission of complex blends of volatile compounds, including green-leaf volatiles and terpenoids, from above-ground portions of maize plants (Tamiru et al., 2011). These volatiles are attractive to both egg and larval parasitoids. Remarkably, feeding on maize roots by larval *Diabrotica virgifera virgifera* also results in the emission of volatile compounds, most prominently the sesquiterpene caryophyllene, from roots, and caryophyllene is attractive to soil-dwelling entomopathogenic nematodes (Rasmann et al., 2005). For both stem borers and *Diabrotica*, qualitative and quantitative variation in volatile release among maize genotypes was demonstrated; in the case of *Diabrotica*, herbivore-induced caryophyllene emission is present in European cultivars but absent in North American cultivars, whereas, in the case of stem borers, volatile emission is present in maize landraces but absent in commercial hybrids. Furthermore, rates of nematode infection were greater, and beetle emergence from soils lower, in cultivars capable of emitting caryophyllene than in cultivars not capable of emitting caryophyllene, and spiking the soil near maize plants incapable of emitting caryophyllene with caryophyllene increased nematode infection rates. Genotypic variation in volatile emission suggests the ability to emit volatiles in response to herbivory is under selection, and thus it may be possible to augment volatile production through breeding. Moreover, the results of Rasmann et al. (2005) suggest that selection for increased volatile emission in some cases may facilitate biological control, which may be useful in crop management programmes.

Finally, our understanding of the physiological responses of plants to herbivore injury is in its infancy, and continued progress in this area should lead to a better understanding of how plants tolerate herbivore injury. For example, Schwachtje et al. (2006), using ^{11}C-photosynthate labelling, found that simulated *Manduca sexta* injury of wild tobacco leaves (mechanical wounding combined with the addition of herbivore regurgitant) led to increased allocation of C (sugars) to roots. This increased allocation of resources to roots was JA-independent and was mediated by the down-regulation of a gene for a SNF1-related kinase following herbivory. Constitutive suppression of this gene provided evidence that increased allocation of resources to roots was accompanied by delayed senescence and prolonged flowering and was thus part of a coordinated tolerance response that allowed plants to sustain seed production after herbivory.

1.6 CASE STUDIES: THE USE OF RESISTANT RICE VARIETIES

Rice is among the world's two or three most important food crops, and is particularly important in tropical Asia. Over 100 arthropod species feed on rice, although far fewer are consistent pests (Stout, 2012). Plant resistance plays a very important role in the management of many of these pests. The successes and failures in the development and use of rice varieties resistant to three important pests – the brown planthopper, *Nilaparvata lugens*, stem-boring lepidopterans, and the rice water weevil, *Lissorhoptrus oryzophilus* – nicely illustrate many of the points made in the preceding discussion.

1.6.1 Brown Planthopper Resistance

The brown planthopper is a sucking insect that, under heavy infestations, can cause the wilting and complete drying of rice plants, a condition known as 'hopperburn' (Bottrell and Schoenly, 2012). The brown planthopper also damages rice by transmitting ragged stunt virus and grassy stunt virus. The insect can complete as many as 12 generations in a single year in tropical areas, where it resides year-round, and fewer generations in temperate areas, where it is a migratory pest. Outbreaks of brown planthopper have occurred throughout the history of rice cultivation, but outbreaks became more frequent and more intense after the introduction of improved rice varieties and input-intensive farming practices during the green revolution of the 1960s. The increased importance of the brown planthopper as a pest prompted efforts to identify sources of planthopper resistance. Large-scale screening efforts at the International Rice Research Institute (IRRI) employing the 'seedbox' method described above identified a number of rice lines with very high levels of resistance to brown planthopper (Khush, 1989; Pathak, 1969). The variety 'Mudgo', for example, showed near-complete immunity to the brown planthopper, with 100% nymphal mortality after 10 days (Pathak, 1969). Because resistance in these lines was conditioned by single major genes, and because phenotyping was a relatively straightforward process, brown planthopper resistance was transferred relatively quickly to cultivars with improved semidwarf plant types and good grain quality. The first brown planthopper-resistant rice variety, 'IR26', which contained the resistance gene *bph-1*, was released by the IRRI in 1973 and was widely adopted by growers throughout Asia. Further screening efforts identified other rice lines with other genes for resistance to the brown planthopper (Khush, 1989).

Unfortunately, release of the highly resistant 'IR26' was followed within 3 years by the development of brown planthopper populations capable of overcoming the plant resistance. Such resistant biotypes apparently developed as quickly as they did because of the high level of resistance present in 'IR26' and because of the simple genetic basis of the resistance. The ensuing decades saw the release of numerous other brown planthopper-resistant lines with other genes for resistance and the development of additional planthopper biotypes (Alam and Cohen, 1998; Bottrell and Schoenly, 2012). The resistance of some varieties has proven to be more stable, notably IR36 and IR64; importantly, the resistance of these two varieties appears be governed by several minor genes in addition to the major gene *Bph-1* (Alam and Cohen, 1998). To date, over 20 *Bph* resistance genes have been identified from cultivated and wild *Oryza* species, although not all have been incorporated into resistant varieties (Bottrell and Schoenly, 2012).

Understanding of the mechanisms of brown planthopper resistance has somewhat lagged behind the development of resistant varieties (Bottrell and Schoenly, 2012), such that there is still 'no clear mechanistic link' between resistance genes and effects on brown planthopper fitness (Horgan, 2009). The amino acid asparagine is a feeding stimulant for the brown planthopper, and levels of sulphur-containing amino acids influence planthopper performance (Horgan, 2009). Constitutive and inducible volatile compounds, including terpenes such as linalool, may be involved in host-plant finding by planthoppers (Xin et al., 2012). Oxalic and silicic acids as well as certain flavonoids and plant sterols can act as feeding inhibitors to brown planthoppers, but causal links between the presence of these compounds and resistance have not been firmly established (Bottrell and Schoenly, 2012). Recent evidence indicates that the differential resistance of rice varieties possessing and lacking *bph* genes is attributable primarily to differential responses of resistant and susceptible varieties to planthopper feeding. Wang et al. (2008) showed that over 100 genes were differentially expressed following brown planthopper feeding on varieties possessing or not possessing the *Bph-14* and *Bph-15* resistance genes. This is consistent with what is known about major gene resistance in plant–pathogen interactions, in which resistance genes function in the recognition of pathogen attack and activation of resistance-related gene expression. The brown planthopper resistance gene *Bph-14* was recently cloned and found to encode a protein with a leucine-rich repeat (LRR) domain similar to those encoded by some pathogen resistance genes (Du et al., 2009). This protein is probably involved in the direct or indirect recognition of attack by planthoppers, leading to the activation of a salicylic acid-dependent pathway and resistance-related biochemical responses such as increased production of trypsin proteinase inhibitors and callose deposition. These responses result in an antibiosis-type resistance that dramatically reduces the feeding, survival, and population growth of planthoppers (Du et al., 2009). Planthoppers on resistant plants spent less time feeding than planthoppers on susceptible plants, an effect that may be directly related to callose synthesis and deposition on sieve plates (Hao et al., 2008).

Although rice varieties containing major genes for resistance to the brown planthopper exhibit high levels of planthopper resistance, other management practices may nonetheless be critical for the sustainable use of these varieties. Heinrichs and colleagues have demonstrated greater effectiveness of both insecticides and generalist predators on planthopper-resistant cultivars than on planthopper-susceptible varieties (Heinrichs, 2009). Furthermore, reducing the use of early-season insecticides, which destroy natural enemy complexes that help regulate brown planthopper populations, is probably critical for the long-term use of resistant varieties (Cohen et al., 1997).

1.6.2 Stem Borer Resistance

Stem-boring insects in rice, nearly all of them lepidopterans in the families Crambidae and

Noctuidae, are found in all important rice-producing regions of the world, and as a group, stem borers probably cause more yield losses than any other type of insect pest (Chaudhary et al., 1984). The life histories of these stem borers in rice are similar. First and second instars feed on leaf blades or in between the leaf sheath and the stem, often leaving feeding scars and characteristic feeding lesions. Later instars bore into rice stalks. Larvae pass through four or five instars and a pupal stage in the stem in 4 to 5 weeks. Feeding severs the growing portion of the plant from the base of the plant. When feeding occurs during the vegetative stage of rice plant development, the tiller in which the larva is present often dies and fails to produce a panicle (deadheart). When feeding occurs after panicle initiation, feeding by a larva within a stem results in drying of the panicle. Affected panicles may not emerge or, if they do, do not produce grains, remain straight, and appear whitish (whitehead). In the mid- to late-1960s, Pathak (1969) and other scientists at IRRI screened over 10,000 rice lines for resistance to *Chilo suppressalis* and identified 20 lines with usable levels of resistance. Resistance was manifested both by reduced oviposition and reduced larval growth and survival on resistant lines, but effects of resistant lines on stem borers were not as dramatic as with the brown planthopper. Importantly, lines and varieties with resistance to *C. suppressalis* also showed resistance to other stem borers ('cross resistance') (Das, 1976). Subsequent research revealed that multiple morphological factors, such as stem hardness, and biochemical factors contributed to resistance (Chaudhary et al., 1984). Consistent with the involvement of multiple plant traits in resistance, stem borer resistance was polygenic and thus more difficult to transfer from donors to high-yielding varieties. Nonetheless, moderate to good resistance was introduced into a large number of varieties released by IRRI in the 1970s (Khush, 1989). Many of these borer-resistant varieties were resistant to multiple other insect and disease pests.

1.6.3 Rice Water Weevil Resistance

Less success has been attained in developing rice varieties resistant to the rice water weevil, the major pest of rice in the United States and an important invasive pest in parts of Asia and Europe. Larvae of this species feed on roots of rice plants, thereby reducing above-ground vegetative growth, tillering, and allocation to grain. A screening programme in place from approximately 1960 to 2000 evaluated over 8000 rice lines for rice water weevil resistance, but no lines possessing high levels of resistance were found. More recent evaluations of commercial varieties and plant introductions also failed to find highly resistant genotypes, although the cultivar 'Jefferson' and several plant introductions show low to moderate resistance (Stout and Riggio, 2003; Stout et al., 2001). Interestingly, results of recent greenhouse experiments suggest that average levels of resistance to the rice water weevil present in commercial varieties released by the Louisiana State University Agricultural Center have decreased over the past 100 years (Figure 1.2). The resistance present in 'Jefferson' appears to be compatible with shallow flooding, a cultural tactic for reducing rice water weevil infestations, and with use of seed treatment insecticides (M.J. Stout and S. Lanka, unpublished data). Crosses of 'Jefferson' with higher-yielding varieties have been made in an effort to develop weevil resistant varieties (Stout, unpublished data).

1.6.4 Induced Resistance in Rice

Induced resistance has not been investigated in rice to the extent it has in other model plants such as tomato, tobacco, and maize, but the pace of research is accelerating (see Karban and Chen, 2007, for a review of earlier

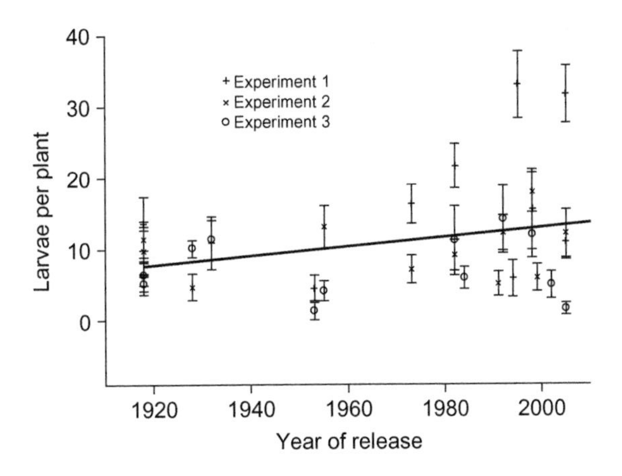

FIGURE 1.2 Relationship between year of variety release and resistance to rice water weevil in varieties released by the Louisiana Agricultural Experiment Station over the past 100 years. Resistance was measured in three greenhouse experiments (pooled for analysis) in which rice water weevil adults were given free access for oviposition to varieties released in different years (choice experiments). Varietal resistance was measured by counting the number of first instars emerging from plants over an approximately 2-week period following infestation. The positive relationship between year of variety release and number of larvae per plant was significant ($P<0.001$).

work). Brown planthopper and caterpillar (*Spodoptera frugiperda*) feeding induce emission of volatiles attractive to various parasitoids, with linalool and caryophyllene among the most important (Xiao et al., 2012; Yuan et al., 2008). Silencing linalool and caryophyllene emission had effects on communities of arthropods in the field as described above (Xiao et al., 2012). Direct induced resistance has also been studied in rice, but to a lesser extent (Stout et al., 2009). The JA- and SA-mediated hormonal pathways are involved in mediating responses to insects, although it is not certain that they play exactly the same roles as they do in dicots. Antisense expression of a key gene in the JA pathway in rice, OsHI-LOX, improved the performance of a stem-boring and leaf-folding species, but enhanced resistance to brown planthoppers (Zhou et al., 2009). In addition, an intriguing recent report showed that applications of the broadleaf herbicide 2,4-dichlorophenoxyacetic acid at low concentrations induced volatile emissions (once again, attractive to parasitoids), proteinase inhibitor activity, resistance to a stem borer, and susceptibility to the brown planthopper (Xin et al., 2012). As with plant resistance in general, understanding the mechanisms of inducible rice resistance to arthropods is probably the key to rapid development of resistant genotypes.

1.7 CONCLUSIONS

Plant resistance is in many ways an ideal management tactic: its effects on the target pest are constant and cumulative, it is usually simple and inexpensive for farmers to implement, it is usually compatible with other tactics, and it does not have the negative environmental impacts of other management tactics. With the need for increased crop production and the growing awareness of the importance of sustainable production practices, it is reasonable to anticipate a greater role for arthropod-resistant varieties in pest management programmes in the future. The traditional approach to host-plant resistance pioneered largely by Painter (1951) has been extremely successful. However, there are limitations to this approach, including barriers to breeding complex, polygenic resistance into agronomically acceptable backgrounds, a conceptual framework that may not encompass the diversity of resistance mechanisms found in plants, and an empirical approach that has historically eschewed mechanistic investigations of plant resistance. The refinement of analytical and molecular genetic tools for characterizing and manipulating plant resistance has profoundly increased the ability to elucidate the causal bases of plant resistance. In the future, elucidation of mechanisms of constitutive and inducible plant resistance will facilitate the identification of novel targets for traditional breeding, marker-assisted selection, and genetic engineering. Moreover, a greater understanding of the natural role and regulation of inducible resistance should lead to novel methods for manipulating resistance phenotypes in crop plants.

Acknowledgments

I thank Dr J. Davis and Dr R. Korada for commenting on an earlier draft of the manuscript. Approved by the Director, Louisiana Agricultural Experiment Station, manuscript number 2012-234-7790.

References

Adkisson, P.L., Dyck, V.A., 1980. Resistant varieties in pest management systems. In: Maxwell, F.G., Jennings, P.R. (Eds.), Breeding Plants Resistant to Insects. John Wiley & Sons, New York, pp. 233–252.

Akbar, W., Ottea, J.A., Beuzelin, J.M., Reagan, T.E., Huang, F., 2008. Selection and life history traits of tebufenozide-resistant sugarcane borer (Lepidoptera: Crambidae). J. Econ. Entomol. 101, 1903–1910.

Alam, S.N., Cohen, M., 1998. Durability of brown planthopper, *Nilaparvata lugens*, resistance in rice variety IR64 in greenhouse selection. Entomol. Exp. Appl. 89, 71–78.

Balkema-Boomstra, A.G., Zijlstra, S., Verstappen, F.W.A., Inggamer, H., Mercke, P.E., Jongsma, M.A., et al., 2003. Role of cucurbitacin C in resistance to spider mite (*Tetranychus urticae*) in cucumber (*Cucumis sativus* L.). J. Chem. Ecol. 29, 225–235.

Barry, B.D., Wiseman, B.R., Davis, F.M., Mihm, J.A., Overman, J.L., 1999. Benefits of insect-resistant maize. In: Wiseman, B.R., Webster, J.A. (Eds.), Economic, Environmental, and Social Benefits of Resistance in Field Crops. Thomas Say Publications in Entomology: Proceedings. Entomological Society of America, Lanham, MD.

Bartlett, R., 2008. Negative interactions between chemical resistance and predators affect fitness in soybeans. Ecol. Entomol. 33, 673–678.

Benheim, D., Rochfort, S., Robertson, E., Potter, I.D., Powell, K.S., 2012. Grape phylloxera (*Daktulosphaira vitifoliae*)–a review of potential detection and alternative management options. Ann. Appl. Biol. 161, 91–115.

Berenbaum, M.R., Zangerl, A.R., 2008. Facing the future of plant-insect interactions research: le retour à la 'Raison d' Être'. Plant Physiol. 146, 804–811.

Berzonsky, W.O., Ding, H., Haley, S.D., Harris, M.O., Lamb, R.J., McKenzie, R.I.H., et al., 2003. Breeding wheat for resistance to insects. Plant Breed. Rev. 22, 221–296.

Black, C.A., Karban, R., Godfrey, L.D., Granett, J., Chaney, W., 2003. Jasmonic acid: a vaccine against leafminers (Diptera: Agromyzidae) in celery. Environ. Entomol. 32, 1196–1202.

Boethel, D.J., 1999. Assessment of soybean germplasm for multiple insect resistance. In: Clement, S.L., Quisenberry, S.S. (Eds.), Global Plant Genetic Resources for Insect Resistant Crops. CRC Press, Boca Raton, FL, pp. 101–130.

Bottrell, D.G., Schoenly, K.G., 2012. Resurrecting the ghost of green revolutions past: the brown planthopper as a recurring threat to highly-yielding rice production in tropical Asia. J. Asia-Pacific Entomol. 15, 122–140.

Buntin, G.D., Raymer, P.L., 1989. Hessian Fly (Diptera: Cecidomyiidae) damage and forage production of winter wheat. J. Econ. Entomol. 82, 301–306.

Campbell, B.C., Duffey, S.S., 1979. Tomatine and parasitic wasps: potential incompatibility of plant antibiosis with biological control. Science 205, 700–702.

Chaudhary, R.C., Khush, G.S., Heinrichs, E.A., 1984. Varietal resistance to rice stem-borers in Asia. Insect Sci. Appl. 5, 447–463.

Clement, S.L., Quisenberry, S.S., 1999. Global Plant Genetic Resources for Insect-Resistant Crops. CRC Press, Boca Raton, FL.

Cohen, M.B., Alam, S.N., Medina, E.B., Bernal, C.C., 1997. Brown planthopper, *Nilaparvata lugens*, resistance in rice cultivar IR64: mechanism and role in successful *N. lugens* management in Central Luzon, Philippines. Entomol. Exp. Appl. 85, 221–229.

Cuong, N.L., Ben, P.T., Phuong, L.T., Chau, L.M., Cohen, M.B., 1997. Effect of host plant resistance and insecticide on brown planthopper *Nilparvata lugens* (Stål) and predator population development in the Mekong Delta, Vietnam. Crop Protect. 16, 707–715.

Das, Y.T., 1976. Cross resistance to stemborers in rice varieties. J. Econ. Entomol. 69, 41–46.

de Vleesschauwer, D., Yang, Y., Vera Cruz, C., Hofte, M., 2010. Abscisic acid-induced resistance against the brown spot pathogen *Cochliobolus miyabeanus* in rice involves MAP kinase-mediated repression of ethylene signaling. Plant Physiol. 152, 2036–2052.

Dervinis, C., Frost, C.J., Lawrence, S.D., Novak, N.G., Davis, J.M., 2010. Cytokinin primes plant responses to wounding and reduces insect performance. J. Plant Growth Regul. 29, 289–296.

Du, B., Zhang, W., Liu, B., Hu, J., Wei, Z., Shi, Z., et al., 2009. Identification and characterization of *Bph14*, a gene conferring resistance to brown planthopper in rice. Proc. Natl. Acad. Sci. U. S. A. 16, 22163–22168.

Duffey, S.S., Stout, M.J., 1996. Antinutritive and toxic components of plant defense against insects. Arch. Insect Biochem. Physiol. 32, 3–37.

El-Wakeil, N.E., Volkmar, C., Sallam, A.A., 2010. Jasmonic acid induces resistance to economically important insect pests in winter wheat. Pest Manage. Sci. 66, 549–554.

Farmer, E.E., Dubugnon, L., 2009. Detritivorous crustaceans become herbivores on jasmonate-deficient plants. Proc. Natl. Acad. Sci. U. S. A. 106, 935–940.

Flanders, K.L., Hawkes, J.G., Radcliffe, E.B., Lauer, F.I., 1992. Insect resistance in potatoes: sources, evolutionary relationships, morphological and chemical defenses, and ecogeographical associations. Euphytica 61, 83–111.

Franzmann, B.A., Hardy, A.T., Murray, D.A.H., Henzell, R.G., 2008. Host-plant resistance and biopesticides: ingredients for successful integrated pest management (IPM) in Australian sorghum production. J. Exp. Agric. 48, 1594–1600.

Hamm, J.C., Riggio, M.R., Stout, M.J., 2010. Herbivore- and elicitor-induced resistance in rice (*Oryza sativa*) to the rice water weevil (*Lissorhoptrus oryzophilus* Kuschel) in the laboratory and field. J. Chem. Ecol. 36, 192–199.

Hao, P., Liu, C., Wang, Y., Chen, R., Tang, M., Du, B., et al., 2008. Herbivore- induced callose deposition on the sieve plates of rice: an important mechanism for host resistance. Plant Physiol. 146, 1810–1820.

Heinrichs, E.A., 2009. Management of rice insect pests. In: Radcliffe, E.B., Hutchison, W.D. (Eds.), Radcliffe's IPM World Textbook. University of Minnesota, St. Paul, MN. URL: <http://ipmworld.umn.edu> (accessed 12.12.12).

Horber, E., 1980. Types and classification of resistance. In: Maxwell, F.G., Jennings, P.R. (Eds.), Breeding Plants Resistant to Insects. John Wiley & Sons, New York, pp. 15–21.

Horgan, F., 2009. Mechanisms of resistance: a major gap in understanding planthopper-rice interactions. In: Heong, K.L., Hardy, B. (Eds.), Planthoppers: New Threats to the Sustainability of Intensive Rice Production Systems in Asia. International Rice Research Institute, Manila, Philippines, pp. 281–302.

Howe, G.A., Jander, G., 2008. Plant immunity to insect herbivores. Annu. Rev. Plant Biol. 59, 46–66.

Hutchison, W.D., Burkness, E.C., Mitchell, P.D., Moon, R.D., Leslie, T.W., Fleischer, S.J., et al., 2010. Areawide suppression of European corn borer with Bt maize reaps savings to non-Bt maize growers. Science 330, 222–225.

Jairin, J., Teangdeerith, S., Leelagud, P., Kothcharerk, J., Sansen, K., Yi, M., et al., 2009. Development of rice introgression lines with brown planthopper resistance and KDML105 grain quality characteristics through marker-assisted selection. Field Crops Res. 110, 263–271.

Kaloshian, I., 2004. Gene-for-gene disease resistance: bridging insect pest and pathogen defense. J. Chem. Ecol. 30, 2419–2438.

Karban, R., Chen, Y., 2007. Induced resistance in rice against insects. Bull. Entomol. Res. 97, 327–335.

Kennedy, G.G., Barbour, J.D., 1992. Resistance variation in natural and managed systems. In: Fritz, R.S., Simms, E.L. (Eds.), Plant Resistance to Herbivores and Pathogens: Ecology, Evolution, and Genetics. The University of Chicago Press, Chicago, pp. 13–41.

Kessler, A., Halitscke, R., Baldwin, I.T., 2004. Silencing the jasmonate cascade: induced plant defenses and insect populations. Science 305, 665–668.

Khush, G.S., 1989. Multiple disease and insect resistance for increased yield stability in rice. Progress in Irrigated Rice Research. International Rice Research Institute, Manila, Philippines, pp. 79–92.

Kim, J., Quaghebeur, H., Felton, G.W., 2011. Reiterative and interruptive signaling in induced plant resistance to chewing insects. Phytochemistry 72, 1624–1634.

Kogan, M., 1986. Plant defense strategies and host-plant resistance. In: Kogan, M. (Ed.), Ecological Theory and Integrated Pest Management Practice. John Wiley and Sons, New York, pp. 83–134.

Kogan, M., Ortman, E., 1978. Antixenosis–a term proposed to define painter's 'Nonpreference' modality of resistance. Bull. Entomol. Soc. Am. 24, 175–176.

McCreight, J.D., Liu, Y.B., 2012. Resistance to lettuce aphid (*Nasonovia ribisnigri*) biotype 0 in Wild Lettuce Accessions PI 49103 and PI 2743378. HortScience 47, 179–184.

Meszaros, A., Beuzelin, J.M., Stout, M.J., Bommireddy, P.L., Riggio, M.R., Leonard., B.R., 2011. Jasmonic acid-induced resistance to the fall armyworm, Spodoptera frugiperda, in conventional and transgenic cottons expressing Bacillus thuringiensis insecticidal proteins. Entomol. Exp. Appl. 140, 226–237.

Nahar, K., Kyndt, T., de Vleesschauwer, D., Hofte, M., Gheysen, G., 2011. The jasmonate pathway is a key player in systemically induced defense against root knot nematodes in rice. Plant Physiol. 157, 305–316.

Omer, A.D., Grannett, J., Karban, R., Villa, M., 2001. Chemically-induced resistance against multiple pests in cotton. Int. J. Pest Manage. 47, 49–54.

Ortega, A., Vasal, S.K., Mihm, J., Hershey, C., 1980. Breeding for insect resistance in maize. In: Maxwell, F.G., Jennings, P.R. (Eds.), Breeding Plants Resistant to Insects. John Wiley and Sons, New York, pp. 371–420.

Painter, R.H., 1951. Insect Resistance in Crop Plants. The University Press of Kansas, Lawrence.

Pathak, M.D., 1969. Stem borer and leafhopper-planthopper resistance in rice varieties. Entomol. Exp. Appl. 12, 789–800.

Quisenberry, S.S., Clement, S.L. (Eds.),, 1999. Global Plant Genetic Resources for Insect-Resistant Crops. CRC Press, Boca Raton, FL.

Quisenberry, S.S., Schotzko, D.J., 1994. Integration of plant resistance with pest management methods in crop production systems. J. Agric. Entomol. 11, 279–290.

Rasmann, S., Agrawal, A.A., 2009. Plant defense against herbivory: progress in identifying synergism, redundancy, and antagonism between resistance. Curr. Opin. Plant Biol. 12, 1–6.

Rasmann, S., Kollner, T.G., Degenhardt, J., Hiltpold, I., Toefper, S., Kuhlman, U., et al., 2005. Recruitment of entomopathogenic nematodes by insect-damaged maize roots. Nature 434, 732–737.

Sarmento, R.A., Lemos, F., Bleeker, P.M., Schuurink, R.C., Pallini, A., Oliveira, M.G.A., et al., 2011. A herbivore that manipulates plant defense. Ecol. Lett. 14, 229–236.

Schoonhoven, L.M., Jermy, T., van Loon, J.J.A., 1998. Insect-Plant Biology: From Physiology to Evolution. Chapman & Hall, London.

Schwachtje, J., Minchin, P.E.H., Jahnke, S., van Dongen, J.T., Schittko, U., Baldwin, I.T., 2006. Snf1-related kinases allow plants to tolerate herbivory by allocating carbon to roots. Proc. Natl. Acad. Sci. U. S. A. 103, 12935–12940.

Shivaji, R., Camas, A., Ankala, A., Engelberth, J., Tumlinson, J.H., Williams, W.P., et al., 2010. Plants on constant alert: elevated levels of jasmonic acid and jasmonate-induced transcripts in caterpillar-resistant maize. J. Chem. Ecol. 36, 179–191.

Simmons, A.T., Gurr, G.M., 2005. Trichomes of *Lycopersicon* species and their hybrids: effects on pests and natural enemies. Agric. For. Entomol. 8, 1–11.

Smith, C.M., 2005. Plant Resistance to Arthropods: Molecular and Conventional Approaches. Springer, Dordrecht, The Netherlands, 423 pp.

Smith, C.M., Clement, S.L., 2012. Molecular bases of plant resistance to arthropods. Annu. Rev. Entomol. 57, 309–328.

Smith, C.M., Khan, Z.R., Pathak, M.D., 1994. Techniques for Evaluating Insect Resistance in Crop Plants. CRC Press, Boca Raton, FL, 320 pp.

Stout, M., Davis., J., 2009. Keys to the increased use of host-plant resistance in integrated pest management. In: Peshin, R., Dhawan, A.K. (Eds.), Integrated Pest Management: Innovation-Development Process. Springer, pp. 163–182.

Stout, M.J., 2012. Rice insects: ecology and control. In: Pimentel, D. (Ed.), Encyclopedia of IPM. Taylor and Francis doi: 10.1081/E-EM-120042889, (last accessed 17.12.12).

Stout, M.J., 2013. Reassessing the conceptual framework for applied host plant resistance research. Insect Sci. 20, 263–272.

Stout, M.J., Riggio, M.R., 2003. Variation in susceptibility of rice lines to infestation by the rice water weevil (Coleoptera: Curculionidae). J. Agric. Urban Entomol. 19, 205–216.

Stout, M.J., Rice, W.C., Linscombe, S.D., Bollich, P.K., 2001. Identification of rice cultivars resistant to *Lissorhoptrus oryzophilus* (Coleoptera: Curculioniade), and their use in an integrated management program. J. Econ. Entomol. 94, 963–970.

Stout, M.J., Zehnder, G.W., Baur, M.E., 2002. Potential for the use of elicitors of plant resistance in arthropod management programs. Arch. Insect Biochem. Physiol. 51, 222–235.

Stout, M.J., Thaler, J.S., Thomma, B.P.H.J., 2006. Plant-mediated interactions between pathogenic microorganisms and arthropod herbivores. Annu. Rev. Entomol. 51, 663–689.

Stout, M.J., Riggio, M.R., Yang, Y., 2009. Direct induced resistance in rice, *Oryza sativa*, to the fall armyworm, *Spodoptera frugiperda*. Environ. Entomol. 38, 1174–1181.

Strauss, S.Y., Rudgers, J.A., Lau, J.A., Irwin, R.E., 2002. Direct and ecological costs of resistance to herbivory. Trends Ecol. Evol. 17, 278–285.

Stuart, J.J., Ming-Shun, C., Shukle, R., Harris, M.O., 2012. Gall midges (Hessian flies) as plant pathogens. Annu. Rev. Phytopathol. 50, 339–357.

Tamiru, A., Bruce, T.J.A., Woodcock, C.M., Caulfield, J.C., Midega, C.A.O., Ogol, C.K.P.O., et al., 2011. Maize landraces recruit egg and larval parasitoids in response to egg deposition by a herbivore. Ecol. Lett. 14, 1075–1083.

Thaler, J.S., Bostock, R.M., 2004. Interactions between abscisic-acid-mediated responses and plant resistance to pathogens and insects. Ecology 85, 48–58.

Thaler, J.S., Stout, M.J., Karban, R., Duffey, S.S., 2001. Jasmonate-mediated induced plant resistance affects a community of herbivores. Ecol. Entomol. 26, 312–324.

Traw, M.B., Bergelson, J., 2003. Interactive effects of jasmonic acid, salicylic acid, and gibberellin on induction of trichomes in *Arabidopsis*. Plant Physiol. 133, 1367–1375.

van Emden, H.V., 2002. Mechanisms of resistance: antibiosis, antixenosis, tolerance, nutrition. In: Pimentel, D. (Ed.), Encyclopedia of Pest Management. CRC Press, Boca Raton, FL, pp. 483–486.

van Eck, L., Schultz, T., Leach, J.E., Scofield, S.R., Peairs, F.B., Botha, A., et al., 2010. Virus-induced gene silencing of WRKY53 and an inducible phenylalanine ammonia-lyase in wheat reduces aphid resistance. Plant Biotechnol. J. 8, 1023–1032.

Walters, D., Heil, M., 2007. Costs and trade-offs associated with induced resistance. Physiol. Mol. Plant Pathol. 71, 3–17.

Walters, D.R., Fountaine, J.M., 2009. Practical application of induced resistance to plant diseases: an appraisal of effectiveness under field conditions. J. Agric. Sci. 147, 523–535.

Wang, Y., Wang, X., Yuan, H., Chen, R., Zhu, L., He, R., et al., 2008. Responses of two contrasting genotypes of rice to brown planthopper. Mol. Plant Microbe Interact. 21, 122–132.

Wearing, C.H., Colhoun, K., McLaren, G.F., Attfield, B., Bus, V.G.M., 2003. Evidence for single gene resistance leafroller, *Ctenopseustis obliquana* and implications for resistance to other New Zealand leafrollers. Entomol. Exp. Appl. 108, 1–10.

Wilde, G., 2002. Arthopod host plant resistant crops. In: Pimentel, D. (Ed.), Encyclopedia of Pest Management. CRC Press, Boca Raton, FL, pp. 33–35.

Willcox, M.C., Khairallah, M.M., Bergvinson, D., Crossa, J., Deutsch, J.A., Edmeades, G.O., et al., 2002. Selection for resistance to southwestern corn borer using marker-assisted and conventional backcrossing. Crop Sci. 42, 1516–1528.

Wiseman, B.R., 1994. Plant resistance to insects in integrated pest management. Plant Dis. 78, 927–932.

Wiseman, B.R., 1999. Successes in plant resistance to insects. In: Wiseman, B.R., Webster, J.A. (Eds.), Economic, Environmental, and Social Benefits of Resistance in Field Crops. Thomas Say Publications in Entomology, Entomological Society of America, Lanham, MD, pp. 3–16.

Worrall, D., Holroyd, G.H., Moore, J.P., Glowacz., M., Croft, P., Taylor, J.E., et al., 2012. Treating seeds with activators of plant defence generates long-lasting priming of resistance to pests and pathogens. New Phytol. 193, 770–778.

Wu, J., Baldwin, I.T., 2010. New insights into plant responses to the attack from insect herbivores. Annu. Rev. Genet. 44, 1–24.

Xiao, Y., Wang, Q., Erb, M., Turlings, T.C.J., Ge, L., Hu, J., et al., 2012. Specific herbivore-induced volatiles defend plants and determine insect community composition in the field. Ecol. Lett. 15, 1130–1139.

Xin, Z., Yu, Z., Erb, M., Turlings, T.C.J., Wang, B., Qi, J., et al., 2012. The broad-leaf herbicide 2,4-dichlorophenoxyacetic acid turns rice into a living trap for a major insect pest and a parasitic wasp. New Phytol. 194, 498–510.

Yuan, J.S., Köllner, T.G., Wiggins, G., Grant, J., Degenhardt, J., Chen, F., 2008. Molecular and genomic basis of volatile-mediated indirect defense against insects in rice. Plant J. 55, 491–503.

Zheng, S.-J., Dicke, M., 2008. Ecological genomics of plant-insect interactions: from gene to community. Plant Physiol. 146, 812–817.

Zhou, G., Qi, J., Ren, N., Cheng, J., Erb, M., Mao, B., et al., 2009. Silencing *OsHI-LOX* makes rice more susceptible to chewing herbivores, but enhances resistance to a phloem feeder. Plant J. 60, 638–648.

CHAPTER

2

Impact of Climate Change on Pest Management and Food Security

Hari C. Sharma and Chandra S. Prabhakar

International Crops Research Institute for the Semi-Arid Tropics (ICRISAT), Andhra Pradesh, India

2.1 INTRODUCTION

The last decade of the 20th century and the first decade of the 21st century have been the warmest periods in the entire global temperature record. The Intergovernmental Panel on Climate Change (IPCC, 2001) has concluded that most of the global warming observed over the last 50 years is attributable to human activities. Climate change, as described by IPCC, refers to '*a change in the state of the climate that can be identified (by using statistical tests) by changes in the mean and/or the variability of its properties that persists for an extended period, typically for decades or longer. It refers to any change in climate over time, whether due to natural variability or as a result of human activity*' (IPCC, 2007). The rise in global climate temperature is mostly due to increased concentrations of greenhouse gases, which include carbon dioxide (CO_2), methane (CH_4), nitrous oxide (N_2O) and chlorofluorocarbons (CFCs). Over the past 200 years, the atmospheric concentration of carbon dioxide has increased by 35%, and is expected to double by the end of this century, i.e. 280 ppm in the preindustrial era vs. 360 ppm at present (Houghton et al., 1995). The global mean surface temperature rose by 0.6 ± 0.2°C during 20th century, and climatic models have predicted an average increase in global temperature of 1.8°C to 4°C over the next 100 years (Collins et al., 2007; Johansen, 2002; Karl and Trenbeth, 2003). The IPCC suggested that if temperatures rise by about 2°C over the next 100 years, negative effects of global warming would begin to extend to most regions of the world, and affect most of the living organisms including humans and plants. Climatic variables such as temperature, rainfall, humidity, and atmospheric gases interact with plants in numerous ways with diverse mechanisms. These changes are affecting plants directly in terms of tissue and organ-specific photosynthetic allocation, and indirectly through change in geographic distribution and population dynamics of the pest species. Experiments have indicated that higher levels of CO_2 generally increase productivity of crop plants (Fuhrer, 2003; Long et al., 2004), as elevated CO_2 increases the photosynthetic rates (Drake et al., 1997; Norby et al., 1999) and

D. P. Abrol (Ed): Integrated Pest Management.
DOI: http://dx.doi.org/10.1016/B978-0-12-398529-3.00003-8

biomass production (Curtis and Wang, 1998; Ledley et al., 1999). However, increase in crop production may be offset through high temperatures and reduced water availability. Global warming and climate changes are having a negative impact on the productivity of cereals and other crops (Anwar et al., 2007; Challinor et al., 2005; Choudhary et al., 2012; Torriani et al., 2007). Increased temperature will cause insect pests to become more abundant (Bale et al., 2002; Cannon, 1998; Patterson et al., 1999) and almost all insects will be affected by changes in temperature. Porter et al. (1991) listed various effects of temperature on insects, including: limitation of geographical range, overwintering, population growth rates, number of generations per annum, crop–pest synchronization, dispersal and migration, and availability of host plants and refugia. Laboratory and modelling experiments with increased temperature support the perception that the biology of agricultural pests is likely to be affected by global warming (Cammell and Knight, 1992; Fleming and Volney, 1995; Fye and McAda, 1972). For example, warming could decrease the occurrence of severe cold events (Diffenbaugh et al., 2005), which in turn might expand the overwintering area for insect pests (Patterson et al., 1999). In-season effects of warming include the potential for increased levels of feeding and growth, including the possibility of additional generations in a given year (Cannon, 1998). This will alter the crop yield, and also influence the effectiveness of insect-pest management practices. Increased global temperature will also influence the phenology of insects including early arrival of insect pests in their agricultural habitats and emergence time of a range of insect pests (Dewar and Watt, 1992; Whittaker and Tribe, 1996, 1998). This will require early and more frequent application of insecticides to reduce the pest damage. Increased temperatures will also increase the pest population, and water stressed plants at times may result in increased insect populations and pest outbreaks. This will affect the crop yield and availability of food grains and threaten food security. Temperature increases associated with climatic changes could result in:

- change in geographical range of insect pests,
- increased overwintering and rapid population growth,
- changes in insect–host plant–natural enemy interactions,
- impact on arthropod diversity and extinction of species,
- changes in synchrony between insect pests and their crop hosts,
- introduction of alternative hosts as green bridges,
- changes in relative abundance and effectiveness of biocontrol agents,
- change in expression of resistance to insects in cultivars with temperature-sensitive genes,
- emergence of new pest problems and increased risk of invasion by migrant pests, and
- reduced efficacy of different components of insect-pest management.

These changes will have major implications for crop protection and food security, particularly in the developing countries, where the need to increase and sustain food production is most urgent. Long-term monitoring of population levels and insect behaviour, particularly in identifiably sensitive regions, may provide some of the first indications of a biological response to climate change. The impact of climate change will vary across regions, crops and species. A large number of models and protocols have been designed to measure the effects of climate change for different species and in different disciplines. There is a need for interdisciplinary cooperation to measure the effects of climate change on the environment and food security. It will be important to keep ahead of undesirable pest adaptations, and consider global warming and climate change for planning research and development efforts for integrated pest management (IPM) in the future.

2.2 IMPACT OF CLIMATE CHANGE ON GEOGRAPHIC DISTRIBUTION AND POPULATION DYNAMICS OF INSECT PESTS

Present and future change in climate will have a significant bearing on the biology and behaviour of insects as insects are poikilothermic (cold-blooded) organisms, and are particularly sensitive to temperature changes. This will change the distribution and severity of infestation of crops through direct effects on the life cycle of insects, and indirectly through climatic effects on hosts, natural enemies, competitors, and insect pathogens (Cammell and Knight, 1992; Dobzhansky, 1965; Fye and McAda, 1972; Harrington and Stork, 1995; Kingsolver, 1989; Mattson and Haack, 1987; Tauber et al., 1986). Low temperatures are often more important than high temperatures in determining the geographical distribution of insect pests (Hill, 1987). Increasing temperatures may result in a greater ability to overwinter in insect species limited by low temperatures at higher latitudes (EPA, 1989; Hill and Dymock, 1989). Recent reports have indicated that the distribution of insects is intensifying at high latitudes and high elevations (Anderson et al., 2008; Hickling et al., 2006; Parmesan and Yohe, 2003; Parmesan et al., 1999; Warren et al., 2001) and diminishing at their low latitudes and low elevations and high-temperature margins (Anderson et al., 2008; Franco et al., 2006; Parmesan, 1996; Wilson et al., 2007). Insect species richness is increasing in cool habitats (Andrew and Hughes, 2005a,b). Butterfly species in the UK are decreasing most rapidly in the south, while species with a southerly distribution are expanding northwards (Breed et al., 2013; Conrad et al., 2004). There is also some evidence that the risk of crop loss will increase due to pole-ward and high-elevation expansion of insect geographical ranges (Bjorkman et al., 2011; Wolf et al., 2008). For all of the insect species, higher temperatures, below the species' upper threshold limit, will result in faster development, resulting in rapid increase in pest populations as the time to reproductive maturity is reduced, and species characterized by high reproduction rates being generally favoured (Southwood and Comins, 1976). Temperature limits geographical range, overwintering, population growth rates, length of crop growing season, crop-pest synchronization, interspecific interactions, dispersal and migration and availability of host plants (Porter et al., 1991). Spatial shifts in the distribution of crops will also influence the distribution of insect pests (Parry and Carter, 1989). However, whether or not an insect species would move with a crop into the new habitats will also depend on the presence of overwintering sites, soil type, and moisture; e.g. corn earworm, *Heliothis zea* (Boddie) might move to higher latitudes/altitudes in North America, leading to greater damage in maize and other crops (EPA, 1989).

Global warming will lead to earlier infestation by *H. zea* in North America (EPA, 1989), and *Helicoverpa armigera* (Hubner) in North India (Sharma, 2010), resulting in increased crop loss. Rising temperatures are likely to result in availability of new niches for insect pests. Temperature has a strong influence on the viability and incubation period of *H. armigera* eggs (Dhillon and Sharma, 2007). Egg incubation period can be predicted based on egg age and storage temperature, and the degree-days required for egg hatching decreased with an increase in temperature from 10 to 27°C, and egg age from 0 to 3 days (Dhillon and Sharma, 2007). An increase of 3°C in mean daily temperature would cause the carrot fly, *Delia radicum*

(L.), to become active a month earlier than at present (Collier et al., 1991), and temperature increases of 5 to 10°C would result in completion of four generations each year, necessitating adoption of new pest control strategies. An increase of 2°C will reduce the generation turnover of the bird cherry aphid, *Rhopalosiphum padi* (L.), by varying levels, depending on the changes in mean temperature (Morgan, 1996). An increase of 1 and 3°C in temperature will cause northward shifts in the potential distribution of the European corn borer, *Ostrinia nubilalis* (Hubner), of up to 1220 km, with an additional generation in nearly all of the regions (Porter et al., 1991). Cottony cushion scale, *Icerya purchasi* Maskell, populations appear to be spreading northwards perhaps as a consequence of global warming; and cottony camellia scale, *Chloropulvinaria floccifera* (West.), has become much more common in the UK, extending its range northwards, and increasing its host range in the last decade in response to climate change. In Sweden, this species was previously only known as a greenhouse species, but is now established as an outdoor species. Warming will allow the cold intolerant pink bollworm, *Pectinophora gossypiella* (Saunders), to expand its range on cotton into formerly inhospitable areas affected by heavy frosts, and damage rates will increase throughout its current range (Gutierrez et al., 2006, 2008). The survival of palm thrips, *Thrips palmi* Karny, is currently limited in the UK due to lack of cold tolerance, but this species may spread to other area in future (McDonald et al., 2000). Fruit flies, *Bactrocera tryoni* (Froggatt), *Bactrocera cucurbitae* (Coquillett) and *Bactrocera latifrons* (Hendel), may be spread into colder areas due to increasing temperature (Prabhakar et al., 2012a,b; Sutherst, 1991; Sutherst et al., 2007). The increased movements of warm air towards high latitudes have caused recent arrivals of diamondback moth, *Plutella xylostella* (L.), on the Norwegian islands of Svalbard in the Arctic Ocean, 800 km north of the edge of its current distribution in the western Russian Federation (Coulson et al., 2002). For a 3°C temperature increase in Japan, Mochida (1991) predicted expanded ranges for tobacco cutworm, *Spodoptera litura* (F.), southern green stink bug, *Nezara viridula* (L.), rice stink bug, *Lagynotomus elongatus* (Dallas), Lima-bean pod borer, *Etiella zinckenella* (Treitschke), common green stink bug, *Nezara antennata* Scott, soybean stem gall midge, *Asphondylia* sp., rice weevil, *Sitophilus oryzae* (L.), and soybean pod borer, *Leguminivora glycinivorella* (Matsumura), but a decreased range for rice leaf beetle, *Oulema oryzae* (Kuwayama), and rice leaf miner, *Agromyza oryzae* (Manukata).

Overwintering of insect pests will increase as a result of climate change, producing larger spring populations as a base for build-up in numbers in the following season. These may be vulnerable to parasitoids and predators if the latter also overwinter more readily. Diamondback moth, *P. xylostella*, overwintered in Alberta (Dosdall, 1994), and if overwintering becomes common, the status of this insect pest will increase dramatically. There will also be increased dispersal of airborne insect species in response to atmospheric disturbances. Many insect species such as *H. armigera* and *H. zea* are migratory, and, therefore, may be well adapted to exploit new opportunities by moving into new areas as a result of climate change (Sharma, 2005).

The effects of precipitation vary with the species as some insects are sensitive to precipitation and are killed or removed from crops by heavy rains, e.g. onion thrips (Reiners and Petzoldt, 2005), cranberry fruit worm and other cranberry insect pests (Vincent et al., 2003). Precipitation has a positive effect on pea aphid (McVean et al., 1999). However, under elevated CO_2 and O_3 in the future, some of the insects may be unaffected as there was no effect on development time, adult weight, embryo number and the weight of nymphs of the aphid, *Cepegillettea betulaefoliae*

Granovsky, feeding on paper birch (Awmack et al., 2004).

2.3 EFFECT OF CLIMATE CHANGE ON THE EFFECTIVENESS OF PEST MANAGEMENT TECHNOLOGIES

2.3.1 Expression of Resistance to Insect Pests

Host-plant resistance to insects is one of the most environmentally friendly components of pest management. However, climate change may alter the interactions between insect pests and their host plants (Sharma et al., 2010). Resistance to sorghum midge, *Stenodiplosis sorghicola* (Coq.), observed in India, breaks down under high humidity and moderate temperatures in Kenya (Sharma et al., 1999). Sorghum midge damage in the midge-resistant lines ICSV 197, TAM 2566 and AF 28 decreased with an increase in open pan evaporation, maximum and minimum temperatures, and solar radiation, while no significant effect was observed on the susceptible cultivars ICSV 112 and CSH 5 (Sharma et al., 2003). There will be an increased impact on insect pests which benefit from reduced host defences as a result of the stress caused by the lack of adaptation to suboptimal climatic conditions. Some plants can change their chemical composition in direct response to insect damage to make their tissues less suitable for growth and survival of insect pests (Sharma, 2002).

Generally, CO_2 impacts on insects are thought to be indirect. Impact on insect damage will result from changes in nutritional quality and secondary metabolites of the host plants. Increased levels of CO_2 will enhance plant growth, but may also increase the damage caused by some phytophagous insects. In the enriched CO_2 atmosphere expected in the 21st century, many species of herbivorous insects will confront less nutritious host plants that will induce both lengthened larval developmental times and greater mortality (Coviella and Trumble, 1999). The effects of climate change on the magnitude of herbivory and direction of response will not only be species-specific, but also specific to each insect–plant system. Bark beetles, wood borers, and sap sucking insects benefit from severe drought (Bjorkman and Larsson, 1999; Huberty and Denno, 2004; Koricheva et al., 1998), while *Spodoptera exigua* (Hub.) exhibited a reduced ability to feed on drought-stressed tomato leaf tissue, which contained higher levels of defence compounds as a result of the abiotic stress (English-Loeb et al., 1997). Severe drought increases the damage by insect species such as spotted stem borer, *Chilo partellus* (Swinhoe), in sorghum (Sharma et al., 2005) and litchi stink bug, *Tessaratoma javanica* (Thunberg), in litchi (Choudhary et al., 2013). However, the effect of drought on leaf miners, leaf defoliators, and gall makers is more uncertain (Jactel et al., 2012).

Although increased CO_2 tends to enhance plant growth rates, the greater effects of increased drought stress will probably result in slower plant growth (Coley and Markham, 1998). In atmospheres experimentally enriched with CO_2, the nutritional quality of leaves declined substantially due to a dilution of nitrogen by 10–30% (Coley and Markham, 1998). Increased CO_2 may also cause a slight decrease in nitrogen-based defences (e.g. alkaloids) and a slight increase in carbon-based defences (e.g. tannins). Lower foliar nitrogen due to CO_2 causes an increase in food consumption by herbivores. Soybeans grown in elevated CO_2 suffered 57% more damage from herbivores (primarily Japanese beetle, potato leafhopper, western corn rootworm and Mexican bean beetle) than those grown in ambient CO_2. Increase in amounts of simple sugars and down-regulation of gene expression for a protease-specific deterrent to coleopteran herbivores may have resulted in greater insect feeding (Hamilton et al., 2005; Zavala et al., 2008). Elevated CO_2

decreases the induction of jasmonic acid and ethylene related transcripts (*lox7*, *aos*, *hpl*, and *acc1*) in soybean plants causing decreased accumulation of defences (polyphenol oxidase, protease inhibitors, etc.) over time compared to plants grown under ambient conditions, suggesting that CO_2 exposure might have resulted in increased insect damage (Casteel, 2010). Problems with new insect pests will occur if climatic changes favour the introduction of non-resistant crops or cultivars into new areas. The introduction of new crops and cultivars could be one of the methods to take advantage of climate change (Parry, 1990; Parry and Carter, 1989).

2.3.2 Transgenic Crops for Pest Management

Transgenic cotton plants expressing the *Bacillus thuringiensis* (*Bt*) (Berliner) insecticidal protein showed a reduction in the level of toxin protein during periods of high temperature, elevated CO_2 levels, or drought, leading to decreased resistance to insect pests (Chen D.H. et al., 2005; Chen F.J. et al., 2005a; Dong and Li, 2007). Cotton bollworm, *Heliothis virescens* (F.), destroyed *Bt* cottons due to high temperatures in Texas, USA (Kaiser, 1996). Similarly, *H. armigera* and *H. punctigera* damaged *Bt*-cotton in the second half of the growing season in Australia because of reduced production of *Bt* toxins in the transgenic crops (Hilder and Boulter, 1999). Cry1Ac levels decrease with plant age, resulting in greater susceptibility of the crop to bollworms during the later stages of crop growth (Adamczyk et al., 2001; Greenplate et al., 2000; Kranthi et al., 2005; Sachs et al., 1998; Sharma, unpublished data). Possible causes for the failure of insect control may be due to inadequate production of the toxin protein, the effect of environment on transgene expression, locally resistant insect populations, and development of resistance due to inadequate management (Sharma and Ortiz, 2000). It is therefore important to understand the effects of climate change on the efficacy of transgenic plants for pest management.

2.3.3 Activity and Abundance of Natural Enemies

The majority of insects are benign to agroecosystems, and there is much evidence to suggest that this is due to population control through interspecific interactions among insect pests and their natural enemies – pathogens, parasites, and predators (Price, 1987). Increases in atmospheric CO_2, low precipitation and increases in temperature will alter plant phenology, influencing herbivore growth and abundance, and indirectly affecting the abundance of prey and insect hosts for natural enemies (Thomson et al., 2010). Relationships between insect pests and their natural enemies will change as a result of climate change, resulting in both increases and decreases in the status of individual pest species. Changes in temperature will also alter the timing of diurnal activity patterns of different groups of insects (Young, 1982), and changes in interspecific interactions could also alter the effectiveness of natural enemies for pest management (Hill and Dymock, 1989). The fitness of natural enemies will decline as the quality of their herbivore hosts decreases (Wang et al., 2007) as has been shown for several groups of predators including spiders (Hvam and Toft, 2005; Toft, 1995), predatory bugs (Butler and O'Neil, 2007) and carabid beetles (Bilde and Toft, 1999). However, a decrease in prey size will not necessarily always lead to a reduction in the success of predators. The number of prey consumed by predators might increase and lead to improved pest control (Chen F.J. et al., 2005b; Coll and Hughes, 2008). For example, the coccinellid predator, *Leis axyridis* Pallas, of cotton aphid,

Aphis gossypii Glover, consumed more prey under higher CO_2 (Chen F.J. et al., 2005b). The pentatomid bug, *Oechalia schellenbergii* Guerin-Meneville, exhibited increased predation of cotton bollworm, *H. armigera*, feeding on peas under elevated CO_2 because the pea plants had reduced nitrogen content when grown under high CO_2, which influenced the size of the cotton bollworm larvae (Coll and Hughes, 2008), whereas a negligible effect was observed in the interactions between *Harmonia axyridis* (Pallas) and its aphid host, *Sitobion avenae* F. (Chen et al., 2007). Quality of insect hosts may also affect parasitoid fitness (Wang et al., 2007), particularly in parasitoids whose hosts continue to feed after parasitization as fecundity of the parasitoid is positively correlated with size and host quality (Harvey et al., 1999). However, increased abundance of the braconid parasitoid, *Aphidius picipes* (Nees), was recorded on *Sitobion avenae* F. parasitism under elevated CO_2 compared to the insects raised under ambient CO_2 (Chen et al., 2007). The oriental armyworm, *Mythimna separata* (Walker), population increases during extended periods of drought (which is detrimental to the natural enemies), followed by heavy rainfall (Sharma et al., 2002). In cassava, parasitism of mealy bugs is reduced under conditions of water stress associated with drought due to improved immune response of mealy bugs on water stressed plants, leading to an increased rate of encapsulation (Calatayud et al., 2002).

Apart from surviving thermal extremes, natural enemies will also need to counter climate change by mating and locating hosts effectively across a wider range of thermal and humidity conditions. Even small changes in thermal conditions might influence the effectiveness of parasitoids in controlling insect pests. Temperatures up to 25°C will enhance the natural control of aphids by coccinellids (Freier and Triltsch, 1996). Temperature not only affects the rate of insect development, but also has a profound effect on fecundity, sex ratio and host location by the parasitoids (Dhillon and Sharma, 2008, 2009; Thomson et al., 2010). Host location of the egg parasitoid, *Trichogramma carverae* Oatman and Pinto, decreases sharply at temperatures above 35°C (Thomson et al., 2001), while fecundity reductions of up to 50% are commonly observed at temperatures >30°C (Naranjo, 1993; Scott et al., 1997). The interactions between insect pests and their natural enemies need to be studied carefully to devise appropriate methods for using natural enemies in pest management programmes under changed climate.

2.3.4 Biopesticides and Synthetic Insecticides

There will be an increased variability in insect damage as a result of climate change. Higher temperatures will make dry seasons drier, and conversely, may increase the amount and intensity of rainfall, making wet seasons wetter than at present. Current sensitivities on environmental pollution, human health hazards, and, pest resurgence are a consequence of improper use of synthetic insecticides. Natural plant products, entomopathogenic viruses, fungi, bacteria, nematodes, and synthetic pesticides are highly sensitive to the environment. Temperature is a major factor affecting insecticide toxicity (DeVries and Georghiou, 1979), and, thus, efficacy (Johnson, 1990; Scott, 1995). The effects of temperature on efficacy can be either positive or negative. The response relationship between temperature and efficacy has been found to vary depending on the mode of action of an insecticide, target species, method of application, and quantity of insecticide ingested or contacted (Johnson, 1990).

Increased temperature will increase the activity of some of the insecticides. Diflubenzuron (an insect growth regulator (IGR)) caused rapid mortality at higher temperatures and was more efficient at 35°C (Amarasekare and Edelson, 2004). This was probably because this IGR is only effective when the insect moults (Ware, 2000), and the insect growth rate and moulting rate increase at higher temperatures (Lactin and Johnson, 1995). However, the biological activity of the entomopathogenic fungus, *Beauveria bassiana* (Balsamo), is reduced at temperatures >25°C (Amarasekare and Edelson, 2004; Inglis et al., 1999).

Increase in temperature and UV radiation, and a decrease in relative humidity, may render many of the pest control tactics to be less effective, and such an effect will be more pronounced on natural plant products and the biopesticides. Entomopathogens used as biocontrol agents suffer from instability after exposure to solar radiation, especially in the ultraviolet (UV) portion of the spectrum (Bullock, 1967; Jaques, 1968; Morris, 1971; Timans, 1982). Several studies have reported a significant decrease in biological activity of entomopathogens, viz. NPV, GV, *Beauveria* and *Bt* (up to 90%) within a few days (Broome et al., 1974; David et al., 1968; Ignoffo et al., 1977; Jones and McKinley, 1986). Another effect of increased temperature and UV radiation may be to slow down the activity even without the loss of activity due to UV radiation; as a result, more time may be required to achieve insect mortality (Moscardi, 1999; Szewczyk et al., 2006). Larvae continue to feed and damage crops until shortly before death. Chen and McCarl (2001) estimated that pest treatment costs under the 2090 projections of climate exhibit increases of 3–10% for corn, soybeans, cotton and potatoes and mixed results for wheat, and show a $200 million per year projected loss to society due to climate change-related pesticide treatment cost effects in the USA. Therefore, there is a need to develop appropriate strategies for pest management that will be effective under situations of global warming in the future. Farmers will need a set of pest control strategies that can produce sustainable yields under climatic change.

2.4 CLIMATE CHANGE AND PEST MANAGEMENT: THE CHALLENGE AHEAD

The greatest challenge facing humanity in this century will be the necessity to double food production to meet the demands of droughts resulting from global warming, and the increasing population, by using less land area, less water, and less soil nutrients. The effects of climate change on pest control will be complex, particularly when new crops are adopted in new areas. As a result, the herbivores will escape the natural enemies, at least temporarily. This will have a major bearing on economic thresholds, as greater variability in climate will result in variable impact of pest damage on crop production. The relationship between the input costs and the resulting benefits will change as a result of changes in plant–insect–natural enemies–environment interactions. Increased temperatures and UV radiation, and low relative humidity, may render many of these control tactics less effective, and therefore, there is a need to: (i) study insect responses to climate change to predict and map the geographical distribution of insect pests and their natural enemies, and understand the metabolic alterations in insects in relation to climate change, (ii) investigate how climatic changes will affect development, incidence, and population dynamics of insect pests, (iii) have a fresh look at the existing economic threshold levels for each crop–pest interaction, as changed feeding habits or increased feeding under high CO_2 will change the economic threshold level for the pest, (iv) study changes in expression of resistance to insect pests and identify stable sources of resistance for use in crop improvement, (v) understand the effect of global warming on the efficacy of transgenic crops in pest management,

(vi) assess the efficacy of various pest management technologies under diverse environmental conditions, and (vii) develop appropriate strategies for pest management to mitigate the effects of climate change.

2.5 CONCLUSIONS

Climate change and global warming will have serious consequences for the diversity and abundance of arthropods, and the extent of losses due to insect pests, which will impact both crop production and food security. Presently, it is estimated that the amount of food that insects consume (pre- and post-harvest) is sufficient to feed more than 1 billion people. By 2050, it is thought that there will be an extra 3 billion people to feed. During this timescale, it is likely that insects will increase in numbers and in pest types. Prediction of changes in geographical distribution and population dynamics of insect pests will be useful for adapting IPM strategies to mitigate the adverse effects of climate change on crop production. Pest outbreaks might occur more frequently, particularly during extended periods of drought, followed by heavy rainfall. Some of the components of pest management such as host-plant resistance, biopesticides, natural enemies, and synthetic chemicals will be rendered less effective as a result of the increase in temperatures and UV radiation, and decrease in relative humidity. Climate change will also alter the interactions between insect pests and their host plants. As result, some of the cultivars that are resistant to insects may exhibit susceptible reactions under global warming. Adverse effects of climate change on the activity and effectiveness of natural enemies will be a major concern in future pest management programmes. The rate of insect multiplication might increase with an increase in CO_2 and temperature. There may be the possibility of evolutionary adaptation in insects to the changing environment. Therefore, climate change might change the population dynamics of insect pests differently in different agro-ecosystem and ecological zones. Therefore, there is a need to take a concerted look at the likely effects of climate change on crop protection and devise appropriate measures to mitigate the effects of climate change on food security.

References

Adamczyk Jr., J.J., Adams, L.C., Hardee, D.D., 2001. Field efficacy and seasonal expression profiles for terminal leaves of single and double *Bacillus thuringiensis* toxin cotton genotypes. J. Econ. Entomol. 94, 1589–1593.

Amarasekare, K.G., Edelson, J.V., 2004. Effect of temperature on efficacy of insecticides to differential grasshopper (Orthoptera: Acrididae). J. Econ. Entomol. 97, 1595–1602.

Anderson, S., Conrad, K., Gillman, M., Woiwod, I., Freeland, J., 2008. Phenotypic changes and reduced genetic diversity have accompanied the rapid decline of the garden tiger moth (*Arctia caja*) in the U.K. Ecol. Entomol. 33, 638–645.

Andrew, N.R., Hughes, L., 2005a. Diversity and assemblage structure of phytophagous Hemiptera along a latitudinal gradient: predicting the potential impacts of climate change. Glob. Ecol. Biogeogr. 14, 249–262.

Andrew, N.R., Hughes, L., 2005b. Herbivore damage along a latitudinal gradient: relative impacts of different feeding guilds. Oikos 108, 176–182.

Anwar, M.R., O'Leary, G., McNeil, D., Hossain, H., Nelson, R., 2007. Climate change impact on rainfed wheat in south-eastern Australia. Field Crops Res. 104, 139–147.

Awmack, C.S., Harrington, R., Lindroth, R.L., 2004. Aphid individual performance may not predict population responses to elevated CO_2 or O_3. Glob. Chang. Biol. 10, 1414–1423.

Bale, J.S., Masters, G.J., Hodkinson, I.D., Awmack, C., Bezemer, T.M., Brown, V.K., et al., 2002. Herbivory in global climate change research: direct effects of rising temperature on insect herbivores. Glob. Chang. Biol. 8, 1–16.

Bilde, T., Toft, S., 1999. Prey consumption and fecundity of the carabid beetle *Calathus melanocephalus* on diets of three cereal aphids: high consumption rates of low-quality prey. Pedobiologia 43, 422–429.

Bjorkman, C., Larsson, S., 1999. Insects on drought-stressed trees: four feeding guilds in one experiment. In: Lieutier, F., Mattson, W.J., Wagner, M.R. (Eds.), Physiology and Genetics of Tree-Phytophage Interactions. International Symposium, Gujan, France, pp. 323–335.

Bjorkman, C., Berggren, A., Bylund, H., 2011. Causes behind insect folivory patterns in latitudinal gradients. J. Ecol. 99, 367–369.

Breed, G.A., Stichter, S., Crone, E.E., 2013. Climate-driven changes in northeastern US butterfly communities. Nature Clim. Chang. 3, 142–145.

Broome, J.R., Sikorowski, P.P., Nee, W.W., 1974. Effect of sunlight on the activity of nuclear polyhedrosis virus from *Malacosoma disstria*. J. Econ. Entomol. 67, 135–136.

Bullock, H.R., 1967. Persistence of *Heliothis* nuclear polyhedrosis virus on cotton foliage. J. Invertebr. Pathol. 9, 434–436.

Butler, C.D., O'Neil, R.J., 2007. Life history characteristics of *Orius insidiosus* (Say) fed diets of soybean aphid, *Aphis glycines* Matsumura and soybean thrips, *Neohydatothrips variabilis* (Beach). Biol. Control 40, 339–346.

Calatayud, P.A., Polania, M.A., Seligmann, C.D., Bellotti, A.C., 2002. Influence of water stressed cassava on *Phenacoccus herreni* and three associated parasitoids. Entomol. Exp. Appl. 102, 163–175.

Cammell, M.E., Knight, J.D., 1992. Effects of climate change on the population dynamics of crop pests. Adv. Ecol. Res. 22, 117–162.

Cannon, R.J.C., 1998. The implications of predicted climate change for insect pests in the UK, with emphasis on non-indigenous species. Global Change Biol. 4, 785–796.

Casteel, C.L., 2010. Impacts of climate change on herbivore induced plant signaling and defenses. Ph.D. Dissertation, University of Illinois, Urbana-Champaign, IL.

Challinor, A.J., Wheeler, T.R., Craufurd, P.Q., Slingo, J.M., 2005. Simulation of the impact of high temperature stress on annual crop yields. Agric. Forest Entomol. 135, 180–189.

Chen, C., McCarl, B., 2001. Pesticide usage as influenced by climate: a statistical investigation. Clim. Change 50, 475–487.

Chen, D.H., Ye, G.Y., Yang, C.Q., Chen, Y., Wu, Y.K., 2005. The effect of high temperature on the insecticidal properties of *Bt* Cotton. Environ. Exp. Bot. 53, 333–342.

Chen, F.J., Wu, G., Ge, F., Parajulee, M.N., Shrestha, R.B., 2005a. Effects of elevated CO_2 and transgenic *Bt* cotton on plant chemistry, performance, and feeding of an insect herbivore, the cotton bollworm. Entomol. Exp. Appl. 115, 341–350.

Chen, F.J., Ge, F., Parajulee, M.N., 2005b. Impact of elevated CO_2 on tritrophic interaction of *Gossypium hirsutum*, *Aphis gossypii* and *Leis axyridis*. Environ. Entomol. 34, 37–46.

Chen, F.J., Gang, W., Megha, N., Parajulee, F.G., 2007. Impact of elevated CO_2 on the third trophic level: a predator *Harmonia axyridis* and a parasitoid *Aphidius picipes*. Biocontrol Sci. Technol. 17, 313–324.

Choudhary, J.S., Shukla, G., Prabhakar, C.S., Maurya, S., Das, B., Kumar, S., 2012. Assessment of local perceptions on climate change and coping strategies in Chotanagpur Plateau of Eastern India. J. Progress. Agric. 3, 8–15.

Choudhary, J.S., Prabhakar, C.S., Moanaro, Das, B., Kumar, S., 2013. Litchi stink bug, Tessaratoma javanica (Thunberg) (Hemiptera: Tessaratomidae) outbreak in Jharkhand (India) on Litchi. Phytoparasitica 41, 73–77.

Coley, P.D., Markham, A., 1998. Possible effects of climate change on plant/herbivore interactions in moist tropical forests. Clim. Change 39, 455–472.

Coll, M., Hughes, L., 2008. Effects of elevated CO_2 on an insect omnivore: a test for nutritional effects mediated by host plants and prey. Agric. Ecosyst. Environ. 123, 271–279.

Collier, R.H., Finch, S., Phelps, K., Thompson, A.R., 1991. Possible impact of global warming on cabbage root fly (*Delia radicum*) activity in the UK. Ann. Appl. Biol. 118, 261–271.

Collins, W., Colman, R., Haywood, J., Manning, R.R., Mote, P., 2007. The physical science behind climate change. Sci. Am. 297, 64–73.

Conrad, K.F., Woiwod, I.P., Parsons, M., Fox, R., Warren, M.S., 2004. Long-term population trends in widespread British moths. J. Insect. Conserv. 8, 119–136.

Coulson, S.J., Hodkinson, I.D., Webb, N.R., Mikkola, K., Harrison, J.A., Pedgley, D.E., 2002. Aerial colonization of high Arctic islands by invertebrates: the diamondback moth *Plutella xylostella* (Lepidoptera: Yponomeutidae) as a potential indicator species. Divers. Distrib. 8, 327–334.

Coviella, C.E., Trumble, J.T., 1999. Effects of elevated atmospheric carbon dioxide on insect-plant interactions. Conserv. Biol. 13, 700–712.

Curtis, P.S., Wang, X., 1998. A meta-analysis of elevated CO_2 effects on woody plant mass, form, and physiology. Oecologia 113, 299–313.

David, W.A.L., Gardiner, B.O.C., Wooner, M., 1968. The effects of sunlight on a purified granulosis virus of *Pieris brassicae* applied to cabbage leaves. J. Invertebr. Pathol. 11, 496–501.

DeVries, D.H., Georghiou, G.P., 1979. Influence of temperature on toxicity of insecticides to susceptible and resistant house flies. J. Econ. Entomol. 72, 48–50.

Dewar, R.C., Watt, A.D., 1992. Predicted changes in the synchrony of larval emergence and budburst under climatic warming. Oecologia 89, 557–559.

Dhillon, M.K., Sharma, H.C., 2007. Effect of storage temperature and duration on viability of eggs of *Helicoverpa armigera* (Lepidoptera: Noctuidae). Bull. Entomol. Res. 97, 55–59.

Dhillon, M.K., Sharma, H.C., 2008. Temperature and *Helicoverpa armigera* food influence survival and development of the ichneumonid parasitoid, *Campoletis chlorideae*. Indian J. Plant Prot. 36, 240–244.

Dhillon, M.K., Sharma, H.C., 2009. Temperature influences the performance and effectiveness of field and laboratory strains of the ichneumonid parasitoid, *Campoletis chlorideae*. BioControl 54, 743–750.

Diffenbaugh, N.S., Pal, J.S., Trapp, R.J., Giorgi, F., 2005. Fine-scale processes regulate the response of extreme events to global climate change. Proc. Natl. Acad. Sci. U.S.A. 102, 15774–15778.

Dobzhansky, T., 1965. 'Wild' and 'domestic' species of *Drosophila*. In: Baker, H.G., Stebbins, G.L. (Eds.), The Genetics of Colonizing Species. Academic Press, New York, pp. 533–546.

Dong, H.Z., Li, W.J., 2007. Variability of endotoxin expression in *Bt* transgenic cotton. J. Agron. Crop Sci. 193, 21–29.

Dosdall, L.M., 1994. Evidence for successful overwintering of diamondback moth, *Plutella xylostella* (L.) (Lepidoptera: Plutellidae), in Alberta. Can. Entomol. 126, 183–185.

Drake, B.G., Gonzalez-Meler, M.A., Long, S.P., 1997. More efficient plants: a consequence of rising atmospheric CO_2? Annu. Rev. Plant Physiol. Plant Mol. Biol. 48, 609–639.

English-Loeb, G., Stout, M.J., Duffey, S.S., 1997. Drought stress in tomatoes: changes in plant chemistry and potential nonlinear consequences for insect herbivores. Oikos 79, 456–468.

Environment Protection Agency (EPA), 1989. The Potential Effects of Global Climate Change on the United States: National Studies, vol. 2, Review of the Report to Congress, US Environmental Protection Agency, Washington, DC, 261 pp.

Fleming, R.A., Volney, W.J., 1995. Effects of climate change on insect defoliator population processes in Canada's boreal forests: some plausible scenarios. Water Air Soil Pollut. 82, 445–454.

Franco, A.M.A., Hill, J.K., Kitschke, C., Collingham, Y.C., Roy, D.B., Fox, R., et al., 2006. Impacts of climate warming and habitat loss on extinctions at species' low-latitude range boundaries. Global Change Biol. 12, 1545–1553.

Freier, B., Triltsch, H., 1996. Climate chamber experiments and computer simulations on the influence of increasing temperature on wheat-aphid-predator interactions. Aspects Appl. Biol. 45, 293–298.

Fuhrer, J., 2003. Agroecosystem responses to combinations of elevated CO_2, ozone, and global climate change. Agric. Ecosyst. Environ. 97, 1–20.

Fye, R.E., McAda, W.C., 1972. Laboratory studies on the development, longevity, and fecundity of six lepidopterous pests of cotton in Arizona. USDA, Washington, DC, p. 75.

Greenplate, J.T., Penn, S.R., Shappley, Z., Oppenhuizen, M., Mann, J., Reich, B., et al., 2000. Bollgard II efficacy: quantification of total lepidopteran activity in a 2-gene product. In: Dugger, P., Richter, D. (Eds.), Proceedings, Beltwide Cotton Conference National Cotton Council of America, Memphis, TN, USA, pp. 1041–1043.

Gutierrez, A.P., Ellis, C.K., d'Oultremont, T., Ponti, L., 2006. Climatic limits of pink bollworm in Arizona and California: effects of climate warming. Acta Oecol. 30, 353–364.

Gutierrez, A.P., Ponti, L., d'Oultremont, T., Ellis, C.K., 2008. Climate change effects on poikilotherm tritrophic interactions. Clim. Change 87, S167–S192. <http://meteora.ucsd.edu/cap/pdffiles/Guiterrez_poikilotherm_jan2008.pdf>.

Hamilton, J.G., Dermody, O., Aldea, M., Zangerl, A.R., Rogers, A., Berenbaum, M.R., et al., 2005. Anthropogenic changes in tropospheric composition increase susceptibility of soybean to insect herbivory. Environ. Entomol. 34, 479–485.

Harrington, R., Stork, N.E., 1995. Insects in a Changing Environment. Academic Press, London, p. 535.

Harvey, J.A., Jervis, M.A., Gols, R., Jiang, N., Vet, L.E.M., 1999. Development of the parasitoid, *Cotesia rubecula* (Hymenoptera: Braconidae) in *Pieris rapae* and *Pieris brassicae* (Lepidoptera: Pieridae): evidence for host regulation. J. Insect Physiol. 45, 173–182.

Hickling, R., Roy, D.B., Hill, J.K., Fox, R., Thomas, C.D., 2006. The distributions of a wide range of taxonomic groups are expanding polewards. Global Change Biol. 12, 450–455.

Hilder, V.A., Boulter, D., 1999. Genetic engineering of crop plants for insect resistance – a critical review. Crop Protect. 18, 177–191.

Hill, D.S., 1987. Agricultural Insects Pests of Temperate Regions and Their Control. Cambridge University Press, Cambridge, UK, p. 659.

Hill, M.G., Dymock, J.J., 1989. Impact of Climate Change: Agricultural/Horticultural Systems. DSIR Entomology Division Submission to the New Zealand Climate Change Program, Department of Scientific and Industrial Research. Auckland, New Zealand, p. 16.

Houghton, J.T., Filho, L.G.M., Callander, B.A., Harris, N., Kattenberg, A., Maskell, K., 1995. Climate Change 1995: The Science of Climate Change, Intergovernmental Panel on Climate Change. Cambridge University Press, Cambridge, GB, pp. 572, <http://www.nysaes.cornell.edu/recommends/>.

Huberty, A.F., Denno, R.F., 2004. Plant water stress and its consequences for herbivorous insects: a new synthesis. Ecology 85, 1383–1398.

Hvam, A., Toft, S., 2005. Effects of prey quality on the life history of a harvestman. J. Arachnol. 33, 582–590.

IPCC, 2001. Climate Change 2001: Impacts, Adaptation, and Vulnerability. Cambridge University Press, Cambridge, UK, p. 1032.

IPCC, 2007. Climate Change 2007: the physical science basis. Summary for policy makers. Report of Working Group I of the Intergovernmental Panel on Climate Change. <http://www.ipcc.ch/pub/spm18-02.pdf>.

Ignoffo, E.M., Hostetter, D.L., Sikorowski, P.P., Sutter, G., Brooks, W.M., 1977. Inactivation of entomopathogenic viruses, a bacterium, fungus, and protozoan by an ultraviolet light source. Environ. Entomol. 6, 411–415.

Inglis, G.D., Duke, G.M., Kawchuck, L.M., Goettel, M.S., 1999. Influence of oscillating temperature on the competitive infection and colonization of the migratory grasshopper by *Beauveria bassiana* and *Metarhizium flavoviride*. Biol. Control 14, 111–120.

Jactel, H., Petit, J., Desprez-Loustau, M., Delzon, S., Piou, D., Battisti, A., et al., 2012. Drought effects on damage by forest insects and pathogens: a meta-analysis. Global Change Biol. 18, 267–276.

Jaques, R.P., 1968. The inactivation of the nuclear polyhedrosis virus of *Trichoplusia ni* by gamma and ultraviolet radiation. Can. J. Microbiol. 14, 1161–1163.

Johansen, B.E., 2002. The Global Warming Desk Reference. Greenwood Press, Westport, CT, p. 353.

Johnson, D.L., 1990. Influence of temperature on toxicity of two pyrethroids to grasshoppers (Orthoptera: Acrididae). J. Econ. Entomol. 83, 366–373.

Jones, K.A., McKinley, D.J., 1986. UV inactivation of *Spodoptera littoralis* nuclear polyhedrosis virus in Egypt: assessment and protection. In: Samson, R.A. Vlak, J.M., Peters, D. (Eds.), Fundamental and Applied Aspects of Invertebrate Pathology. Proceedings, IV International Colloqium on Invertebrate Pathology, Wageningen, The Netherlands, p. 155.

Kaiser, J., 1996. Pests overwhelm *Bt* cotton crop. Nature 273, 423.

Karl, T.R., Trenbeth, K.E., 2003. Modern global climate change. Science 302, 1719–1723.

Kingsolver, J.G., 1989. Weather and the population dynamics of insects: integrating physiological and population ecology. Physiol. Ecol. 62, 314–334.

Koricheva, J., Larsson, S., Haukioja, E., 1998. Insect performance on experimentally stressed woody plants: a meta-analysis. Annu. Rev. Entomol. 43, 195–216.

Kranthi, K.R., Naidu, S., Dhawad, C.S., Tatwawadi, A., Mate, K., Patil, E., et al., 2005. Temporal and intraplant variability of *Cry1Ac* expression in *Bt*-cotton and its influence on the survival of the cotton bollworm, *Helicoverpa armigera* (Hubner) (Noctuidae: Lepidoptera). Curr. Sci. 89, 291–298.

Lactin, D.J., Johnson, D.L., 1995. Temperature dependent feeding rates of *Melanoplus sanguinipes* nymphs (Orthoptera: Acrididae) in laboratory trials. Environ. Entomol. 24, 1291–1296.

Ledley, T.S., Sundquist, E.T., Schwartz, S.E., Hall, D.K., Fellows, J.D., Killeen, T.L., 1999. Climate change and greenhouse gases. EOS 80 (39), 453 <http://www.ecd.bnl.gov/steve/pubs/LedleyAGUgreenhousegasEOS99.pdf>.

Long, S.P., Ainsworth, E.A., Rogers, A., Ort, D.R., 2004. Rising atmospheric carbon dioxide: plants FACE the future. Annu. Rev. Plant Biol. 55, 591–628.

Mattson, W.J., Haack, R.A., 1987. The role of drought in outbreaks of plant-eating insects. BioScience 37, 110–118.

McDonald, J.R., Head, J., Bale, J.S., Walters, K.F.A., 2000. Cold tolerance, overwintering and establishment potential of *Thrips palmi*. Physiol. Entomol. 25, 159–166.

McVean, R.I.K., Dixon, A.F.G., Harrington, R., 1999. Causes of regional and yearly variation in pea aphid numbers in eastern England. J. Appl. Entomol. 123, 495–502.

Mochida, O., 1991. Impact of CO_2-climate change on pests distribution. Agric. Hortic. 66, 128–136.

Morgan, D., 1996. Temperature changes and insect pests: a simulation study. Aspects Appl. Biol. 45, 277–283.

Morris, O.N., 1971. The effect of sunlight, ultraviolet and gamma radiation and temperature on the infectivity of a nuclear polyhedrosis virus. J. Invertebr. Pathol. 18, 292–294.

Moscardi, F., 1999. Assessment of the application of baculoviruses for control of Lepidoptera. Annu. Rev. Entomol. 44, 257–289.

Naranjo, S.E., 1993. The life-history of *Trichogrammatoidea bactrae* (Hymenoptera: Trichogrammatidae) an egg parasitoid of pink-bollworm (Lepidoptera: Gelechidae) with emphasis on performance at high temperatures. Environ. Entomol. 22, 1051–1059.

Norby, R.J., Wullschleger, S.D., Gunderson, C.A., Johnson, D.W., Ceulemans, R., 1999. Tree responses to rising CO_2 in field experiments: implications for the future forest. Plant Cell Environ. 22, 683–714.

Parmesan, C., 1996. Climate and species' range. Nature 382, 765–766.

Parmesan, C., Yohe, G., 2003. A globally coherent fingerprint of climate change impacts across natural systems. Nature 421, 37–42.

Parmesan, C., Ryrholm, N., Stefanescu, C., Hill, J.K., Thomas, C.D., Descimon, H., et al., 1999. Poleward shifts in geographical ranges of butterfly species associated with regional warming. Nature 399, 579–583.

Parry, M.L., 1990. Climate Change and World Agriculture. Earthscan, London, UK, 157 pp.

Parry, M.L., Carter, T.R., 1989. An assessment of the effects of climatic change on agriculture. Clim. Change 15, 95–116.

Patterson, D.T., Westbrook, J.K., Joyce, R.J.V., Lingren, P.D., Rogasik, J., 1999. Weeds, insects and diseases. Clim. Change 43, 711–727.

Porter, J.H., Parry, M.L., Carter, T.R., 1991. The potential effects of climate change on agricultural insect pests. Agric. Forest Meteorol. 57, 221–240.

Prabhakar, C.S., Mehta, P.K., Sood, P., Singh, S.K., Sharma, P., Sharma, P.N., 2012a. Population genetic structure of the melon fly, *Bactrocera cucurbitae* (Coquillett) (Diptera:

Tephritidae) based on *mitochondrial cytochrome oxidase (COI)* gene sequences. Genetica 140, 83–91.

Prabhakar, C.S., Sood, P., Mehta, P.K., 2012b. Fruit fly (Diptera: Tephritidae) diversity in cucurbit fields and surrounding forest areas of Himachal Pradesh, a North-Western Himalayan state of India. Arch. Phytopathol. Plant Protect. 45, 1210–1217.

Price, P.W., 1987. The role of natural enemies in insect populations. In: Barbosa, P., Schultz, J.C. (Eds.), Insect Outbreaks. Academic Press, San Diego, CA, USA, pp. 287–312.

Reiners, S., Petzoldt, C., 2005. Integrated Crop and Pest Management Guidelines for Commercial Vegetable Production. Cornell Cooperative Extension Publication #124VG. Cornell University, Comstock Hall, Ithaca, New York.

Sachs, E.S., Benedict, J.H., Stelly, D.M., Taylor, J.F., Altman, D.W., Berberich, S.A., et al., 1998. Expression and segregation of genes encoding *Cry1A* insecticidal proteins in cotton. Crop Sci. 38, 1–11.

Scott, J.G., 1995. Effects of temperature on insecticide toxicity In: Roe, M. Kuhr, R.J. (Eds.), Reviews in Pesticide Toxicology, vol. 3. Toxicology Communications Inc., Raleigh, NC, pp. 111–135.

Scott, M., Berrigan, D., Hoffmann, A.A., 1997. Costs and benefits of acclimation to elevated temperature in *Trichogramma carverae*. Entomol. Exp. Appl. 85, 211–219.

Sharma, H.C., 2002. Host plant resistance to insects: principles and practices. In: Sarath Babu, B. Varaprasad, K.S. Anitha, K. Prasada Rao, R.D.V.J. Chandurkar, P.S. (Eds.), Resources Management in Plant Protection, Vol. 1. Plant Protection Association of India, Rajendranagar, Hyderabad, Andhra Pradesh, India, pp. 37–63.

Sharma, H.C., 2005. *Heliothis/Helicoverpa* Management: Emerging Trends and Strategies for Future Research. Oxford & IBH Publishers, Inc., New Delhi, India, 469 pp.

Sharma, H.C., 2010. Effect of climate change on IPM in grain legumes. In: Fifth International Food Legumes Research Conference (IFLRC V), and the Seventh European Conference on Grain Legumes (AEP VII), 26–30 April 2010, Anatalaya, Turkey.

Sharma, H.C., Ortiz, R., 2000. Transgenics, pest management, and the environment. Curr. Sci. 79, 421–437.

Sharma, H.C., Mukuru, S.Z., Manyasa, E., Were, J., 1999. Breakdown of resistance to sorghum midge, *Stenodiplosis sorghicola*. Euphytica 109, 131–140.

Sharma, H.C., Sullivan, D.J., Bhatnagar, V.S., 2002. Population dynamics of the Oriental armyworm, *Mythimna separata* (Walker) (Lepidoptera: Noctuidae) in South-Central India. Crop Protect. 21, 721–732.

Sharma, H.C., Venkateswarlu, G., Sharma, A., 2003. Environmental factors influence the expression of resistance to sorghum midge, *Stenodiplosis sorghicola*. Euphytica 130, 365–375.

Sharma, H.C., Dhillon, M.K., Kibuka, J., Mukuru, S.Z., 2005. Plant defense responses to sorghum spotted stem borer, *Chilo partellus* under irrigated and drought conditions. Int. Sorghum Millets Newsl. 46, 49–52.

Sharma, H.C., Srivastava, C.P., Durairaj, C., Gowda, C.L.L., 2010. Pest management in grain legumes and climate change. In: Yadav, S.S., McNeil, D.L., Redden, R., Patil, S.A. (Eds.), Climate Change and Management of Cool Season Grain Legume Crops. Springer Science + Business Media, Dordrecht, The Netherlands, pp. 115–140.

Southwood, T.R.E., Comins, H.N., 1976. A synoptic population model. J. Anim. Ecol. 45, 949–965.

Sutherst, R.W., 1991. Pest risk analysis and the greenhouse effect. Rev. Agric. Entomol. 79, 1177–1187.

Sutherst, R.W., Baker, R.H.A., Coakley, S.M., Harrington, R., Kriticos, D.J., Scherm, H., 2007. Pests under global change – meeting your future landlords? In: Canadell, J.G., Pataki, D.E., Pitelka, L.F. (Eds.), Terrestrial Ecosystems in a Changing World. Springer-Verlag, Berlin, Heidelberg.

Szewczyk, B., Hoyos-Carvajal, L., Paluszek, M., Skrzecz, I., Lobo de Souza, M., 2006. Baculoviruses re-emerging biopesticides. Biotechnol. Adv. 24, 143–160.

Tauber, M.J., Tauber, C.A., Masaki, S., 1986. Seasonal Adaptations of Insects. Oxford University Press, New York, 411 pp.

Thomson, L.J., Robinson, M., Hoffmann, A.A., 2001. Field and laboratory evidence for acclimation without costs in an egg parasitoid. Funct. Ecol. 15, 217–221.

Thomson, L.J., Macfadyen, S., Hoffmann, A.A., 2010. Predicting the effects of climate change on natural enemies of agricultural pests. Biol. Control 52, 296–306.

Timans, U., 1982. Zur Wirkung von UV-Strahlen auf des Kempolyedevirus des Schwammspinners, *Lymantria dispar* L. (Lepidoptera: Lymantriidae). Z. Angew. Entomol. 94, 382–401.

Toft, S., 1995. Value of the aphid *Rhopalosiphum padi* as food for cereal spiders. J. Appl. Ecol. 32, 552–560.

Torriani, D.S., Calanca, P., Schmid, S., Beniston, M., Fuhrer, J., 2007. Potential effects of changes in mean climate and climate variability on the yield of winter and spring crops in Switzerland. Clim. Res. 34, 59–69.

Vincent, C., Hallman, G., Panneton, B., Fleurat-Lessardú, F., 2003. Management of agricultural insects with physical control methods. Annu. Rev. Entomol. 48, 261–281.

Wang, X.Y., Yang, Z.Q., Wub, H., Gould, J.R., 2007. Effects of host size on the sex ratio, clutch size, and size of adult *Spathius agrili*, an ectoparasitoid of emerald ash borer. Biol. Control 44, 7–12.

Ware, G.W., 2000. The Pesticide Book, Fifth ed. Thompson Publication, Fresno, CA.

Warren, M.S., Hill, J.K., Thomas, J.A., Asher, J., Fox, R., Huntley, B., et al., 2001. Rapid responses of British butterflies to opposing forces of climate and habitat change. Nature 414, 65–69.

Whittaker, J.B., Tribe, N.P., 1996. An altitudinal transect as an indicator of responses of a spittlebug (Auchenorrhyncha: Cercopidae) to climate change. Eur. J. Entomol. 93, 319–324.

Whittaker, J.B., Tribe, N.P., 1998. Predicting numbers of an insect (*Neophilaenus lineatus*: Homoptera) in a changing climate. J. Anim. Ecol. 67, 987–991.

Wilson, R.J., Gutierrez, D., Gutierrez, J., Monserrat, V., 2007. An elevational shift in butterfly species richness and composition accompanying recent climate change. Global Change Biol. 13, 1873–1887.

Wolf, A., Kozlov, M.V., Callaghan, T.V., 2008. Impact of non-outbreak insect damage on vegetation in northern Europe will be greater than expected during a changing climate. Clim. Change 87, 91–106.

Young, A.M., 1982. Population Biology of Tropical Insects. Plenum Press, New York, USA, 511 pp.

Zavala, J.A., Cast, C.L., DeLucia, E.H., Berenbaum, M.R., 2008. Anthropogenic increase in carbon dioxide compromises plant defense against invasive insects. Proc. Natl. Acad. Sci. U.S.A. 105, 5129–5133.

CHAPTER

3

Application of Remote Sensing in Integrated Pest Management

J.L. Willers[1], *S.L. DeFauw*[2], *P.J. English*[3] *and J.N. Jenkins*[1]

[1]Genetics and Precision Agriculture Research Unit, MS, USA, [2]Pennsylvania State University, PA, USA, [3]Mississippi State University, MS, USA

Disclaimer: Mention of trade names or commercial products in this publication is solely for providing specific information and does not imply recommendations or endorsement by the US Department of Agriculture.

3.1 INTRODUCTION

Analysing the dispersion of insect pests in any ecosystem is a challenging problem for many applied ecology investigations, especially when considering the spatio-temporal dynamics of habitat conditions and pest preferences. Pest dispersion impacts sampling efforts (Davis, 1994; Trumble, 1985), rate of habitat colonization (Southwood et al., 1983) and the establishment of economic thresholds or injury levels (Byerly et al., 1978; Stern et al., 1959; Wilson, 1994) as well as a host of population ecology-related analyses (Banerjee, 1976; Dalthorp et al., 2000; Fleischer et al., 1999). Previously, Willers et al. (1999, 2005, 2009) and Willers and Riggins (2010) described some site-specific sampling strategies for tarnished plant bug (TPB) (*Lygus lineolaris* [P. de B.] (Heteroptera: Miridae)) detection aided by remotely-sensed imagery of commercial cotton fields. Other complementary works have developed the initial capacity for site-specific pesticide applications (Dupont et al., 2000; Seal et al., 2001). As these efforts expand and become more mainstream in the management of commercial cropping systems, it is necessary to make continued improvements in methodology.

One important issue with image-based sampling is focused on understanding how sample unit size choices and habitat-related differences in pest density affect a field scout's assessment of pest dispersion in commercial cotton fields. To examine this question, two general courses of investigation are available: (i) conduct a detailed field-scale study or (ii) conduct a simulation study. Under commercial conditions, influences due to discrepancies in observer ability, physiographic location of arable parcels and cropland-hedgerow

D. P. Abrol (Ed): Integrated Pest Management.
DOI: http://dx.doi.org/10.1016/B978-0-12-398529-3.00004-X

adjacencies, field heterogeneity, management practices, sampling error, and/or lack of sufficient time (cost efficiencies) to conduct thorough field sampling efforts heavily influence outcomes, interpretations, conclusions, and subsequent management actions. However, a simulation model can generate large numbers of observations useful for discerning patterns and trends to make better informed choices about the most economically efficient sample unit size as related to pest infestation rates and dispersion under heterogeneous field conditions.

Various methods utilized in the analysis of species-specific spatial patterns have been described in the literature (Davis, 1994; Ludwig and Reynolds, 1988; Pielou, 1960, 1977, 1978). With field data, the goodness of fit test (Davis, 1994; Poole, 1974; Steel and Torrie, 1960) is traditionally employed. In this investigation, Lloyd's mean crowding and patchiness indices (Lloyd, 1967) are used to examine relationships among pest density, dispersion pattern and sample unit size. These indices developed by Lloyd (1967) were based on the use of quadrats (a cell or small sized unit of area) and, thus, appear to correspond well with the characteristics/conditions of the two choices of simulation models used in this study.

Previous work by Willers et al. (2005) assumed that the dispersion of TPBs in various cotton habitat classes was random. The main objective of the current study is to further examine this assumption by application of additional simulation modelling efforts. If a parametric random pattern is established as a condition, the null hypothesis is that there is no relationship between sample unit size and pest density. Estimates for mean crowding or patchiness (Lloyd, 1967) are used to test this hypothesis. The practical utility of the computer simulations is underscored by other simulation experiments with dice, and a real-world example involving several adjacent cotton fields during the 2006 production season.

3.2 METHODS

The simulation model is comprised of two parts. The first part models a simple random sample (SRS) obtained from a simulated habitat. The second part models the characteristics of a randomly dispersed pest insect population set to only one of the several choices 'available' for an infestation rate describing the mean number of insects per plant (Willers et al., 2005) in a simulated habitat.

3.2.1 Simulation of a Simple Random Sample Design

The quadrat-based SRS design modelled here has been previously described (Willers and Akins, 2000; Willers et al., 1999, 2005). The simulation model employs two different sampling unit sizes.

The smallest sampling unit is called the quadrat, whose dimensions are a crop row length of 0.914 m (which emulates one drop cloth sample from that row) and a width set equal to two row spacings (as typically used in actual Mid-South (USA) cotton fields and is either 0.762 m or 1.016 m). For the simulation model, this single quadrat is assumed to bisect these two drills of each crop row. These quadrat units provide the main link between the simulated system and actual field conditions. Therefore, 0.4047 ha (or 1 ac) of field area will contain 4356.3 units of this size (at 0.9290 m^2 when the row spacing is 1.016 m). It is important to emphasize that the total number of quadrat units in a cotton field constitute a countable number of sampling units (Thompson, 1992). These units are nested within different sizes, selected at random and assessed for counts of a particular insect. If these quadrats are apportioned amongst one or more habitats of crop growth and development (Willers et al., 2005) using remote sensing information, a SRS plan by habitat class is constructed. Since the spatial resolution of individual pixels (often 1 m^2) in a geo-registered multi-spectral image of a cotton field conceptually corresponds to the size of one of these quadrats, the practical linkage between the computer simulation model and field applications with remote sensing is demonstrable.

The larger sized sampling unit, referred to as a belt transect, consists of a sequential arrangement of n quadrats joined together (Willers et al., 1999) for a variable length (L). For different runs, each quadrat outputs a random variate of insect counts for a simulated habitat class assigned one of several simulated insect infestation rates. An additional programming module collects the counts of simulated insects from each quadrat of a belt transect sample of a particular size and summarizes the total number of insects found in each simulated sample.

Under field conditions, each belt transect sample is selected by consultant/producer-determined preferences within meso-scale habitat maps derived from classified imagery (Richards and Jia, 1999) of the cotton field (Willers et al., 1999, 2005, 2009, 2012). To represent these field practices in the simulated system, different pest densities for various habitat classes were modelled by changing the infestation rate parameter (λ_c, as explained in the following section) and by aggregating (or stacking) adjacent belt transects to create larger areas that approximate the size of individual sprayer polygons contained in a field grid. For spatial pesticide applications in a commercial cotton field, each cell of the field grid can be assigned an application rate to apply a spatial pesticide prescription. These very large-sized sample units matched to sprayer traits represent another construct that enables the simulation model to be applicable to real-world field conditions. Since the boom width of the sprayer determines the length of each belt transect (L) in a stack of transects, the breadth of the simulated stack of transects is determined by the variable-rate controller's response time. However, the model does not account for other spatial relationships (e.g. the distance, direction or proximity to edges or differences in the sharpness of gradient effects) among simulated samples within a simulated habitat during a given simulation run.

3.2.2 Insect Infestation of Habitats Simulation Model

The basic approach to model infestation rate was to employ the negative binomial distribution (NBD) (Anscombe, 1949; Davis, 1994). Other detailed modifications have been described in Willers et al. (1990). The model generates integer values of counts to simulate numbers of insects per quadrat for simulated belt transects of different sizes and infestation rates (λ_c), while setting the dispersion parameter (k) to a constant value of 50. (Note: If k is set to very small values (e.g. 1, 2 or 3), the model would generate clustered or aggregated variates for a simulated belt transect sample, while for increasingly larger values of k, the NBD converges to the Poisson (or random) distribution.) The number of insects (or events (counts)) per quadrat for a belt transect of size L was subsequently generated by the inverse transformation method (Pritsker and Pegden, 1979) using the probability values presented in Table 3.1. The simulation model was programmed in SAS® (SAS Institute, 1990).

The primary objective of this simulation study is to investigate how the assessment of insect dispersion differs with changes in (i) the pest density (or infestation rate, where $\lambda_c = 0.01, 0.04, 0.08, 0.24$ and 0.40) and (ii) belt transect sizes within a simulated cotton habitat class. The various lengths of belt transects employed were $L = 4, 8, 16, 24$ crop rows (for 1.016 m row spacings) or $L = 315$ crop rows in a stack (for 0.7620 m row spacings). The infestation rate parameter (λ_c) is linked to assumptions about the number of plants contained in a quadrat (Willers et al., 1990). For simplicity, the plant density was 'fixed' at a value of 10 cotton plants/quadrat in the simulation model, although other values could be specified.

TABLE 3.1 Probability of Observing Various Counts of Insects Per Quadrat (Simulated Sample Units)

Insects/SU	$\lambda_c = 0.01$	$\lambda_c = 0.04$	$\lambda_c = 0.08$	$\lambda_c = 0.16$	$\lambda_c = 0.24$	$\lambda_c = 0.40$
0	0.9049278	0.6713877	0.4521834	0.2070213	0.0959259	0.0213212
1	0.0903122	0.2664237	0.3560499	0.3209633	0.2196776	0.0789675
2	0.0045967	0.0539191	0.1429807	0.2537850	0.2565701	0.1491609
3	0.0001590	0.0074174	0.0390288	0.1364012	0.2036892	0.1915152
4		0.0007800	0.0081438	0.0560408	0.1236129	0.1879686
5		0.0000669	0.0013851	0.0187672	0.0611459	0.1503749
6		0.0000049	0.0001999	0.0053343	0.0256720	0.1021064
7			0.0000252	0.0013232	0.0094065	0.0605075
8			0.0000028	0.0002923	0.0030697	0.0319345
9				0.0000584	0.0009061	0.0152445
10				0.0000107	0.0002448	0.0066624
11				0.0000018	0.0000612	0.0026919
12					0.0000142	0.0010136
13					0.0000031	0.0003581
14					0.0000006	0.0001194
15						0.0000377
16						0.0000114
17						0.0000033
18						0.0000009

The plant sample size per quadrat is 10 cotton plants (0.9144 m length of row for different infestation rates (λ_c) (see text)).

3.2.3 Dispersion Analyses of Simulated Conditions

Lloyd's mean crowding and index of patchiness (Lloyd, 1967) was used to assess the random dispersion assumption for various simulated combinations of infestation rate (λ_c) and belt transect size (L) under the conditions of the model to generate a parametric random dispersion pattern. Lloyd's mean crowding index (Davis, 1994; Lloyd, 1967) is estimated (without correction for bias (Pielou, 1978, p. 151)) as

$$\overset{*}{x} = \overline{x} + \frac{s^2}{\overline{x}} - 1 \tag{3.1}$$

The index of mean crowding describes the mean number of individuals occupying the same habitat space (here, a quadrat sample unit of a particular size) (Lloyd, 1967; Pielou, 1977, 1978). Lloyd's patchiness index (or the ratio of mean crowding to the mean) is expressed as

$$\overset{*}{x}/\overline{x} \tag{3.2}$$

where: <1 corresponds to regular dispersion, =1 random dispersion, and >1 aggregated dispersion.

The index of patchiness is derived from the mean crowding index (Lloyd, 1967; Southwood, 1978) and is dependent upon quadrat size (Davis, 1994). Choices for sample unit size are known to influence the assessment of dispersion, particularly when artificial units as opposed to natural units are employed (Pielou, 1977, 1978; Poole, 1974). See discussion in Ludwig and Reynolds (1988) for additional details.

The index of patchiness was estimated for the larger belt transect sample units of different sizes. Patterns of these statistics, generated over 10,000 simulation runs for each combination of infestation rate and belt transect size, were summarized by histograms. A single run represents a single belt transect sample at a particular combination of L (i.e. aggregated quadrat units) and λ_c (i.e. pest density).

3.2.4 Simulation Experiments with Dice

The purpose of the dice simulations is to enable the reader to better grasp the 'geographical' and 'statistical' correspondences between the computer simulation model and field applications. The particular aim is to understand how the sample size (or sample time) may be reduced without compromising precision for management decisions about insect control. The first sampling 'universe' (Ash, 1993) assigned one of five colours (red, orange, yellow, green, and blue) to each face of each die. The colour assignment of the countable events (the integers 1–6) (Table 3.2) corresponds to a SRS scheme for each colour. Each face of a die represents the count of insects found for adjacent quadrats of different belt transect lengths assigned to a colour (i.e. habitat class) where the mean infestation rate of each colour is similar. To add some variability to this system, the event 3 was assigned twice to the colour 'Red' and was not assigned to the colour 'Blue' (Table 3.3). Table 3.4 presents the frequency count of the outcomes of the events for 30 rolls by colour.

The second sampling 'universe' was established from 10 new dice, where the events 1–6 were assigned to one of five colours.

TABLE 3.2 Event List for Sampling Universe with Dice with Similar Mean Values among the Colours

Colour	Lists of Face Values per Die per Colour
Red	(5) (3, 4) (1, 2, 3, 6)
Orange	(4) (6) (1, 3) (2, 5)
Yellow	(1) (3, 6) (2, 4, 5)
Green	(5) (6) (1, 3) (2, 4)
Blue	(1) (6) (2, 4, 5)

TABLE 3.3 Table of Frequencies for 30 Rolls of Five Dice for the Sampling Universe Presented in Table 3.2

	Face Value					
Colour	**1**	**2**	**3**	**4**	**5**	**6**
Red	3	6	5	6	5	7
Orange	5	8	2	6	3	5
Yellow	6	7	6	4	6	4
Green	2	6	6	4	6	5
Blue	6	5	0	5	5	5

(Diagramming of this sampling universe is not presented here.) These colour assignments correspond to a SRS scheme where the mean infestation rate is not equivalent over the colours, since the mean ranks of events are Red < Orange < Yellow < Green < Blue. The belt transect length is determined by the number of times a particular colour occurred on a roll of the 10 dice. All colour and event combinations were not observed in every roll. This corresponds to the situation under field conditions when the sampler chooses to *not* sample a habitat class. The outcomes observed by colour for seven rolls of these 10 dice are summarized in Table 3.4.

3.2.5 Field Data Illustrations

The field data originally used to validate the model consisted of insect counts obtained from

TABLE 3.4 Outcomes for Seven Rolls of 10 Dice for Another Sampling Universe Where the Expected Mean by Colour is Ranked as Red < Orange < Yellow < Green < Blue

	Colour				
Roll	Red	Orange	Yellow	Green	Blue
1	1, 1	–	1, 3	3, 4, 4	5, 5, 6
2	–	3	1, 3	4, 4, 5, 6	6, 6, 6
3	1	2, 2	3, 3	3, 4, 4, 5	5
4	1	2, 2	3, 3, 3, 4	4, 4	5
5	1, 1	2	3, 3	5, 5, 6	6, 6
6	1, 1	–	3	4, 4, 5, 5, 5, 6	6
7	1, 1	1, 2, 2	1, 3, 4	4	6

48 transect lines across a large commercial cotton field in Bolivar County, MS. Quadrat-based samples (0.9144 m) were collected across eight successive rows (L = 8.128 m) on 17 July and 22 July 1997. For each row at each sample site, the number of plants and total number of TPB were recorded. A partial listing of these data is included in Table 3.5; the complete dataset may be examined in Willers et al. (1999).

The sample locations of the 2006 field datasets (collected in June, July and August) from a farm in Noxubee County, MS, were consolidated onto a single classified map (Figure 3.9). Row spacing for each cotton field was 0.7620 m (30 in). Coordinates for insect collection loci (n = 104 sites) were obtained using a Garmin Model 12GPS unit (Olathe, KS) and overlain on a classified image using ESRI® ArcMap (Ver. 9.1 (Redlands, CA)) using various geoprocessing techniques similar to those described by Nelson et al. (2005). The classified image layer (accomplished with ERDAS® Imagine software (NorCross, GA)) represents the categorical change in vegetation vigour of the cotton crop, including other surrounding features (i.e. corn, grasses, trees, pond, soybean and roads) detected between early June and July 2006. Higher class values represent the most change in cotton plant vigour between these two months. The details of each scouting site, how the imagery of cotton fields was processed and classified into habitats, and the subsequent preparation of any pesticide prescription maps (based on sprayer polygons built using a custom application programmed for ERDAS® Imagine, Ver. 8.7) are not described here. (Interested readers may consult Dupont et al., 2000; Frigden et al., 2002; McKinion et al., 2009; Richards and Jia, 1999; Seal et al., 2001; Theobald, 2003; Willers and Riggins, 2010 and Willers et al., 1999, 2005, 2009, 2012 for details.)

3.3 RESULTS

This study improves prior simulation efforts (Willers et al., 1990, 2000, 2005) to further explore emergent details of relationships among different infestation rates and patterns of dispersion as sampling unit size changes. Simulated results were first compared with actual TPB counts obtained from belt transects sampled in commercial cotton fields to test the model's ability to generate variates by quadrat similar to field data (see Willers et al., 1999). Only one of these comparisons (for 1.016 m row spacing) is presented here, for reasons of brevity.

These simulation runs were based on an 8-row long belt transect using an infestation rate of λ_c = 0.20 distributed across a uniform stand of 10 plants per row. Examining the counts by quadrat (or crop row) between the simulated runs and the actual field sample, it is observed that the simulation runs provide similar estimates of insect abundance. On occasion, however, the number of insects 'observed' per quadrat in the simulated runs was much larger than that obtained from field data for a comparable density (e.g. run 4593, with eight insects counted in quadrat 5).

TABLE 3.5 Comparison of a Sample Transect from a Cotton Field of Tarnished Plant Bug Counts per Quadrat with Seven Random Selections from a Total of 10,000 Simulation Runs

	Crop row (Quadrat)									
	1	2	3	4	5	6	7	8	Density/0.405ha[1]	Infestation rate[2]
Totals	2	2	2	0	2	0	2	3	7085	0.20
Stand	8	10	6	7	11	11	6	7	35,937	–
ILLUSTRATIVE SIMULATION RUNS										
Run 328	2	1	1	3	1	2	2	1	7085	0.16
Run 607	1	2	2	3	1	0	0	2	5995	0.14
Run 2880	5	1	0	1	4	0	3	3	9265	0.21
Run 4593	0	0	3	3	8	2	2	0	9810	0.22
Run 5378	1	1	4	0	3	3	4	0	8720	0.20
Run 7627	1	3	2	0	1	2	3	1	7085	0.16
Run 8573	2	2	2	1	4	3	4	3	11,445	0.26
Stand (Model)	10	10	10	10	10	10	10	10	43,600	–

Equivalent densities/0.405 ha, infestation rates (near 0.20) and numbers of insects per quadrat are obtained irrespective of the variability of numbers of plants per quadrat for the field data vs. the fixed number of 10 plants per quadrat used by the model (see text).
[1]*These values were determined using a line-intercept sampling estimator (Willers et al., 1999; Williams et al., 1995).*
[2]*Infestation rate is determined by dividing the insect density estimate by the stand density estimate.*

Patterns of dispersion using estimates of Lloyd's index of patchiness at infestation rates of λ_c = 0.01, 0.08 and 0.24, simulated for several relatively short transect lengths, are summarized in Figures 3.1–3.3. Only a few of the many possible graphs from these simulation runs are shown. As belt transect length and pest density increased, the patchiness index decreased in its range and centred about a value of 1.0. These trends indicated increasing 'opportunities' for random dispersion and decreased capacity for aggregated dispersion pattern tendencies within a homogeneous habitat. Therefore, this index is sensitive to choices for belt transect length, but not to values for infestation rate, once the rate departs from small values close to zero. Despite this sensitivity, the convergence of results about a mode of 1.0 supports the assumption that a random dispersion pattern is plausible for most pest densities estimated with larger sample unit sizes. Consistency in estimating this index is most doubtful with a pest density close to zero (Figure 3.1) and if too short a belt transect has been employed (see the uppermost panels in Figures 3.2 and 3.3).

Mean crowding tended to be a noisy parameter when belt transects shorter than 24 crop rows were used and lower infestation rates (<8%) were modelled. Mean crowding results are presented for a partial collection of sample unit size and pest density combinations (Figures 3.4–3.6). Emergent trends are most clear if very long transects (e.g. 315 quadrats or crop rows) are utilized (as shown in Figures 3.7 and 3.8); however, sampling these very long (stacked) belt transects would require an excessive amount of scouting time in commercial field settings. The practicality of the mean crowding trends detected does provide an

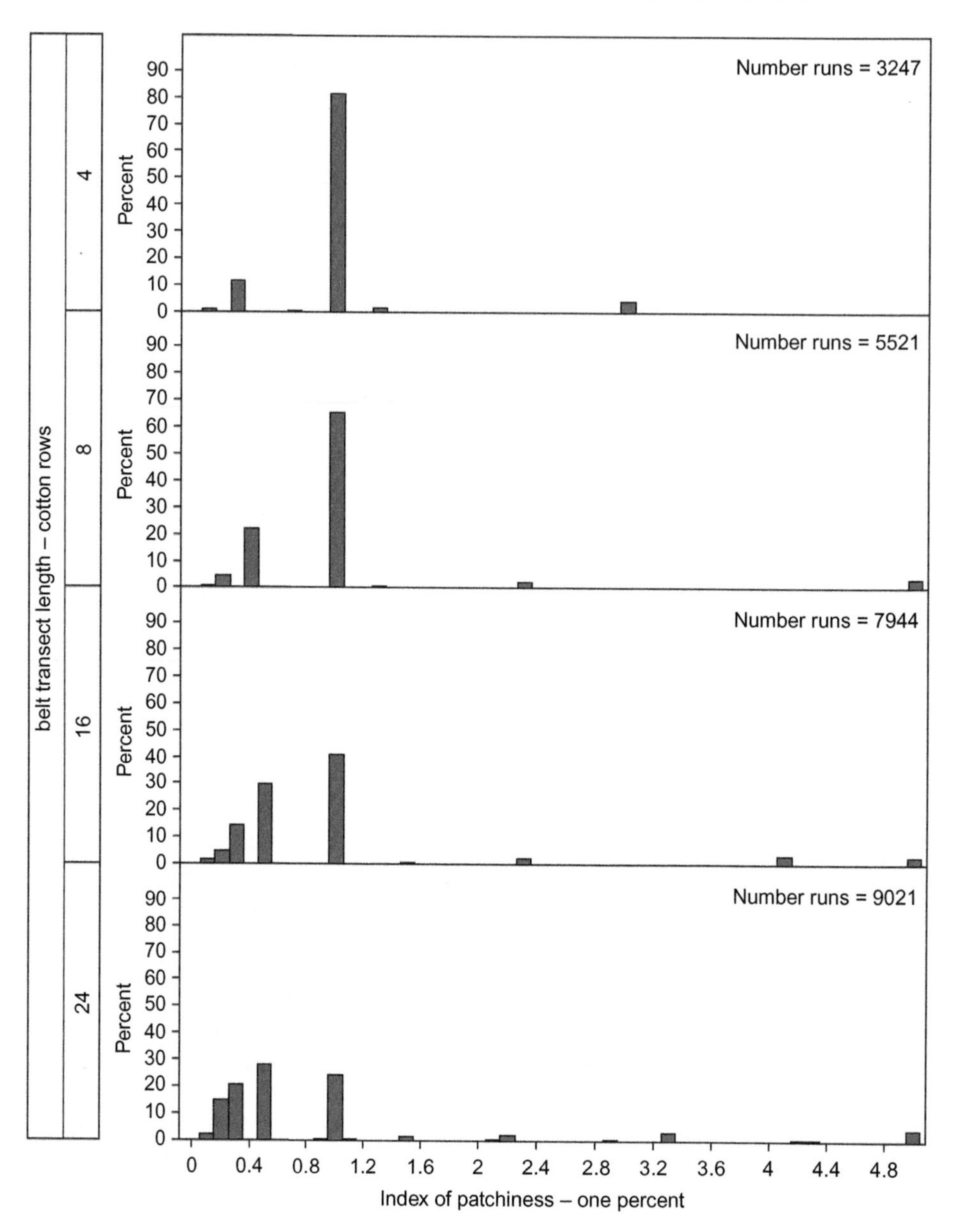

FIGURE 3.1 Histograms of results for Lloyd's (1967) index of patchiness for four different transect lengths at a simulated infestation rate of 1%. The index has a value <1 for regular dispersion patterns.

advantage when preparing site-specific pesticide prescriptions for application by a variable-rate sprayer. Thus, patterns of simulated dispersion responses at selected pest densities (λ_c = 0.01, 0.04, 0.16, 0.24, and 0.40) while using a large stacked belt transect are very consistent in contrast to the shorter transects of 24 or fewer quadrats. This stacked transect

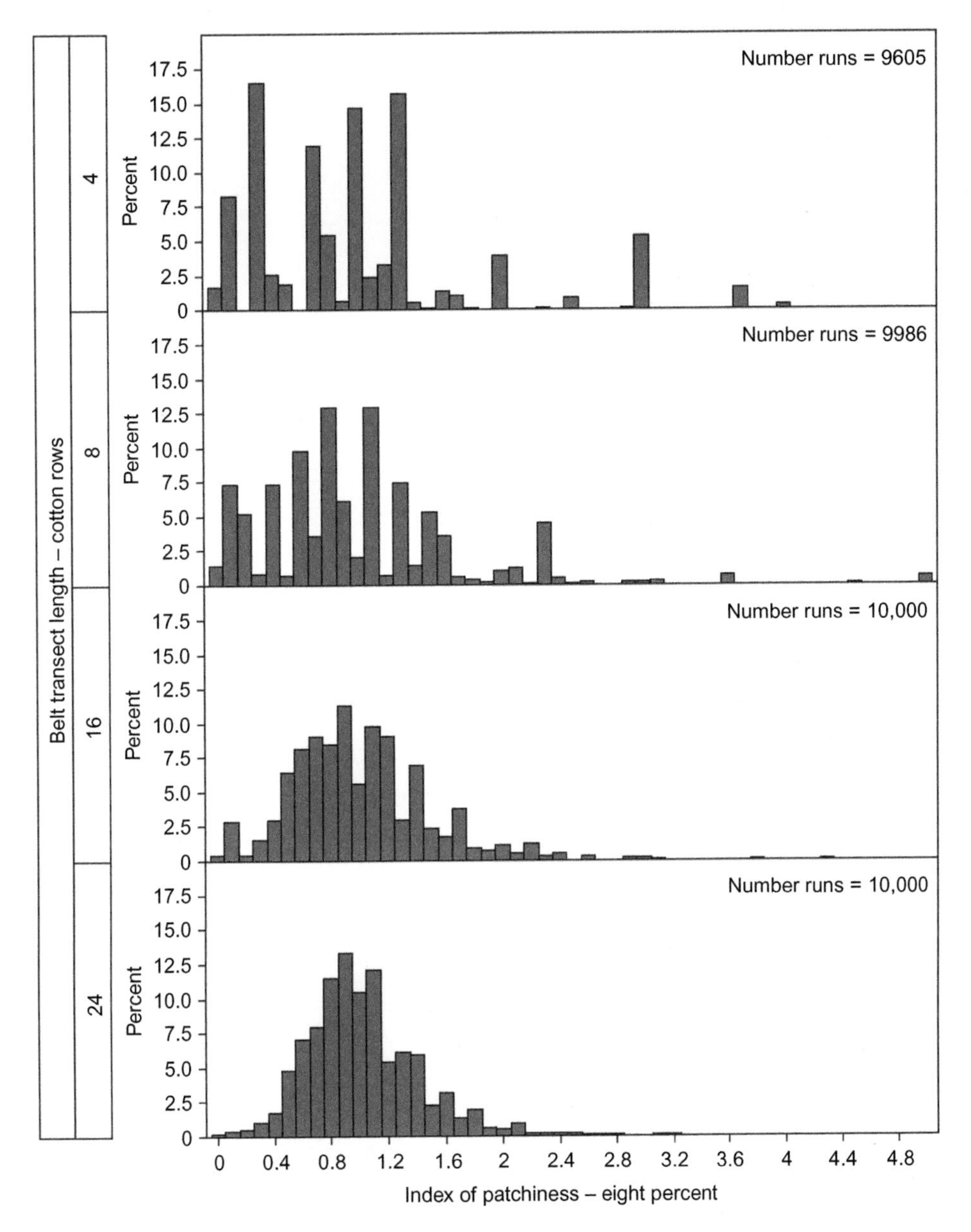

FIGURE 3.2 Histograms of results for the index of patchiness for four different transect lengths at a simulated infestation rate of 8%. The index has a value >1 for aggregated dispersion patterns.

(approximately 219.456 m^2) corresponds to a practical polygon size that matches the characteristics of a variable-rate sprayer (i.e. its boom length and controller response time).

Additional analysis of the simulated outcomes for this large belt transect revealed an interesting insight about increases in pest numbers per unit area. Unlike results shown

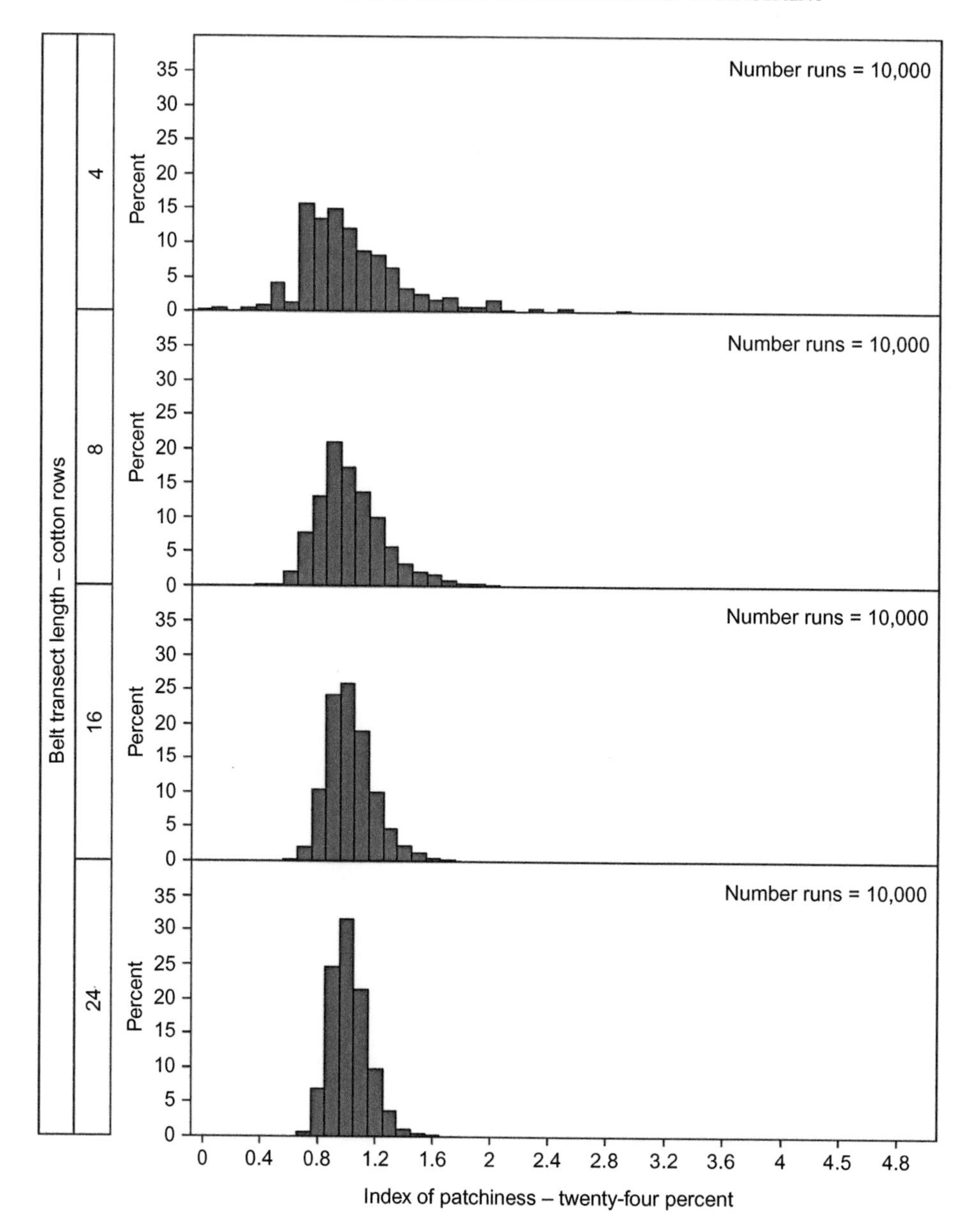

FIGURE 3.3 Histograms of results for the index of patchiness for four different transect lengths at a simulated infestation rate of 24%. The index is equivalent to 1 for random dispersion patterns.

for transect samples smaller than 24 quadrats (Figures 3.4–3.6), mean crowding estimates for a large stacked transect (L = 315 rows) were well separated as pest density increased (Figure 3.7). However, even for a very large sample unit size, if the pest density is low (see Figure 3.8, top panel), wide discrepancies in the patchiness index still occurred. For several runs when

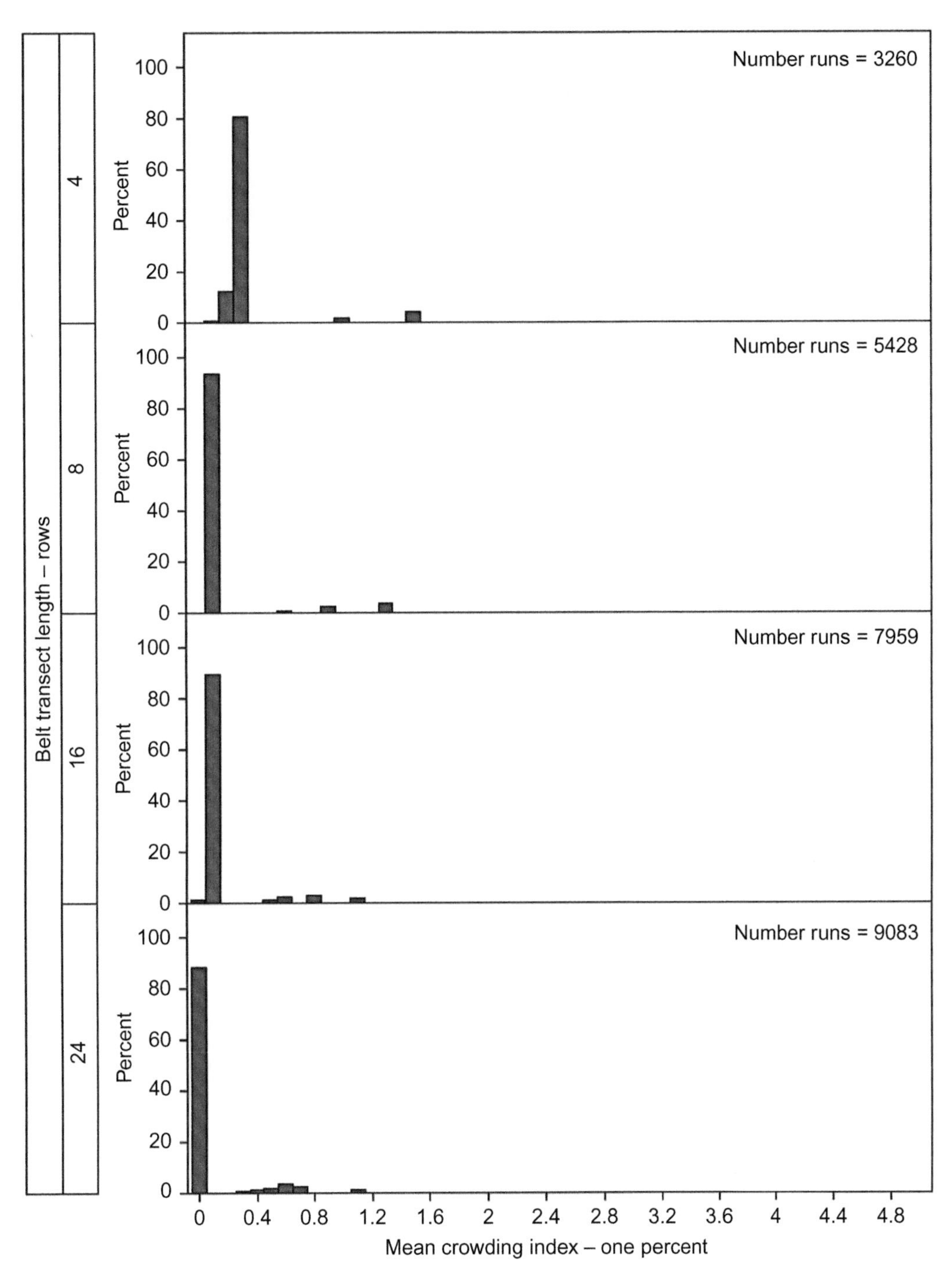

FIGURE 3.4 Histograms of results for mean crowding for four different transect lengths at a simulated infestation rate of 1%. Compare to Figure 3.7 for a sample unit size of 315 units.

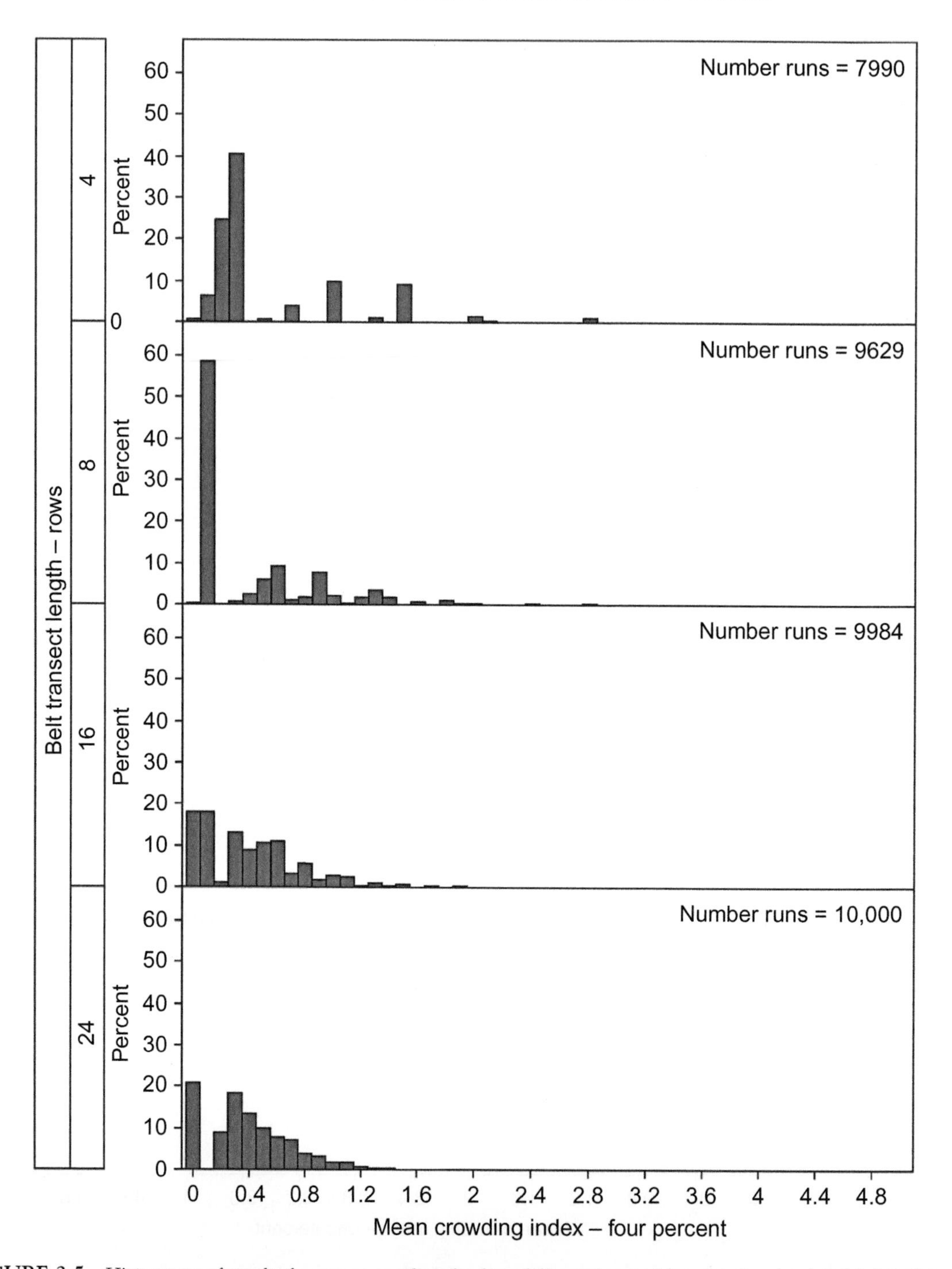

FIGURE 3.5 Histograms of results for mean crowding for four different transect lengths at a simulated infestation rate of 4%. Compare to Figure 3.7 for a sample unit size of 315 units.

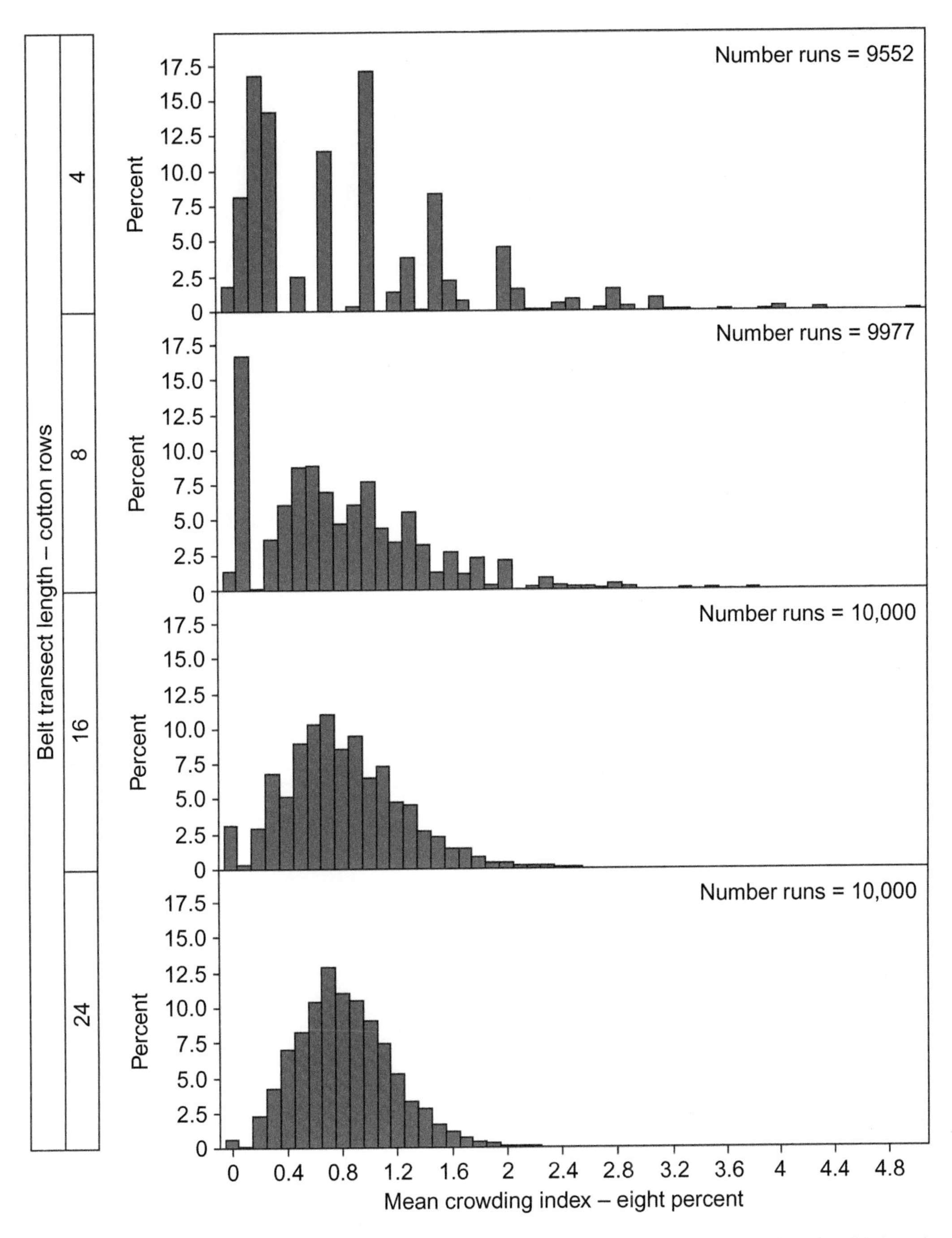

FIGURE 3.6 Histograms of results for mean crowding for four different transect lengths at a simulated infestation rate of 8%. Infestation rates larger than 8%, but less than 40% (not shown) would show less variability for different transect sizes. Compare to Figure 3.7 for a sample unit size of 315 units.

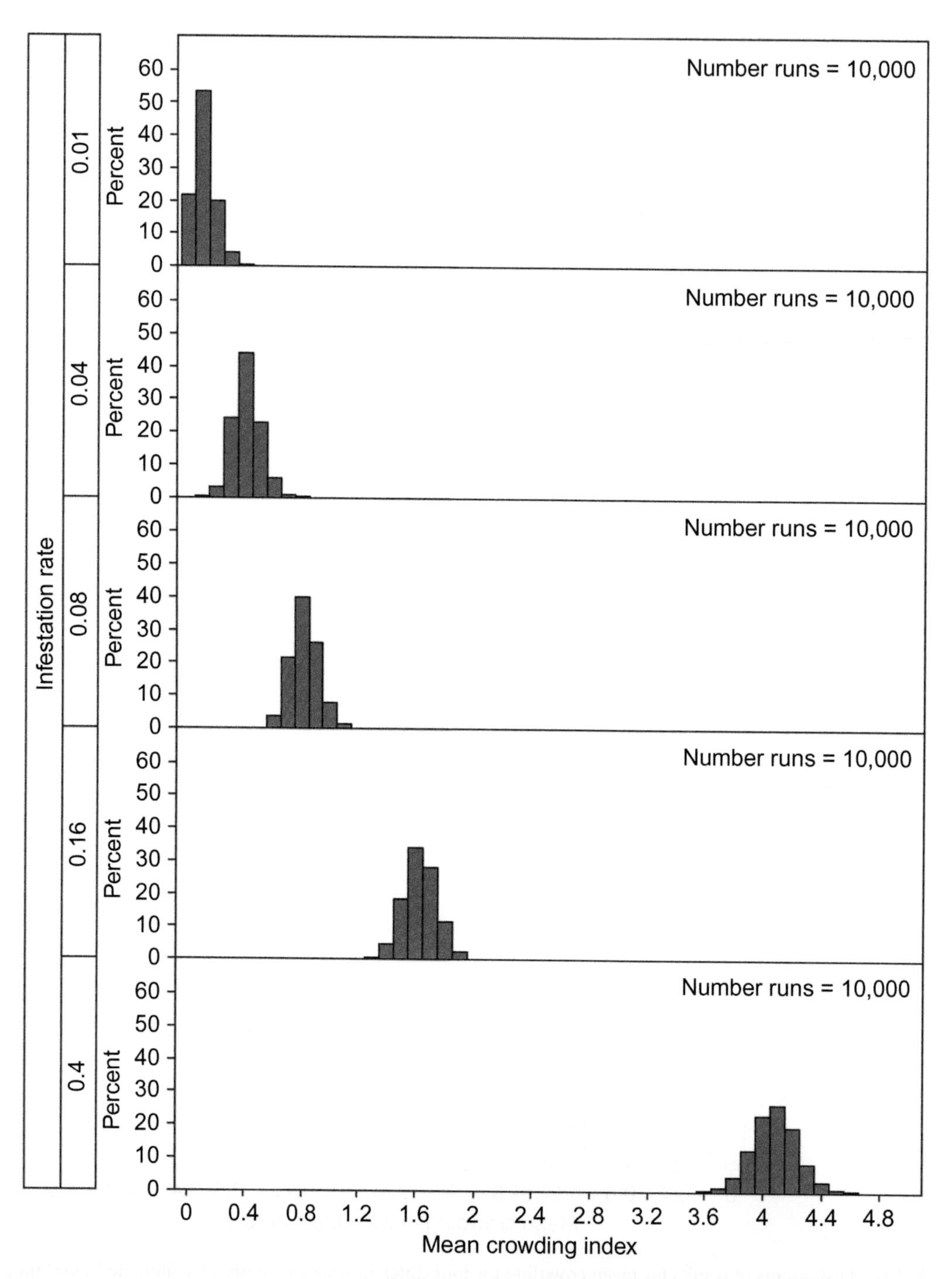

FIGURE 3.7 Histograms of results for Lloyd's (1967) mean crowding index for five different simulated pest densities where the belt transect length is 315 crop rows.

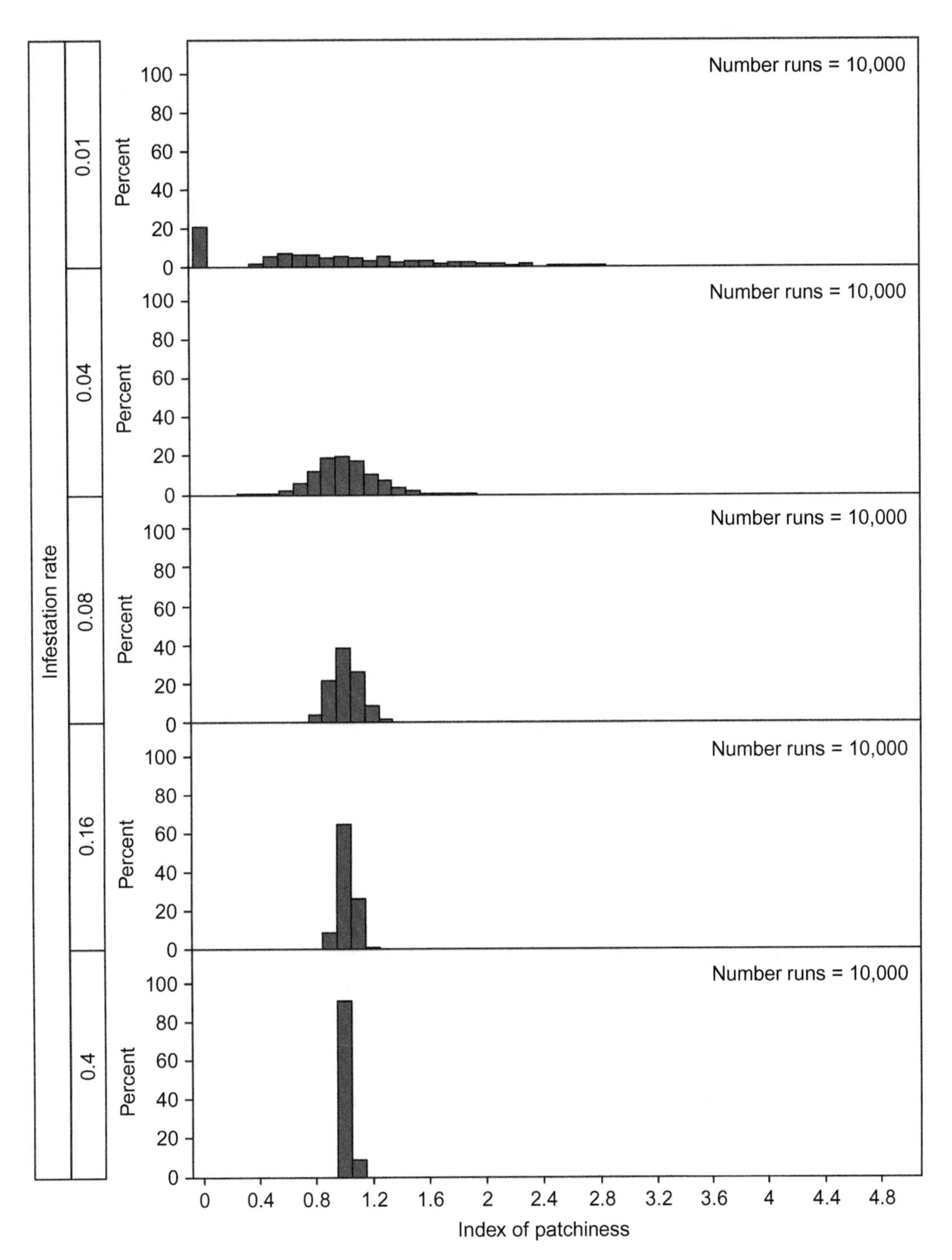

FIGURE 3.8 Histograms of results for the index of patchiness for five different simulated pest densities across a belt transect length of 315 crop rows.

the pest density was 1%, the index of patchiness was estimated to be >1, indicative of an aggregated dispersion pattern. The lack of stability at very low pest densities ($\lambda_c = 0.01$ as shown in the top panel of Figure 3.8) was previously reported by Byerly et al. (1978), and was attributed to the proportionately large effect that sampling errors contribute to estimation of standard error for a mean based on a small sample size. Similar influences are also likely to be at work on the outcomes observed for Lloyd's index of patchiness when a short belt transect is used to sample a habitat class having a low pest density. At low population densities, there may be insufficient numbers of pests to occupy all 'available habitat units' within a given class (or state of plant vigour) captured by the classified image. This scenario would also result in deviations from a random dispersion index of 1.0. However, the patchiness index is independent of population density once there is a departure from low infestation rates and when the sample unit size is very large. (Compare Figures 3.7 and 3.8 at infestation rates of 0.04, 0.08, 0.16, and 0.40.) This decoupling of Lloyd's index of patchiness from the mean crowding index was first reported by Myers (1978).

Convergent behavior about a patchiness value of 1.0 for the largest belt transect length (315 quadrats) used in the simulation study reinforces the validity of a random dispersion pattern over a large, consistent, spatial extent, particularly when pest density (or infestation rate) is large enough to take action. A map of sample site allocations based upon a classified, remotely-sensed image of several nearby cotton fields during the 2006 season typifies a real-world application of these simulation results. While it is impractical to sample very long belt transects in a commercial field setting, it is practical to apply SRS within homogeneous habitat classes (as determined by classified colour infrared (CIR) aerial imagery), select a few widely-spaced sampling locations (Figure 3.9) within a particular field over the course of a production season (Willers et al., 2005), and use belt transect lengths shorter than 4–8 crop rows (McKinion et al., 2009).

If the estimate of pest density is similar at the widely spaced sites within a particular habitat class, it is reasonable to infer that other unsampled locations between these sites are also infested at a similar rate. This pattern in the selection of scouting sites is strongly evident in Figure 3.9. The different habitats were repeatedly sampled at similar locations over time while other regions of the various habitat classes were never sampled during the 2006 production season. A habitat classification map derived from crop imagery (and based on some measure of plant vigour – such as Normalized Difference Vegetation Index (NDVI)) has been effectively used by field scouts to select sites for estimating infestation severity in commercial cotton fields without resorting to large sample sizes (e.g. Willers et al., 2005). Once the scout has learned that a particular habitat class (Figure 3.9) is occupied by the pest above the economic threshold, he/she can conclude that the entire habitat class needs treatment because, in all likelihood, the pests will be randomly dispersed throughout. He/she will come to know from sampling other habitat classes whether they demonstrate tendencies to be above or below an economic threshold. However, if the conclusion is that pest insect abundance is low (say, 1%), localized collections of pests in that habitat may occur, because simulation results indicated that, for low pest densities, clustered dispersion patterns are probable at times for even the largest sample unit size (recall top panel, Figure 3.8). For very low pest densities, the pattern of dispersion is unimportant because the cost of control would exceed the probable benefits of an application.

For each sampling date shown in the case study (Figure 3.9), only a small number of scouting sites (3–7) were required to make decisions 'not to spray', 'wait and see one more

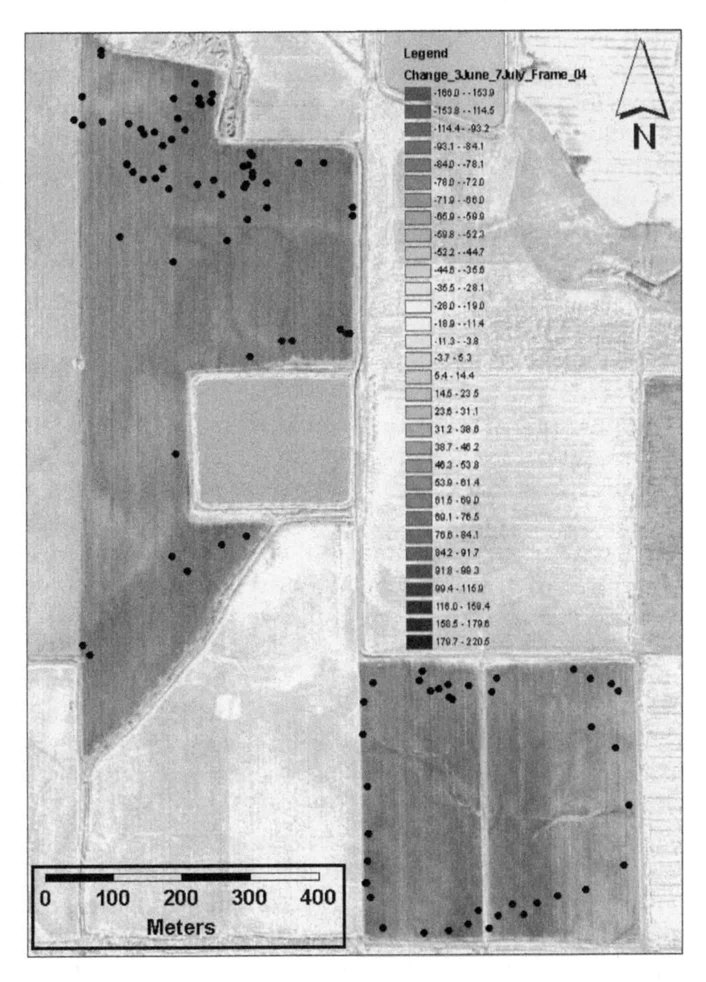

FIGURE 3.9 Example of scouting sites selection by a field scout for the entire 2006 production season for three cotton fields in Noxubee County, MS. This figure is reproduced in colour in the colour plate section.

week', or 'spray now here' for TPB management during any given week of the production season. Likewise, the histogram series (Figures 3.1–3.8) reflects this phenomenon of random dispersion because as sample unit size increased and pest density increased, the required sample unit size could become smaller. This concept learned from a simulation analysis provides information highly useful for refining pest sampling efforts in commercial cotton fields.

3.4 DISCUSSION

Stern et al. (1959) first presented the concept of an economic threshold (ET) or economic injury level (EIL). They elaborated on several key ecological concepts and discussed the impact of changes in pest density in both space and time (Fleischer et al., 1999). The simulation results presented here indicate that when a pest population exceeds an ET as low as four insects per 100 plants, it is plausible and practical to conclude that the pest persists at that density until the 'edge' of another habitat class is encountered. Across this edge or habitat boundary, pest density will either decrease or increase in response to shifts in environmental conditions encountered within these adjacent areas. These effects are stable during a small increment of time so that crop management decisions can be made. From this perspective, the chief value of remote sensing is the delineation of edges of cotton habitats (Willers et al., 1999, 2005; Willers, unpublished data) that are due to variability in edaphic or hydropedologic conditions as well as nutrient availability (Daubenmire, 1974; DeFauw et al., 2006; Gish et al., 2005).

In this study, Lloyd's indices evaluated the plausibility of an assumption (see also Willers et al., 2005, p. 438) that a random dispersion pattern occurs within a habitat class for many fruit feeding cotton insect pests if sufficiently sized sampling units are geographically nested. To link this finding to field applications, other small experiments, based upon discrete probability applications with dice, were also conducted. These transitional experiments convey how the linkages between the spatial resolution of image pixels and the spatial resolution of a variable-rate equipped ground sprayer combine to create sample units of various sizes that represent the entities actually sampled. Consequently, the perspective that pests follow a random dispersion pattern, in homogeneous habitat classes, whenever the infestation rate is large enough to require action, leads to simplicity in field applications. Here are the salient points. First, the field can be apportioned into distinctive habitat classes, similar to the use of colour in the experiments with dice. Second, these habitat classes can be apportioned into discrete sample units according to characteristics of variable-rate sprayer equipment. This is the same as the face of each side of any one die. Third, the intersections of these sprayer unit polygons with the pixel ground spatial distances of the classified remote sensing imagery comprise a population of countable sampling units (Figure 3.10) from which a SRS scheme can be employed to assess the risk of pest infestation in the habitat classes. Once a location in a field is sampled and its locus mapped by GPS, the collection of counts of insects using a quadrat-based construct generates a list of values from each location, just as lists can be generated from the simulations using dice. If the insect is ubiquitous throughout the field, then the lists of counts will behave similarly to those found in Tables 3.3 and 3.4. On the other hand, if the insect is only ubiquitous in some habitat classes and not others, then the list of sample counts from the field will behave similarly to that found in Table 3.5. And, if a count does not arise (for example, the absent 3 for the colour blue (Table 3.4)), it will show up promptly as occurrences of zeros in just a few samples ($n < 30$). Interestingly, however, for a count such as the 3, which was doubled up for the red colour, it requires a very large number of samples to assess a subtle pattern involving nonzero counts. But, if there is a pattern of counts strongly associated with a spatial structure, such as colour (Table 3.5), then that reality is quickly noticed and can be determined with very, very small sample sizes (for example, seven rolls or less). These points, for example, are further confirmation of the idea that there are patterns in the counts from a geo-referenced SRS plan as first investigated by a sensitivity analysis using resampling methods (Willers et al., 2000) and a second work comparing resampling and count model regression approaches of field samples obtained by two persons (Willers et al., 2009). Thirdly, the effects of sample unit size on these patterns is also strengthened as described in Willers et al. (1990, 2005). Therefore, all of these sources of evidence indicate that it is not necessary to sample all patches of habitat in any given field, as far too many people believe necessary to do with a site-specific sampling plan employing classified imagery products.

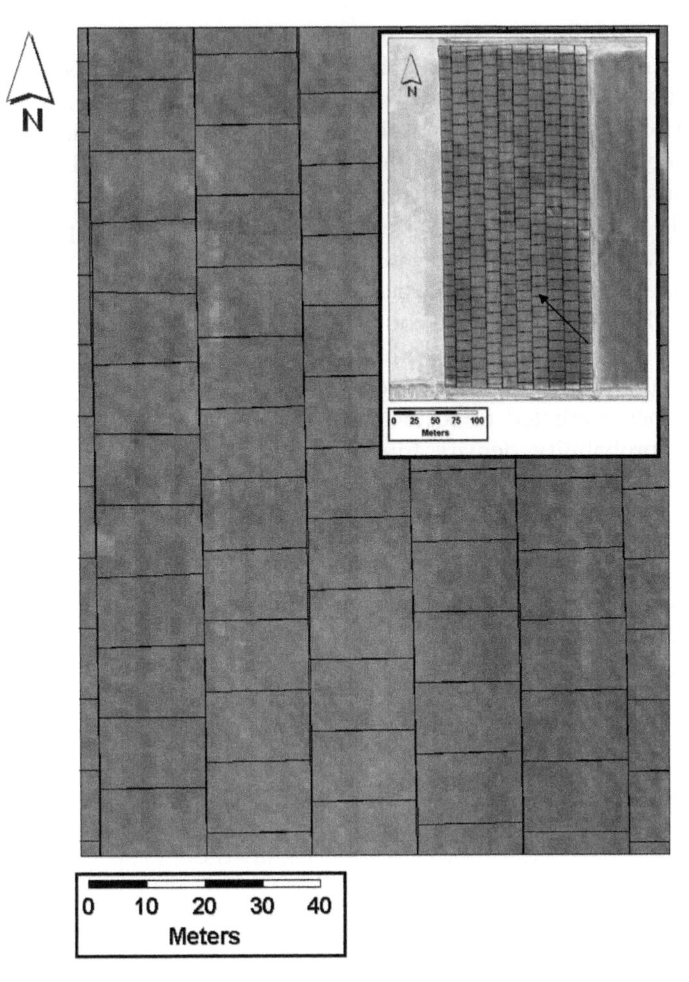

FIGURE 3.10 Example spray grid showing the polygons within sprayer paths that are the basis of assignments for different pesticide rates to build a spatial prescription. Each polygon of the grid is equivalent in area to a stacked belt transect sample of 315 units (or 219.456 m^2). This figure is reproduced in colour in the colour plate section.

In commercial applications, it is not possible to satisfy the sample number sizes often required by theory (Karandinos, 1976); therefore, discovering new estimators accurate at low pest densities while using small sample sizes is advantageous. Armed with an understanding of the relationships gleaned from this and previous simulation efforts (Willers et al., 2005), numerous seasons of fieldwork confirm that it is possible to use smaller numbers of field samples. The samples can be obtained more effectively and better interpreted if remotely-sensed imagery is available (Willers and Riggins, 2010). Information derived from timely and frugal sampling methods linked to remote sensing imagery of row crop landscapes provides the template for site-specific applications of pesticides, reducing costs and providing environmental benefits (Dupont et al., 2000; Frigden et al., 2002; Seal et al., 2001). This additional benefit of classified imagery for field sampling was also demonstrated in the experiments with coloured dice.

Opportunities exist to further refine the computer simulation model. For instance, the model may be too heavily influenced by values of abundance from the upper tail of the NBD. It may be more realistic to assume that biological populations within a homogeneous habitat (Willers et al., 1999, 2005) class are less likely to exhibit the higher count values with respect to their population mean for a particular habitat class. In other words, sample count distributions may be more restrictive within habitat classes under field conditions, such that the higher extremes predicted to occur by a fit to a particular probability density function may actually not occur. Other possibilities to consider are that the classification procedure (e.g. Backoulou et al., 2013; Willers et al., 2012) applied to the imagery may not have been the most appropriate choice for a particular pest species or that the information captured by a particular sensor type is erroneous due to inadequate calibration or sensor failure.

3.5 CONCLUSIONS

The simulation analyses indicate that the assumption of random dispersion of a pest within crop habitats (based on remotely-sensed phenologic indicators that correlate well with pest density differences) is practical and prudent for commercial scouting and decision-making purposes. The assumption strengthens continued use of SRS or line-intercept (LIS) estimators for pest densities within habitat classes established by the geoprocessing of remote sensing images acquired during key crop-specific stages of the growing season. The findings reported here, when linked to outcomes presented in our earlier works (Willers and Akins, 2000; Willers and Riggins, 2010; Willers et al., 1990, 1999, 2005, 2009, 2012) and other investigators (e.g. Carrière et al., 2006; Dammer and Adamek, 2012; Karimzadeh et al., 2011; van Helden, 2010) build up a body of knowledge helpful for assessing pest dispersion and abundance in large, remotely-sensed, commercial production fields. These concepts and methods on image-based, geographical SRS sampling procedures are expected to change the future of insect pest control once further research is accomplished.

Acknowledgments

We thank Drs Eric Villavaso and Wes Burger for their initial comments on an earlier draft of this manuscript. Thanks are also due to other anonymous reviewers who have prompted the lead author to make considerable changes compared to an initial manuscript. The Advanced Spatial Technologies for Agriculture Program (ASTA) of Mississippi State University has provided partial funding for this investigation. The cooperative assistance rendered by Mr Kenneth Hood, Mr John Freeman, Mr John Bassie, Sr, and Mr M. Doug Cauthen and other staff of Perthshire Farms, Gunnison, MS and Mr Paul Good, Mr Dale Weaver and Mr Bert Falkner, Good's Longview Farm, Macon, MS, is also appreciated.

References

Anscombe, F.J., 1949. The statistical analysis of insect counts based on the negative binomial distribution. Biometrics 5, 165–173.

Ash, C., 1993. The probability tutoring book. An Intuitive Course for Engineers and Scientists. IEEE Press, New York.

Backoulou, G.F., Elliott, N.C., Giles, K.L., Rao, M.N., 2013. Differentiating stress to wheat fields induced by *Diuraphis noxia* from other stress causing factors. Comput. Electron. Agric. 90, 47–53.

Banerjee, B., 1976. Variance to mean ratio and the spatial distribution of animals. Experientia 72, 993–994.

Byerly, K.F., Gutierrez, A.P., Jones, R.E., Luck, R.F., 1978. A comparison of sampling methods for some arthropod populations in cotton. Hilgardia 46, 257–282.

Carrière, Y., Ellsworth, P.C., Dutilleul, P., Ellers-Kirk, C., Barkley, V., Antilla, L., 2006. A GIS-based approach for areawide pest management: the scales of *Lygus hesperus* movements to cotton from alfalfa, weeds, and cotton. Entomol. Exp. Appl. 118, 203–210.

Dalthorp, D., Nyrop, J., Villani, M.G., 2000. Foundations of spatial ecology: The reification of patches through quantitative description of patterns and pattern repetition. Entomol. Exp. Appl. 96, 119–127.

Dammer, K.-H., Adamek, R., 2012. Sensor-based insecticide spraying to control cereal aphids and preserve lady beetles. Agron. J. 104 (6), 1694–1701.

Daubenmire, R.R., 1974. Plants and Environment: A Textbook of Autecology, third ed. John Wiley and Sons, New York.

Davis, P.M., 1994. Statistics for describing populations. In: Pedigo, L.P., Buntin, G.D. (Eds.), Handbook of Sampling Methods for Arthropods in Agriculture. CRC Press, Boca Raton, FL, pp. 33–54.

DeFauw, S.L., English, P.J., Harris, F.A., Willers, J.L., 2006. Field-scale stability assessment of cotton productivity zones using NDVI imagery, soils, TDR, Veris and yield data. Proceedings of the Beltwide Cotton Conference, vol. 8, pp. 2092–2099d.

Dupont, J.K., Campenella, R., Seal, M.R., Willers, J.L., Hood, K.B., 2000. Spatially variable insecticide applications through remote sensing. Proceedings of the Beltwide Cotton Conference, vol. 2, pp. 426–429.

Fleischer, S.J., Blom, P.E, Weisz, R., 1999. Sampling in precision IPM: when the objective is a map. Phytopathology 89, 1112–1118.

Frigden, J.J., Seal, M.R., Lewis, M.D., Willers, J.L., Hood, K.B., 2002. Farm level spatially-variable insecticide applications based on remotely sensed imagery. Proceedings of the Beltwide Cotton Conference. 13 pp. (unpaginated CD).

Gish, T.J., Walthall, C.L., Daughtry, C.S.T., Kung, K.-J.S., 2005. Using soil moisture and spatial yield patterns to identify subsurface flow patterns. J. Environ. Qual. 34, 274–286.

van Helden, M., 2010. Spatial and temporal dynamics of arthropods in arable fields. In: Oerke, E., Gerhards, R., Menz, G., Sikora, R.A. (Eds.), Precision Crop Protection – the Challenge and Use of Heterogeneity, first ed. Springer, pp. 51–64.

Karandinos, M.G., 1976. Optimum sample size and comments on some published formulae. Bull. Entomol. Soc. Am 22, 417–421.

Karimzadeh, R., Hejazi, M.J., Helali, H., Iranipour, S., Mohammadi, S.A., 2011. Assessing the impact of site-specific spraying on control of Eurygaster integriceps (Hemiptera: Scutellieridae) damage and natural enemies. Prec. Agric. 12, 576–593.

Lloyd, M., 1967. Mean crowding. J. Anim. Ecol. 36, 1–30.

Ludwig, J.A., Reynolds, J.F., 1988. Statistical Ecology: A Primer on Methods and Computing. John Wiley and Sons, New York.

McKinion, J.M., Jenkins, J.N., Willers, J.L., Zusmanis, A., 2009. Spatially variable insecticide applications for early season control of cotton insect pests. Comp. Electron. Agric. 67, 71–79.

Myers, J.H., 1978. Selecting a measure of dispersion. Environ. Entomol. 7, 619–621.

Nelson, T., Wilson, H.G., Boots, B., Wulder, M.A., 2005. Use of ordinal conversion for radiometric normalization and change detection. Int. J. Rem. Sens. 26, 535–541.

Pielou, E.C., 1960. A single mechanism to account for regular, random and aggregated populations. J. Anim. Ecol. 48, 574–584.

Pielou, E.C., 1977. Mathematical Ecology. John Wiley and Sons, New York.

Pielou, E.C., 1978. Population and Community Ecology: Principles and Methods. Gordon and Breach, New York.

Poole, R.W., 1974. An Introduction to Quantitative Ecology. McGraw-Hill, New York.

Pritsker, A.B., Pegden, C.D., 1979. Introduction to Simulation and SLAM. Halsted Press, John Wiley and Sons, New York.

Richards, J.A., Jia, X., 1999. Remote Sensing Digital Image Analysis. An Introduction, third ed. Springer-Verlag, Berlin.

SAS Institute, 1990. SAS® Language: Reference, Ver. 6, first ed. SAS Institute, Cary, NC.

Seal, M., Dupont, K., Bethel, M., Lewis, D., Johnson, J., Willers, J., et al., 2001. Utilization of remote sensing technologies in the development and implementation of large-scale spatially variable insecticide experiments in cotton. Proceedings of the Beltwide Cotton Conference, Vol. 2, 1010–1018.

Southwood, T.R.E., 1978. Ecological Methods with Particular Reference to the Study of Insect Populations. Halsted Press, John Wiley and Sons, New York.

Southwood, T.R.E., Brown, V.K., Reader, P.M., 1983. Continuity of vegetation in space and time: a comparison of insects' habitat templet in different successional stages. Res. Popul. Ecol. (Suppl. 3), 61–74.

Steel, R.G.D., Torrie, R.H., 1960. Principles and Procedures of Statistics. McGraw-Hill, New York.

Stern, V.M., Smith, R.F., van den Bosch, R., Hagen, K.S., 1959. The integrated control concept. Hilgardia 29, 81–101.

Theobald, D.M., 2003. GIS Concepts and ArcGIS® Methods. Conservation Planning Technologies, Fort Collins, CO.

Thompson, S.K., 1992. Sampling. Wiley-Interscience, New York.

Trumble, J.T., 1985. Implications of changes in arthropod distribution following chemical applications. Res. Popul. Ecol. 27, 277–285.

Willers, J.L., Akins, D.S., 2000. Sampling for tarnished plant bugs in cotton. Southwest. Entomol. (Suppl. 23), 39–57.

Willers, J.L., Riggins, J.J., 2010. Geographical approaches for integrated pest management of arthropods in forestry and row crops. In: Oerke, E., Gerhards, R., Menz, G., Sikora, R.A. (Eds.), Precision Crop Protection – the Challenge and Use of Heterogeneity, first ed. Springer, pp. 183–202.

Willers, J.L., Boykin, D.L., Hardin, J.M., Wagner, T.L. Olson, R.L., Williams, M.R., 1990. A simulation study on the relationship between the abundance and spatial distribution of insects and selected sampling schemes. Proceedings of the Conference on Applied Statistics in Agriculture, 29 April–1 May 1990, Kansas State University, Manhattan, KS, pp. 33–45.

Willers, J.L., Seal, M.R., Luttrell, R.G., 1999. Remote sensing, line-intercept sampling for tarnished plant bugs (Heteroptera: Miridae) in Mid-south cotton. J. Cotton Sci. 3 (4), 160–170.

Willers, J.L., Ladner, W.L., McKinion, J.M., Cooke, W.H., 2000. Application of computer intensive methods to evaluate the performance of a sampling design for use in cotton insect pest management. Proceedings of the Conference on Applied Statistics in Agriculture, 30 April–2 May 2000, Kansas State University, Manhattan, KS, pp. 119–133.

Willers, J.L., Jenkins, J.N., Ladner, W.L., Gerard, P.D., Boykin, D.L., Hood, K.B., et al., 2005. Site-specific approaches to cotton insect control. Sampling and remote sensing analysis techniques. Prec. Agric. 6, 431–452.

Willers, J.L., Jenkins, J.N., McKinion, J.M., Gerard, P., Hood, K.B., Bassie, J.R., et al., 2009. Methods of analysis for georeferenced sample counts of tarnished plant bugs in cotton. Prec. Agric. 10, 189–212.

Willers, J.L., Wu, J., O'Hara, C., Jenkins, J.N., 2012. A categorical, improper probability method for combining NDVI and LiDAR elevation information for potential cotton precision agricultural applications. Comput. Electron. Agric. 82, 15–22.

Williams, M.R., Wagner, T.L., Willers, J.L., 1995. Revised protocol for scouting arthropod pests of cotton in the Midsouth. Miss. Agric. For. Exp. Stn. Tech. Bull. 206.

Wilson, L.T., 1994. Estimating abundance, impact, and interactions among arthropods in cotton agroecosystems. In: Pedigo, L.P., Buntin, G.D. (Eds.), Handbook of Sampling Methods for Arthropods in Agriculture. CRC Press, Boca Raton, FL, pp. 475–514.

CHAPTER

4

Weather-based Pest Forecasting for Efficient Crop Protection

Rabiu Olatinwo[1] *and Gerrit Hoogenboom*[2]

[1]**Louisiana State University, LA, USA,** [2]**Washington State University, WA, USA**

4.1 INTRODUCTION

4.1.1 Crop Protection and Current Challenges

Although insects, pathogens, mites, nematodes, weeds, vertebrates, and arthropods are different in many ways, they are regarded as pests. They are a major constraint to crop productivity and profitability around the world caused by direct and indirect damage to valuable crops. Insect pests, pathogens, and weeds account for an estimated 45% of pre- and post-harvest losses worldwide (Pimentel, 1991), in addition to losses caused by vertebrate pests (Strand, 2000). Each year, farmers are confronted with several questions and uncertainties on how best to manage potential threats posed by pests to valuable crops, particularly when a significant amount of resources is committed to cultivation and production process, in expectation of profitable yields. Finding ways to address these problems has led to changes in agriculture production systems over the years, with an increase in the use of chemical pesticides to minimize pest damage. However, some unintended consequences, such as emergence of pest resistance due to repeated use of pesticides, mean resurgence of pests and crop damage are still of great concern, in addition to lingering negative effects of pesticide residues on the environment.

These uncertainties have led to pertinent questions such as: What can be done to mitigate the risks caused by disease and pests to valuable crops? Can weather forecasts reliably help predict the risk of disease and pest outbreaks? To what extent can we have forewarning for effective management of pests and diseases with minimal inputs? Are the incidences or severity of these diseases and pests avoidable or predictable? If predictable, can farming practices be improved by incorporating weather information into existing management strategies, for instance to improve the effectiveness of pesticides through minimal and timely applications, to manage pests, but also reduce risks?

Faced with huge risks of crop damage and many uncertainties, farmers need effective and sustainable solutions every season to ensure profitability. This is especially so in many parts of the world where expanding agricultural

D. P. Abrol (Ed): Integrated Pest Management.
DOI: http://dx.doi.org/10.1016/B978-0-12-398529-3.00005-1

productivity depends on timely, effective, and accurate use of information gathered from multiple sources (including, for example weather forecasts). Weather information is especially critical for making management decisions to avoid or mitigate potential disease and pest outbreaks, improve crop development, and achieve profitable yields.

For a pest attack or disease outbreak to occur, three basic factors must be present; a susceptible host plant, a virulent pathogen or pest, and favourable environmental factors that facilitate disease initiation or pest attack. Favourable weather factors (e.g. temperature, rainfall, wind, relative humidity) may exist within the canopy, on a local scale within the field, or on a regional scale, across several farms. Seasonal variability in weather patterns influenced by preceding or prevailing climatic conditions not only creates a conducive environment for pest population development and distribution, but also influences crop growth and development, and ultimately final yield. Therefore, understanding the delicate balance between host and pest sensitivity to environmental factors such as weather is critical for survival of vulnerable host crops, or a successful attack by aggressive pests or virulent pathogens. Infestations mostly occur when environmental conditions are favourable for initial attack and subsequent interactions with the host. Conducive weather conditions or lack there of, are therefore critical for both pest the population and crop development.

However, since no two growing seasons are the same, extreme weather events driven by climate variability, in addition to increasing global demand for crops, and productivity pressure, have pushed cultivations into regions where conditions are becoming more favourable for invasive pest development, making crops in those regions more predisposed to non-native pest attacks. The ever-expanding worldwide trade and globally increasing demands for food and plant products have also led to crop production pressure, an increase in pests' resistance to pesticides that in the long term may increase pest activities even further, due to intensification of cropping, reduced crop rotation, and increased monoculture (Rosenzweig et al., 2001).

4.1.2 Weather, Pest, and Crop Interactions

Whatever the nature of interactions between pests and host crops, weather factors create an additional layer of uncertainty to already complex dynamic interactions between a pest and its host plant. Understanding the nature of this complex interaction requires an interdisciplinary approach to identify critical components needed to develop management tools to address the pest and disease concerns of a farmer. The relationship between two or more organisms within the immediate ecosystem of a crop, in many ways can facilitate the extent of damage caused by a pest, or symptoms observed on host tissues. For example, a warmer than usual condition that favours pest attack may equally favour a competitor, or be conducive for a crop variety to resist attack, whereas reverse conditions such as stress may predispose the same variety to successful attacks by pests or pathogens. However, unfavourable dry conditions may actually be detrimental to both crop growth and pests (e.g. fungal pathogen sporulation).

The dynamic nature of sequences of ecological processes is hard to predict due to many uncertainties inherent in such complex interactions. This complexity and uncertainties, therefore, create opportunities for scientists across different fields to study, understand, and develop management strategies in solving emerging pest problems faced by farmers. Achieving comprehensive integrated management through a multidisciplinary approach is a viable option to mitigate the potential risks of disease or pest epidemics. However, any integrated approach starts with examining the three key components individually, i.e. crop, pests, and weather, followed by understanding how the delicate interactions that exist among them could be exploited in mitigating potential threats of pest and disease attacks.

4.2 WEATHER

In the past, favourable weather conditions such as warm weather that boosts the pest population, or mild winter temperatures that increase the chance of pest survival through the winter, are known to increase the use of agricultural chemical pesticides, thereby heightening health risks and increasing ecological and economic costs. In the future, extreme weather events from climate variability are expected to contribute directly and indirectly even more to a potential increase in pest damage and the use of chemical pesticides to control the increased pest pressure (Rosenzweig et al., 2001, 2000; Yang and Scherm, 1997).

4.2.1 Weather Factors and Derived Variables

Weather variables including temperature, rainfall, and relative humidity, have been tested and reported on extensively in many disease studies (Bailey et al., 1994; Nokes and Young, 1991; Wharton et al., 2008; Olatinwo et al., 2008, 2009, 2010). In some studies, individual computer programs have been developed based on various weather parameters to make predictions, while others studies have incorporated computer programs into commercial advisory equipment (Cu and Phipps, 1993; Grichar et al., 2005; Jensen and Boyle, 1965, 1966; Linvill and Drye, 1995; Parvin et al., 1974; Shew et al., 1988; Wu et al., 1999).

Whether excessive, optimal, or insufficient, temperature and rainfall are perhaps the most important variables affecting crop–pest interactions. For example, many pest species favour warm and humid conditions, while moisture stress may cause direct or indirect effects to crop development, making crops more vulnerable to damage by pests, especially at the early stages. Pest infestations often coincide with favourable climatic conditions or weather patterns, such as early or late rains, drought, or increases in humidity, which in themselves can reduce yield.

In most cases, favourable temperature is critical for pest development, population growth, pest epidemics, the extent of damage caused to crops, and the overall crop yield. Cold-blooded pests (i.e. insects) are sensitive to temperature, and therefore insects typically respond to higher temperature, which increases the rate of development and reduces the time between generations. However, very high temperatures may also reduce insect longevity. Warmer winters (mild winters) reduce winterkill of pests through the winter, thereby allowing a greater number of pests to survive through a normally expected harsh winter season, and consequently, increase insect populations in subsequent growing seasons. Rosenzweig et al. (2001) noted that drought resulting from extreme high temperatures and reduced rainfall, changes the physiology of host species, leading to changes in the insects that feed on them, and can reduce populations of friendly insects (such as predators or parasitoids), spiders and birds, and ultimately influences the impact of pest infestations. In addition, abnormally cool, wet conditions can also bring on severe insect and plant pathogen infestations, although excessive soil moisture may drown soil-residing insects.

In addition to temperature and precipitation, relative humidity is another weather variable that has been shown in many studies to be related to the development of fungal pathogens (Damicone et al., 1994; Jensen and Boyle, 1965, 1966; Jewell, 1987; Olatinwo et al., 2008, 2009, 2011; Shew et al., 1988; Wu et al., 1999). In monitoring the likelihood of infection initiation through sporulation of fungal spores, the available moisture on a leaf surface can be estimated using relative humidity, since it correlates with wetness of a leaf surface within a canopy. It is a critical component for estimating the likelihood of successful infection initiation and foliar disease development by fungal pathogens (Jensen and Boyle, 1965, 1966). Generally, a relative humidity of ≥95%, equivalent to saturation, is assumed to indicate a level of leaf wetness or moisture on the leaf surface sufficient for sporulation and infection initiation on leaf tissue. Although leaf wetness as a weather parameter is rarely measured, a few empirical methods (Matra et al., 2005) have been used to derive leaf wetness durations from meteorological parameters. Dew is another important weather parameter that also influences leaf wetness duration and plays a significant role in facilitating germination of spores and entrance of disease spores into crop tissues (Das et al., 2007).

The infection process of a disease such a Downey mildew (*Bremia lectucae*) may occur rapidly (i.e. within 48h when the leaf wetness requirement is met) to the extent that sufficient time is unavailable for fungicide application or for any meaningful control measure to be taken (Strand, 2000; Scherm and van Bruggen, 1993). Strand (2000) noted that, for such diseases, obtaining leaf wetness and the period of wetting information from weather forecast can provide farmers with sufficient lead-time to take adequate control measures for preventing disease outbreak.

In developing disease and pest models, input variables are not limited to only air temperature, rainfall, and relative humidity, but also include other variables such as wind speed, wind direction, soil temperature, soil moisture, and solar radiation. Depending on pest model needs and knowledge about the biology of a pest of interest or the corresponding host plant, additional weather variables may be required for developing a predictive model. Therefore, measurements of other weather parameters can be obtained by using different techniques and equipment at the field level.

4.2.2 Critical Weather Variables for Pest Forecasting

Whether complex or simple, a disease or pest model mostly requires essential environmental variables as inputs to be operational, depending on individual pathogen or pest sensitivity to environmental factors. Access to weather data and derived variables from temperature, rainfall, humidity, and other measurements, is essential for developing, testing, and evaluating these models. For example, models that are based on insect phenology, using derived variables from degree-days accumulation, are more applicable in most environments, since they utilize knowledge about individual pest species and its sensitivity to baseline temperature that correlates with pest population growth. This is usually determined from prior laboratory experiments, field trials, and specific information about the pest biology. For example, Dawidziuk et al. (2012) found that higher winter temperatures could increase the ability of *Leptosphaeria maculans* and *L. biglobosa* pseudothecia to discharge ascospores into the air, causing damaging and early plant infections in oilseed rape.

Rainfall patterns (i.e. frequency and intensity) are among the commonly used weather information that is needed for timely scheduling of pesticide applications to prevent pest development or protect crops with signs of early symptoms of a pest attack. In the past, farmers have also used prevailing weather information to modify the microclimate conditions within the canopy (i.e. by lowering the humidity) to reduce the likelihood of infection initiation and disease development (Strand, 2000).

For example, the significance of weather parameters in the development of the thrips population has been reported in several studies, including *Thrips palmi* (McDonald et al., 1998), onion thrips *Thrips tabaci* (Edelson and Magaro, 1988; Morsello et al., 2008), tobacco thrips (Morsello et al., 2008), and western flower thrips *F. occidentalis* (Katayama, 1997). The tobacco thrips (*Frankliniella fusca* Hinds) and the western flower thrips (*Frankliniella occidentalis* Pergande) are particularly important thrips species that have a significant economic impact on several crops in the southeastern United States (Olatinwo et al., 2008, 2009, 2011). The population peaks of important thrips in field crops and vegetables in the southeastern US mostly occur during the first and second week of May (McPherson et al., 1999; Riley and Pappu, 2000, 2004). Therefore, management decisions such as scouting for pests at weekly (McPherson et al., 1999) or biweekly intervals may be expensive and particularly time consuming, apart from when weather information is available for monitoring population progressions (Olatinwo et al., 2011).

After repeated fungicide applications, many diseases such as leaf spots of peanut may develop resistance to fungicides (Culbreath et al., 2002; Woodward et al., 2010). Monitoring environmental conditions such as rainfall, relative humidity, leaf surface wetness, and temperature to optimize pesticide applications is critical for reducing infection initiation and disease development (Alderman and Beute, 1986; Jensen and Boyle, 1965, 1966; Shew et al., 1988; Wu et al., 1999). Although chemical pesticides are effective tools for managing diseases and pests, they can be inefficient methods for managing pests due to unintended negative impacts on the ecosystems. Constantly monitoring these weather variables are important for delivering an effective pest management strategy, through timely pesticide applications that minimize the overall negative impacts on the environment.

Compared to only 5 or 10 years ago, access to readily available weather information makes many farm management decisions less complicated, especially for disease or pest models that require historical weather data, prevailing weather conditions, and weather forecasts for predicting potential pest risks. However, some gaps still exist in terms of weather data reliability and in translating complex weather information to timely warnings that may significantly reduce the risks associated with pest attacks. With increasing global access to mobile phones, electronic text messages, emails, and dynamic internet website information (i.e. that incorporates weather forecasts into existing pest models) are becoming effective means of instantly communicating pest risk information to farmers. To earn the trust of end-users, the uncertainty inherent in pest models and weather forecasts used in generating a risk alert must be addressed and concisely presented to users, since the accuracy and reliability of disseminated risk information are critical for management decisions and adoption of such products.

Apart from the general weather forecasts, which are limited to the meteorological elements and factors such as maximum and minimum temperature, type, duration and amount of precipitation, cloudiness, and wind speed and direction, there are other types of forecast (Das et al., 2007) that might be useful depending on the required range of the forecast. These include the *nowcasting and very short-range forecasts*, the *short- and medium-range forecasts*, and the *long-range forecasts*. Whatever the type or source of weather data, reliability is crucial.

4.2.3 Sources of Weather Data and Reliability

Weather information is available from several sources, including from national governments such as in the US, the National Weather Service (NWS; http://www.weather.gov/) or automated weather networks that are managed by Land Grant Universities, such as the Georgia Automated Environmental Monitoring

Network (AEMN; www.Georgiaweather.net), one of the largest automated weather station networks in the southeastern USA (Hoogenboom, 2000, 2001; Hoogenboom et al., 2003) and AgWeatherNet, managed by Washington State University (www.weather.wsu.edu). In some cases, weather data from some regional networks have been integrated with pest prediction models, as demonstrated by the NSCU-APHIS Plant Pest Forecasting (NAPPFAST) system (Magarey et al., 2007). Therefore, it is not surprising to see automated weather stations becoming more available to complement other sources of weather data for implementing pest predictive models. However, spatial resolutions of the weather data sources for most of the existing pest forecast models are coarse due to insufficient coverage by the monitoring stations. With the rapid advancement of technology in agriculture, there is a growing need for customized local weather forecasts to enhance pest and crop production decisions at a local farm level.

Strand (2000) noted that improved technology has made automated weather stations more accessible; in particular, the availability of internet services, mobile phone applications, and other portable hand-held devices is making data from such stations even more accessible to farmers and stakeholders. In fact, several regional networks of automated weather stations including AgWeatherNet (http://weather.wsu.edu/awn.php), Georgia Automated Environmental Monitoring Network (http://www.georgiaweather.net/), North Dakota Agricultural Weather Network (http://ndawn.ndsu.nodak.edu/), and many others are now available. However, some limitations still exist on the type of weather data available. Also, due to limitations in spatial coverage by existing networks, it is impossible to obtain farm-specific forecasts or measurements for monitoring biological processes (pests and diseases) on a specific field scale or within the crop canopy at every farm in a region, if such parameters are need for making pest predictions.

The distance from a farm location where weather information is needed for making management decisions, to the nearest station on the weather network may play a critical role in pest prediction accuracy. Available data through the station may not accurately represent the current weather conditions on a farm that is perhaps 10–20 miles away, yet it could be the nearest and only source of weather information available for making meaningful management decisions at the field level. However, the accuracy of on-farm measurements could be improved by developing correlations or statistical relationships between measurements in the field and data from the nearest stations of an automated weather station network. This type of technique may enhance pest prediction accuracy and integration of weather information into farm management schemes (Strand, 2000; Weiss, 1990). Weather variables also play a crucial role as inputs in models for predicting insect vector/pathogen population dynamics. For instance, weather patterns were shown to have a significant but indirect effect on the incidence of Tomato spotted wilt virus transmitted by thrips in peanut (Olatinwo et al., 2008, 2009). Several studies have also used weather parameters as a management tool for monitoring pests/vectors and diseases/pathogens in valuable crops (De Wolf and Isard, 2007; Magarey et al., 2007; Wharton et al., 2008). However, innovative approaches to weather forecasts are also emerging.

The Weather Research and Forecasting (WRF: http://www.wrf-model.org) model is a next-generation mesoscale numerical weather prediction system designed to serve both operational forecasting and atmospheric research needs. The WRF model developed by the National Center for Atmospheric Research (NCAR) features multiple dynamical cores, a three-dimensional variational (3DVAR) data assimilation system, and a software architecture allowing for computational parallelism and system extensibility. The WRF model is suitable for a broad spectrum of applications across scales ranging from metres to thousands of kilometres. Potential applications of WRF in plant disease and insect vectors were recently evaluated (Olatinwo et al., 2011, 2012), of which two examples are discussed as case studies later in this chapter.

Implementing a disease or pest model requires easy access to reliable sources of weather data, and knowledge of the pest and host crop. Although several weather parameters may be required, as mentioned earlier, the key inputs from weather measurements mostly include temperature, rainfall, and relative humidity. Since not all inputs needed for developing a model are available through the standard weather station data, other variables that are not measured are either calculated, computed, or derived from actual weather measurements. Usually, this is done by using tested algorithms, statistical analyses, and mathematical functions in calculating new derived variables. A good example is *leaf wetness* (i.e. the wetness of a leaf surface) that can be estimated from the relative humidity as mentioned earlier. Derived variables are extremely important when instruments for measuring the variables are limited or impossible to deploy for collecting reliable data. Therefore, in this case, the leaf wetness can be estimated from relative humidity measurements from the local weather station to help determine the likelihood of sporulation of fungal pathogens on a leaf surface. The data from regional scale weather monitoring networks obtained from different sources is as important in monitoring pest outbreaks as the on-farm measurements of weather parameters for monitoring disease or pest development within the canopy. A regional-scale weather forecast is a useful source of data input for disease and pest models, and is needed for monitoring disease epidemiology and pest population dynamics on a larger scale.

4.3 PESTS

4.3.1 Sensitivity and Vulnerability to Weather Factors – Extreme Events and Prevailing Climate

Sensitivity of pests to temperature and rainfall usually varies by species. Extreme weather conditions such as high temperatures, low temperatures, a decrease in precipitation, or extreme flooding could have direct effects on pests and crops, while host crops (depending on individual variety) may be indirectly affected through weather influences on soil processes, nutrient dynamics, and abiotic stressors that predispose crops to disease and pest attacks. Ultimately, variability in temporal and spatial weather conditions due to short- and long-term climate variability could have an impact on soil conditions, water availability, agricultural yield, and susceptibility of crops to pest and pathogen infestations (Rosenzweig et al., 2001).

The USPEST.org (http://uspest.org/wea/) is an Integrated Pest Management (IPM) model and forecasting web resource for agricultural, pest management, and plant biosecurity decision support in the US. The internet site provides over 78 degree-day and 18 hourly weather-driven models serving many IPM, regulatory, and plant biosecurity uses in the United States and specializes in IPM needs for the Pacific Northwest, according to available information on the site. Degree-day data presented on the site are very useful for monitoring pest developments and prevailing weather conditions, and in evaluating different options available for management through the growing season. Models developed based on a simple technique of degree-days, may utilize air or soil temperatures to describe the phenology of an individual pest, and helps determine when they reach a pre-determined population threshold that would warrant pest management actions. The information may also be useful for scheduling pesticides application based on known biology of the pest. Strand (2000) noted that the degree-days technique has been useful for controlling insect pest populations, such as the European corn borer, rice water weevil, and pink bollworm, particularly in tree, vegetable, and field crops, where pesticide applications may be accurately timed using phenology models.

According to Rosenzweig et al. (2001, 2000), and Yang and Scherm (1997), mild winter weather or other extreme events such as abnormally high summer temperatures are expected to increase in frequency, and may directly or indirectly contribute to increase the risk of pest damage in the near future. Currently, there are few examples of known pests of valuable crops other than soybean cyst nematode (*Heterodera glycines*) and sudden death syndrome (*Fusarium solani f. sp. glycines*) that have recently expanded their geographical ranges due to more favorable conditions for development (Hartman et al., 1995; Rosenzweig et al., 2000; Roy et al., 1997). The dynamic nature (i.e. expanding or shrinking) of geographical ranges of several important insects may accelerate with changing global climate, resulting in gradual expansion of the reach of pests beyond the traditional ranges we currently know. Matching this shift with early detection methods and effective management strategies to deal with potential threats from invasive pests shifting beyond known ranges into a new geographical region will be critical. Seasonal pest monitoring efforts may be strengthened by using weather forecasts that provide forewarning information and a scouting guide for locating areas where favourable conditions are met and impending pest population increases or emergencies are expected.

4.3.2 Weather Forecasts for Early Warning/Scouting of Pest

Detecting an impending disease outbreak or pest attack early enough by itself serves as a strong management tool. According to Das et al. (2007), 'the projections for optimum flight periods from daily synoptic weather forecasts facilitate the detection of invasions of pest and disease vectors and also the timing of pesticide applications to intercept and eliminate pest infestations during displacement from breeding areas.' Since the cost of pesticide application constitutes a sizeable amount of a farmer's total overall cost during a given crop production season, minimizing the use of agrochemicals will likely make more cash resources available to a farmer by reducing the overall costs needed to increase the acreage that is protected against pests or diseases. It will also free-up resources to provide additional plant nutrition needed to increase crop productivity, while reducing environmental contamination from chemical residues. In view of this, the use of weather forecasts in predictive models for early warning of an impending attack, scouting of insects, or early detection of diseases and weeds, can not only help minimize the volume of agrochemicals applied, but also make the applications more effective. It will prevent overuse of chemical pesticides and reduce the development of chemo-resistant strains of pests and pathogens (Das et al., 2007).

A carefully evaluated disease model coupled with the weather forecasts from WRF output as discussed later in this chapter could provide an approach for routine spatiotemporal predictions of potential threats for many diseases of valuable crops, especially those for which IPM can play an important role in the long term. The easy to understand spatiotemporal distribution map would provide growers with a simple to understand warning and ample time to take preventative measures in protecting high value crops.

4.4 CROPS

4.4.1 Agronomic Dependence on Weather Factors – Planting Days, Phenology, and Host Maturity

Genotypic and phenotypic traits of a crop can make it either vulnerable or resistant to pest attack. Susceptibility of a crop to weather-induced stresses, and infestations or infections caused by pests or diseases, vary among crops, among different varieties within the same crop, and among different growth stages within the same crop variety (Das et al., 2007). Over the years, crop breeders have selected several traits in breeding programmes (depending on the crop), to meet consumer expectation, to address crop vulnerability to pests and diseases, to improve productivity, and increase profitability. These needs led to intermittent releases of improved crop varieties that differ in many ways, ranging from attributes such as maturity (early or late), yields (high or low), how sensitive or tolerant they are to environmental factors (such as drought), and how susceptible they are to pathogens and pest attack (resistance or susceptible).

Availability of different crop varieties with varying levels of sensitivity to pests makes variety selection decisions by farmers a critical component of any IPM approach. It provides farmers with a decision tool for pest management. For example, selecting an early maturing variety may be uniquely suitable for cultivating a crop at a specific period, to avoid diseases or insect pest attacks during the latter part of the growing season as a management strategy. In view of this, for an IPM strategy, weather forecasts (i.e. temperature) serve as a useful means to monitor crop phenology effectively, estimate crop growth and development, and quantify changes under varying environmental conditions through the growing season. Accumulation of average daily temperature above a pre-determined base temperature (degree-days; unique to individual species) is a common and simple technique that has been used in the past for monitoring crop development and insect pest phenology.

The *cropping systems model* (Strand, 2000) is a more comprehensive approach that utilizes complex mathematical equations, incorporated with weather parameters (air and soil temperature, rainfall, etc.,) and derived variables including degree-days, to generate information on the status of crops, their pests, and potential threats under multiple scenarios, and probable management options. Although there are few examples of cropping system models, some have been developed into products with pest models as an optional module in the management decision process (Boote et al., 1983; Jones et al., 2003). Overall, an important benefit of this type of model is that it allows simultaneous evaluations of interactions between crop and pest components, potentially providing a farmer with more in-depth information needed to improve overall crop-pest management decisions (Tsuji et al., 1998).

4.4.2 Synchronization of Pest Emergence and Host Development; Avoidance and Planting Dates

Whether it is traditional or genetically improved crop varieties, both are critically sensitive to environment factors such as changing weather patterns. Although drought-tolerant varieties of various crops are mostly available to farmers, many varieties are still prone to infections or infestations under stressful environmental conditions such as drought or flooding. Manipulating planting dates, i.e. planting early or late, or planting an early- or late-maturing variety, coupled with weather forecasts, is an approach that farmers can exploit to avoid population peaks of pests or insect vectors, thereby lowering the probability of host vulnerability and risk of potential attack (Strand, 2000).

Olatinwo et al. (2008, 2009) described the synchronization of peanut planting date and early populations of thrips as critical components in managing *Tomato spotted wilt virus* in the southeastern US. Like several other soil-borne pathogens, development of sugar beet cyst nematode (*Heterodera schachtii*) is quite sensitive to changes in soil temperature, and therefore its population in the soil is usually measured using degree-days based on soil temperature. Studies (Olatinwo et al., 2006a,b,c), have exploited this sensitivity in monitoring progression of sugar beet cyst nematode generations in greenhouse experiments and in vegetable fields during the growing season. Roberts and Thomason (1981) indicated how early plantings of sugar beet could take advantage of cooler temperature when *H. schachtii* nematode is inactive and unable to attack due to temperature conditions that are below the required base threshold for development. Therefore, selecting the most suitable variety (i.e. at the beginning of the season) based on variety phenology and seasonal weather forecasts or prevailing weather conditions may strengthen a farmer's ability to manage pest attack effectively within an IPM approach.

Weather factors influence insect occurrence and govern the general distribution and numbers of insects, and, therefore, can either foster or suppress insect life. Das et al. (2007) noted that temperature and relative humidity control the time interval between successive generations of insects as well as the numbers produced in each generation, while wind patterns are an important factor for the migration of insect pests. Strand (2000) also noted that frequent and heavy rainfall characterized by run-off and flooding could serve as an impetus for conducive and suitable habitats for locust survival and population growth. A study by Prior and Streett (1997) on strategies for the use of entomopathogens in the control of the desert locust found that, although preventing locust outbreaks by destroying flightless nymphs (which can be monitored using weather information) might be desirable, emergency measures are usually preferred to control the destructive swarms of desert locust adults.

Interactions involving crops, pests and the environment can be very complex to untangle. However, almost all components of an IPM strategy including variety, biological control agents, planting date, crop rotation and other cultural practices are either directly or indirectly affected by environmental factors such as changing weather patterns or climate variability. For example, a crop variety might not do well under extreme weather conditions, or a biological control agent might be less effective and out-competed by targeted pest if the environmental conditions are unfavourable for it to establish. A planting date might be too early or too late if soil temperature is not suitable for planting. Even the amount of moisture on plant surfaces and wind speed/direction might affect uptake and coverage of pesticide applications, respectively. Hence, environmental limitations to any of these factors could have serious constraints for effective and successful implementation of IPM methods, and, thus, efforts to control diseases and pests of valuable crops at critical periods during the growing season.

4.5 EFFICIENT CROP PROTECTION PRODUCT

4.5.1 Weather-Based Forecasts and IPM

IPM is a crop production technique that is generally accepted as an effective strategy for balancing between management of pests using a minimum amount of chemical pesticides and reducing the negative impacts of pesticide applications on the environment. It is fast becoming a favoured approach in many regions across the world because it combines multiple management techniques including use of resistant varieties, a natural enemy, and biological control agents, improved cropping practices such as crop rotation, tillage, and irrigation methods, and a minimal amount of pesticide use, based on weather forecasts and timely applications of pesticides. As part of an IPM, weather information is critical for selecting the most suitable variety (i.e. early or late maturing, according to phenology), and the best planting date(s) to avoid diseases or pest pressure at an early stage of crop development. Hence, IPM provides farmers with a variety of choices to plan and take preventative measures against disease development and pest attacks from pre-planting throughout the growing season until final harvest.

Scientific and technological advances in biotechnology, agro-meteorology, and computer science are complementing traditional methods of pest management with newer and more efficient techniques that have produced transgenic resistant varieties, high-resolution weather forecasts, and accurate pest predictive models. Where IPM has been implemented, it has demonstrated that both emerging technologies and traditional pest management methods can be complementary to each other in preventing threats posed by pests to valuable crops. Profitable and efficient crop production is achievable with IPM techniques, while minimizing the impacts of chemical pesticides and fertilizer inputs on the environment.

In addition to the use of chemical pesticides to improve crop productivity, selecting a resistant variety that can tolerate or resist pathogens or pest attacks is not only good for the environment, but also relatively inexpensive compared to pesticide applications alone. Generally, resistance expressed by a crop variety is a product of genetic traits of that variety, the virulence of the corresponding pathogen or pest, and how they interact with other components such as weather, soils, and cultural practices within the environment. Genetic engineering and biotechnology have so far played a significant role as tools in crop improvement, turning out new cultivars each year to address crop production needs. They have improved and complemented the traditional crop breeding methods, through relatively quick gene transfer techniques that have resulted in new varieties with desirable traits against potential impacts of pests and diseases. A good example is genetically engineered corn carrying bacteria (*Bacillus thuringiensis*) endotoxin gene, which makes it tolerant to insect attacks.

Strand (2000) alluded to other examples such as genetically engineered carrots with antifungal genes from tobacco that offer protection against powdery mildew. In some cases, crops are engineered to tolerate herbicide applications, while competitive weeds at the target site are killed. This type of crop improvement (resistant host crop) provides farmers with another tool to fight pest attacks, in addition to the use of weather information for short-term monitoring of pests. However, a sudden breakdown in crop resistance to a pathogen or a pest due to frequent use of pesticides may be catastrophic for overall crop production. Resistance to a single pest or disease may be less effective when multiple pests are involved, but perhaps more effective when used in combination with other techniques and as one of the components of an IPM approach.

Apart from increasing pest resistance to pesticides due to repeated use, there are other negative impacts of chemical residues on the ecosystems, including contamination of surface and groundwater, and beneficial organisms. Hence, strategies for managing pests by farmers will continue to shift to an integrated management approach that reduces both the frequency and the amount of chemical applications released into the environment. IPM is a viable option the farmer can explore and be improved upon every year through precise and timely applications of pesticides. With significant progress made in the field of science and technology, pest models have been coupled with high-resolution weather forecast data to predict the risk of an impending increase in pest populations or favourable conditions for infection initiation and disease development. Olatinwo et al. (2011, 2012) recently examined the application of this promising scientific approach for managing insect vectors and a peanut disease in the southeastern United States. The disease and insect vector models were coupled with the weather predictions or forecasts provided by the WRF model. A part of this study (Olatinwo et al., 2012) evaluated a potential short-range forewarning concept that triggers an alert and generates a 3×3km grid high-resolution map when a favourable condition for the potential onset of a disease is met.

4.5.2 Existing Products

How farmers consume weather forecast information is by itself important. Several web-based pest models that are driven by weather information can now deliver location-specific risk alerts using simple web graphics. Most existing internet-based interactive systems/models can output spatial and temporal distribution maps to depict the potential level of pest risks which is easy enough for farmers or stakeholders to understand and incorporate into a quick decision process for monitoring the likelihood of a pest outbreak. Measurements and observations from regional weather networks are coupled with disease and pest prediction models, while online maps are generated and frequently updated on a regional scale, for monitoring likelihood of infections or outbreaks in valuable crops. Examples of internet-based interactive systems include the potato late blight in Michigan (http://www.lateblight.org/forecasting.php) described by Wharton et al. (2008); the AWIS Weather Services, Inc (http://awis.com); the Oklahoma mesonet peanut leaf spot advisor (http://www.mesonet.org/index.php/agriculture/category/crop/peanut/leaf_spot_advisor) based on a model described by Damicone et al. (1994); North American Plant Disease Forecast Center, North Carolina State University (http://cdm.ipmpipe.org/); HortPlus (http://www.hortplus.com/Brochure/MetWatch/MWSoftware.htm); and a web-based tool for Fusarium Head Blight risk assessment (http://www.wheatscab.psu.edu). AgWeatherNet also incorporates several disease models for cherry and grass for the state of Washington, USA (www.weather.wsu.edu). These web-based IPM risk assessment tools usually generate distribution maps that are very easy to understand and useful for monitoring the potential level of risks across a given area. Generally, they are models that have been developed based on an in-depth understanding of how weather factors affect biology and development of a particular pest and evaluated with local data.

The University of California, Davis also developed an online database of IPM models (http://www.ipm.ucdavis.edu/WEATHER/index.html) from a large collection of research summaries of phenology models for insects, mites, diseases, plants, and beneficial organisms. This internet-based interactive system is a useful source of information on key weather parameters that are needed as input for designing a successful IPM strategy that improves crop productivity. In addition to local weather conditions, some systems also include parameters such as type of soil, type of crop, and phenological stages, as well as level and type of insect pest infestation. Usually, a combination of these parameters is considered in offering advisories for decision-making on sowing, harvesting, irrigation, nutrient management, and chemical application (Dacom, 2003).

A typical internet weather-based pest forecasting web site frequently updates predictions or risk assessments using the most recent weather data (depending on parameters) available as input. The output/predictions, in most cases, are translated into simple management recommendations that farmers can use in deciding what action is needed, if any. Seeley

(2002) noted that insect and disease control, pheromone release, irrigation, freeze prevention, maturity indices, and fruit damage have benefited from weather database prediction programs. This is largely due to the significant improvement in computer technologies that deliver new tools, and increasing accessibility to information disseminated through media such as the internet and cell phones using push technology.

4.5.3 Case Studies

We examined two case studies on the significance of weather-based pest forecasting for efficient peanut protection in Georgia, USA. The early leaf spot disease of peanut caused by a fungal pathogen, and the *Tomato spotted wilt virus* (TSWV) of peanut transmitted by two major thrips vectors, were examined as examples of complex pathogen–vector interactions. The two case studies explored the potential application of the high-resolution WRF model, which is a next-generation mesoscale numerical weather prediction system designed to serve both operational forecasting and atmospheric research needs (Prabha and Hoogenboom, 2008).

4.5.3.1 Case Study 1: WRF model and Early Leaf Spot in Peanut

Cercospora arachidicola S. Hori is a major fungal pathogen that causes early leaf spot in peanut (*Arachis hypogaea L.*), a devastating foliar disease of peanut that can result in complete defoliation of susceptible peanut cultivars. The disease accounts for significant yield losses in the absence of fungicide applications (Cantonwine et al., 2006), and it is a major problem for peanut production in the southeastern United States, mostly resulting from inadequate and untimely applications of fungicides. In Georgia, losses due to peanut leaf spot diseases were approximately $42 million in 2005 (Kemerait, 2006). Generally, symptoms of infection typically appear in the lower canopy and later progress to the upper canopy. Economic losses can increase significantly from ineffective monitoring where timely management of leaf spot is required (Jacobi et al., 1995a,b; Woodward et al., 2010).

Although applications of fungicide remain an effective tool for managing leaf spot in peanut, Culbreath et al. (2002) and Woodward et al. (2010) noted that repeated applications of fungicides can lead to risks of fungal pathogen resistance. Therefore, timely and effective management of the disease rely on good monitoring of environmental conditions, i.e. rainfall, relative humidity, leaf wetness, and temperature, which are required for infection to occur (Alderman and Beute, 1986; Jensen and Boyle, 1965, 1966; Shew et al., 1988; Wu et al., 1999).

For early leaf spot, weather information is crucial for developing prediction models (Cu and Phipps, 1993; Linvill and Drye, 1995), especially for monitoring favourable conditions for disease development on host crops during the growing season. Jewell (1987) identified a strong correlation between early leaf spot incidence and cumulative hours of relative humidity (RH $\geq$ 95). The *Oklahoma peanut leaf spot* model described by Damicone et al. (1994) calculates the daily 'infection hours' based on 24 h of temperature, and leaf wetness or relative humidity. According to Grichar et al. (2005), other systems have used a similar combination of relative humidity/leaf wetness and temperature to forecast favourable conditions for disease development and scheduling of fungicide applications.

Olatinwo et al. (2012) demonstrated the possibility of coupling the high-resolution WRF data output with a leaf spot disease model, i.e. *Oklahoma peanut leaf spot* model in predicting favourable conditions for early leaf spot infection. The spatial–temporal distribution maps of infection threats generated from the coupled models highlighted the usefulness of the approach. Maps of areas identified as having favourable conditions for the disease could complement field scouting for early leaf spot symptoms and for timely applications of management measures. The coupled model output (risks distribution maps) is particularly useful at the beginning of the growing season when management decisions are taken at critical peanut development phases.

The spatial and temporal distribution probability of leaf spot occurrence (Figure 4.1) showed that peanut fields in southeast Georgia and the coastal areas would be more vulnerable to leaf spot risk due to favourable weather conditions during the period evaluated. The infection hours required for leaf spot development from the coupled models increased along the coastal areas where the required optimum number of infection hours (36h) was met earlier, compared to southwest and central Georgia. The probability of favourable conditions for infection was 0.8–0.9 along the coastal areas. After several days, the trend extended to the southwest and central parts of Georgia during the evaluated period of this study.

It is important to note that the disease model used the prevailing variability in weather conditions for each grid location, i.e. 3×3km grid to produce the distribution map, which could assist farmers with timely applications, rather than using a pre-determined traditional spray calendar, which does not take into account the prevailing weather conditions during the growing season. The alert of favourable conditions for the potential onset of early leaf spot at a resolution of 3×3km grid was demonstrated in the study (Figure 4.1). Developing an early warning tool based on the approach would be useful for locations where weather stations are currently not available. The spatiotemporal distribution could be produced by coupling the two models, i.e. WRF and the *Oklahoma peanut leaf spot* model. This could be useful by itself or complement existing tools for disease management activities such as scheduling of fungicide applications.

4.5.3.2 Case Study 2: WRF Model and Thrips-Vector Populations

Several million dollars in crop damage are reported annually due to infestations of tobacco thrips (*Frankliniella fusca* Hinds) and western flower thrips (*Frankliniella occidentalis* Pergande), which are economical pests of valuable crops such as cotton (*Gossypium hirsutum* L.) and peanut (*A. hypogaea* L.). Both thrips are also important vectors that transmit TSWV to field crops, ornamentals, and vegetables (Olatinwo et al., 2008). As with many insect pests, the populations of both thrips species are sensitive to changes in seasonal temperature. The population growth relies on favourable weather conditions such as prolonged temperatures above a minimum developmental threshold (i.e. base temperature) during the season. Therefore, access to accurate weather information is critical for predicting thrips' population dynamics during early spring when thrips' population information could assist farmers in mitigating damage to crops.

Among many factors, Lewis (1997) noted that the thrips' infestation of a crop depends on favourable weather conditions for population growth, while several studies (Brown et al., 2005a,b; Chaisuekul and Riley, 2005; Harding, 1961; McDonald et al., 1998) have linked rainfall patterns and temperature to thrips vectors and spotted wilt development. Heavy rainfall was reported to have a negative effect on thrips' larvae survival (Kirk, 1997) and adult flight (Lewis, 1997), while increased temperatures during the spring were associated with greater thrips' activity and population growth (Kirk, 1997; Lewis, 1997; Lowry et al., 1992; Pearsall and Myers, 2001). Harding (1961) suggested that cool temperatures and rains are detrimental to thrips' colonization on onions in south Texas. Thrips often migrate into cropping fields during the spring after overwintering on uncultivated plants or alternative hosts (Groves et al., 2002, 2003; Kirk, 1997; Lewis, 1997; Pearsall

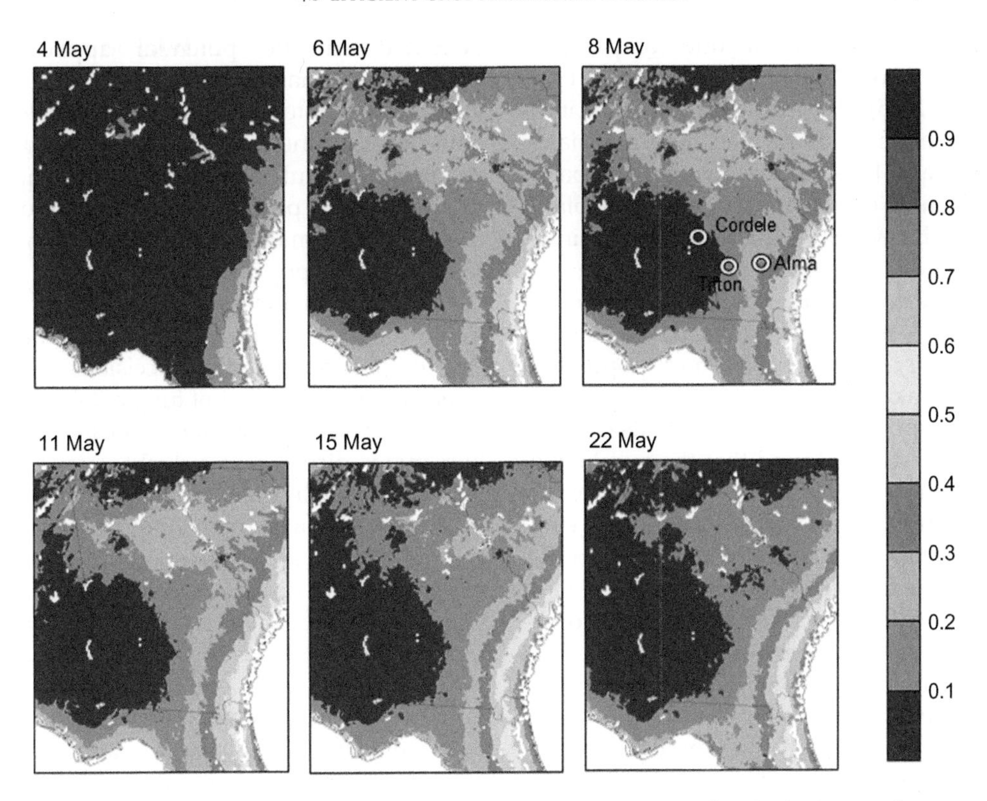

FIGURE 4.1 Spatiotemporal distribution of the probability of occurrence of early leaf spot of peanut (*Arachis hypogaea* L.), a disease caused by *Cercospora arachidicola* S. Hori, during the period from 4 May to 22 May 2007 for Georgia, USA (Olatinwo et al., 2012). This figure is reproduced in colour in the colour plate section.

and Myers, 2001). Hence, the timing of peanut emergence in relation to the movement of viruliferous thrips vectors into a cultivated field can significantly affect the incidence of TSWV for the remainder of the season (Culbreath et al., 2003).

Olatinwo et al. (2008, 2009) noted a high probability of spotted wilt if the number of rain days during March was greater than or equal to 10 days and planting was before 11 May or after 5 June. The total evapotranspiration in April and the average daily minimum temperature in March similarly increased the risk of spotted wilt. Knowing in advance the level of spotted wilt risk expected in a peanut field could assist growers with evaluating management options and significantly improve the impact of their decisions against spotted wilt risk in peanut (Brown et al., 2005a, 2008; Olatinwo et al., 2010).

Stormy weather conditions have been linked to mass flights of thrips. Weather fronts and incipient thunderstorms are reported to discourage the mass flight of thrips, thereby resulting in high densities above the soil surface due to the landing attempts of thrips (Kirk, 2004; Lewis, 1964, 1965, 1973, 1997), while Morsello et al. (2008) also found that the number of thrips captured in flight has a positive relationship with the number of wet days or days with precipitation. In peanut, populations of adult thrips vectors *F. occidentalis* and *F. fusca* were reported to be greater for early planting in April or late planting in June compared to planting in May (Mitchell and Smith, 1991; Todd et al., 1995). Field observations also indicate a higher level of spotted wilt associated with early- and late-planted peanuts compared to those planted during the middle of the planting season (Brown et al., 2005b, 2008; Olatinwo et al., 2008).

Studies (Kirk, 1997; Lewis, 1997; Lowry et al., 1992; Pearsall and Myers, 2001) have shown that a higher thrips activity and population growth are linked to an increase in temperature during the spring, while Morsello (2007) and Morsello et al. (2008) found that the numbers of *F. fusca* captured in flight was positively related to degree-days. Olatinwo et al. (2008, 2009) evaluated potential application of the WRF model in developing high-resolution spatial and temporal distribution maps of favourable conditions for thrips' development. Results based on degree-day models showed that southwestern Georgia is more favourable for thrips' development during the early part of the growing season examined, with a varied rate of development according to thrips species (Figure 4.2). The high-resolution forecasts map of favourable conditions could serve as a scouting guide in places where weather information is limited, thereby assisting growers in pest management decisions and timely application of pesticides.

4.5.4 Accuracy, Limitations, and Uncertainties

Initially, computational resource requirements for running the high-resolution weather WRF model were a great challenge in the implementation of this approach. However, computational limitations and access to high-resolution weather data are no longer major constraints compared to earlier years. A 3-day forecast range of high-resolution disease prediction based on WRF forecast data is achievable and now possible (Olatinwo et al., 2011, 2012). For thrips, the degree-day accumulation demonstrated the potential application of WRF in pest management, although Olatinwo et al. (2011) noted that degree-day calculation alone does not necessarily translate to the exact changes in population of thrips. Depending on the individual pest model, several biotic and abiotic factors may still be necessary to be able to estimate the population accurately.

Growers can incorporate model predictions into a decision support system for routine disease or pest management decisions. The traditional spray scheduling of fungicide applications by growers as a preventative means of controlling many diseases of valuable crops such as leaf spots, usually calls for intermittent pesticide applications at a regular intervals (e.g. 15 days through the season), irrespective of incidence or severity of a disease or pest population pressure. Olatinwo et al. (2011) observed that the traditional approach generally does not consider weather factors in scheduling pesticide applications, except for avoiding rainfall or other factors that could hinder the pesticide application process. However, spraying at intervals irrespective of disease or pest biology, may lead to unnecessary sprays that have the potential to leave behind a high amount of chemical pesticide residue harmful to the environment.

Developing weather-based forecasting for efficient crop protection is, therefore, not only necessary, but the implementation of the approach is obviously dependent on several factors for it to be successful and operational. Some crops require an in-depth understanding of biological processes and extensive studies on pests of interest, to identify and accurately measure the parameters needed for quantitative forecasts, while additional knowledge and research may be required for others. Overall, there are uncertainties inherent in the biotic or abiotic parameters that are needed for qualitative forecasts or accurate predictions. The extent of these uncertainties could significantly affect the accuracy of a model, creating an unreliable assessment of pest populations and potential risks

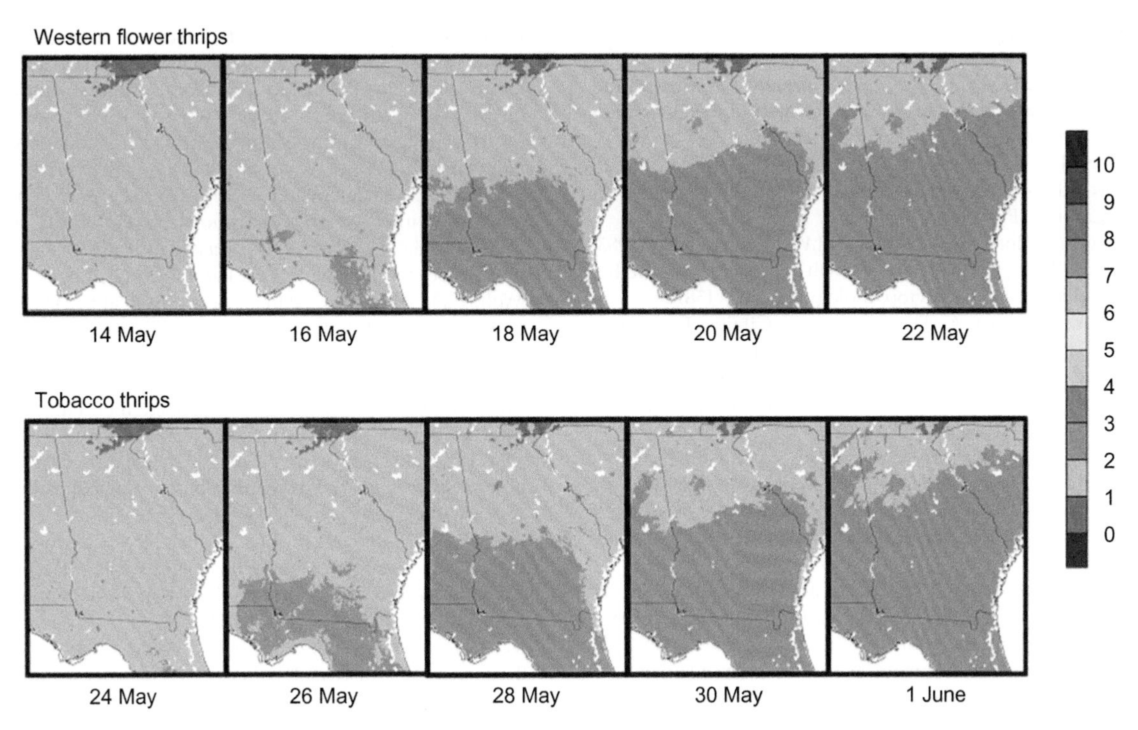

FIGURE 4.2 Spatial and temporal distribution of potential thrips accumulated generations in the southeastern United States using predictions from the Weather Research and Forecasting (WRF) model and the base temperature requirement for tobacco thrips (*Frankliniella fusca*) and western flower thrips (*Frankliniella occidentalis*) for 2007 (Olatinwo et al., 2011). This figure is reproduced in colour in the colour plate section.

of crop damage, which could complicate farmers' decisions on suitable management options. Therefore, in addition to reliable weather forecasts and thorough knowledge of a disease or pest required to develop predictive models, evaluating predictions from such models with actual field observations is a cautious and critical step required in the process, as it provides an added level of confidence to end-users in terms of accuracy of the model.

4.6 CONCLUSIONS

The threat of a disease epidemic or pest outbreak is real; hence constant monitoring is required to avert risks of significant damage to valuable crops from one year to the next. Therefore, accurate and reliable weather-based pest forecasting remains a critical component in current and emerging IPM strategies. It is not only important to protect valuable crops, improve crop productivity, or increase economic returns for farmers, but also vital for efficient use of pesticides and overall protection of the environment. Obviously, there are several uncertainties inherent in using weather parameters for disease/pest forecasts. However, as our knowledge on the biology of individual diseases or pests improves through new scientific findings and application of emerging technologies in fields such as computing and statistics, the accuracy of weather-based pest forecasts is expected to become a lot more reliable in the future, thereby enhancing successful implementation of long-term IPM strategies.

References

Alderman, S.C., Beute, M.K., 1986. Influence of temperature and moisture on germination and germ tube elongation of *Cercospora arachidicola*. Phytopathology 76, 715–719.

Bailey, J.E., Johnson, G.L., Toth, S.J., 1994. Evolution of weather-based peanut leaf spot spray advisory in North Carolina. Plant Dis. 78, 530–535.

Boote, K.J., Jones, J.W., Mishoe, J.W., Berger, R.D., 1983. Coupling pests to crop growth simulators to predict yield reductions. Phytopathology 73, 1581–1587.

Brown, S., Csinos, A., Díaz-Pérez, J.C., Gitaitis, R., LaHue, S.S., Lewis, J., et al. 2005a. Tospoviruses in solanaceae and other crops in the coastal plain of Georgia. College of Agriculture and Environmental Sciences, University of Georgia Research Report (ISSN 0072-128X) 704:19.

Brown, S.L., Culbreath, A.K., Todd, J.W., Gorbet, D.W., Baldwin, J.A., Beasley, J.P., 2005b. Development of a method of risk assessment to facilitate integrated management of spotted wilt of peanut. Plant Dis. 89, 348–356.

Brown, S.L., Todd, J.W., Culbreath, A.K., Baldwin, J., Beasley, J., Kemerait, B., et al., 2008. Minimizing disease of peanut in the southeastern United States. In: 2008 Peanut Update. University of Georgia, Cooperative Extension Service, College of Agriculture and Environmental Sciences. CSS-08-0114, pp. 36–52.

Cantonwine, E.G., Culbreath, A.K., Stevenson, K.L., Kemerait Jr., R.C., Brenneman, T.B., Smith, N.B., et al., 2006. Integrated disease management of leaf spot and spotted wilt of peanut. Plant Dis. 90, 493–500.

Chaisuekul, C., Riley, D.G., 2005. Host plant, temperature, and photoperiod effects on ovipositional preference of *Frankliniella occidentalis* and *Frankliniella fusca* (Thysanoptera: Thripidae). J. Econ. Entomol. 98, 2107–2113.

Cu, R.M., Phipps, P.M., 1993. Development of a pathogen growth response model for the Virginia peanut leaf spot advisory program. Phytopathology 83, 195–201.

Culbreath, A.K., Stevenson, K.L., Brenneman, T.B., 2002. Management of late leaf spot of peanut with benomyl and chlorothalonil: A study in preserving fungicide utility. Plant Dis. 86, 349–355.

Culbreath, A.K., Todd, J.W., Brown, S.L., 2003. Epidemiology and management of tomato spotted wilt in peanut. Annu. Rev. Phytopathol. 41, 54–75.

Dacom, 2003. Crop management with plant-plus. Emmen, Dacom Plant Service.

Damicone, J.P., Jackson, K.E., Sholar, J.R., Gregory, M.S., 1994. Evaluation of a weather-based spray advisory for management of early leaf spot of peanut in Oklahoma. Peanut Sci. 21, 115–121.

Das, H.P., Doblas-Reyes, F.J., Garcia, A., Hansen, J.W., Mariani, L., Nain, A., et al., 2007. Weather and climate forecasts for agriculture. Ch. 4 In: Stigter, K. et al. (Ed.), Guide to Agricultural Meteorological Practices (GAMP), Draft third ed. WMO-No.134. World Meteorological Organization, Rome. <http://www.wmo.ch/pages/prog/wcp/agm/gamp/gamp_en.html/> (accessed November 2012).

Dawidziuk, A., Kaczmarek, J., Jedryczka, M., 2012. The effect of winter weather conditions on the ability of pseudothecia of *Leptosphaeria maculans* and *L. biglobosa* to release ascospores. Eur. J. Plant Pathol. 134 (2), 329–343.

De Wolf, E.D., Isard, S.A., 2007. Disease cycle approach to plant disease prediction. Annu. Rev. Phytopathol. 45, 203–220.

Edelson, J.V., Magaro, J.J., 1988. Development of onion thrips, *Thrips tabaci* Lindeman, as a function of temperature. Southwestern Entomol. 13, 171–176.

Grichar, W.J., Jaks, A.J., Besler, B.A., 2005. Response of peanuts (*Arachis hypogaea*) to weather-based fungicide advisory sprays. Crop Prot. 24, 349–354.

Groves, R.L., Walgenbach, J.F., Moyer, J.W., Kennedy, G.G., 2002. The role of weed hosts and tobacco thrips, *Frankliniella fusca*, in the epidemiology of Tomato spotted wilt virus. Plant Dis. 86, 573–582.

Groves, R.L., Walgenbach, J.F., Moyer, J.W., Kennedy, G.G., 2003. Seasonal dispersal patterns of *Frankliniella fusca* (Thysanoptera: Thripidae) and tomato spotted wilt virus occurrence in central and eastern North Carolina. J. Econ. Entomol. 96, 1–11.

Harding, J.A., 1961. Effect of migration, temperature, and precipitation on thrips infestations in south Texas. J. Econ. Entomol. 54, 77–79.

Hartman, G.L., Noel, G.R., Gray, L.E., 1995. Occurrence of soybean sudden death syndrome in east-central Illinois and associated yield losses. Plant Dis. 79, 314–318.

Hoogenboom, G., 2000. The Georgia automated environmental monitoring network. Reprints of the 24th Conference on Agricultural and Forest Meteorology. American Meteorological Society, Boston, pp. 24–25.

Hoogenboom, G., 2001. Weather monitoring for management of water resources. In: Hatcher, K.J. (Ed.), Proceedings of the 2001 Georgia Water Resources Conference. Institute of Ecology, The University of Georgia, Athens, GA, pp. 778–781.

Hoogenboom, G., Coker, D.D., Edenfield, J.M., Evans, D.M., Fang, C., 2003. The Georgia automated environmental monitoring network: 10 years of weather information for water resources management. In: Hatcher, K.J. (Ed.), Proceedings of the 2003 Georgia Water Resources Conference, The University of Georgia, Athens, GA, pp. 896–900.

Jacobi, J.C., Backman, P.A., Davis, D.P., Brannen, P.M., 1995a. AU-Pnuts advisory II: modification of the rule-based leaf spot advisory system for a partially resistant peanut cultivar. Plant Dis. Rep. 79, 672–676.

Jacobi, J.C., Backman, P.A., Davis, D.P., Brannen, P.M., 1995b. AU-Pnuts advisory I: development of a rule-based system for scheduling peanut leaf spot fungicide applications. Plant Dis. Rep. 79, 666–671.

Jensen, R.E., Boyle, L.W., 1965. The effect of temperature, relative humidity and precipitation on peanut leaf spot. Plant Dis. Rep. 49, 810–814.

Jensen, R.E., Boyle, L.W., 1966. A technique for forecasting leafspot on peanut. Plant Dis. Rep. 50, 810–814.

Jewell, E.L., 1987. Correlation of early leaf spot disease in peanut with a weather-dependent infection index. M.S. thesis. Virginia Polytechnic Institute and State University, Blacksburg, VA, p. 55.

Jones, J.W., Hoogenboom, G., Porter, C.H., Boote, K.J., Batchelor, W.D., Hunt, L.A., et al., 2003. The DSSAT cropping system model. Eur. J. Agron. 18, 235–265.

Katayama, H., 1997. Effect of temperature on development and oviposition of western flower thrips *Frankliniella occidentalis* (Pergande). Jpn. J. Appl. Entomol. Zool. 41, 225–231.

Kemerait, R.C., 2006. Peanut. In: (2005) Georgia Plant Disease Loss Estimate, compiled by Alfredo Martinez. The University of Georgia, College of Agricultural and Environmental Sciences, Cooperative Extension Special Bulletin 41-08, 11.

Kirk, W.D.J., 1997. Distribution, abundance, and population dynamics. In: Lewis, T. (Ed.), Thrips as Crop Pests CAB Wallingford, Oxon, UK, pp. 217–258.

Kirk, W.D.J., 2004. The link between cereal thrips and thunderstorms. Acta Phytopathol. Entomol. Hung. 39, 13–16.

Lewis, T., 1964. The weather and mass flights of Thysanoptera. Ann. Appl. Biol. 53, 165–170.

Lewis, T., 1965. The species, aerial density and sexual maturity of Thysanoptera caught in mass flights. Ann. Appl. Biol. 55, 219–225.

Lewis, T., 1973. Thrips: Their Biology, Ecology and Economic Importance. Academic Press, London, 349 pp.

Lewis, T., 1997. Flight and dispersal. In: Lewis, T. (Ed.), Thrips as Crop Pests. CAB, Oxon, UK.

Linvill, D.E., Drye, C.E., 1995. Assessment of peanut leaf spot disease control guidelines using climatological data. Plant Dis. 79, 876–879.

Lowry, V.K., Smith Jr., J.W., Mitchell, F.L., 1992. Life fertility tables for *Frankliniella fusca* and *F. occidentalis* on peanut. Ann. Entomol. Soc. Am. 85, 744–754.

Magarey, R.D., Fowler, G.A., Borchert, D.M., Sutton, T.B., Colunga-Garcia, M., Simpson, J.A., 2007. NAPPFAST: An internet system for the weather-based mapping of plant pathogens. Plant Dis. 91, 336–345.

Matra, A.D., Magarey, R.D., Orlandini, S., 2005. Modelling leaf wetness duration and downy mildew simulation on a grapevine in Italy. Agric. For. Meteorol. 132, 84–95.

McDonald, J.R., Bale, J.S., Walters, K.F.A., 1998. Effect of temperature on development of the western flower thrips, *Frankliniella occidentalis* (Thysanoptera: Thripidae). Eur. J. Entomol. 95, 301–306.

McPherson, R.M., Pappu, H.R., Jones, D.C., 1999. Occurrence of five thrips species on flue-cured tobacco and impact on spotted wilt disease incidence in Georgia. Plant Dis. 83, 765–767.

Mitchell, F.L., Smith, J.W. Jr., 1991. Epidemiology of tomato spotted wilt virus relative to thrips populations. Virus–Thrips–Plant Interactions of Tomato Spotted Wilt Virus. Proceedings of a USDA Workshop. USDA-ARS Bulletin ARS-87:46–52.

Morsello, S.C., 2007. The Role of Temperature and Precipitation on Thrips Populations in Relation to the Epidemiology of Tomato Spotted Wilt Virus. Ph.D. Thesis. Department of Entomology, North Carolina State University, Raleigh, NC.

Morsello, S.C., Groves, R.L., Nault, B.A., Kennedy, G.G., 2008. Temperature and precipitation affect seasonal patterns of dispersing tobacco thrips, *Frankliniella fusca*, and onion thrips, *Thrips tabaci* (Thysanoptera: Thripidae) caught on sticky traps. Environ. Entomol. 37, 79–86.

Nokes, S.E., Young, J.H., 1991. Simulation of the temporal spread of leafspot and the effect on peanut growth. Trans. ASAE 34, 653–662.

Olatinwo, R., Becker, J.O., Borneman, J., 2006a. Suppression of *Heterodera schachtii* populations by *Dactylella oviparasitica* in four soils. J. Nematol. 38, 345–348.

Olatinwo, R., Borneman, J., Becker, J.O., 2006b. Induction of beet-cyst nematode suppressiveness by the fungi *Dactylella oviparasitica* and *Fusarium oxysporum* in field microplots. Phytopathology 96, 855–859.

Olatinwo, R., Yin, B., Becker, J.O., Borneman, J., 2006c. Suppression of the plant-parasitic nematode *Heterodera schachtii* by the fungus *Dactylella oviparasitica*. Phytopathology 96, 111–114.

Olatinwo, R.O., Paz, J.O., Brown, S.L., Kemerait, R.C., Culbreath, A.K., Beasley Jr, J.P., et al., 2008. A predictive model for spotted wilt epidemics in peanut based on local weather conditions and the *Tomato spotted wilt virus* risk index. Phytopathology 98, 1066–1074.

Olatinwo, R.O., Paz, J.O., Brown, S.L., Kemerait Jr., R.C., Culbreath, A.K., Hoogenboom, G., 2009. Impact of early spring weather factors on the risk of tomato spotted wilt in peanut. Plant Dis. 93, 783–788.

Olatinwo, R.O., Paz, J.O., Kemerait, R.C., Culbreath, A.K., Hoogenboom, G., 2010. El Niño-Southern Oscillation (ENSO): Impact on tomato spotted wilt intensity in peanut and the implication on yield. Crop Prot. 29, 448–453.

Olatinwo, R.O., Prabha, T.V., Paz, J.O., Riley, D.G., Hoogenboom, G., 2011. The weather research and forecasting (WRF) model: application in prediction of TSWV-vectors populations. J. Appl. Entomol. 135 (1–2), 81–90.

Olatinwo, R.O., Prabha, T.V., Paz, J.O., Hoogenboom, G., 2012. Predicting favorable conditions for early leaf spot of peanut using output from the weather research and forecasting (WRF) model. Int. J. Biometeorol. 56, 259–268.

Parvin, D.W., Smith, D.H., Crosby, F.L., 1974. Development and evaluation of a computerized forecasting method for Cercospora leaf spot of peanuts. Phytopathology 64, 385–388.

Pearsall, I.A., Myers, J.H., 2001. Spatial and temporal patters of dispersal of western flower thrips (Thysanoptera: Thripidae) in nectarine orchards in British Columbia. J. Econ. Entomol. 94, 831–843.

Pimentel, D., 1991. CRC Handbook of Pest Management in Agriculture, Vol. 1 second ed. CRC Press, Boca Raton, FL.

Prabha, T., Hoogenboom, G., 2008. Evaluation of the weather research and forecasting model for two frost events. Comput. Electron. Agric. 64, 234–247.

Prior, C., Streett, D.A., 1997. Strategies for the use of entomopathogens in the control of the desert locust and other acridoid pests. Mem. Entomol. Soc. Can. 171, 5–25.

Riley, D.G., Pappu, H.R., 2000. Evaluation of tactics for management of thrips-vectored Tomato spotted wilt virus in tomato. Plant Dis. 84, 846–852.

Riley, D.G., Pappu, H.R., 2004. Tactics for management of thrips (Thysanoptera: Thripsidae) and tomato spotted wilt virus in tomato. J. Econ. Entomol. 97, 1648–1658.

Roberts, P.A., Thomason, I.J., 1981. Sugarbeet Pest Management: Nematodes. University of California Division of Agriculture Sciences Publications, 3272. University of California, Oakland, CA.

Rosenzweig, C., Iglesias, A., Yang, X.B., Epstein, P.R., Chivian, E., 2000. Implications of Climate Change for U.S. Agriculture: Extreme Weather Events, Plant Diseases, and Pests. Cambridge, MA: Center for Health and the Global Environment, Harvard Medical School. Cambridge, MA, 56 pp.

Rosenzweig, C., Iglesias, A., Yang, X.B., Epstein, P.R., Chivian, E., 2001. Climate change and extreme weather events. Implications for food production, plant diseases, and pests. Glob. Chang. Hum. Health 2, 90–104.

Roy, K.W., Rupe, J.C., Hershman, D.E., Abney, T.S., 1997. Sudden death syndrome. Plant Dis. 81, 1100–1111.

Scherm, H., van Bruggen, A.H.C., 1993. Sensitivity of simulated dew duration to meteorological variations in different climatic regions of California. Agric. For. Meteorol. 66, 229–245.

Seeley, S.D., 2002. Reducing Chemical Inputs in Arid Climates through Sustainable Orchard Management. Department of Plants, Soils, and Biometeorology, Utah State University, Logan, UT.

Shew, B.B., Beute, M.K., Wynne, J.C., 1988. Effects of temperature and relative humidity on expression of resistance to *Cercosporidium personatum* in peanut. Phytopathology 78, 493–498.

Strand, J.F., 2000. Some agrometeorological aspects of pest and disease management for the 21st century. Agric. For. Meteor. 103, 73–82.

Todd, J.W., Culbreath, A.K., Chamberlin, J.R., Beshear, R.J., Mullinix, B.J., 1995. Colonization and population dynamics of thrips in peanuts in the southern United States. In: Parker, B.L., Skinner, M., Lewis, T. (Eds.), Thrips Biology and Management. Plenum, New York, pp. 453–460.

Tsuji, G.Y., Hoogenboom, G., Thornton, P.K., 1998. Understanding Options for Agricultural Production. Kluwer Academic Publishers, Dordrecht, The Netherlands.

Weiss, A., 1990. The role of climate-related information in pest management. Theor. Appl. Climatol. 41, 87–92.

Wharton, P.S., Kirk, W.W., Baker, K.M., Duynslager, L., 2008. A web-based interactive system for risk management of potato late blight in Michigan. Comput. Electron. Agric. 61, 136–148.

Woodward, J.E., Brenneman, T.B., Kemerait Jr., R.C., Culbreath, A.K., Smith, N.B., 2010. Management of peanut diseases with reduced input fungicide programs in fields with varying levels of disease risk. Crop Prot. 29, 222–229.

Wu, L., Damicone, J.P., Duthie, J.A., Melouk, H.A., 1999. Effects of temperature and wetness duration on infection of peanut cultivars by *Cercospora arachidicola*. Phytopathology 89, 653–659.

Yang, X.B., Scherm, H., 1997. El Niño and infectious disease. Science 275, 739.

CHAPTER

5

Forecasting of Colorado Potato Beetle Development with Computer Aided System SIMLEP Decision Support System

Paolo Racca[1], *Beate Tschöpe*[1], *Kristina Falke*[1], *Benno Kleinhenz*[1] *and Dietmar Rossberg*[2]

[1]**Central Institution for Decision Support Systems in Crop Protection, Bad-Kreuznach, Germany**
[2]**Julius Kühn Institute – Federal Research Centre for Cultivated Plants JKI, Kleinmachnow, Germany**

5.1 INTRODUCTION

5.1.1 Brief History of the Colorado Potato Beetle

In 1811, Thomas Nuttal discovered the Colorado potato beetle (CPB) for the first time, but the first scientific description of *Leptinotarsa decemlineata* was recorded 13 years later by Thomas Say, who collected the beetle in the Colorado Rocky mountains from buffalo-bur (*Solanum rostratum*, Ramur) plant. It was believed to have originated in Central Mexico (Arnett et al., 2002).

Since 1867, the beetle has had a series of names including the 'ten-striped spearman', 'ten-lined potato beetle', 'potato bug' and 'new potato bug'. The State of Colorado was not connected with the insect until, in 1865, Walsh received reports from colleagues of large numbers of the insect in the territory of Colorado feeding on buffalo-bur. This convinced him that it was native to Colorado and in 1867, Riley first used the term 'Colorado potato beetle' (Gauthier et al., 1981).

The association with the potato plant (*Solanum tuberosum* L.), was detected in 1840 and the first severe damage to crops was reported in 1859 at Omaha, Nebraska. The insect spread rapidly eastwards and reached the Atlantic coast in 1874. CPB is today well distributed in all US federal states and in Canada. It is also present in the southern American Continent region, Mexico, and in Central American Costa Rica, Guatemala, and Cuba (EPPO, 2012).

D. P. Abrol (Ed): Integrated Pest Management.
DOI: http://dx.doi.org/10.1016/B978-0-12-398529-3.00006-3

CPB migrated from the USA to France (Bordeaux region) in 1922 and spread rapidly over the European mainland. In 1935, it was found in Belgium, Romania, Russia, Slovakia, and Spain; a year later in Italy, Latvia, Lithuania, Luxembourg, Czech Republic, Estonia, and Germany; in 1937, in Moldova, the Netherlands, and Switzerland; in 1941, in Austria; in 1943, in Portugal; in 1946, in Poland; in 1947, in Hungary; in 1958, in Bulgaria; and in 1963, in Greece. The insect has been reported from but is not actually established in Denmark, Finland, Norway, Sweden, and the UK (EPPO, 2012).

There are some reports of the appearance of CPB in western China and Iran (Jolivet, 1991). Potentially, CPB could occupy large areas of China and Asia Minor, and spread to Korea, Japan, and certain areas of the Indian subcontinent, parts of North Africa, and the temperate Southern Hemisphere (Worner, 1988).

5.1.2 Biology and Life Cycle

CPB belongs to the family Chrysomelidae, the leaf beetles, and the subfamily Chrysomelinae.

The adult beetles are oval and robust and are convexly rounded with average size of 6–11 mm in length and about 3–5 mm in width. The elytra on the dorsum of the thorax and abdomen display 10 characteristic black stripes (Figure 5.1, A). The eggs are oblong, yellow-orange in colour, and normally grouped in clusters of 20–30 (Figure 5.1, E). The four larval instars are characterized by their large abdomen and arched back. The orange pink larvae have black spots and are up to 15 mm in length in their last instar (Figure 5.1, L1–L4). CPB pupae are oval and orange in colour (Figure 5.1, P) (Capinera, 2001; Wilkerson et al., 2005).

Adult beetles of CPB overwinter in the soil, with the majority aggregating in areas adjacent to potato fields (Weber and Ferro, 1993). The emergence of post-diapause beetles is more or less synchronous with potatoes. In extensive potato growing areas, the new potato fields are

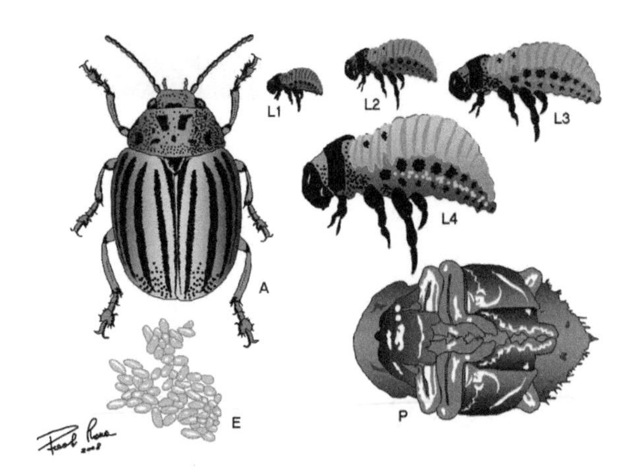

FIGURE 5.1 *Leptinotarsa decemlineata* life stages (original hand drawing by Racca).

colonized by overwintered adults that walk to the field from their overwintering sites or emerge from the soil within the field (Voss and Ferro, 1990). The beetles are able to fly up to several kilometres to find a new host habitat (Ferro et al., 1999).

After feeding, the beetles mate. They must feed before mating; food intake is zero at 10°C and maximum at 25°C. Oviposition follows within a day or two, females laying their eggs (from 15 to 30°C), 10–30 at a time, in several orderly rows on the lower leaf surface. Egg laying usually continues over a period of several weeks, until midsummer, with each female laying up to 2000 eggs. The eggs hatch in 4–12 days (provided temperatures are above 12°C) and the emerging larvae start to feed immediately. Molting occurs four times during the course of 2–3 weeks (optimum 30°C). Larvae are hardy and resistant to unfavourable weather, though heavy rain and strong winds may lead to high mortality, especially in the earlier instars (Hurst, 1975).

The mature larvae fall to the ground and bury themselves in the soil at varying depths (a few centimetres) according to conditions. Pupation, in smoothly lined cells, lasts for 10–20 days, after which the first generation adult beetles emerge (Hurst, 1975). Development from the time of oviposition to adult emergence from pupae takes between 14 and 56 days (Logan et al., 1985).

Reproduction continues until the hibernation diapause is induced by photoperiod and temperature (Lefevere and Kort, 1989). The mean mortality during hibernation is about 30%, but could be as high as 83%, due mainly to fungal and bacterial infections (Hurst, 1975; Koval, 1984). Soil temperature determines the length of diapause and the emergence from the soil (Hurst, 1975; Mailloux et al., 1988).

The number of yearly generations is mostly a function of temperature and varies between about four in the hottest areas of the CPB habitat (cycle completed in 30 days) to one full and one partial generation near the colder extremes (Hurst, 1975).

5.1.3 Economic Impact and Control Measures

CPB is the most destructive potato pest in Europe; both adults and larvae feed on this host, and often cause complete defoliation of potato plants, with yield losses up to 50% (EPPO, 2012). Under favourable weather conditions, the populations are liable to expand dramatically. Owing to the warm weather conditions during the 1990s, severe losses occurred in Germany and Poland. Consequently, insecticide use increased considerably (Pruszynski and Węgorek, 1991). In Germany, on average, 2–3 treatments are carried out per year (Roßberg et al., 2002).

In order to solve the problems with CPB, the decision support system (DSS) SIMLEP was developed by the Central Institution for Decision Support Systems in Crop Protection in Germany (German acronym: ZEPP). ZEPP was founded in October 1997 on the basis of an administrative agreement of the Federal States. The mission of ZEPP is to collect and examine existing predictive and simulation models for important agricultural and horticultural pests and diseases and to develop these models for practical use. Moreover, it initiates the development of predictive models for further pests and diseases not yet considered. More than 60 meteorologically based predictive models for pests and diseases have been successfully developed and introduced for practical use by governmental crop protection services within recent years. The plant protection services of the Federal States provide the prediction information to farmers and horticulturists. Calculation of pest attack and disease predictions is based on more than 570 meteorological stations using the latest information technology and media (Kleinhenz et al., 1996; Racca et al., 2010).

5.2 SIMLEP DSS

SIMLEP consists basically of two modules, SIMLEP1-Start for the prediction of the first occurrence of the beetle in a region and SIMLEP3 for the simulation of further development in the potato fields; this is a plot-specific model.

5.2.1 SIMLEP1-Start

5.2.1.1 Model Description and Development

The model predicts the hatching of the overwintering CPB from soil depending on the sum of the mean daily temperatures (8°C base level) from 1 March.

The basis of the model was overwintering data collected from special trials from season 2002–2003 to 2005–2006 done by the ZEPP in two locations (Mainz and Bad Kreuznach) in Rhineland-palatinate in Germany.

Vertically open wooden cages (1m height × 1.5m width) were buried in the soil and filled with new sandy, loam and sandy-loam soil from normal potato fields. Each wooden cage was considered as a single experimental replication, and four replications were carried out for each soil type. The wooden cages were covered with a fine plastic grid. At the end of August, 100 CPB adults were placed in each cage. They were randomly chosen from almost 500 adults collected in a normal potato growing field. To maintain the same genetic homogeneity of the CPB population, the adults were collected in one restricted potato growing area in Rhineland-palatinate.

Up to two surveys were performed in winter at the end of October to mid-November to check the buried rate and, consequently, the winter mortality. In the following spring (beginning of March), the cages were checked daily and the number of overwintering CPBs were recorded. The surveys ended at the start of May to mid-May, when no new CPB adults came out from the soil. The winter mortality was expressed as follows:

$$\text{Wm} = 100 - \text{Au} - \text{S\_Ow} \tag{5.1}$$

where Wm = winter mortality; Au = autumn mortality; S_Ow = sum of the hatching adults (March–May).

Since the aim of this model was to predict the start and course of CPB hatching from the soil and not the size of the overwintering population, the survey data were percent-transformed as follows:

$$\text{Ha}(\%) = \text{Ow}/\text{S\_Ow} \tag{5.2}$$

where Ha(%) = percentage of hatching CPB (for each survey date); Ow = overwintering CPB (for each survey date); S_Ow = sum of the hatching adults (March–May).

Since the sandy-loam soil was the most common soil in the German potato growing region, only the results of these trials were considered for model development. The results are illustrated in Figure 5.2.

Buried rate varied from 0.57 (Mainz, 2002–2003) to 0.98 (Bad Kreuznach, 2005–2006). Autumn mortality rate was low compared to winter mortality; in both cases, the variation was high within the trials and statistically significant for all values (Tuckey, $\alpha = 0.05$). As a consequence, the survival rate was only >0.5 in one case (Bad Kreuznach, 2005–2006), and on average, only 22% of the CPB adults overwintered.

The CPB hatched from the soil, expressed as a percentage of population, was correlated with a soil temperature (sensor at 20 cm depth) sum from 1 March with 8°C base level. To perform a statistical comparison of the data, they were linearly transformed with Logit (Equation 5.3) and Richard transformations (Equations 5.4 and 5.5) (Richards, 1959) as follows:

$$\text{LOGIT\_Ha} = \text{LN}[\text{Ha}/(1 - \text{Ha})] \tag{5.3}$$

where LOGIT_Ha = Logistical transformed hatching CPB (for each survey date); Ha = percentage of hatching CPB (for each survey date).

$$\text{Richard\_1\_Ha} = \text{LN}[1/(\text{Ha}^{(1-m)}) - 1] \tag{5.4}$$

where Richard_1_Ha = Richards' transformed hatching CPB (for each survey date); Ha = percentage of hatching CPB (for each survey date); m = Richards' formula coefficient ($m > 1$).

$$\text{Richard\_2\_Ha} = \text{LN}\{1/[1 - (\text{Ha}^{(1-m)})]\} \tag{5.5}$$

where Richard_2_Ha = Richards' transformed hatching CPB (for each survey date); Ha = percentage of hatching CPB (for each survey date); m = Richards' formula coefficient ($m<1$).

The transformed values obtained from Equations 5.3, 5.4, and 5.5 were correlated with the sum of soil temperature by means of a simple linear regression.

$$\text{TV} = a + b \times T\text{sum} \tag{5.6}$$

where TV = transformed values (LOGIT_Ha, Richard_1_Ha and Richard_2_Ha); a = constant of the linear equation; b = slope of the linear regression.

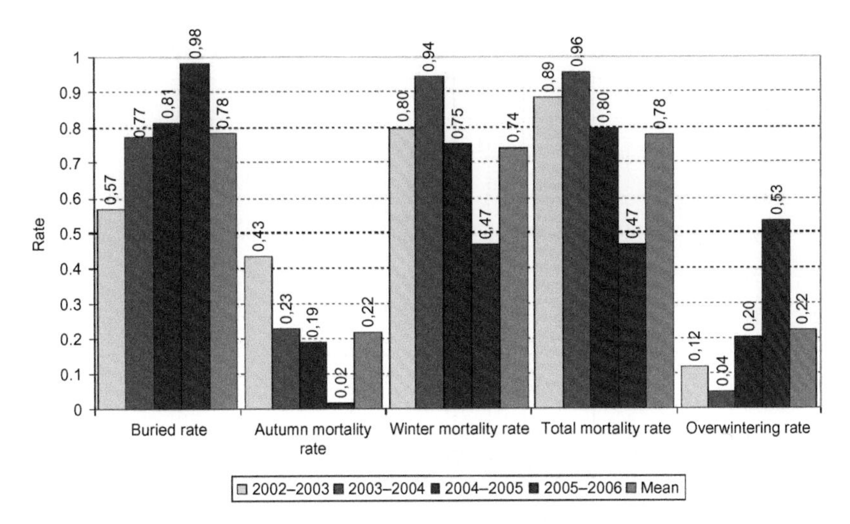

FIGURE 5.2 Buried rate, autumn, winter and total mortality rate, and overwintering rate of CPB adult population in trials 2002–2003 to 2005–2006. This figure is reproduced in colour in the colour plate section.

No significant differences were found between the course of hatching for the different years according to both parametric (significance of linear regression coefficients according to Armitage (1980)) and non-parametric tests (Kolmogorov-Smirnov) for the transformed data. The final model, Richard-2, was chosen using the maximized linear coefficient of determination (Table 5.1).

To transform the linear correlation into a typical 's' curve, the final model chosen, Richard-2, was rewritten as:

$$\text{Ha} = [1 - \text{EXP}(-(a + b^{*}T\text{sum})]^{[1/(1-m)]} \quad (5.7)$$

where Ha = percentage of hatching CPB; Tsum = sum of temperature (8°C base level) from 1 March; a = constant; b = slope; m = Richard's formula coefficient ($m < 1$).

The data and interpolated model are described in Figure 5.3.

TABLE 5.1 Values and Significance of the Parameters of the Tested Models

Model	Parameter	Value	Significance	r^2
Logistic (Equation 5.3)	a	−2.6905	<0.001	0.85
	b	0.0120	<0.001	
Richard-1 (Equation 5.4)	a	−1.8739	<0.001	0.87
	b	0.0112	<0.001	
	m	1.68	<0.001	
Richard-2 (Equation 5.5)	a	0.3165	<0.001	0.96
	b	0.0093	<0.001	
	m	0.98	<0.001	

5.2.1.2 Model Validation

The model was validated using validation methods proposed for praxis simulation models (Racca et al., 2010, 2011). Data for observed hatching of CPB recorded from 2008 to 2012 (n = 28 localities and year) by the German governmental crop protection services in the most important potato growing area of Germany were transformed with Equation 5.2 as a percentage of the overwintering population and then compared with the simulation results. In particular, three simulation results were tested: 10%, 50% and 80% occurrence dates for the overwintering population. The results are displayed in Figure 5.4.

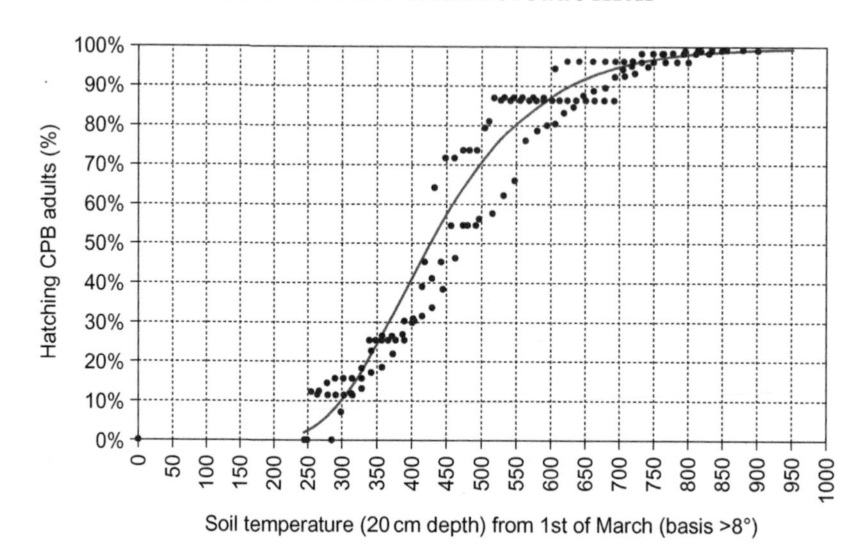

FIGURE 5.3 Scatter plot of the percentage of CPB adults hatching from the soil, expressed as a percentage of the overwintering population, depending on the sum of temperature from 1 March (8°C base level) and the SIMLEP1-Start interpolating Richards function. This figure is reproduced in colour in the colour plate section.

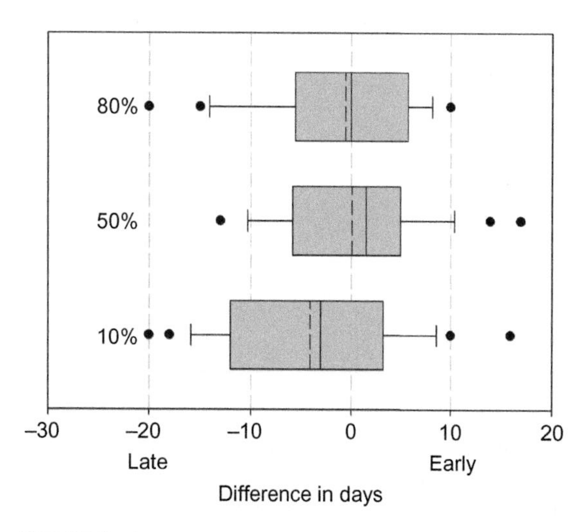

FIGURE 5.4 Box plot of the results of the validation of the SIMLEP1-Start model. Difference in days between simulation and observation for 10%, 50% and 80% occurrence of the overwintering CPB population (n = 28, dashed vertical line in box = mean, unbroken vertical line in box = median, points = outliers, negative and positive values on x axis represent late and early simulated occurrence date compared with the observation, respectively).

The results were satisfactory and, on average, the 10% occurrence date was about 4 days late for the simulated compared to the observed date. The model was more accurate for the 50% and 80% occurrence dates for the overwintering CPB, in which case the model simulated, on average, a slightly early (0.1) and a slightly late (−0.5) occurrence, respectively. The interquartile ranges (50% of the data) were −12 to 3, −5 to 4, and −5 to 6 for the 10%, 50%, and 80% occurrence dates, respectively.

Since, in extensive potato growing regions, identification of the beginning of hatching of the overwintering CPB with monitoring was very difficult and was not useful for a practical approach like the beginning of egg laying, a correlation was performed between the beginning of egg laying and the SIMLEP1-Start simulated percentage of hatching overwintering CPB. The correlation was done using monitoring data obtained from the governmental crop protection services in extensive potato growing regions where overwintering beetles migrate from the previous year's potato

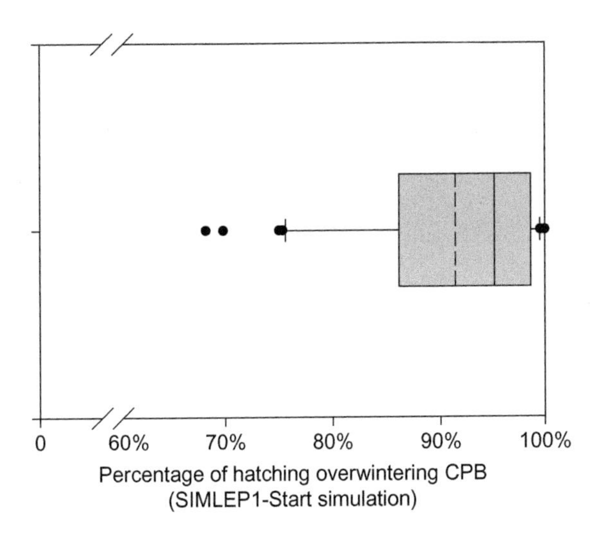

FIGURE 5.5 Correlation between the beginning of egg laying (detected in the field) and the percentage of the hatching overwintering CPB simulated by SIMLEP1-Start model displayed as a box plot (n = 45, dashed vertical line in box = mean, unbroken vertical line in box = median, points = outliers).

fields to the new potato fields in a short time span. For each observation (n = 45, years 1994 to 2008), the date of the beginning of egg laying was correlated with the respective percentage of the hatching CPB overwintering population simulated by SIMLEP1-Start. The results are illustrated in Figure 5.5. No egg laying was observed at hatching percentages lower than 50%, and the mean start of egg laying was detected at 91.6% of the hatching overwintering CPB.

5.2.1.3 *Practical Approach of the Model*

SIMLEP1-Start was implemented on the homepage of www.isip.de (ISIP: German acronym for Information System for Integrated Plant production), an official government website for the German governmental crop protection services and partner of the ZEPP (Racca et al., 2009). Simulation automatically begins on 1 March for each weather station with soil temperature measured at 20 cm depth. The model results are clearly and simply displayed in a geographical chart representing the region (Rhineland-palatinate in the example in Figure 5.6). Each weather station is represented by a cloud symbol with different colours: grey; no available data; yellow, beginning of hatching of the overwintering CPB population (10%); red, beginning of egg laying (hatching overwintering CPB population more than 91%).

Additionally, the results are also summarized in a simple table that shows both the dates of the beginning of hatching from soil and the beginning of egg laying (Figure 5.7). This second date is now used as the starting point for the SIMLEP3 model (see next section).

5.2.2 SIMLEP 3

5.2.2.1 *Model Description and Development*

SIMLEP3 simulates the development of CPB from the beginning of egg laying to the occurrence of old larvae at a field-specific scale.

Original functions derived from data from the 1960s (Kittlaus, 1961) and originally used in the former German Democratic Republic (Kurth, 1980; Kurth and Roßberg, 1983; Weber et al., 1988) for the creation of a model called SIMLEP2, were further elaborated in the SIMLEP3 model (Jörg et al., 2007; Roßberg et al., 1999). The model requires the following input parameters:

1. air temperature (2 m height) on an hourly basis (from 1 April)
2. the date of the beginning of egg laying on the potato field
3. the date of the previous assessment without egg laying (in cases where the date is unavailable, e.g. assessment started too late, then formally 1 January has to be inserted as the 'default value')
4. number of eggs laid (recorded on the date of their first observation; recorded on a potato field on at least five plants taken randomly in five locations).

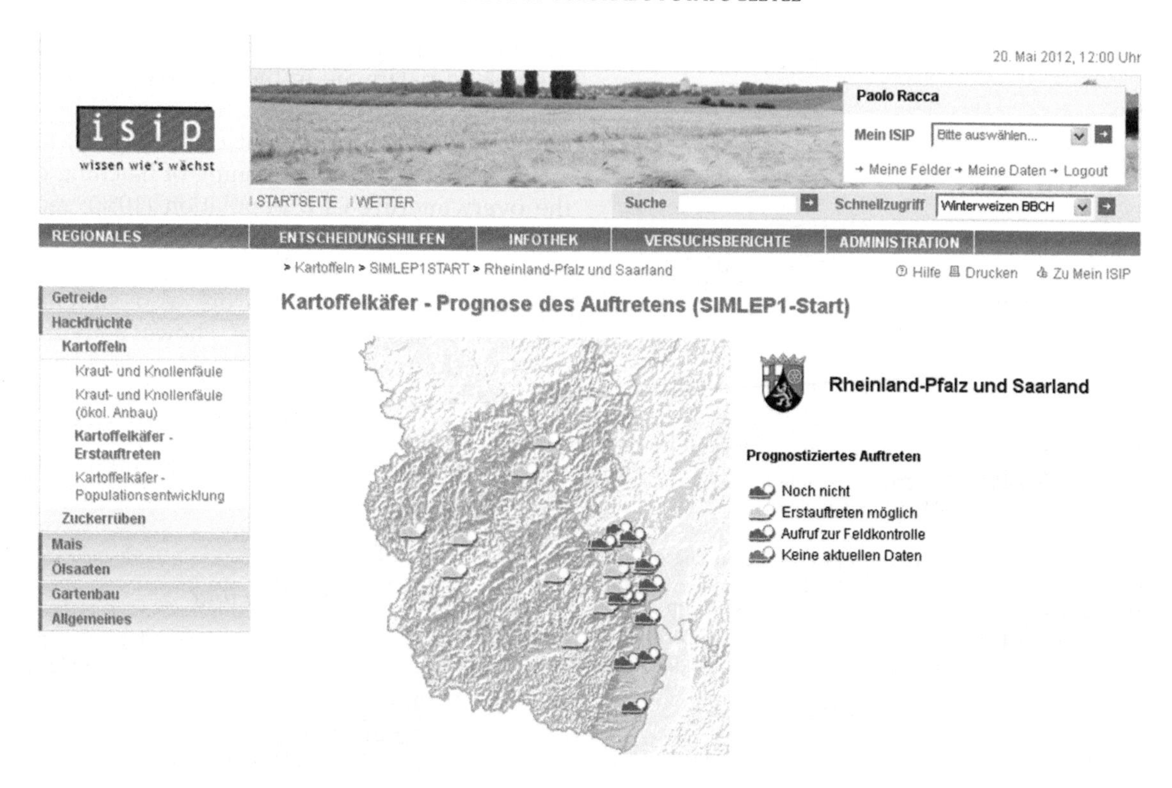

FIGURE 5.6 Output of the SIMLEP1-Start model from the ISIP website for the Rhineland-palatinate region. Clouds indicate weather stations. Cloud colours indicate: green, CPB overwintering population still in diapause; yellow, beginning of hatching of the overwintering population (10% CPB overwintering population have emerged from the soil); red, beginning of the egg laying in potato fields (>91% CPB overwintering population have emerged from the soil). This figure is reproduced in colour in the colour plate section.

The simulation starts on the date that egg laying is observed in the field. Based on the data inserted and the meteorological data provided, SIMLEP3 calculates an internal date for the start of egg laying by employing a complex algorithm. All calculations and forecasts refer to that date.

The model simulates the date of first occurrence of young larvae (L1/L2), the date of first occurrence of old larvae (L3/L4), the period of maximum egg laying (= period for optimal assessment of population density with respect to decision-making on insecticide use; advance warning of about 6–9 days), and the period of maximum abundance of young larvae (= optimal period for insecticide application; advance warning of about 4–7 days).

5.2.2.2 *Model Validation*

SIMLEP3 was validated in Germany and in several European countries in the years 1999–2004. According to Racca et al. (2010, 2011), the method used for the validation was only subjective, comparing the forecasting dates of the maximum abundance of egg clusters and young larvae with field observations. The model output was considered correct when the forecast was within an interval of 1 week of the observed date (Table 5.2).

In general, SIMLEP3 results were satisfactory. The first occurrence of young larvae

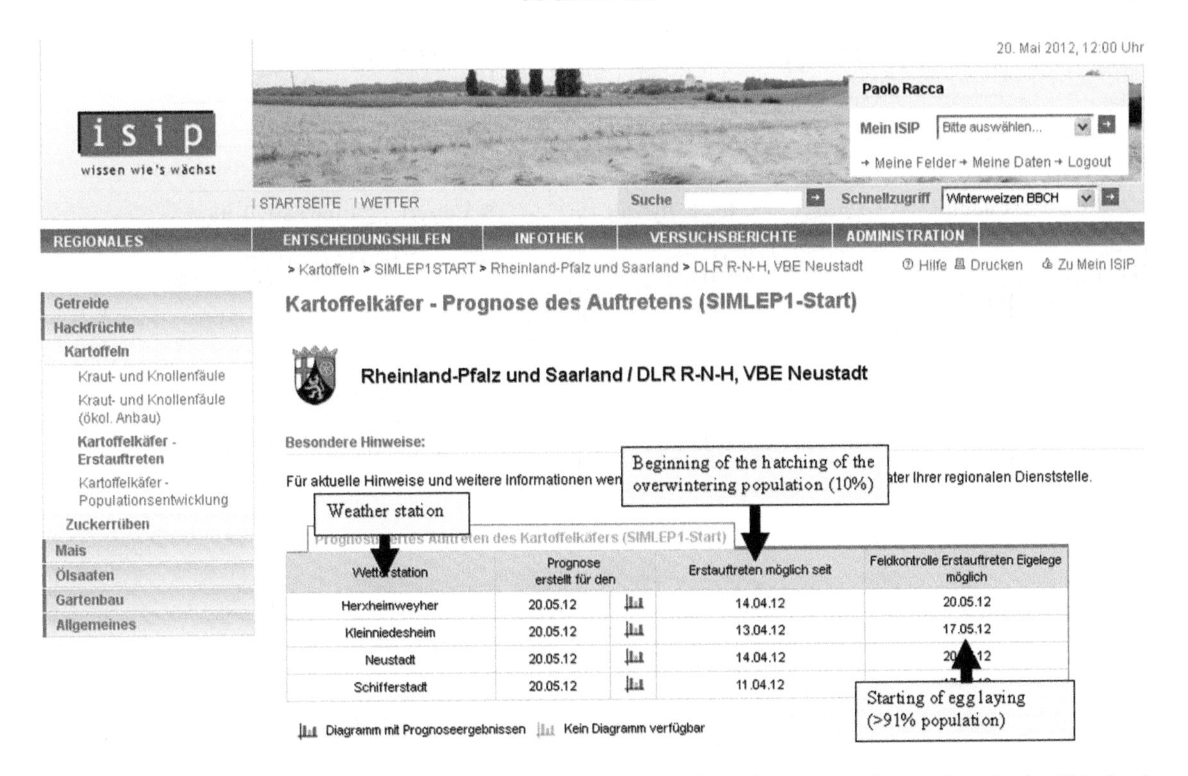

FIGURE 5.7 Output of the SIMLEP1-Start model from the ISIP website for two weather stations in the Rhineland-palatinate region. The first results column shows the beginning of hatching of the overwintering population (10%), and the second column indicates the start of egg laying. This figure is reproduced in colour in the colour plate section.

in most of the cases was predicted correctly. Nevertheless, differences between forecasting and observed date were registered ranging from 18 days too early up to 10 days too late. Good results were also obtained for the prediction of maximum egg cluster occurrence. Throughout Germany, Poland, Austria, and Italy, the mean share of correct forecasts given from SIMLEP3 (both egg clusters and young larvae) amounted to about 92%. In Austria, the share of correct predictions was the lowest (approx. 70%) and in Germany the share of correct predictions exceeded 90%. Maximum occurrences of young larvae predictions were correct in about 93% of cases on the European scale. Again, optimum results were obtained in Italy and Poland. In Austria and Germany, the share of correct forecasts exceeded 85%.

5.2.2.3 *Practical Approach of the Model*

Since, in Germany, action thresholds for insecticide applications are based on numbers of egg clusters (Jörg and Beck, 2000), assessments (whether or not action thresholds are overridden) should be done when SIMLEP3 identifies the period of maximum egg density.

As with SIMLEP1-Start, the SIMLEP3 results are presented on the ISIP website in a simple and clear way (Figure 5.8).

First, the occurrence date of young (L1/L2) and old (L3/L4) larvae is displayed. Additionally, for the period of maximum abundance of young larvae, two forecasts are

TABLE 5.2 Results of SIMLEP3 Subjective Validation in Several European Countries (1999–2004): Share of Correct Forecasts (%)

	Maximum Abundance of					
	Egg Clusters			Young Larvae		
Country	% Correct	% Too Early/Late	*n*	% Correct	% Too Early/Late	*n*
Germany	91	9	33	87	13	38
Italy	100	0	6	100	0	6
Austria	71	29	7	86	14	7
Poland	100	0	2	100	0	2
Mean	90.5	7.5		93.25	6.75	

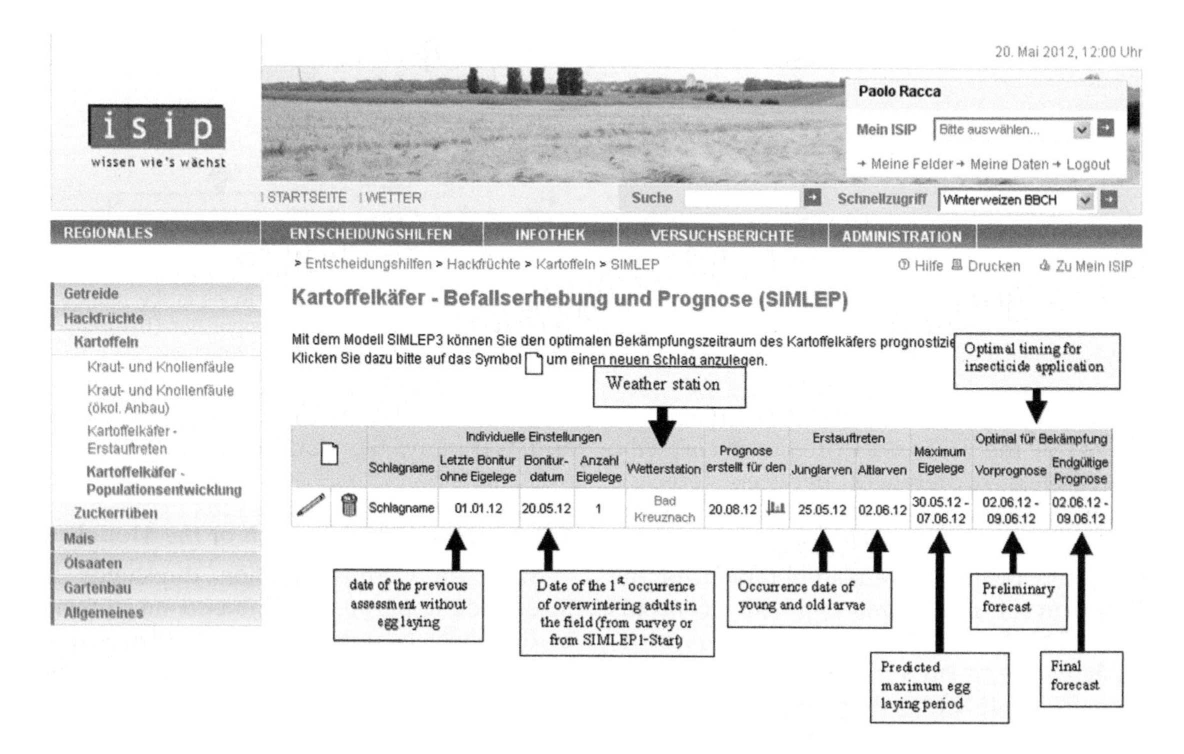

FIGURE 5.8 SIMLEP3 results from the ISIP website. This figure is reproduced in colour in the colour plate section.

calculated: a 'preliminary forecast' and a 'final forecast'. The 'preliminary forecast' is done using statistical calculations from the previous year's observations and calculating long year mean temperatures, the 'final forecast' with actual temperature data.

In most cases, the 'preliminary forecast' and 'final forecast' are identical. In years when the temperature drops severely after the calculation of the 'preliminary forecast', the final forecast period of maximum abundance of young larvae may be postponed. Conventional insecticides should be applied when maximum young larvae (L1/L2) abundance is forecast by the model because the control efficacy is at a maximum during this stage. If a biological insecticide (e.g. *Bacillus thuringiensis*) should be applied, the most appropriate date is the date of first occurrence of young larvae (Jörg et al., 2007). SIMLEP3 gives a prediction of these dates with an advance warning so that control measures can be planned in time. Time spans of about 5–6 days' duration are identified during which assessments or sprayings have to be carried out.

5.2.3 Planning Insecticide Strategy with SIMLEP-DSS

For both farmers and advisors, the control strategy for CPB begins in the spring when, at the beginning of March, the SIMLEP1-Start module begins to calculate the percentage of the population of overwintering CPB adults emerging from the soil. A first warning is given when the threshold of 10% of the hibernating population emerging is reached. The second and more important warning is given when reaching the threshold of 90% of the hibernating population emerging, which corresponds to the beginning of egg laying on potato plants in the field. At this point, this date can be used to initialize the module SIMLEP3 that will calculate the phase of maximum presence of L1/L2 larvae, which is the optimum date to ensure maximum efficacy of an insecticide treatment.

5.2.4 Outlook: the Developing Model SIMLEP-Res, a New Module for Anti-Resistance Strategy

From the mid-1990s onward, reports of reductions in the efficacy of insecticide applications became numerous. In part, these reductions in efficacy were due to inappropriate application dates or conditions (Jörg, 1998). In addition, resistance of CPB to organophosphates, carbamates, and pyrethroids was detected (Jörg et al., 2003; Preiß et al., 2004; Richerzhagen et al., 2010; Tschöpe et al., 2012).

A new module called SIMLEP-Res is now in the development phase in a 3-year project (2011–2014) (Tschöpe et al., 2012). The aim of this project is to develop an expert system that can be used to plan a strategy against potato beetle. The existing prediction models of population dynamics, SIMLEP1-Start and SIMLEP3 described above, used together with action thresholds, agronomic measures, and monitoring results, will be combined with the new module SIMLEP-Res to make the SIMLEP-DSS more complete and able not only to provide predictions for insecticide scheduling, but also to facilitate both farmers and advisors in the choice of appropriate active ingredients to minimize the risk of development of insecticide resistance by CPB.

5.3 CONCLUSIONS

Validation efforts have shown that SIMLEP1-Start is able to give correct predictions of the emergence of the overwintering CPB adult population. The analysis of monitoring data also showed a good correlation between the emergence of 91% of the hibernating CPB adult population and the beginning of egg laying on potato plants in the field. The SIMLEP3 model was also extensively validated in Germany and other European countries; results are satisfactory and demonstrate good prediction by the model for the most important developmental stages of CPB. Both models have been grouped in the SIMLEP-DSS and integrated on the ISIP governmental crop protection services official website and have been widely introduced into agricultural practice in Germany.

SIMLEP-DSS application has led to some improvements in crop protection efforts by farmers. The control efficacy of conventional as well as biological insecticides is higher because they are sprayed on the most susceptible larval stages of CPB (young larvae). Farmers have stopped the practice of overdosing insecticides to increase control efficacy. Another benefit of SIMLEP-DSS is the increase in acceptance of action thresholds because labour for field inspections can be minimized.

References

Armitage, P., 1980. Statistical Methods in Medical Research. Blackwell Scientific Publications, Oxford, UK.

Arnett Jr., R., Thomas, M., Skeppey, P., Frank, J., 2002. American Beetles. CRC Press, Boca Raton, Florida, USA.

Capinera, J., 2001. Handbook of Vegetable Pests. Academic Press, San Diego, CA.

EPPO, 2012. PQR – EPPO Database on Quarantine Pests. European and Mediterranean Plant Protection Organization (EPPO), Paris.

Ferro, D.N., Alyokhin, A.V., Tobin, D.B., 1999. Reproductive status and flight activity of the overwintered Colorado potato beetle. Entomol. Exp. Appl. 91, 443–448.

Gauthier, N., Hofmaster, R., Semel, M., 1981. History of Colorado potato beetle control. In: Casagrande, R., Lashomb, J.H. (Eds.), Advances in Potato Pest Management, Hutchinson Ross Publishing Company, Stroudsburg, PA.

Hurst, G.W., 1975. Meteorology and the Colorado Potato Beetle. World Meterological Organization.

Jolivet, P., 1991. The Colorado beetle menaces Asia (Leptinotarsa decemlineata Say) (Coleoptera: Chrysomelidae). L'Entomologiste 47, 29–48.

Jörg, E., 1998. Kartoffelkäferbekämpfung – Minderwirkungen beim Insektizideinsatz. Kartoffelbau 49, 172–174.

Jörg, E., Beck, W., 2000. Schadwirkung und Bekämpfung des Kartoffelkäfers. Kartoffelbau 51, 202–204.

Jörg, E., Węgorek, P., Racca, P., 2003. Kartoffelkäfer – Insektizidresistenz in Deutschland und Polen. Kartoffelbau 6, 235–237.

Jörg, E., Racca, P., Preiß, U., Butturini, A., Schmiedl, J., Wójtowicz, A., 2007. Control of colorado potato beetle with the SIMLEP decision support system. EPPO Bull., 353–358.

Kittlaus, E., 1961. Die Embryonalentwicklung von *Leptinotarsa decemlineata* SAY, *Epilancha sparsa* HERBST und *Epilancha vigintioctomaculata* MOTSCH. var. *niponica* LEWIS in Abhängigkeit von der Temperatur. Dtsch. Entomol. Z. N.F. 8, 41–61.

Kleinhenz, B., Jörg, E., Gutsche, V., Kluge, E., Roßberg, D., 1996. PASO-computer-aided models for decision making in plant protection. OEPP Bull./EPPO Bull. 26, 461–468.

Koval, Y.V., 1984. Characteristics of overwintering of the Colorado beetle. Zashchita Rastenii 5, 34.

Kurth, H., 1980. Untersuchungen zur anwendbarkeit von effektivtemperatursummen für die terminbestimmung beim kartoffelkäfer, *Leptinotarsa decemlineata* Say. Arch. Phytopathol. Pflanzensch. 16, 45–50.

Kurth, H., Roßberg, D., 1983. Ein modellgestütztes verfahren zur prognose des kartoffelkäfers (ISSN 0323-5912). Nachrichtenblatt Pflanzenschutz DDR 37, 49–51.

Lefevere, K.S., Kort, C.A.D., 1989. Adult diapause in the Colorado potato beetle Leptinotarsa decemlineata: effects of external factors on maintenance, termination and post-diapause development. Physiol. Entomol. 14, 299–308.

Logan, P.A., Casagrande, R.A., Faubert, H.H., Drummond, F.A., 1985. Temperature-dependent development and feeding of immature Colorado potato beetles, Leptinotarsa decemlineata Say (Coleoptera: Chrysomelidae). Environ. Entomol. 14, 275–283.

Mailloux, G., Richard, M.A., Chouinard, C., 1988. Spring, summer and autumn emergence of the colorado potato beetle, *Leptinotarsa decemlineata* (Say) (Coleoptera: Chrysomelidae). Agric. Ecosyst. Environ. 21, 171–179.

Preiß, U., Racca, P., Jörg, E., 2004. Insektizidresistenzentwicklung beim Kartoffelkäfer. In: Węgorek, P. (Ed.) Biologische Bundesanstalt für Land- und Forstwirtschaft, Biologische Bundesanstalt für Land- und Forstwirtschaft, Berlin–Dahlem, Hamburg, pp. 364.

Pruszynski, S., Węgorek, W., 1991. Control of Colorado beetle (*Leptinotarsa decemlineata*) in Poland. OEPP. Bull/EPPO Bull. 21, 11–16.

Racca, P., et al., 2009. In Proceedings of the 4th Conference on Statistical, Mathematical and Computer Methods on Plant Pathology and Forestry – Research and Application. Aracne, Viterbo.

Racca, P., Zeuner, T., Jung, J., Kleinhenz, B., 2010. Model validation and use of geographic information systems in crop protection warning service. In: Oerke, E.-C. (Ed.), Precision Crop Protection – the Challenge and Use of Heterogeneity. Springer, Netherlands, pp. 259–276.

Racca, P., Kleinhenz, B., Zeuner, T., Keil, B., Tschöpe, B., Jung, J., 2011. Decision support systems in agriculture: Administration of weather data, use of geographic information systems (GIS) and validation methods in crop protection warning service. In: Jao, C. (Ed.), Efficient Decision Support Systems – Practice and Challenges From Current to Future, vol. 1. InTech, Rijeka, pp. 331–354.

Richards, F.J., 1959. A flexible growth function for empirical use. J. Exp. Bot. 10 (2), 290–301. ISSN 0022-0957 10.

Richerzhagen, D., Falke, K., Racca, P., 2010. Julius Kühn-Institut, Untersuchungen zur Insektizidresistenz des Kartoffelkäfers (Leptinotarsa decemlineata (SAY)), 57. Deutsche Pflanzenschutztagung, Berlin, pp. 200.

Richerzhagen, D., Heibertshausen, D., Racca, P., Zeuner, T., Kleinhenz, B., Hau, B., 2010b. Einsatz regionaler Klimaprojektionen zur Untersuchung des Auftretens von Blattkrankheiten an Zuckerrüben (Poster), pp. 429. In: Pflanzenschutztagung, D., Kühn-Institut, J. (Eds.) 57. Deutsche Pflanzenschutztagung. 6–9 September 2010. Humboldt-Universität zu Berlin; Gesunde Pflanze, gesunder Mensch. Julius Kühn-Inst., Bundesforschungsinst. für Kulturpflanzen, Berlin.

Roßberg, D., Jörg, E., Kleinhenz, B., 1999. SIMLEP2 – ein Modell zur schlagspezifischen Prognose des Kartoffelkäferauftretens. Eugen Ulmer GmbH & CO., Stuttgart 51, 81–87: ISSN 0027-7479.

Roßberg, D., Gutsche, V., Enzian, S., Wick, M., 2002. NEPTUN 2000 – Erhebung von Daten zum tatsächlichen Einsatz chemischer Pflanzenschutzmittel im Ackerbau Deutschlands 98. BBA Braunschweig, Braunschweig.

Tschöpe, B., Breckheimer, B., Richerzhagen, D., Racca, P., 2012. Aktuelle Untersuchungen zur Insektizidresistenz des Kartoffelkäfers (*Leptinotarsa decemlineata* (SAY)) (Poster) 58. Deutsche Pflanzenschutztagung, Braunschweig, p. 463.

Voss, R.H., Ferro, D.N., 1990. Phenology of flight and walking by Colorado potato beetle (Coleoptera: Chrysomelidae) adults in western Massachusetts. Environ. Entomol. 19, 117–122.

Weber, B., Roßberg, D., Kurth, H., Otto, D., 1988. Simulation of insecticide influences on the population dynamics of the Colorado beetle. OEPP Bull./EPPO Bull. 18, 163–172.

Weber, D.C., Ferro, D.N., 1993. Distribution of overwintering Colorado potato beetle in and near Massachusetts potato fields. Entomol. Exp. Appl. 66, 191–196.

Wilkerson, J., Webb, S., Capinera, J., 2005. Vegetable Pests 1: Coleoptera, Diptera, Hymenoptera. UF/IFAS Publications, University of Florida, Gainesville, FL.

Worner, S.P., 1988. Ecoclimatic assessment of potential establishment of exotic pests. J. Econ. Entomol. 81, 973–983.

CHAPTER

6

Role of Semiochemicals in Integrated Pest Management

L.E. Smart, G.I. Aradottir and T.J.A. Bruce

Biological Chemistry and Crop Protection Department, Rothamsted Research, Harpenden, Hertfordshire, UK

6.1 INTRODUCTION

Semiochemicals are signalling chemicals used to carry information between living organisms and which cause changes in their behaviour (Dicke and Sabelis, 1988; Nordlund and Lewis, 1976). They are emitted by one individual and cause a response in another. Most invertebrates rely on olfaction as the principal sensory modality for sensing their external environment (Krieger and Breer, 1999). Attraction of insects to plants and other host organisms involves detection of specific semiochemicals or specific ratios of semiochemicals (Bruce et al., 2005a). Avoidance of unsuitable hosts can involve the detection of specific semiochemicals, or mixtures of semiochemicals, associated with non-host taxa (Agelopoulos et al., 1999; Bruce and Pickett, 2011; Hardie et al., 1994). For integrated pest management, there is an opportunity to develop non-toxic interventions using semiochemicals that influence the behaviour of pest insects. Attractants can be used in baited traps to monitor pest populations. Furthermore, semiochemicals that repel pests or attract their natural enemies could be used to keep pest populations below damaging levels.

Semiochemicals are divided into pheromones, which act within the same species, and allelochemicals, which act between species. Pheromones consist of sex, alarm, aggregation or territory marking signals and have evolved for communication purposes. Allelochemicals can be divided into signals that benefit the receiver (kairomones), the emitter (allomones), or both (synomones) (Nordlund et al., 1981). Semiochemicals can have multiple roles, being used for different purposes at different trophic levels. For example, herbivore-induced volatiles often repel plant-feeding insects while at the same time attracting their natural enemies. This terminology is somewhat limited since the same chemical compound may have several functions, e.g. a pheromone that also acts as a kairomone for another species, hence Dicke and Sabelis (1988) proposing the use of the term 'infochemical', which may be particularly appropriate in situations where tritrophic interactions are being considered. There is strong selection pressure on insects to evolve sophisticated means for detecting food resources as

D. P. Abrol (Ed): Integrated Pest Management.
DOI: http://dx.doi.org/10.1016/B978-0-12-398529-3.00007-5

their survival and reproduction depend on finding them. Indeed, even generalist insects have mechanisms to avoid alighting on non-host plants.

There is an environmental case and public demand for reduction in the use of toxic insecticides for pest control. Semiochemicals have great potential to provide alternative solutions, because they are relatively non-toxic to vertebrates and to beneficial insects, are generally used in small amounts, and are often species-specific. This chapter describes some of the ways in which semiochemicals have been used in integrated pest management (IPM) to date, using selected examples, some based on research by the authors at Rothamsted. Limitations to their use are also considered along with possible ways in which these limitations may be overcome in the future.

6.2 SEMIOCHEMICALS FOR MONITORING PEST POPULATIONS

One of the most widespread and successful practical applications of semiochemicals is in detection and monitoring of pest populations (Witzgall et al., 2010). To rationalize pesticide use, monitoring systems are used to time treatments so that they are only applied when economic thresholds are exceeded. Crop scouting by direct inspection of crops is often labour intensive and not feasible for large-scale agriculture. Semiochemical baited insect monitoring traps can provide a solution to this problem. Sex pheromones are good for this purpose because they are very strong attractants and are species-specific although they usually only attract males. One of the first pheromone monitoring traps was for the pea moth, *Cydia nigricana* (Wall et al., 1987). Currently, pheromone lures are used in traps to monitor many different crop pest species (Witzgall et al., 2010). Such monitoring systems allow farmers to time insecticide applications, which reduces economic and environmental costs of insecticide application. Poorly targeted and unnecessary insecticide sprays can have a negative impact on natural enemies of pests.

At Rothamsted, the authors recently developed a pheromone trap-based monitoring system for the orange wheat blossom midge (OWBM), *Sitodiplosis mosellana* (Bruce et al., 2007). OWBM is a common and increasingly important pest of wheat in the Northern Hemisphere, causing severe yield losses in years of high infestation. Larval feeding on the developing seeds causes shrivelling and pre-sprouting damage and also facilitates secondary fungal attack by *Fusarium graminearum* and *Septoria nodorum*. This affects both the yield and quality of grain harvested. Due to difficulties in detection of OWBM before pheromone traps were developed, the actual degree of damage to crops was often not realized. However, in an outbreak in the UK in 2004, crop losses were estimated at 6% (1 million tonnes) nationally, which was compounded by reductions in grain quality, despite insecticide application to around 500,000 ha of wheat. OWBM has a very patchy spatial distribution and varies from year to year depending on climatic conditions. In the UK, precipitation causing moist soil conditions at the end of May, followed by warm still weather in late May/early June, can lead to serious OWBM outbreaks. The ovipositing female is a small insect which can remain well hidden in the crop canopy. The larvae are also hidden within the wheat ear, which is a cryptic position as well as a difficult spray target. Thus, to achieve effective control, any insecticide application has to be applied promptly before larvae burrow in-between the lemma and palea.

The female produced sex pheromone of OWBM has been identified as (2S,7S)-nonanediyl dibutyrate (Gries et al., 2000). The authors synthesized the pheromone at Rothamsted and tested different formulations of the pheromone, with different release profiles, in a series of field trials, and effective trap and dispenser designs were determined. Observations of variability in trap catch, and how it related to subsequent infestations, were used to develop a decision support model (Bruce and Smart, 2009). This model is a distillation of some complicated data obtained during several years of research but has been framed in terms of what it means for the farmers when using the traps. With this in mind, it has been kept as simple and user-friendly as possible, being based on a stepwise decision tree involving yes/no answers to questions (Figure 6.1).

6.3 MASS TRAPPING

Mass trapping is an extension of the use of species-specific semiochemical baited monitoring traps, with the aim of reducing or eradicating populations of target pests by capturing as many individuals as possible. The lure can be a synthetic pheromone, a food or host attractant, or a combination of the two, which is sufficiently effective when deployed in an optimally efficient trap design at a suitable density to suppress the pest and reduce economic damage to the target crop. To achieve this, traps have to capture a large proportion of the population in an area, before mating or oviposition, and retain or kill captured individuals. The lure must be more effective than natural sources of attraction such as mates or food/oviposition sites and ideally retain efficacy over the entire period of adult insect reproductive activity to reduce damage to a minimum. In addition, the yield benefits and the cost of traps, and the manual labour required to deploy them, must be economically comparable to alternative control methods for this approach to be feasible. The biology and ecology of the pest can also be an obstacle to the use of this approach, the best targets being small or isolated populations with low immigration rates, univoltine species or those with a limited life cycle, host range or flight period (El-Sayed et al., 2006).

Despite the above-mentioned constraints, there are many examples of the attempted use of mass trapping to control a range of pest Coleoptera, Lepidoptera, Diptera and Homoptera, and classic case studies have been reviewed extensively by El-Sayed et al. (2006) and Witzgall et al. (2010). Apart from practical aspects such as achieving optimal trap and dispenser life/design and an appropriate trap density, the constraints to successful mass trapping vary according to the semiochemical attractant used. For example, some species of coleopteran pests produce aggregation pheromones, which are equally attractive to both sexes and thereby provide an opportunity to reduce the local population as a whole. Examples include the more injurious bark beetles, *Ips* spp. and *Dendroctonus* spp. (Curculionidae: Scolytinae) (Byers et al., 1990; Silverstein et al., 1968) and pest weevils, e.g. the boll weevil, *Anthonomus grandis*, the palm weevils, *Rhynchophorus* spp. (Curculionidae) (Rochat et al., 1991; Tumlinson et al., 1969) and the banana weevil, *Cosmopolites sordidus* (Dryophthoridae) (Beauhaire et al., 1995). Some of the earliest and currently most successful attempts at mass trapping, both for pest management and eradication, have been against Coleoptera using optimized aggregation pheromone blends (Alpizar et al., 2012; Oehlschlager et al., 2002; Reddy et al., 2009; Schlyter et al., 2001; Smith, 1998; Witzgall et al., 2010). The reason these worked well was mainly because these lures were powerful attractants for female insects, the sex which lays eggs from which the damaging larvae emerge.

In contrast, Lepidoptera predominantly use species-specific female produced sex pheromones that attract only males. Due to the males' capacity for multiple matings, a very large proportion of the male population has to be removed before female fecundity is reduced. The underlying behavioural mechanism is competitive attraction between calling females and discrete pheromone point sources, the latter

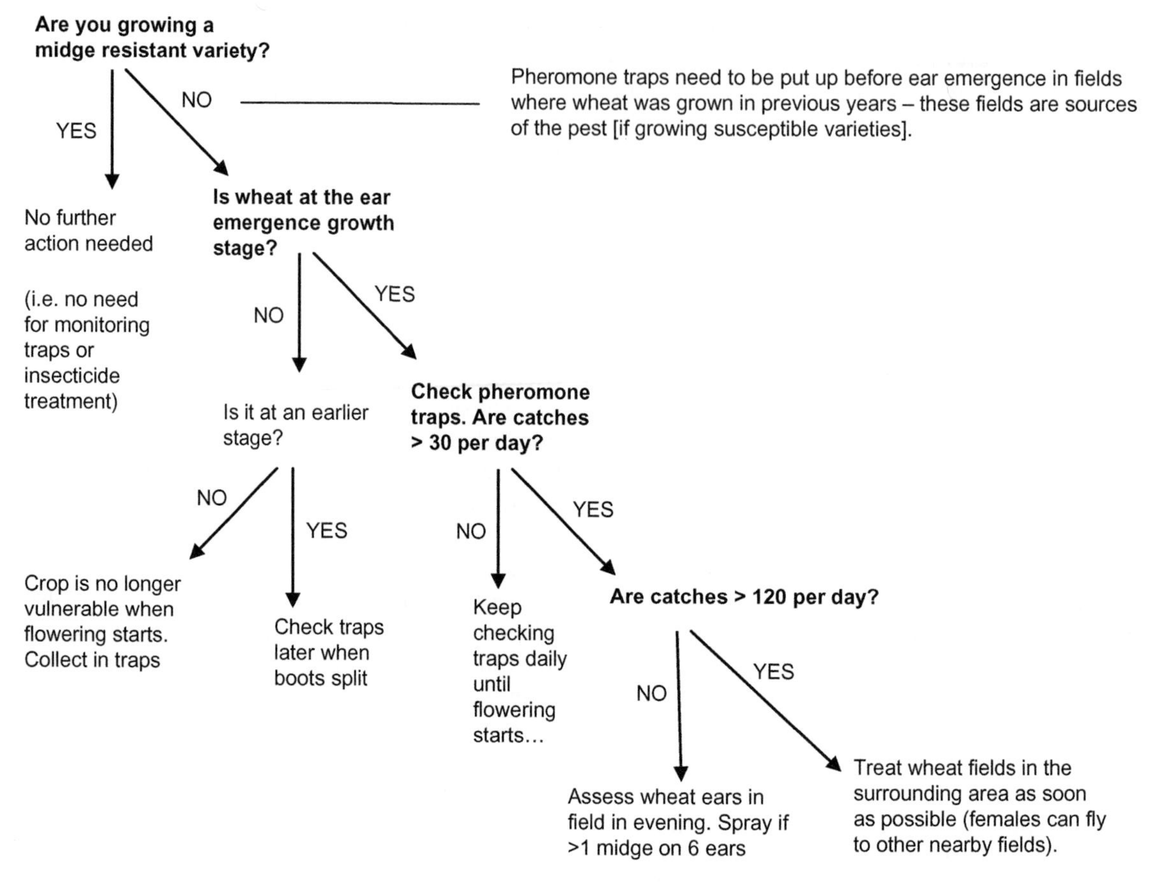

FIGURE 6.1 Decision support tree devised for wheat growers for orange wheat blossom midge management.

resulting in the permanent removal of potential mates (Miller et al., 2010). Initial attempts at control or eradication of pest Lepidoptera such as gypsy moth, *Lymantria dispar* L., codling moth, *Cydia pomonella* L. and more recently dogwood borer, *Synanthedon scitula*, using mass trapping techniques did not provide adequate or economical control (Hagley, 1978; Leskey et al., 2009; Myers et al., 1998; Sharov et al., 1998) and subsequently mating disruption techniques have provided more promising results (Brunner et al., 2002; Leskey et al., 2009; Tcheslavskaia et al., 2005; Witzgall et al., 2008). Control of the brinjal fruit and shoot borer moth, *Leucinodes orbonalis*, has been achieved, due to a combination of a reduction in the pest population and a greater impact of natural enemies, numbers of which increased after cessation of the use of insecticides (Cork et al., 2005). In other cases, where intervention with insecticides or biological control has proved expensive, ineffective or impossible, mass trapping has provided a reasonably effective alternative, e.g. for controlling the leopard moth, *Zeuzera pyrina* L., in olive orchards (Hegazi et al., 2009), the cerambycid beetle, *Prionus californicus*, attacking hop yards (Maki et al., 2011) and the microlepidopteran tomato leaf miner, *Tuta absoluta*, in solanaceous crops (Chermiti and Abbes, 2012). In many cases, the addition of semiochemical attractants to pheromone traps, which enable trapping and removal of females as well, will have a more direct effect on population growth than reliance on reduced mating through male removal (Camelo et al., 2007).

Volatile kairomones from food or host plant sources have been used alone, or in combination with pheromones, to trap both male and female pests in search of food or oviposition sites and have targeted predominantly dipteran and some coleopteran species (El-Sayed et al., 2009). However, caution is needed as some plant volatiles such as floral odours attract non-target beneficial insects and the level of attraction of the target pest may not be enough for effective control. The banana weevil, *C. sordidus*, was controlled by mass trapping using baits made from host plant pseudostems, until the aggregation pheromone was identified and proved to be more efficacious (Alpizar et al., 2012). Food odours including hydrolysed proteinaceous baits were developed to trap a wide range of tephritid fruit fly species and are still in use in lure and kill strategies (El-Sayed et al., 2009, and see below). The natural product, methyl eugenol (4-allyl-1,2-dimethoxybenzene), which is a male pheromone precursor extremely attractive to males of the genus *Bactrocera*, has been used in IPM programmes to eradicate these flies in areas of the USA (Witzgall et al., 2010). Parapheromones, e.g. Trimedlure (tert-butyl 4(or 5)-chloro-2-methylcyclohexanecarboxylate), Ceralure (ethyl-*cis*-5-iodo-*trans*-2-methylcyclohexane-1-carboxylate), some plant volatiles, and essential oils (Cunningham et al., 1990; El-Sayed et al., 2009) have also been used widely to control fruit flies, including male Mediterranean fruit fly, *Ceratitis capitata*, and the food attractants ammonium acetate, putrescine and trimethylamine have been shown to be attractive to female *C. capitata* (Katsoyannos and Papadopoulos, 2004).

Mass trapping usually requires a killing agent such as a dichlorvos strip inside the trap. However, the addition of an insecticide is not always possible or acceptable. Aurelian et al. (2012) demonstrated that a trap baited with grape juice was as effective at capturing male *Synanthedon myopaeformis* (Lepidoptera: Sesiidae), a recent arrival in organic apple growing areas of Canada, as sex pheromone baited traps, with the additional benefit of attracting female moths. At higher densities, there was interference between pheromone traps, which restricted the total male catch, but there was no such interaction between juice baited traps and summed catches of males and females were greater, indicating that a mass trapping approach could be effective at suppressing the pest while populations are still small and isolated. Another promising target for mass trapping with female attractant kairomones is the blow fly *Calliphora vicina*, which attacks stockfish production in Norway. Chemical treatments and physical methods to protect the fish, a traditional source of protein, are undesirable or difficult. The fly is univoltine in this region and occurs in isolated areas of fish production, so is an ideal target for mass trapping. Aak et al. (2011) showed that damage to drying fish could be reduced by 63% after a 4-year trapping programme that included traps baited with attractive odours.

There is an increased interest in the use of mass trapping as an alternative to pest control with conventional insecticides. The development of simulation models, e.g. of insect searching behaviour in association with male or female produced sex pheromones, and pheromone trap and point source density (Byers, 2007, 2012), and individual-based models (IBMs) that integrate key behaviours such as habitat selection and dispersal with spatial heterogeneity (Vinatier et al., 2012), have led to a better understanding of the variables affecting the efficacy of mass trapping programmes. Recent large-scale studies have demonstrated that mass trapping, using pheromone baited micro sticky traps, has the potential to be more effective in controlling codling moth and the oblique banded leafroller, *Choristoneura rosaceana*, in orchards, than mating disruption techniques (Reinke et al., 2012). In addition, the same group, having identified the behavioural mechanism underlying mating disruption and mass trapping of three different lepidopteran pests of cherry and peach orchards, demonstrated the possibility of using mass trapping to target the three species using just one trap and bait system (Teixeira et al., 2010). The cost of this technique is dependent on the efficacy of the attractant lure and the number of traps required to provide effective control. Ultimately, as pheromone components become more readily available and with the development of cheap biodegradable traps, there will be greater opportunities to utilise mass trapping for pest control.

6.4 LURE AND KILL

The most common instances of the combination of mass trapping with insecticides are found in 'lure and kill' or 'attracticide' technology (Jones, 1998). Lure and kill approaches, consisting of specific formulations of attractants and insecticides, have been used in pest management for several decades and many case studies have been reviewed comprehensively by El-Sayed et al. (2009). The pest is attracted to the semiochemical lure, but instead of/as well as being trapped, it is killed by a toxicant, which is usually an insecticide, but can be an insect pathogen ('lure and infect'). Pest population reduction by lure and kill is cost effective compared to mating disruption since generally smaller amounts of pheromone are required and the insecticide component is limited to very small areas, which reduces crop contamination, and is therefore more environmentally benign. Nevertheless, the inclusion of an insecticide does affect public acceptance of this method and some formulation technologies will be susceptible to pesticide regulatory processes, especially within the European Union (Regulation EU 1107/2009). There are also some concerns about the possible attraction of non-target and beneficial species to toxic baits (Michaud, 2003).

The success of the approach encompasses similar principles to those of mass trapping in general such as deployment density and placement, but with the additional complication of devising the correct insecticide dose and formulation which has sufficient longevity, is compatible with and does not affect the efficacy of the semiochemical or the target pest's interaction with it, e.g. when in competition with natural sources of pheromone. Efficacy is also dependent on the target pest contacting the insecticide, which is formulated with the attractant lure or applied adjacent to it. If the formulation is not in a trap then the pest must receive a sufficient dose before leaving the lure, which leads to death or disablement and reduces the population substantially (El-Sayed et al., 2009). Lure and kill technology has attracted considerable commercial development, since formulations are patentable, and many products against a wide range of targets are now available in some countries. Simple products include an insecticide within a standard semiochemical baited trap, e.g. the Magnet™ trap range (Agrisense) for the tephritid pests olive fruit fly, *Bactrocera oleae*, which is baited with the female produced sex pheromone, a spiroacetal, and Mediterranean fruit fly, *C. capitata*, using the male parapheromone Trimedlure. Mass trapping, using an attractant and a toxic lure, has been used widely in the Mediterranean regions to control tephritid flies, in particular *C. capitata* in citrus groves, and the technique is often reviewed and improved (Martinez-Ferrer et al., 2012; Navarro-Llopis et al., 2008, 2011).

Synthetic food-based attractant protein hydrolysates (e.g. Solbait and GF-120) mixed with insecticides have been used successfully in female biased mass trapping strategies against multiple tephritid fly species (Mangan et al., 2006) and against *C. capitata* in Mallorca (Leza et al., 2008). More advanced formulations usually incorporate the semiochemical attractant with the insecticide in a paste, gel or wax, e.g. SPLAT, a wax which is also used for mating disruption (Stelinski et al., 2007) and Magnet, a mix of attractant plant volatiles and an insecticide that is ingested by and kills *Helicoverpa* spp. (Lepidoptera: Noctuidae) (Del Socorro et al., 2010; Gregg et al., 2010). These have the advantage in that they can be applied, at specifically defined numbers of droplets per hectare, directly by a commercial applicator. Other formulations use microencapsulated semiochemical and insecticide applied by hand-held sprayer to provide a known number of lures per hectare, or hollow fibres containing pheromone that are mixed with an adhesive containing the insecticide just before application, e.g. Nomate for management of pink bollworm, *Pectinophora gossypiella* (Conlee and Staten, 1986). For a full list of products, see El-Sayed et al. (2009). As with traps, the life of the semiochemical lure and the density per unit area are crucial to successful control, but high densities of point sources are more easily achieved with the sprayable formulations and labour costs are lower. However, too high a density of lures may result in interference between them thereby reducing efficacy. A high density could also give rise to increased immigration of pests into the treated area reducing overall control. A clear understanding of insect behaviour in these contexts is essential (El-Sayed et al., 2009).

6.5 MATING DISRUPTION WITH PHEROMONES

Mating disruption aims to disrupt chemical communication by organisms and interrupt normal mating behaviour by dispensing synthetic sex pheromone, thereby affecting the organism's chance of reproduction (Cardé and Minks, 1995). This can be done by using both attractive and non-attractive pheromone blends. Mating disruption with sex pheromones can be effective if the edge effect of mated females flying in from outside the treated area can be avoided. This can be done if very large areas are treated or if the area treated is isolated such as in a mountain valley. The use of pheromones in IPM has recently been reviewed by Witzgall et al. (2010) and mating disruption was reviewed by Rodriguez-Saona and Stelinski (2009). The area under mating disruption has increased almost exponentially from the 1990s, and it is reported that the crop area being managed for specific pests using mating disruption worldwide was 770,000 ha in 2010 (Ioriatti et al., 2011; Witzgall et al., 2010). The three species with the highest land area under mating disruption were the gypsy moth (*Lymantria dispar*) in North American forests, the codling moth (*Cydia pomonella*) in apple and pear trees worldwide and the grapevine moth (*Lobesia botrana*) in grape in the EU and Chile (Witzgall et al., 2010).

The European grapevine moth, *Lobesia botrana*, is the principal native pest of grape in the Palearctic. A coordinated strategy of mating disruption in vineyards in Northern Italy has been successful in area-wide reduction of *L. botrana* populations, and the reduction in insecticide use has improved the quality of life for growers and consumers, as well as the public (Ioriatti et al., 2011), and now pheromone treated vineyards in Europe are estimated at 100,000 ha (Ioriatti et al., 2008), which includes 0.03% of the approximately 3 million hectares under vineyards in the EU. Many plant-feeding midges are important as pests and can cause substantial crop losses in forestry and field crops, as well as horticultural and fruit crops. Progress in use of pheromones for the direct control of midge pests has been slow, in part due to the expense of producing midge pheromones and the sporadic nature of outbreaks

(Hall et al., 2012). A recent study has, however, shown the effectiveness of pheromone-based mating disruption of a member of the dipteran family, the swede midge *Contarinia nasturtii*, in field evaluations of small-scale plots with Brussels sprouts, and in commercial-scale fields with broccoli and cauliflower. Crop damage was reduced by 59% in broccoli and by an average of 91% in the large-scale experiments (Samietz et al., 2012).

6.6 SEMIOCHEMICALS TO REPEL PESTS AND ATTRACT NATURAL ENEMIES

As they influence the behaviour of insects, it has been suggested that semiochemicals could play a role in repelling pest insects and attracting natural enemies of the pests (Agelopoulos et al., 1999; Foster and Harris, 1997). Laboratory-based bioassays have shown that semiochemicals have such activity and importantly the same compounds which repel pests often also attract their natural enemies. This is because key semiochemicals are produced by plants only when they are attacked by insects and the herbivorous insects associate these compounds with a poorer quality host which has defence induced whereas the predators and parasitoids associate the semiochemicals with a plant that has a supply of prey items (e.g. Du et al., 1998). Bioassays where compounds are tested against clean air often give very promising results but translating this activity into field performance is more challenging. Under real field conditions, insects are not exposed to semiochemicals in a vacuum but against a background of naturally occurring semiochemicals produced by plants in the habitat. In an agricultural ecosystem, there is a very large area of host plant and associated volatiles that are attractive to pest species which are adapted to that particular crop. It is thus a considerable challenge to artificially release repellent semiochemical to counteract the natural semiochemical emission by the crop.

Methyl salicylate, a winter host volatile, is repellent to summer forms of the bird-cherry oat aphid, *Rhopalosiphum padi*, and has been shown to reduce cereal aphid infestation levels in small plot field trials (Pettersson et al., 1994). The best reduction obtained was approximately 50%. Formulation of artificial dispensers requires development for large-scale use. Oil of wintergreen contains methyl salicylate and formulations of this could be used by organic growers to obtain some measure of crop protection. Effects on natural enemies have also been demonstrated; for example, deployment of aphid sex pheromone lures in wheat in small plot trials resulted in a doubling in the number of parasitized aphids and earlier parasitism than in control plots (Powell and Pickett, 2003). Aphid sex pheromones are not attractive to the aphids themselves during the summer because they are in an asexual form that undergoes parthenogenetic reproduction. Another approach involved use of the essential oil of *Hemizygia petiolata*, which contains the aphid alarm pheromone, (*E*)-β-farnesene. Dispensers containing *H. petiolata* oil caused significant reductions in aphid settlement in field trials both with pea aphids in beans and cereal aphids in wheat (Bruce et al., 2005b) but the reduction obtained was not sufficient for this to be used as a standalone method of crop protection. Use of semiochemical repellents would need to be part of an integrated package and constraints associated with formulation and registration need to be overcome.

6.7 COMPANION PLANTS RELEASING SEMIOCHEMICALS IN PUSH-PULL SYSTEMS

Companion plants can be used to deliver semiochemicals in the field if plants with suitable phytochemical release profiles can be found. This approach is appropriate for smallholder farming in Africa but may not be feasible for larger scale agriculture because of the high labour requirement. Simultaneously deploying a repellent and an attractant semiochemical can increase efficacy in insect population management in 'Push–pull' systems. Push-pull involves use of intercrops and trap crops in a mixed cropping system (Khan et al., 2010). These companion plants release semiochemicals to manipulate the distribution and abundance of stemborers and beneficial insects for management of stemborer pests (Figure 6.2). The system relies on an in-depth understanding of chemical ecology, agrobiodiversity, and plant–plant and insect–plant interactions and is well suited to African socio-economic conditions. The main cereal crop is planted with a repellent intercrop such as *Desmodium* (push) and an attractive trap plant such as Napier grass (pull) planted as a border crop around this intercrop. Gravid stemborer females are repelled from the main crop and are simultaneously attracted to the trap crop (Khan et al., 2010). Companion crops are valuable themselves as high quality animal fodder.

Napier grass trap crop produces significantly higher levels of volatile cues (chemicals), used by gravid stemborer females to locate host plants, than maize or sorghum (Birkett et al., 2006). There is also an increase of approximately 100-fold in the total amounts of these compounds produced in the first hour of nightfall (scotophase) by Napier grass (Chamberlain et al., 2006), the period at which stemborer moths seek host plants for oviposition, causing the differential oviposition preference. However, most of the stemborer larvae, about 80%, do not survive (Khan et al., 2006, 2007) as Napier grass tissues produce sticky sap in response to feeding by the larvae which traps them causing their mortality. The intercrop, legumes in the *Desmodium* genus (silverleaf, *D. uncinatum* and greenleaf, *D. intortum*), on the other hand, produce repellent volatile chemicals that push away the stemborer moths. These include (*E*)-β-ocimene and (*E*)-4,8-dimethyl-1,3,7-nonatriene, semiochemicals produced during damage to plants by herbivorous insects and are responsible for the repellency of *Desmodium* to stemborers (Khan et al., 2000).

6.8 USING SEMIOCHEMICALS AS ACTIVATORS OF PLANT DEFENCES

Semiochemical treatments can be used to switch on plant defence against pests because they cause upregulation of defence traits and alter plant secondary metabolism. Direct defence involves production of antibiotic or antinutritive compounds or repellents whereas indirect defence involves production of attractants for natural enemies of pests. Both direct and indirect defence can be enhanced when plants are treated with certain semiochemicals that function as activators of plant defence. The ability of plants to respond to semiochemicals that are associated with insect or pathogen attack allows them to fine tune their metabolism according to the likelihood of exposure to biotic stress factors. For example, emission of herbivore induced volatiles from neighbouring plants can lead to activation of defence pathways that make a plant more resistant to insect attack (Baldwin et al., 2006; Farmer and Ryan, 1990; Karban et al., 2000).

Many chemical activators of induced defences against biotic attackers are known (Paré et al., 2005) and some of these have been commercialized for crop protection (Vallad and Goodman, 2004; von Rad et al., 2005). Furthermore, artificial compounds can also have activity as plant activators. Some plant activators may be phytotoxic and this aspect should be considered when developing them. Another problem is that sustained activation of defence may be costly in terms of resources, and long-term activation of induced

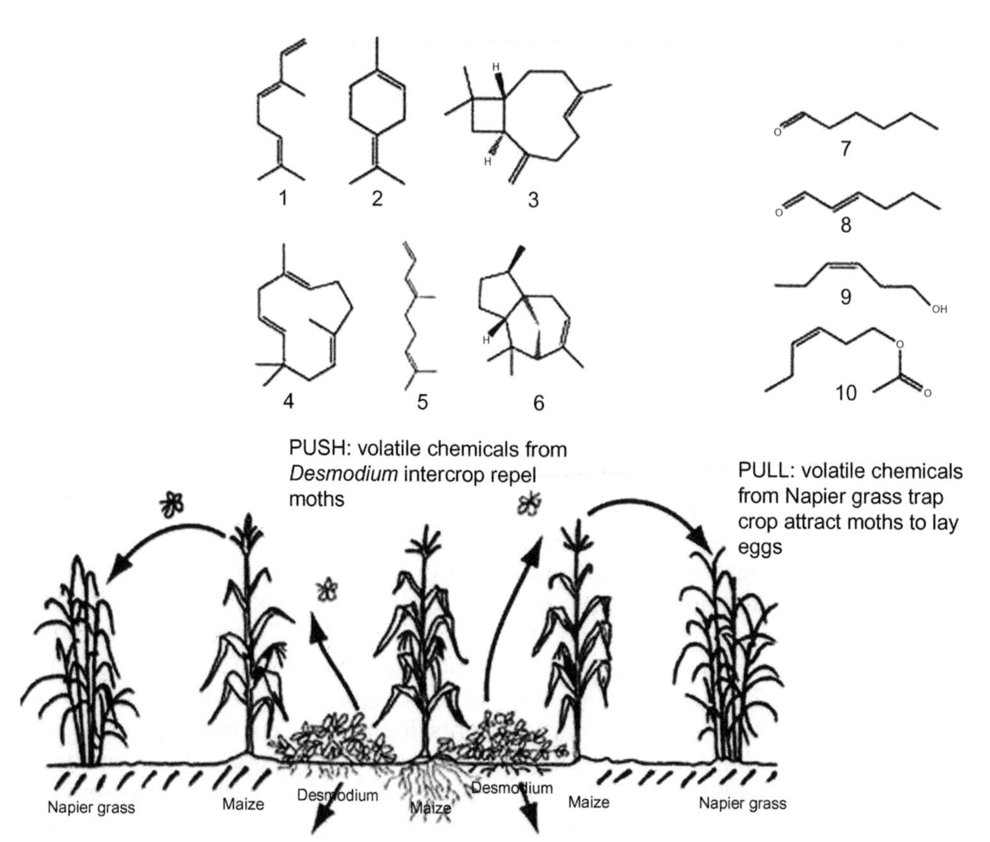

1
2
3
4
5
6
7
8
9
10
PUSH: volatile chemicals from *Desmodium* intercrop repel moths
PULL: volatile chemicals from Napier grass trap crop attract moths to lay eggs
Napier grass
Maize
Desmodium
Maize
Desmodium
Maize
Napier grass
ALLELOPATHY: chemicals exuded by *Desmodium* roots inhibit attachment of *Striga* to maize roots and cause suicidal germination of *Striga*

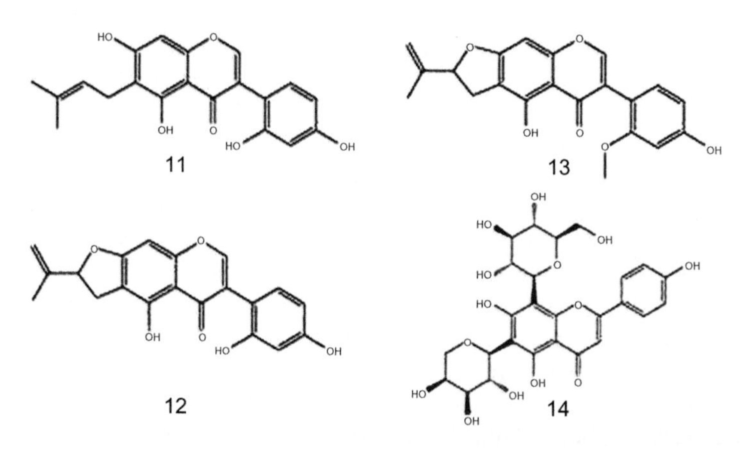

11
13
12
14

FIGURE 6.2 How the Push-Pull system works. Semiochemicals that repel pests and attract their natural enemies are released by the intercrop: 1 = (*E*)-β-ocimene; 2 = α-terpinolene; 3 = β-caryophyllene; 4 = humulene; 5 = (*E*)-4,8-dimethyl-1,3,7-nonatriene; 6 = α-cedrene. Semiochemicals that attract pests are released by the trap crop: 7 = hexanal; 8 = (*E*)-2-hexenal; 9 = (Z)-3-hexen-1-ol; 10 = (Z)-3-hexen-1-yl acetate. The intercrop also releases root exudates supressing striga weed: 11 = uncinanone A; 12 = uncinanone B; 13 = uncinanone C, and 14 = di-C-glycosylflavone 6-C-α-L-arabinopyranosyl-8-C-β-D-glucopyranosylapigenin. *Figure obtained from Khan et al., Exploiting phytochemicals for developing a 'push-pull' crop protection strategy for cereal farmers in Africa, J. Exp. Bot., 2010, 61: 4185, by permission of Oxford University Press.*

defences can result in yield penalties (Vallad and Goodman, 2004; van Hulten et al., 2006) although this was not found with jasmonic acid induced tomato plants (Thaler, 1999) and the release of volatiles is not necessarily very costly (Aharoni et al., 2005). Induced defence occurs when a plant becomes more resistant to insect pests or pathogens after a signal causes a change in its metabolism (Karban and Kuc, 1999). An alternative to direct activation of defence is 'priming'. The process of priming occurs when prior exposure to a biotic or an abiotic stimulus sensitizes a plant to express a more efficient defence response to future a biotic stress (Beckers and Conrath, 2007; Bruce et al., 2007; Conrath et al., 2006). Primed plants display either faster and/or stronger activation of the various defence responses that are induced following pathogen or insect attack or exposure to abiotic stress.

One plant activator that the authors have investigated is *cis*-jasmone, or (Z)-jasmone. Its activity was first discovered at Rothamsted when components of blackcurrant volatiles that repelled the summer form of lettuce aphid, *Nasonovia ribis-nigri*, were being identified. Due to structural similarities with jasmonic acid, it was tested as a plant treatment and was found to have intricate effects on interactions between pest insects and crop plants (Birkett et al., 2000; Pickett et al., 2007). It occurs naturally as a component of flower volatiles, but is also released from cotton leaves and flowers upon feeding by lepidopterous larvae (Loughrin et al., 1995) and there is evidence that *cis*-jasmone has a role in plant defence. This was first shown by placing low levels of *cis*-jasmone over bean plants contained in bell jars. The plants were tested for residual *cis*-jasmone, which was found to be completely absent after 48h, and these and control plants were then placed in a wind tunnel and the effect on an aphid parasitoid, *Aphidius ervi*, was investigated. In both dual and single choice experiments, there were, respectively, three-fold and two-fold increases in oriented flight to the *cis*-jasmone treated plant, with both results being highly significant statistically (Birkett et al., 2000). One of the compounds showing induced release as a consequence of the *cis*-jasmone treatment was (*E*)-ocimene, which is known to be partly responsible for the response by *A. ervi*. Although this compound was also induced by methyl jasmonate, the effect was short-lived and had disappeared 48h after the initial treatment. However, the effect with *cis*-jasmone remained for 8 days (Birkett et al., 2000).

Studies with *cis*-jasmone then focused on the interaction between the grain aphid *Sitobion avenae* and wheat, *Triticum aestivum*. Wheat plants sprayed with low levels of *cis*-jasmone as an aqueous emulsion became less attractive to aphids but more attractive to their parasitoids in laboratory bioassays. In the field, similarly treated plants had lower aphid infestations (Bruce et al., 2003). Field plots of wheat were sprayed hydraulically with *cis*-jasmone, at a rate equivalent to $50\,g\,ha^{-1}$, in mid-May and early June in four consecutive seasons, and aphid counts were made at weekly intervals. It was consistently found that aphid infestations were reduced in *cis*-jasmone treated plots (Bruce et al., 2003). It appears that part of this effect is due to increased parasitism of aphids by parasitoids. In simulated field trials on wheat seedlings treated with *cis*-jasmone, it has been shown that there is a statistically significant increase in foraging by *A. ervi* on treated plants. 6-Methyl-5-hepten-2-one is an important foraging cue for *A. ervi* (Du et al., 1998) and, in certain elite wheat cultivars, there is upregulation of the production of 6-methyl-5-hepten-2-one with *cis*-jasmone. As a consequence of this and other effects, there is repellency to the cereal aphid *S. avenae* when the wheat cultivar is treated with *cis*-jasmone. In addition to the behavioural effects, it was observed that there were reductions in aphid development. These involved statistically significant reductions in the mean intrinsic rate of population increase and in nymph production by *S. avenae* on certain wheat varieties previously treated with *cis*-jasmone. This appears to relate to induction of antibiotic secondary metabolites such as 2,4-dihydroxy-7-methoxy-1,4-benzoxazin-3-one (DIMBOA) from the hydroxamic acid pathway.

6.9 ALTERING EMISSION OF SEMIOCHEMICALS FROM CROPS

There is genetic variation in the profile of volatiles released by plants and thus potential for selecting lines that release appropriate volatiles. For example, Tamiru et al. (2011) showed that certain maize landraces emit semiochemicals that attract natural enemies of stemborer pests when the insects lay their eggs on the plant but this trait is absent in mainstream commercial maize lines. Emission of such semiochemicals could be bred into crops through conventional plant breeding and marker assisted selection. Conventional breeding can take a long time because the trait needs to be crossed in without bringing other undesirable traits with it in a phenomenon known as linkage drag (Bruce, 2012). There is further scope to deliver altered volatile production in plants using genetic engineering approaches (Aharoni et al., 2005; Degenhardt et al., 2003; Dudareva and Negre, 2005).

Progress has been made with this with initial studies in *Arabidopsis* and more recent studies in crop plants. Schnee et al. (2006) overexpressed a terpene synthase gene, *TPS10*, in *Arabidopsis* and found that transformed plants emitted a mixture of sesquiterpenes and were more attractive to the parasitic wasp *Cotesia marginiventris* than wild-type plants. Another study in *Arabidopsis* showed that increasing green leaf volatile biosynthesis and emission led to increased attractiveness of plants to *C. glomerata* parasitic wasps and increased resistance to grey mould fungal infection (Shiojiri et al., 2006). A terpene synthase gene for production of the aphid alarm pheromone (*E*)-β-farnesene has been cloned into *Arabidopsis* and transformed plants were less attractive to the aphid *M. persicae* but more attractive to the aphid parasitoid *Aphidius ervi* (Beale et al., 2006). Recent studies at Rothamsted have demonstrated that this compound is also released by transformed wheat and has similar behavioural effects on cereal aphids. The authors are currently testing wheat plants with this trait in field trials.

Another sesquiterpene, (*E*)-β-caryophyllene, is emitted by maize leaves in response to attack by lepidopteran larvae such as *Spodoptera littoralis* and released from roots after damage by larvae of the coleopteran *Diabrotica virgifera virgifera*. It is synthesized by maize terpene synthase 23 (TPS23) and can attract natural enemies of both herbivores: entomopathogenic nematodes below ground and parasitic wasps (Köllner et al., 2008). The gene encoding TPS23 is active in teosinte species and European maize lines but not in most North American lines. Köllner et al. (2008) suggested that the (*E*)-β-caryophyllene defence signal was lost during breeding of the North American lines and that its restoration might help to increase the resistance of these lines. This was confirmed experimentally by transformation of non-emitting maize with a (*E*)-β-caryophyllene synthase gene from oregano to produce plants which suffered significantly less root damage in field trials (Degenhardt et al., 2009). Similarly, the rice (*E*)-β-caryophyllene synthase (OsTPS3) plays an important role in inducible volatile sesquiterpene biosynthesis and the parasitoid *A. nilaparvatae* was attracted to plants overexpressing this gene (Cheng et al., 2007).

6.10 CONCLUSIONS AND FUTURE OUTLOOK

To date, the main application of semiochemicals in IPM has been in pest monitoring systems where semiochemicals are used as attractive baits to lure insects into traps. Achieving reductions in pest populations with semiochemicals has been more challenging. Volatile semiochemicals are difficult to apply over large areas of crop and in a way that their release is extended over the whole crop season and, even with the best formulations, reductions in insect infestations are often not strong enough to prevent pests from entering the large areas of host plants present in agricultural systems. However, release from the plants seems a promising way to deliver semiochemicals in the field and is done successfully by the push-pull system which uses companion plants. Use of plant activators has also been found to provide limited and variable effects in reducing pest populations.

Much remains to be learnt about plant defence processes. In the future, it is likely that crop breeding will allow development of improved crop cultivars that are able to respond to pest attack by switching on appropriate defence metabolism which may include production of antibiotic secondary metabolites as well as volatile semiochemicals. It is known that plants can respond to pest attack and most plants are non-hosts to most insects. The process that makes incompatible interactions incompatible requires attention because crop plants could benefit from introgression of defence traits from their wild relatives.

References

Aak, A., Birkemoe, T., Knudsen, G.K., 2011. Efficient mass trapping: catching the pest, *Calliphora vicina*, (Diptera: Calliphoridae), of Norwegian stockfish production. J. Chem. Ecol. 37, 924–931.

Agelopoulos, N., Birkett, M.A., Hick, A.J., Hooper, A.M., Pickett, J.A., Pow, E.M., et al., 1999. Exploiting semiochemicals in insect control. Pestic. Sci. 55, 225–235.

Aharoni, A., Jongsma, M.A., Bouwmeester, H.J., 2005. Volatile science? Metabolic engineering of terpenoids in plants. Trends Plant Sci. 10, 594–602.

Alpizar, D., Fallas, M., Oehlschlager, A.C., Gonzales, L.M., 2012. Management of *Cosmopolites sordidus* and *Metamasius hemipterus* in banana by pheromone-based mass trapping. J. Chem. Ecol. 38, 245–252.

Aurelian, V.M., Evenden, M.L., Judd, G.J.R., 2012. Small-plot studies comparing pheromone and juice baits for mass-trapping invasive *Synanthedon myopaeformis* in Canada. Entomol. Exp. Appl. 145, 102–114.

Baldwin, I.T., Paschold, A., Von Dahl, C.C., Halitschke, R., Preston, C.A., 2006. Volatile signaling in plant–plant interactions: 'Talking trees' in the genomics era. Science 311, 812–815.

Beale, M.H., Birkett, M.A., Bruce, T.J.A., Chamberlain, K., Field, L.M., Huttly, A.K., et al., 2006. Aphid alarm pheromone produced by transgenic plants affects aphid and parasitoid behavior. Proc. Natl. Acad. Sci. U.S.A. 103, 10509–10513.

Beauhaire, J., Ducrot, P-H., Malosse, C., Rochat, D., Ndiege, I.O., Otieno, D.O., 1995. Identification and synthesis of sordidin, a male pheromone emitted by *Cosmopolites sordidus*. Tetrahedron Lett. 36, 1043–1046.

Beckers, G.J., Conrath, U., 2007. Priming for stress resistance: from the lab to the field. Curr. Opin. Plant Biol. 10, 425–431.

Birkett, M.A., Campbell, C.A.M., Chamberlain, K., Guerrieri, E., Hick, A.J., Martin, J.L., et al., 2000. New roles for cis-jasmone as an insect semiochemical and in plant defense. Proc. Natl. Acad. Sci. U.S.A. 97, 9329–9334.

Birkett, M.A., Chamberlain, K., Khan, Z.R., Pickett, J.A., Toshova, T., Wadhams, L.J., et al., 2006. Electrophysiological responses of the lepidopterous stemborers Chilo partellus and Busseola fusca to volatiles from wild and cultivated host plants. J. Chem. Ecol. 32, 2475–2487.

Bruce, T.J.A., 2012. GM as a route for delivery of sustainable crop protection. J. Exp. Bot. 63, 537–541.

Bruce, T.J.A., Pickett, J.A., 2011. Perception of plant volatile blends by herbivorous insects–Finding the right mix. Phytochemistry 72, 1605–1611.

Bruce, T.J.A., Smart, L.E., 2009. Orange wheat blossom midge, Sitodiplosis mosellana, management. Outlooks Pest Manag. 20, 89–92.

Bruce, T.J.A., Martin, J.L., Pickett, J.A., Pye, B.J., Smart, L.E., Wadhams, L.J., 2003. cis-Jasmone treatment induces resistance in wheat plants against the grain aphid, Sitobion avenae (Fabricius) (Homoptera: Aphididae). Pest Manag. Sci. 59, 1031–1036.

Bruce, T.J.A., Wadhams, L.J., Woodcock, C.M., 2005a. Insect host location: a volatile situation. Trends Plant Sci. 10, 269–274.

Bruce, T.J.A., Birkett, M.A., Blande, J., Hooper, A.M., Martin, J.L., Khambay, B., et al., 2005b. Response of economically important aphids to components of Hemizygia petiolata essential oil. Pest Manag. Sci. 61, 1115–1121.

Bruce, T.J.A., Hooper, A.M., Ireland, L., Jones, O.T., Martin, J.L., Smart, L.E., et al., 2007. Development of a pheromone trap monitoring system for orange wheat blossom midge, Sitodiplosis mosellana, in the UK. Pest Manag. Sci. 63, 49–56.

Brunner, J., Welter, S., Calkins, C., Hilton, R., Beers, E., Dunley, J., et al., 2002. Mating disruption of codling moth: a perspective from the Western United States. IOBC WPRS Bull. 25 (9), 11–19.

Byers, J.A., 2007. Simulation of mating disruption and mass trapping with competitive attraction and camouflage. Environ. Entomol. 36, 1328–1338.

Byers, J.A., 2012. Modelling female mating success during mass trapping and natural competitive attraction of searching males or females. Entomol. Exp. Appl. 145, 228–237.

Byers, J.A., Schlyter, F., Birgersson, G., Francke, W., 1990. *E*-myrcenol in *Ips duplicatus:* an aggregation pheromone component new for bark beetles. Experientia 45, 1209–1211.

Camelo, L., Landolt, P.J., Zack, R.S., 2007. A kairomone based attract-and-kill system effective against alfalfa looper (Lepidoptera: Noctuidae). J. Econ. Entomol. 100, 366–374.

Cardé, R.T., Minks, A.K., 1995. Control of moth pests by mating disruption: successes and constraints. Annu. Rev. Entomol. 40, 559–585.

Chamberlain, K., Khan, Z.R., Pickett, J.A., Toshova, T., Wadhams, L.J., 2006. Diel periodicity in the production of green leaf volatiles by wild and cultivated host plants of stemborer moths, Chilo partellus and Busseola fusca. J. Chem. Ecol. 32, 565–577.

Cheng, A.X., Lou, Y.G., Mao, Y.B., Lu, S., Wang, L.J., Chen, X.Y., 2007. Plant terpenoids: Biosynthesis and ecological functions. J. Integr. Plant Biol. 49, 179–186.

Chermiti, B., Abbes, K., 2012. Comparison of pheromone lures used in mass trapping to control the tomato leafminer *Tuta absoluta* (Meyrick, 1917) in industrial tomato crops in Kairouan (Tunisia). Bull. OEPP/EPPO Bull. 42, 241–248.

Conlee, J.K., Staten, R.T., 1986. Device for insect control. U.S. Patent 4,671,010. 6 September 1987.

Conrath, U., Beckers, G.J.M., Flors, V., García-Agustín, P., Jakab, G., Mauch, F., et al., 2006. Priming: Getting ready for battle. Mol. Plant Microbe Interact. 19, 1062–1071.

Cork, A., Alam, S.N., Rouf, F.M.A., Talekar, N.S., 2005. Development of mass trapping technique for control of brinjalshoot and fruit borer, Leucinodes orbonalis (Lepidoptera: Pyralidae). Bull. Entomol. Res. 95, 589–596.

Cunningham, R.T., Kobayashi, R.M., Miyashita, D.H., 1990. The male lures of tephritid fruit flies. In: Ridgway, R.L., Silverstein, R.M., Inscoe, M.N. (Eds.), Behavior Modifying Chemicals for Insect Management: Applications of Pheromones and Other Attractants. Marcel Dekker, New York, pp. 113–129.

Degenhardt, J., Gershenzon, J., Baldwin, I.T., Kessler, A., 2003. Attracting friends to feast on foes: engineering terpene emission to make crop plants more attractive to herbivore enemies. Curr. Opin. Biotechnol. 14, 169–176.

Degenhardt, J., Hiltpold, I., Kollner, T.G., Frey, M., Gierl, A., Gershenzon, J., et al., 2009. Restoring a maize root signal that attracts insect-killing nematodes to control a major pest. Proc. Natl. Acad. Sci. U. S. A. 106, 13213–13218.

Del Socorro, A.P., Gregg, P.C., Hawes, A.J., 2010. Development of a synthetic plant volatile based attracticide for female noctuid moths. III. Insecticides for adult Helicoverpa armigera (Hübner) (Lepidoptera: Noctuidae). Aust. J. Entomol. 49, 31–39.

Dicke, M., Sabelis, M.W., 1988. Infochemical terminology: based on cost-benefit analysis rather than origin of compounds? Funct. Ecol 2, 131–139.

Du, Y.J., Poppy, G.M., Powell, W., Pickett, J.A., Wadhams, L.J., Woodcock, C.M., 1998. Identification of semiochemicals released during aphid feeding that attract parasitoid Aphidius ervi. J. Chem. Ecol. 24, 1355–1368.

Dudareva, N., Negre, F., 2005. Practical applications of research into the regulation of plant volatile emission. Curr. Opin. Plant Biol. 8, 113–118.

El-Sayed, A.M., Suckling, D.M., Wearing, C.H., Byers, J.A., 2006. Potential of mass trapping for long-term pest management and eradication of invasive species. J. Econ. Entomol. 99, 1550–1564.

El-Sayed, A.M., Suckling, D.M., Byers, J.A., Jang, E.B., Wearing, C.H., 2009. Potential of 'lure and kill' in long-term pest management and eradication of invasive species. J. Econ. Entomol. 102, 815–835.

Farmer, E.E., Ryan, C.A., 1990. Interplant communication–airborne methyl jasmonate induces synthesis of proteinase-inhibitors in plant-leaves. Proc. Natl. Acad. Sci. U.S.A. 87, 7713–7716.

Foster, S.P., Harris, M.O., 1997. Behavioral manipulation methods for insect pest-management. Annu. Rev. Entomol. 42, 123–146.

Gregg, P.C., Del Socorro, A.P., Henderson, G., 2010. Development of a synthetic plant volatile based attracticide for female noctuid moths. II. Bioassays of synthetic plant volatiles as attractants for the adults of the cotton bollworm, Helicoverpa armigera (Hübner) (Lepidoptera: Noctuidae). Aust. J. Entomol. 49, 1–9.

Gries, R., Gries, G., Khaskin, G., King, S., Olfert, O., Kaminski, L.A., et al., 2000. Sex pheromone of orange wheat blossom midge, Sitodiplosis mosellana. Naturwissenschaften 87, 450–454.

Hagley, E.A.C., 1978. Sex pheromones and suppression of the codling moth (Lepidoptera: Olethreutidae). Can. Entomol. 110, 781–783.

Hall, DR, Amarawardana, L, Cross, JV, Francke, W, Boddum, T, Hillbur, Y, 2012. The chemical ecology of cecidomyiid midges (Diptera: Cecidomyiidae). J. Chem. Ecol. 38, 2–22.

Hardie, J., Isaacs, R., Pickett, J.A., Wadhams, L.J., Woodcock, C.M., 1994. Methyl salicylate and (−)-(1r,5s)-myrtenal are plant-derived repellents for black bean aphid, Aphis-fabae Scop (Homoptera, Aphididae). J. Chem. Ecol. 20, 2847–2855.

Hegazi, E., Khafagi, W.E., Konstantopoulou, M., Ratptopoulos, D., Tawfik, H., Abd El-Aziz, G.M., et al., 2009. Efficient mass-trapping as an alternative tactic for suppressing populations of leopard moth (Lepidoptera: Cossidae). Ann. Entomol. Soc. Am. 102, 809–818.

Ioriatti, C, Lucchi, A, Bagnoli, B, 2008. Grape areawide pest management in Italy. In: Koul, O., Cuperus, G., Elliot, N. (Eds.), Areawide Pest Management: Theory and Implementation CABI, Wallingford, pp. 208–225.

Ioriatti, C, Anfora, G, Tasin, M, De Cristofaro, A, Witzgall, P, Lucchi, A, 2011. Chemical ecology and management of Lobesia botrana (Lepidoptera: Tortricidae). J. Econ. Entomol. 104, 1125–1137.

Jones, O.T., 1998. Practical applications of pheromones and other semiochemicals (section II. Lure and kill). In: Howse, P., Stevens, I., Jones, O.T. (Eds.), Insect Pheromones and Their Use in Pest Management. Chapman and Hall, London, UK, pp. 280–300.

Karban, R, Kuc, J, 1999. Induced resistance against pathogens and herbivores: an overview. In: Agrawal, AA, Tuzun, S, Bent, E (Eds.), Induced Plant Defenses against Pathogens and Herbivores. APS Press, St. Paul, Minnesota, pp. 1–15.

Karban, R., Baldwin, I.T., Baxter, K.J., Laue, G., Felton, G.W., 2000. Communication between plants: induced resistance in wild tobacco plants following clipping of neighboring sagebrush. Oecologia 125, 66–71.

Katsoyannos, B.I., Papadopoulos, N.T., 2004. Evaluation of synthetic female attractants against Ceratitis capitata (Diptera: Tephritidae) in sticky coated spheres and McPhail type traps. J. Econ. Entomol. 97, 21–26.

Khan, Z.R., Pickett, J.A., van den Berg, J., Wadhams, L.J., Woodcock, C.M., 2000. Exploiting chemical ecology and species diversity: stem borer and striga control for maize and sorghum in Africa. Pest Manag. Sci. 56, 957–962.

Khan, Z.R., Midega, C.A.O., Hutter, N.J., Wilkins, R.M., Wadhams, L.J., 2006. Assessment of the potential of Napier grass (Pennisetum purpureum) varieties as trap plants for management of Chilo partellus. Entomol. Exp. Appl. 119, 15–22.

Khan, Z.R., Midega, C.A.O., Wadhams, L.J., Pickett, J.A., Mumuni, A., 2007. Evaluation of Napier grass (Pennisetum purpureum) varieties for use as trap plants for the management of African stemborer (Busseola fusca) in a push-pull strategy. Entomol. Exp. Appl. 124, 201–211.

Khan, Z.R., Midega, C.A.O., Bruce, T.J.A., Hooper, A.M., Pickett, J.A., 2010. Exploiting phytochemicals for developing a 'push–pull' crop protection strategy for cereal farmers in Africa. J Exp. Bot. 61, 4185–4196.

Köllner, T.G., Held, M., Lenk, C., Hiltpold, I., Turlings, T.C., Gershenzon, J., et al., 2008. A maize (E)-β-caryophyllene synthase implicated in indirect defense responses against herbivores is not expressed in most American maize varieties. Plant Cell 20, 482–494.

Krieger, J., Breer, H., 1999. Olfactory reception in invertebrates. Science 286, 720–723.

Leskey, T.C., Bergh, J.C., Walgenbach, J.F., Zhang, A., 2009. Evaluation of pheromone-based management strategies for dogwood borer (Lepidoptera: Sesiidae) in commercial apple orchards. J. Econ. Entomol. 102, 1085–1093.

Leza, M.M., Juan, A., Capllonch, M., Alemany, A., 2008. J. Appl. Entomol. 132, 753–761.

Loughrin, J.H., Manukian, A., Heath, R.R., Tumlinson, J.H., 1995. Volatiles emitted by different cotton varieties damaged by feeding beet armyworm larvae. J. Chem. Ecol. 21, 1217–1227.

Maki, E.C., Millar, J.G., Rodstein, J., Hanks, L.M., Barbour, J.D., 2011. Evaluation of mass trapping and mating disruption for managing Prionus californicus (Coleoptera: Cerambycidae) in hop production yards. J. Econ. Entomol. 104, 933–938.

Mangan, R.L., Moreno, D.S., Thompson, G.D., 2006. Bait dilution, spinosad concentration and efficacy of GF-120 based fruit fly sprays. Crop Prot. 25, 125–133.

Martinez-Ferrer, M.T., Campos, J.M., Fibla, J.M., 2012. Field efficacy of Ceratitis capitata (Diptera: Tephritidae) mass trapping technique on clementine groves in Spain. J. Appl. Entomol. 136, 181–190.

Michaud, J.P., 2003. Toxicity of fruit fly baits to beneficial insects in citrus. J. Insect Sci. 3, 1–9.

Miller, J.R., McGhee, P.S., Siegert, P.Y., Adams, C.G., Huang, J., Grieshop, M.J., et al., 2010. General principles of attraction and competitive attraction as revealed by large-cage studies of moths responding to sex pheromone. Proc. Natl. Acad. Sci. U. S. A. 107, 22–27.

Myers, J.H., Savoie, A., van Renden, E., 1998. Eradication and pest management. Annu. Rev. Entomol. 43, 471–491.

Navarro-Llopis, V., Alfaro, F., Dominguez, J., Sanchis, J., Primo, J., 2008. Evaluation of traps and lures for mass trapping of Mediterranean fruit fly in citrus groves. J. Econ. Entomol. 101, 126–131.

Navarro-Llopis, V., Alfaro, F., Primo, J., Vacas, S., 2011. Response of two tephritid species, Bactrocera oleae and Ceratitis capitata, to different emission levels of pheromone and parapheromone. Crop Prot. 30, 913–918.

Nordlund, D.A., Lewis, W.J., 1976. Terminology of chemical releasing stimuli in intraspecific and interspecific interactions. J. Chem. Ecol. 2 (2), 211–220.

Nordlund, D.A., Jones, R.L., Lewis, W.J., 1981. Semiochemicals. Their Role in Pest Control. John Wiley and Sons, New York.

Oehlschlager, A.C., Chinchilla, C., Castillo, G., Gonzalez, L., 2002. Control of red ring disease by mass trapping of *Rhynchophorus palmarum* (Coleoptera: Curculionidae). Fla. Entomol. 85, 507–513.

Paré, P.W., Krishnamachari, V., Zhang, H., Farag, M.A., Ryu, C.-M., Kloepper, J.W., 2005. Elicitors and priming agents initiate plant defense responses. Photosynth. Res. 85, 149–159.

Pettersson, J., Pickett, J.A., Pye, B.J., Quiroz, A., Smart, L.E., Wadhams, L.J., et al., 1994. Winter host component reduces colonization by bird-cherry oat aphid, Rhopalosiphum-Padi (L) (Homoptera, Aphididae), and other aphids in cereal fields. J. Chem. Ecol. 20, 2565–2574.

Pickett, J.A., Birkett, M.A., Moraes, M.C.B., Bruce, T.J.A., Chamberlain, K., Gordon-Weeks, R., et al., 2007. cis-Jasmone as allelopathic agent in inducing plant defence. Allelopath. J. 19, 109–117.

Powell, W., Pickett, J.A., 2003. Manipulation of parasitoids for aphid pest management: progress and prospects. Pest Manag. Sci. 59, 149–155.

Reddy, G.V.P., Cruz, Z.T., Guerrero, A., 2009. Development of an efficient pheromone-based trapping method for the banana root borer Cosmopolites sordidus. J. Chem. Ecol. 35, 111–117.

Reinke, M.D., Miller, J.R., Gut, L.J., 2012. Potential of high-density pheromone-releasing microtraps for control of codling moth Cydia pomonella and oblique banded leafroller Choristoneura rosaceana. Physiol. Entomol. 37, 53–59.

Rochat, D., Malosse, C., Lettere, M., Ducrot, P.-H., Zagatti, P., Renou, M., et al., 1991. Male-produced aggregation pheromone of the American palm weevil, Rhynchophorus palmarum (L.) (Coleoptera: Curculionidae): collection, identification, electrophysiological activity and laboratory bioassay. J. Chem. Ecol. 17, 1221–1230.

Rodriguez-Saona, CR, Stelinski, L.L., 2009. Behavior-modifying strategies in IPM: theory and practice. In: Peshin, R.D., Dhawan, A.K. (Eds.), Integrated Pest Management: Innovation-Development Process. Springer, pp. 263–315.

Samietz, J, Baur, R, Hillbur, Y, 2012. Potential of synthetic sex pheromone blend for mating disruption of the Swede midge, Contarinia nasturtii. J. Chem. Ecol. 38, 1171–1177.

Schlyter, F., Zhang, Q.-H., Liu, G.-T., Ji, L.-Z., 2001. A successful case of pheromone mass trapping of the bark beetle *Ips duplicatus* in a forest island, analyzed by 20-year time-series data. Integr. Pest Manag. Rev. 6, 185–196.

Schnee, C., Köllner, T.G., Gershenzon, J., Degenhardt, J., Held, M., Turlings, T.C.J., 2006. The products of a single maize sesquiterpene synthase form a volatile defense signal that attracts natural enemies of maize herbivores. Proc. Natl. Acad. Sci. U. S. A. 103, 1129–1134.

Sharov, A.A., Liebhold, A.M., Roberts, E.A., 1998. Optimizing the use of barrier zones to slow the spread of gypsy moth (Lepidoptera: Lymantriidae) in North America. J. Econ. Entomol. 91, 165–174.

Shiojiri, K., Kishimoto, K., Ozawa, R., Kugimiya, S., Urashimo, S., Arimura, G., et al., 2006. Changing green leaf volatile biosynthesis in plants: an approach for improving plant resistance against both herbivores and pathogens. Proc. Natl. Acad. Sci. U.S.A. 103, 16672–16676.

Silverstein, R.M., Brownlee, R.G., Bellas, T.E., Wood, D.L., Browne, L.E., 1968. Brevicomin: principal sex attractant in the frass of the female Western pine beetle. Science (Wash., DC) 159, 889–891.

Smith, J.W., 1998. Boll weevil eradication: area-wide pest management. Ann. Entomol. Soc. Am. 91, 239–247.

Stelinski, L.L., Miller, J.R., Ledebuhr, R., Siegert, P., Gut, L.J., 2007. Season-long mating disruption of *Grapholita molesta* (Lepidoptera: Tortricidae) by one machine application of pheromone in wax drops (SPLAT-OFM). J. Pest Sci. 80, 109–117.

Tamiru, A., Bruce, T.J.A., Woodcock, C.M., Caulfield, J.C., Midega, C.A.O., Ogol, C.K.P.O., et al., 2011. Maize landraces recruit egg and larval parasitoids in response to egg deposition by a herbivore. Ecol. Lett. 14, 1075–1083.

Tcheslavskaia, K.S., Thorpe, K.W., Brewster, C.C., Sharov, A.A., Leonard, D.S., Reardon, R.C., et al., 2005. Optimization of pheromone dosage for gypsy moth mating disruption. Entomol. Exp. Appl. 115, 355–361.

Teixeira, L.A.F., Miller, J.R., Epstein, D.L., Gut, L.J., 2010. Comparison of mating disruption and mass trapping with Pyralidae and Sesiidae moths. Entomol. Exp. Appl. 137, 176–183.

Thaler, J.S., 1999. Induced resistance in agricultural crops: effects of jasmonic acid on herbivory and yield in tomato plants. Environ. Entomol. 28, 30–37.

Tumlinson, J.H., Guelder, R.C., Hardee, D.D., Thompson, A.C., Hedin, P.A., Minyard, J.P., 1969. Sex pheromones produced by male boll weevil: isolation, identification and synthesis. Science 166, 1010–1012.

Vallad, G.E., Goodman, R.M., 2004. Systemic acquired resistance and induced systemic resistance in conventional agriculture. Crop Sci. 44, 1920–1934.

Van Hulten, M., Pelser, M., Van Loon, L.C., Pieterse, C.M.J., Ton, J., 2006. Costs and benefits of priming for defense in Arabidopsis. Proc. Natl. Acad. Sci. U.S.A. 103, 5602–5607.

Vinatier, F., Lescourret, F., Duyck, P-F., Tixier, P., 2012. From IBM to IPM: Using individual-based models to design the spatial arrangement of traps and crops in integrated pest management strategies. Agric. Ecosys. Environ. 146, 52–59.

von Rad, U., Mueller, M.J., Durner, J., 2005. Evaluation of natural and synthetic stimulants of plant immunity by microarray technology. New Phytol. 165, 191–202.

Wall, C., et al., 1987. The efficacy of sex-attractant monitoring for the pea moth, Cydia nigricana, in England, 1980–1985. Ann. Appl. Biol. 110 (2), 223–229.

Witzgall, P., Stelinski, L., Gut, L., Thomson, D., 2008. Codling moth management and chemical ecology. Annu. Rev. Entomol. 53, 503–522.

Witzgall, P., Kirsch, P., Cork, A., 2010. Sex pheromones and their impact on pest management. J. Chem. Ecol. 36, 80–100.

CHAPTER

7

Pesticides Applied for the Control of Invasive Species in the United States

David Pimentel
Cornell University, NY, USA

7.1 INTRODUCTION

In the history of the United States, approximately 50,000 alien invasive (non-native) species are estimated to have been introduced (Pimentel, 2011). Introduced crop species, such as corn, wheat, rice, and other food crops now provide more than 98% of the US food system at a value of approximately $800 billion per year (USCB, 2007). Other exotic species have been used for landscape restoration, biological pest control, sport, pets, and food processing. Some non-indigenous species, however, have caused major economic losses in agriculture, forestry, and several other segments of the US economy, in addition to harming the environment. One recent study reported approximately $100–$200 billion in damages from exotic species per year (Pimentel, 2011).

Estimating the full extent of the environmental damage caused by exotic species and the number of species extinctions they have caused is difficult because little is known about the estimated 750,000 species in the United States; note, half of these have not been described as yet (Raven and Johnson, 1992). Nonetheless, about 400 of the 958 species that are listed as threatened or endangered under the Endangered Species Act are considered to be at risk primarily because of competition with and predation by non-indigenous species (Wilcove et al., 1998). In other regions of the world, as many as 80% of the endangered species are threatened and at risk due to the pressures of non-native species (Armstrong, 1995). In this article, an assessment is made of the use of pesticides for the control of a large number of invasive species in the United States.

7.2 ENVIRONMENTAL DAMAGE AND ASSOCIATED CONTROL COSTS

Most plant and vertebrate animal introductions have been intentional, whereas most invertebrate animal and microbe introductions have been accidental. In the past 40 years, the rate of and risk associated with biotic invaders have increased enormously because of human

D. P. Abrol (Ed): Integrated Pest Management.
DOI: http://dx.doi.org/10.1016/B978-0-12-398529-3.00008-7

population growth, rapid movement of people, and alteration of the environment. In addition, more goods and materials are being traded among nations than ever before, thereby creating opportunities for unintentional introductions (Bryan, 1996; USCB, 2007).

Some of the approximately 50,000 species of plants and animals that have invaded the United States cause a wide array of damage to managed and natural ecosystems (Table 7.1). Some of the $120 billion in damage and control costs are assessed below.

7.2.1 Plants

Most alien plants now established in the United States were introduced for food, fibre, and/or ornamental purposes. An estimated 5000 plant species have escaped and now exist in US natural and managed ecosystems (Beers, 2003; Morse et al., 1995), compared with a total of about 17,000 species of native US plants (Morin, 1995). In Florida, of the approximately 25,000 alien plant species imported mainly as ornamentals for cultivation, more than 900 have escaped and become established in surrounding natural ecosystems (Frank and McCoy, 1995a; Frank et al., 1997; Simberloff et al., 1997). More than 1800 plant species have been introduced into California (California Invasive Plant Council, 2006).

Most of the 5000 alien plants established in US natural ecosystems have displaced several native plant species (Beers, 2003; Morse et al., 1995). Alien weeds are spreading and invading approximately 700,000 ha/year of the US wildlife habitat (Babbitt, 1998). One of these pest weeds is the European purple loosestrife (*Lythrum salicaria*), which was introduced in the early 19th century as an ornamental plant (Malecki et al., 1993). It has been spreading at a rate of 115,000 ha/year and is changing the basic structure of most of the wetlands it has invaded (Thompson et al., 1987). Competitive stands of purple loosestrife have reduced the biomass of 44 native plants and endangered wildlife, such as the bog turtle and several duck species, which depend on these native

TABLE 7.1 Estimated Annual Costs Associated with Some Alien Species Introduced into the United States

		Non-Indigenous Losses and Control		
Category	**Species**	**Damage (million $)**	**Costs (million $)**	**Total (million $)**
PLANTS	25,000			
Purple loosestrife		–	–	45
Aquatic weeds		10	100	110
Melaleuca tree		NA	3–6	3–6
Crop weeds		17,500	3000	20,500
Weeds in pastures		1000	5000	6000
Weeds in lawns, gardens, golf courses		NA	1500	1500
MAMMALS	20			
Wild horses and burros		5	NA	5
Feral pigs		1000	0.5	1000.5

(Continued)

TABLE 7.1 (Continued)

Category	Species	Non-Indigenous Losses and Control		
		Damage (million $)	Costs (million $)	Total (million $)
Mongooses		50	NA	50
Rats		19,000	NA	19,000
Cats		17,000	NA	18,000
Dogs		620	NA	620
BIRDS	97			
Pigeons		2200	NA	2200
Starlings		800	NA	800
REPTILES & AMPHIBIANS	53			
Brown tree snake		1	11	12
FISH	138	5400	NA	5400
ARTHROPODS	4500			
Imported fire ant		1200	800	2000
Formosan termite		1000	NA	1000
Green crab		44	NA	44
Gypsy moth		NA	11	11
Crop pests		10,400	500	10,900
Pests in lawns, gardens, golf courses		NA	1500	1500
Forest pests		2100	NA	2100
MOLLUSCS	88			
Zebra mussel		–	–	1000
Asian clam		1000	NA	1000
Shipworm		205	NA	205
MICROBES	20,000			
Crop plant pathogens		18,000	400	18,400
Plant pathogens in lawns, gardens, golf courses		NA	2000	2000
Forest plant pathogens		2100	NA	2100
Dutch elm disease		NA	100	100
LIVESTOCK DISEASES		9000	200	9200
HUMAN DISEASES		NA	200	200
TOTAL				**$125,005.50**

See text for details and sources.

plants (Gaudet and Keddy, 1988). Loosestrife now occurs in 48 states and costs $45 million per year in pest control costs and forage losses (ATTRA, 1997).

Many introduced plant species established in the wild are having an effect on US national parks (Hiebert and Stubbendieck, 1993). In Great Smoky Mountains National Park, 400 of approximately 1500 vascular plant species are exotic, and 10 of these are currently displacing and threatening other native species in the park (Hiebert and Stubbendieck, 1993).

Hawaii has a total of 2690 plant species, 946 of which are alien species (Eldredge and Miller, 1997). About 800 native species are endangered and more than 200 endemic species are believed to be extinct because of alien species (Endangered Species, 2008).

Sometimes one non-indigenous plant species competitively overruns an entire ecosystem. For example, in California, yellow star thistle (*Centaurea solstitialis*) now dominates more than 4 million ha of northern California grassland, resulting in the total loss of this once productive grassland (Campbell, 1994).

European cheatgrass (*Bromus tectorum*) is dramatically changing the vegetation and fauna of many natural ecosystems in the US west. This annual grass has invaded and spread throughout the shrub-steppe habitat of the Great Basin in Idaho and Utah, predisposing the invaded habitat to fires (Kurdila, 1995; Vitousek et al., 1997). Before the invasion of cheatgrass, fire burned once every 60–110 years, and shrubs had a chance to become well established. Now, fires occur about every 3–5 years; shrubs and other vegetation are diminished and competitive monocultures of cheatgrass now exist on 5 million ha in Idaho and Utah (Invasive.org, 2010). The animals dependent on the shrubs and other original vegetation have been reduced or eliminated.

An estimated 138 alien tree and shrub species have invaded native US forest and shrub ecosystems (Campbell, 1998). Introduced trees include salt cedar (*Tamarix pentandra*), eucalyptus (*Eucalyptus* spp.), Brazilian pepper (*Schinus terebinthifolius*), and Australian melaleuca (*Melaleuca quinquenervia*) (Miller, 1995; OTA, 1993; Randall, 1996). Some of these trees have displaced native trees, shrubs, and other vegetation types, and populations of some associated native animal species have been reduced in turn (OTA, 1993). For example, the melaleuca tree is competitively spreading at a rate of 11,000 ha/year throughout the vast forest and grassland ecosystems of the Florida Everglades (Campbell, 1994), where it damages the natural vegetation and wildlife (OTA, 1993).

Exotic aquatic weeds in the Hudson River basin of New York number 53 species (Mills et al., 1997). In Florida, exotic aquatic plants, such as hydrilla (*Hydrilla verticillata*), water hyacinth (*Eichhornia crassipes*), and water lettuce (*Pistia stratiotes*), are altering fish and other aquatic animal species, choking waterways, altering nutrient cycles, and reducing recreational use of rivers and lakes. Active control measures of aquatic weeds have become necessary (OTA, 1993).

For instance, Florida spends about $14.5 million each year on hydrilla control (Center et al., 1997). Despite this large expenditure, hydrilla infestations in just two Florida lakes have prevented their recreational use, causing $10 million annually in losses (Center et al., 1997). In the United States, a total of $100 million is invested annually in alien species aquatic weed control mostly with herbicides (OTA, 1993).

7.2.2 Mammals

About 20 species of mammals have been introduced into the United States; these include dogs, cats, horses, burros, cattle, sheep, pigs, goats, and deer (Layne, 1997). Several of these species have escaped or were released into the wild; many have become pests by preying on native animals, grazing on vegetation, or intensifying soil erosion.

Many small mammals have also been introduced into the United States. These species include a number of rodents: the European [black or tree] rat (*Rattus rattus*), Asiatic (Norway or brown) rat (*Rattus norvegicus*), house mouse (*Mus musculus*), and the European rabbit (*Oryctolagus cuniculus*) (Layne, 1997).

Some introduced rodents have become serious pests on farms, in industries, and in homes (Layne, 1997). On farms, rats and mice are particularly abundant and destructive. On poultry farms, there is approximately 1 rat per 5 chickens (D. Pimentel, unpublished data, 1951; Smith, 1984). Using this ratio, the total rat population on US poultry farms may easily number more than 1.4 billion (USDA, 2001). Assuming that the number of rats per chicken has declined because of improved rat control since these observations were made, it is estimated that the number of rats on poultry and other farms is approximately 1 billion. With an estimated 1 rat per person in the United States (Wachtel and McNeely, 1985), there are an estimated 250 million rats in US urban and suburban areas (USCB, 2007).

If we assume, conservatively, that each adult rat consumes and/or destroys stored grains (Ahmed et al., 1995; Chopra, 1992) and other materials valued at $15/year, then the total cost of destruction by introduced rats in the United States is more than $19 billion per year. In addition, rats cause fires by gnawing electric wires, pollute foodstuffs, and act as vectors of several diseases, including salmonellosis and leptospirosis, and, to a lesser degree, plague and murine typhus (Richards, 1989). They also prey on some native invertebrate and vertebrate species such as birds and bird eggs (Amarasekare, 1993).

7.2.3 Birds

About 97 of the 1000 bird species in the United States are exotic (Exotic Birds in Urban Environments, 2007). Of the 97 introduced bird species, only 5% are considered beneficial, while most (56%) are pests (Temple, 1992). However, several species, including chickens and pigeons, were introduced into the United States for agricultural purposes.

In Hawaii, 35 of 69 alien bird species introduced between 1850 and 1984 are still extant on the islands (Moulton and Pimm, 1983; Pimm, 1991). The common myna (*Acridotheres tristis*), introduced into Hawaii, helped in the control of pest cutworms and armyworms in sugarcane (Kurdila, 1995). However, it became the major disperser of seeds of the introduced pest-weed, *Lantana camara*. To cope with the weed problem, Hawaii resorted to the use of herbicides and the introduction of insects as biocontrol agents (Kurdila, 1995).

The English or house sparrow (*Passer domesticus*) was introduced into the United States intentionally in 1853 to control the canker worm (Laycock, 1966; Roots, 1976). By 1900, the birds were considered pests because they damage plants around homes and public buildings and consume wheat, corn, and the buds of fruit trees (Laycock, 1966). Furthermore, English sparrows harass robins, Baltimore orioles, yellow-billed cuckoos, and black-billed cuckoos, and displace native bluebirds, wrens, purple martins, and cliff swallows (English Sparrow, 2005). They are also associated with the spread of about 29 diseases of humans and livestock (Weber, 1979).

The exotic common pigeon (*Columba livia*) exists in most cities of the world, including those in the United States (Robbins, 1995). Pigeons are considered a nuisance because they foul buildings, statues, cars, and sometimes people, and feed on grain (Long, 1981; Smith, 1992). The chemical and physical control costs of pigeons are at least $9 per pigeon per year (Haag-Wackernagel, 1995). Assuming there is 1 pigeon per ha in urban areas (Johnston and Janiga, 1995) or approximately 0.5 pigeons per person in urban areas, and using potential chemical and physical control costs as a surrogate for losses, pigeons cause an estimated $1.1 billion/year in damage. These damage costs do not include the environmental damage associated with their serving as reservoirs and vectors for over 50 diseases, including parrot fever, ornithosis, histoplasmosis and encephalitis (Long, 1981; Weber, 1979).

7.2.4 Arthropods

Approximately 4500 arthropod species (2582 species in Hawaii and more than 2000 in the continental United States) have been introduced. More than 95% of these introductions were accidental, with many species gaining entrance via plants or through soil and water ballast from ships.

The introduced balsam woolly adelgid (*Adelges piceae*) inflicts severe damage in balsam-fir natural forest ecosystems (Jenkins, 1998). According to Alsop and Laughlin (1991), this aphid is destroying the old-growth spruce-fir forest in many regions. Over about a 20-year period, it has spread throughout the southern Appalachians and has destroyed up to 95% of the Fraser firs (personal communication, H.S. Neufeld, Appalachian State University, 1998).

Other introduced insect species have become pests of livestock and wildlife. For example, the red imported fire ant (*Solenopsis invicta*) kills poultry chicks, lizards, snakes, and ground nesting birds (Vinson, 1994). A 34% decrease in swallow nesting success as well as a decline in the northern bobwhite quail populations was reported due to these ants (Allen et al., 1995). The estimated damage to livestock, wildlife, and public health caused by fire ants in Texas is estimated to be $300 million/year. Two people were killed by fire ants in Mississippi in 2002. An additional $200 million is invested in pesticide controls per year (Vinson, 1992; TAES, 1998). Assuming equal damage in other infested southern states, the fire ant damage and pesticide use total approximately $1 billion per year.

7.3 CROP, PASTURE, AND FOREST LOSSES AND ASSOCIATED PESTICIDE USE

Many weeds, pest insects, and plant pathogens are biological invaders causing several billion dollars in losses to crops, pastures, and forests annually in the United States. In addition, several billion dollars are spent on pest control.

7.3.1 Weeds

In crop systems, including forage crops, an estimated 500 introduced plant species have become weed pests; many of these were actually introduced as crops and then became pests (Pimentel et al., 1989). Most of these weeds were accidentally introduced with crop seeds, from ship-ballast soil, or from various imported plant materials; among them were yellow rocket (*Barbarea vulgaris*) and Canada thistle (*Cirsium arvense*).

In US agriculture, weeds cause a reduction of 12% in crop yields. In economic terms, this represents about $24 billion in lost crop production annually, based on the crop potential value of all US crops of more than $200 billion/year (Pimentel, 2011). Based on the estimate that about 73% of the weeds are alien (Pimentel, 1993), it follows that about $17.5 billion of these crop losses are due to introduced weeds. Note, alien invasive weeds are more serious pests than native weeds; thus, this is likely to be a conservative estimate. In addition, approximately $4 billion in herbicides are applied to US crops (Pimentel, 2011), of which about $3 billion is used for control of alien invasive weeds. Therefore, the total costs of introduced weeds to the US economy is about $20.5 billion annually.

In pastures, 45% of weeds are alien species (Pimentel, 1993). US pastures provide about $10 billion in forage crops annually (USDA, 1998), and the estimated losses due to weeds is approximately $2 billion (Pimentel, 1993). Since about 45% of the weeds are alien invasives (Pimentel, 1993), the forage losses due to these non-indigenous weeds are nearly $1 billion/year. An estimated $250 million is invested in herbicides per year for weed control in pastures.

Some introduced weeds are toxic to cattle and wild ungulates, such as leafy spurge (*Euphoria esula*) (Trammel and Butler, 1995). In addition, several alien thistles have replaced desirable native plant species in pastures, rangelands, and forests, thus reducing cattle grazing (Invasive Plants of California's Wildland, 2008). According to Interior Secretary Bruce Babbitt (1998), ranchers spend about $5 billion (mostly herbicides) each year to control invasive alien weeds in pastures and rangelands, yet these weeds continue to spread.

Management of weed species in lawns, gardens, and golf courses is a significant proportion of their total management costs of about $40 billion/year (USCB, 2007). In addition, Templeton et al. (1998) estimated that about $1.3 billion of the $40 billion is spent just on residential pesticide weed, insect, and disease pest control each year. Because a large proportion of these weeds are exotics, the estimate is that $500 million is spent on residential exotic weed control and an additional $1 billion is invested in alien invasive weed control on golf courses.

7.3.2 Insect and Mite Pests

Approximately 500 alien insect and mite species are pests in crops. Hawaii has 5246 identified native insect species, and an additional 2582 introduced insect species (Eldredge and Miller, 1997; Frank and McCoy, 1995a; Howarth, 1990). Introduced insects account for 98% of the pest insects in the state (Beardsley, 1991). In addition to Florida's 11,500 native insect species, 949 introduced species have invaded the state (42 species were introduced for biological control) (Frank and McCoy, 1995b). In California, the 600 introduced species are responsible for 67% of all crop losses (Dowell and Krass, 1992).

Each year, pest insects destroy about 13% of potential crop production representing a value of about $18 billion in US crops (USBC, 2001). Considering that about 40% of the pests were introduced (Pimentel, 1993), we estimate that these pests cause about $10.4 billion in crop losses each year. In addition, about $1.2 billion in pesticides are applied for all insect control each year in the US (Pimentel, 1997). The portion applied against introduced pest insects is approximately $500 million/year. Therefore, the total cost for introduced invasive insect pests is approximately $10.9 billion/year. In addition, based on the earlier discussion of management costs of lawns, gardens, and golf courses, the pesticide control costs of pest insects and mites in lawns, gardens, and golf courses are estimated to be at least $1.5 billion/year.

About 360 alien insect species have become established in American forests (Liebold et al., 1995). Approximately 30% of these are now serious pests. Insects cause the loss of approximately 9% of forest products, amounting to a cost of $7 billion per year (Hall and Moody, 1994; USCB, 2007). Because 30% of the pests are alien pests, annual losses attributed to alien invasive species is about $2.1 billion per year.

The gypsy moth (*Lymantria dispar*), intentionally introduced into Massachusetts in the 1800s, has developed into a major pest of US forests and ornamental trees, oaks in particular (Campbell and Schlarbaum, 1994). The US Forest Service currently spends about $11 million annually on pesticides for gypsy moth control (Campbell and Schlarbaum, 1994).

7.3.3 Plant Pathogens

There are an estimated 50,000 parasitic and non-parasite diseases of plants in the United States and most of these are caused by fungi species (Baker et al., 2005; USDA, 1960). In addition, there are more than 1300 species of viruses that are plant pests in the United States (Baker et al., 2005; USDA, 1960). Many of these microbes were introduced inadvertently with seeds and other parts of host plants and have become major crop pests in the United States (Pimentel, 1993). Including the introduced plant pathogens plus other microbes, more than 20,000 species of microbes are estimated to have invaded the United States.

US crop losses to all plant pathogens total approximately $18 billion per year (Pimentel, 2011; USCB, 2007). Approximately 65% (Pimentel, 1993), or an estimated $11.7 billion per year of losses, are attributable to alien plant pathogens. In addition, $0.72 billion is spent annually for fungicides (Pimentel, 1997), with approximately $0.47 billion for the pesticidal control of alien plant pathogens. This brings the costs of damage and control of alien invasive plant pathogens to about $12.17 billion/year. In addition, based on the earlier discussion of pests in lawns, gardens, and golf courses, the control costs of plant pathogens in lawns, gardens, and golf courses are estimated to be at least $2 billion/year.

In forests, more than 20 alien species of plant pathogens attack woody plants (Liebold et al., 1995). Two of the most serious plant pathogens are the chestnut blight fungus (*Cryphonectria parasitica*) and Dutch elm disease (Ophiostoma ulmi). Before the introduction of chestnut blight, approximately 25% of eastern US deciduous forest consisted of American chestnut trees (Campbell, 1994). Elm tree removal costs about $100 million/year (Campbell and Schlarbaum, 1994).

Approximately 9%, or $7 billion, of forest products are lost each year due to plant pathogens (Hall and Moody, 1994; USCB, 2007). Assuming that the proportion of introduced plant pathogens in forests is similar to that of introduced insects (about 30%), approximately $2.1 billion in forest products are lost each year to alien invasive plant pathogens in the United States.

7.4 LIVESTOCK PESTS

Similarly to crops, exotic microbes (e.g. calf-diarrhoea-rotavirus) and parasites (e.g. face flies, *Musca autumnalis*) were introduced when livestock were brought to the United States (Drummond et al., 1981; Morgan, 1981). In addition to the hundreds of microbes and parasites that have already been introduced and are pests, there are more than 60 microbes and parasites that could invade and become serious pests to US livestock (USAHA, 1984). A conservative estimate of the losses to US livestock from exotic microbes and parasites is approximately $9 billion/year (personal conversation, Kelsey Hart, College of Veterinary Medicine, Cornell University, 2001). The author's estimate of the cost of pesticide used in pest control of livestock is $200 million per year.

7.5 HUMAN DISEASES

The alien diseases now having the greatest impact are Acquired Immune Deficiency Syndrome (AIDS), syphilis, and influenza (Newton-John, 1985; Pimentel et al., 2011). West Nile virus and pest mosquitoes are serious problems and an estimated $200 million is spent on pesticidal control each year.

7.6 CONCLUSIONS

With more than 50,000 alien invasive species in the United States, the fraction that is harmful does not have to be large to inflict significant damage to natural and managed ecosystems and cause public health problems. There is a suite of ecological factors that may cause alien invasive species to become abundant and persistent. These include: the lack of controlling natural enemies (e.g. purple loosestrife and imported fire ant); the development of new associations between alien parasite and host (e.g. West Nile virus in humans and gypsy moth in US oaks); effective predators in new ecosystems (e.g. feral cats); artificial and/or disturbed habitats that provide favourable invasive ecosystems for the aliens (e.g. weeds in crop and lawn habitats); and invasion by some highly adaptable and successful alien species (e.g. water hyacinth).

Although specific economic damage and associated control costs have been estimated at $125 billion, precise economic costs associated with some of the most ecologically damaging exotic species are not available. If we had been able to assign monetary values to species extinctions and losses in biodiversity, ecosystem services, and aesthetics, the costs of destructive alien invasive species would undoubtedly be several times higher than $125 billion/year. Yet even this understated economic loss indicates that alien invasive species are exacting a significant toll.

We recognize that nearly all our crop and livestock species are alien and have proven essential to the viability our agriculture and economy. Although certain alien crops (e.g. corn and wheat) are vital to agriculture and the US food system, this does not diminish the enormous negative impacts of other non-indigenous species (e.g. exotic weeds and insects).

The true challenge lies not in determining the precise costs of the impacts of exotic species, but in preventing further damage to natural and managed ecosystems. Formulation of sound prevention policies needs to take into account the means through which alien species gain access to and become established in the United States. Since the invasions vary widely, we should expect that a variety of strategies would be needed for prevention programmes. For example, public education, sanitation, and effective prevention programmes at airports, seaports, and other ports of entry will help reduce the chances of biological invaders becoming established in the United States.

While these policies and practices may help prevent accidental and intentional introduction of potentially harmful exotic species, we have a long way to go before the resources devoted to the problem are in proportion to the risks. We hope that this environmental and economic assessment will advance the argument that investments made now to prevent future introductions will be returned many times over in the form of preservation of natural ecosystems, diminished losses to agriculture and forestry, and lessened threats to public health.

Acknowledgments

We thank the following people for reading an earlier draft of this article and for their many helpful suggestions: D. Bear, Council on Environmental Quality, Executive Office of the President, Washington, DC; J.W. Beardsley, University of Hawaii; A.J. Benson, U.S. Geological Survey, Gainesville, FL; B. Blossey, Cornell University; C.R. Bomar, University of Wisconsin, Stout; F.T. Campbell, Western Ancient Forest Campaign, Springfield, VA; R. Chasen, Editor, *BioScience*; P. Cloues, Geologic Resources Division, Natural Resource Program Center, Lakewood, CO; W.R. Courtenay, Florida Atlantic University; R.H. Cowie, Bishop Museum, Honolulu, HI; D. Decker, Cornell University; R.V. Dowell, California Department of Food and Agriculture; T. Dudley, University of California, Berkeley; H. Fraleigh, Colorado State University; H. Frank, University of Florida; T. Fritts, U.S. Geological Survey, Washington, DC; E. Groshoz, University of New Hampshire; J. Jenkins, Forest Service, USDA, Radnor, PA; J.N. Layne, Archbold Biological Station, Lake Placid, FL; J. Lockwood, University of Tennessee; J.D. Madsen, U.S. Army Corps of Engineers, Vicksburg, MS; R.A. Malecki, N.Y. Cooperative Fish & Wildlife Research Unit, Ithaca, NY; E.L. Mills, Cornell University; S.F. Nates, University of Southwestern Louisiana; H.S. Neufeld, Appalachian State University; P.J. O'Connor, Colorado State University; B.E. Olson, Montana State University; E.F. Pauley, Coastal Carolina University; M. Pimentel, Cornell University; S. Pimm, University of Tennessee; W.J. Poly, Southern Illinois University; W. Roberts, Rainbow Beach, Australia; M. Sagoff, Institute for Philosophy and Public Policy, University of Maryland, College Park; B. Salter, Maryland Department of Natural Resources; D.L. Scarnecchia, University of Idaho; D. Simberloff, University of Tennessee; G.S. Rodrigues, Empresa Brasilerira de Pesquisa Agropecuaria, Brazil; J.N. Stuart, University of New Mexico; S.B. Vinson, Texas A & M University; L.A. Wainger, University of Maryland; J.K. Wetterer, Columbia University; and C.E. Williams, Clarion University of Pennsylvania. This research was supported in part from a grant from the Podell Emertii award at Cornell University.

References

Ahmed, E., Hussain, I., Brooks, J.E., 1995. Losses of stored foods due to rats at grain markets in Pakistan. Int. Biodeter. Biodegrad. 36 (1–2), 125–133.

Allen, C.R., Lutz, R.S., Demarais, S., 1995. Red imported fire ant impacts on northern bobwhite populations. Ecol. Appl. 5 (3), 632–638.

Alsop, F.J., Laughlin, T.F., 1991. Changes in the spruce-fir avifauna of Mt Guyot, Tennessee, 1967–1985. J. Tennessee Acad. Sci. 66 (4), 207–209.

Amarasekare, P., 1993. Potential impact of mammalian nest predators on endemic forest birds of western Mauna Kea, Hawaii. Conserv. Biol. 7 (2), 316–324.

Armstrong, S., 1995. Rare plants protect Cape's water supplies. New Scientist 11, 8.

ATTRA, 1997. Purple Loosestrife: Public Enemy #1 on Federal Lands. ATTRA Interior Helper Internet, Washington, DC, <http://refuges.fws.gov/NWRSFiles/HabitatMgmt/PestMgmt/LoosestrifeProblem.html/>.

Babbitt, B., 1998. Statement by Secretary of the Interior on invasive alien species. Proceedings, National Weed Symposium, BLM Weed Page. April 8–10, 1998, <http://www.nps.gov/plants/alien/pubs/bbstat.htm>.

Baker, R., Cannon, R., Bartlett, P., Baker, I., 2005. Novel strategies for assessing and managing the risks posed by invasive alien species to global crop production and biodiversity. Ann. Appl. Biol. 146 (2), 177–191.

Beardsley, J.W., 1991. Introduction of arthropod pests into the Hawaiian Islands. Micronesia Suppl. 3, 1–4.

Beers, J., 2003. Invasive Species, Part I. <http://www.landrights.org/invasive_species/invasive1.htm>. (Last accessed 28.06.13).

Bryan, R.T., 1996. Alien species and emerging infectious diseases: past lessons and future applications. In: Sandlund GT, Schel PJ, Viken A, (Eds.), Proceedings of the Norway/UN Conference on Alien Species, 1–5 July 1996, Trondheim, Norway: Norwegian Institute for Nature Research, pp. 74–80.

California Invasive Plant Council, 2006. <http://www.cal-ipc.org/ip/inventory/pdf/Inventory2006.pdf>. (Last accessed 28.06.13).

Campbell, F.T., 1994. Killer pigs, vines, and fungi: alien species threaten native ecosystems. Endanger. Spec. Tech. Bull. 19 (5), 3–5.

Campbell, F.T., 1998. 'Worst' Invasive Plant Species in the Conterminous United States. Report. Western Ancient Forest Campaign, Springfield, VA.

Campbell, F.T., Schlarbaum, S.E., 1994. Fading Forests: North American Trees and the Threat of Exotic Pests. Natural Resources Defense Council, New York.

Center, T.D., Frank, J.H., Dray, F.A., 1997. Biological control. In: Simberloff, D., Schmitz, D.C., Brown, T.C. (Eds.), Strangers in Paradise Island Press, Washington, DC, pp. 245–266.

Chapman, L.E., Tipple, M.A., Schmeltz, L.M., Good, S.E., Regenery, H.L., Kendal, A.P., et al., 1992. Influenza – United States, 1989–90 and 1990–91 seasons. Mortal. Morbid. Weekly Rep. Surveill. Sum. 41 (SS-3), 35–46.

Chopra, G., 1992. Poultry farms. In: Prakash, I., Ghosh, P.K. (Eds.), Rodents in Indian Agriculture Scientific Publishers, Jodhpur, India, pp. 309–330.

Colburn, D., 1999. Dogs take a big bite out of health care costs. The Washington Post, 2 February 1999, p. z5.

Dill, W.A., Cordone, A.J., 1997. History and Status of Introduced Fishes in California, 1871–1996. Fish Bulletin 178. The Resources Agency, Department of Fish and Game, State of California.

Dowell, R.V., Krass, C.J., 1992. Exotic pests pose growing problem for California. Calif. Agric. 46 (1), 6–10.

Drummond, R.O., Lambert, G., Smalley, H.E., Terrill, C.E., 1981. Estimated losses of livestock to pests. In: Pimentel, D. (Ed.), Handbook of Pest Management in Agriculture. CRC Press, Inc., Boca Raton, FL, pp. 111–127.

Dunn, E.H., Tessaglia, D.L., 1994. Predation of birds at feeders in winter. J. Field Ornithol. 65 (1), 8–16.

Eldredge, LG, Miller, SE., 1997. Numbers of Hawaiian species: supplement 2, Including a review of freshwater invertebrates. Bishop Museum Occas. Pap. 48, 3–32.

Endangered Species, 2008. Why Save Endangered Species? <http://www.empowermentzone.com/species.txt>. (Last accessed 28.06.13).

English Sparrow, 2005. House Sparrows. Internet Center for Wildlife Damage Management. Cornell University et al. <http://icwdm.org/handbook/birds/HouseSparrows.asp>. (Last accessed 28.06.13).

Exotic Birds in Urban Environments, 2007. <http://thedrinkingbirds.blogspot.com/2007/11/exotgic-birds-in-urban-environments-part_27.html/>. (Last accessed 14.01.09.)

Frank, J.H., McCoy, E.D., 1995a. Introduction to insect behavioral ecology: the good, the bad and the beautiful: non-indigenous species in Florida. Fla. Entomol. 78 (1), 1–15.

Frank, J.H., McCoy, E.D., 1995b. Precinctive insect species in Florida. Fla. Entomol. 78 (1), 21–35.

Frank, J.H., McCoy, E.D., Hall, H.G., O'Meara, F., Tschinkel, W.R., 1997. Immigration and introduction of insects. In: Simberloff, D., Schmitz, D.C., Brown, T.C. (Eds.), Strangers in Paradise. Island Press, Washington, DC, pp. 75–100.

Gaudet, C.L., Keddy, P.A., 1988. Predicting competitive ability from plant traits: a comparative approach. Nature 334, 242–243.

Haag-Wackernagel, D., 1995. Regulation of the street pigeon in Basel. Wildl. Soc. Bull. 23 (2), 256–260.

Hall, J.P., Moody, B., 1994. Forest Depletions Caused by Insects and Diseases in Canada 1982–1987. Forest Insect and Disease Survey Information Report ST-X-8, Forest Insect and Disease Survey, Canadian Forest Service, Natural Resources Canada, Ottawa, Canada.

Hiebert, R.D., Stubbendieck, J., 1993. Handbook for Ranking Exotic Plants for Management and Control. U.S. Department of Interior, National Park Service, Denver, CO.

Howarth, F.G., 1990. Hawaiian terrestrial arthropods: an overview. Bishop Museum Occas. Pap. 30, 4–26.

Invasive.org, 2010. Cheatgrass. Invasive.org. <http://www.invasive.org/browse/subinfo.cfm?sub=5214>. (Last accessed 28.06.13).

Invasive Plants of California's Wildland, 2008. <http://www.cal-ipc.org/ip/management/ipcw/>. (Last accessed 01.07.13).

Jenkins, J.C., 1998. Measuring and Modeling Northeaster Forest Response to Environmental Stresses. Ph.D. Dissertation Submitted to the University of New Hampshire, Durham, N.H.

Johnston, R.F., Janiga, M., 1995. Feral Pigeons. Oxford University Press, New York.

Keniry, T., Marsden, J.E., 1995. Zebra mussels in Southwestern Lake Michigan. In: LaRoe, E.T., Farris, G.S., Puckett, C.E., Doran, P.D., Mac, M.J. (Eds.), Our Living Resources: a Report to the Nation on the Distribution, Abundance, and Health of U.S. Plants, Animals and Ecosystems. U.S. Department of the Interior, National Biological Service, Washington, DC, pp. 445–448.

Kent, J.H., Chapman, L.E., Schmeltz, L.M., Regnery, H.L., Cox, N.J., Schonberger, L.B., 1992. Influenza surveillance – United States, 1991–92. Mortal. Morbid. Weekly Rep. Surveill. Sum. 41 (SS-5), 35–43.

Khalanski, M., 1997. Industrial and ecological consequences of the introduction of new species in continental aquatic ecosystems: the zebra mussel and other invasive species. Bull. Franc. Peche Pisciculture 0 (344–345), 385–404.

Kotanen, P.M., 1995. Responses of vegetation to a changing regime of disturbance: effects of feral pigs in a California coastal prairie. Ecography 18, 190–197.

Kurdila, J., 1995. The introduction of exotic species into the United States: there goes the neighborhood. Environ. Aff. 16, 95–118.

Lafferty, K.D., Kuris, A.M., 1996. Biological control of marine pests. Ecology 77 (7), 1989–2000.

Lafferty, K.D., Page, C.J., 1997. Predation of the endangered tidewater goby, Eucyclogobius newberryi, by the introduced African Clawed frog, Xenopus laevis, with notes on the frog's parasites. Copeia 3, 589–592.

Laycock, G., 1966. The Alien Animals. Natural History Press, New York.

Layne, J.N., 1997. Nonindigenous mammals. In: Simberloff, D., Schmitz, D.C., Brown, T.C. (Eds.), Strangers in Paradise. Island Press, Washington, DC, pp. 157–186.

Liebold, A.M., MacDonald, W.L., Bergdahl, D., Mastro, V.C., 1995. Invasion by exotic forest pests: a threat to forest ecosystems. For. Sci. 41 (2), 1–49.

Long, J.L., 1981. Introduced Birds of the World: the Worldwide History, Distribution, and Influence of Birds Introduced to New Environments. Universe Books, New York.

Malecki, R.A., Blossey, B., Hight, S.D., Schroeder, D., Kok, D.T., Coulson, J.R., 1993. Biological control of purple loosestrife. BioScience 43 (10), 680–686.

McCoid, M.J., Kleberg, C., 1995. Non-native reptiles and amphibians. In: LaRoe, E.T., Farris, G.S., Puckett, C.E., Doran, P.D., Mac, M.J. (Eds.), Our Living Resources: a Report to the Nation on the Distribution, Abundance, and Health of U.S. Plants, Animals and Ecosystems. U.S. Department of the Interior, National Biological Service, Washington, DC, pp. 433–437.

McKay, G.M., 1996. Feral cats in Australia: origin and impacts. Unwanted Aliens? Australia's Introduced Animals. Nature Conservation Council of NSW, The Rocks, NSW, Australia.

Miller, J.H., 1995. Exotic plants in southern forests: their nature and control. Proc. South. Weed Sci. Soc. 48, 120–126.

Mills, E.L., Scheuerell, M.D., Carlton, J.T., Strayer, D.L., 1997. Biological Invasions in the Hudson River Basin. New York State Museum Circular No. 57. The University of the State of New York, State Education Department.

Morgan, N.O., 1981. Potential impact of alien arthropod pests and vectors of animal diseases on the U.S. livestock industry. In: Pimentel, D. (Ed.), Handbook of Pest Management in Agriculture. CRC Press, Inc., Boca Raton, FL, pp. 129–135.

Morin, N., 1995. Vascular plants of the United States. In: LaRoe, E.T., Farris, G.S., Puckett, C.E., Doran, P.D., Mac, M.J. (Eds.), Our Living Resources: a Report to the Nation on the Distribution, Abundance, and Health of U.S. Plants, Animals and Ecosystems. U.S. Department of the Interior, National Biological Service, Washington, DC, pp. 200–205.

Morse, L.E., Kartesz, J.T., Kutner, L.S., 1995. Native vascular plants. In: LaRoe, E.T., Farris, G.S., Puckett, C.E., Doran, P.D., Mac, M.J. (Eds.), Our Living Resources: a

Report to the Nation on the Distribution, Abundance, and Health of U.S. Plants, Animals and Ecosystems. U.S. Department of the Interior, National Biological Service, Washington, DC, pp. 205–209.

Moulton, M.P., Pimm, S.L., 1983. The introduced Hawaiian avifauna: biogeographic evidence for competition. Am. Nat. 121 (5), 669–690.

Newton-John, H., 1985. Exotic human diseases. In: Gibbs, A.J., Meischke, H.R.C. (Eds.), Pests and Parasites as Migrants. Cambridge University Press, Sydney, pp. 23–27.

OTA, 1993. Harmful Non-Indigenous Species in the United States. Office of Technology Assessment, United States Congress, Washington, DC.

Pimentel, D., 1993. Habitat factors in new pest invasions. In: Kim, K.C., McPheron, B.A. (Eds.), Evolution of Insect Pests – Patterns of Variation. John Wiley & Sons, New York, pp. 165–181.

Pimentel, D., 1997. Techniques for Reducing Pesticides: Environmental and Economic Benefits. John Wiley & Sons, Chichester, UK.

Pimentel, D., 2011. Environmental and economic costs associated with alien invasive species in the United States. In: Pimentel, D. (Ed.), Biological Invasions: Economic and Environmental Costs of Alien Plant, Animal, and Microbe Species. 2nd Edition. CRC Press (Taylor & Francis Group), Boca Raton, Florida, pp. 411–430.

Pimentel, D., Hunter, M.S., LaGro, J.A., Efronymson, R.A., Landers, J.C., Mervis, F.T., et al., 1989. Benefits and risks of genetic engineering in agriculture. BioScience 39, 606–614.

Pimm, S.L., 1991. The Balance of Nature? The University of Chicago Press, Chicago.

Randall, J.M., 1996. Weed control for the preservation of biological diversity. Weed Technol. 10, 370–381.

Raven, P.H., Johnson, G.B., 1992. Biology, third ed. Mosby Year Book, St. Louis, MO.

Richards, C.G.J., 1989. The pest status of rodents in the United Kingdom. In: Putman, R.J. (Ed.), Mammals as Pests. Chapman and Hall, London, pp. 21–33.

Robbins, C.S., 1995. Non-native birds. In: LaRoe, E.T., Farris, G.S., Puckett, C.E., Doran, P.D., Mac, M.J. (Eds.), Our Living Resources: a Report to the Nation on the Distribution, Abundance, and Health of U.S. Plants, Animals and Ecosystems. U.S. Department of the Interior, National Biological Service, Washington, DC, pp. 437–440.

Roots, C., 1976. Animal Invaders. Universe Books, New York.

Simberloff, D., Schmitz, D.C., Brown, T.C., 1997. Strangers in Paradise. Island Press, Washington, DC.

Smith, R., 1984. Producers need not pay startling 'rodent tax' losses. Feedstuffs 56 (22), 13–14.

Smith, R.H., 1992. Rodents and birds as invaders of stored-grain ecosystems. In: Jayas, D.S., White, N.D.G., Muir, W.E. (Eds.), Books in Soils, Plants, and the Environment: Stored-Grain Ecosystems. Marcel Dekker, Inc., New York, pp. 289–323.

Stone, C.P., Cuddihy, L.W., Tunison, T., 1992. Response of Hawaiian ecosystems to removal of pigs and goats. In: Stone, C.P., Smith, C.W., Tunison, J.T. (Eds.), Alien Plant Invasions on Native Ecosystems in Hawaii: Management and Research University of Hawaii Cooperative National Park Studies Unit, Honolulu, pp. 666–702.

TAES, 1998. Texas Imported Fire Ant Research & Management Plan. Report. Texas Agricultural Extension Service, Texas A & M University, College Station, TX.

Temple, S.A., 1992. Exotic birds, a growing problem with no easy solution. The Auk 109, 395–397.

Templeton, S.R., Zilberman, D., Yoo, S.J., 1998. An economic perspective on outdoor residential pesticide use. Environ. Sci. Technol. 32 (17), 416A–423A.

Thompson, D.G., Stuckey, R.L., Thompson, E.B., 1987. Spread, impact, and control of purple loosestrife (Lythrum salicaria) in North American wetlands. U.S. Fish and Wildlife Service, Fish and Wildlife Research, Washington, DC, 2, 55 pp.

Trammel, M.A., Butler, J.L., 1995. Effects of exotic plants on native ungulate use of habitat. J. Wildl. Manag. 59 (4), 808–816.

USAHA, 1984. Foreign Animal Diseases: Their Prevention, Diagnosis and Control. Committee on Foreign Animal Diseases of the United States Animal Health Association, Richmond, VA.

USBC, 2001. Statistical Abstract of the United States 2001. U.S. Bureau of the Census, U.S. Government Printing Office, Washington, DC.

USCB, 2007. Statistical Abstract of the United States: 2008. (127th Edition). U.S. Census Bureau, Washington, DC.

USDA, 1960. Index of Plant Diseases in the United States. Crop Research Division, ARS. U.S. Department of Agriculture, Washington, DC.

USDA, 1998. Agricultural Statistics 1998. U.S. Government Printing Office, Washington, DC.

USDA, 2001. Agricultural Statistics 2001. U.S. Government Printing Office, Washington, DC.

Vilella, F.J., Zwank, P.J., 1993. Ecology of the small Indian mongoose in a coastal dry forest of Puerto Rico where sympatric with the Puerto Rican nightjar. Caribb. J. Sci. 29 (1–2), 24–29.

Vinson, S.B., 1992. The economic impact of the imported fire ant infestation on the State of Texas. Report. Texas A & M University, College Station, TX.

Vinson, S.B., 1994. Impact of the invasion of Solenopsis invicta (Buren) on native food webs. In: Williams,

D.F. (Ed.), Exotic Ants: Biology, Impact, and Control of Introduced Species. Westview Press, Boulder, CO, pp. 241–258.
Vitousek, P.M., D'Antonio, C.M., Loope, L.L., Rejmanek, M., Westerbrooks, R., 1997. Introduced species: a significant component of human-caused global change. New Zealand J. Ecol. 21 (1), 1–16.
Wachtel, S.P., McNeely, J.A., 1985. Oh rats. Int. Wildl. 15 (1), 20–24.
Weber, W.J., 1979. Health Hazards from Pigeons, Starlings and English Sparrows: Diseases and Parasites Associated with Pigeons, Starlings, and English Sparrows which Affect Domestic Animals. Thomson Publications, Fresno, CA.
Wilcove, D.S., Rothstein, D., Bubow, J., Phillips, A., Losos, E., 1998. Quantifying threats to imperilled species in the United States. BioScience 48 (8), 607–615.

CHAPTER

8

Potential and Utilization of Plant Products in Pest Control

R.T. Gahukar

Arag Biotech Pvt. Ltd., Nagpur, India

8.1 INTRODUCTION

Agricultural crops are attacked by several species with potential losses as a result estimated at 16.9% (Oerke, 2006). In the era of global food security, this loss cannot be permitted. Therefore, nearly 3.5 tons of pesticides including 35.2% insecticides are applied globally in 2010 to save the crops from pests, the maximum use (25.7%) being on fruit and vegetables (David and Shankar, 2012). Generally, synthetic pesticides of various chemical groups are applied to food crops, fibre crops, fruit trees, root crops, ornamental and flowering plants, and industrial crops. However, large scale application of incorrect doses of active ingredient and faulty equipment have resulted in several undesirable effects such as acute and chronic poisoning of applicators and farmworkers, food poisoning and health hazards to consumers, danger to non-target organisms (pollinators, honey bees, natural enemies of insect pests), destruction of fish, birds and wildlife, disruption of natural equilibrium, pollution of the environment and groundwater, severe outbreaks of pests, minor pests assuming the status of serious pests, and increased pest resistance to one or more groups of pesticides. Pesticide residues have been detected in daily foods including cow's milk and milk products. Apart from these adverse effects, changes in the odour of plant parts occur as nitrogen oxides in air react with degrading hydrocarbons (McFrederick et al., 2008). All of these factors degrade the quality of food products.

In recent years, application of insecticides has been reduced to some extent but not totally (David and Shankar, 2012). There is thus the possibility to further restrict pesticide use by searching for alternative tools. Amongst the available options, plant parts and plant-derived products have recently gained importance because plant biodiversity has provided an excellent source of biologically active constituents or allelochemicals for use in traditional crop protection. Studies on plant-derived biopesticides/botanical pesticides tested under field and laboratory conditions have been published but these are mostly local or regional publications that cannot be easily procured.

D. P. Abrol (Ed): Integrated Pest Management.
DOI: http://dx.doi.org/10.1016/B978-0-12-398529-3.00009-9

Nowadays, plant products (crude or formulated) are mixed generally with other plant products. In this chapter, emphasis is therefore given to combinations of plant products with other plant protection measures. In this review, recent information is compiled and the bioefficacy of plant products is discussed for formulating an effective and economic integrated pest management (IPM) strategy in agricultural crops in different agro-ecosystems.

8.2 POTENTIAL OF PLANT PRODUCTS

Various plant parts are used for preparing crude or aqueous extracts in water or for extracting oil while commercial products are formulated by chemical processes. In reality, all products act in the same way since active substances play a major role. However, this can be related to structure–activity. For example, dihydrofuran acetal present in azadirachtin (AZ) imparts antifeeding activity whereas decalin moieties disrupt insect growth and development. Inert materials (spreading agent, emulsifier, stabilizer, etc.) are added to extend shelf life and for a better, more uniform coverage of plant surfaces. Degradation of plant products due to high temperatures or direct sunlight, water, pH, microorganisms, etc. is also minimized.

Plant products are mainly stomach poisons; their contact toxicity reported against soft bodied insects and larvae is rather slow and reduced probably because plant products must be ingested by pests to be effective. Systemic action is exerted when plant products are incorporated into soil or applied to plant parts and seedlings absorb and accumulate the products. This action enhances product bioefficacy and persistence in the field and makes the plant pest-resistant. Essential oils act as fumigants and contact poisons.

Generally, plant products act as antifeedants, repellents, growth regulators, sterilants, or oviposition deterrents, or have toxic/pesticidal actions. These actions depend on the dose/concentration of the product and are related to the behaviour and physiology of different systems of insects. Plant products may exhibit any single action or many actions on any life stage of the pest, which may not die immediately by a knock-down effect but may die after a few hours of treatment. The impact may be greater on one pest species than on others. Also, it is difficult to ascertain which mode of action is predominant when there is interspecific variation in behavioural responses of a pest species; e.g. a substance deterring feeding in one pest species can act as an attractant, growth inhibitor or stimulant for other pests or substances initially acting as feeding deterrents and may lack toxicity if ingested rending antifeedant action ineffective due to reduction in proteins and nucleic acid content (Isman, 2002). In the case of crude extracts, the detrimental effect can occur due to modulating the nature of extracts and alteration of the biochemical pathways in reproductive organs, macromolecules, mineral levels in the alimentary canal and detoxification enzyme levels in the fat body and intestine (Sahayaraj and Antony, 2006; Sahayaraj and Shoba, 2012).

The stimulation of gustatory receptors is inhibited or receives a negative response and therefore the pest is unable to recognize suitable food. Consequently, biochemical changes occur in the digestive system affecting consumption and utilization of food because of a reduction in the activity of midgut enzymes and disturbance of the neural regulation of gut peristalsis. Nevertheless, rapid desensitization to a feeding deterrent makes the pest tolerant upon repeated or continuous exposure or feeding deterrence is not produced at lower doses of pesticides (Isman, 2002). With antifeedant action, insect feeding is deterred and the pest dies due to starvation.

Insect mating and sexual communication are disrupted and oviposition is deterred. A sterilant effect is produced when yolk deposition is reduced, the ovarian sheath and inter-follicular tissues are disrupted and the follicular epithelium is damaged. Eggs with malformed chorion are sterile. Low ovarian weight is due to a reduction in proteins and nucleic acid content. Concentration of vitellogenin and vitellin is reduced affecting ovarian development. A very few eggs are laid and the proportion of eggs hatching is quite low. These effects may be carried over to the next generation. Normal growth and development are regulated by growth hormone systems. The synthesis and release of neurosecretions are delayed which acts on the secretion of juvenile hormone and moulting enzymes (ecdysone) and the product is known as an insect growth regulator (IGR). Often intermediates of two life stages or instars are produced or life stages are malformed interrupting the life cycle.

8.3 UTILIZATION OF PLANT PRODUCTS

8.3.1 Plant Species

In the tropics, plant biodiversity has provided an excellent source of allelochemicals for crop protection in traditional agriculture for centuries and crude local products were applied to crops and stored food grains to protect them from an array of pest species. Globally, more than 6000 plant species have been screened and >2500 species belonging to 235 families contain toxins (Saxena, 1998). Indigenous plants in Africa possess a wide range of biological activities against pests, but only some of them have been scientifically validated (Odeyemi et al., 2008).

Neem (*Azadirachta indica* A. Juss.) has been extensively exploited for about two decades for both crude/traditional preparations and commercial products because of its high content and effectiveness of allelochemicals, viz. azadirachtin (AZ), salanin, nimbecidine, nimbinin, nimbin, nimbocinol, nimbocinone, meliantriol, azadirachtol, azadirone, slannolide and other isomers. Other plants exploited for commercial production include *Chrysanthemum cinerifolium* Vis. for pyrethrins, *Derris* spp., *Tephrosia* spp. and *Lonchocarpus* spp. for rotenone, *Ryania speciosa* Vahl for ryanodine, *Schoenocaulon officinale* Schltdl. & Cham. for sabadilla, *Anacyclus pyrethrum* (L.) Link for pellitorine, *Heliopsis longipes* Blake for affinine, *Piper nigram* L. for pipericide, *Quassia amara* L. for quassin, *Nicotiana tabacum* L. for nicotine sulphate, *Pongamia pinnata* (L.) Pierre for karanjin, *Melia azedarach* L. for limonoids, and *Vitex negundo* L. for norditerpene alkaloids. Essential oils are extracted from plants mostly of the Lamiaceae family, such as *Eucalyptus* spp., *Rosmarinus officinalis* L., *Syzygium aromaticum* (L.) Merrill & Perry, *Mentha* spp. and *Thymus vulgaris* L. More plant species will be added to the list as and when research on allelochemicals reveals their toxic properties. In fact, many indigenous plants found abundantly in the tropics have not been studied probably because some species are protected by regulations against exploitation and some of them may even be endangered species.

Plants are natural sources of allelochemicals used in plant protection in agricultural crops (Farooq et al., 2011). These are water soluble secondary plant metabolites or by-products of the principal metabolic pathways in plants. They are non-nutritional and can be synthesized in any plant part but are easily and rapidly biodegradable due to shorter half-life and thus residual effects on the biological environment are negligible. Most of them are volatiles of short-chain alcohols and aldehydes, ketones, esters, aromatic phenols, mono- and sesquiterpenes and include alkaloids, terpenoids, flavonoids, steroids, phenolic compounds, glycosides, lipids, sitosterols, tannins, monoterpenes, clerodane diterpenes, sugar esters, acetogenins, and light-activated allelochemicals such as thiophenes, acetylenic thiophenes, quinines and furanocoumarins.

With advances in research, extraction/isolation, identification and synthesis of allelochemicals have been possible. Plant essential oils are complex mixtures of mono- and sesquiterpenes and related phenyl propenes. The content of allelochemicals differs significantly per plant species, plant part, ecotype, climatic condition and genetic diversity (Gupta et al. 2010; Kaur et al., 2005; Sidhu et al., 2004; Vir 2007).

8.3.2 Plant Preparations/Products

8.3.2.1 Traditional Crude Products

Traditionally, farmers prepare extracts of plant leaves, bark, seeds, kernels or seed cake in cold or hot water by a simple soaking method. Oil is extracted from leaves, seeds, berries or kernels with local wooden rotating wheels. The water preparations (5–10%) and oils (2–5%) are sprayed by knapsack or hand sprayers @ 500 l/ha so as to obtain better plant coverage. Seed treatment with neem oil (NO) or neem products at 25 ml/kg seed, and incorporation of neem cake (NC) @ 2 t/ha or neem seed kernel powder (NSKP) @ 20 kg/ha are effective treatments against soil insects.

Baits containing plant products have been successfully used in agricultural crops. For example, a mixture of ripe banana fruit pulp and water extract of sacred basil, *Ocimum sanctum* L. leaves attracts fruit flies, *Bactrocera cucurbitae* (Coq.), in cucurbits (Satpathy and Rai 2002). Farmers can use these products as preventive measures without knowing the economic threshold of the pest populations but are reluctant to use them as home-made products due to their slow effects, poor contact action and reduced residual toxicity.

8.3.2.2 Commercial Formulations

Oil mills extract oil in industrial expellers by cold pressing. Also, organic solvents such as ethanol, methanol, acetone, and petroleum ether are used to extract the maximum quantity of allelochemicals and essential oils from selected plant parts. Over 100 protolimonoids, limonoids or tetranor triterpenoids, pentanortriterpenoids, hexanortriterpenoids and some nonterpenoid constituents have been isolated from various parts of the neem tree. But isolation, synthesis or formulation of allelochemicals is time consuming and expensive. Therefore, research on simplified, rapid and cheaper technologies should be carried out. For example, non-destructive near infrared spectroscopy and isocratic high performance liquid chromatography (Highi and Hatami, 2010), high performance thin layer chromatography (Verma et al., 2011) or supercritical fluid extraction and hydro-distillation (Pavela et al., 2012) have been proved to be useful for simultaneous accurate quantification of allelochemicals. These sophisticated facilities are not available in developing and less-developed countries mainly due to limited financial allocations to scientific research.

In the case of neem, over 50 commercial products (mostly emulsifiable concentrates based on oil or extracts in organic solvents) are now available in the market in India. Rotenone dust (1–5% a.i.) and liquid (8% a.i.) formulations are available in Latin America. Recent products include neem kernel-based pellets, neem cake in flakes, and emulsifiable concentrates containing AZ at 3000, 5000, 10,000, 20,000, 30,000, 50,000 or 65,000 ppm.

8.4 PEST MANAGEMENT

The quantity of active ingredients in crude preparations or formulated products is an important criterion for bioefficacy. Mixtures of plant products give better pest mortality than individual products due to synergism but accurate determination of toxicity values and the mode of action poses difficulties. The quantity of active ingredients in crude preparations is unknown and the use of such products may result in inadequate pest control or may induce partial resistance in target pests.

Water extracts, oils or commercial products based on pure allelochemicals have been found to be as effective as synthetic pesticides or often more effective in controlling all kinds of pests, viz. soil dwellers (termites, white grubs, root borers), sap sucking pests (aphids, thrips, white flies, mites), defoliators (semiloopers, hairy caterpillars, armyworms), boll worms and gram caterpillars, flower and earhead feeding pests (gall midge, blister beetles), stem borers, and fruit feeding insects (fruit flies, fruit borers) infesting cereals (Gahukar, 2007a), grain legumes (Gahukar, 2005), vegetables (Gahukar, 2007b), oilseed crops (Gahukar, 2008b), fruit trees (Gahukar, 2008c), floriculture (Gahukar, 2011a), medicinal plants (Gahukar, 2012), cotton (Gahukar, 2000), sugarcane (Gahukar, 2008a), plantation crops (Gahukar, 2010a), spices and condiments (Gahukar, 2011b) and forest trees (Gahukar, 2010b). In some instances, plant products could not control insect pests but they were more effective than water sprays (control). This ineffectiveness was probably due to dose and quality of product, time of application and type of equipment resulting in inadequate plant coverage.

It is common practice to mix several plant products together (using oil of *P. pinnata* or *Sesamum indicum* L.), along with chemical insecticides or fungicides (deltamethrin, endosulfan, cypermethrin, carbendazim), biopesticides (*Bacillus thuringiensis* Berl. (Bt), *Beauveria bassiana* (Bals.) Vuill., nuclear polyhedrosis virus (NPV) or adjuvants (teepol, soap water) (Murugan et al. 1998). Several pesticide companies have also formulated commercial products containing 2–3 plant products; these have been registered and are sold in the market. Some of the combinations are shown in Table 8.1 and how they could be used for effective and practical IPM is discussed below.

Generally, mixtures are more toxic than individual constituents because they are more difficult to detoxify than a single molecule (Koul, 2008). Other possible advantages of mixtures are that the dose of the synthetic pesticide can be halved and application costs can be reduced. Mixtures are generally more toxic than single components or individual natural active constituents due to synergism and compatibility of products, and inhibition of an insect's ability to employ detoxifying enzymes against a compound complex. Plant products such as cake, compost or decomposing residues of crops releasing allelochemicals can deter plant pests and improve soil fertility (Panthi et al., 2008). Plant products sprayed on trap crops or weeds containing allelochemicals keep the pests away (Kong, 2010). Another advantage is that mixtures can lower pest resistance. For example, non-edible oils extracted from rosemary (*Rosemarinus officinalis* L.) and lemon grass (*Cymbopogon citratus* Spreng.) when applied at 100 ppm, suppressed the resistance in diamond backmoth, *Plutella xylostella* (L.) to monocrotophos, quinalphos, carbosulfan and fenvalerate (Manoharan et al., 2010).

The major objective is to reduce the cost of plant protection by increasing the bioefficacy of plant products, mixing plant products or by alternating them with synthetics in a compatible manner. The following examples illustrate how the combinations are used in pest control. Neem, *Curcuma zedoaria* Roscoe, *Phyllanthus emblica* L., *Allium cepa* L., *Allium sativum* Linn., *Calotropis procera* (Aiton), *Lycopersicon esculentum* L., and *Ocimum canum* Sims have been used in an indigenous formulation that could control 70–80% of tomato fruit borer, *Helicoverpa armigera* (Hb.) and other pests, and doubled the fruit yield (Arora et al., 2012). A month after transplanting eggplants, and soil incorporation of NC @ 250 kg/ha followed by 3–4 sprays of neem seed extract (NSE, 5%), seven field releases of *Trichogramma chilonis* Ishii and one application of synthetic pesticide could control the shoot and fruit borer, *Leucinodes orbonalis* Guen. in eggplant (Sardana et al., 2004).

TABLE 8.1 Examples of Plant Products Used in Combination in the Management of Crop Pests

Combination	Pests Controlled	Crop	References
Prosopis juliflora extract (1%) + endosulfan (1.1%)	*Spodoptera litura*	Chick pea	Murugesan et al., 2004
Nimbecidine® (2 ml/l)/tobacco decoction (13.3 g/l) + Bt (1 kg a.i/ha) + HNPV(200 LE/ha)	*Helicoverpa armigera*	Chick pea	Bhat et al., 2002
AZ (0.1%) + monocrotophos (0.02%)	*Toxoptera citricidus*	Citrus	Chaterjee and Mondal, 2006
Nimbecidine® (3 ml/l) + Biocatch® (5 g/l)	*Aceria guerreronis*	Coconut	Ramarethinam et al., 2000
NSKE (5%)/Neemgold® (3 ml/l)/NO (19 ml/l + cow urine (30 ml/l)	*Antigastra catalaunalis*	Sesamum	Gupta, 2003
NO (2%) + monocrotophos (0.05%)	Lepidopteran pests	Peanut	Sahayaraj and Amalraj, 2005
Oil of *Pongamia pinnata* (0.2%) + acephate (1.5 g/l)	Thrips	Flowering plants	IIHR, 2008
Oil of *Pongamia pinnata* (0.2%) + fenazaquin (0.015%)	Mites	Flowering plants	IIHR, 2008
NO/Oil of *Pongamia pinnata* (0.2%) + imidacloprid (0.5 ml/l)	*Scirtothrips dorsalis*	Chilli	IIHR, 2008
NO/Oil of *Pongamia pinnata* (3%) + dimethoate (0.05%)	*Aleurodicus dispersus*	Mulberry	Sakthivel et al., 2011
NSE (5%)/NO (1%)/AZ (1500 ppm) + HNPV (250 LE/ha)/Bt (1 kg/ha)/spinosad® (0.01%)	Bollworm complex	Cotton	Borkar and Sarode, 2011
Nimbecidine® (0.5%) + endosulfan (0.07%) + carbosulfan (0.05%)/spinosad® (0.05%)/Halt® (0.3%)	*Acrocercops cramerella*	Litchi	Singh et al., 2009b
NSKE (5%) + imidacloprid (4 g a.i./kg seed)/thiamethoxam (1 g a.i./kg seed)	*Chilo partellus*	Corn	Ahad et al., 2012
NSKE (5%) + imidacloprid (0.05%)	*Chilo partellus*	Corn	Amjad et al., 2001
NSKE (5%) + imidacloprid (5 g a.i./kg seed)/thiamethoxam (5 g a.i./kg seed)	*Sesamia inferens*	Corn	Reddy et al., 2004
NO (3%) + endosulfan (0.07%)	*Chilo partellus* + *Atherigona soccata*	Sorghum	Ameta and Kumar, 2003

AZ = azadirachtin; Biocatch = commercial product containing fungus, Hirsutellia thompsonii (1×10^{-7} CFU/g); Bt = Bacillus thuringiensis, commercial products containing 55,000 SU applied at 1 kg a.i./ha or 2 ml/l; HNPV = Helicoverpa nuclear polyhedrosis virus containing 1×10^{-9} polyhedral occlusion bodies/ml = 500 larval equivalent (LE)/ha (0.10%); NO = neem oil; NSKE = neem seed kernel extract in water.

Panickar et al. (2003) reported lower insect populations by up to 50% and higher fruit yield by spraying the okra crop twice with endosulfan (0.05%) followed by two applications of Achook®, a neem-based product (3 ml/l). Lakshmi Narayana and Savitri (2003) managed infestations of citrus (5% butterfly, *Papilio demoleus* (L.)) with a single spray of Bt (0.005%) followed by three sprays of neem products (NSKE 5%, NO 0.5%, Neemazal® 0.005%) that resulted in 60–90% pest mortality in sweet orange.

Prabhakar et al. (2003) combined the installation of 5 bird perches/ha, two sprays of NSKE (5%) and one spray of Bt (0.2%) or granulosis virus at 250 Larval equivalent (LE)/ha against *Achaea janata* (L.) on castor crop. Verma et al. (2010) adopted IPM for fruit fly, *B. cucurbitae*, in round gourd (*Citrullus vulgaris* Schrad. var. *fistulosus* Stocks) that consisted of a first spray of acephate (0.03%) at the fruiting stage followed by a single spray of NSKE (10%) and malathion (0.05%) at 10-day intervals.

An IPM schedule consisting of Multineem® (4 ml/l) sprayed three times, starting from the fruit initiation stage and subsequent sprays at 10-day intervals was significantly effective against *H. armigera* and *Spodoptera litura* (Fb.) attacking tomato crop (Sudharani and Rath, 2011a). This treatment resulted in a lower larval population density (5 larvae/5 plants versus 11–12 larvae in control), fruit damage (17–23% versus 44–53% in control) and higher yield (153 q/ha versus 98 q/ha in control). These workers also suggested alternation of insecticides against *H. armigera* in tomato, e.g. Multineem (4 ml/l) followed first by endosulfan (0.05%) and later by Multineem. Alternatively, three sprays of Multineem were applied at an interval of 10 days starting from fruit development (Sudharani and Rath, 2011b). Patel et al. (2002) sprayed a pigeon pea crop with NSKE (5%) at 50% flowering, *Helicoverpa* NPV (HNPV) @ 250 LE/ha on young larvae, and monocrotophos (0.04%) at 50% pod setting followed by endosulfan (0.07%) during pod development and installed 10 pheromone traps for birds. This package doubled the grain yield over control.

In Bt cotton (cv. Bunny), Prasad and Rao (2009) experimented with NSKE (5%) in an IPM schedule with other control measures (seed treatment and stem application with imidacloprid, intercropping, trap crops, bird perches, pheromone traps). Pest incidence was low, natural enemies (NE) were in abundance and CB ratio was 1:2.26 compared to conventional control (1:1.57). In groundnut IPM, along with other components, incorporation of NC @500 kg/ha before sowing and spraying with NSKE (5%) at seedling stage resulted in 51.62% higher yield when compared to farmers' practice in India (Singh et al. 2009a). In IPM paddy fields, Dhawan et al. (2011) reported a lower pest population (stem borer, brown plant hopper) and a greater number of predators (coccinellids, spiders) than in non-IPM fields. Consequently, the number of sprays was only 2.78 versus 3.74 sprays in non-IPM, and additional income of US $252/ha over the non-IPM fields was obtained. To achieve effective and economic control of two major insect pests of cabbage, Bana and Jat (2012) evaluated nine modules integrating different groups of chemicals, IGR and botanicals and recommended a combination of NSKE (5%)+lufenuron (300 ppm)+endosulfan (0.07%) against aphid, *Lipaphis erysimi* (Kalt.) with a 76.2–79.2% reduction in pest population, and a combination of spinosad (0.05%)+Bt (1 kg a.i./ha)+endosulfan (0.07%) against diamond backmoth with an 85–88% reduction in the larval population.

8.5 CONSTRAINTS AND OPPORTUNITIES IN RESEARCH AND DEVELOPMENT

8.5.1 Phytotoxicity

Toxic effect is generally expressed as leaf injury (burning, yellowing or reddening on the tips and leaf surface, vein clearing or necrosis), necrosis or plant wilting. Often these visual observations are rated on a suitable scale, viz. 1–5 or 1–10. In the case of plant products, phytotoxicity has not been reported on agricultural crops (Nagar et al., 2012).

8.5.2 Persistence on Crops

The residual activity of plant products depends upon the storage period of raw materials or finished products, climatic conditions, and type of preparation or formulation and stabilizers used while spraying. The period after treatment and the stage of the insect against which the applications are directed are also important for residual activity. Generally, crude extracts in water are to be applied as soon as possible after preparation and residual persistence is limited to 5–8 days as allelochemicals dissolve in water and degrade quickly. On the other hand, AZ in NO retains its potency much longer when stored at 40°F in low light, oil can be used even after 3 months of storage, and commercial products have an expiry of 1 year.

Sprays of water extracts can be washed off plants due to heavy rain and susceptibility to hydrolysis, and deposits are sensitive to high temperatures and break down easily due to sunlight (Gahukar, 2012). From the viewpoint of applicability, they have minimal residual activity because of rapid volatility. Consequently, crops cannot be sprayed with them in hot weather and repeated applications are needed to achieve maximum pest mortality and as a result, costs increase. Thus, toxicity values are low when AZ and other constituents degrade due to chemical changes in the presence of high temperatures and organic solvents. Similarly, microorganisms are responsible for degradation of plant products applied to soil. Pyrethrins are labile in UV rays, and rotenone residue disappears due to photo-degradation (Cabras et al., 2002). In the case of essential oils, they have rapid volatility and minimum residual activity.

By adding stickers, stabilizers and antioxidants, the residual activity can be extended. Farmers can use cheap and locally available stickers such as gum *arabic*, white yolk of chicken egg, and UV ray-protectants (soap/detergent powder in water (0.5%) or commercial products (0.05%) such as Sandovit®, Saver® or APSA-80®, and antioxidants such as hydroquinone, resorcinol.)

8.5.3 Toxic Action

Quick pest mortality cannot be achieved with plant products as their action is slow and with a lack of residual toxicity and therefore some plant damage can occur even after treatment. Also, farmers are often reluctant to use plant products as they are accustomed to the 'quick or knock-down effect' of synthetic pesticides. However, in some instances, plant products even with slow action, have been found to be either equally effective or more effective than synthetics.

8.5.4 Awareness

If plant material is not cleaned and stored in a dry and well-ventilated place, it can be attacked by saprophytic bacteria and fungi. For example, moist or damaged neem seeds can become heavily infected by a fungus, *Aspergillus flavus* Link and such seeds become useless for planting. Farmers are not trained in the proper methods of collection and storage of plant material used for crude (home-made) preparations.

Apart from plant protection, neem is used for soap making, traditional medicines and pharmaceuticals. Essential oils are used for flavouring and fragrances in food recipes and beverages. These multiple uses make the plant material competitive and result in a rise in the market price making their use in agriculture costly. Taking advantage of this situation, illegal trading is encouraged or quality norms of raw materials are not respected. Apart from raw materials, competition occurs between plant products and new pesticides including microorganisms. In developing and less-developed countries, farmers have to decide whether to purchase ecofriendly products to safeguard beneficial and non-target organisms but they are not aware of their bioefficacy, expiry period for use, active substances, dosages and spraying instructions for products sold in the local market. Furthermore, new pesticides including synthetics (neonicotinoids, pyrethroids, growth regulators) are being preferred because they are more effective than botanicals. In reality, farmers are not aware of the economic and ecological aspects. Also, awareness of health hazards and particularly the possible effects of technological factors on the proportion of pesticide transferred from the raw material to the end product is needed in future.

8.5.5 Patenting

Patenting of plant products used in crop protection is rarely performed in less-developed and developing countries though it has worldwide importance probably because the legislation is not respected to the expected level or this system does not exist for any traditional preparations (Gahukar, 2003). Each country should have internal mechanisms to protect their useful indigenous plant species. The law must ensure equitable sharing of the profits arising from the utilization of these plants with local communities. For example, a patent has been granted in India for products derived from neem, *Vitex negundo* L., *Zanthoxylum alatum* Roxb. and *Annona squamosa* L. There are instances of incorrect recognition and also biopiracy. For example, the USDA granted a patent for turmeric (*Curcuma domestica* Val.) to the University of Mississippi Medical Center in 1995 which was revoked in 1997; a patent for NO given to W.R. Grace and USDA was challenged and revoked in May 2000 by NGOs and environmentalists. In fact, with globalization and privatization, there is scope for obtaining a patent for geographical indications as well as because the content of allelochemicals varies considerably as per agro-climatic zones. To ease the current procedure, products with a 'green label' should be exempt from lengthy and tedious regulatory procedures.

8.5.6 Standardization

The content of active substances in crude products (water extracts, seed/kernel oil and seed cake) needs to be verified. Other inert ingredients (emulsifier, stabilizer, sticker, UV-protectant) are added to enhance the bioefficacy of the products. Quality maintenance through product standardization is necessary to achieve maximum pest mortality. The process involving isolation/extraction, synthesis and standardization is expensive and needs sophisticated equipment. Therefore, recent less costly techniques should be made readily available for screening. Of course, the procedure is cheaper and quicker for plant essential oils than plant parts.

8.5.7 Effects on Human Health, Beneficial Insects and Other Non-Target Organisms

For each commercial plant product, details on the acute toxicity and chronic effects of exposure, expiry date, warnings and antidotes are given on the container (Trumble, 2002). These data are nowhere indicated for handling of crude preparations but sometimes, precautionary measures are informed by concerned organizations. Despite these measures, accidental poisoning of farm workers and consumers has been reported from remote villages probably because villagers in poor countries often use plant products with other unauthorized toxic materials (local wine, herbs, etc.). Also, suitable protective clothing, face masks and equipment are not available in villages or sometimes these materials are not used properly due to illiteracy. The available protective clothing/accessories should be field tested for appropriateness (Srivastava et al., 2012). Consequently, persons engaged in spraying are exposed to inhalation, ingestion or dermal contact. Poisoning is possible when recommended methods of collection and storage of plant material are not followed. In the case of crude preparations of insect-infested or disease-infected plant parts, it is most likely that farm workers are exposed to toxic substances. Symptoms of acute or chronic toxicity are often not diagnosed properly due to poor health services and similar symptoms caused by several poisons. For example, ingestion of nicotine causes symptoms of poisoning similar to chemicals and rapid dermal absorption is possible (Isman 2002, 2006). Moderate poisoning or low mammalian toxicity due to ingestion of unrefined neem oil or purified terpenoid constituents of plant essential oils is possible (Isman, 2000; Dhongade et al., 2008).

Plant products are environmentally friendly. They have no hazardous effects when mixed in water or, incorporated in soil or when their vapours combine with air. Being biodegradable, residual toxicity or environmental persistence are negligible or minimal. Some are more toxic than others. For example, for the parasitoid *Diaeretiella rapae* (McIntosh), a decoction (2–5%) of *Lantana camara* L., *A. sativum* or *Ipomoea purpurea* (L.) Roth is safe in comparison to a decoction of *N. tabacum*, neem or *Ocimum gratissimum* L. (Barasker et al., 2012). In the laboratory bioassay, Kaushik and Shankarganesh (2009) recorded that survival of *Coccinella septempunctata* Linn. and *Chrysoperla carnea* (Stephens), two major predators of mustard aphid, was greater (94%) in Bollcure 30EC (a eucalyptus leaf extract formulation) than in Neemban® (a neem-based formulation containing 1500 ppm of AZ) (61% survival). Nath and Singh (2003) reported greater toxicity of neem products to predators of coriander aphid, *Hyadaphis coriandri* Das, in comparison with extracts of *Lagerstroemia indica* L. or *P. pinnata*.

It is generally observed that plant products are safe to natural enemies and other non-target organisms (Nagar et al., 2012). For example, when Neemazal® (5%) or Econeem® (0.3% AZ) at 2 ml/l, were sprayed against *Earias vittella* (Fb.) in okra fields, neither affected the adult population of the predators (coccinellids, spiders) (Mohanasundaram and Sharma, 2011). In the laboratory, Nimbecidine® (300 ppm of AZ) was found to be safe to the larvae and pupae of chrysopid *Mallada boninensis* (Okomoto) and did not affect the larval and pupal development period and pupal weight (More et al., 2011). However, with higher doses and improper application techniques, plant products can affect survival, conservation and augmentation of predators and parasitoids and the pathogenicity of pathogens of the crop pests. Coccinellid beetles exposed to sub-lethal effects of *Melia volkensii* Gurke show morphogenetic abnormalities and prolongation of larval period (Reveling and Ely, 2006). The development period of a chrysopid, *C. carnea*, larva is prolonged by NSKE (5%) or Neemark® containing 1500 ppm of AZ applied at 5 ml 2.5 l/ha (More et al., 2005), or water extract (5%) of *Catharanthus roseus* L. or *Eucalyptus tereticornis* Sm. applied at 500 l/ha (Alagar and Sivasubramanian, 2007). In paddy ecosystems in India, water extract (2%) of *Ocimum gratissimum* L. was most harmful to coccinellids and spiders (*Lycosa pseudoannulata* Boesenberg & Strand, *Argiope catenulata* (Doleschall) and *Clubiona japonicola* L. Koch.) followed by NSE (2%) and *Cymbopogon citratus* (DC) Stapf (2%) (Firake et al., 2010). Similarly, water extract (2%) of *Datura stramonium* L. was more harmful than NO (1%) (Rao et al., 2005). A water extract (>5%) of *A. squamosa* seeds or the neem product nimbecidine® (>0.003%) significantly reduced the level of parasitism and adult emergence of a trichogrammatid *T. chilonis* and a braconid *Cotesia flavipes* (Cameron), which are major parasitoids of stem borers of sugarcane and cereals in India (Singh, 2007). In the laboratory, NO (1%) reduced the level of parasitism on *H. armigera* eggs (35–39%, compared with 87–93% in control) (Fand et al., 2009).

Apart from a few examples of toxic effects on beneficial organisms, the plant products are safe to at least generalist predators such as spiders, mites, insects (coccinellid beetles, chrysopids, anthocorid bugs), generalist hymenoptera parasitoids (trichogrammatids, braconids, scelionids, ichneumonids, chalcids, aphelinids, eulophids) (Mishra and Mishra, 2002; Ramanjaneyulu et al., 2004), and generalist disease pathogens (*B. bassiana, M. anisopliae, Vairimorpha* spp.), and growth regulator viruses (nuclear polyhedrosis) (Gahukar, 2008c; Gupta et al., 2002). Therefore, direct or indirect toxic effects on beneficial insects present in habitats other than agricultural crops need intensive studies. In future, it would be necessary to recommend only those plant products which are effective against pests and diseases and also safe to the environment and natural enemies. By manipulating herbivore-induced plant volatile (HIPV) signals, predators and parasitoids can become desensitized and confused or may perceive those cues as repellents in the absence of prey if timed inappropriately. How to employ HIPV effectively with plant products is a subject for further study.

8.5.8 Regulations and Vigilance

From the categories of labelling of plant products, those with green labels are generally safe to the beneficial organisms and human health. However, vigilance and execution of legislation are of utmost importance. In fact, rules and regulations for preventing the hazardous effects of pesticides have been established to safeguard farm workers and consumers of farm produce, and the amendments are regularly published in gazettes by government departments (Trumble, 2002). The current difficulty is that similar tests are imposed for registration of synthetic pesticides and plant products. Therefore, several plant products are not registered but used illegally by farmers at their own risk. Application of such material may result in partial pest control or may not serve any purpose. Amongst developing countries, Brazil leads in registered products based on pyrethrins, rotenone, neem, garlic and nicotine. Pyrethrins are registered in South Africa and Australia; and neem, *P. pinnata* and *Madhuca indica* J.F. Gmel in India, USA and Latin America. Israel started technology for the gradual release of essential oils and natural compounds derived from *Capsicum annuum* L, pyrethrins and methylene di-oxy compounds in Japan (Isman, 2006). Other products such as neem cake powder, NO, and commercial products including neem emulsion, and extract of *A. squamosa*, are sold under several brand names in the open market (Gahukar, 2012).

8.5.9 Treatment Costs

Plant biomass (particularly leaves and fruits) is readily available in many villages. Removal of plant parts does not affect plant growth and regeneration or existence and multiplication of plant species. Farmers can prepare water extract decoctions with little technical knowledge, and with limited financial resources and labour. These preparations, being cheaper, reduce the cost of plant protection considerably. Therefore, commercial production based on local plant material is certainly possible even in villages. For example, plant oils and extracts are produced on a small scale and are distributed to farmers in Latin America and Africa. Furthermore, the neem ecotypes with higher content of AZ have been identified and recommended for large scale plantations. The laboratory production of such clones through tissue culture may require further efforts to make this technique profitable or feasible (Allan et al., 2002). However, a few spices from condiments and medicinal plants are used as botanical pesticides. Large scale exploitation may therefore result in decreased availability.

There is a need to establish a mechanism of quality control for plant products that are best suited for use in organic cultivation and post-harvest protection. Since plant-derived products can be used for industrial production of pesticides, plant species of proved bioefficacy can be integrated into forest ecosystems to conserve biodiversity and adopt sound ecological pest management.

In most field trials, data on crop yields and resulting net profit are lacking. This information is of utmost importance to farmers as they must be convinced of profit using plant products. On a global basis, many indigenous plant species have not been exploited for extracting allelochemicals or testing bioefficacy. With improved analytical methods and infrastructure facilities, these studies should be undertaken by agricultural universities or regional research centres. As per an old/traditional saying, 'prevention is better than cure', pest attack can be avoided by spraying with plants which will kill pests and save natural enemies.

8.5.10 Resistance to Compounds

Generally, pest resistance to plant products has not been reported even though compound mixtures reduce pest resistance better than single compounds (Koul and Walia, 2009). Therefore, pest outbreak/resurgence in crops may not occur.

8.6 CONCLUSIONS

Apart from insects and mites, plant products have been found to be effective against plant disease pathogens, viruses, nematodes, and snails, and are recommended as an alternative to synthetics to prevent further development of pest resistance and pest resurgence. As a preventive measure, pest monitoring may not be needed and estimation of Economic Threshold Level (ETL)/Economic Injury Level (EIL) becomes optional. In such case, feasibility of utilization of plant products is an important criterion for applicability under the farmer's economic circumstances since the yield data, cultivation cost and cost/benefit ratio show whether the IPM schedule can be recommended to farmers.

Plant-derived pesticides are environmentally friendly, are not toxic to non-target organisms, are not persistent in nature and do not promote pest resistance. Thus, knowing the bioefficacy of plant products, concerted efforts are needed by government agencies to promote wide-scale application of plant products. This initiative would help farmers to reduce the cost of crop production, and produce foods without chemical residues. In the future, marketing networks, and education programmes/awareness campaigns for farmers and shopkeepers and consumers, need to be strengthened by establishing close links through extension. On the global front, efforts are needed to collaborate possibly through a close link which would facilitate exchange of information and experimental data to finally formulate a pest management strategy in each agro-ecological zone. Wherever the regulatory mechanism is well established, plant products based on pyrethrum, rotenone, and neem are being used globally for commercial purposes. In developing and less-developed countries, farmers are unable to afford costly synthetic pesticides. Infrastructure facilities for storing, cleaning and timely preparations are lacking at the village level (Morse et al., 2002). Nevertheless, preservation of natural sources, development of quality norms and adoption of standardization would be helpful for large scale adoption.

Plant products are best suited for use in organic farming, which is gaining momentum even in developed countries. The movement of organic farming using plant products initiated by the International Federation of Organic Agricultural Movements (IFOAM) should be strengthened by every means such as infrastructure, financial aid, technical know-how and international collaboration. In Africa, the African Dryland Alliance for Pesticidal Plant Technologies is associated with development and testing of products derived from local plants. In India, the Neem Foundation organizes international seminars.

Acknowledgement

I am grateful to Dr A. S. Sohi, Retired Entomologist (PAU), Ludhiana, India, for his valuable suggestions.

References

Ahad, I., Bhagat, R.M., Ahmad, H., 2012. Efficacy of insecticides and some biopesticides against maize stem borer, *Chilo partellus* Swinhoe. Indian J. Entomol. 74, 99–102.

Alagar, M., Sivasubramanian, P., 2007. Influence of botanicals and pesticides on predatory potential and biology of *Chrysopa carnea* Stephens. Indian J. Entomol. 69, 117–121.

Allan, E.J., Eeswara, J.P., Jarvis, A.P., Mordue Luntz, A.J., Morgan, E.D., Stuchbury, T., 2002. Introduction of hairy root cultures of *Azadirachta indica* A. Jusss. and their production of azadirachtin and other important insect bioactive metabolites. Plant Cell Res. 21, 374–379.

Ameta, O.P., Kumar, A., 2003. Relative efficacy of botanicals against shoot borers and midges infesting sorghum. Indian J. App. Entomol. 17 (1), 12–14.

Amjad, M., Afzal, M., Khalid, M., 2001. New synthetic and bio-insecticides against maize stem borer, *Chilo partellus* (Swinhoe), on golden maize. Online J. Biol. Sci. 1 (1), 38–39.

Arora, S., Kanojia, A.K., Kumar, A., Mogha, N., Sahu, V., 2012. Biopesticide formulation to control tomato lepidopteran pest menace. Curr. Sci. 102, 1051–1057.

Bana, J.K., Jat, B.L., 2012. Bioefficacy and economics of IPM modules against major insect pests of cabbage. Indian J. Entomol. 74, 256–260.

Barasker, D., Pachori, R.K., Panse, R., 2012. Bioefficacy of green pesticides against mustard aphid, *Lipaphis erysimi* (Kalt.) and its parasitoid, *Diaeretiella rapae* (McIntosh). Pestology 36 (1), 39–42.

Bhat, N.J., Patel, R.K., Patel, R.M., 2002. Bioefficacy of various insecticides against *Helicoverpa armigera* on chickpea. Indian J. Entomol. 64, 27–34.

Borkar, S.L., Sarode, S.V., 2011. Evaluation of botanicals and biopesticides in the management of bollworm complex in cotton. Indian J. Entomol. 73, 263–269.

Cabras, P., Caboni, P., Cabras, M., Angioni, A., Russo, M., 2002. Rotenone residues on olives and in olive oil. J. Agric. Food Chem. 50, 2576–2580.

Chaterjee, H., Mondal, P., 2006. Efficacy of some neem-based formulations against citrus brown aphid, *Toxoptera citricidus* Kirkaldy in Darjeeling hills. J. Insect Sci. 19, 103–104.

David, B.V., Shankar, G., 2012. Crop production products in India: The present and future trends. Pestology 36 (4), 8–20.

Dhawan, A.K., Matharu, K.S., Kumar, V., 2011. Impact of integrated pest management strategies on pest complex and economics of *basmati* rice. J. Insect Sci. 24, 219–224.

Dhongade, R.K., Kawade, S.G., Damle, R.S., 2008. Neem oil poisoning. Indian Pediatr. 45, 56–58.

Fand, B.B., Satpute, N.S., Dadmal, S.M., Bag, R.P., Sarode, S.V., 2009. Effect of some newer insecticides and biopesticides on parasitization and survival of *Trichogramma chilonis* Ishii. Indian J. Entomol. 71, 105–109.

Farooq, M., Jabran, K., Chemma, Z.A., Wahid, A., Siddiq, K.H.M., 2011. The role of allelopathy in agricultural pest management. Pest Manag. Sci. 67, 493–506.

Firake, D.M., Pande, R., Karnatak, A.K., 2010. Influence of medicinal plant extracts on predatory fauna in rice ecosystem. Indian J. Entomol. 72, 192–193.

Gahukar, R.T., 2000. Use of neem products/pesticides in cotton pest management. Int. J. Pest Manag. 46, 149–160.

Gahukar, R.T., 2003. Issues relating to the patentability of biotechnological subject matter in Indian agriculture. J. Intell. Property Rights 8 (1), 9–22.

Gahukar, R.T., 2005. Plant-derived products against insect pests and plant diseases of tropical grain legumes. Int. Pest Contr. 47, 315–318.

Gahukar, R.T., 2007a. Indigenous plant-derived products for pest management in cereal crops in India. J. Entomol. Res. 31, 129–136.

Gahukar, R.T., 2007b. Botanicals for use against vegetable pests and diseases: a review. Int. J. Veg. Sci. 13, 41–60.

Gahukar, R.T., 2008a. Pest management in sugarcane using indigenous plant products in India. Indian J. Sugarcane Technol. 23, 51–55.

Gahukar, R.T., 2008b. Management of pests and diseases of oilseed crops in India using indigenous plant products. Outlook Agric. 37, 225–232.

Gahukar, R.T., 2008c. Role of plant products in pest management of tropical fruit crops. J. Insect Sci. 21, 1–10.

Gahukar, R.T., 2010a. Organic production of tea, coffee and cocoa in India: potential of botanicals in pest management. J. Insect Sci. 23, 351–358.

Gahukar, R.T., 2010b. Bioefficacy of indigenous plant products against pests and diseases of Indian forest trees: a review. J. For. Res. 21, 231–238.

Gahukar, R.T., 2011a. Use of neem and plant-based biopesticides in floriculture: current challenges and perspectives – a review. J. Hortic. Sci. Biotechnol. 86, 203–209.

Gahukar, R.T., 2011b. Use of indigenous plant products for management of pests and diseases of spices and condiments: Indian perspective. J. Spices Arom. Crops 20, 1–8.

Gahukar, R.T., 2012. Evaluation of plant-derived products against pests and diseases of medicinal plants – a review. Crop Prot. 42, 202–209.

Gupta, M.P., 2003. Comparative efficacy of neem products and endosulfan against major insect pests of sesame. Indian J. Plant Prot. 31, 96–97.

Gupta, R.B.L., Sharma, S., Yadava, C.P.S., 2002. Compatibility of two entomofungi, *Metarhizium anisopliae* and *Beauveria bassiana* with certain fungicides, insecticides and organic manures. Indian J. Entomol. 64, 48–52.

Gupta, V.K., Ahlawat, S.P., Kumar, R.V., Datta, A., 2010. Effect of season and year on azadirachtin A and oil content in neem (*Azadirachta indica* A. Juss.) seeds and relationship of azadirachtin A and oil content with rainfall, temperature and humidity. Curr. Sci. 99, 953–956.

Highi, G., Hatami, A., 2010. Simultaneous quantification of flavonoids and phenolic acids in plant materials by a newly developed isocratic high performance liquid chromatography approach. J. Agric. Food Chem. 58, 10812–10816.

IIHR, 2008. Annual Report for 2007-08. Indian Institute of Horticultural Research, Bangalore, India, 125 pp.

Isman, M.B., 2000. Plant essential oils for pest and disease management. Crop Prot. 19, 603–608.

Isman, M.B., 2002. Insect antifeedants. Pestic. Outlook 13, 152–157.

Isman, M.B., 2006. Botanical insecticides, deterrents, and repellents in modern agriculture and an increasingly regulated world. Annu. Rev. Entomol. 51, 45–66.

Kaur, V.D., Singh, G., Singh, D., 2005. Effect of feeding *Melia* extracts from different locations to adults of *Plutella xylostella* (Linnaeus) on their fecundity. J. Entomol. Sci. 18, 81–82.

Kaushik, N., Shankarganesh, K., 2009. Effect of bollcure (Eucalyptus leaf extract formulation) on mustard aphid, *Lipaphis erysimi* Kalt. and its predator complex. Indian J. Entomol. 71, 359–360.

Kong, C.H., 2010. Ecological pest management and control by using allelopathic weeds (*Ageratum conyzoides*, *Ambrosia trifida* and *Lantana camara*) and their allelochemicals in China. Weed Biol. Manag. 10, 73–80.

Koul, O., 2008. Phytochemicals and insect control: an antifeedant approach. Crit. Rev. Plant Sci. 27, 1–24.

Koul, O., Walia S., 2009. Comparing Impacts of Plant Extracts and Pure Allelochemicals and Implications for Pest Control. CAB Rev. no. 49, CAB International, London.

Lakshmi Narayana, V., Savitri, P., 2003. Evaluation of biopesticides against citrus butterfly, *Papilio demoleus* L. on sweet orange. Indian J. Plant Prot. 31, 105–106.

Manoharan, T., Muralitharan, V., Preetha, G., 2010. Effect of edible and non-edible oils as synergists in suppressing insecticide resistance in diamondback moth, *Plutella xylostella* (L.). Indian J. Entomol. 72, 11–22.

McFrederick, Q.S., Kathilankal, J.C., Fuentes, J.D., 2008. Air pollution modifies floral scent trails. Atmos. Environ. 42, 2336–2338.

Mishra, N.C., Mishra, S.N., 2002. Impact of biopesticides on insect pests and defenders of okra. Indian J. Plant Prot. 30, 99–101.

Mohanasundaram, A., Sharma, R.K., 2011. Effect of newer pesticide schedules on the population of *Earias vittella* (Fabricius) and its predators on okra. J. Insect Sci. 24, 280–290.

More, S.A., Kabre, G.B., Saindane, Y.S., 2005. Effect of various insecticides on larval period of *Chrysopa carnea* (Stephens). J. Insect Sci. 18, 111–113.

More, S.A., Patil, P.D., Shinde, B.D., Lad, S.K., Gharge, C.P., 2011. Effect of insecticides on larval and pupal stages of potential predator, *Mallada boninensis* (Okomoto) under laboratory conditions. Pestology 35 (3), 29–31.

Morse, S., Ward, A., McNamara, N., Denholm, I., 2002. Exploring the factors that influence the uptake of botanical insecticides by farmers: a case study of tobacco-based products in Nigeria. Exp. Agric. 38, 469–479.

Murugan, K., Sivaramakrishnan, S., Senthil Kumar, N., Jayabalan, d., Senthil Nathan, S., 1998. Synergistic interactions of botanicals and biocide nuclear polyhedrosis virus on pest control. J. Sci. Ind. Res. 57, 732–739.

Murugesan, S.S., Baskaran, S., Mahadevan, N.R., 2004. Individual and combined effect of endosulfan with leaf extract of *Prosopis julifera* L. on biochemical contents and fecundity of *Spodoptera lituta* (Fabricius). Pestology 28 (12), 23–27.

Nagar, A., Singh, S.P., Singh, Y.P., Singh, R., Meena, H., Nagar, R., 2012. Bioefficacy of vegetable and organic oils, cakes and plant extracts against mustard aphid, *Lipaphis erysimi* (Kalt.). Indian J. Entomol. 74, 114–119.

Nath, L., Singh, A.K., 2003. Safety of some plant extracts and a neem formulation to predators of the coriander aphid, *Hyadaphis coriandri* Das, under field conditions. Pest Manag. Econ. Zool. 11, 159–162.

Odeyemi, O.O., Masika, P., Afolayan, A.J., 2008. A review of the use of phytochemicals for insect pest control. Afr. Plant Prot. 14, 1–7.

Oerke, E.C., 2006. Crop losses to pests. J. Agric. Sci. 144, 31–43.

Panickar, B.T., Bharpoda, T.M., Patel, J.R., Patel, J.J., 2003. Evaluation of various schedules based on botanical and synthetic insecticides in okra ecology. Indian J. Entomol. 65, 344–346.

Panthi, B.B., Devkota, B., Devkota, J.U., 2008. Effect of botanical pesticides on soil fertility of coffee orchards. J. Agric. Environ. 9, 16–22.

Patel, M.G., Bharpoda, T.M., Patel, J.J., Chavda, A.J., Patel, J.R., 2002. Evaluation of various modules for IPM in pigeon pea. Indian J. Entomol. 64, 39–43.

Pavela, R., Sajfrtova, M., Sovova, H., Barnet, M., Karban, J., 2012. The insecticidal activity of *Tanacetum parthenium* (L.) Schultz Bip. extracts obtained by supercritical fluid extraction and hydrodistillation. In: Pascual-Villalobos, M.J. (Ed.), Essential Oils as Natural Insecticides. Elsevier, Amsterdam, Netherlands, pp. 449–454.

Prabhakar, M., Srinivasa Rao, M., Prasad, Y.G., 2003. Evaluation of bio-intensive integrated pest management modules against castor semilooper, *Achaea janata* Linn. Indian J. Plant Prot. 31, 56–58.

Prasad, N.V.V.S.D., Rao, N.H.P., 2009. Evaluation of Bt cotton under integrated pest management. Indian J. Entomol. 71, 279–283.

Ramanjaneyulu, K.V., Arjun Rao, P., Vijayalakshmi, K., 2004. Relative toxicity of lufenuron and certain insecticides against the natural enemies of *Spodoptera litura* (Fab.) infesting groundnut. Pestology 28 (5), 19–22.

Ramarethinam, S., Marimuthu, S., Murugesan, N.V., 2000. Studies on the effect of *Hirsutella thompsonii* K. and a neem derivative in the control of coconut eriophyid mite, *Aceria (Eriophytes guerreronis* (K.). Pestology 24 (2), 3–8.

Rao, N.B.V., Singh, V.S., Chander, S., 2005. Studies on compatibility between predator and insecticides in management of rice leaf folder, *Cnaphalocrocis medinalis* (Guenee). Indian J. Plant Prot. 33, 72–74.

Reddy, M.L., Babu, T.R., Reddy, D.D.R., Sreeramalu, M., 2004. Bioefficacy of insecticides against pink borer, *Sesamia inferens* (Walker) in maize. Indian J. Entomol. 66, 209–211.

Reveling, G., Ely, S.O., 2006. Side effects of botanical insecticides derived from *Meliaceae* on coccinellid predators of the date palm scale. Crop Prot. 25, 1253–1258.

Sahayaraj, K., Amalraj, A., 2005. Impact of monocrotophos and neem oil mixture on defoliator management in groundnut. J. Food Agric. Environ. 3, 313–315.

Sahayaraj, K., Antony, N., 2006. Impact of five plant extracts on the digestive and detoxification enzymes of *Spodoptera litura* (Fab.) (Lepidoptera: Noctuidae). Hexapoda 13, 53–57.

Sahayaraj, K., Shoba, J., 2012. Toxic effects of *Tephrosia purpurea* (Linn.) and *Acalypha indica* (Linn.) aqueous extracts impact on the mortality, macromolecules, intestinal electrolytes and detoxification enzymes of *Dysdercus cingulatus* (Fab.). Asian J. Biochem. 7, 112–122.

Sakthivel, N., Punithavathy, G., Qadri, S.M.H., 2011. Evaluation of different insecticides and botanicals against spiraling whitefly infesting mulberry. Indian J. Seric. 50, 98–102.

Sardana, H.R., Arora, S.S., Singh, D.K., Kadu, L.N., 2004. Development and validation of adoptable IPM in eggplant, *Solanum melongena* L. in a farmer's participation approach. Indian J. Plant Prot. 32, 123–128.

Satpathy, S., Rai, S., 2002. Luring ability of indigenous food baits for fruit fly, *Bactrocera cucurbitae* (Coq.). J. Entomol. Res. 26, 249–252.

Saxena, R.C., 1998. Botanical pest control. In: Dhaliwal, G.S., Heinrichs, E.A. (Eds.), Critical Reviews in Insect Pest Management. Commonwealth Publishers, New Delhi, India, pp. 155–179.

Sidhu, O.P., Vishal, K., Behl, H.M., 2004. Variability in triterpenoids (nimbin and salanin) composition of neem among different provenances of India. Ind. Crops Prod. 19, 69–75.

Singh, M.R., 2007. Effect of biopesticides on *Trichogramma chilonis* Ishii and *Cotesia flavipes* Cameron: parasitoids of borer pests of sugarcane. Indian J. Entomol. 69, 218–220.

Singh, S., Gaur, R.B., Singh, S.K., Ahuja, D.B., 2009a. Development and evaluation of farmers-participatory integrated pest management technology in groundnut. Indian J. Entomol. 71, 160–164.

Singh, S.S., Yadav, S.K., Rai, M.K., Singh, V.B., 2009b. Efficacy of some insecticides against fruit borer, *Acrocercops cramerella* Snellen in litchi. Indian J. Entomol. 71, 152–154.

Srivastava, M., Udawat, P., Kumar, A., 2012. Design development and field testing of protective clothing/accessories to combat health hazards. Pestology 36 (10), 46–53.

Sudharani, D., Rath, L.K., 2011a. Biopesticides against fruit borer complex in tomato. Indian J. Plant Prot. 39, 316–317.

Sudharani, D., Rath, L.K., 2011b. Studies on efficacy of biopesticides in controlling the American bollworm (*Helicoverpa armigera*) infesting tomato. Pestology 36 (6), 25–27.

Trumble, J.T., 2002. Caveat emptor: safety considerations for natural products used in arthropod control. Am. Entomol. 48, 7–13.

Verma, H., Singh, S., Jat, B.L., Ahuja, D.B., 2010. Evaluation of ecofriendly IPM modules against fruit fly, *Bactrocera cucurbitae* (Coquillett) on round gourd. Indian J. Entomol. 72, 185–187.

Verma, S.C., Nigam, S., Jain, S.K., Pant, P., Padhi, M.M., 2011. Microwave assisted extraction of gallic acid in leaves of *Eucalyptus x hybrida* Maiden and its quantitative determination by HPTLC. Der Chem. Sinica 2, 268–277.

Vir, S., 2007. Neem genetic diversity in India and its use as biopesticide and biofertilizer. Indian J. Plant Prot. 35, 185–193.

CHAPTER

9

Use of Pheromones in Insect Pest Management, with Special Attention to Weevil Pheromones

Sunil Tewari[1,3], *Tracy C. Leskey*[2], *Anne L. Nielsen*[3], *Jaime C. Piñero*[4] *and Cesar R. Rodriguez-Saona*[1,3]

[1]Rutgers PE Marucci Center for Blueberry & Cranberry Research & Extension, NJ, USA

[2]USDA-ARS Appalachian Fruit Research Station, WV, USA

[3]Department of Entomology, Rutgers University, NJ, USA

[4]Cooperative Research and Extension, Lincoln University, MO, USA

9.1 INTRODUCTION

In the natural world, chemical cues produced by either plants or animals can elicit behavioural or physiological responses in other organisms. Termed semiochemicals, these compounds can mediate interactions either between individuals of the same species (pheromones), or across varied biological entities (allelochemicals). Allelochemicals can be further classified based on whether they favour the receiver (kairomones), or the sender (allomones). In the last few decades, the pest management paradigm has undergone an intentional shift from calendar-based, broad-spectrum insecticide applications to using more holistic, integrated, and high-efficacy approaches. Food safety, environmental conservation, higher input costs and resistance management are some of the key factors guiding current pest management policies and practices in commercial agriculture (Witzgall et al., 2010). As a result, the integration of conventional, behaviourally based, biological, and cultural pest management strategies into integrated pest management (IPM) systems has become increasingly important in the twenty-first century.

Meeting the demands of the rapidly expanding global population while incorporating sustainability and ecological stewardship are the major challenges facing modern agriculture (Kogan and Jepson, 2007). Thus, the impetus to develop and integrate alternative pest management strategies has gained considerable interest and significant effort has been directed towards the conceptualization and

D. P. Abrol (Ed): Integrated Pest Management.
DOI: http://dx.doi.org/10.1016/B978-0-12-398529-3.00010-5

validation of novel tools and approaches. A better understanding of the chemical ecology of many insects has led to the incorporation of behaviour-modifying compounds into existing IPM programmes (Pickett et al., 1997). The application of pheromones and/or allelochemicals as behavioural manipulation tools can supplant or complement existing management programmes (Witzgall et al., 2008), leading to a reduction in the use of broad-spectrum insecticides. This tactic affords a high degree of specificity and ensures that non-target and beneficial insects in and around the operation area are not adversely affected. Furthermore, their relatively lower costs and reduced toxicity may make semiochemical-based approaches ideal for economically depressed regions (Cork et al., 2005) and for high value crops with export potential (Yongmo et al., 2005). Semiochemicals, including insect pheromones, now form the cornerstone of pest management strategies against a broad category of insects worldwide.

The topics covered in the chapter include an overview of insect pheromones and their use in pest management programmes. Within the context of integrated pest management (IPM), pheromone-based approaches can be used to either monitor the target pest, or play a more direct role in population suppression. Mating disruption, mass trapping, and attract-and-kill are some of the most common direct pest control tactics that depend on the use of semiochemicals, some of which will be discussed in the first part of the chapter. The second part of the chapter discusses the use of aggregation pheromones in management of economically important weevil pests (Coleoptera: Curculionidae). Lures and traps baited with pheromones have been particularly successful in the monitoring and management of weevils in various agroecosystems. For many insects, the attractiveness of pheromone-based lures has been further enhanced by combining them with blends of volatiles derived from the host plants, a phenomenon termed synergism. Finally, case studies are presented in which pheromones have been used as part of management strategies against weevil pests. These include the boll weevil, plum curculio, cranberry weevil, and pepper weevil.

9.2 INSECT PHEROMONES

Insect pheromones are volatile organic molecules of low molecular weight that elicit a behavioural response from individuals of the same species and can be used to communicate between members of the same or the opposite sex (Phillips, 1997). Pheromones are generally produced by specialized exocrine glands associated with the cuticle (Ayasse et al., 2001; Billen and Morgan, 1998; Law and Regnier, 1971). Hall et al. (2002a,b) suggested that the aggregation pheromone components of two bark beetle species, *Dendroctonus jeffreyi* Hopkins and *Ips pini* Say (Coleoptera: Scolytidae), were produced in the midgut tissue. Similarly, for the boll weevil, *Anthonomus grandis grandis* Boheman (Coleoptera: Curculionidae), it was suggested that biosynthesis of aggregation pheromone occurred in the gut tissue (Taban et al., 2006). The production and release of insect pheromones is governed by a variety of environmental factors and physiological mechanisms. Furthermore, the amount of pheromones that insects release is extremely low and varies from a few nanograms to micrograms per unit of time, depending on the species (Piñero and Ruiz-Montiel, 2012). For example, it was reported that the release rate of the main pheromonal component in the agave weevil, *Scyphophorus acupunctatus* Gyllenhaal (Coleoptera: Curculionidae), was between 0.2 and 2.1 ng/24 h (Ruiz-Montiel et al., 2009). Factors that may have an impact on the release of pheromones include circadian rhythm, temperature, presence of food sources, and age of the insects.

Chemoreceptor cells present in the exoskeleton of insects, also called sensilla, mediate the perception of pheromones present in the environment (Schneider, 1969). For example, sensilla found in insect antennae and palps contain pheromone receptive olfactory receptor neurons (ORNs) (Dickens 1990; Keil, 1999). The ability to detect different pheromones or physiological specificity is determined by receptor proteins present in the ORNs (Elmore et al., 2003; Keller and Vosshall, 2003). The effectiveness of pheromones in insect communication is affected by multiple factors including chemical nature, volatility, solubility, and persistence in the environment (Heuskin et al., 2011). Pheromones modulate critical activities such as mate and host location in insects and are primarily classified on the basis of their effects.

9.2.1 Types of Insect Pheromones

Pheromones are subdivided into several types based on the nature of the interactions between emitters and receivers. Furthermore, releaser pheromones (e.g. alarm pheromone) bring about immediate changes in the behaviour of receivers whereas primer pheromones (e.g. 9-keto-2-decenoic acid or queen honeybee substance) cause relatively slow and longer-term physiological changes (Ginzel, 2010; Law and Regnier, 1971).

9.2.1.1 Sex Pheromones

Sex pheromones act as a signal to attract potential mates over long distances (e.g. moths). Sensitive chemoreceptive sensilla in insects facilitate the detection of very low concentrations of sex pheromones in the environment (Regnier and Law, 1968). Release of sex pheromones may be governed by factors such as time of day, weather, and the availability of host plants (Law and Regnier, 1971). Furthermore, both the immature and adult stages of insects can sequester chemicals from host plants and use them as precursors for sex pheromones (Landolt and Phillips, 1997).

9.2.1.2 Alarm Pheromones

Some insects (e.g. aphids) release alarm pheromones in response to attack by natural enemies. Alarm pheromones serve as a trigger for dispersal and avoidance behaviour among the conspecifics. However, some social insects may respond aggressively to alarm pheromones (e.g. bees in genus *Apis* and leaf-cutting ants) (Ginzel, 2010).

9.2.1.3 Aggregation Pheromones

These can be defined as intraspecific signals that facilitate group formation and mating at a food source (Tinzaara et al., 2002). For example, aggregation pheromones released by some species of bark beetles (Scolytidae: Coleoptera) result in the recruitment of other individuals of either sex to the feeding site (Blomquist et al., 2010).

9.2.1.4 Anti-Aggregation Pheromones

These compounds result in the dispersal of individuals (both sexes) and help maintain optimum spacing in a resource-limited environment.

9.2.1.5 Oviposition-Deterring or Epideictic Pheromones

These compounds help females of certain insect species avoid egg deposition on hosts that have already been utilized by conspecifics and thus reduce intraspecific competition (Stelinski et al., 2007). Females of numerous species of fruit flies (Diptera: Tephritidae), e.g. Mediterranean fruit fly, *Ceratitis capitata* Wiedemann, deposit an oviposition-deterring fruit-marking pheromone during ovipositor dragging after egg-laying (Prokopy et al., 1978). Similarly, pepper weevil, *Anthonomus eugenii* Cano (Coleoptera: Curculionidae) females deposit an oviposition plug that deters egg-laying (Addesso et al., 2007).

9.2.1.6 Trail Pheromones

Social insects (e.g. ants and termites) use trail pheromones to mark feeding or nest sites to guide members of their colony.

9.2.2 Use of Pheromones in IPM

The characterization of the silk moth sex pheromone (*E*, *Z*)-10,12-hexadecadien-1-ol (Butenandt et al., 1959), which is released by the female silkworm moth, *Bombyx mori* (Lepidoptera: Bombycidae) to attract mates, opened doors to the establishment of chemical ecology as a scientific discipline (Heuskin et al., 2011). Focused on elucidating the origins, functions, and significance of semiochemicals mediating interactions between organisms, chemical ecology has played a vital role in the development of sustainable pest management strategies over the past 50 years (Pickett et al., 1997; Witzgall et al., 2010). Post World War II agriculture was facilitated by a dramatic increase in the use of synthetic broad-spectrum insecticides. However, frequent use of the chemicals resulted in toxic food residues, environmental degradation, and development of resistance (Kirsch, 1988). The need to develop an integrated approach to pest management, combined with advances in analytical chemistry and insect behaviour, sparked the initial interest in exploring the use of insect pheromones as biorational pesticides (Gut et al., 2004).

Currently, pheromones and other semiochemicals are being used to monitor and control pests in millions of hectares of land (Witzgall et al., 2010). Highly relevant for IPM programmes of tephritid fruit flies is the use of parapheromones. These are chemical compounds of anthropogenic origin, not known to exist in nature but are structurally related to natural pheromone components, that in some way affect physiologically or behaviourally the insect pheromone communication system, eliciting a similar response to that of a true pheromone (Renou and Guerrero, 2000). For example, males of many *Bactrocera* and *Dacus* species are strongly attracted to specific chemical compounds, which either occur naturally in plants (e.g. methyl eugenol) or are synthetic analogues of plant-borne substances (e.g. cue lure) (Cunningham, 1989; Fletcher, 1987). Parapheromones are very powerful lures used in current programmes aimed at detecting, monitoring, and controlling (through the Male Annihilation Technique) invasive tephritid pests (Vargas et al., 2008).

9.2.2.1 Monitoring

Monitoring is an effective way to determine the population trends of insects and plays a critical role in pest management programmes (Binns and Nyrop, 1992; Cohnstaedt et al., 2012). Pheromone-based behavioural manipulation has been developed as a monitoring tool for many pests. It involves the use of a synthetically-derived pheromone formulated into a dispenser and trap to selectively attract and intercept the target insect. However, the technique can be modified to meet the requirements of specific agroecosystems. For example, a 'trap tree' approach has been developed to monitor plum curculio, *Conotrachelus nenuphar* Herbst (Coleoptera: Curculionidae) in apple orchards (Prokopy et al., 2003, 2004). It involves baiting the branches of a perimeter-row apple tree with grandisoic acid (aggregation pheromone) and benzaldehyde (synthetic fruit volatile) to attract plum curculio adults, monitoring then focuses on sampling just a few fruit from that perimeter-row odor-baited trap tree.

The advantages of using pheromones for monitoring pests include lower costs, specificity, ease of use, and high sensitivity (Laurent and Frérot, 2007; Wall, 1990). The data collected from monitoring traps can be utilized in an IPM programme that includes components such as species detection, early warning, timing of control treatments, population trends, and dispersion of target pests (Nealis et al., 2010; Wall, 1990). Some common types of pheromone dispensers currently in use include hollow fibres, plastic laminates, impregnated ropes, twist ties, wax formulations, polyethylene vials, sol-gel polymers, and rubber septa (Vacas et al., 2009; Zada et al., 2009). Biodegradability, low cost, and maintenance of appropriate pheromone release rates during the flight period of a pest are some of the key attributes of an ideal dispenser (Vacas et al., 2009). Similarly, the design, colour, and placement (including vertical height) may influence the monitoring efficiency of pheromone-baited traps (Athanassiou et al., 2004; Bergh et al., 2006; Diaz-Gomez et al., 2012; Hoddle et al., 2011; Isaacs and Van Timmeren, 2009; Knight, 2007; Knight and Fisher, 2006; Leskey et al., 2012; Roubos and Liburd, 2008; Sanders, 1986a; Strong et al., 2008). For example, in monitoring adults of jasmine moth, *Palpita unionalis* Hubner (Lepidoptera: Pyralidae), funnel traps were more efficient compared to adhesive traps and a higher number of males were caught along the edge than in the interior of the groves (Athanassiou et al., 2004). It was easier to identify, count, and remove the moths from funnel traps compared to the adhesive traps. Furthermore, funnel traps were less contaminated by plant debris. The same study also found that white funnel traps were ≈2.5 times more attractive to the male moths than brown funnel traps. Sticky traps become saturated at relatively low population density and are thus not suitable to monitor pests whose density fluctuates significantly (Houseweart et al., 1981; Sanders, 1986a,b). In apple trees, monitoring traps placed at 4m captured more male codling moths, *Cydia pomonella* L. (Lepidoptera: Tortricidae), than those placed at 2m (Epstein et al., 2011). A higher number of male codling moths were caught at traps placed along the border of orchards compared to traps 30 or 50m inside (Knight, 2007). A higher number of male obliquebanded leafroller, *Choristoneura rosaceana* Harris (Lepidoptera: Tortricidae), were caught in traps placed on top of apple trees compared to the middle and lower position traps (Agnello et al., 1996). The bucket trap captured more pickleworm (*Diaphania nitidalis* Stoll (Lepidoptera: Pyralidae)) males than three other trap types and also more males were caught at heights of 80 and 150cm compared to 30 and 180cm (Valles et al., 1991). Furthermore, factors such as wind direction and topography may also influence catches at pheromone-baited traps (Knight, 2007). The protocols for monitoring pests will evolve as more data become available on the different variables affecting catch and modifications can be made to improve current practices (Knight, 2007). Among others, female-produced sex pheromones have been used to monitor gypsy moth, *Lymantria dispar* L. (Lepidoptera: Lymantriidae) (Kolodny-Hirsch and Schwalbe 1990), *Heliothis* spp. (Lepidoptera: Noctuidae) populations (Lopez et al., 1990), and codling moth (Knight et al., 2005). Similarly, aggregation pheromones have been used to monitor the boll weevil and plum curculio (Kroschel and Zegarra, 2010; Leskey and Wright, 2004; Piñero et al., 2011; Ridgway et al., 1990).

Monitoring pests using pheromone lures can benefit management decisions such as insecticide application timing (Leskey et al., 2012; Peng et al., 2012). For example, it was demonstrated that the diamondback moth, *Plutella xylostella* L. (Lepidoptera: Yponomeutidae), was more effectively controlled when an insecticide was applied based on pheromone trap catches compared to a calendar application approach (Reddy and Guerrero, 2001). Monitoring traps can also be used to develop quantitative relationships between adult capture and injury-causing life stages of a pest (Allen et al., 1986; Bacca et al., 2012; Evenden et al., 1995; Jones et al., 2009; Knight and Light, 2005; Kolodny-Hirsch and Schwalbe, 1990; McBrien et al., 1994; Reddy and Guerrero, 2001; Reddy et al., 2012). For example, a positive relationship was found between the trap capture of adult mullein bug, *Campylomma verbasci* Meyer (Hemiptera: Miridae), and the density of first-generation nymphs in the following year in conventional apple orchards (McBrien et al., 1994). Similarly, correlations predicting future population outbreaks or damage risk on the basis of adult capture have also been reported (Sanders, 1988; Weslien et al., 1989). A strong correlation was found between annual catches of male spruce budworm, *Choristoneura fumiferana* Clemens (Lepidoptera: Tortricidae), and larval population densities in the following year (Sanders, 1988). Based on data from the monitoring traps, warning of extensive defoliation by spruce budworm can be given 6 years in advance (Sanders, 1988). Similarly, a strong linear correlation was reported between adult capture of spruce bark beetle, *Ips typographus* L., and log-transformed tree mortality (Weslien et al., 1989). However, the utility of pheromone-based monitoring may be limited in some cases due to the lack of precise quantitative relationship between trap catches and the presence of damaging life stages of the pest (Campbell et al., 1992; Latheef et al., 1991; Shepherd et al., 1985). For instance, the number of males caught at the pheromone trap may not reflect the oviposition activity of females (Latheef et al., 1991).

Attempts have also been made to combine the pheromones of two or more overlapping pests of a common host to monitor the different species simultaneously (Allison et al., 2012; Jones and Evenden, 2008; Jones et al., 2009; Miller et al., 2005; Waterworth et al., 2011; Wong et al., 2012). In western Canada, a combined blend of *Malacosoma disstria* Hubner (Lepidoptera: Lasoicampidae) and *Choristoneura conflictana* Walker (Lepidoptera: Tortricidae) pheromones was used to monitor the populations of the two defoliators simultaneously in trembling aspen *Populus tremuloides* Michenaux (Jones et al., 2009). One of the advantages of monitoring several species simultaneously is that the method is cost-effective (Miller et al., 2005). However, there can also be considerable disadvantages as significantly more insects can be captured, leading to frequent need for servicing traps. Alternatively, the pheromone component of one species can serve as a behavioural antagonist of another species. Indeed, multiple-component pheromone blends, produced by conspecific females, and behavioural antagonists, produced by heterospecific females, contribute to reproductive isolation by creating a very specific communication channel (Linn and Roelofs, 1995). Such is the case for dogwood borer, *Synanthedon scitula* Harris (Lepidoptera: Sesiidae); the main component of the sex pheromone of the lesser peach tree borer, *S. pictipes* Grote & Robinson, is a powerful behavioural antagonist for the dogwood borer (Zhang et al., 2005).

The performance of pheromone-based monitoring tools may be affected by variations in biotic and abiotic factors. Variables also reported to affect catches of corn earworm moths, *Helicoverpa zea* Boddie (Lepidoptera: Noctuidae), at pheromone traps included temperature variation, moth behaviour, moth age, pheromone release rate, moonlight, crop phenology, and crop cover (Hartstack and Witz, 1981; Hartstack et al., 1979; Latheef et al., 1991). Traps constructed with a sticky material and deployed for extended durations are particularly vulnerable to deterioration as the surface may become saturated with insects and debris (Sanders, 1986b). Furthermore, variables such as pheromone release rate may fluctuate when a standard dispenser is used to monitor a pest across a wide geographical area with varying climatic conditions and thus the performance may be adversely affected (Tobin et al., 2011).

Pheromones can be combined with attractive plant-derived kairomones to increase the efficiency of monitoring traps (Knight et al., 2005). For example, traps baited with a 3.0/3.0 mg pheromone/kairomone blend caught significantly more codling moth males and total moths (including females) than traps baited with either compound alone in apple orchards (Knight et al., 2005). Furthermore, traps with pheromone/kairomone blends are attractive to both males and females and may improve the predictive correlations between adult capture and variables such as egg density and timing of egg hatch (Knight et al., 2005).

Identification and optimization of pheromones for monitoring pests is an active field of research and new formulations are being introduced globally (Heath et al., 2006; Leskey et al., 2012; Peng et al., 2012; Zhu et al., 2006). Other developments include attempts to integrate information technology with monitoring by providing real time data on trap catches via the Internet (Kim et al., 2011).

9.2.2.2 Mating Disruption

During mate location, sex pheromones are commonly used as long-range cues to orient insect species toward potential mates. Synthetic blends of sex pheromones can be used to permeate the environment and disrupt the orientation of males to females, thereby inhibiting the mating process (Byers, 2007; Witzgall et al., 2008). Through mating disruption, male search behaviour is diverted due to competition between females and synthetic pheromone sources, sensory adaptation and habituation of the males, and camouflage of the female plume (Baker et al., 1988; Byers, 2007; Byers, 2011; Cardé, 1990; Cardé and Minks, 1995; Daly and Figueredo, 2000; Stelinski et al., 2013; Teixeira et al., 2010). Prolonged exposure to a high concentration of a synthetic pheromone blend may render males insensitive to pheromone plumes produced by females (Witzgall et al., 2008). Mating disruption in moths works mainly through competitive attraction compared to the non-competitive mechanisms (camouflage, desensitization, and sensory imbalance) (Miller et al. 2006a,b). Specifically, it involves the deployment of multiple synthetic pheromone point sources that act to divert, arrest, and possibly deactivate males seeking mating partners (Miller et al., 2010). This behavioural manipulation in turn reduces the frequency with which males encounter calling females because of preoccupation with more proximate synthetic pheromone dispensers (Miller et al., 2010).

In contrast to traditional insecticide-dependent pest management programmes, mating disruption does not affect non-target organisms and is environmentally benign and approved for organic and biorational production systems. Furthermore, it does not affect the efficacy of biological control agents making it well suited for sustainable IPM programmes (Kirsch, 1988; Vacas et al., 2012). Mating disruption has been implemented against multiple lepidopteran pests of field crops, orchards, forests, and vineyards. These include the pink bollworm (*Pectinophora gossypiella* Saunders) (Lepidoptera: Gelechiidae), oriental fruit moth (*Grapholita molesta* Busck) (Lepidoptera: Tortricidae), obliquebanded leafroller (*Choristoneura rosaceana* Harris) (Lepidoptera: Tortricidae), grape berry moth (*Endopiza viteana* Clemens) (Lepidoptera: Tortricidae), European vine moth (*Lobesia botrana* Den. & Schiff.) (Lepidoptera: Tortricidae), honey dew moth (*Cryptoblabes gnidiella* Mill.) (Lepidoptera: Pyralidae), rice striped stem borer (*Chilo suppressalis* Walker) (Lepidoptera: Pyralidae), gypsy moth, and codling moth (Atanassov et al., 2002; Cardé and Minks, 1995; Harari et al., 2007; Knight et al., 1998; Kolodny-Hirsch et al., 1990; Sharov et al., 2002; Thorpe et al., 2007; Tollerup et al., 2012; Trimble, 1993). Non-lepidopteran pests targeted by mating disruption include oriental beetle, *Anomala orientalis* Waterhouse, and *Prionus californicus* Motschulsky (Maki et al., 2011; Wenninger and Averill, 2006).

Factors such as pheromone application rate (trap density), optimum dispenser design, and dispenser height are important for the proper and cost-effective implementation of mating disruption technique (Alfaro et al., 2009; Epstein et al., 2006, 2011). For example, mating disruption was more effective against codling moth when pheromone dispenser density was high (Epstein et al., 2006). The lowest proportion of mated female codling moth on apple trees was recorded when pheromone dispensers were placed simultaneously at 2 and 4m height, compared to both dispensers at the same height (2 or 4m) (Epstein et al., 2011). Reduction in trap density from standard 51 dispensers/ha to 31, 25, or 16 dispensers/ha did not affect the performance of mating disruption for rice striped stem borer, allowing the cost of deploying the traps to be reduced from €66/ha to €43, €36, and €24/ha, respectively (Alfaro et al., 2009). The determination of appropriate trap density for satisfactory mating disruption of different pests has historically depended on a trial and error approach (Byers, 2011). However, there is potential to improve the process by employing simulation models (Byers, 2011).

Efficiency of mating disruption may be improved either through supplemental insecticidal applications or under area-wide management. Mating disruption can also be combined with conventional pest management programmes or an IPM programme to reduce insecticide applications, particularly when the pest density is high (Atanassov et al., 2002; Knight et al., 1998; Vickers et al., 1998). It was observed that unintentional drift of insecticide into a high pest density mating disruption plot throughout the growing season provided excellent control of codling moth (Vickers et al., 1998). Immigration of mated females and ballooning larvae to pheromone-treated plots may undermine the efficacy of mating disruption as a pest management strategy (Agnello et al., 1996; Cardé and Minks, 1995; Knight et al., 1998; Vickers et al., 1998; Wenninger and Averill, 2006). For example, immigration of mated females has been suggested as a potential problem in mating disruption targeted against obliquebanded leafroller, codling moth, and oriental beetle (Agnello et al., 1996; Knight et al., 1998; Vickers et al., 1998; Wenninger and Averill, 2006). However, adoption of this technique on an area-wide basis may alleviate this issue (Cardé and Minks, 1995; McGhee et al., 2011). For example, it was demonstrated that area-wide implementation of mating disruption for codling moth in apple orchards reduced both fruit injury and insecticide use, and resulted in average savings of $55–65/ha (McGhee et al., 2011). Mating disruption may not provide a satisfactory level of control when the pest population density is high (Cardé and Minks, 1995; Gut and Brunner, 1998; Trimble, 1995; Vickers et al., 1998; Webb et al., 1990). A relatively high number of insects per unit area increases the probability of random encounters between males and females (Cardé and Minks, 1995). Since high moth density and immigration from untreated areas can have an adverse impact on the success of mating disruption, precise monitoring of the adult population in orchards treated with pheromones is critical (Cardé and Minks, 1995).

Furthermore, cost effectiveness, uniformity of application, and ease of use are some of the factors that can be considered when several mating disruption dispensers and formulations are available for a single pest (Trimble, 2007). For example, a sprayable formulation of a pheromone is preferable to a hand-applied dispenser in some systems (Agnello et al., 1996; Trimble, 2007). Recent developments in the management of codling moth using mating disruption include attempts to increase its efficiency by co-releasing sex pheromone (codlemone) with attractive plant-derived kairomones (e.g. pear ester) (Ansebo et al., 2005; Knight and Light, 2012; Knight et al., 2012; Stelinski et al., 2013). However, the results from several field studies on the efficacy of this technique have been mixed (Knight and Light, 2012; Knight et al., 2012; Stelinski et al., 2013).

9.2.2.3 *Mass Trapping*

This strategy involves the use of attractive semiochemicals (synthetic aggregation and sex pheromones, host volatiles, etc.) involved in the mate finding and/or foraging behaviours of the target pest with the purpose of bringing them to killing devices. Insects are removed from the population using small amounts of insecticides, adhesives, water, or other physical structures. Competitiveness of traps with wild calling females, pest density, biology and ecology of the target pest, operational costs, and mated female immigration risk are the important considerations in devising a mass trapping pest management programme (El-Sayed et al., 2006; Kroschel and Zegarra, 2010). Mass trapping is usually more effective against pest populations that are isolated and occur at low-density (El-Sayed et al., 2006). Some other key factors involved in successful implementation of mass trapping include the optimization of lures and traps so that the target pest population is reduced below the threshold for economic injury (El-Sayed et al., 2006). Pests of field crops, orchards, and forestry in the orders Lepidoptera, Coleoptera, Diptera, and Homoptera have been targeted using the mass trapping approach (Alpizar et al., 2002; El-Sayed et al., 2006; Oehlschlager et al. 1995a; Reddy et al., 2005; Ross and Daterman, 1997; Sallam et al., 2007; Suckling et al., 2007). Some examples include codling moth, pink bollworm, bark beetles, and palm weevils (El-Sayed et al., 2006; Flint et al., 1974, 1976; Oehlschlager et al. 1995a, 2002). Mass trapping has also played an important role in the management programmes targeted against gypsy moth, boll weevil, and tephritid fruit flies (El-Sayed et al., 2006; Martinez-Ferrer et al., 2012).

As a management strategy, the track record of mass trapping is better for coleopteran and dipteran pests compared with Lepidoptera (Cork et al., 2003). For many lepidopteran pests, both mass trapping and mating disruption may be feasible control options (Teixeira et al., 2010). The efficiency of the lure in attracting male moths and operational costs can aid in determining which method is the most appropriate (Leskey et al., 2009; Yamanaka, 2007). For example, farmers in China prefer mass trapping to mating disruption for the control of diamondback moth owing to lower costs and ease of use (Dai et al., 2008). Like mating disruption, the utility of mass trapping as a pest management tool can be enhanced under certain circumstance by combining it with practices such as sanitation and limited insecticide use (Cork et al., 2005; James et al., 1996). Instances where mass trapping has not yielded satisfactory results include attempts to manage codling moth under moderate to high population pressure (El-Sayed et al., 2006). Competitive attraction of calling female moths relative to pheromone traps, inadequate number of traps per unit area, and the polygamous nature of male codling moths were identified as some of the factors responsible for inadequate control (El-Sayed et al., 2006). The success of mass trapping as a pest control tactic may also be hindered by factors such as inefficient trap design, trap saturation, high costs, and immigration of pests from outside the treated areas (Cox, 2004; El-Sayed et al., 2006; Kroschel and Zegarra, 2010; Leskey et al., 2009). Moreover, traps designed for pests may also inadvertently attract and remove their natural enemies (Dahlsten et al., 2003), which may have evolved to locate their hosts through kairomones (Raffa, 1991; Raffa and Dahlsten, 1995). The catch potential of traps can be increased by physical modifications such as treatment with surface lubricants (Allison et al., 2011; Graham and Poland 2012). Treatment of funnel traps with Fluon, a surface conditioner that makes the surface more slippery, increased the number of cerambycid beetles collected in the traps (Graham and Poland, 2012). Improving the trap design, while simultaneously lowering its cost and maintenance, can also increase the efficiency of mass trapping (Reinke et al., 2012). For example, low cost micro-traps applied at high density had the potential to be more effective than mating disruption against two lepidopteran pests of apples (Reinke et al., 2012). Lures that attract both males and females can also enhance the utility of mass trapping for pest control (Dai et al., 2008). Furthermore, using one trap to capture multiple pests can also improve the efficiency of mass trapping (Hallett et al., 1999).

9.2.2.4 *Attract-and-Kill*

The attract-and-kill method is one type of behavioural manipulation method that combines a long-distance olfactory stimulus to attract a particular pest in combination with some type of killing agent. This approach is similar to mass trapping but does not require the physical entrapment of the target pests (El-Sayed et al., 2009). The use of target-specific pheromones minimizes the impact on beneficial non-target insects, while overall insecticide use is also reduced (Kroschel and Zegarra, 2010). Both mass trapping and attract-and-kill approaches work best when pest density is relatively low (El-Sayed et al., 2006, 2009). For certain lepidopteran pests, attract-and-kill may be a better option than mating disruption since it minimizes the risk that males will recover and subsequently mate (Suckling, 2000). The aggregation pheromone of dried fruit beetles, *Carpophilus* spp., was successfully incorporated with a co-attractant (ripening fruit) and insecticide to prevent the infestation of ripening peaches (Hossain et al., 2006). High costs may hinder the adoption of attract-and-kill as a pest management option and efforts are needed to make it more affordable for farmers (Hossain et al., 2010). As with other pheromone-based systems, attract-and-kill may not be a feasible option if the initial pest density is relatively high or if there is potential for immigration from surrounding areas (Charmillot et al., 2000). El-Sayed et al. (2009) provide a detailed discussion on the use of attract-and-kill as a long-term pest management strategy for economically important pests and also for the eradication of invasive species.

For some pests, all three pheromone-based direct control tactics have been tested to help select the most appropriate one. For example, in Switzerland, separate implementation of mass trapping and mating disruption did not result in satisfactory suppression of codling moth in apple orchards (Charmillot et al., 2000). However, attract-and-kill provided effective control of the pest in 14 of the 15 orchards tested (Charmillot et al., 2000).

9.2.2.5 Push, Pull, and Push-Pull Approaches

Semiochemicals can also be used to 'push' pests away from a valuable resource. The anti-aggregation pheromone verbenone has been used to reduce the attack rates on pines by the mountain pine beetle, *Dendroctonus ponderosae* Hopkins (Coleoptera: Curculionidae) (Bentz et al., 2005; Borden et al., 2006; Gillette et al., 2012a,b; Progar, 2005). Similarly, treatment of cherry trees with the synthetic host-marking pheromone of the European cherry fruit fly, *Rhagoletis cerasi* L., reduced the infestation of fruit (Aluja and Boller, 1992). The pull approach is similar in principle and results in the aggregation of pests in pre-determined zones. This technique was demonstrated in apple orchards for the management of plum curculio (Leskey et al., 2008). Perimeter-row trap trees baited with grandisoic acid (aggregation pheromone) and benzaldehyde (fruit volatile) attracted plum curculio adults. Compared with standard full block insecticide applications for the pest, limiting treatment to pheromone-baited 'trap' trees provided satisfactory suppression of fruit injury. It was estimated that $\approx$93% fewer trees were treated with insecticide in the trap tree approach (Leskey et al., 2008). Pheromones can also be used to increase the attractiveness of trap crops. For example, it was demonstrated that application of synthetic aggregation pheromone increased the number of Colorado potato beetle (*Leptinotarsa decemlineata* Say) adults in the trap crop perimeter (Kuhar et al., 2006).

The attractive and repellent stimuli corresponding to the olfactory and/or visual communications of a pest can be manipulated simultaneously to devise a more potent 'push-pull' strategy. Using multiple stimuli, pests can be 'shepherded' into pre-determined zones and then be targeted for elimination by conventional and/or biological control methods. Push-pull was tested against *D. ponderosase* using anti-aggregation pheromones (push) in combination with perimeter traps baited with its aggregation pheromone (pull) (Gillette et al., 2012b). Other pests that have been targeted using the push and pull approach include the onion fly, *Delia antiqua* Meigen (Diptera: Anthomyiidae), and German cockroach, *Blattella germanica* L. (Dictyoptera: Blattellidae) (Miller and Cowles, 1990; Nalyanya et al., 2000). Successful implementation of the pull and push-pull approach in agroecosystems can lower input costs and lighten the ecological footprint as targeted pest elimination dramatically decreases insecticide use (Cook et al., 2007; Leskey et al., 2008).

9.2.2.6 Other Uses

The principles of behavioural manipulation can be applied to attract the natural enemies of pests and enhance biological control services in managed agroecosystems (Rodriguez-Saona et al., 2012). For example, alarm pheromones of some aphids can attract their natural enemies to the fields (Bruce et al., 2005). Aggregation pheromones of natural enemies can be used for mass trapping and inundative releases into crops. For example, it was suggested that the aggregation volatiles released by seven-spot ladybird beetle, *Coccinella septempunctata* L. (Coleoptera: Coccinellidae), could be used in the biological control of aphids (Al Abassi et al., 1998). Furthermore, pheromones can also be utilized to monitor biological control agents and help in the detection of exotic or invasive species at ports of entry (Allison et al., 2004; DeLury et al., 1999; Graham and Poland, 2012; Suckling et al., 2002, 2006). Another application of pheromone baited traps is to monitor insecticide resistance in pest populations (Haynes et al., 1987; Sauphanor et al., 2000; Shearer and Riedl, 1994; Varela et al., 1993). The utility of pheromones in enhancing the spread of entomopathogenic control agents has also been investigated (Baverstock et al., 2010; Kreutz et al., 2004; Roditakis et al., 2000; Tinzaara et al., 2007; Yasuda, 1999). For example, the aggregation pheromone of banana weevil, *Cosmopolites sordidus* Germar, was used to augment the spread of *Beauveria bassiana* (Balsamo) Vuillemin, its biological control agent (Tinzaara et al., 2007). Nishisue et al. (2010) reported that the foraging activity of Argentine ant, *Linepithema humile* Mayr (Hymenoptera: Formicidae), was inhibited by the application of synthetic trail pheromone.

9.3 SYNERGISM WITH PLANT VOLATILES

The addition of host-plant volatiles can enhance the attractiveness of some insect pheromones (Landolt and Phillips, 1997). The phenomenon, termed synergism, is observed when the behavioural response to a mixture of pheromone and plant volatiles is greater than the sum of responses to the separate stimuli (Reddy and Guerrero, 2004). Insects in the order Coleoptera have received considerable attention for research related to the interactions between pheromones and host-plant volatiles. The attractiveness of boll weevil traps was greatly enhanced when the aggregation pheromone of *A. grandis* was combined with green leaf volatiles (trans-2-hexen-1-ol, cis-3-hexen-1-ol, or 1-hexanol) from cotton plants (Dickens, 1989). Increases in trap captures were also observed when benzaldehyde, a fruit volatile, was used in combination with the plum curculio aggregation pheromone (Piñero and Prokopy, 2003; Piñero et al., 2001). Pheromone trap capture has also been synergized by plant volatiles for the mountain pine beetle (Borden et al., 2008). For example, traps baited with pheromone and host-plant volatiles captured approximately 5 to 13 times more adults than traps with pheromone alone (Borden et al., 2008). Similarly, host-plant volatiles synergized the response of Asian palm weevil, *Rhynchophorus ferrugineus* (Coleoptera: Curculionidae), adults to traps baited with the aggregation pheromone ferrugineol (Hallett et al., 1999). Examples of some other insect species for which synergistic responses have been reported include flea beetle, palmetto weevil (*Rhynchophorus cruentatus* F.), codling moth, corn earworm, and oriental fruit moth (Light et al., 1993; Ochieng et al., 2002; Soroka et al., 2005; Weissling et al., 1994; Yang et al., 2004).

9.4 WEEVIL PHEROMONES IN PEST MANAGEMENT

Weevils (Coleoptera: Curculionidae) constitute a diverse group of phytophagous insects and many species are important pests of field and orchard crops. Pheromones (aggregation and sex) and attractive plant volatiles have been identified for 29 and 36 weevil species, respectively. Furthermore, synergistic interactions between pheromones and plant volatiles have been demonstrated for 13 of the approximately 61 weevil species studied (Piñero and Ruiz-Montiel, 2012). The aggregation pheromones of insects, and particularly in Curculionidae, represents a powerful yet sensitive tool for early detection of infestations (Piñero and Ruiz-Montiel, 2012). Monitoring and control strategies based specifically on aggregation pheromones and their synergistic enhancement by host-plant volatiles have been used extensively in the management of many economically important weevil pests, including those in genus *Anthonomus*. For example, grandisoic acid, a male-produced aggregation pheromone, has played an important role in the monitoring and control of the plum curculio in apple orchards (Leskey et al., 2008). Similarly, aggregation pheromone is an important component of the boll weevil eradication programme in the United States.

The aggregation pheromones of several weevil species consist of multiple behaviourally active components. For example, the seven common aggregation pheromone components of genus *Anthonomus* include (*Z*)-2-isopropenyl-1-methylcyclobutaneethanol (grandlure I); (*Z*)-2-(3,3-dimethyl-cyclohexylidene) ethanol (*Z* grandlure II); (*E*)-2-(3,3-dimethyl-cyclohexylidene) ethanol (*E* grandlure II); (*Z*)-(3,3-dimethylcyclohexylidene) acetaldehyde (grandlure III); (*E*)-(3,3-dimethylcyclohexylidene) acetaldehyde (grandlure IV); (*E*)-3,7-dimethyl-2,6-octadien-1-ol) (geraniol); and (*E*)-3,7-dimethyl-2,6-octadienoic acid (geranic acid) (Figure 9.1). The distribution and relative abundance of these compounds in the aggregation pheromone varies among the different *Anthonomus* species; these include the boll weevil, pepper weevil, strawberry blossom weevil (*Anthonomus rubi* Herbst.), and cranberry weevil (*Anthonomus musculus* Say). For example, grandlure I, *Z* grandlure II, grandlure III, and grandlure IV are the main components of boll weevil aggregation pheromone, whereas *Z* grandlure II, grandlure III, grandlure IV, and geraniol are the principal constituents

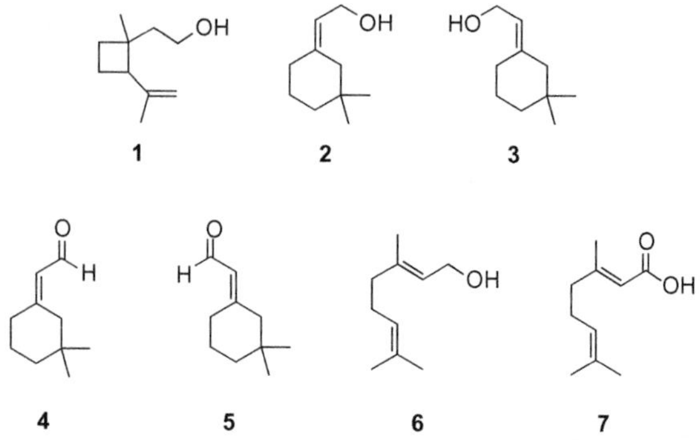

OH OH HO
1 2 3
O O
H H OH OH
4 5 6 7

FIGURE 9.1 The chemical structures of seven common aggregation pheromone components in *Antonomus* spp. weevils (Coleoptera: Curculionidae). **1**: (*Z*)-2-isopropenyl-1-methylcyclobutaneethanol (grandlure I); **2**: (*Z*)-2-(3,3-dimethyl-cyclohexylidene) ethanol (*Z* grandlure II); **3**: (*E*)-2-(3,3-dimethyl-cyclohexylidene) ethanol (*E* grandlure II); **4**: (*Z*)-(3,3-dimethylcyclohexylidene) acetaldehyde (grandlure III); **5**: (*E*)-(3,3-dimethylcyclohexylidene) acetaldehyde (grandlure IV); **6**: (*E*)-3,7-dimethyl-2,6-octadien-1-ol) (geraniol); and **7**: (*E*)-3,7-dimethyl-2,6-octadienoic acid (geranic acid).

of cranberry weevil aggregation pheromone (Szendrei et al., 2011). Aggregation pheromones have also been used in the management of other weevil pests including the American palm weevil (*Rhynchophorus palmarum* L.), Asian palm weevil (*Rhynchophorus ferrugineus* Olivier), banana corm weevil (*Cosmopolites sordidus* Germar), strawberry blossom weevil, and pecan weevil (*Curculio caryae* Horn) (Alpizar et al., 2002, 2012; Cross et al., 2006; Hallett et al., 1999; Hedin et al., 1997; Leskey et al., 2008; Oehlschlager et al., 1995a).

There can be geographical variations in the production and behavioural significance of aggregation pheromone components (Giblin-Davis et al., 2000). For example, the Hawaiian and Australian populations of the New Guinea sugarcane weevil, *Rhabdoscelus obscurus* Boisduval, have different communication ecology. Both the populations produce male-specific 2-methyl-4-octanol. However, the compound was able to enhance the attractiveness of sugarcane bait to male and female weevils for the Hawaiian population only. Furthermore, in the Australian population, 2-methyl-4-octanol had to be combined with another male-specific component (*E*2)-6-methyl-2-hepten-4-ol (rhynchophorol) to increase the attractiveness of sugarcane. It was hypothesized that the observed differences between the Hawaiian and Australian populations could be due to either the founder effect or because the two populations represented sibling species (Giblin-Davis et al., 2000). The above-mentioned study demonstrates that multiple factors can influence the chemical ecology of insects and careful studies are needed to untangle the complex interactions between organisms and their pheromones.

Table 9.1 lists some of the economically important weevil species and the identity of their aggregation pheromone components. Information on the host plants and distribution range of the weevils is also provided. Examples of traps commonly used to test weevil pheromones are shown in Figure 9.2.

9.5 CASE STUDIES

Research on the pheromones of four weevil species: boll weevil, plum curculio, cranberry weevil, and pepper weevil, is discussed below. These weevils are of economic importance in the US, and two of them (plum curculio and cranberry weevil) are the focus of ongoing research programmes of the authors.

9.5.1 Boll Weevil

Boll weevil, *A. grandis*, a historically devastating pest of cotton, was first reported from the US (Texas) in 1894 (Burke et al., 1986). For the next 30 years, approximately 87% of the growing area was infested and the cotton industry was decimated (Smith, 1998). Early insecticides targeting boll weevil, including arsenates and chlorinated hydrocarbons, were effective but resistance was reported by 1960 (Perkins, 1980). The next stage of the boll weevil management programme was launched in 1962 with the establishment of the Boll Weevil Research Laboratory at Mississippi State University.

A major breakthrough in the management of boll weevils came with the release of its synthetic aggregation pheromone, grandlure (Hardee et al., 1972, 1974) (Table 9.1). Grandlure proved to be an effective monitoring tool (Figure 9.2B), with the potential for playing a significant role in the control and eradication programme targeted against the boll weevil (Mitchell and Hardee, 1974). A pilot eradication trial was initiated in 1971 and involved the use of pheromone traps, trap crops, sterile male releases, and insecticides (Perkins, 1980). Subsequently, a second eradication trial was carried out to incorporate the latest research on the use of pheromone traps and to also address issues such as weevil immigration from non-treated areas. It was concluded that the two trials were successful in demonstrating the utility of large-scale, area-wide programmes for

TABLE 9.1 Weevil Species, Including Crops Impacted and Geographical Distribution, for Which Aggregation Pheromones Have Been Reported

Weevil species	Subfamily	Crops impacted	Distribution	Aggregation pheromone components	References
Anthonomus grandis Boheman (boll weevil)	Curculioninae	Cotton	Native to southern Mexico and central America, target of eradication programme in the United States	Grandlure I, *Z* grandlure II, grandlure III, and grandlure IV	Tumlinson et al., 1971
Anthonomus eugenii Cano (pepper weevil)	Curculioninae	Pepper	Southern United States, Hawaii, Puerto Rico, Central America, Caribbean, Mexico, Honduras, Guatemala, El Salvador	*Z* grandlure II, *E* grandlure II, grandlure III, grandlure IV, geraniol, geranic acid	Eller et al., 1994
Anthonomus musculus Say (cranberry weevil)	Curculioninae	Blueberry, Cranberry	Northeastern United States, Michigan, Wisconsin	*Z* grandlure II, grandlure III, grandlure IV, geraniol	Szendrei et al., 2011
Anthonomus rubi Herbst. (strawberry blossom weevil)	Curculioninae	Strawberry	UK, Continental Europe	Grandlure I, *Z* grandlure II, lavandulol	Innocenzi et al., 2001
Curculio caryae Horn (pecan weevil)	Curculioninae	Pecan	Southern United States	Grandlure I, *Z* grandlure II, grandlure III, grandlure IV	Hedin et al., 1997
Conotrachelus nenuphar Herbst. (plum curculio)	Molytinae	Apple, Plum, Peach	United States, Canada	Grandisoic acid	Eller and Bartelt, 1996
Rhynchophorus palmarum L. (American palm weevil)	Dryophthorinae	Palm, Coconut plantations in Brazil, Mexico, and the Caribbean	South and central America, Mexico	Rhynchophorol	Oehlschlager et al., 1992
Rhynchophorus ferrugineus Olivier (Asian palm weevil)	Dryophthorinae	Palm	Native to Asia, introduced to many countries in Africa and Europe	Ferrugineol	Hallett et al., 1993
Rhynchophorus bilineatus Montr. (black palm weevil)	Dryophthorinae	Palm	Indonesia, Papua New Guinea, Solomon Islands	Ferrugineol	Oehlschlager et al., 1995b

(*Cotinued*)

TABLE 9.1 (Continued)

Weevil species	Subfamily	Crops impacted	Distribution	Aggregation pheromone components	References
Rhabdoscelus obscurus Boisduval (New Guinea sugarcane weevil)	Dryophthorinae	Sugarcane, Ornamental palms, Coconut, occasionally Papaya	Native to Austromalaya. Present in Micronesia, Hawaii, Queensland, southern Japan, and Indonesia	2-Methyl-4-octanol, rhynchophorol	Giblin-Davis et al., 2000
Rhynchophorus cruentatus Fabricius (palmetto weevil)	Dryophthorinae	Palm	Florida and the southeastern US	Cruentol	Weissling et al., 1994
Rhynchophorus phoenicis Fabricius (African palm weevil)	Dryophthorinae	Palm	Africa	Phoenicol	Perez et al., 1994
Scyphophorus acupunctatus Gyllenhaal (agave weevil)	Dryophthorinae	Agavaceae and Dracaenaceae	Southern United States to Brazil, the Caribbean, Hawaii, Borneo, Java, Australia, East Africa	2-Methyl-4-heptanol, 2-methyl-4-octanol, 2-methyl-4-heptanone, 2-methyl-4-octanone	Ruiz-Montiel et al., 2008
Dynamis borassi Fabricius (palm weevil)	Dryophthorinae	Palm	South America	4-Methyl-5-nonanol	Giblin-Davis et al., 1997
Metamasius hemipterus sericeus Oliv. (West Indian sugarcane weevil)	Dryophthorinae	Banana, Pineapple, Palms, Sugarcane	Florida (US), Central and South America, Caribbean, Africa	3-Pentanol, 2-methyl-4-heptanol, 2-methyl-4-octanol, 4-methyl-5-nonanol (ferrugineol)	Perez et al. 1997
Metamasius spinolae Gyllenhaal (cactus weevil)	Dryophthorinae	Cactus	Mexico	2-Methyl-4-heptanone, 6-methyl-2hepten-4-one, 2-hydroxy-2-methyl-4-heptanone	Tafoya et al., 2007
Metamasius hemipterus L. (West Indian sugarcane borer)	Dryophthorinae	Sugarcane, Banana, Palm	Florida (US), West Indies, Uruguay, northern Argentina, Africa	4-Methyl-5-nonanol (ferrugineol), 2-methyl-4-heptanol, 2-methyl-4-octanol, 5-nonanol, 3-hydroxy-4-methyl-5-nonanone	Ramirez-Lucas et al., 1996
Sitona lineatus L. (pea leaf weevil)	Entiminae	Pea	Europe, Canary Islands, northern Africa, Israel, North America	4-Methyl-3,5-heptanedione	Blight et al., 1984
Cosmopolites sordidus Germar (banana corm weevil)	Calendrinae	Banana, Plantain	Central America, Africa	Sordidin	Beauhaire et al., 1995

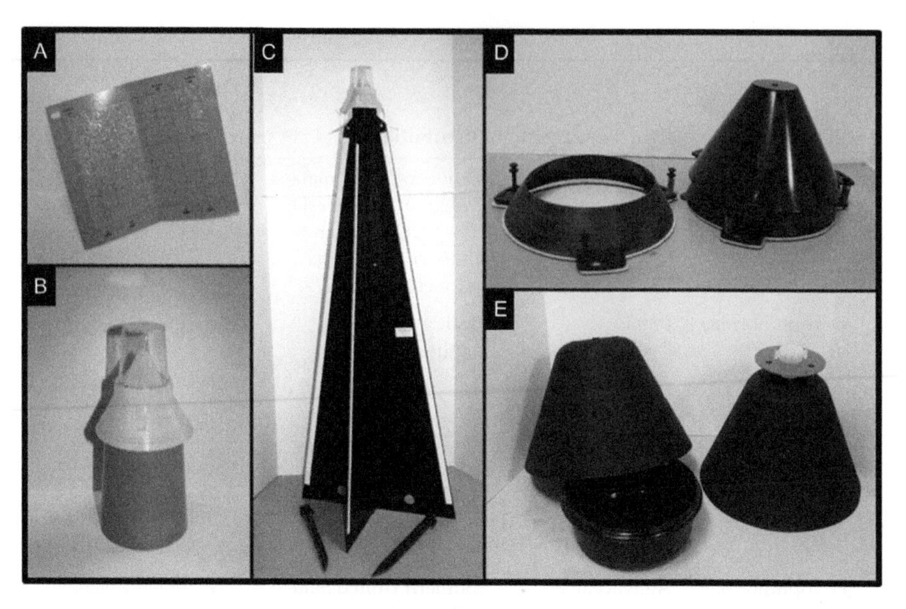

FIGURE 9.2 Types of trap used for testing weevil pheromones. A) Sticky card (Trécé, Adair, OK, USA), B) Boll weevil trap (Great Lakes IPM, Inc., Vestaburg, MI, USA), C) Plum curculio pyramid trap (Great Lakes IPM, Inc., Vestaburg, MI, USA), D) Dome trap with opening at the bottom (AgBio Inc., Westminster, CO, USA), E) Dome trap with opening at the top (ISCA Technologies, Riverside, CA, USA).

the management of boll weevil (Smith, 1998). In 1983, an eradication programme was initiated in the southeastern Cotton Belt (North and South Carolina) and was later expanded to parts of Georgia, Alabama, and all of Florida. The main thrust of the programme was to prevent boll weevil diapause and reproduction, combined with control during the growing season (Smith, 1998). In 1985, the programme was expanded to the southwestern US and by 1993, boll weevil eradication had been achieved in California, Arizona, and northwest Mexico (Smith, 1998).

In the boll weevil eradication programme, pheromone-based traps are used for detection, population estimation, mass trapping, and guiding insecticide application decisions (El-Sayed et al., 2006; Suh et al., 2009). Furthermore, insecticide-impregnated kill strips can also be incorporated into the pheromone traps to induce mortality and thus prevent escape (Suh et al., 2009). Villavaso et al. (1998) demonstrated that an attract-and-kill strategy using insecticide treated sticky bait sticks was ≈3 times more effective than conventional pheromone traps in removing boll weevils from the population.

9.5.2 Plum Curculio

The plum curculio, *C. nenuphar*, is one of the most serious pests of stone and pome fruits in eastern and central North America (Leskey and Wright, 2004; Vincent et al., 1999). It is also considered a serious pest of peaches in the eastern US (Akotsen-Mensah et al., 2010). Effective monitoring of adults was identified as a key factor for the successful management of plum curculio (Akotsen-Mensah et al., 2010). Research into the behavioural manipulation of plum curculio has focused on the monitoring and control strategies using the synthetic aggregation pheromone, grandisoic acid (Table 9.1), and fruit volatiles. Adults are also attracted to a number of fruit volatiles including (*E*)-2-hexenal, hexyl acetate, ethyl isovalerate, limonene, benzaldehyde, benzyl alcohol, decanal, and geranyl propionate (Leskey et al., 2001; Prokopy et al., 2001). Traps (Figure 9.2C) baited with benzaldehyde, in particular, in combination with grandisoic acid attracted significantly more adults than those baited with grandisoic acid alone or the unbaited traps (Piñero et al., 2001). Only benzaldehyde was found to synergize the response of plum curculio to its aggregation pheromone in apple orchards (Piñero and Prokopy, 2003).

Historically, the management of plum curculio was dependent on calendar-based full block application of broad-spectrum insecticides in apple orchards (Leskey et al., 2008). This was in part because effective and reliable monitoring techniques were not available after petal fall. In response to this problem, a trap tree technique was developed to monitor the oviposition activity of plum curculio in which one perimeter-row tree was baited with a synergistic two-component lure (grandisoic acid + benzaldehyde) (Prokopy et al., 2003, 2004). The technique was effective in determining oviposition injury and allowed growers to determine the appropriate spray timings (Piñero et al., 2011). The strategy was further refined when Leskey et al. (2008) demonstrated that plum curculio could be managed by applying insecticides to a few perimeter-row trap trees baited with a synergistic blend of grandisoic acid and benzaldehyde.

9.5.3 Cranberry Weevil

The cranberry weevil, *A. musculus*, is a key pest of highbush blueberries and cranberries in the USA, in Massachusetts, Michigan, New Jersey, and Wisconsin (Long and Averill, 2003; Szendrei et al., 2009). The economic injury is due to the larvae, which feed and develop inside the flower buds and prevent the formation of fruit (Szendrei et al., 2011). Management strategies are targeted against the mobile adults and monitoring is usually done by using beat trays or through visual assessment of blossom damage in blueberries (Szendrei et al., 2009). In cranberries, sweep-nets are used to monitor this pest (Averill and Sylvia, 1998). However, clumped spatial distribution of adults makes the aforementioned monitoring techniques unreliable and expensive (Szendrei et al., 2009).

In order to develop behaviourally-based monitoring and management tools for the cranberry weevil, the response of adults to volatiles collected from blueberry buds and open flowers was investigated. Four compounds collected from both the buds and open flowers [hexanol, (*Z*)-3-hexenyl acetate, hexyl acetate, and (*Z*)-3-hexenyl butyrate] elicited a significant antennal response from cranberry weevil adults (Szendrei et al., 2009). Furthermore, a significantly higher number of cranberry weevil adults were captured on traps baited with the aggregation pheromone of *A. eugenni* (pepper weevil) than on traps baited either with cinnamyl alcohol (a major blueberry floral component) or unbaited traps (Szendrei et al., 2011). This result indicated that the two *Anthonomus* species shared common aggregation pheromone components. Subsequent headspace analysis confirmed that cranberry weevil and pepper weevil adults did share four aggregation pheromone components (*Z* gradlure II, gradlure III, gradlure, IV, geraniol) (Szendrei et al., 2011) (Table 9.1).

Field evaluations demonstrated that yellow sticky traps (Figure 9.2A) baited with the blend of four aggregation pheromone components trapped significantly more cranberry weevil adults than eight other blends (including aggregation pheromones of boll and pepper weevil), and unbaited controls (Szendrei et al., 2011). These results indicate that a low cost and reliable monitoring tool can be developed for the cranberry weevil in blueberry and cranberry production systems. There is also potential to develop an attract-and-kill strategy for the cranberry weevil, similar to the one used for boll weevil (McKibben et al., 1990). Current research efforts are directed at optimizing the aggregation pheromone lure and trap parameters (colour, design, and placement) to develop an effective monitoring and control strategy for the cranberry weevil.

9.5.4 Pepper Weevil

The pepper weevil, *A. eugenii*, is a pest of cultivated peppers in the southern US, Mexico, Central America, and the Caribbean region (Addesso et al., 2011; Eller et al., 1994). Damage caused by the feeding of larvae and adults can range from contamination of fruit with frass, oviposition and feeding punctures on blooms and fruits, defoliation, to fruit drop (Bottenberg and Lingren, 1998). Wild nightshade plants (*Solanum* spp.) can be used as an alternative host for feeding and reproduction when pepper is not in production (Addesso et al., 2011).

Two of the most common sampling methods for pepper weevils include use of sticky yellow cards (Figure 9.2A) or whole-plant visual inspections (Riley and Schuster, 1994). However, it was suggested that the use of sticky cards could be made more economical by combining them with a pheromone attractant (Riley and Schuster, 1994). The aggregation pheromone of pepper weevil was identified (Table 9.1) and sticky traps baited with the synthetic blend captured more adults than the unbaited traps (Eller et al., 1994). The adoption of early pheromone-based monitoring traps was hindered by high production costs and relatively short field longevity (Bottenberg and Lingren, 1998). However, an improved lure with a longer activity period was released subsequently (Bottenberg and Lingren, 1998).

Pepper weevil adults can also orient to constitutive host-plant volatiles and prefer damaged plants over undamaged ones (Addesso and McAuslane, 2009; Addesso et al., 2011). Furthermore, adults preferred plants with actively feeding weevils compared to plants with previous damage (Addesso et al., 2011). Based on these results, it was suggested that a more effective monitoring tool could be developed for pepper weevils by combining host-plant volatiles with the male aggregation pheromone (Addesso et al., 2011).

9.6 CONCLUSIONS

Pheromones and other behaviour-modifying semiochemicals are now an integral part of numerous pest management programmes and are expected to play an important role in high-tech crop protection of the future (Zijlstra et al., 2011). These will help provide a sustainable and environmentally friendly replacement to the broad-spectrum insecticides, either as monitoring or management tools of critical IPM programmes. Potentially useful interactions occur not only within one sensory modality such as olfaction, but can derive from different sensory modalities such as vision and olfaction. Since the interactions within or between cues can increase the chances of host and mate location in nature, more reliable control of the target insect pest might be accomplished under variable environmental conditions using pheromone-based systems that exploit such interactions. Therefore, interactions among modalities of host- and mate-finding and visual cues offer a wide field of opportunities for future research and development of pheromone-based systems (Dorn and Piñero, 2009).

Other promising areas of research include the use of multiple pheromones to monitor several pests simultaneously, use of pheromones for behavioural manipulation of natural enemies, understanding the mechanisms underlying mating disruption and other pheromone-based control approaches, and the use of multiple pheromones for controlling several pests simultaneously in a mating disruption scenario. The ultimate challenge will be to increase the adoption of pheromone-based pest management technologies by making them cost-effective, but without sacrificing efficacy.

ACKNOWLEDGEMENTS

We thank the editor for inviting us to write this chapter. Thanks also to Robert Holdcraft for providing the figures. This work was supported by the USDA Pest Management Alternatives Program (grant no. 2012-34381-20108) to C.R.R-S., A.L.N. and T.C.L.

References

Addesso, K.M., McAuslane, H.J., 2009. Pepper weevil attraction to volatiles from host and nonhost plants. Environ. Entomol. 38, 216–224.

Addesso, K.M., McAuslane, H.J., Stansly, P.A., Schuster, D.J., 2007. Host-marking by female pepper weevils, *Anthonomus eugenii*. Entomol. Exp. Appl. 125, 269–276.

Addesso, K.M., McAuslane, H.J., Alborn, H.T., 2011. Attraction of pepper weevil to volatiles from damaged pepper plants. Entomol. Exp. Appl. 138, 1–11.

Agnello, A.M., Reissig, W.H., Spangler, S.M., Charlton, R.E., Kain, D.P., 1996. Trap response and fruit damage by obliquebanded leafroller (Lepidoptera: Tortricidae) in pheromone-treated apple orchards in New York. Environ. Entomol. 25, 268–282.

Akotsen-Mensah, C., Boozer, R., Fadamiro, H.Y., 2010. Field evaluation of traps and lures for monitoring plum curculio (Coleoptera: Curculionidae) in Alabama peaches. J. Econ. Entomol. 103, 744–753.

Al Abassi, S., Birkett, M.A., Pettersson, J., Pickett, J.A., Woodcock, C.M., 1998. Ladybird beetle odour identified and found to be responsible for attraction between adults. Cell. Mol. Life Sci. 54, 876–879.

Alfaro, C., Navarro-Llopis, V., Primo, J., 2009. Optimization of pheromone dispenser density for managing the rice striped stem borer, *Chilo suppressalis* (Walker), by mating disruption. Crop Prot. 28, 567–572.

Allen, D.C., Abrahamson, L.P., Eggen, D.A., Lanier, G.N., Swier, S.R., Kelley, R.S., et al., 1986. Monitoring spruce budworm (Lepidoptera: Tortricidae) populations with pheromone-baited traps. Environ. Entomol. 15, 152–165.

Allison, J.D., Borden, J.H., Seybold, S.J., 2004. A review of the chemical ecology of the Cerambycidae (Coleoptera). Chemoecology 14, 123–150.

Allison, J.D., Johnson, C.W., Meeker, J.R., Strom, B.L., Butler, S.M., 2011. Effect of aerosol surface lubricants on the abundance and richness of selected forest insects captured in multiple-funnel and panel traps. J. Econ. Entomol. 104, 1258–1264.

Allison, J.D., McKenney, J.L., Miller, D.R., Gimmel, M.L., 2012. Role of ipsdienol, ipsenol, and cis-verbenol in chemical ecology of *Ips avulsus*, *Ips calligraphus*, and *Ips grandicollis* (Coleoptera: Curculionidae: Scolytinae). J. Econ. Entomol. 105, 923–929.

Alpizar, D., Fallas, M., Oehlschlager, A.C., Gonzalez, L.M., Chinchilla, C.M., Bulgarelli, J., 2002. Pheromone mass trapping of the West Indian sugarcane weevil and the American palm weevil (Coleoptera: Curculionidae) in palmito palm. Fla. Entomol. 85, 426–430.

Alpizar, D., Fallas, M., Oehlschlager, A.C., Gonzalez, L.M., 2012. Management of *Cosmopolites sordidus* and *Metamasius hemipterus* in banana by pheromone-based mass trapping. J. Chem. Ecol. 38, 245–252.

Aluja, M., Boller, E.F., 1992. Host marking pheromone of *Rhagoletis cerasi*: field deployment of synthetic pheromone as a novel cherry fruit-fly management strategy. Entomol. Exp. Appl. 65, 141–147.

Ansebo, L., Ignell, R., Lofqvist, J., Hansson, B.S., 2005. Responses to sex pheromone and plant odours by olfactory receptor neurons housed in *sensilla auricillica* of the codling moth, *Cydia pomonella* (Lepidoptera: Tortricidae). J. Insect Physiol. 51, 1066–1074.

Atanassov, A., Shearer, P.W., Hamilton, G., Polk, D., 2002. Development and implementation of a reduced risk peach arthropod management program in New Jersey. J. Econ. Entomol. 95, 803–812.

Athanassiou, C., Kavallieratos, N., Mazomenos, B., 2004. Effect of trap type, trap color, trapping location, and pheromone dispenser on captures of male *Palpita unionalis* (Lepidoptera: Pyralidae). J. Econ. Entomol. 97, 321–329.

Averill, A.L., Sylvia, M.M., 1998. Cranberry Insects of the Northeast: A Guide to Identification, Biology, and Management. Cranberry Experiment Station Publication, University of Massachusetts, Amherst, 112 pp.

Ayasse, M., Paxton, R., Tengo, J., 2001. Mating behavior and chemical communication in the order hymenoptera. Annu. Rev. Entomol. 46, 31–78.

Bacca, T., Saraiva, R.M., Lima, E.R., 2012. Capture of *Leucoptera coffeella* (Lepidoptera: Lyonetiidae) in sex pheromone traps and damage intensity. Rev. Colomb. Entomol. 38, 42–49.

Baker, T.C., Hansson, B.S., Lofstedt, C., Lofqvist, J., 1988. Adaptation of antennal neurons in moths is associated with cessation of pheromone-mediated upwind flight. Proc. Natl. Acad. Sci. U.S.A. 85, 9826–9830.

Baverstock, J., Roy, H.E., Pell, J.K., 2010. Entomopathogenic fungi and insect behaviour: from unsuspecting hosts to targeted vectors. Biocontrol 55, 89–102.

Beauhaire, J., Ducrot, P., Malosse, C., Rochat, D., Ndiege, I.O., Otieno, D.O., 1995. Identification and synthesis of sordidin, a male pheromone emitted by *Cosmopolites sordidus*. Tetrahedron Lett. 36, 1043–1046.

Bentz, B.J., Kegley, S., Gibson, K., Thier, R., 2005. A test of high-dose verbenone for stand-level protection of lodgepole and whitebark pine from mountain pine beetle (Coleoptera: Curculionidae: Scolytinae) attacks. J. Econ. Entomol. 98, 1614–1621.

Bergh, J.C., Leskey, T.C., Sousa, J.M., Zhang, A., 2006. Diel periodicity of emergence and premating reproductive behaviors of adult dogwood borer (Lepidoptera: Sesiidae). Environ. Entomol. 35, 435–442.

Billen, J., Morgan, E.D., 1998. Pheromone communication in social insects: Sources and secretions. In: Vander Meer, R.K., Breed, M.D., Espelie, K.E., Winston, M.L. (Eds.), Pheromone Communications in Social Insects. Westview Press, Boulder, Colorado, pp. 3–33.

Binns, M.R., Nyrop, J.P., 1992. Sampling insect populations for the purpose of IPM decision making. Annu. Rev. Entomol. 37, 427–453.

Blight, M.M., Pickett, J.A., Smith, M.C., Wadhams, L.J., 1984. An aggregation pheromone of *Sitona lineatus*: identification and initial field studies. Naturwissenschaften 71, 480.

Blomquist, G.J., Figueroa-Teran, R., Aw, M., Song, M., Gorzalski, A., Abbott, N.L., et al., 2010. Pheromone production in bark beetles. Insect Biochem. Mol. Biol. 40, 699–712.

Borden, J.H., Birmingham, A.L., Burleigh, J.S., 2006. Evaluation of the push-pull tactic against the mountain pine beetle using verbenone and non-host volatiles in combination with pheromone-baited trees. For. Chron. 82, 579–590.

Borden, J.H., Pureswaran, D.S., Lafontaine, J.P., 2008. Synergistic blends of monoterpenes for aggregation pheromones of the mountain pine beetle (Coleoptera: Curculionidae). J. Econ. Entomol. 101, 1266–1275.

Bottenberg, H., Lingren, B., 1998. Field performance of a new pepper weevil pheromone formulation. Proc. Fla. State Hort. Soc. 111, 48–50.

Bruce, T.J.A., Birkett, M.A., Blande, J., Hooper, A.M., Martin, J.L., Khambay, B., et al., 2005. Response of economically important aphids to components of *Hemizygia petiolata* essential oil. Pest Manag. Sci. 61, 1115–1121.

Burke, H.R., Clark, W.E., Cate, J.R., Fryxell, P.A., 1986. Origin and dispersal of the boll weevil. Bull. Entomol. Soc. Am. 32, 228–238.

Butenandt, A., Beckmann, R., Stamm, D., Hecker, E., 1959. Uber den Sexual-Lockstoff des Seidenspinners *Bombyx mori*. Reindarstellung und Konstitution. Z. Naturforsch. B. 14, 283.

Byers, J.A., 2007. Simulation of mating disruption and mass trapping with competitive attraction and camouflage. Environ. Entomol. 36, 1328–1338.

Byers, J.A., 2011. Analysis of vertical distributions and effective flight layers of insects: Three-dimensional simulation of flying insects and catch at trap heights. Environ. Entomol. 40, 1210–1222.

Campbell, C.D., Walgenbach, J.F., Kennedy, G.G., 1992. Comparison of black light and pheromone traps for monitoring helicoverpa-zea (boddie) (Lepidoptera: Noctuidae) in tomato. J. Agric. Entomol. 9, 17–24.

Cardé, R.T., 1990. Principles of mating disruption. In: Ridgway, R.L., Silverstein, R.M., Inscoe, M.N. (Eds.), Behavior-Modifying Chemicals for Insect Management. Marcel Dekker, Inc., New York, New York, pp. 47–71.

Cardé, R.T., Minks, A.K., 1995. Control of moth pests by mating disruption: successes and constraints. Annu. Rev. Entomol. 40, 559–585.

Charmillot, P.-J., Hofer, D., Pasquier, D., 2000. Attract and kill: A new method for control of the codling moth *Cydia pomonella*. Entomol. Exp. Appl. 94, 211–216.

Cohnstaedt, L.W., Rochon, K., Duehl, A.J., Anderson, J.F., Barrera, R., Su, N., et al., 2012. Arthropod surveillance programs: Basic components, strategies, and analysis. Ann. Entomol. Soc. Am. 105, 135–149.

Cook, S.M., Khan, Z.R., Pickett, J.A., 2007. The use of push-pull strategies in integrated pest management. Annu. Rev. Entomol. 52, 375–400.

Cork, A., Alam, S.N., Rouf, F.M.A., Talekar, N.S., 2003. Female sex pheromone of brinjal fruit and shoot borer, *Leucinodes orbonalis* (Lepidoptera: Pyralidae): Trap optimization and application in IPM trials. Bull. Entomol. Res. 93, 107–113.

Cork, A., Alam, S., Rouf, F., Talekar, N., 2005. Development of mass trapping technique for control of brinjal shoot and fruit borer, *Leucinodes orbonalis* (Lepidoptera: Pyralidae). Bull. Entomol. Res. 95, 589–596.

Cox, P.D., 2004. Potential for using semiochemicals to protect stored products from insect infestation. J. Stored Prod. Res. 40, 1–25.

Cross, J.V., Hall, D.R., Innocenzi, P.J., Hesketh, H., Jay, C.N., Burgess, C.M., 2006. Exploiting the aggregation pheromone of strawberry blossom weevil Anthonomus rubi (Coleoptera: Curculionidae): Part 2. Pest monitoring and control. Crop Prot. 25, 155–166.

Cunningham, R.T., 1989. Parapheromones. In: Robinson, A.S., Hooper, G. (Eds.), World Crop Pests 3(A). Fruit Flies; Their Biology, Natural Enemies and Control. Elsevier, Amsterdam, Netherlands, pp. 221–230.

Dahlsten, D.L., Six, D.L., Erbilgin, N., Raffa, K.F., Lawson, A.B., Rowney, D.L., 2003. Attraction of *Ips pini* (Coleoptera: Scolytidae) and its predators to various enantiomeric ratios of ipsdienol and lanierone in California: Implications for the augmentation and conservation of natural enemies. Environ. Entomol. 32, 1115–1122.

Dai, J., Deng, J., Du, J., 2008. Development of bisexual attractants for diamondback moth, *Plutella xylostella* (Lepidoptera: Plutellidae) based on sex pheromone and host volatiles. Appl. Entomol. Zool. 43, 631–638.

Daly, K.C., Figueredo, A.J., 2000. Habituation of sexual response in male *Heliothis* moths. Physiol. Entomol. 25, 180–190.

DeLury, N.C., Gries, G., Gries, R., Judd, G.J.R., Brown, J.J., 1999. Sex pheromone of *Ascogaster quadridentata*, a parasitoid of *Cydia pomonella*. J. Chem. Ecol. 25, 2229–2245.

Diaz-Gomez, O., Malo, E.A., Patino-Arrellano, S.A., Rojas, J.C., 2012. Pheromone trap for monitoring *Copitarsia decolora* (Lepidoptera: Noctuidae) activity in cruciferous crops in Mexico. Fla. Entomol. 95, 602–609.

Dickens, J., 1989. Green leaf volatiles enhance aggregation pheromone of boll weevil, *Anthonomus grandis*. Entomol. Exp. Appl. 52, 191–203.

Dickens, J., 1990. Specialized receptor neurons for pheromones and host plant odors in the boll weevil, *Anthonomus grandis* Boh. (Coleoptera: Curculionidae). Chem. Senses 15, 311–331.

Dorn, S., Piñero, J.C., 2009. How do key tree-fruit pests detect and colonize their hosts: Mechanisms and application for IPM. In: Aluja, M., Leskey, T.C., Vincent, C. (Eds.), Biorational Tree-Fruit Pest Management. CABI Publishers, Wallingford, UK, pp. 85–109.

Eller, F.J., Bartelt, R.J., Shasha, B.S., Schuster, D.J., Riley, D.G., Stansly, P.A., et al., 1994. Aggregation pheromone for the pepper weevil, *Anthonomus eugenii* Cano (Coleoptera: Curculionidae): Identification and field activity. J. Chem. Ecol. 20, 1537–1555.

Eller, F.J., Bartelt, R.J., 1996. Grandisoic acid, a male-produced aggregation pheromone from the plum curculio, *Conotrachelus nenuphar*. J. Nat. Prod. 59, 451–453.

Elmore, T., Ignell, R., Carlson, J., Smith, D., 2003. Targeted mutation of a *Drosophila* odor receptor defines receptor requirement in a novel class of sensillum. J. Neurosci. 23, 9906–9912.

El-Sayed, A.M., Suckling, D.M., Wearing, C.H., Byers, J.A., 2006. Potential of mass trapping for long-term pest management and eradication of invasive species. J. Econ. Entomol. 99, 1550–1564.

El-Sayed, A.M., Suckling, D.M., Byers, J.A., Jang, E.B., Wearing, C.H., 2009. Potential of 'lure and kill' in long-term pest management and eradication of invasive species. J. Econ. Entomol. 102, 815–835.

Epstein, D.L., Stelinski, L.L., Reed, T.P., Miller, J.R., Gut, L.J., 2006. Higher densities of distributed pheromone sources provide disruption of codling moth (Lepidoptera: Tortricidae) superior to that of lower densities of clumped sources. J. Econ. Entomol. 99, 1327–1333.

Epstein, D.L., Stelinski, L.L., Miller, J.R., Grieshop, M.J., Gut, L.J., 2011. Effects of reservoir dispenser height on efficacy of mating disruption of codling moth (Lepidoptera: Tortricidae) in apple. Pest Manag. Sci. 67, 975–979.

Evenden, M.L., Borden, J.H., Van Sickle, G.A., 1995. Predictive capabilities of a pheromone-based monitoring-system for western hemlock looper (Lepidoptera: Geometridae). Environ. Entomol. 24, 933–943.

Fletcher, B., 1987. The biology of dacine fruit flies. Annu. Rev. Entomol. 32, 115–144.

Flint, H.M., Kuhn, S., Horn, B., Sallam, H.A., 1974. Early season trapping of pink bollworm with gossyplure. J. Econ. Entomol. 67, 738–740.

Flint, H.M., Smith, R.L., Bariola, L.A., Horn, B.R., Forey, D.E., Kuhn, S.J., 1976. Pink bollworm: trap tests with gossyplure. J. Econ. Entomol. 69, 535–538.

Giblin-Davis, R.M., Gries, R., Gries, G., Pena-Rojas, E., Pinzon, I., Pena, J.E., et al., 1997. Aggregation pheromone of palm weevil, *Dynamis borassi*. J. Chem. Ecol. 23, 2287–2297.

Giblin-Davis, R.M., Gries, R., Crespi, B., Robertson, L.N., Hara, A.H., Gries, G., et al., 2000. Aggregation pheromones of two geographical isolates of the New Guinea sugarcane weevil, *Rhabdoscelus obscurus*. J. Chem. Ecol. 26, 2763–2780.

Gillette, N.E., Hansen, E.M., Mehmel, C.J., Mori, S.R., Webster, J.N., Erbilgin, N., et al., 2012a. Area-wide application of verbenone-releasing flakes reduces mortality of whitebark pine *Pinus albicaulis* caused by the mountain pine beetle *Dendroctonus ponderosae*. Agric. For. Entomol. 14, 367–375.

Gillette, N.E., Mehmel, C.J., Mori, S.R., Webster, J.N., Wood, D.L., Erbilgin, N., et al., 2012b. The push-pull tactic for mitigation of mountain pine beetle (Coleoptera: Curculionidae) damage in lodgepole and whitebark pines. Environ. Entomol. 41, 1575–1586.

Ginzel, M.D., 2010. Olfactory signals. In: Breed, M. Moore, J. (Eds.), Encyclopedia of Animal Behavior, vol. 2. Elsevier Ltd., Oxford, UK, pp. 584–588.

Graham, E.E., Poland, T.M., 2012. Efficacy of fluon conditioning for capturing cerambycid beetles in different trap designs and persistence on panel traps over time. J. Econ. Entomol. 105, 395–401.

Gut, L.J., Brunner, J.F., 1998. Pheromone-based management of codling moth (Lepidoptera: Tortricidae) in Washington apple orchards. J. Agric. Entomol. 15, 387–406.

Gut, L.J., Stelinski, L.L., Thompson, D.R., Miller, J.R., 2004. Behaviour-modifying chemicals: prospects and constraints in IPM. In: Koul, O., Dhaliwal, G.S., Cuperus, G.W. (Eds.), Integrated Pest Management: Potential, Constraints and Challenges. CABI Publishing, Cambridge, MA, USA, pp. 73–121.

Hall, G.M., Tittiger, C., Blomquist, G.J., Andrews, G.L., Mastick, G.S., Barkawi, L.S., et al., 2002a. Male jeffrey pine beetle, *Dendroctonus jeffreyi*, synthesizes the pheromone component frontalin in anterior midgut tissue. Insect Biochem. Mol. Biol. 32, 1525–1532.

Hall, G.M., Tittiger, C., Andrews, G.L., Mastick, G.S., Kuenzli, M., Luo, X., et al., 2002b. Midgut tissue of male pine engraver, *Ips pini*, synthesizes monoterpenoid pheromone component ipsdienol de novo. Naturwissenschaften 89, 79–83.

Hallett, R.H., Gries, G., Gries, R., Borden, J.H., Czyzewska, E., Oohlschlager, A.C., et al., 1993. Aggregation pheromones of two Asian palm weevils, *Rhynchophorus ferrugineus* and *R. vulneratus*. Naturwissenschaften 80, 328–331.

Hallett, R.H., Oehlschlager, A.C., Borden, J.H., 1999. Pheromone trapping protocols for the Asian palm weevil, *Rhynchophorus ferrugineus* (Coleoptera: Curculionidae). Int. J. Pest Manag. 45, 231–237.

Harari, A.R., Zahavi, T., Gordon, D., Anshelevich, L., Harel, M., Ovadia, S., et al., 2007. Pest management programmes in vineyards using male mating disruption. Pest Manag. Sci. 63, 769–775.

Hardee, D.D., McKibben, G.H., Gueldner, R.C., Mitchell, J.H., Tumlinson, J.H., Cross, W.H., 1972. Boll weevils in nature respond to grandlure, a synthetic pheromone. J. Econ. Entomol. 65, 97–100.

Hardee, D.D., Graves, T.M., McKibben, G.H., Johnson, W.L., Gueldner, R.C., Olsen, C.M., 1974. Slow-release formulation of grandlure synthetic pheromone of the boll weevil. J. Econ. Entomol. 67, 44–46.

Hartstack, A.W., Witz, J.A., 1981. Estimating field populations of tobacco budworm moths (Lepidoptera: Noctuidae) from pheromone trap catches. Environ. Entomol. 10, 908–914.

Hartstack, A.W., Witz, J.A., Buck, D.R., 1979. Moth traps for the tobacco budworm. J. Econ. Entomol. 72, 519–522.

Haynes, K.F., Miller, T.A., Staten, R.T., Li, W.-G., Baker, T.C., 1987. Pheromone trap for monitoring insecticide resistance in the pink-bollworm moth (Lepidoptera: Gelechiidae): new tool for resistance management. Environ. Entomol. 16, 84–89.

Heath, R.R., Teal, P.E.A., Epsky, N.D., Dueben, B.D., Hight, S.D., Bloem, S., et al., 2006. Pheromone based attractant for males of *Cactoblastis cactorum* (Lepidoptera: Pyralidae). Environ. Entomol. 35, 1469–1476.

Hedin, P.A., Dollar, D.A., Collins, J.K., Dubois, J.G., Mulder, P.G., Hedger, G.H., et al., 1997. Identification of male pecan weevil pheromone. J. Chem. Ecol. 23, 965–977.

Heuskin, S., Verheggen, F.J., Haubruge, E., Wathelet, J., Lognay, G., 2011. The use of semiochemical slow-release devices in integrated pest management strategies. Biotechnol. Agron. Soc. Environ. 15, 459–470.

Hoddle, M.S., Millar, J.G., Hoddle, C.D., Zou, Y., McElfresh, J.S., Lesch, S.M., 2011. Field optimization of the sex pheromone of *Stenoma catenifer* (Lepidoptera: Elachistidae): evaluation of lure types, trap height, male flight distances, and number of traps needed per avocado orchard for detection. Bull. Entomol. Res. 101, 145–152.

Hossain, M.S., Williams, D.G., Mansfield, C., Bartelt, R.J., Callinan, L., Il'ichev, A.L., 2006. An attract-and-kill system to control *Carpophilus* spp. in Australian stone fruit orchards. Entomol. Exp. Appl. 118, 11–19.

Hossain, M.S., Hossain, M.A.B.M., Williams, D.G., Chandra, S., 2010. Potential to reduce the spatial density of attract and kill traps required for effective control of *Carpophilus* spp. (Coleoptera: Nitidulidae) in stone fruit in Australia. Aust. J. Entomol. 49, 170–174.

Houseweart, M.W., Jennings, D.T., Sanders, C.J., 1981. Variables associated with pheromone traps for monitoring spruce budworm populations (Lepidoptera: Tortricidae). Can. Entomol. 113, 527–537.

Innocenzi, P.J., Hall, D.R., Cross, J.V., 2001. Components of male aggregation pheromone of strawberry blossom weevil, *Anthonomus rubi* Herbst. (Coleoptera: Curculionidae). J. Chem. Ecol. 27, 1203–1218.

Isaacs, R., Van Timmeren, S., 2009. Monitoring and temperature-based prediction of the whitemarked tussock moth (Lepidoptera: Lymantriidae) in blueberry. J. Econ. Entomol. 102, 637–645.

James, D.G., Bartelt, R.J., Moore, C.J., 1996. Mass-trapping of *Carpophilus* spp. (Coleoptera: Nitidulidae) in stone fruit orchards using synthetic aggregation pheromones and a coattractant: Development of a strategy for population suppression. J. Chem. Ecol. 22, 1541–1556.

Jones, B.C., Evenden, M.L., 2008. Ecological applications of pheromone trapping of *Malacosoma disstria* and *Choristoneura conflictana*. Can. Entomol. 140, 573–581.

Jones, B.C., Roland, J., Evenden, M.L., 2009. Development of a combined sex pheromone-based monitoring system for *Malacosoma disstria* (Lepidoptera: Lasiocampidae) and *Choristoneura conflictana* (Lepidoptera: Tortricidae). Environ. Entomol. 38 (2), 459–471.

Keil, T.A., 1999. Morphology and development of the peripheral olfactory organs. In: Hansson, B.S. (Ed.), Insect Olfaction. Springer, Berlin, pp. 5–48.

Keller, A., Vosshall, L., 2003. Decoding olfaction in *Drosophila*. Curr. Opin. Neurobiol. 13, 103–110.

Kim, Y., Jung, S., Kim, Y., Lee, Y., 2011. Real-time monitoring of oriental fruit moth, *Grapholita molesta*, populations using a remote sensing pheromone trap in apple orchards. J. Asia-Pacific Entomol. 14, 259–262.

Kirsch, P., 1988. Pheromones: their potential role in the control of agricultural insect pests. Am. J. Altern. Agric. 3, 83–97.

Knight, A.L., Fisher, J., 2006. Increased catch of codling moth (Lepidoptera: Tortricidae) in semiochemical-baited orange plastic delta-shaped traps. Environ. Entomol. 35, 1597–1602.

Knight, A.L., Light, D.M., 2005. Developing action thresholds for codling moth (Lepidoptera: Tortricidae) with pear ester and codlemone baited traps in apple orchards treated with sex pheromone mating disruption. Can. Entomol. 137, 739–747.

Knight, A.L., Light, D.M., 2012. Monitoring codling moth (Lepidoptera: Tortricidae) in sex pheromone-treated orchards with (E)-4,8-dimethyl-1,3,7-nonatriene or pear ester in combination with codlemone and acetic acid. Environ. Entomol. 41, 407–414.

Knight, A.L., Thomson, D.R., Cockfield, S.D., 1998. Developing mating disruption of obliquebanded leafroller (Lepidoptera: Tortricidae) in Washington state. Environ. Entomol. 27, 1080–1088.

Knight, A.L., Hilton, R., Light, D.M., 2005. Monitoring codling moth (Lepidoptera: Tortricidae) in apple with blends of ethyl (E, Z)-2,4-decadienoate and codlemone. Environ. Entomol. 34, 598–603.

Knight, A.L., Stelinski, L.L., Hebert, V., Gut, L., Light, D., Brunner, J., 2012. Evaluation of novel semiochemical dispensers simultaneously releasing pear ester and sex pheromone for mating disruption of codling moth (Lepidoptera: Tortricidae). J. Appl. Entomol. 136, 79–86.

Kogan, M., Jepson, P., 2007. Ecology, sustainable development and IPM: the human factor. In: Kogan, M., Jepson, P. (Eds.), Perspectives in Ecological Theory and Integrated Pest Management. Cambridge University Press, New York, USA, pp. 1–44.

Kolodny-Hirsch, D.M., Schwalbe, C.P., 1990. Use of disparlure in the management of the gypsy moth. In: Ridgway, R.L., Silverstein, M. Inscoe, M.N. (Eds.), Behavior-Modifying Chemicals For Insect Management. Marcel Dekker, Inc., New York, USA, pp. 363–385.

Kolodny-Hirsch, D.M., Webb, R.E., Olsen, R., Venables, L., 1990. Mating disruption of gypsy moth (Lepidoptera: Lymantriidae) following repeated ground application of racemic disparlure. J. Econ. Entomol. 83, 1972–1976.

Kreutz, J., Zimmermann, G., Vaupel, O., 2004. Horizontal transmission of the entomopathogenic fungus *Beauveria bassiana* among the spruce bark beetle, *Ips typographus* (Col., Scolytidae) in the laboratory and under field conditions. Biocontrol Sci. Technol. 14, 837–848.

Knight, A.L., 2007. Influence of within orchard trap placement on catch of codling moth (Lepidoptera: Tortricidae) in sex pheromone-treated orchards. Environ. Entomol. 36, 425–432.

Kroschel, J., Zegarra, O., 2010. Attract-and-kill: A new strategy for the management of the potato tuber moths *Phthorimaea operculella* (Zeller) and *Symmetrischema tangolias* (Gyen) in potato: Laboratory experiments towards optimising pheromone and insecticide concentration. Pest Manag. Sci. 66, 490–496.

Kuhar, T.P., Mori, K., Dickens, J.C., 2006. Potential of a synthetic aggregation pheromone for integrated pest management of Colorado potato beetle. Agric. For. Entomol. 8, 77–81.

Landolt, P.J., Phillips, T.W., 1997. Host plant influences on sex pheromone behavior of phytophagous insects. Annu. Rev. Entomol. 42, 371–391.

Latheef, M.A., Witz, J.A., Lopez, J.D., 1991. Relationships among pheromone trap catches of male corn earworm moths (Lepidoptera: Noctuidae), egg numbers, and phenology in corn. Can. Entomol. 123, 271–281.

Laurent, P., Frérot, B., 2007. Monitoring of European corn borer with pheromone-baited traps: Review of trapping system basics and remaining problems. J. Econ. Entomol. 100, 1797–1807.

Law, J., Regnier, F., 1971. Pheromones. Annu. Rev. Biochem. 40, 533–548.

Leskey, T.C., Wright, S.E., 2004. Monitoring plum curculio, *Conotrachelus nenuphar* (Coleoptera: Curculionidae), populations in apple and peach orchards in the mid-Atlantic. J. Econ. Entomol. 97, 79–88.

Leskey, T.C., Prokopy, R.J., Wright, S.E., Phelan, P., Haynes, L.W., 2001. Evaluation of individual components of plum odor as potential attractants for adult plum curculios. J. Chem. Ecol. 27, 1–17.

Leskey, T.C., Pinero, J.C., Prokopy, R.J., 2008. Odor-baited trap trees: A novel management tool for plum curculio (Coleoptera: Curculionidae). J. Econ. Entomol. 101, 1302–1309.

Leskey, T.C., Bergh, J.C., Walgenbach, J.F., Zhang, A., 2009. Evaluation of pheromone-based management strategies for dogwood borer (Lepidoptera: Sesiidae) in commercial apple orchards. J. Econ. Entomol. 102, 1085–1093.

Leskey, T.C., Wright, S.E., Short, B.D., Khrimian, A., 2012. Development of behaviorally-based monitoring tools for the brown marmorated stink bug (Heteroptera: Pentatomidae) in commercial tree fruit orchards. J. Entomol. Sci. 47, 76–85.

Light, D.M., Flath, R.A., Buttery, R.G., Zalom, F.G., Rice, R.E., Dickens, J.C., et al., 1993. Host-plant green-leaf volatiles synergize the synthetic sex pheromones of the corn earworm and codling moth (Lepidoptera). Chemoecology 4, 145–152.

Linn Jr., C.E., Roelofs, W.L., 1995. Pheromone communication in moths and its role in the speciation process. In: Lambert, D.M., Spencer, H.G. (Eds.), Speciation and the Recognition Concept: Theory and Application. Johns Hopkins University Press, Baltimore, MD, USA, pp. 263–300.

Long, B.B., Averill, A.L., 2003. Compensatory response of cranberry to simulated damage by cranberry weevil (*Anthonomus musculus* Say) (Coleoptera: Curculionidae). J. Econ. Entomol. 96, 407–412.

Lopez, J.D., Shaver, T.N., Dickerson, W.A., 1990. Population monitoring of *Heliothis.* spp. using pheromones. In: Ridgway, R.L., Silverstein, R.M., Inscoe, M.N. (Eds.),

Behavior-Modifying Chemicals For Insect Management. Marcel Dekker, Inc., New York, NY, pp. 473–496.

Maki, E.C., Millar, J.G., Rodstein, J., Hanks, L.M., Barbour, J.D., 2011. Evaluation of mass trapping and mating disruption for managing *Prionus californicus* (Coleoptera: Cerambycidae) in hop production yards. J. Econ. Entomol. 104, 933–938.

Martinez-Ferrer, M.T., Campos, J.M., Fibla, J.M., 2012. Field efficacy of *Ceratitis capitata* (Diptera: Tephritidae) mass trapping technique on clementine groves in Spain. J. Appl. Entomol. 136, 181–190.

McBrien, H.L., Judd, G.J.R., Borden, J.H., 1994. *Campylomma verbasci* (Heteroptera: Miridae): Pheromone-based seasonal flight patterns and prediction of nymphal densities in apple orchards. J. Econ. Entomol. 87, 1224–1229.

McGhee, P.S., Epstein, D.L., Gut., L.J., 2011. Quantifying the benefits of areawide pheromone mating disruption programs that target codling moth (Lepidoptera: Tortricidae). Am. Entomol. 57, 94–100.

McKibben, G.H., Smith, J.W., McGovern, W.L., 1990. Design of an attract-and-kill device for the boll weevil (Coleoptera: Curculionidae). J. Entomol. Sci. 25, 581–586.

Miller, D.R., Asaro, C., Berisford, C.W., 2005. Attraction of southern pine engravers and associated bark beetles (Coleoptera: Scolytidae) to ipsenol, ipsdienol, and lanierone in southeastern United States. J. Econ. Entomol. 98, 2058–2066.

Miller, J.R., Cowles, R.S., 1990. Stimulo-deterrent diversion: a concept and its possible application to onion maggot control. J. Chem. Ecol. 16, 3197–3212.

Miller, J.R., Gut, L.J., de Lame, F.M., Stelinski, L.L., 2006a. Differentiation of competitive vs. non-competitive mechanisms mediating disruption of moth sexual communication by point sources of sex pheromone (part I): Theory. J. Chem. Ecol. 32, 2089–2114.

Miller, J.R., Gut, L.J., de Lame, F.M., Stelinski, L.L., 2006b. Differentiation of competitive vs. non-competitive mechanisms mediating disruption of moth sexual communication by point sources of sex pheromone (part 2): Case studies. J. Chem. Ecol. 32, 2115–2143.

Miller, J.R., McGhee, P.S., Siegert, P.Y., Adams, C.G., Huang, J., Grieshop, M.J., et al., 2010. General principles of attraction and competitive attraction as revealed by large-cage studies of moths responding to sex pheromone. Proc. Natl. Acad. Sci. U.S.A. 107, 22–27.

Mitchell, E.B., Hardee, D.D., 1974. In-field traps: A new concept in survey and suppression of low populations of boll weevils. J. Econ. Entomol. 67, 506–508.

Nalyanya, G., Moore, C.B., Schal, C., 2000. Integration of repellents, attractants, and insecticides in a 'push-pull' strategy for managing German cockroach (Dictyoptera: Blattellidae) populations. J. Med. Entomol. 37, 427–434.

Nealis, V.G., Silk, P., Turnquist, R., Wu, J., 2010. Baited pheromone traps track changes in populations of western blackheaded budworm (Lepidoptera: Tortricidae). Can. Entomol. 142, 458–465.

Nishisue, K., Sunamura, E., Tanaka, Y., Sakamoto, H., Suzuki, S., Fukumoto, T., et al., 2010. Long-term field trial to control the invasive argentine ant (Hymenoptera: Formicidae) with synthetic trail pheromone. J. Econ. Entomol. 103, 1784–1789.

Ochieng, S.A., Park, K.C., Baker, T.C., 2002. Host plant volatiles synergize responses of sex pheromone-specific olfactory receptor neurons in male *Helicoverpa zea*. J. Comp. Physiol. A Neuroethol. Sens. Neural Behav. Physiol. 188, 325–333.

Oehlschlager, A.C., Pierce, H.D., Morgan, B., Wimalaratne, P.D.C., Slessor, K.N., King, G.G.S., 1992. Chirality and field activity of rhynchophorol, the aggregation pheromone of the American palm weevil. Naturwissenschaften 79, 134–135.

Oehlschlager, A.C., McDonald, R.S., Chinchilla, C.M., Patschke, S.N., 1995a. Influence of a pheromone-based mass-trapping system on the distribution of *Rhynchophorus palmarum* (Coleoptera: Curculionidae) in oil palm. Environ. Entomol. 24, 1005–1012.

Oehlschlager, A.C., Prior, R.N.B., Perez, A.L., Gries, R., Gries, G., Pierce, H.D., et al., 1995b. Structure, chirality, and field testing of a male-produced aggregation pheromone of asian palm weevil *Rhynchophorus bilineatus* (Montr.) (Coleoptera: Curculionidae). J. Chem. Ecol. 21, 1619–1629.

Oehlschlager, A.C., Chinchilla, C., Castillo, G., Gonzalez, L., 2002. Control of red ring disease by mass trapping of *Rhynchophorus palmarum* (Coleoptera: Curculionidae). Fla. Entomol. 85, 507–513.

Peng, C., Gu, P., Li, J., Chen, Q., Feng, C., Luo, H., et al., 2012. Identification and field bioassay of the sex pheromone of *Trichophysetis cretacea* (Lepidoptera: Crambidae). J. Econ. Entomol. 105, 1566–1572.

Perez, A.C., Gries, G., Gries, R., Giblin-Davis, R.M., Oehlschlager, A.C., 1994. Pheromone chirality of African palm weevil, *Rhynchophorus phoenicis* (F.) and palmetto weevil, *Rhynchophorus cruentatus* (F.) (Coleoptera: Curculionidae). J. Chem. Ecol. 20, 2653–2671.

Perez, A.L., Campos, Y., Chinchilla, C., Oehlschlager, A., Gries, G., Gries, R., et al., 1997. Aggregation pheromones and host kairomones of West Indian sugarcane weevil, *Metamasius hemipterus sericeus*. J. Chem. Ecol. 23, 869–888.

Perkins, J.H., 1980. Boll weevil eradication. Science 207, 1044–1050.

Phillips, T., 1997. Semiochemicals of stored-product insects: Research and applications. J. Stored Prod. Res. 33, 17–30.

Pickett, J., Wadhams, L., Woodcock, C., 1997. Developing sustainable pest control from chemical ecology. Agric. Ecosyst. Environ. 64, 149–156.

Piñero, J.C., Prokopy, R.J., 2003. Field evaluation of plant odor and pheromonal combinations for attracting plum curculios. J. Chem. Ecol. 29, 2735–2748.

Piñero, J.C., Ruiz-Montiel, C., 2012. Ecología Química y Manejo de Picudos (Coleóptera: Curculionidae) de Importancia Económica. In: Rojas, J.C., Malo, E.A. (Eds.), Temas Selectos en Ecología Química de Insectos. El Colegio de la Frontera Sur, México, pp. 361–400.

Piñero, J.C., Wright, S.E., Prokopy, R.J., 2001. Response of plum curculio (Coleoptera: Curculionidae) to odor-baited traps near woods. J. Econ. Entomol. 94, 1386–1397.

Piñero, J.C., Agnello, A.M., Tuttle, A., Leskey, T.C., Faubert, H., Koehler, G., et al., 2011. Effectiveness of odor-baited trap trees for plum curculio (Coleoptera: Curculionidae) monitoring in commercial apple orchards in the northeast. J. Econ. Entomol. 104, 1613–1621.

Progar, R.A., 2005. Five-year operational trial of verbenone to deter mountain pine beetle (*Dendroctonus ponderosae*; Coleoptera: Scolytidae) attack of lodgepole pine (*Pinus contorta*). Environ. Entomol. 34, 1402–1407.

Prokopy, R., Ziegler, J., Wong, T., 1978. Deterrence of repeated oviposition by fruit-marking pheromone in *Ceratitis capitata* (Diptera: Tephritidae). J. Chem. Ecol. 4, 55–63.

Prokopy, R.J., Phelan, P.L., Wright, S.E., Minalga, A.J., Barger, R., Leskey, T.C., 2001. Compounds from host fruit odor attractive to adult plum curculios (Coleoptera: Curculionidae). J. Entomol. Sci. 36, 122–134.

Prokopy, R.J., Chandler, B.W., Dynok, S.A., Piñero, J.C., 2003. Odor-baited trap trees: A new approach to monitoring plum curculio (Coleoptera: Curculionidae). J. Econ. Entomol. 96, 826–834.

Prokopy, R.J., Jacome, I., Gray, E., Trujillo, G., Ricci, M., Pinero, J.C., 2004. Using odor-baited trap trees as sentinels to monitor plum curculio (Coleoptera: Curculionidae) in apple orchards. J. Econ. Entomol. 97, 511–517.

Raffa, K.F., 1991. Temporal and spatial disparities among bark beetles, predators, and associates responding to synthetic bark beetle pheromones: *Ips pini* (Coleoptera: Scolytidae) in Wisconsin. Environ. Entomol. 20, 1665–1679.

Raffa, K.F., Dahlsten, D.L., 1995. Differential responses among natural enemies and prey to bark beetle pheromones. Oecologia 102, 17–23.

Ramirez-Lucas, P., Malosse, C., Ducrot, P., Lettere, M., Zagatti, P., 1996. Chemical identification, electrophysiological and behavioral activities of the pheromone of *Metamasius hemipterus* (Coleoptera: Curculionidae). Bioorg. Med. Chem. 4, 323–330.

Reddy, G.V.P., Guerrero, A., 2001. Optimum timing of insecticide applications against diamondback moth *Plutella xylostella* in cole crops using threshold catches in sex pheromone traps. Pest Manag. Sci. 57, 90–94.

Reddy, G.V.P., Guerrero, A., 2004. Interactions of insect pheromones and plant semiochemicals. Trends Plant Sci. 9, 253–261.

Reddy, G.V.P., Cruz, Z.T., Bamba, J., Muniappan, R., 2005. Development of a semiochemical-based trapping method for the New Guinea sugarcane weevil, *Rhabdoscelus obscurus* in Guam. J. Appl. Entomol. 129, 65–69.

Reddy, G.V.P., Shi, P., Mann, C.R., Mantanona, D.M.H., Dong, Z., 2012. Can a semiochemical-based trapping method diminish damage levels caused by *Rhabdoscelus obscurus* (Coleoptera: Curculionidae)? Ann. Entomol. Soc. Am. 105, 693–700.

Regnier, F.E., Law, J.H., 1968. Insect pheromones. J. Lipid Res. 9, 541–551.

Reinke, M.D., Miller, J.R., Gut, L.J., 2012. Potential of high-density pheromone-releasing microtraps for control of codling moth *Cydia pomonella* and obliquebanded leafroller *Choristoneura rosaceana*. Physiol. Entomol. 37, 53–59.

Renou, M., Guerrero, A., 2000. Insect parapheromones in olfaction research and semiochemical-based pest control strategies. Annu. Rev. Entomol. 45, 605–630.

Ridgway, R.L., Inscoe, M.N., Dickerson, W.L., 1990. Role of the boll weevil pheromone in pest management. In: Ridgway, R.L., Silverstein, R.M., Inscoe, M.N. (Eds.), Behavior-Modifying Chemicals for Insect Management. Marcel Dekker, Inc., New York, NY, pp. 437–471.

Riley, D.G., Schuster, D.J., 1994. Pepper weevil adult response to colored sticky traps in pepper fields. Southwestern Entomol. 19, 93–107.

Roditakis, E., Couzin, I.D., Balrow, K., Franks, N.R., Charnley, A.K., 2000. Improving secondary pick up of insect fungal pathogen conidia by manipulating host behaviour. Ann. Appl. Biol. 137, 329–335.

Rodriguez-Saona, C., Blaauw, B.R., Issacs, R., 2012. Manipulation of natural enemies in agroecosystems: Habitat and semiochemicals for sustainable insect pest control. In: Integrated Pest Management and Pest Control – Current and Future Tactics, pp. 89–126.

Ross, D., Daterman, G., 1997. Using pheromone-baited traps to control the amount and distribution of tree mortality during outbreaks of the douglas fir beetle. For. Sci. 43, 65–70.

Roubos, C.R., Liburd, O.E., 2008. Effect of trap color on captures of grape root borer (Lepidoptera: Sesiidae) males and non-target insects. J. Agric. Urban Entomol. 25, 99–109.

Ruiz-Montiel, C., Garcia-Coapio, G., Rojas, J.C., Malo, E.A., Cruz-Lopez, L., del Real, I., et al., 2008. Aggregation pheromone of the agave weevil, *Scyphophorus acupunctatus*. Entomol. Exp. Appl. 127, 207–217.

Ruiz-Montiel, C., Rojas, J.C., Cruz-Lopez, L., Gonzalez-Hernandez, H., 2009. Factors affecting pheromone release by *Scyphophorus acupunctatus* (Coleoptera: Curculionidae). Environ. Entomol. 38, 1423–1428.

Sallam, M.N., Peck, D.R., McAvoy, C.A., Donald, D.A., 2007. Pheromone trapping of the sugarcane weevil borer, *Rhabdoscelus obscurus* (Boisduval) (Coleoptera: Curculionidae): An evaluation of trap design and placement in the field. Aust. J. Entomol. 46, 217–223.

Sanders, C., 1986a. Evaluation of high-capacity, nonsaturating sex-pheromone traps for monitoring population densities of spruce budworm (Lepidoptera: Tortricidae). Can. Entomol. 118, 611–619.

Sanders, C., 1986b. Accumulated dead insects and killing agents reduce catches of spruce budworm (Lepidoptera: Tortricidae) male moths in sex-pheromone traps. J. Econ. Entomol. 79, 1351–1353.

Sanders, C., 1988. Monitoring spruce budworm population density with sex pheromone traps. Can. Entomol. 120, 175–183.

Sauphanor, B., Brosse, V., Bouvier, J.C., Speich, P., Micoud, A., Martinet, C., 2000. Monitoring resistance to diflubenzuron and deltamethrin in french codling moth populations (*Cydia pomonella*). Pest Manag. Sci. 56, 74–82.

Schneider, D., 1969. Insect olfaction: Deciphering system for chemical messages. Science 163, 1031–1037.

Sharov, A.A., Leonard, D., Liebhold, A.M., Clemens, N.S., 2002. Evaluation of preventive treatments in low-density gypsy moth populations using pheromone traps. J. Econ. Entomol. 95, 1205–1215.

Shearer, P.W., Riedl, H., 1994. Comparison of pheromone trap bioassays for monitoring insecticide resistance of *Phyllonorycter elmaella* (Lepidoptera: Gracillariidae). J. Econ. Entomol. 87, 1450–1454.

Shepherd, R.F., Gray, T.G., Chorney, R.J., Daterman, G.E., 1985. Pest-management of douglas fir tussock moth, *Orgyia pseudotsugata* (Lepidoptera: Lymantriidae) – monitoring endemic populations with pheromone traps to detect incipient outbreaks. Can. Entomol. 117, 839–848.

Smith, J.W., 1998. Boll weevil eradication: area-wide pest management. Ann. Entomol. Soc. Am. 91, 239–247.

Soroka, J.J., Bartelt, R.J., Zilkowski, B.W., Cosse, A.A., 2005. Responses of flea beetle *Phyllotreta cruciferae* to synthetic aggregation pheromone components and host plant volatiles in field trials. J. Chem. Ecol. 31, 1829–1843.

Stelinski, L.L., Oakleaf, R., Rodriguez-Saona, C., 2007. Oviposition deterring pheromone deposited on blueberry fruit by the parasitic wasp, *Diachasma alloeum*. Behaviour 144, 429–445.

Stelinski, L.L., Gut, L.J., Miller, J.R., 2013. An attempt to increase efficacy of moth mating disruption by co-releasing pheromones with kairomones and to understand possible underlying mechanisms of this technique. Environ. Entomol. 42, 158–166.

Strong, W.B., Millar, J.G., Grant, G.G., Moreira, J.A., Chong, J.M., Rudolph, C., 2008. Optimization of pheromone lure and trap design for monitoring the fir coneworm, *Dioryctria abietivorella*. Entomol. Exp. Appl. 126, 67–77.

Suckling, D.M., 2000. Issues affecting the use of pheromones and other semiochemicals in orchards. Crop Prot. 19, 677–683.

Suckling, D.M., Gibb, A.R., Burnip, G.M., Delury, N.C., 2002. Can parasitoid sex pheromones help in insect biocontrol? A case study of codling moth (Lepidoptera: Tortricidae) and its parasitoid *Ascogaster quadridentata* (Hymenoptera: Braconidae). Environ. Entomol. 31, 947–952.

Suckling, D.M., Gibb, A.R., Johnson, T., Hall, D.R., 2006. Examination of sex attractants for monitoring weed biological control agents in Hawaii. Biocontrol Sci. Technol. 16, 919–927.

Suckling, D.M., Walker, J.T.S., Shaw, P.W., Manning, L., Lo, P., Wallis, R., et al., 2007. Trapping *Dasinuera mali* (Diptera: Cecidomyiidae) in apples. J. Econ. Entomol. 100, 745–751.

Suh, C.P.-C., Armstrong, J.S., Spurgeon, D.W., Duke, S., 2009. Comparisons of boll weevil (Coleoptera: Curculionidae) pheromone traps with and without kill strips. J. Econ. Entomol. 102, 183–186.

Szendrei, Z., Malo, E., Stelinski, L., Rodriguez-Saona, C., 2009. Response of cranberry weevil (Coleoptera: Curculionidae) to host plant volatiles. Environ. Entomol. 38, 861–869.

Szendrei, Z., Averill, A., Alborn, H., Rodriguez-Saona, C., 2011. Identification and field evaluation of attractants for the cranberry weevil, *Anthonomus musculus* Say. J. Chem. Ecol. 37, 387–397.

Taban, A.H., Fu, J., Blake, J., Awano, A., Tittiger, C., Blomquist, G.J., 2006. Site of pheromone biosynthesis and isolation of HMG-CoA reductase cDNA in the cotton boll weevil, *Athonomus grandis*. Arch. Insect Biochem. Physiol. 62, 153–163.

Tafoya, F., Whalon, M.E., Vandervoot, C., Coombs, A.B., Cibrian-Tovar, J., 2007. Aggregation pheromone of *Metamasius spinolae* (Coleoptera: Curculionidae): Chemical analysis and field test. Environ. Entomol. 36, 53–57.

Teixeira, L.A.F., Miller, J.R., Epstein, D.L., Gut, L.J., 2010. Comparison of mating disruption and mass trapping with Pyralidae and Sesiidae moths. Entomol. Exp. Appl. 137, 176–183.

Thorpe, K.W., Tcheslavskaia, K.S., Tobin, P.C., Blackburn, L.M., Leonard, D.S., Roberts, E.A., 2007. Persistent effects of aerial applications of disparlure on gypsy moth: trap catch and mating success. Entomol. Exp. Appl. 125, 223–229.

Tinzaara, W., Dicke, M., van Huis, A., Gold, C.S., 2002. Use of infochemicals in pest management with special reference to the banana weevil, *Cosmopolites sordidus*

(Germar) (Coleoptera: Curculionidae). Insect Sci. Appl. 22, 241–261.

Tinzaara, W., Gold, C.S., Dicke, M., Van Huis, A., Ragama, P.E., Nankinga, C.M., et al., 2007. The use of aggregation pheromone to enhance dissemination of *Beauveria bassiana* for the control of the banana weevil in Uganda. Biocontrol Sci. Technol. 17, 111–124.

Tobin, P.C., Zhang, A., Onufrieva, K., Leonard, D.S., 2011. Field evaluation of effect of temperature on release of disparlure from a pheromone-baited trapping system used to monitor gypsy moth (Lepidoptera: Lymantriidae). J. Econ. Entomol. 104, 1265–1271.

Tollerup, K.E., Rucker, A., Shearer, P.W., 2012. Whole farm mating disruption to manage *Grapholita molesta* (Lepidoptera: Tortricidae) in diversified New Jersey orchards. J. Econ. Entomol. 105, 1712–1718.

Trimble, R.M., 1993. Efficacy of mating disruption for controlling the grape berry moth, *Endopiza viteana* (Clemens) (Lepidoptera: Tortricidae), a case study over three consecutive growing seasons. Can. Entomol. 125, 1–9.

Trimble, R.M., 1995. Mating disruption for controlling the codling moth, Cydia pomonella (L) (Lepidoptera: Tortricidae), in organic apple production in southwestern Ontario. Can. Entomol. 127, 493–505.

Trimble, R.M., 2007. Comparison of efficacy of pheromone dispensing technologies for controlling the grape berry moth (Lepidoptera: Tortricidae) by mating disruption. J. Econ. Entomol. 100, 1815–1820.

Tumlinson, J.H., Gueldner, R.C., Hardee, D.D., Thompson, A.C., Hedin, P.A., Minyard, J.P., 1971. Identification and synthesis of the four compounds comprising the boll weevil sex attractant. J. Org. Chem. 36, 2616–2621.

Vacas, S., Alfaro, C., Navarro-Llopis, V., Zarzo, M., Primo, J., 2009. Study on the optimum pheromone release rate for attraction of *Chilo suppressalis* (Lepidoptera: Pyralidae). J. Econ. Entomol. 102, 1094–1100.

Vacas, S., Vanaclocha, P., Alfaro, C., Primo, J., Jesus Verdu, M., Urbaneja, A., et al., 2012. Mating disruption for the control of *Aonidiella aurantii* Maskell (Hemiptera: Diaspididae) may contribute to increased effectiveness of natural enemies. Pest Manag. Sci. 68, 142–148.

Valles, S.M., Capinera, J.L., Teal, P.E.A., 1991. Evaluation of pheromone trap design, height, and efficiency for capture of male *Diaphania nitidalis* (Lepidoptera: Pyralidae) in a field cage. Environ. Entomol. 20, 1274–1278.

Varela, L.G., Welter, S.C., Jones, V.P., Brunner, J.F., Riedl, H., 1993. Monitoring and characterization of insecticide resistance in codling moth (Lepidoptera: Tortricidae) in four western states. J. Econ. Entomol. 86, 1–10.

Vargas, R.I., Stark, J.D., Hertlein, M., Neto, A.M., Coler, R., Pinero, J.C., 2008. Evaluation of SPLAT with spinosad and methyl eugenol or cue-lure for 'attract-and-kill' of oriental and melon fruit flies (Diptera: Tephritidae) in Hawaii. J. Econ. Entomol. 101, 759–768.

Vickers, R.A., Thwaite, W.G., Williams, D.G., Nicholas, A.H., 1998. Control of codling moth in small plots by mating disruption: alone and with limited insecticide. Entomol. Exp. Appl. 86, 229–239.

Villavaso, E.J., McGovern, W.L., Wagner, T.L., 1998. Efficacy of bait sticks versus pheromone traps for removing boll weevils (Coleoptera: Curculionidae) from released populations. J. Econ. Entomol. 91, 637–640.

Vincent, C., Chouinard, G., Hill, S.B., 1999. Progress in plum curculio management: a review. Agric. Ecosyst. Environ. 73, 167–175.

Wall, C., 1990. Principle of monitoring. In: Ridgway, R.L., Silverstein, R.M., Inscoe, M.N. (Eds.), Behavior-Modifying Chemicals for Insect Management. Marcel Dekker, Inc., New York, NY, pp. 9–23.

Waterworth, R.A., Redak, R.A., Millar, J.G., 2011. Pheromone baited traps for assessment of seasonal activity and population densities of mealybug species (Hemiptera: Pseudococcidae) in nurseries producing ornamental plants. J. Econ. Entomol. 104, 555–565.

Webb, R.E., Leonhardt, B.A., Plimmer, J.R., Tatman, K.M., Boyd, V.K., Coehn, D.L., et al., 1990. Effect of racemic disparlure released from grids of plastic ropes on mating success of gypsy moth (Lepidoptera: Lymantriidae) as influenced by dose and by population-density. J. Econ. Entomol. 83, 910–916.

Weissling, T.J., Giblin-Davis, R.M., Gries, G., Gries, R., Perez, A.L., Pierce, H.D., et al., 1994. Aggregation pheromone of palmetto weevil, *Rhynchophorus cruentatus* (F.) (Coleoptera: Curculionidae). J. Chem. Ecol. 20, 505–515.

Wenninger, E.J., Averill, A.L., 2006. Mating disruption of oriental beetle (Coleoptera: Scarabaeidae) in cranberry using retrievable, point-source dispensers of sex pheromones. Environ. Entomol. 35, 458–464.

Weslien, J., Annila, E., Bakke, A., Bejer, B., Eidmann, H.H., Narvestad, K., et al., 1989. Estimating risks for spruce bark beetle (*Ips typographus* (L.)) damage using pheromone-baited traps and trees. Scand. J. For. Res. 4, 87–98.

Witzgall, P., Stelinski, L., Gut, L., Thomson, D., 2008. Codling moth management and chemical ecology. Annu. Rev. Entomol. 53, 503–522.

Witzgall, P., Kirsch, P., Cork, A., 2010. Sex pheromones and their impact on pest management. J. Chem. Ecol. 36, 80–100.

Wong, J.C.H., Mitchell, R.F., Striman, B.L., Millar, J.G., Hanks, L.M., 2012. Blending synthetic pheromones of cerambycid beetles to develop trap lures that simultaneously attract multiple species. J. Econ. Entomol. 105, 906–915.

Yamanaka, T., 2007. Mating disruption or mass trapping? Numerical simulation analysis of a control strategy for lepidopteran pests. Popul. Ecol. 49, 75–86.

Yang, Z., Bengtsson, M., Witzgall, P., 2004. Host plant volatiles synergize response to sex pheromone in codling moth, *Cydia pomonella*. J. Chem. Ecol. 30, 619–629.

Yasuda, K., 1999. Auto-infection system for the sweet potato weevil, *Cylas formicarius* (Fabricius) (Coleoptera: Curculionidae) with entomopathogenic fungi, *Beauveria bassiana* using a modified sex pheromone trap in the field. Appl. Entomol. Zool. 34, 501–505.

Yongmo, W., Feng, G., Xianghui, L., Feng, F., Lijun, W., 2005. Evaluation of mass-trapping for control of tea tussock moth *Euproctis pseudoconspersa* (Strand) (Lepidoptera: Lymantriidae) with synthetic sex pheromone in south China. Int. J. Pest Manag. 51, 289–295.

Zada, A., Falach, L., Byers, J.A., 2009. Development of sol-gel formulations for slow release of pheromones. Chemoecology 19, 37–45.

Zhang, A., Leskey, T.C., Bergh, J.C., Walgenbach, J.F., 2005. Sex pheromone of the dogwood borer, *Synanthedon scitula*. J. Chem. Ecol. 31, 2463–2479.

Zhu, J., Zhang, A., Park, K., Baker, T., Lang, B., Jurenka, R., et al., 2006. Sex pheromone of the soybean aphid, *Aphis glycines* Matsumura, and its potential use in semiochemical-based control. Environ. Entomol. 35, 249–257.

Zijlstra, C., Lund, I., Justesen, A.F., Nicolaisen, M., Jensen, P.K., Bianciotto, V., et al., 2011. Combining novel monitoring tools and precision application technologies for integrated high-tech crop protection in the future (a discussion document). Pest Manag. Sci. 67, 616–625.

CHAPTER

10

Role of Entomopathogenic Fungi in Integrated Pest Management

Margaret Skinner[1], *Bruce L. Parker*[1] *and Jae Su Kim*[2]

[1]The University of Vermont, VT, USA, [2]Chonbuk National University, Jeollabuk-do, Republic of Korea

10.1 INTRODUCTION

For centuries, agrochemicals have been used to protect crops from pests and diseases. They have been responsible for maintaining and increasing the quality and quantity of food and fibre worldwide. However, their extensive use has resulted in pest resistance, resurgence of secondary pests, and a disruption or elimination of natural enemy complexes reducing the efficacy of natural control processes. These factors, combined with concerns about environmental impacts and human safety, have provided the momentum to develop more environmentally safe strategies that are cost-effective and reliable. Integrated pest management (IPM) is a comprehensive approach to crop production, combining a broad array of compatible techniques such as sanitation, survey and detection, use of resistant varieties, cultural manipulations, trap and companion cropping, biological control, and even agricultural chemicals when necessary, to maintain pests below economic injury levels. This is a shift from the traditional individual pest-centred strategies that relied heavily on chemical pesticides to a more holistic approach, viewing the entire crop production system together to manage rather than eradicate the pests.

Several microbial agents have been developed to manage insect pests, including fungi, bacteria and viruses. Some such as *Bacillus thuringiensis* (Bt) have shown dramatic success, and are integral components of pest management, whereas fungi remain underutilized microbial agents in IPM. Generally speaking, this group of microorganisms specifically infects insects, not plants or other animals. Over 700 different fungal species from at least 90 genera are known to be pathogenic to insects (Khachatourians and Sohail, 2008). However, a fairly select few fungal genera are well-recognized as entomopathogens, including *Beauveria*, *Metarhizium*, *Isaria*, *Lecanicillium*, *Hirsutella* and *Entomophthorales*. Various fungal-based products containing *Beauveria bassiana*, *Metarhizium anisopliae*, *Isaria* spp. and *Lecanicillium* spp. have been developed for use against a wide variety of pests in forest, field, and greenhouse environments, and against structural and household pests. At this

D. P. Abrol (Ed): Integrated Pest Management.
DOI: http://dx.doi.org/10.1016/B978-0-12-398529-3.00011-7

time, commercial formulations of a range of entomopathogenic fungi are available to farmers in most parts of the world.

Though it is common in nature to find insects infected with a fungus, and epizootics are observed having a significant impact on insect pest populations, mortality from fungal infection rarely occurs naturally at sufficiently high levels or early enough in a pest outbreak to prevent economic damage. Therefore, despite their great potential, many advantages and extensive research, few entomopathogenic fungi are commonly used by growers. However, in recent years due to greater commercialization, their use has expanded. Among the microbial biological control agents, fungal pathogens have received particular interest because of their effective management of pests with piercing and sucking mouthparts (Wraight et al., 2001). Experience has taught growers, pest managers, researchers, and administrators that rarely is there a silver bullet that will solve any pest problem on a long-term basis. A multi-faceted approach combining all available IPM tools and strategies together into a holistic compatible strategy has a far better likelihood of suppressing pests. Entomopathogenic fungi are poised to become a more significant component of IPM. This chapter provides a review of these promising microbials, and factors that contribute to their success and failure under field conditions. It also describes two agricultural scenarios where entomopathogenic fungi have been developed or are being evaluated as an integral and essential component of IPM.

10.2 AN OVERVIEW OF ENTOMOPATHOGENIC FUNGI

10.2.1 How They Work

Entomopathogenic fungi comprise a diverse group of microorganisms that collectively can be found in a wide range of environmental conditions (including arid to tropical settings, terrestrial to aquatic habitats and arctic to temperate climates) and infecting a broad array of insects (Goettel et al., 2000; Meyling et al., 2012; Scholte et al., 2004; Tanada and Kaya, 1993).

Though an insect can become infected by ingesting infective propagules of entomopathogenic fungi, it is more common for spores (conidia, zoospores, ascospores, etc.) to attach to the cuticle of the host insect. Attachment can be passive in spores that are covered with a sticky or slimy substance (e.g. *Lecanicillium* spp., *Entomophthorales* spp. and *Hirsutella* spp.). Alternatively, deuteromycetous fungi produce dry conidia, which have special structures (rodlets) that attach to the cuticle. A complex group of factors interact to stimulate the spores to germinate, including ambient humidity and temperature, nutritional and chemical cues and cuticular extracts from the host (Tanada and Kaya, 1993). The germinating spore produces a germ tube with a penetration peg or appressorium which uses both enzymatic and physical pressure to penetrate the insect cuticle. Fungi can also enter through openings in the insect's body, such as spiracles, sensory pores or wounds. Once inside, the fungus multiplies, feeding on the insect's internal contents. The host is killed by one or more factors, including nutritional deficiency, tissue destruction or disruption of normal biological functions through clogging of the vessels with blastospores, or by toxic substances from the fungus that are released into the insect (Goettel et al., 2000; Tanada and Kaya, 1993). After the insect host is killed and all nutrition has been consumed, hyphae grow out of the cadaver, particularly at the margins of the intersegmental regions, and produce resting or infective spores that promote the spread of the fungus. Several means of spore dispersal are known, including passive dispersal through water, wind or secondary agents (insects or other adjacent arthropods, etc.), or by forcible discharge from the sporophores.

Entomopathogenic fungi fall into two general categories. Many are facultative saprophytes, such as *Beauveria* and *Metarhizium* spp., which are parasitic microbes that attack and grow on or in a living organism, but have the capacity to survive and reproduce on non-living substances. This group of fungi can readily be mass-produced on artificial media or a solid substrate, which makes them highly desirable for commercialization. Others are obligate parasites, such as *Entomophthora* spp., which require a suitable live host to survive, and are often highly specific, making them ideal for IPM in that they are unlikely to negatively impact non-target organisms. However, mass production of this group of fungi is complicated by the fact that they require a live host. Suppression of gypsy moth (*Lymantria dispar*), a forest pest in the northeastern US with *Entomophthora maimaiga*, is a dramatic example of the effectiveness of this fungal group. It is believed to be largely responsible for the sustained suppression of this pest that for over 125 years caused intermittent severe defoliation of hardwood trees over a wide geographical area. Because of the obligate nature of *E. maimaiga*, gypsy moth larvae infected with a Japanese strain of the fungus were released in a small area in New York and Virginia in 1985 and 1986. It was never recovered from the release areas during the following 3–4 years (Reardon and Hajek, 1998). Surprisingly, widespread infection of gypsy moth larvae by *E. maimaiga* was observed in 1989 in many northeastern states and since then, gypsy moth outbreaks in many areas have been relatively rare.

10.2.2 Advantages and Disadvantages

Entomopathogenic fungi possess several characteristics that make them excellent candidates for use in IPM. They are relatively harmless to beneficial insects and have minimal impact on the natural biodiversity of the ecosystem in which they exist. Though exceptions have been observed, in general they are specifically pathogenic to arthropods, not plants or mammals. They leave no toxic residues on crops, as some chemical pesticides do. They are somewhat host-specific, which enables them to fit into an IPM programme that draws on other beneficials, such as predators and parasitoids. In addition, most are considered safe for humans and do not pose a hazard for disposal, particularly when compared with chemical insecticides (Laird et al., 1990). Many fungi are active over a range of environmental conditions year-round and are unaffected by day length, which can inhibit performance of other natural enemies. Although infection through the gut lining occurs, entomopathogenic fungi are more likely to infect the host by penetrating the cuticle, eliminating the need for a target pest to ingest it to become infected. Of particular importance in terms of their commercialization is that most entomopathogenic fungi are facultative saprophytes and they can be mass-produced on a solid substrate in large quantities at low cost compared with many parasitoids and predators. Many can be dried and if held under proper conditions, they can be stored with minimal loss of viability or efficacy. A variety of simple techniques have been developed to apply fungal-based products, but most commonly the spore powder solutions are sprayed on the crops similar to a conventional chemical pesticide formulation.

Despite the many advantages of entomopathogenic fungi, several abiotic factors detract from their effectiveness, and have contributed to their limited use in agricultural production. The spores of many entomopathogenic genera are damaged or killed by direct exposure to UV-B radiation for only a few hours (Braga et al., 2001, 2002; Fargues et al., 1996; Goettel et al., 2000). UV-A has also been found to inactivate and delay germination of conidia of some fungi (Braga et al., 2002). Temperature also influences fungal efficacy. While most entomopathogenic fungi tolerate a wide range of temperatures (commonly 0–40°C), the optimal temperatures for germination, growth and sporulation are generally 20–30°C (Goettel et al., 2000). Historically, moisture has been considered one of the most significant factors limiting their effectiveness, and low humidity has been implicated in failures of field trials. However, it is now recognized that ambient humidity levels may not accurately reflect moisture conditions in the microhabitat around the insect where the spore germinates. Thus, issues of low moisture can be addressed by timing the application of fungi when humidity levels are naturally higher (e.g. early morning or late afternoon). In addition, oil formulations have been shown to protect spores from the negative impact of low humidity. Rainfall can have a negative effect on fungal efficacy by washing off propagules before they are able to germinate and enter the insect. Research has been done on formulation technology to minimize this problem. Although several environmental factors are known to inhibit fungal efficacy, wide variation in the susceptibility to individual abiotic factors has been observed among and within fungal species and genera. Careful strain selection can minimize these disadvantages.

A range of biotic factors can negatively impact the efficacy of an entomopathogenic fungus, such as the host stage of the pest, competitive microbial organisms, and antagonistic enzymes and compounds on the plant or host surface (Butt et al., 2001; Goettel, 1992; O'Callaghan and Brownbridge, 2009). For example, soft-bodied immature stages of the host tend to be more susceptible to fungal infection than the egg stage. Differential susceptibility among the instars of the greenhouse whitefly, *Trialeurodes vaporariorum*, has been reported, though the relationship between insect age and susceptibility is unclear. Insect behaviour can reduce the effectiveness of a fungal treatment. Some termites avoid coming in contact with infected individuals within the colony, and thereby escape infection (Chouvenc et al., 2008). The termite *Coptotermes lacteus* showed an avoidance response, walling off tunnels with *M. anisopliae*, and thus protecting the colony from infection. Aphids and mites are sometimes able to escape infection by moulting before the fungus enters the body (Alavo et al., 2002).

10.3 ENTOMOPATHOGENIC FUNGI AS A SUCCESSFUL COMPONENT OF IPM

IPM is built on the concept that pest populations can be cost-effectively maintained below damaging levels by combining a broad array of compatible management tactics that have as little negative impact on the environment as possible. Each individual IPM component exists within a complex interrelated ecosystem and thus, to be a successful contributor to pest suppression, it must be assessed for its impact alone and together with other management practices. Several critical factors must be assessed when considering an entomopathogenic fungus for IPM, including: virulence; mass production potential; compatibility with chemical insecticides, fungicides and natural enemies; persistence; shelf life; and ease of application.

10.3.1 Virulence

For centuries, research has been done to assess the ability of a fungus to kill a target pest. Though these studies have focused on a wide range of agricultural, human and household pests, including those attacking field crops, forests and greenhouse ornamentals, and stored product insects, most have focused on a fairly limited group of fungi, in particular, species in the genera *Beauveria*, *Metarhizium*, *Paecilomyces*, *Lecanicillium*, and *Isaria*.

10.3.2 Mass Production

Though the virulence of a fungal strain against a target pest is an important characteristic, its ability to be mass-produced in large quantities at low cost may be more important. Growers are dependent on the commercial availability of fungal-based products, and those that are relatively easy to produce are more likely to be marketable. Extensive research has been and continues to be conducted on developing, refining and improving fungal mass production techniques. Fungal mass production is a complex subject involving the effects of substrate, additives and other factors on the virulence, viability and thermotolerance of fungal spores (Feng et al., 1994; Kassa et al., 2008; Machado et al., 2010; Sahayaraj and Namasivayam, 2008).

10.3.3 Compatibility with Other Components of IPM

Compatibility is a critical issue that growers must factor into their decision when considering if, when and how to use an entomopathogenic fungus. When designing an effective IPM plan, one must consider the relationships between these diverse inputs and organisms, both those that occur naturally and those released or applied to the crop or pest. Depending on the crop ecosystem, naturally occurring biological control agents, such as parasitoids and predators, may play an important role in pest suppression. The application of an insecticide, whether it is a chemical or fungal-based material, may upset the balance of natural enemies, resulting in the outbreak of a secondary pest that had previously been maintained below damaging levels. Commercially produced natural enemies – parasitoids, predators, nematodes, etc. – are commonly released as components of IPM. A wide array of agrochemicals, among them insecticides, fertilizers and fungicides, are sometimes needed to address other aspects of successful plant production. Therefore, a full understanding of the impact of an entomopathogenic fungus on other components of IPM is essential. Similarly, the impact of other IPM components on the entomopathogenic fungus is equally important.

10.3.4 Natural Enemies

Some fungi only kill one particular pest; others are generalists. Many of the *Entomophagous* fungi only infect one or a few closely related species, such as *Entomophaga maimaiga*, a specific pathogen of gypsy moth, *Lymantria dispar*. In contrast, *Beauveria bassiana* infects a wide variety of arthropods over a broad range of environments, including fields, forests and greenhouses. In general, entomopathogenic fungi are compatible with most parasitoids and predators (Copping, 2001; Goettel and Hajek, 2001; Sterk et al., 2003). However, a blanket statement on compatibility cannot be assumed given that variation occurs among fungal species and isolates, the specific natural enemies and the environment in which they are co-existing. To accurately assess compatibility, tests must be conducted under the agricultural conditions in which they will occur. Results obtained in laboratory bioassays may indicate that a fungus is pathogenic to a natural enemy, though in the agricultural setting, mortality of the non-target may not occur (Brown and Khan, 2009; Sterk et al., 2003).

10.3.5 Plant Extracts and Botanical Products

While agricultural chemicals have dominate the world's pesticide market, in recent years there has been an increased interest in the use of plant-based materials, such as neem oil, as a biorational approach to pest management (Rosell et al., 2008). Several of these products are now available commercially and have been adopted by growers. Though these products offer a safer way to manage pests than conventional chemical insecticides, they are not benign, and their effect on other IPM components must be considered. Many studies have been done assessing the impact of various botanically based insecticides on the growth, germination, and efficacy of entomopathogenic fungi (Islam et al., 2011; Islam and Omar, 2012; Rosell et al., 2008; Sahayaraj et al., 2011). A wide array of plant products is available, each with their unique effects. Similarly, there are many different entomopathogenic fungi, and each may respond differently to a plant extract. For example, Sahayaraj et al. (2011) conducted Petri dish tests on the effect of several neem products and other plant extracts on *B. bassiana*, *I. fumosoroseus* and *L. lecanii*, and most were compatible with the fungi. In contrast, Hirose et al. (2001) found that neem oil and various biofertilizers had a negative effect on germination, colony growth and spore production of several *M. anisopliae* and *B. bassiana* isolates. This demonstrates that the effect of a plant product may vary between strains of the same fungal species. It is impossible to make broad assumptions and as the availability of these biorational products increases, their compatibility with entomopathogenic fungi must be assessed. While *in vitro* trials of compatibility are easy to do, they do not answer the more important question of the effect of a botanical on a fungus under field conditions.

10.3.6 Agrochemicals

Compared with the many chemical insecticides that are on the market, relatively few fungal-based products are available. Though alternatives to chemical insecticides and fungicides are becoming more available to growers, these conventional compounds remain a primary means of managing pests and diseases, despite the decline in their efficacy. For non-organic farmers, it is essential that an IPM programme integrates the judicious use of select compatible chemical pesticides (biorationals), to handle a persistent pest outbreak that for whatever reason is not responding to non-chemical approaches. Extensive research assessing the compatibility of entomopathogenic fungi with agrochemicals has been reported (Asi et al., 2010; McCoy et al., 1988). As for botanicals, the compatibility of entomopathogenic fungi and agrochemicals depends on the compound and the fungus. It is impossible to make broad assumptions with regard to fungal tolerance. Logically, fungicides are generally considered to be less compatible with entomopathogenic fungi than insecticides, but even this is not consistently true. As is true for the compatibility of fungi and natural enemies, variation in sensitivity has been found among fungal species and strains and various chemical pesticides, and it is difficult to make broad generalizations (Asi et al., 2010; Islam et al., 2011; Malekan et al., 2012; Neves et al., 2001; Sharififard et al., 2011; Sterk et al., 2003). Adjusting the timing of applications of either the fungus or the pesticide can reduce the negative impact of agrochemicals (Copping, 2001; Jaros-Su et al., 1999; Trissi et al., 2012).

10.4 FUNGAL FORMULATIONS AND APPLICATION TECHNOLOGY

10.4.1 Formulations

Hypocrealean entomopathogenic fungi are available commercially as formulated biological control mycoinsecticides for managing agricultural pests (Roberts and Hajek, 1992). A formulation refers to substances that are added to enhance the viability, efficacy/virulence or shelf life of the active ingredient, in this case fungal conidia or propagules. The formulation ingredients include carriers, spreaders, stickers, etc., that can improve conidial thermotolerance or persistence, increase the wettability or adhesive properties of the material, or extend the shelf life (Tanada and Kaya, 1993). Several types of fungal-based products containing conidia are available commercially in a variety of formulations, including granulars, wettable

TABLE 10.1 Commercial Mycoinsecticides

Fungus	Product	Formulation	Shelf life (at recommended temperature)
Beauveria bassiana	BotaniGard, Naturalis-L, Mycotrol, Bio-Power, Beauverin, Boverol, Proecol	WP, SC	1 year (≤20°C)
Beauveria brongniartii	Betel, Schweizer Beauveria	WP	1 year (≤2°C)
Lecanicillium lecanii	Mycotal, Bio-Catch, Vertalec	WP	0.5 year (≤4°C)
Metarhizium anisopliae	Bio-Catch -M	WD	1 year (≤4°C)
	Green Muscle, BioCane	G	
Metarhizium flavoviride var. flavoviride	BioGreen	WP	0.5 year (≤4°C)
Isaria fumosorosea	Preferal, Priority, Futureco, NoFly	WP, WDG	0.5 year (≤4°C)

Revised from Copping (2004).
WP, wettable powder; WDG, water-disposable powder; SC, suspension concentrate; G, granular.

powders, water-disposable powders, liquids (emulsifiable and suspension concentrates) dusts or baits (Tables 10.1 and 10.2; Copping, 2004; Faria and Wraight, 2007; Kabaluk and Gazdik, 2005; Tanada and Kaya, 1993). Several factors have been cited as reasons why fungal products make up a small percentage of the total insecticide market worldwide (Yatin et al., 2006). Fungal infection and death of the pest are gradual processes in contrast with the quick knockdown effect that is common for most chemical insecticides. This delay allows the pest time to continue to damage the crop and leads some growers to have less confidence in the efficacy of fungal products. In addition, there can be significant variation in the efficacy of a commercial fungal application, for many of the reasons discussed previously, such as adverse biotic and abiotic conditions. Because fungi are living organisms, care must be taken to store them properly. Their shelf life can be shortened significantly if held under adverse conditions (Bateman and Alves, 2000; Inglis et al., 1997). A wide range of strategies have been devised to address these issues through the development of special formulations that include adjuvants, carriers and buffers.

10.4.1.1 Granular Formulations and Baits

Fungal-based granules contain inert carriers, such as clay minerals, and ground plant residues, etc., that hold fungal propagules together (Burges, 1998). They are commonly used to apply the fungus to the soil or in potting mix. This type of formulation is relatively easy to apply and can result in a sustained source of fungal inoculum, that not only persists, but spreads throughout the target area. The effectiveness of this type of formulation depends on biotic factors such as the occurrence of other antagonistic microbial organisms, and abiotic factors such as temperature, humidity and chemical factors. For example, the persistence of *B. bassiana* was shown to decrease after application of fresh cow manure, but was improved when compost was added (Rosin et al., 1996). This demonstrates the importance of considering multiple factors when incorporating different production components into IPM. Several types of granules have been developed, some including different clays (e.g. bentonite, attapulgite), while others are grain-based.

A variety of grains are commonly used as solid substrates to produce entomopathogenic fungi, including rice, barley, whole and cracked

TABLE 10.2 Main Target Pests of Commercially Available Mycoinsecticides

Fungus	Product	Target pests
Beauveria bassiana	BotaniGard, Boverol, Naturalis-L, Proecol Mycotrol, Beauverin Bio-Power	Lepidoptera (diamondback moth, beet armyworm, cabbage looper, cutworm, etc.), Coleoptera (scarab beetle grubs, weevils, coffee berry borer, cutworms, etc.), Heteroptera (psyllids, stinkbug, plant bugs, leafhoppers, mealybugs, aphids, whitefly, etc.), Thysanoptera (western flower thrips)
Beauveria brongniartii	Betel, Schweizer Beauveria	Lepidoptera (diamondback moth, beet armyworm, cabbage looper, cutworm, etc.)
Lecanicillium lecanii	Mycotal, Bio-Catch Vertalec	Heteroptera (stinkbug, aphids, whitefly, plant and leafhoppers, mealybugs), Thysanoptera (western flower thrips, onion thrips)
Metarhizium anisopliae	Bio-Magic	Coleoptera (scarab beetle grubs, weevils), Blattodea (termites), Heteroptera (leafhoppers), Orthoptera (grasshoppers), Lepidoptera (cutworms)
Metarhizium flavoviride var. flavoviride	BioGreen, BioCane	Coleoptera (scarab beetles, weevils), Orthoptera (grasshoppers and locusts), Blattodea (termites)
Isaria fumosorosea	Preferal, Priority, FuturEcoNofly	Heteroptera (whiteflies, aphids, etc.), Lepidoptera (tomato moth), Acari (rust mites, spider mite, etc.)
Paecilomyces lilacinus	Bio-Nematon	Nematodes (Root knot, cyst, lesion burrowing)

Revised from Copping (2004).

wheat, corn, and millet; and can be dried after the production phase to produce simple formulated granules. The suitability of a grain depends on the fungus being produced and the setting in which it will be used. For example, millet and cracked wheat are ideal because their shape allows for multiple surfaces on which the fungus will grow. The size of the grain particles also allows the granule to drop through the litter layer and into crevices in the soil, where the pest occurs.

A millet-based fungal production system has been reported previously (Bartlett and Jaronski, 1988; Gouli et al., 2008; Jenkins et al., 1998; Li and Feng, 2005). The millet provides nutrition to support fungal growth in the soil in the absence of an insect host. This type of formulation is relatively simple to produce. Millet grains are placed in a polyvinyl bag and soaked in water containing citric acid and boiled at 90°C for 1 h, after which the mixture is autoclaved at 121°C for 30 min. The bag is inoculated with a 3-day old liquid culture and incubated at 25°C with a 16:8 h (L/D) photoperiod for 3 weeks. The culture is air dried until it reaches a moisture content of less than 5%. This process commonly produces a granule with a concentration of 1.1×10^8 *B. bassiana* conidia per gram of grain and a germination rate of 98.2% at 20°C after 24 h.

Research has shown that the type of substrate on which a fungus is grown can affect the thermotolerance of the conidia produced. For example, a study was conducted to compare the thermotolerance of conidia produced on agar made from millet, whey permeate or quarter strength Sabouraud dextrose agar (¼ SDAY) (Kim et al., 2011). *B. bassiana* conidia were then exposed to 45°C for 90 min and stored for 30 days at 25°C. Conidia produced on whey permeate or millet-based agar exhibited significantly greater thermotolerance than those grown on ¼ SDAY (Kim et al., 2011).

Fungal-based baits are commonly produced on a solid substrate that is attractive to the target insect. Insects with an active feeding stage in the soil (soil-dwelling termites and ants, black vine weevil, etc.) or household pests (cockroaches, carpenter ants, etc.) are often targets for baits. Baits have also been tested for use against field pests such as grasshoppers and locusts. Some insects demonstrate an ability to detect the presence of entomopathogenic fungi and avoid them. For example, termites are generally susceptible to infection by *B. bassiana* and *M. anisopliae*, but it is difficult to achieve contact between the pest and sufficient concentrations of the fungus to obtain successful management. Some termite species, when they detect *M. anisopliae* in the colony, seal off the contaminated tunnel to prevent it from contacting them and others in the colony (Staples and Milner, 2000). However, when *M. anisopliae* was produced as a cellulose bait, repellency was overcome (Wang and Powell, 2004). However, the response of healthy termites to fungal-contaminated termites may further limit the effectiveness of this strategy. Within 24 min of coming in contact with a contaminated termite, uninfected termites have been observed to initiate several defensive behaviours, including grooming, biting, defecation, and burial of the infected termite (Myles, 2002). A bait made from agar, sugarcane molasses and cellulose powder and treated with *M. anisopliae* has been lab-tested against the termite *Microcerotermes diversus* (Cheraghi et al., 2013). No evidence of repellency was observed and a high level of mortality was obtained, suggesting differences among termite species in their response to fungi. Combining a chemical pesticide bait with exposure to *M. anisopliae* to target German cockroaches, *Blattella germanica*, has also shown promise. When cockroaches were exposed to a solid imidacloprid bait, they displayed signs of toxicity, but most recovered over time (Kaakeh et al., 1997). However, cockroaches died faster and at higher rates when they fed on the pesticide-laced bait and were treated topically with the fungus, suggesting a synergistic effect of the two treatments. Given the secretive habits of cockroaches, it is difficult to contact them with the entomopathogen, and thus it may be possible to combine both ingredients in a bait to achieve positive results.

10.4.1.2 Wettable Powders

Entomopathogenic fungi are most commonly produced as wettable powder formulations (WP), which contain 50–80% technical powder, 15–45% filler, 1–10% dispersant and 3–5% surfactant (Burges, 1998). These are mixed with water and applied to the foliage as a standard insecticidal spray with ultra-low volume (ULV) or hydraulic applicators. They can also be applied to the soil as a drench. Formulations have been developed using a wide array of compounds, each with unique properties that affect particular factors to enhance efficacy or spore survival. For example, additives can protect the spores from UV light, enhance the ability of the spores to stick to the foliage (reduce spores from washing off the leaf due to watering or rain) or increase humidity around the spore to promote germination under adverse environmental conditions (Burges, 1998). To improve the efficacy of an *I. fumosorosea* (SFP-198) WP for use against greenhouse whitefly, *Trialeurodes vaporariorum*, in greenhouse tomatoes, Kim et al. (2010b) tested the addition of five photoactive dyes. They found that, among the dyes tested, Phloxine B conferred the greatest benefit in terms of efficacy. The most suitable dose of Phloxine B was $0.005\,g\,l^{-1}$, based on the dosage-dependent control efficacy and negative aspects of the dye, including its phytotoxicity and effect on conidial germination.

Moisture can significantly impact conidial stability and viability during storage, and is an important factor that affects shelf life of a fungal-based product. To maintain low moisture levels, the potential of incorporating moisture absorbents (e.g. calcium chloride, silica gel, magnesium sulfate, white carbon, and sodium sulfate) has been investigated in 10% WP conidial powder formulations (J.S. Kim, unpublished data). Of the materials tested, white carbon was superior to the others, maintaining conidial viability at room temperature for 60 days. Tests were also conducted in which fresh rather than dried conidia were combined with white carbon in a standard WP formulation. The stability of conidia in the WP formulation with white carbon during storage was significantly better than those without this absorbent.

10.4.1.3 Oil Formulations

Whereas most formulation technology information is protected by intellectual property restrictions, it is well known that a variety of oils can be added to formulations to improve the shelf life of fungal products and increase their field efficacy in dry climates. The use of oil as a carrier helps to wet the waxy hydrophobic or lipophilic surfaces of insects and leaves. This improves the survival of spores in arid conditions and facilitates the spreading of spores over the leaf surface, thereby raising the potential for the pest to come in contact with the fungus. Oils also facilitate spore adhesion to the insect, stimulating germination and assisting with penetration by disrupting the waxy layer of the cuticle (Bateman et al., 1993).

It has been shown that conidia will break dormancy when they come in contact with even a small amount of water or moisture, which will reduce the viability of the conidia (Jenkins et al., 1998). The simple addition of oil to spore powder increases the survival and viability of conidia (Moore et al., 1995). Isoparaffinic hydrocarbon solvents, such as paraffin oil and mineral oil, have been used as carriers for oil-based formulations. However, other oils, such as vegetable oils, are also suitable for *M. anisopliae* and *B. bassiana* (Morley-Davis et al., 1995). For example, Naturalis-L® (*B. bassiana*) and Green Muscle® (*M. anisopliae*), have been formulated with soybean oil as a carrier (McClatchie et al., 1994). An oil-based suspension concentrate (SC) can be exploited to improve the shelf life of *M. anisopliae* conidia (McClatchie et al., 1994).

Kim et al. (2011) investigated the use of methyl oleate (a wetting agent and emulsifier), and corn, cottonseed and paraffin oils as carriers. They found that corn oil was superior to the other substances in terms of promoting the heat tolerance of *Isaria fumosorosea* (strain SFP-198) after exposure to 50°C for 2h. Even when the corn oil-based suspension was exposed to 50°C for 8h the germination rate of conidia was 91.6% compared to only 28.4% for the conidial powder. The long-term storage of the corn oil-based conidial suspension and conidial powder was also compared at 25°C for 24 months. The viability of conidia in corn oil was over 98% for up to 9 months of storage at 25°C, and 23% at 21 months. However, the viability of the conidial powder was only 34% after 3 months at 25°C, after which its viability rapidly decreased. This further supports the fact that oil suspensions greatly enhance the thermotolerance and stability of conidia during storage.

10.4.2 Application Technology

The method of application used to apply an entomopathogenic fungus influences the effectiveness of a treatment. Several factors must be considered, including spore distribution and coverage to ensure that the pest comes in contact with the fungus and the environmental conditions that favour survival and germination of the fungus before, during and after application. The four primary methods of fungal application are dipping the plant or roots, spraying the foliage, treating the soil and indirect transmission by vectors.

10.4.2.1 *Dipping*

Application of entomopathogenic fungal spores by dipping of plant roots or cuttings into a spore suspension is not commonly recommended, but has gained favour in recent years to manage western flower thrips, *Frankliniella occidentalis*. It is thought that the fungus targets the larval stage in the soil, though research supporting the effectiveness of this strategy is lacking.

10.4.2.2 *Foliar Sprays*

Foliar sprays are historically the most common way to apply fungi, probably because growers are familiar with this technology, having used it for chemical pesticides. Extensive research has been done to maximize the efficacy of fungi by spraying the foliage. A wide array of applicators can be used to apply fungi from the ground. Application from the air has also been used to treat large acreages to combat locust and grasshopper outbreaks in Africa. Care must be taken to select the right type of sprayer for the particular agricultural setting and the right formulation for the conditions. For example, in greenhouses, it is commonly recommended to make fungal sprays in the late afternoon, to maximize on the higher relative humidity that occurs at that time, which facilitates fungal germination.

Ensuring that the fungal spores are distributed evenly to the site where the target insect occurs is probably the most important factor affecting successful control, yet it is challenging to achieve. Kim et al. (2010a) described the use of supernatant from the liquid culture of *Beauveria bassiana* which contained a thermotolerant chitinase, and an enzyme linked to pathogenicity, to enhance the efficacy of a fungal spray against aphids. The supernatant fraction was incorporated into the fungal preparation with Attagel® (an inert powder used as a thickener made by BSAF Corp.) at 0.5% (*w/v*) and mixed with 0.01% polyoxyethylene-(3)-isotridecyl ether (TDE-3) as a spreading agent. This demonstrates just one of many options for enhancing fungal efficacy through advanced application and formulation technology.

10.4.2.3 *Granular Soil Treatments*

Given the conditions under which fungi thrive, soil treatment of fungal spores is probably the most reasonable application strategy, though it is only suitable if the target insect has a susceptible soil phase. In general, natural soil moisture levels are sufficient to promote conidial germination and mycelial growth, allowing fungal inoculum to be sustained over time. Many agricultural settings irrigate their crops regularly, ensuring that suitable moisture conditions for fungal growth are maintained. Due to concerns about the potential for spores to be leached out of potting medium during watering, Kim et al. (2010c) investigated the downward movement of *B. bassiana* in potting medium with different levels of moisture and top-watering. They found that the concentration of *B. bassiana* was greatest in the upper 3cm of the potting medium, and no significant movement of the spores occurred after 18 days of regular watering.

Another benefit of applying fungi to soil rather than on foliage is that the conidia are protected from damaging UV light within the soil. Temperature is also an important factor affecting fungal efficacy. Soil helps to moderate large fluctuations in ambient temperature, which often favours fungal growth and infectivity. One significant potential negative aspect of treating soil with fungi is the occurrence of antagonistic fungi, bacteria and fungal-feeding arthropods and microorganisms.

Skinner et al. (2012) tested a granular fungal application system to control western flower thrips, *F. occidentalis* (WFT). Mycotized millet grains with entomopathogenic fungi applied to soil of potted marigold plants were tested to target pupating thrips. Two experimental fungal isolates, (*B. bassiana* [ARS7060] and *M. anisopliae* [ERL1171]), were compared with the registered *B. bassiana* strain GHA [commercialized as BotaniGard®] and untreated controls in greenhouse caged trials. Mycotized millet grains were mixed into the upper surface of the potting soil in pots of flowering 'Hero Yellow' marigolds (4g/pot). At 8 weeks post-infestation, the mean total number of thrips per plant was 81% and 90% less in the ERL1171 and ARS 7060 treatments, respectively, than in the controls. Plant damage was 60% less on plants treated with the experimental fungi than with the control and GHA treatments. At 10 weeks post-application, 75–90% of WFT collected from the treatments were infected with the experimental isolates.

10.4.2.4 Indirect Vector Transmission

Entomopathogenic fungi are readily transmitted in nature by non-pest species to locations where a pest occurs. For example, collembolans, which generally reside in the soil, are known to transport fungal conidia which stick to their cuticle. Research demonstrated that some species were able to vector enough fungal propagules to cause mortality of mealworms (Dromph, 2003). Other examples of indirect vectoring in nature have been reported (Baverstock et al., 2010), but the impact of this random transfer on pest populations is uncertain. In caged greenhouse trials, Kapongo et al. (2008) assessed the potential of using bumble bees, *Bombus impatiens*, a common pollinator in greenhouse-grown vegetables, to transmit *B. bassiana* to greenhouse whitefly and tarnished plant bug, *Lygus lineolaris*. Depending on the spore concentration tested, mortality rates of up to 56% among whiteflies and 67% for the tarnished plant bug were obtained following fungal dispersal by the bees. Mortality among the bees exposed to the fungus was observed, demonstrating the delicate balance between the positive and negative effects that must be achieved to make a system such as this work.

10.5 TWO CASE STUDIES OF THE USE OF ENTOMOPATHOGENIC FUNGI FOR IPM

10.5.1 Sunn Pest

A major constraint to wheat and barley production in West and Central Asia is the Sunn Pest, *Eurygaster integriceps*, an insect that damages the above-ground parts of the plant. It is found from North Africa through the Middle East, and eastern Europe to the independent states of the former Soviet Union, infesting over 15 million hectares (37.1 million acres) (Figure 10.1). In addition to reducing crop yield, Sunn Pest also affects quality by injecting a chemical during feeding on the grain that destroys gluten. Bread made with flour from damaged grains fails to rise and burns easily and imparts an off-flavour (Figure 10.2). If as little as 2% of the grains in a lot have been fed upon by Sunn Pest, the value of the entire lot is reduced and may be unsalable. In addition, if populations are high, yield can be reduced by 50–90%. Though much of the damage is done during feeding on the wheat spikes, in the spring it also feeds on the vegetative stage of the plant, constricting the stem which reduces sap flow and deforming the foliage. Before 1996, no standardized IPM strategies for Sunn Pest existed, though research on various management components (parasitoids, predators, etc.) had been done. In areas where Sunn Pest was a particular problem, management decisions were generally made at the discretion of federal agencies and area-wide applications of chemical insecticides from the air (paid for by the government) was the norm. Over US$45–50 million was commonly spent annually for insecticide treatments. This resulted in higher Sunn Pest levels, in part due to development of insecticide-resistant populations, and destruction of the natural enemy complex. An intensive region-wide initiative to develop IPM for Sunn Pest was started in 1996 led by the

FIGURE 10.1 Wheat spike heavily infested with Sunn Pest. *(Photo by ICARDA).*

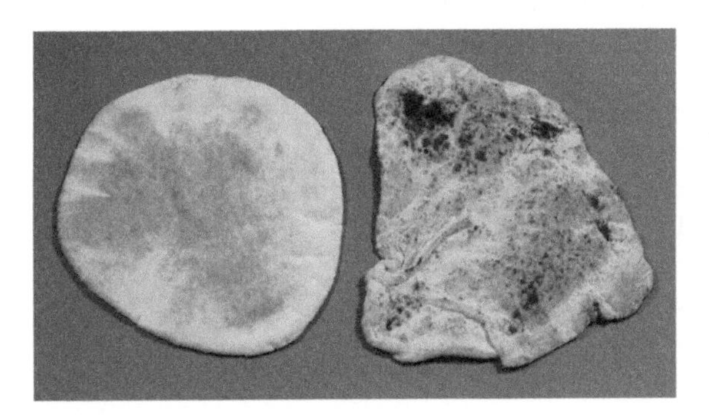

FIGURE 10.2 Pita bread made with flour from undamaged grains (left) and ones damaged by Sunn Pest (right). *(Photo by ICARDA).*

International Center for Agricultural Research in the Dry Areas (ICARDA) in collaboration with the University of Vermont and several national agricultural specialists.

10.5.1.1 Life Cycle

To develop suitable IPM approaches, knowledge of the pest's life cycle and assessment of its bioecology are critical to identify 'windows of opportunity' to reduce populations to tolerable levels, minimizing damage and production costs while increasing yield, quality and economic returns.

Sunn Pest has one generation per year; it is in cereal fields causing damage for only about 2.5 months per year (Figure 10.3). For 9–10 months of the year, the adults overwinter in the foothills and mountains around cereal fields resting beneath the leaf litter around trees or under bushes, where they are protected from animal predators, the hot sun and dry conditions in summer, and cold temperatures in winter. Typically, they are found at elevations of 900 to 2000m. It is common to find hundreds of Sunn Pest under a single bush resting at the interface of the litter and soil surface. In the early spring, adults fly from overwintering sites to grassy areas around wheat fields, where they begin to feed and mate. Over time, they migrate into the fields where they reproduce, lay eggs and die. The next generation completes development and as temperatures rise and the wheat matures, these new adults fly back to the overwintering sites. In years when populations are high, large numbers of adults are commonly caught in harvesting equipment, which causes problems for grain processing and storage.

10.5.1.2 IPM Components

10.5.1.2.1 CULTURAL CONTROLS

Several production practices are recommended to minimize Sunn Pest damage, such as planting the crop early or growing early-maturing wheat varieties which allows harvesting to be done early before significant damage occurs. Research is also under way to identify wheat varieties resistant to Sunn Pest; some have been identified that show resistance to Sunn Pest at the vegetative stage (Abdullah,

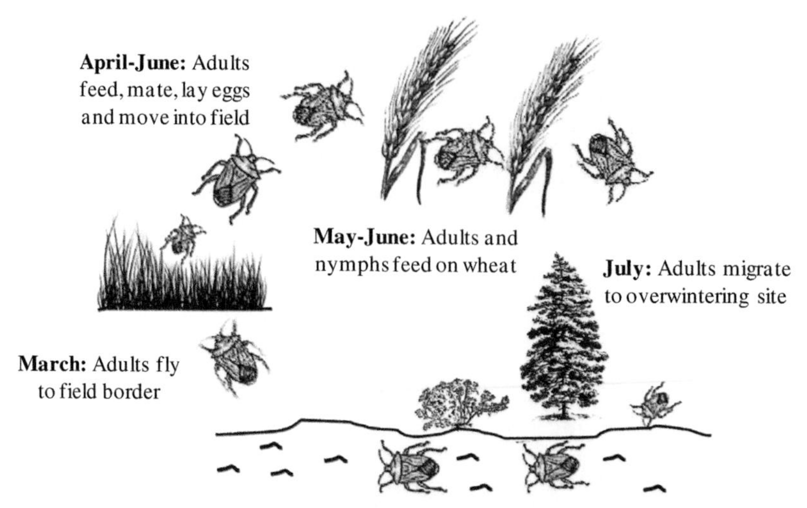
April-June: Adults feed, mate, lay eggs and move into field
May-June: Adults and nymphs feed on wheat
July: Adults migrate to overwintering site
March: Adults fly to field border
July-March: Adults in diapause under litter in foothills

FIGURE 10.3 Generalized life cycle of Sunn Pest; specific timing varies with location and weather.

2007; El Bouhssini et al., 2007; Yildirim et al., 2007). Because Sunn Pest congregate in large numbers in overwintering sites, some governments have initiated programmes to support hand-collection of adult Sunn Pest. Anecdotal reports suggest that this tactic, when done on an area-wide basis, significantly reduced subsequent damage in the field. It is possible that these insects could be used as feed for poultry, though further research is needed to assess the effect on the flavour of the eggs or meat.

10.5.1.2.2 ACTION THRESHOLDS AND SAMPLING

A key first step towards reducing unnecessary pesticide use is to determine the appropriate action threshold for Sunn Pest, i.e. the number of insects that would reduce production revenues more than the cost of control. Standardized sampling procedures were developed and validated for use in overwintering sites and crop fields. Sunn Pest thresholds based on field sampling were increased, which has reduced the number of pesticide treatments made.

10.5.1.2.3 AGRICULTURAL CHEMICALS

Government policy changes have been made in several countries prohibiting most aerial pesticide sprays, and requiring that farmers make applications from the ground only on fields in which pest populations exceeded the threshold (Erman et al., 2007; Gul et al., 2007). This reduced exposure of the natural enemy complex that commonly resides in the hedgerows. In addition, many farmers who welcomed the government sprays were less inclined to take the time to make the applications themselves, which further reduced chemical treatments.

10.5.1.2.4 PARASITOIDS AND PREDATORS

Several species of indigenous egg and adult parasitoids of Sunn Pest have been identified (Abdulhai et al., 2007; Al-Izzi et al., 2007; Trissi et al., 2007). Rearing methods have been developed for *Trissolcus grandis* and, in some countries, mass releases have been made to reestablish them around Sunn Pest infested areas (Amir-Maafi, 2007; Kodan and Gurkan,

2007). In addition, an effort is being made to preserve or create refuges for natural enemies, such as hedgerows and habitat belts around fields, to provide suitable sites to sustain their populations throughout the year. Because aerial spraying of chemical pesticides has been reduced and the timing of applications has been adjusted to minimize their negative impacts on the ecosystem, the natural enemy complex is being restored.

10.5.1.2.5 ENTOMOPATHOGENIC FUNGI

Historical references reported evidence of high levels of natural mortality of Sunn Pest by infection with entomopathogenic fungi. A concerted effort was initiated in 1997 to develop indigenous fungi for Sunn Pest IPM. Inspection at overwintering sites revealed that many of the current year's Sunn Pest adults were dead and showed signs of fungal infection (Figure 10.4). Fungi were isolated from these cadavers, and a collection of over 200 different isolates from throughout the range of Sunn Pest are maintained at ICARDA and the University of Vermont (Aquino de Muro et al., 2005; Parker et al., 2003). Through laboratory bioassays, several fungal isolates were found to be highly pathogenic to immature and adult Sunn Pest (Parker et al., 2003). Research has also been done to assess the compatibility of *B. bassiana* with the parasitoid *T. grandis*, and no evidence of a negative impact from the fungus was observed (Trissi et al., 2010). Several approaches for incorporating fungi into an IPM programme were investigated taking into consideration the Sunn Pest life cycle and conditions that favour fungal efficacy.

The traditional approach would be to spray fungal-based suspensions on the crop as a mycopesticide. Field trials were conducted testing seven strains of *B. bassiana* applied as an oil formulation (kerosene and sunflower oil) with an ultra-low volume sprayer (Edgington et al., 2007). Two fungal applications were made, one in mid-April, and another in late-April. A significant treatment effect was not observed among the adults after the first spray, but after the second spray, over 93% mortality of the nymphs was obtained compared to 35% in the controls, suggesting the potential of managing the summer population. Despite the promising results, the persistence of this type of treatment is short lived because fungal spores are rapidly killed when exposed to high temperature, UV light, and low humidity. Specialized formulations would be needed to enhance fungal persistence over time to eliminate the need for multiple reapplications.

An alternative approach investigated was application of fungal-based granular formulations to the ground around bushes and trees in Sunn Pest overwintering sites. The hypothesis of this strategy was that conditions for fungal survival and persistence are ideal in these locations. The soil tends to remain moist throughout the year, and temperature is moderated by leaf litter and shade from the plants. In addition, adults remain relatively inactive for 9 months allowing time for infection to occur. Because Sunn Pest tend to return to the same hillside sites each year, fungal-infected

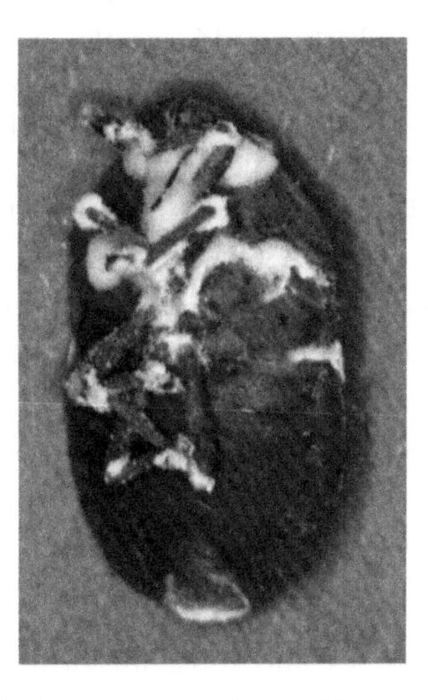

FIGURE 10.4 Sunn Pest cadaver with signs of infection. *(Photo by ICARDA).*

cadavers from previous years may serve as a persistent source of inoculum. Research demonstrated an elevated level of Sunn Pest mortality in fungal plots, and evidence of persistence of fungal inoculum for at least 2 years following application. However, because Sunn Pest tend to move around within the overwintering sites when temperatures rise, it was difficult to fully assess the efficacy of this management strategy. Alternatively, fungal granules could be applied to the soil within the wheat field. Sunn Pest commonly drop off the plants and crawl around on the soil during the middle of the day, where it is cooler and they are protected from bird predation. A fungal treatment applied to the soil is protected from UV light by the foliage and high moisture levels are maintained naturally. Yet another option that has been considered is to apply a fungal treatment in the spring to the foliage or soil of plants growing around field borders, where the Sunn Pest usually land first when migrating from the overwintering sites, and before they move gradually into the field. A trap crop of a highly susceptible variety could be planted there to further enhance its attractiveness to the pest. This strategy would reduce the area to be treated, which would greatly reduce costs. Research is needed to further assess these novel yet promising fungal-based approaches which could be combined with other tactics to achieve an effective multi-faceted IPM programme.

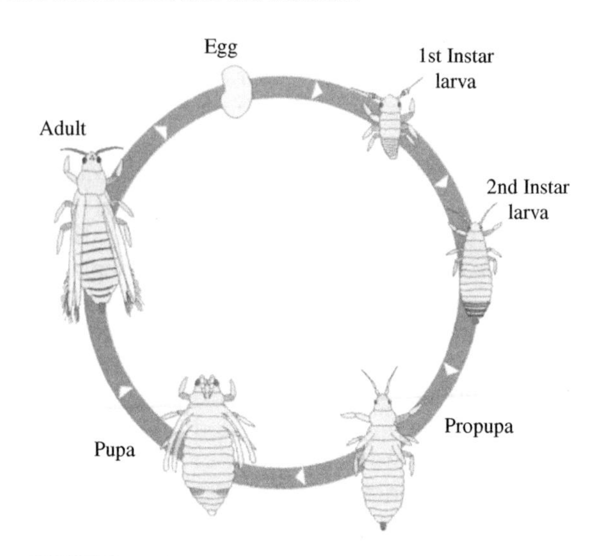

FIGURE 10.5 Life cycle of western flower thrips.

10.5.2 Western Flower Thrips in Greenhouse Ornamentals

Western flower thrips (WFT) is one of the most serious pests of greenhouse ornamentals worldwide (Reitz, 2009). Native to the western US, WFT has spread throughout the world because of the global trade in ornamentals. It feeds on a wide range of crops, including fruit, vegetables and most importantly greenhouse ornamentals. It causes substantial economic loss by causing cosmetic damage during feeding and transmitting viral diseases (tomato spotted wilt virus [TSWV] and impatiens necrotic spot virus [INSV]), which can spread quickly within a greenhouse and destroy the entire crop. Its cryptic behaviour, rapid reproductive rate and potential to develop resistance to insecticides make WFT particularly difficult to control (Broadbent and Pree, 1997; Reitz, 2009; Robb and Parrella, 1995). Historically, growers have relied on chemical pesticides to control WFT, and this pest is a primary reason greenhouse ornamentals are rarely grown as certified organic. Extensive research has been and continues to be done to develop effective strategies to address this persistent pest problem.

10.5.2.1 Life Cycle

At ambient temperatures of around 20°C, WFT can complete one generation in 7–10 days, and there are several generations annually (Figure 10.5) (Reitz, 2009). Females lay eggs individually in the plant's leaf, stem or flower tissue, with a specialized ovipositor, laying 150–300 over its 30-day life span. Eggs hatch into 1st instars after 2–4 days, then moult into 2nd instars in 2 days. For 2–4 days, they feed on the undersides of foliage or in flowers, after which they pupate. Some pupate in the crevices of foliage or flowers, but most drop to the soil to pupate (Berndt et al., 2004; Broadbent et al., 2003). They undergo two non-feeding pupal stages over 2–5 days, before emerging as adults.

10.5.2.2 IPM Components

10.5.2.2.1 CULTURAL AND PHYSICAL CONTROLS

The first line of defence for WFT is to try to keep them out of the crop. Growers have reported a 50–90% reduction in pesticide use by placing fine-mesh thrips-proof screening over vents, doorways and other openings to exclude them from greenhouses (Robb and Parrella, 1995). However, in recent years, some growers have removed the screening because it prevents entry of beneficial predators and parasitoids from outside. Good sanitation practice in and around the greenhouse also reduces the build-up of WFT populations and migration into the crop. Growers are encouraged to remove weeds and flowering plants that grow inside and outside within 3 m of their greenhouses, and dispose of infested plant debris away from the production area. When feasible, fallowing the greenhouse for a month or more is recommended. Research has shown that high temperatures can kill WFT, and some growers close down their greenhouses in autumn to allow internal temperatures to rise. Greenhouses with concrete floors or floors covered with weed-cloth barriers may minimize WFT problems, perhaps because of a reduction in weed growth and pupal survival.

10.5.2.2.2 SCOUTING

Early detection of a pest problem is fundamental to IPM. Growers are advised to check incoming plants for WFT by tapping plug trays over white paper. Standardized crop scouting procedures have been devised, including placing sticky traps throughout the greenhouse. Whereas yellow traps are commonly used, research has shown that blue traps are especially attractive to WFT. Because WFT pupate in the soil, sticky cards are sometimes placed close to the soil under benches to capture emerging adults. Several types of plants have been found to be highly attractive to WFT, such as marigolds, chrysanthemums, ornamental peppers or gerbera daisy (Buitenhuis et al., 2005; Skinner, unpublished data). These are used either as indicator plants for early detection, or to lure WFT adults out of the crop. Some varieties of petunia (e.g. Summer madness, Blue wave) and fava beans are particularly sensitive to TSWV and INSV, displaying symptoms of infection early in the disease cycle (Allen and Matteoni, 1991; Stack et al., 2013).

10.5.2.2.3 CHEMICAL INSECTICIDES

The conventional strategy for managing WFT outbreaks is to apply chemical pesticides. A wide array of chemical insecticides is registered for thrips, but most require direct contact with the pests to be effective (Stack et al., 2013). Historically, the foliage has been the primary target for pesticide applications, overlooking the importance of the soil phase of WFT (Berndt et al., 2004; Broadbent et al., 2003; Deligeorgidis and Ipsilandis, 2004; Wiethoff et al., 2004). Because eggs are embedded in plant tissue and pupae in the soil escape exposure, multiple reapplications at 4-day intervals are needed to reach these stages as they develop. Resistance to most available chemical insecticides has been observed (Gao et al., 2012). In addition, there is evidence that thrips are repelled by some chemical sprays which diminishes treatment efficacy. All of these factors contribute to reducing the effectiveness of conventional chemical control, and most growers realize IPM is the only way to protect their crops from WFT.

10.5.2.2.4 PARASITOIDS AND PREDATORS

Several natural enemies are commercially available for the above and below ground stages of WFT, and are the subject of extensive research to assess their efficacy. Several species of predatory mites are used to combat WFT in the above-ground parts of the plant (Berndt et al., 2004; de Courcy Williams, 2001; van den Meiracker and Sabelis, 1999; Jacobson et al., 2001). The anthocorid bug, *Orius laevigatus*, has been shown to prey on WFT larvae and adults in the plant canopy (van den Meiracker and Sabelis, 1999; Weintraub et al., 2011). Both nematodes and predatory mites are recommended for use against the soil phases of WFT (Arthurs and Heinz, 2006; Ebssa et al., 2006; Wiethoff et al., 2004). While all of these natural enemies demonstrated an ability to reduce WFT populations, none of them alone has been found to be sufficiently effective to suppress WFT at levels acceptable to growers.

FIGURE 10.6 Marigold guardian plant in greenhouse of ornamentals. A predatory mite sachet is placed in the plant canopy and a fungal-based granular material is incorporated into the potting mix.

10.5.2.2.5 ENTOMOPATHOGENIC FUNGI

Various entomopathogenic fungal species have been studied for WFT management, and preparations based on *B. bassiana*, *M. anisopliae* and *L. lecanii* have significantly reduced WFT populations in greenhouse vegetable and floral crops under research conditions. However, in commercial greenhouse settings, fungi have provided inconsistent results (Bradley et al., 1998; Butt and Brownbridge, 1997). Todd et al. (2007) reported a 40% reduction in WFT populations using a clay-based wettable powder formulation of *B. bassiana* applied 3–4 times over 1 month. In contrast, Jacobson et al. (2001) reported that WFT populations were reduced by 87% with three consecutive high volume sprays or low volume mist applications of Naturalis-L® or BotaniGard WP®, both *B. bassiana*-based products, applied at 6-day intervals. Good spray coverage is essential for reliable pest control with mycopesticides, and spray techniques commonly used in greenhouse crops are unlikely to provide the coverage needed to achieve success. In addition, as for conventional pesticides, efficacy is reduced because WFT pupating in the soil escape contact with fungal inoculum. Developing a fungal formulation targeting the soil stage could enhance efficacy by reducing the number of emerging adults, thereby breaking the WFT reproductive cycle.

Research is under way to combine several IPM components for WFT into what has been called a guardian plant (Figure 10.6). It uses a flowering marigold (var. Hero Yellow) as a trap plant to attract thrips out of the crop. A fungal-based granular formulation is incorporated into the potting mix of the marigold to target the pupae. A slow-release sachet containing a predatory mite, *Neoseiulus cucumeris*, and grain mites (on which the predators are reared), is hung within the canopy of the marigold. The predatory mites move onto the plant where they feed on WFT immatures. This system is based on the concept that adult WFT are attracted out of the crop to the flowering marigolds, where they reproduce. The immatures serve as prey for the predatory mites, sustaining them and encouraging their dispersal throughout the crop. In the absence of prey, the mites are sustained on marigold pollen. Thrips escaping predation will drop to the soil to pupate, where they become infected with the fungus. The granular formulation enables the fungus to colonize the potting mix, eliminating the need for reapplication. This system represents a low-cost, easy-to-use, non-chemical pesticide approach, suppressing WFT populations through a holistic system. Guardian plants are positioned throughout the crop at 120 sq. m (1000 sq. ft.) intervals. Because fungal treatments and mite releases are applied to the guardian plant rather than the entire crop, management costs are reduced, while control is maximized.

Trials testing this system were conducted in 2012 in Vermont commercial greenhouses containing spring ornamentals, testing two *B. bassiana* strains (GHA, the fungus in BotaniGard®; and an experimental isolate). Marigolds with and without fungal treatments and predators were monitored bi-weekly for 12 weeks. Yellow sticky traps were also positioned throughout the crop. Three plants located within 1 meter of each marigold were inspected for WFT, predatory mites and damage. Marigold flowers were collected and dissected to quantify WFT and mites. In general, six times more WFT were detected on marigolds than on randomly selected crop plants, and damage was greater on the marigolds than on the crop plants. Toward the end of the experiment, some of the marigolds had a high level of foliar damage whereas nearby crop plants had little or no damage, suggesting that the marigolds were more attractive to WFT than the crop, and that they did not migrate away from the trap plant.

From week 6 to 12, higher numbers of thrips were found on marigolds with no mites or fungal treatments than on those treated with mites and fungi. At week 12, thrips populations were significantly less in marigolds containing the experimental fungus and mites than in those with the commercial isolate and mites, demonstrating variability in efficacy among strains. Low numbers of predatory mites were detected throughout the experimental period during plant tapping, demonstrating that they were sustained on pollen and/or prey. However, they were not observed on the randomly inspected crop plants adjacent to the marigolds, suggesting limited mite dispersal occurred.

Both fungal strains persisted within the soil throughout the 12-week test period. The granular fungal formulations were applied at a rate of 1.0×10^8 (13.2 g/pot). The number of spores per gram of soil was $4–5\times10^5$ at week 0 (post-application) and $5–8\times10^3$ after 12 weeks. This shows that the fungal-based granular formulation can provide viable inoculum for over 12 weeks. This experiment demonstrates the potential of combining multiple IPM strategies to cost-effectively reduce WFT populations and damage on crop plants. In addition, the guardian plant demonstrated an ability to maintain WFT populations at relatively low levels for up to 12 weeks in greenhouses.

References

Abdulhai, M., Canhilal, R., El Bouhssini, M., Reid, W., Rihawi, F., 2007. Survey of Sunn Pest adult parasitoids in Syria. In: Parker, B.L., Skinner, M., El Bouhssini, M., Kumari, S.G. (Eds.), Sunn Pest Management: A Decade of Progress 1994-2004. Arab Society for Plant Protection, Beirut, Lebanon, pp. 315–318.

Abdullah, S.I., 2007. Screening of wheat varieties in Nineveh Province, Iraq. In: Parker, B.L., Skinner, M., El Bouhssini, M., Kumari, S.G. (Eds.), Sunn Pest Management: A Decade of Progress 1994-2004. Arab Society for Plant Protection, Beirut, Lebanon, pp. 357–361.

Al-Izzi, M.A.J., Amin, A.M., Al-Assadi, H.S., 2007. Role of biocontrol agents in decreasing populations of Sunn Pest in northern Iraq. In: Parker, B.L., Skinner, M., El Bouhssini, M., Kumari, S.G. (Eds.), Sunn Pest Management: A Decade of Progress 1994-2004. Arab Society for Plant Protection, Beirut, Lebanon, pp. 265–271.

Alavo, T.B.C., Sermann, H., Bochow, H., 2002. Virulence of strains of the entomopathogenic fungus *Verticillium lecanii* to aphids: strain improvement. Arch. Phytopathol. Plant Prot. 34, 379–398.

Allen, W.R., Matteoni, J.A., 1991. Petunia as an indicator plant for use by growers to monitor for thrips carrying the tomato spotted wilt virus in greenhouses. Plant Dis. 75, 78–82.

Amir-Maafi, M., 2007. Mass rearing of the Sunn Pest egg parasitoid, *Trissolcus granis* Thompson (Hymenoptera: Scelionidae): a demographic framework. In: Parker, B.L., Skinner, M., El Bouhssini, M., Kumari, S.G. (Eds.), Sunn Pest Management: A Decade of Progress 1994-2004. Arab Society for Plant Protection, Beirut, Lebanon, pp. 303–307.

Aquino de Muro, M., Elliott, M.S., Moore, D., Skinner, M., Reid, W., Parker, B.L., et al., 2005. Molecular characterisation of *Beauveria bassiana* isolates obtained from overwintering sites of Sunn Pests (*Eurygaster* and *Aelia* spp). Mycol. Res. 109, 294–306.

Arthurs, S., Heinz, K., 2006. Evaluation of the nematodes *Steinernema feltiae* and *Thripinema nicklewoodi* as biological control agents of western flower thrips *Frankliniella occidentalis* infesting chrysanthemum. Biol. Sci. Technol. 16, 141–155.

Asi, M.R., Bashir, M.H., Afzal, M., Ashfaq, M., Sahi, S.T., 2010. Compatibility of entomopathogenic fungi, *Metarhizium anisopliae* and *Paecilomyces fumosoroseus* with select insecticides. Pak. J. Bot. 42, 4207–4214.

Bartlett, M.C., Jaronski, S.T., 1988. Mass production of entomopathogenous fungi for biological control of insects. In: Burge, M.N. (Ed.), Fungi in Biological Control Systems. Manchester University Press, Manchester, UK, pp. 61–85.

Bateman, R.P., Alves, R.T., 2000. Delivery systems for mycoinsecticides using oil-based formulations. Aspect. Appl. Biol. 57, 163–170.

Bateman, R.P., Carey, M., Moore, D., Prior, C., 1993. The enhanced infectivity of *Metarhizium flavoviride* in oil formulations to desert locusts at low humidities. Ann. Appl. Biol. 12, 145–152.

Baverstock, J., Roy, H.E., Pell, J.K., 2010. Entomopathogenic fungi and insect behaviours: from unsuspecting hosts to targeted vectors. Biocontrol 55, 89–102.

Berndt, O., Poehling, H.-M., Meyhöfer, R., 2004. Predation capacity of two predatory laelapid mites on soil-dwelling thrips stages. Entomol. Exp. Appl. 112, 107–115.

Bradley, C.A., Lord, J.C., Jaronski, S.T., Gill, S.A., Dreves A.J., Murphy, B.C., 1998. Mycoinsecticides in thrips management. Proceedings of the Brighton Crop Protection Conference – Pests and Diseases 35, 177–181.

Braga, G.U., Flint, S.D., Miller, C.D., Anderson, A.J., Roberts, D.W., 2001. Both solar UVA and UVM radiation impart conidial culturability and delay germination in the entomopathogenic fungus *Metarhizium anisopliae*. Photochem. Photobiol. 74, 734–739.

Braga, G.U., Rangel, D.E.N., Flint, S.D., Miller, C.D., Anderson, A.J., Roberts, D.W., 2002. Damage and recovery from UV-B exposure in conidia of the entomopathogens *Verticillium lecanii* and *Aphanocladium album*. Mycologia 94, 912–920.

Broadbent, A.B., Pree, D.J., 1997. Resistance to insecticides in population of *Frankliniella occidentalis* (Pergande) (Thysanoptera: Thripidae) from greenhouses in the Niagara region of Ontario. Can. Entomol. 129, 907–913.

Broadbent, A.B., Rhainds, M., Shipp, L., Murphy, G., Wainman, L., 2003. Pupation behaviour of western flower thrips (Thysanoptera: Thripidae) on potted chrysanthemum. Can. Entomol. 135, 741–744.

Brown, H.A., Khan, A., 2009. Pathogenicity and virulence of four isolates of *Metarhizium anisopliae* on selected natural enemies: *Cryptolaemus montrouzieri, Anagyrus kamali, Lysiphlebus testaceipes* and *Bracon thurberiphagae*. J. Biopestic. 2, 199–203.

Buitenhuis, R., Shipp, L., Jandricic, S., Short, M. 2005. New control strategy for thrips in chrysanthemum. Greenhouse Canada. http://www.greenhousecanada.com/content/view/952/38/.

Burges, H.D., 1998. Formulation of mycoinsecticides. In: Burges, H.D. (Ed.), Formulation of Microbial Biopesticides: Beneficial Microorganisms, Nematodes and Seed Treatments. Kluwer Academic Publishing, Dordrecht, NL, pp. 131–186.

Butt, T.M., Brownbridge, M., 1997. Fungal pathogens of thrips. In: Lewis, T. (Ed.), Thrips as Crop Pests. CAB International, Wallingford, UK, pp. 399–433.

Butt, T.M., Jackson, C., Magan, N., 2001. Introduction – Fungal biological control agents: Progress, problems and potential. In: Butt, T.M., Jackson, C., Magan, N. (Eds.), Fungi as Biocontrol Agents. CABI Publishing, Wallingford, UK, pp. 1–8.

Cheraghi, A., Habibpour, B., Mossadegh, M.S., 2013. Application of bait treated with the entomopathogenic fungus *Metarhizium anisopliae* (Metsch.) Sorokin for the control of Microcerotermes diversus Silv. Psyche. http://dx.doi.org/10.1155/2013/865102.

Chouvenc, T., Su, N.Y., Elliott, M.L., 2008. Interaction between the subterranean termite *Reticulitermes flavipes* (Isoptera: Rhinotermitidae) and the entomopathogenic fungus *Metarhizium anisopliae* in foraging arenas. J. Econ. Entomol. 101, 885–893.

Copping, L.G., 2001. The BioPesticide Manual, second ed. British Crop Protection Council, Surrey, UK.

Copping, L.G., 2004. The Manual of Biocontrol Agents, third ed. British Crop Protection Council, Surrey, UK.

de Courcy Williams, M.E., 2001. Biological control of thrips on ornamental crops: interactions between the predatory mite *Neoseiulus cucumeris* (Acari: Phytoseiidae) and western flower thrips, *Frankliniella occidentalis* (Thysanoptera: Thripidae), on cyclamen. Biocontrol Sci. Technol. 11, 41–55.

Deligeorgidis, P.N., Ipsilandis, C.G., 2004. Determination of soil depth inhabited by *Frankliniella occidentalis* (Pergande) and *Thrips tabaci* Lindeman (Thysan., Thripidae) under greenhouse cultivation. J. Appl. Entomol. 128, 108–111.

Dromph, K.M., 2003. Collembolans as vectors of entomopathogenic fungi. Pedobiologia 47, 245–256.

Ebssa, L., Borgemeister, C., Poehling, H.-M., 2006. Simultaneous application of entomopathogenic nematodes and predatory mites to control western flower thrips, *Frankliniella occidentalis*. Biol. Control 39, 66–74.

Edgington, S., Moore, D., Kutuk, H., Satar, H., El Bouhssini, M., 2007. Progress in the development of a mycoinsecticide for biological control of Sunn Pest. In: Parker, B.L., Skinner, M., El Bouhssini, M., Kumari, S.G. (Eds.), Sunn Pest Management: A Decade of Progress 1994-2004. Arab Society for Plant Protection, Beirut, Lebanon, pp. 237–243.

El Bouhssini, M., Nachit, M., Valkoun, J., Moussa, M., Ketata, H., Abdallah, O., et al., 2007. Evaluation of wheat and its wild relatives for resistance to Sunn Pest under artificial infestation. In: Parker, B.L., Skinner, M., El Bouhssini, M., Kumari, S.G. (Eds.), Sunn Pest Management: A Decade of Progress 1994-2004. Arab Society for Plant Protection, Beirut, Lebanon, pp. 363–368.

Erman, A., Sabahoglu, Y., Babaroglu, N., Canhilal, R., 2007. Aerial vs ground applications for Sunn Pest Management. In: Parker, B.L., Skinner, M., El Bouhssini, M., Kumari, S.G. (Eds.), Sunn Pest Management: A Decade of Progress 1994-2004. Arab Society for Plant Protection, Beirut, Lebanon, pp. 209–213.

Fargues, J., Goettel, M.S., Smits, N., Ouedraogo, A., Vidal, C., Lacey, L.A., et al., 1996. Variability in susceptibility to simulated sunlight of conidia among isolates of entomopathogenic Hyphomycetes. Mycopathologia 135, 171–181.

Faria, M.R., Wraight, S.P., 2007. Mycoinsecticides and mycoacaricides: a comprehensive list with worldwide coverage and international classification of formulation types. Biol. Control 43, 237–256.

Feng, M.G., Poprawski, T.J., Khachatourians, G.G., 1994. Production, formulation and application of the entomopathogenic fungus *Beauveria bassiana* for insect control: current status. Biocontrol Sci. Technol. 4, 3–34.

Gao, Y., Lei, Z., Reitz, S.R., 2012. Western flower thrips resistance to insecticides: detection, mechanisms and management strategies. Pest Manag. Sci. 68, 1111–1121.

Goettel, M.S., 1992. Fungal agents for biocontrol. In: Lomer, C.J., Prior, C. (Eds.), Biological Control of Locusts and Grasshoppers. CABI International and International Institute of Tropical Agriculture, Wallingford, Oxon, UK, pp. 122–132.

Goettel, M.S., Hajek, A.E., 2001. Evaluation of non-target effects of pathogens used for management of arthropods. In: Wainberg, E., Scott, J.K., Quimby, P.C. (Eds.), Evaluating indirect effects of biological control. CAB International, Wallingford, UK, pp. 81–97.

Goettel, M.S., Inglis, G.D., Wraight, S.P., 2000. Fungi. In: Lacey, L.A., Kaya, H.K. (Eds.), Field Manual of Techniques in Invertebrate Pathology. Kluwer Academic Publishers, Dordrecht, NL, pp. 255–282.

Gouli, V., Gouli, S., Skinner, M., Shternshis, M.V., 2008. Effect of the entomopathogenic fungi on mortality and injury level of western flower thrips, *Frankliniella occidentalis*. Arch. Phytopathol. Plant Prot. 4 (2), 37–47.

Gul, A., Aw-Hassan, A., Kutuk, H., Canhilal, R., Mazid, A., Hoghaddam, M.H., et al., 2007. Shifting from aerial to ground spraying for Sunn Pest control: Farmers' perceptions and problems. In: Parker, B.L., Skinner, M., El Bouhssini, M., Kumari, S.G. (Eds.), Sunn Pest Management: A Decade of Progress 1994-2004. Arab Society for Plant Protection, Beirut, Lebanon, pp. 331–339.

Hirose, E., Neves, P.M.O.J., Zequi, J.A.C., Martins, L., Peralta, C.H., Moino Jr., A., 2001. Effect of biofertilizers and neem oil on the entomopathogenic fungi *Beauveria bassiana* (Bals.) Vuill. and *Metarhizium anisopliae* (Metsch.) Sorok. Braz. Arch. Biol. Technol. 44, 419–423.

Inglis, G.D., Johnson, D.L., Goettel, M.S., 1997. Effects of temperature and sunlight on mycosis of *Beauveria bassiana* (Hyphomycetes: Sympodulosporae) of grasshoppers under field conditions. Environ. Entomol. 26, 400–409.

Islam, M.T., Omar, D.B., 2012. Combined effect of *Beauveria bassiana* with neem on virulence of insect in case of two application approaches. J. Anim. Plant Sci. 22, 77–82.

Islam, M.T., Omar, D.B., Latif, M.A., Morshed, M.M., 2011. The integrated use of entomopathogenic fungus, *Beauveria bassiana* with botanical insecticide, neem against *Bemisia tabaci* on eggplant. J. Microbiol. Res. 5, 3409–3413.

Jacobson, R.J., Chandler, D., Fenlon, J., Russell, K.M., 2001. Compatibility of *Beauveria bassiana* (Balsamo) Vuillemin with *Amblyseius cucumeris* Oudemans (Acarina: Phytoseiidae) to control *Frankliniella occidentalis* Pergande (Thysanoptera: Thripidae) on cucumber plants. Biocontrol Sci. Technol. 11, 391–400.

Jaros-Su, J., Groden, E., Zhang, J., 1999. Effects of selected fungicides and the timing of fungicide application on *Beauveria bassiana*-induced mortality of Colorado potato beetle (Coleoptera: Chrysomelidae). Biol. Control 15, 259–269.

Jenkins, N.E., Heviefo, G., Langewald, J., Cherry, A.J., Lomer, C.J., 1998. Development of a mass production technology for aerial conidia of mitosporic fungi for use as mycopesticides. Biocontrol News Inform. 19, 21–31.

Kaakeh, W., Reid, B.L., Bohnert, T.J., Bennett, G.W., 1997. Toxicity of imidacloprid in the German cockroach (Dictyoptera: Blattellidae), and the synergism between imidacloprid and *Metarhizium anisopliae* (Imperfect Fungi: Hyphomycetes). J. Econ. Entomol. 90, 473–482.

Kabaluk, T., Gazdik, K., 2005. *Directory of Microbial Pesticides for Agricultural Crops in OECD Countries*. Agriculture and Agri-Food Canada., pp. 1–242.

Kapongo, J.P., Shipp, L., Kevan, P., Broadbent, B., 2008. Optimal concentration of *Beauveria bassiana* vectored by bumble bees in relation to pest and bee mortality in greenhouse tomato and sweet pepper. Biocontrol 53, 799–812.

Kassa, A., Brownbridge, M., Parker, B.L., Skinner, M., Gouli, V., Gouli, S., et al., 2008. Whey for mass production of *Beauveria bassiana* and *Metarhizium anisopliae*. Mycol. Res. 112, 583–591.

Khachatourians, G.G., Sohail, S.Q., 2008. Entomopathogenic Fungi. In: Brakhage, A.A., Zipfel, P.F. (Eds.), Biochemistry and Molecular Biology, Human and Animal Relationships, 2nd edn. The Mycota VI. Springer-Verlag, Berlin, Heidelberg.

Kim, J.S., Roh, J.Y., Choi, J.Y., Wang, Y., Shim, H.J., Je, Y.H., 2010a. Correlation of the aphicidal activity of *Beauveria bassiana* SFB-205 supernatant with enzymes. Fungal Biol. 114 (1), 120–128.

Kim, J.S., Je, Y.H., Choi, J.Y., 2010b. Complementary effect of Phloxine B on the insecticidal efficacy of *Isaria fumosorosea* SFP-198 wettable powder against greenhouse whitefly, *Trialeurodes vaporariorum* West. Pest Manag. Sci 66, 1337–1343.

Kim, J.S., Skinner, M., Gouli, S., Parker, B.L., 2010c. Influence of top-watering on the movement of *Beauveria bassiana*, GHA (Deuteromycota: Hyphomycetes) in potting medium. Crop Prot. 29 (6), 631–634.

Kim, J.S., Je, Y.H., Woo, E.O., Park, J.S., 2011. Persistence of *Isaria fumosorosea* (Hypocreales: Cordycipitaceae) SFP-198 conidia in corn oil-based suspension. Mycopathologia 171, 67–75.

Kodan, M., Gurkan, M.O., 2007. Mass production and storage of *Trissolcus grandis* (Thomson) (Hymenoptera: Scelionidae). In: Parker, B.L., Skinner, M., El Bouhssini, M., Kumari, S.G. (Eds.), Sunn Pest Management: A Decade of Progress 1994-2004. Arab Society for Plant Protection, Beirut, Lebanon, pp. 295–301.

Laird, M., Lacey, L.A., Davidson, E.W. (Eds.), 1990. Safety of Microbial Insecticides. CRC Press, Boca Raton, FL.

Li, H., Feng, M.G., 2005. Broomcorn millet grains cultures of the entomophthoralean fungus *Zoophthora radicans*: sporulation capacity and infectivity to *Plutella xylostella*. Mycol. Res. 109, 319–325.

Machado, A.C.R., Monteiro, A.C., Belasco de Almeida, A.M., Martins, M.I.E.G., 2010. Production technology for entomopathogenic fungus using a biphasic culture system. Pesq. Agropec. Bras. 45, 1157–1163.

Malekan, M., Hatami, B., Ebadi, R., Akhavan, A., Radjabi, R., 2012. The singular and combined effects of entomopathogenic fungi, *Beauveria bassiana* (Bals) and *Lecanicillium muscarium* (Petch) with insecticide imidacloprid on different nymphal stages of *Trialeurodes vaporariorum* in the laboratory conditions. Adv. Environ. Biol. 6 (1), 423.

McClatchie, G.V., Moore, D., Bateman, R.P., Prior, C., 1994. Effect of temperature on the viability of the conidia of *Metarhizium flavoviride* in oil formulations. Mycol. Res. 98, 749–756.

McCoy, C.W., Samson, R.A., Boucias, D.G., 1988. Entomogenous fungi. In: Ignoffo, C.M., Mandava, N.B. (Eds.), Handbook of Natural Pesticides. Vol. V: Microbial Insecticides, Part A: Entomogenous Protozoa and Fungi. CRC Press, Boca Raton, FL, pp. 151–236.

Meyling, N.V., Schmidt, N.M., Eilenberg, J., 2012. Occurrence and diversity of fungal entomopathogens in soils of low and high Arctic Greenland. Polar. Biol. 35, 1439–1445.

Moore, D., Bateman, R.P., Carey, M., Prior, C., 1995. Long-term storage of *Metarhizium flavoviride* conidia in oil formulation for the control of locusts and grasshoppers. Biocontrol Sci. Technol. 5, 193–199.

Morley-Davis, J., Moore, D., Prior, C., 1995. Screening of *Metarhizium* and *Beauveria* spp. conidia with exposure to stimulated sunlight and a range of temperatures. Mycol. Res. 100, 31–38.

Myles, T.G., 2002. Alarm, aggregation, and defense by *Reticulitermes flavipes* in response to a naturally occurring isolate of *Metarhizium anisopliae*. Sociobiology 40, 243–255.

Neves, P.M.O.J., Hirose, E., Tchujo, P.T., Moino Jr., A., 2001. Compatibility of entomopathogenic fungi with neonicotinoid insecticides. Neotrop. Entomol. 30, 263–268.

O'Callaghan, M., Brownbridge, M., 2009. Environmental impacts of microbial control agents used for control of invasive pests In: Hajek, A.J. Glare, T. O'Callaghan, M. (Eds.), Progress in Biological Control: Use of Microbials for Control and Eradication of Invasive Arthropods, Vol. 6. Springer, pp. 305–330.

Parker, B.L., Skinner, M., Costa, S.D., Gouli, S., Reid, W., El Bouhssini, M., 2003. Entomopathogenic fungi of *Eurygaster integriceps* Puton (Hemiptera: Scutelleridae): Collection and characterization for development. Biol. Control 27, 260–272.

Reardon, R.C., Hajek, A.E., 1998. The gypsy moth fungus *Entomophaga maimaiga* in North America. US Dept. of Agric. Forest Service, FHTET-97-11. Morgantown, WV, USA. 22 pp.

Reitz, S.R., 2009. Biology and ecology of the western flower thrips (Thysanoptera: Thripidae): The making of a pest. Fla. Entomol. 92, 7–13.

Robb, K.L., Parrella, M.P., 1995. IPM of western flower thrips. In: Parker, B.L., Skinner, M., Lewis, T. (Eds.), Thrips Biology and Management. Plenum Press, NY, pp. 365–370.

Roberts, D.W., Hajek, A.E., 1992. Entomopathogenic fungi as bioinsecticides. In: Leatham, G.F. (Ed.), Frontiers of Industrial Mycology. Chapman and Hall, New York, pp. 114–159.

Rosell, G., Quero, C., Coll, J., Guerrero, A., 2008. Biorational insecticides in pest management. J. Pest Sci. 33, 601–606.

Rosin, R., Shapiro, D.I., Lewis, L.C., 1996. Effects of fertilisers on the survival of *Beauveria bassiana*. J. Invertebr. Pathol. 68, 194–195.

Sahayaraj, K., Namasivayam, S.K.R., 2008. Mass production of entomopathogenic fungi using agricultural products and by products. Afr. J. Biotechnol. 7, 1907–1910.

Sahayaraj, K., Namasivayam, S.K.R., Martin, J.R., 2011. Compatibility of entomopathogenic fungi with extracts of plants and commercial botanicals. Afr. J. Biotechnol. 10, 933–938.

Scholte, E.-J., Knols, B.G.J., Samson, R.A., Takken, W., 2004. Entomopathogenic fungi for mosquito control: a review. J. Insect Sci. 4, 19.

Sharififard, M., Mossadegh, M.S., Vazirianzadeh, B., Zarei-Mahmoudabadi, A., 2011. Interactions between entomopathogenic fungus, *Metarhizium anisopliae* and sublethal doses of spinosad for control of house fly, *Musca domestica*. Iran J. Arthropod-Borne Dis. 5, 28–36.

Skinner, M., Gouli, S., Frank, C.E., Parker, B.L., Kim, J.S., 2012. Management of *Frankliniella occidentalis* (Thysanoptera: Thripidae) with granular formulations of entomopathogenic fungi. Biol. Control 63, 246–252.

Stack, L.B., Cloyd, R., Dill, J., McAvoy, R., Pundt, L., Smith, C., et al., 2013. New England Greenhouse Floriculture Guide. New England Floriculture, Inc.

Staples, J.A., Milner, R.J., 2000. A laboratory evaluation of the repellency of *Metharizium anisopliae* conidia to *Coptotermes lacteus* (Isoptera: Rhinotermitidae). Sociobiology 36, 1–16.

Sterk, G., Heuts, F., Merck, N., Bock, J., 2003. Sensitivity of non-target arthropods and beneficial fungal species to chemical and biological plant protection products: Results of laboratory and semi-field trials. First International Symposium on Biological Control of Arthropods. USDA Forest Service FHTET-03-05. <www.insectscience.org/4.19>; <http://www.bugwood.org/arthropod/day4/sterk.pdf>.

Tanada, Y., Kaya, H.K., 1993. Fungal infections. In: Tanada, Y., Kaya, H.K. (Eds.), Insect Pathology. Academic Press, NY, USA, pp. 318–387.

Todd, A., Ugine, W.S., John, S., 2007. Effect of manipulating spray-application parameters on efficacy of entomopathogenic fungus *Beauveria bassiana* against western flower thrips, *Frankliniella occidentalis* infesting greenhouse impatiens crops. Biocontrol Sci. Technol. 17, 193–219.

Trissi, A.N., El Bouhssini, M., Ibrahem, J., Abdulhai, M., Reid, W., 2007. Survey of egg parasitoids of Sunn Pest in northern Syria. In: Parker, B.L., Skinner, M., El Bouhssini, M., Kumari, S.G. (Eds.), Sunn Pest Management: A Decade of Progress 1994-2004. Arab Society for Plant Protection, Beirut, Lebanon, pp. 309–314.

Trissi, A.N., El-Bouhssini, M., Al-Salti, N., Parker, B.L., Skinner, M., Massri, A., 2010. Efficiency of *Beauveria bassiana* on Sunn Pest, *Eurygaster integriceps* Puton (Hem: Scutelleridae) and effects on the egg parasite, *Trissolcus grandis* Thomson. Arab J. Plant Prot. 28, 163–168.

Trissi, A.N., El Bouhssini, M., Al-Salti, N., Abdulhai, M., Skinner, M., Parker, B.L., 2012. Virulence of *Beauveria bassiana* against Sunn pest, *Eurygaster integriceps* Puton (Hemiptera: Scutelleridae) at different time periods of application. J. Entomol. Nematol. 4, 49–53.

van den Meiracker, R.A.F., Sabelis, M.W., 1999. Do functional responses of predatory arthropods reach a plateau? A case study of *Orius insidiosus* with western flower thrips as prey. Entomol. Exp. Appl. 90, 323–329.

Wang, C.L., Powell, J.E., 2004. Cellulose bait improves the effectiveness of *Metarhizium anisopliae* as a microbial control of termites (Isoptera: Rhinotermitidae). Biol. Control 30, 523–529.

Weintraub, P.G., Pivonia, S., Steinberg, S., 2011. How many *Orius laevigatus* are needed for effective western flower thrips, *Frankliniella occidentalis*, management in sweet pepper. Crop Prot. 30, 1443–1448.

Wiethoff, J., Poehling, H.M., Meyhöfer, R., 2004. Combining plant- and soil-dwelling predatory mites to optimise biological control of thrips. Exp. Appl. Acarol. 34, 239–261.

Wraight, S.P., Jackson, M.A., de Kock, S.L., 2001. Production, stabilization and formulation of fungal biological agents. In: Butt, T.M., Jackson, C., Magan, N. (Eds.), Fungi as Biocontrol Agents. CABI, Wallingford, UK, pp. 253–287.

Yatin, B.T., Venkataraman, N.S., Parija, T.K., Panneerselvam, D., Govindanayagi, P., Geetha, K., 2006. The New Biopesticide Market. Business Communications Research, Denver.

Yildirim, A.F., Kinaci, E., Uysal, M., 2007. Sunn Pest overwintering sites, parasitoids and effects on cereal lines and varieties. In: Parker, B.L., Skinner, M., El Bouhssini, M., Kumari, S.G. (Eds.), Sunn Pest Management: A Decade of Progress 1994-2004. Arab Society for Plant Protection, Beirut, Lebanon, pp. 273–277.

CHAPTER

11

Potential of Entomopathogenic Nematodes in Integrated Pest Management

S.S. Hussaini

National Bureau of Agriculturally Important Insects, Bangalore, Karnataka State, India

11.1 INTRODUCTION

Constraints on the use of chemical insecticides have limited the availability of control measures against soil-borne insect pests. An overall integrated pest management (IPM) programme involves, in part, biological control methods to reduce the environmental and safety hazards of chemicals. This may be a more economical alternative to some insecticides. Biological controls are often very specific for a particular pest unlike most insecticides with least danger of impact on the environment and water quality. Concern over chemical pesticides in respect of groundwater contamination, residues on food, and resistance development prompted safer alternatives. A Biological Control programme may range from choosing a pesticide least harmful to beneficial insects, to raising and releasing one insect to attack another like a 'living insecticide'. Successful use thus requires a thorough understanding of the biology of both the pest and its antagonists.

The annual growth increase in microbial insecticides is 10-fold compared to that of chemical insecticides. The entomopathogenic nematodes (EPN) are potential insect parasites used in biological control. The third-stage infective juveniles (IJs) serve as agents with their associated bacterium. EPN possess impressive attributes of parasitoids/predators and pathogens such as quick kill, broad host range, high virulence, chemoreceptors, easy culture *in vitro*, and safety to vertebrates, plants and non-targets. Standard equipment is used for application, and the technique is compatible with many chemical pesticides, in addition to being amenable to genetic selection. Commercialization still has to address issues such as mass production, formulation and use. An attempt has been made to analyse some critical factors in the successful utilization of EPN in IPM for insect control.

D. P. Abrol (Ed): Integrated Pest Management.
DOI: http://dx.doi.org/10.1016/B978-0-12-398529-3.00012-9

The IJs locate host insects in soil and invade through natural openings gaining access to the haemocoel, and releasing a symbiotic bacterium (*Xenorhabdus/Photorhabdus*) that multiplies rapidly, killing the host within 24–48h. Juveniles feed on bacteria and disintegrated host tissues, grow to adults, and produce 2–3 generations. On depletion of nutrients, the nematodes produce next generation juveniles which leave the host cadaver seeking new hosts (Poinar, 1990). Limited shelf-life of nematodes is a major obstacle (Grewal and Georgis, 1998). Since the discovery of infection of Japanese beetle with *Steinernema glaseri* in 1932, biological control of insects has progressed significantly and currently, nematodes are produced commercially worldwide and are mostly used in niche markets.

11.2 SPECIES AND STRAIN

The importance of inter-specific and intra-specific variation has been recognized and demonstrated leading to differences in efficacy among species and strains of a number of entomopathogens. Accurate, rapid and easy methods are required for identification of entomopathogenic nematodes. Very few morphological differences are discernible due to their specialized feeding remaining inside the hosts, and species identification has been difficult and has been based on host/geographical area. For a species to be distinct, morphological/physiological/ecological differences and reproductive isolation need to be detectable compared to closely related species. Morphological characteristics are helpful but often fail. DNA sequence and hybridization data are supplemented to confirm distinctiveness. Novel diagnostic characteristics have been developed based on molecular methods in taxonomy where morphology lacked discriminatory power. Protein or isozyme phenotype differences, affinity for antibody, DNA-RFLP, and sequence data for any gene/region of the genome are considered important in diagnosis.

Cross-breeding to produce fertile offspring is important for identifying species and for showing species divergence. PCR techniques have widespread application in molecular taxonomy replacing DNA analysis by other techniques. PCR amplifies DNA from small quantities and amplifies taxonomically useful genome areas. In India, Poinar et al. (1992) detected *Heterorhabditis indica* in sugarcane fields at Coimbatore. Extensive surveys by the Project Directorate of Biological Control, Bangalore since 1996 detected *Steinernema carpocapsae* from Bangalore, Madurai, and Rajahmundry, *S. bicornutum* from Delhi, and *H. indica* from Bangalore, Coimbatore, Chidambaram, Kanyakumari, and Aligarh. Detailed RFLP analysis showed the presence of *S. tami* from Jorhat, *S. abbasi* from Delhi and unidentified *Steinernema* spp. SSL2 from Aligarh and Coimbatore (Hussaini et al., 2001a). In addition, *S. thermophilum* was recorded from IARI, New Delhi (Ganguly and Singh, 2000).

11.3 VIRULENCE AND INFECTIVITY

Entomopathogenic nematodes have been tested against a large number of insect pest species with results varying from poor to excellent control (Georgis et al., 2006). They have been most efficacious in habitats that provide protection from environmental extremities, especially in the soil, which is their natural habitat, and in cryptic habitats. Excellent control has been achieved against plant-boring insects because their cryptic habitats are favourable for nematode survival and infectivity. Nematode infectivity consists of a sequence of events initiated with host location, host attachment, host penetration, bacterial release, bacterial proliferation and host death. Infectivity is influenced by several biotic and abiotic factors including insect and nematode species, production method (Georgis and Gaugler, 1991), bacterial strain, temperature (Hussaini et al., 2005b; Molyneux, 1986) and bioassay methods (Grewal et al., 1994b).

Infectives respond differently to unparasitized hosts or hosts parasitized by conspecific/heterospecific nematodes and thus may reduce inter-/intra-specific competition. Apart from *S. carpocapsae*, other *Steinernema* spp. were attracted to unparasitized hosts. *S. carpocapsae* IJs were repelled from hosts infected for 4h with all heterospecific infections apart from *S. anomali*, whereas *S. glaseri* were repelled from *S. riobrave*-infected hosts. *S. glaseri* were attracted to 4 of 5 heterospecific infections and *S. anomali* and *S. riobrave* were attracted to 2 of 5 heterospecific infections. *S. anomali* and *S. glaseri* were more attracted to hosts infected with the out-group *H. bacteriophora* than those infected by conspecific nematodes. *S. carpocapsae, S. anomali*, and *S. glaseri* IJs were more attracted to insects colonized by conspecific nematodes than to un-infested insects. Bacteria are a source of active volatiles and odour-mediated host recognition by IJ may reduce inter- and intra-specific competition among steinernematids (Grewal et al., 1997). Pre-infection of *S. carpocapsae* and *S. glaseri* on subsequent invasion at 25°C did not change 6h post-infection, but invasion decreased by 50% compared to previously uninfected hosts. Pre-infection with *S. glaseri* promptly suppressed subsequent invasion of the two nematode species. At the early stage of the subsequent invasion, the sex ratio (% male) of the invading *S. carpocapsae* in the hosts with and without the pre-infections was 5–12%. Insects infected with a *Steinernema* complex may produce suppressive substance(s) against invaders (X.D. Wang et al., 1998). The penetrative rate of *S. carpocapsae, S. ceratophorum* and *S. cubanum* on *G. mellonella* was influenced in degrees by emergence time; the penetrative rate of those emerging on the 1st day was higher than that of those emerging on the 7th day, especially *S. cubanum*. The highest penetrative rate for *S. carpocapsae/S. ceratophorum* occurred 7 days after storage, and for *S. cubanum* after 15 days, and for 10 days, *S. carpocapsae* and *S. ceratophorum* were more active than newly emerged (Yang et al., 1999).

Virulence and infectivity of native *Steinernema* spp. and *H. indica* alone and in combination against *A. ipsilon* were studied in sand and sandy loam soil columns. *H. indica* PDBC EN 13.22 out-competed *Steinernema* spp. in all soil types, depth and time for mortality. *S. abbasi* PDBC EN 3.1 was promising among steinernematids. The performance of all isolates was better in sandy loam soil than in sand at 5cm depth. The mortality of *A. ipsilon* increased with increase in time of exposure. Combination of *S. carpocapsae* PDBC EN 6.11 and *H. indica* PDBC EN 13.22 had an additive effect over their individual populations in both soil types at 10cm depth (Hussaini et al. 2000b). Nematode isolates differ in virulence between the stages of the insect and among the species (Glazer and Navon, 1990). Stimulation of *Steinernema* sp. and *H. indica* PDBC isolates, exposed to Mg and Mn ions, caused increased pathogenicity and progeny production (Hussaini et al., 2001b). Similar results were obtained by Menti et al. (2000).

Another important factor affecting efficacy is temperature. The temperature range for infectivity was greater than that for development with the optimal temperature for infection and development at 23°C. Infectivity of *Steinernema* stored for 12 weeks was unaffected, as in *H. megidis*, which were less infective than fresh IJ. Similar differences were found by Boff et al. (2000). Time to first emergence and body length of IJ were affected by time, storage temperature and inoculum level. Highest infectivity and optimal development were observed in IJs stored at 10°C and 15°C. Temperature and host age influenced the biocontrol potential against sugarcane rootstalk borer weevil, *Diaprepes abbreviatus*. Virulence and reproductive potential were compared among *S. riobrave*, *H. bacteriophora* and *H. indica* in moist sand, soil temperature regimes and larval ages. Older larvae (100-day-old) were less susceptible to infection than younger larvae. They were less virulent at 21°C than at 24°C or 27°C. Virulence of *H. indica* was greater than that of *H. bacteriophora* in 50-day-old larvae. *H. bacteriophora* was more virulent than *S. riobrave* in 20-day-old larvae and *S. riobrave* more virulent than *H. bacteriophora* at 21°C in 50-day-old larvae. Reproductive potential was greatest in *H. indica*. Potential for recycling is inherent in high reproduction, temperature and host age, critical factors for field application (Shapiro et al., 1999a). *S. riobrave* was superior to *S. abbasi* in virulence to 4th-instar *S. littoralis*, at 3 days post-treatment with LD_{50} values of 49.6 and 60.3 IJs/larva, respectively, and highly virulent to full-grown larvae, pre-pupae and adult moths in sand. Six-day-old pupae were resistant to attack (Abbas and Saleh, 1998).

Environmental cues influence the responses of IJs and their oral or cuticle penetration (X.D. Wang et al. 1998). *S. glaseri* migrated 5.54 and 3.20 mm towards whole and wounded grass roots, respectively, while *H. bacteriophora* migrated 7.46 and 3.24 mm, in 30 min. Wounded roots were attractive. *S. glaseri* and *H. bacteriophora* migrated 4.15 and 16.43 mm towards gut fluid of *Popillia japonica*. *S. glaseri* migrated towards host haemolymph at 4.76 mm, but *H. bacteriophora* was not attracted to haemolymph. The presence of grass roots enhanced penetration by *S. glaseri*. In oral injection tests, *S. glaseri* penetrated haemocoel better via the gut. The proportion of penetration sites for *H. bacteriophora* on cuticle were higher than on gut. *S. carpocapsae* IJs entered *Liriomyza trifolii* through oviposition punctures made by the female during egg laying, or through an unnatural tear in the mine surface. Nematodes were unable to enter mines by penetrating the intact leaf cuticle. Infection through the anus was higher and they were unable to enter via the larval or puparial spiracles because of small size and morphology. Susceptible stages were larval stages, prepuparium and 'early puparium'. Mortality was highest in 2nd-instar larvae (93.3%/1-h exposure). First- and 2nd-instar larvae died 0.25 and 0.66 h, respectively after penetration. Rapid death was due to internal mechanical damage. Pre-puparia died 15 h post penetration due to mechanical and bacterial factors. The maximum production of IJs was 250/large 3rd-instar larva (LeBeck et al., 1993). *S. feltiae* invaded *S. litura* via routes other than the alimentary canal, larvae, pupae and adult spiracles of *G. mellonella* but not the last-instar larva. The firm framework of the head may be responsible for mechanical invasion through the cuticular membrane (Kondo and Ishibashi, 1989).

Host-plant background affects the susceptibility of the pest to EPN. *Sitona lineatus* infection by *S. carpocapsae* was greater for larvae from peas than from faba beans. Adults from pea-fed larvae were more susceptible and larvae from beans appeared favourable hosts for nematode multiplication (Jaworska and Ropek, 1994). In lab assays, the mortality of *Diabrotica undecimpunctata howardi* and progeny production of *S. carpocapsae* and *H. bacteriophora*, varied according to the host plant on which the pest had fed. Mortality from *S. carpocapsae* was lower when the pest had fed on groundnut roots than when on squash or maize roots. Mortality from *H. bacteriophora* was lower when the pest had fed on maize roots than on groundnut or squash roots. Progeny production from larvae fed on squash roots was lower than from those fed on maize or groundnut (Barbercheck, 1993). Similarly effects were found in other crops (Barbercheck et al., 1995; Hussaini et al., 2001c).

11.4 BIOEFFICIENCY

Bioefficiency tests in India have been confined mostly to lab studies. In field trials of *S. carpocapsae* (strain DD 136), *H. bacteriophora* (strain Burliar) and *Heterorhabditis* spp. (strains Chekkanurnai and Melur) against 4th-instar larvae of *Amsacta albistriga* on groundnut, *S. carpocapsae* was the most effective followed by *Heterorhabditis* sp. (Chekkanurnai). *Heterorhabditis* spp. did not develop in the larvae but *S. carpocapsae* did (Bhaskaran et al., 1994). Indigenous isolates of EPN were tested against *Leucinodes orbonalis* on brinjal in the lab and field. *S. carpocapsae* and *H. indica* @ 0.5–2.0 billion/ac spray reduced the borer holes on fruits and increased yields comparable with sprays of neem seed kernel extract. *S. carpocapsae* PDBC EN 6.11 was found to be more effective (Hussaini et al., 2002). Similarly, *H. bacteriophora* (NC strain) reduced populations of *Popillio japonica* in turf grass in Ohio up to 60%, 34 days after treatment in autumn. This increased to 96% before pupation the following spring, and was 93–99% in the next larval generation. *S. carpocapsae* (All strain) and *H. bacteriophora* HP88 yielded 51% and 100% of control after 34 days and 20 days, 90% and 93% the next spring, and 0% after 386 days, respectively. No adverse effects were observed on non-target mites or Collembola (Klein and Georgis, 1992). Application of *Heterorhabditis* and *Steinernma* species controlled the white grub, *Holotrichia longipennis,* population in turf grass in Srinagar (Hussaini et al., 2005b). In Florida Citrus orchards, *S. carpocapsae* as BioVector at 5 million IJ/tree and *H. bacteriophora* as Otinem at 1, 2 and 5 million IJ/tree controlled *Diaprepes abbreviatus* and *Pachnaeus litus*. Adult weevils were monitored as they emerged from the soil. Otinem resulted in a reduction compared to untreated trees, with the lowest rate giving the best level of control. BioVector also reduced adult emergence by 45% (Downing et al., 1991). Paunikar et al. (2011) evaluated the infectivity of two native *Steinernema* species/strains (TFRI-EPN-49 and TFRI-EPN-56) isolated from tropical forest areas of Madhya Pradesh, India in the lab against wax moth, *G. mellonella*. Dose-dependent mortality was observed and the lowest dose of 3IJs/larva gave 19.04% and 42.85% mortality with TFRI-EPN-49 and TFRI-EPN-56, respectively. Paunikar et al. (2010) investigated the susceptibility of teak skeletonizer, *Eutectona machaeralis*, to *S. carpocapsae* in the lab and found that larvae were susceptible to the nematode indicating their future role in the management of these forest insect pests. Susceptibility of target insect varied based on nematode species and strain. Lab studies carried out by Shakeela and Hussaini (2006) showed varying degrees of virulence of *S. abbasi, S. carpocapsae, S. feltiae, S. riobrave, H. indica* and *H. bacteriophora* against larvae of tobacco cut worm *Spodoptera litura.* Besides insects, EPN have been utilized against plant-parasitic nematodes. The effect of *S. abbasi, S. carpocapsae, S. feltiae* and *H. indica* has been documented against *Meloidogyne incognita* in tomato (Hussaini et al., 2008).

The field bioefficiency of selected target pests is presented in Table 11.1.

The thermal niche breadth for different species is variable. Grewal et al. (1994a) found that thermal niche breadths for establishment within hosts was the widest for *S. glaseri* (10–37°C) and the narrowest for *S. feltiae* (8–30°C). *S. riobrave*-infected *G. mellonella* larvae at a wider range (10–39°C) and *S. feltiae* at the narrowest (8–30°C), and reproduction was widest for *S. glaseri* (12–32°C) and narrowest for *S. carpocapsae* (20–30°C). *S. scapterisci* (20–32°C), *S. riobrave* (20–35°C), and *Steinernema* sp. (20–32°C) were more adapted to warm temperature reproduction, and *S. feltiae* to cooler temperatures (10–25°C). Heterorhabditids endemic to warmer climates had the upper thermal limits and temperatures. Thermal niche breadths did not differ between conspecific populations isolated from different localities, and were different for different species from the same locality. Thus, EPN have well-defined thermal niches unaffected by their locality. Hussaini et al. (2005a) observed absolute mortality of *G. mellonella* and *A. ipsilon* with *S. carpocapsae* PDBCEN 6.11,

TABLE 11.1 Bioefficacy of EPN Against Targeted Pests Worldwide (1990–2012)

Crop	Pest	Nematode sp.	Dosage	% mortality	Reference
Brinjal	Shoot and fruit borer, *Leucinodes orbonalis* Guen	*S. carpocapsae, H. indica*		Effective	Hussaini et al. (2002); Ganga Visalakshi et al. (2009)
Tobacco	Cutworm, *Spodoptera litura*	*S. carpocapsae* (Talc)	1–4 lakh/sq.m.	Effective ETL	Sitaramaiah et al. (2002)
Citrus	*Papilio* spp.				Singh (1993)
Groundnut	*Amsacta albistriga*	*H. bacteriophora*			Bhaskaran et al. (1994)
Rice	*Scirpophaga incertulas* Wlk, *Sesamia inferens* Wlk, *Chilo suppressalis, Cnaphalocrosis medinalis* G., *Orseolia oryzae*	*S. carpocapsae*		Effective	Srinivas and Prasad (1991)
Chrysanthemum	Serpentine leafminer, *Liriomyza trifolii*	*S. carpocapsae*	50×10^8/ha	64	Harris et al. (1990)
Vegetables	Serpentine leafminer, *L. trifolii*	*S. carpocapsae, H. bacteriophora*	50×10^8/ha	93	LeBeck et al. (1993)
Cabbage	Diamondback moth, *Plutella xylostella*	*Steinernema* sp., *H. indica*	100 IJ/larva	73–100	Mason and Wright (1997)
Cabbage	Diamondback moth, *P. xylostella*	*S. carpocapsae., S. riobrave, S. feltiae*	100 IJ/larva	67–100	Ratnasinghe and Hague (1997)
	Conorrhynchus mendicus	*S. carpocapsae, S. feltiae*		*Sc* 89.5, 59.5% *S.f* 68.5, 28%	Akalach and Wright (1995)
Maize	Earworm, *Heliothis zea*	*S. feltiae*	40 IJ/ml	100	Cabanillas and Raulston (1994)
Paddy	Stem borer, *Chilo suppressalis*	*S. carpocapsae, H. bacteriophora*	2000 IJ/ml 800 IJ/ml	69 91–100	Choo et al. (1991)
	Filbertworm, *Cydia latiferreana*	*S. carpocapsae*	40–200/cm^2	90–92% larvae without hibernacula (80–95%), pupal 50–75%, larvae 65–75%	Chambers et al. (2010)
Sweet potato (*Ipomoea batatas* (L.))	*Cylas puncticollis* Boheman (Coleoptera: Apionidae)	*S. karii/H. indica*			Nderitu et al. (2009)
Litchi (*Litchi chinensis*), longan (*Dimocarpus longan*)	Metarbelid borer *Arbela dea* (*Indarbela dea*)	*S. carpocapsae*	8–10,000/ml injection/spraying	89–100%	Xu and Yang (1992)

(*Continued*)

TABLE 11.1 (Continued)

Crop	Pest	Nematode sp.	Dosage	% mortality	Reference
Banana	Root weevil, *Cosmopolites sordidus*	*S. carpocapsae, S. glaseri, S. bibionis*	400–40,000	100	Figueroa (1990)
Sweet potato	Weevil, *Cylas formicarius*	*S. carpocapsae, S. feltiae, H. bacteriophora*			Jannson et al. (1991)
Pear	*Haplocampa brevis* Hym: Tentherinidae	*S. carpocapsae, S. feltiae, H. bacteriophora*	250,000/m^2 or 50,000/m^2	10–25	Curto et al. (2007)
Apple	Apple borer, *Zeuzera pyrina*	*S. riobravae, S. abbasi, S. carpocapsae* S2, *Heterorhabditis* sp. SAA1 *S. feltiae*	1000–2000 IJ/ml spray/injection	*S.c* S2 14.58, 39.39, 66.6% spray/injection *S.f* 81.8%	Saleh and Abbas (1998)
Turf	White grub, *Holotrichia longipennis*	*S. carpocapsae, H. bacteriophora, S. abbasi, H. indica*	5×10^9 IJ/ha soil	*S.c-55; H.i-42 S.a-53; H.b-39*	Hussaini et al. (2005b)
Hazelnut	Hazelnut borer, *Curculio nucum*	*H. bacteriophora/ H. megidis*			Blum et al. (2009)
Strawberry	*Cydia pomonella*	*S. feltiae*	0.75×10^9	40–50%	Kienzle et al. (2010)
Sugarbeet	*Cassida vittata*	*S.c; H. bacteriophora; S. feltiae; Heterorhabditis* sp.	500–400	*S.c* larvae 65% P 92 Adult 53%	Saleh et al. (2009)
Cauliflower	*Pieris brassicae*	*Steinernema* sp. Ats; At4; *Heterorhabditis* sp. *B20,H44*		Significant	Atwa et al. (2009)
Cabbage	*Delia radicum*, cabbage maggot				Leger and Riga (2009)
Strawberry	*Tymnorrhynchus*	*S. glaseri*	20,000 IJ/m^2	Highly effective	Atwa et al. (2009)
Cauliflower	*Pieris brassicae*	*S. carpocapsae (PDBC); H. indica (PDBC); S.c (JMU)*	0.5, 1.0, 2.0 billion/ha	41.8%	Gupta et al. (2009)
Oak	*Caliroa varipes* (Klug, 1814) (Hymenoptera, Tenthredinidae)	*H. bacteriophora, S. feltiae*	5×10^9 IJ/ha	Highly effective	Curto et al. (2008)
Groundnut	*Amsacta albistriga*	*H. indica; S. carpocapsae*	100,000/ml	Highly effective	Prabhu and Sudheer (2008)
	S. cretica	*H. bacteriophora* (BA1) *S. carpocapsae* (BA2) spray		97–100% 1 week post	El-Wakeil and Hussein (2009)

(Continued)

TABLE 11.1 (Continued)

Crop	Pest	Nematode sp.	Dosage	% mortality	Reference
Olives	*Bactrocera oleae*	*S. feltiae* sprayed over infested fallen olives		67.9%	Sirjani et al. (2009)
Litchi (*Litchi chinensis* Sonn.)	Bark-feeding moth, *Indarbela dea* (Swinhoe)	*S. carpocapsae*	100% in both trials	Mortality (100% and 95%) 3 weeks after application of *S.c* alone 2 weeks after combined applcn of *B.b* and *S.c*	Schulte et al. (2009)
Asparagus (*Asparagus officinalis)*	Asparagus beetle (*Crioceris asparagi)*	*H. megidis, S. carpocapsae, H. bacteriophora* and *S. feltiae*	2500 nemas/ml, 250 ml/pl	54, 58, 66, 96%	Schelt and Hoogerbrooks (2008)
					Nyasani et al. (2008)
Stone fruit	Flat-headed rootborer *Capnodis tenebrionis* (Coleop: Buprestidae)	*S. feltiae* (strain Bpa)	Drench and injection 1 million IJs/tree/week 4 or 8 weeks, 4 × 10^6 IJs/tree and 8 × 10^6 IJs/tree	88.3–97%	Morton and Garcia-del-Pino (2008)
Maize	Western corn rootworm (*Diabrotica virgifera virgifera* LeConte, Coleop: Chrysomelidae)	*S. scarabaei, H. bacteriophora, H. zealandica*	2.5 × 10^9 IJ/ha 5 × 10^9 IJ/ha	57%	Toefer et al. (2008)
	Leek moth, *Acrolepiopsis assectella*	*S. feltiae*	30,000 IJs/leek	Field (80–100%)	Garcia-del-Pino and Morton (2008)
	Xanthogaleruca luteola (Coleop: Chrysomelidae)	*S. feltiae* and *S. carpocapsae, H. bacteriophora*	–	*S.c* 57%	Triggiani and Tarasco (2007)
	Phyllophaga georgiana (Coleop: Scarabaeidae)	*S. scarabaei, H. bacteriophora H. zealandica*	2.5 × 10^9 IJ/ha 5 × 10^9 IJ/ha	76–100%	Koppenhöffer et al. (2008)
Stone fruit and seed fruit	Mediterranean flat-headed rootborer, *Capnodis tenebrionis*	*S. feltiae* (strain Bpa)	1 million IJs per tree/week, 4–8 weeks	88.3–97%	Morton and Garcia-del-Pino (2008)

(*Continued*)

TABLE 11.1 (Continued)

Crop	Pest	Nematode sp.	Dosage	% mortality	Reference
Turfgrass	*Popellia quadrigutta*	*S. glaseri* and *S. carpocapsae*, *Heterorhabditis* sp.		30–97%	Lee et al. (2005)
S. exitiosa	Peachtree borer, *Synanthedon exitiosa*,	*S. carpocapsae* strains (All and Hybrid) *H. bacteriophora*	150,000–300,000 IJ/tree		Shapiro-Ilan et al. (2009b)
Apple	*Cydia pomonella*	*S. carpocapsae*, *S. feltiae*, *H. bacteriophora*, *H. megidis*		42%	Züger et al. (2005)
Walnut	Leopard moth, *Zeuzera pyrina* L. (Lep.: Cossidae)	*S. carpocapsae*, *H. bacteriophora*	Injection into galleries 2000 IJs/larva	Significant control	Ashtari et al. (2011)
Cut grassland	White grubs June beetle (*Amphimallon solstitialis*), margined vine chafer (*Anomala dubia*), garden chafer (*Phyllopertha horticola*)	*Beauveria brongniartii*, *B. bassiana*, *B. t. kurstaki*, *H. bacteriophora*	Water suspension and infested grain	One application in April, abundance of overwintered white grubs reduced	Laznik et al. (2012)
	Third-stage larvae of cockchafer (*Melolontha melolontha*)	*S. feltiae* Entonem and indigenous strain C76	20 and 25°C 250,000 IJs/m^2, 500,000 1,000,000	27–53% C76 20% entonem	Laznik et al. (2009)
	C. elephas (Coleop: Curculionidae)	*H. bacteriophora*, *H. megidis*, *S. feltiae*	4000 IJ/1000 IJ	*H.b* 92% *H. megidis* (48%). H.b 87.5%; *S. feltiae* (86.6%)	

S. abbasi PDBCEN 3.1, *S. tami* PDBCEN 2.1, *H. indica* PDBCEN 6.71, and PDBCEN 13.3 after 48 h at 32°C whereas at 25°C, absolute mortality was observed after 72 h by all isolates apart from *S. glaseri*; and in *A. ipsilon*, the mortality ranged from 66.7% to 93.3% after 120 h whereas at 15°C, the mortality of both insects was drastically reduced. An increase in temperature from 15°C to 32°C was favourable for infection and progeny production. Many *S. feltiae* actively nictated at 25°C on soil particles RH 10–50% per unit wt, and invaded *S. litura* on soil for 24 h. They had higher survival rates at RH 10, 15 and 50% than at 25–40%, Nictating larvae did not aggregate under any of the soil moisture conditions (Kondo and Ishibashi, 1985). Diapausing cocooned larvae of Codling moth, *Cydia pomonella* L., overwinter in cryptic habitats in the soil or in the bark of infested trees. Cocooned larvae were more susceptible than non-cocooned larvae (Navaneethan et al., 2010). UV radiation of 302 nm inactivated *H. bacteriophora* juveniles and not *S. carpocapsae*, resulted in delayed progeny emergence at 1.5 min, and caused declined reproductive capacity at 2 min and loss of pathogenicity at 4 min. Negative effects were noted after 6 min for *S. carpocapsae*. The IJ sheath did not provide protection (Gaugler et al., 1992). Crop residues afforded protection from desiccation or UV light, enhanced persistence of *S. carpocapsae* in soil and hence reduced tillage leads to increased insect pest suppression (Shapiro et al. 1999b).

The survival and behaviour mechanisms of specific *Steinernema/Heterorhabditis* sp. in the soil environment have to be understood to develop an effective biological insecticide (Kaya, 1990). Soil texture affects nematode dispersal and infectivity and dense roots adversely affect movement and infectivity. Soil type and texture are other important factors that influence efficacy and persistence of applied EPN. Infectivity and persistence of *S. carpocapsae* and *H. indica* were affected by soil type and texture with lighter soil types favouring infectivity and heavier soil types with higher silt fraction favouring persistence (Shakeela and Hussaini, 2009). The host-finding ability varied in humus, clay, loam and sand with and without roots (Choo and Kaya, 1991). *H. bacteriophora* infected *G. mellonella* larvae more in soils with roots; infectivity was greatest in humus followed by sandy loam and clay, and was influenced by pore size and presence of roots. Root metabolites served as cues. The activity and infectivity of indigenous EPN isolates in sand and the soil column were compared in relation to soil type and depth. *H. indica* out-competed *Steinernema* spp. irrespective of soil depth, type and time for mortality of *A. ipsilon* populations and together had an additive effect over a single population. The presence of *S. glaseri* enhanced the performance of *S. carpocapsae* when applied in combination. *S. carpocapsae* and *S. glaseri* together had additive effect against *A. ipsilon* over a single population (Hussaini et al. 2000b). Survival, movement and infectivity of EPN were affected by several factors in soil. Absolute mortality of *G. mellonella* was obtained with *H. indica* PDBCEN 13.22 in both soil types. The activity of nematodes was higher in sandy loam than in sand. *H. indica* PDBCEN 13.22 population recorded 100% mortality irrespective of soil depth and the infectivity of *Steinernema* sp. was higher at 5 cm than at 10 cm depth (80–91.3%) apart from *S. carpocapsae* PDBCEN 13.1 which recorded higher infectivity at both depths (Hussaini and Sanakaranarayanan, 2001).

Survival of *H. bacteriophora* decreased linearly with time (4–70 days) and quadratically with increasing bulk density, whereas *S. glaseri* decreased linearly with time, but increased quadratically with increasing bulk density. Survival of *S. carpocapsae* was unaffected by bulk density. *H. bacteriophora* and *S. glaseri* infected larvae of *G. mellonella* (L.) for up to 10 weeks after soil inoculation, and infection incidence was not related to bulk density or time. Movement differed among species and soil textures: *H. bacteriophora* was the least restricted, and *S. carpocapsae* the most (Portillo-Aguilar et al., 1999). There was increased movement in sandy loam, compared with loam or silty clay loam, and movement decreased at relatively high bulk densities. Soil pore space regulated movement and infection within dimensions of the nematodes. *H. bacteriophora* moved 18 cm within 4 days in a sandy loam soil, and *S. carpocapsae* moved 9 cm and less at higher soil density. *S. glaseri* showed intermediate levels.

Maximum survival for *H. indica* and *S. carpocapsae* was at pH 5 and pH 7 and least survival at pH 2 and pH 3, respectively (Hussaini et al., 2004). *S. carpocapsae* tolerated a broader range of pH values (4–9) than did *H. indica*. Infectivity and progeny production were not affected by varying pH levels apart from at pH 2 and pH 3. *S. carpocapsae* PDBCEN 6.11, PDBCEN 6.61, *S. abbasi* PDBCEN 3.1, *S. bicornutum* PDBCEN 3.2, *S. tami* PDBCEN 2.1, *H. indica* PDBCEN 6.71, and PDBCEN 13.3 survived storage at room temperature for 8 weeks in distilled water. *S. bicornutum* and *S. carpocapsae* stored better than *H. indica* (Hussaini et al., 2000). At 15°C, distilled water was found to be better than glycerin and liquid paraffin, and liquid paraffin most suitable at 8°C and 24°C. Steinernematids stored better than heterorhabditids. Osmotic stress reduced the viability of fresh IJs of *S. carpocapsae* exposed to 45°C/12h in a high-temperature assay (HTA). Potassium chloride caused low mortality and survival was 0% in HTA. Glycerol (2.2–3.8) for 72h, PEG-300 (1.2–1.6) and PEG-600 (0.8) increased heat tolerance in an 8-h HTA. Survival rate was high in nematodes stored for 72 days with a gradual increase in glycerol concentration from 0.7 to 3.5 over 72h; survival was lower in nematodes exposed to 3.5 glycerol or stored in distilled water for 36 days. *S. carpocapsae*, desiccated by gradual osmotic pressure, infected *Tenebrio molitor* similarly to fresh nematodes stored for 54 days at 25°C (Glazer and Salame, 2000).

11.5 APPLICATION TECHNOLOGY

11.5.1 Soil

Many factors affect the ability to place quantities of nematodes on or in close proximity to the target host to produce optimal results economically. Inundation at high concentration overcomes the impact of abiotic and biotic factors on nematode efficacy and persistence, and has been used as a primary control strategy targeted at the soil and against cryptic habitats (Parkman and Frank, 1992). Soil applications need to coincide with the life cycle of the target insect. Application through furrow irrigation or post irrigation was better than spraying onto soil before irrigation (Cabanillas and Raulston, 1996). *S. riobrave* was effective in the field of high temperature with irrigation. Host susceptibility, pest population levels, host behaviour, species and host cues affect nematode behaviour. Environmental extremes reduce field efficacy (Kaya and Gaugler, 1993). One or two pre and post application irrigations provide sufficient moisture. Spraying nematodes directly onto the soil surface is the most common method, provides good coverage and is simple and quick. Pesta-pelletized *S. carpocapsae* (All strain) were used in soil treatments in the greenhouse against larvae of Western corn rootworm (*Diabrotica virgifera virgifera*) and pre-pupae of Colorado potato beetle (*Leptinotarsa decemlineata*). The pesta-pellets delivered 100,000 living nematodes/g IJs and bacteria survived the pesta-pellet process, emerged from the pellets in large numbers in the soil, and reduced adult emergence of pests by 90% (Nickle et al., 1994).

Oxygen concentration and pH influence IJ survival in soil. *S. carpocapsae* and *S. glaseri* populations declined over 16 weeks as the soil pH decreased from 8 to 4. Survival dropped sharply after 1 week at pH 10. It was similar at pH 4, 6 and 8 during the first 4 weeks, but *S. carpocapsae* survived better than *S. glaseri* at pH 10 for 16 weeks. Both nematodes were stored at pH 4, 6 and 8 for 16 weeks, and storing at pH 10 for one or more weeks was not infective to *G. mellonella*. *S. carpocapsae* survived better than *S. glaseri* at oxygen/nitrogen ratios of 1:99, 5:95, and 10:90 during the first 2 weeks, and declined sharply to less than 20% after 4 weeks. Survival decreased after 8 weeks as the oxygen concentrations decreased from 20% to 1%, and none survived at 16 weeks. *S. carpocapsae* pathogenicity was greater in the first 2 weeks. No nematode pathogenicity was observed at oxygen concentrations of 1%, 5%, and 10% after 2 weeks and at 20% after 16 weeks (Kung et al., 1990).

11.5.2 Foliar

Conventional equipment such as high volume sprayers are used to apply treatments, in addition to drip irrigation and food baits or through sound traps (Parkman and Frank, 1992). Antidesiccants were used to retard evaporation of the nematode suspension on foliage, and reduce desiccation (Glazer and Navon, 1990). Glycerin 10% was the most effective adjuvant increasing survival/activity. Excellent control of Japanese beetle, *P. japonica*, and the chaffer, *Cyclocephala borealis*, was obtained with irrigation before application and again within 24 h after treatment (Downing, 1994). TX 7719, Rodspray oil, and Nufilm P provided the best antidesiccant activity in the lab, and TX 7719 + Blankophor BBH in the greenhouse increased persistence on watercress leaves and efficacy against *P. xylostella* (Baur et al., 1997). Phagostimulants enhanced the efficacy of *S. carpocapsae* against the 4th instar *S. litura* larvae on sunflower head (Sezhian et al., 1996). Improvements were achieved through optimization of spray conditions using standard hydraulic equipment and spinning disc (SD) sprayers. The quality of the spray produced should be optimum for nematode carriage to compete directly with chemicals. Improvement in the operation of the disc prevented clumping of nematodes (Piggott et al., 1999).

Mortality of *S. carpocapsae* IJs on bean foliage was related to the RH, and a gradual reduction in survival was found during a 6-h exposure period at 80% and 60% RH, and at 45% RH, high mortality was observed within 2 h. The antidesiccant 'Folicote' (6% w/w) added to the nematode suspension ensured survival at 60% RH, with 38–60% increased viability. At 80% RH, 'Folicote' affected a 10–20% increase in viability, and at 45% RH, no increase in survival was recorded. Survival on tomato and soybean leaves was 30–35% higher than on cotton, pepper (*Capsicum frutescens*) and bean. At 60% RH, IJ movement ceased within 45–60 min and the nematode body shrank without alteration in pathogenicity (Glazer, 1992). Several antidesiccants and adjuvants were tested for survival and pathogenicity of indigenous isolates of EPN without adverse effects (Hussaini et al., 2005c, 2005d). Richter and Fuxa (1990) evaluated *S. feltiae* against *S. frugiperda* and *H. zea* in field maize; spraying resulted in 33–43% infection of *S. frugiperda* larvae in autumn. Spraying at 0.70 kg/cm^2 pressure gave better grade marketable ears, compared to 1.76 kg/cm^2. Adjuvants should not be toxic to IJs or larvae of the pest; the proportion of droplets did not affect infectivity to *G. mellonella*. SD sprayers such as Micron Ulva+ and Micron Herbaflex improved the deposition per cm^2 (Mason et al., 1998).

Fluorescent brighteners were compared with standard tinopal LPW as solar radiation protectants for *S. carpocapsae* (All). Blankophor BBH and tinopal LPW were the most successful UV screens, with 95% of the original nematode infectivity to larvae of *G. mellonella* retained after 4 h of exposure to direct sunlight. The Blankophors HRS and DML preserved 80% and 85% infectivity, and P167 preserved 70% infectivity after exposure. Blankophors (RKH, LPG, and BSU) were not effective (Nickle and Shapiro, 1994). Foliar persistence of *H. indica* and *S. glaseri*, on cotton leaves at RH 57–69% and temperature 27–30°C for 12 h was evaluated and IJ survival was 2% and 6%, respectively. Maximum mortality of 89% and 82% of *H. indica* and *S. glaseri* was noticed within 7 h (Subramanian and Rajeswari Sundarababu, 2002). Application and delivery of EPNs require improvement for their widespread acceptance as biopesticides, possibly being directly delivered through commercial potting medium and garden soil used in nurseries and greenhouses. Survival of commercially *in vitro*-produced *S. carpocapsae, S. feltiae* and *H. zealandica* stored at 22°C for 2 months was 32% and 23% after 1 month in garden soil and potting medium, respectively. Survival declined below 5% with 2 months of storage. *In vivo*-produced *S. carpocapsae* also declined below 5% after 2 months of storage. *In vivo*-produced *S. carpocapsae* showed higher virulence (Deol et al., 2006).

11.6 GENETIC MANIPULATION

Limitations such as susceptibility to environmental extremes, and host-finding behaviour prevent the EPN from reaching their full biocontrol potential. Most research has centred on areas related to behaviour, ecology, commercial production and field application. Genetic manipulation for strain improvement in EPN has received little attention. Now much of the focus in EPN genetics is on strain improvement in areas such as environmental tolerance, infectivity to target pests, and host searching behaviour. Strategies for genetic manipulation include artificial selection, hybridization and genetic transformation. Molecular methods were mostly used for species identification (Curran, 1993; Hashmi et al., 1996; Joyce et al., 1994). Genetic variability within species provides insight into their phylogenetic relationships and population dynamics. Conventional techniques have provided nematode enhancement (Rahimi et al., 1993; Zioni et al., 1992), and the introduction of specific characteristics required biotechnological approaches. The close relationship of *Caenorhabditis elegans*, the model organism in molecular genetics, provides an extraordinary opportunity to exploit this new technology. Progress was made in genetic transformation, and modifying the functionality of *C. elegans* genes in *H. bacteriophora* (Hashmi et al., 1995). The approaches require genetic diversity for the beneficial traits (Glazer et al., 1991) and are overcome by surveying the natural population for the desired traits. Selection is employed to find genetically superior species. A heat tolerant strain of *H. bacteriophora* designated IS 5 was discovered in the Negev desert, Israel in 1996 and was found to be superior to HP 88, a commercial strain. Selection was also used to enhance the resistance of EPN to nematicides (Glazer et al., 1997).

Szalanski et al. (2000) used DNA sequence analysis to characterize the ribosomal DNA ITS1 region and a portion of the COII and 16S rDNA genes of the mitochondrial genome from *Steinernema* spp. Nuclear ITS1 nucleotide divergence ranged from 6% to 22%, and mtDNA divergence among five species ranged from 12% to 20%. No intra-specific variation was observed among three *S. feltiae* strains. Phylogenetic analysis of nuclear and mitochondrial DNA sequences confirmed the existing morphological relationships of several *Steinernema* species. Both the rDNA ITS1 and mtDNA sequences were useful for resolving relationships among *Steinernema* taxa.

A hybridized foundation population of *S. feltiae* was bidirectionally selected for enhanced and diminished host-finding ability, with no response for the latter; selection for enhancement produced a 20- to 27-fold increase. IJs with positive chemotaxis were increased to 80%. Relaxation of selection pressure produced a gradual decrease in host-finding. This regression, coupled with the high realized heritability for enhanced host-finding (0.64), suggests that wild-type populations take a passive approach to host-finding (Gaugler et al., 1989). An HaeIII satellite DNA family was cloned from *S. carpocapsae*. This repeated sequence appears to be an unusually abundant satellite DNA, since it constitutes about 62% of the *S. carpocapsae* genome. The nucleotide sequences of 13 monomers were determined. This satellite DNA family is represented by two sub-families: one with monomeric units of 170 bp and the other with monomeric units of 182 bp. These monomers are homogeneous in sequence, showing an intermonomeric variability of 6% from the consensus sequence. These results indicate that some homogenizing mechanism is acting to maintain the homogeneity of this satellite DNA. After hybridization with the genomic DNA of several other *Steinernema* species, this DNA sequence appears to be specific to the *S. carpocapsae* genome. Therefore, the species specificity and high copy number of the HaeIII satellite DNA sequence should provide a rapid and powerful tool for *Steinernema* identification (Grenier et al., 1997).

Tomalak (1994) devised a selection method for *S. feltiae* that was effective against *Lycoriella solani*. Efficacy of this strain was evaluated. Preference was given to nematodes with the greatest ability to search effectively for larvae of the target insect in their natural habitat, infect quickly, and to reproduce in the haemocoel of the target insect. After 34 rounds, a 4-fold improvement in nematode ability to find and parasitize 3rd- and 4th-instar larvae of *L. mali* in the mushroom substrate was achieved. No difference was found between the efficacy of selected nematodes applied at 1–3 × 10^6 IJs/m^2, and unselected strains performed better at the higher concentration. All strains persisted in the mushroom casing, due to recycling. The success in genetic transformation of EPN opens the way for generating transgenic nematodes carrying genes conferring resistance to various environmental extremes, most notably heat shock genes.

11.7 DESICCATION TOLERANCE

Desiccation limits the utilization of EPN as biocontrol agents. Nematode migration vertically is more evident than horizontally in the presence/absence of the host insect. IJs follow a geotropic movement along with the gravity. Foraging strategy varies with species as some cruise through soil following host cues (Campbell et al., 2003; Grewal et al., 1994b). Cruisers spend more time moving in soil following cues increasing the probability of locating sedentary and cryptic hosts (Lewis et al., 2006). Fresh IJs cover up to 8 cm in the presence of hosts. *H. indica* responds to CO_2, temperature, faeces, cuticle, electromagnetic fields, and vibration (Lewis et al., 1993; Shapiro et al. 2009a; Torr et al., 2004). Tyson et al. (2012) reported that *P. superbus* utilized a strategy of combined constitutive and inducible gene expression in preparation for anhydrobiosis where the late embryogenesis abundant genes (LEA) are important components of the anhydrobiotic protection repertoire. Hallem et al. (2011) reported a CO_2 response in host-finding behaviour in free living and plant-parasitic nematodes. *H. bacteriophora* (cruiser) and *S. carpocapsae* (ambusher) were attracted to wax moth, *G. mellonella*, superworms, *Zophobas morio*, meal worms, *T. molitor*, and crickets, *Acheta domesticus*. Odours stimulated jumping by *S. carpocapsae*. Such BAG genes are sensitive to environmental stress and may not function after rehydration of stressed insects. Anhydrobiosis is important for storage stability. It is a reversible, physiologically arrested, state of dormancy induced by dehydration (Barrett, 1991), a characteristic feature being biosynthesis of low molecular carbohydrates, proteins and glycerol (Crowe, 2002; Wharton, 2003) and a high concentration of reducing sugars such as trehalose (Hoekstra et al., 2001). It occurs naturally in nematodes/other invertebrates, a survival strategy under drought stress. Trehalose protects membranes and proteins from desiccation damage, and replacing these with structural water also contributes to the formation of intracellular glass. Other proteins are synthesized by anhydrobiotes that are essential for survival (Browne et al., 2002; Solomon et al., 2000). Anhydrobiotes can lose up to 95–98% of their body water and lower their metabolism to below detectable levels if dehydration persists for a longer period leading to cryptobiosis. Many genes have been identified which play a role in stress acclimation and survival. Nematodes are bound by a cuticle, an extracellular protein matrix composed of collagen, and a highly crosslinked external envelope of proteins called cuticlin. The cuticle exhibits an extremely restricted profile that may be species-/stage-specific (Cookson et al., 1992). Biochemical and physical changes occur in cuticle properties on drying, reducing the secretion of total proteins in anhydrobiotic IJs (Wharton et al., 2008). The desiccation stressed IJs respond well to the challenge of host-finding in soil. The stress response induces resistance to cold-heat desiccation, UV radiation, and pH changes. Structural alterations are extremely slow during anhydrobiosis as the cytoplasm is viscous and diffusion of cellular molecules is virtually arrested achieving stability. Coiling and aggregation into clumps is a unique response to desiccation to reduce the surface area of exposure and slow the rate of drying (Womersley et al., 1998). Solomon et al. (2000) identified a novel heat-stable, water-stress-related protein (Desc47) with molecular mass of 47 kDa in *S. feltiae* IS-6, which accumulated 10-fold in dehydrated clumps of IJs, with loss of 34% initial water content at 97% RH. Trehalose accumulated 300–600 mg/g protein during induction of a quiescent anhydrobiotic state. Desc47 retained its high RCL for 3 days in rehydrated IJs. No homology to other known proteins was found. The five sequences obtained from the protein (11–21 amino acids), the 21-amino-acid peptide N V A S D A V E T V G N A A G Q A G (D/T) A V, showed excellent homology (74% identity). In the *Caenorhabditis elegans* predicted proteome database search, the N21 yielded the first-best identity score (59% identity in 17 amino acids) to the CE-LEA homologue protein (g2353333). Reduced metabolism during storage prolongs shelf-life by desiccation of the IJs. Tolerance is increased by adaptation to moderate desiccation conditions. Heritability of the desiccation tolerance is high, justifying a genetic selection for enhanced tolerance. In *H. bacteriophora*, dehydrating conditions produced by treating IJ with polymer PEG 600 were measured as water activity. Inter-specific variations existed between strains and species. The mean tolerated aw value (MW50) was 0.90–0.95 for non-adapted and 0.67–0.99 for adapted populations. The lowest aw value tolerated by 10% of a population (MW10) used for selective breeding was 0.845–0.932 for non-adapted and 0.603–0.950 for adapted populations. Adaptation increased the desiccation tolerance (Mukuka et al., 2010).

11.8 COMPATIBILITY

11.8.1 Pesticides

EPN may contribute better if integrated with other methods of control than when use separately. Compatibility is important for any IPM strategy. Pesticides in aqueous solutions tested in the lab at Project Directorate of Biological Control, Bangalore (PDBC) varied in their effects on sensitivity (Hussaini et al., 2001d). *Heterorhabditis* spp. was more compatible than *Steinernema* spp. Neem was the best followed by endosulfan and fenvalerate. Quinalphos was deleterious for nematode survival and progeny production. EPN are compatible with most of the commonly used pesticides and so could be included as part of a biointensive IPM schedule. The bioefficiency of tolerant strains of *Steinernema* sp. (SSL2) PDBC EN 13.21, PDBC EN 14.1, *H. indica* 13.22, PDBC EN 14.3, and PDBC EN 6.71 was tested in the lab against *G. mellonella* and was not impaired by any exposure. *S. bicornutum* PDBC EN 3.2, *S. tami* PDBCEN 2.1, *Steinernema* sp. PDBCEN 13.1 and *H. indica* PDBC EN 13.3 tolerated twice the recommended dose of endosulfan, malathion, mancozeb and carbofuran. Progeny production of pesticide tolerant strains was on par with control for fenvalerte and endosulfan tolerant populations and it was higher than control for neem tolerant populations. Quinalphos tolerant populations lost their virulence and did not produce any progeny. *Steinernema* sp. PDBC EN 14.1 and *H. indica* PDBC EN 13.22, EN 14.3, and EN 6.71 produced normal progeny with neem. Vainio (1994) found that isofenphos plus thiram, permethrin, trifluralin, fluazifop butyl, or iprodion applied every season along with one inundative release of *S. feltiae* did not have any adverse effect on nematodes and resulted in increased numbers of nematodes in a turnip field. *H. bacteriophora* and *S. feltiae* IJs showed no adverse reaction to diazinon EC 25, dimethoate 37%, bromfenvinfos 50% and bromophos EC 40 at 0.2% to 0.0001% conc. The mortality rate of *S. feltiae* varied from 12.3% (dimethoate) to 14.7% (bromophos) and for *H. bacteriophora* from 14.3% (diazinon) to 16.7% (dimethoate). They were lower at 10°C (Jarowska, 1990). *Heterorhabditis* spp. was less tolerant to some pesticides than *Steinernema* spp., which showed higher survival and greater ability to recover, and maintained infectivity after exposure. Most pesticides affected normal development. *S. glaseri* was the most tolerant and most suitable for integration with pesticides (Sirjusingh et al., 1991). *S. feltiae* were adversely affected by mevinphos, fenamiphos, trichlorfon, methomyl and oxamyl causing partial paralysis with curled or coiled posture, could not infect *S. exigua* (Hb.) larvae, but recovered when washed in water (Hara and Kaya, 1982). Survival and infectivity of *S. glaseri* and *H. indica* were unaffected with dimethoate, carbosulfan, imidacloprid, endosulfan, carbofuran and phorate at 500, 1000 and 2000 ppm. *S. glaseri* was not compatible with endosulfan at 2000 ppm. *H. indica* was not compatible with phorate or dimethoate at all concentrations and endosulfan at 2000 ppm (Priya and Subramanian, 2008). Toxic effects of organophosphates, carbamates, synthetic pyrethroid insecticides, cartap and imidacloprid to IJs of *S. carpocapsae* were present in insecticide solutions. Cartap, profenofos and pyraclofos were the most toxic causing 83–48% mortality at 100 μg/ml/48 h. Diazinon, dichlorvos, fenthion, malathion, trichlorfon, propetamphos and prothiofos showed weak toxicity. OPs (apart from acephate, malathion and temephos), methomyl, permethrin, ethofenprox, and cartap inhibited the pathogenicity of IJs (Zhang et al., 1994).

On sweet potato, *S. carpocapsae* provided significant protection from damage by *Conoderus* spp., *Diabrotica balteata, D. undecimpunctata howardi, Systena blanda, S. elongata* and *S. frontalis* in the field. Fonofos reduced damage caused by these insect pests, particularly when applied in combination with the nematode. Use of resistant cultivars in combination with nematode and/or fonofos treatment has been advocated (Schalk et al., 1993). *S. carpocapsae* and *S. feltiae* tolerated most of the 75 commercial pesticides tested. Only dodine (fungicide) and alachlor and paraquat (herbicides) seriously affected the juveniles. Parathion, aldicarb, methomyl, flubenzimine, metam sodium and phenamiphos were the most toxic. Overall, the feasibility of an integrated use of these nematode species with chemical pesticides in crop protection is indicated (Rovesti and Deseo, 1990).

Vainio and Hokkanen (1990) found no effect of bentazone, ioxynil, hexaconazole, cyromazine and buprofezin on *S. feltiae*, but quizalofop-ethyl, tralkoxydim, sulfur and potassium soap were toxic. Pirimicarb, cypermethrin, diazinon, simazine and metalaxyl plus mancozeb did not affect *Metarhizium anisopliae* and *Beauveria bassiana* and glyphosate, dimethoate, MCPA, vinclozolin, trifluralin, thiram and propiconazole inhibited at least one species. Soil characteristics such as organic matter (chlorpyrifos, bendiocarb and *S. glaseri*) and pH (carbaryl and isazofos) affected efficacy. *S. glaseri* was the most effective followed by chlorpyrifos, and bendiocarb under turf conditions. Irrigation improved insecticide efficacy with higher soil organic matter (Cowles and Villani, 1994). Kulkarni and Paunikar (2009) found tolerance of *H. indica* with seven common chemical pesticides viz., endosulfan, monocrotophos, chlorpyrifos, dimethoate and imidacloprid. Survival of IJs of *H. indica* revealed good tolerance to most of the agrochemicals even at the highest range of recommended concentrations tested, apart from dimethoate, with 100% mortality at 0.20%. Kulkarni et al. (2009) also reported the compatibility of *S. carpocapsae* with biopesticides such as Agropest *Bt.*, Conserve and neem formulation. Radovi (2010) studied the effect of selected pesticides on the virulence of *S. feltiae* after exposure to eight insecticides (a.i. kinoprene, lufenuron, methomyl, metoxyfenozide, oxamyl, piperonyl-butoxide, pyriproxyfen, tebufenozide), seven acaricides (a.i. azocyclotin, clofentezin, diafenthiuron, etoxazole, fenbutatinoxide, fenpyroximate, tebufenpyrad) and four fungicides (a.i. captan, fenhexamid, kresoxim-methyl, nuarimol) under laboratory conditions. It was found that *S. feltiae* was tolerant to all tested pesticides and mortality over 72h varied from 2.26% to 18.68% and from 7.04% to 8.86%, respectively. Fetoh et al. (2009) found that the combined effect of *S. carpocapsae* and *H. bacteriophora* with two biopesticides – spinosad and proclaim – was more evident than when the nematodes were used separately; spinosad was more effective than proclaim. The effects of nematodes and biopesticides were higher in the lab compared to the field. Sankar et al. (2009) observed the tolerance of *H. indica* with fungi, *M. anisopliae, B. bassiana* and *T. viride*, bacteria, *P. fluorescence*, and neem-based biopesticides, Neem and Nimor, under lab conditions. When tested in isolation, *B. bassiana* caused 40% mortality of larvae after 24h of storage while the combination of *P. fluorescens* and *H. indica* caused 100% mortality. Progeny produced by *H. indica* on single *G. mellonella* was found to be greater (140,108 IJs/larva) in the combination treatment with *T. viride*. The pathogenic influence of *H. indica* when exposed with other biopesticides on host larva, proved to be more virulent and compatible. Paunikar et al. (2012) examined the compatibility of *H. indica* to biopesticide products viz., four botanical pesticides Neem oil, Agropest *Bt*., Derisome and Ozomite and two microbial pesticides, Bioprahar and Conserve and one insect growth regulator, Cigna, at different concentrations. All of the biopesticides were compatible with *H. indica* at all concentrations.

Karunakar et al. (2002) reported that *S. feltiae, S. glaseri* and *H. indica* were highly compatible with urea and muriate of potash (400ppm). *Steinernema* spp. was highly compatible with superphosphate (100ppm); *H. indica* was also highly compatible with superphosphate (200ppm). *S. glaseri* and *H. indica* were highly compatible with atrazine (10,000ppm). *S. feltiae* was also highly compatible with atrazine (5000ppm). All were compatible with 312, 625, and 1250ppm of carbofuran; 312ppm of phorate; 2500, 5000, and 1250ppm of quinalphos; and 2500, >312, and >312ppm of aldrin, respectively. Animal manure such as poultry, swine, and beef cattle manure were inhibitory to *S. carpocapsae* (Hsiao and All, 1997).

11.8.2 Natural Enemies

An important aspect in the utilization of EPN in IPM is whether EPN have an effect on beneficial insects such as natural enemies of pests. In general, if the natural enemies of an insect pest live above the ground, they are not affected by nematodes. If they are living in the soil there may be a possibility of infection, although in practice, this did not seem to occur. Exposing *Agrotis ipsilon* larvae to *S. carpocapsae* 24h after parasitism by *Meteorus rubens* prevented the emergence of *M. rubens* from host larvae, at concentrations of 15 IJs. The effect of *S. carpocapsae* decreased when exposure was delayed to 7 and 12 days after parasitism by *M. rubens*. The survival of *A. ipsilon* larvae was shorter when exposure was delayed from 24h to 7 and 12 days (Zaki et al., 1997). El-Wakeil and Hussein (2009) tested EPN and an egg parasitoid for controlling three corn borers, *Sesamia cretica, Chilo agamemnon* and *Ostrinia nubilalis* in corn fields. *H. bacteriophora* (BA1) and *S. carpocapsae* (BA2) were applied to control *S. cretica* (40 days post planting) on different planting dates. Three releases of *Trichogramma evanescens* were performed at 2-week intervals to control *C. agamemnon* and *O. nubilalis* starting at tasseling time for the three planting dates; release levels, 20 and 30 cards (1000 parasitized eggs/card/acre) were used. During the first planting period, spraying of EPN resulted in 97% and 100% mortality of *S. cretica* larvae with *H. bacteriophora* (BA1) and *S. carpocapsae* (BA2), respectively, 1 week post spraying. After 2 weeks, there was 100% mortality of *S. cretica* with both EPN species. During the second planting period, infestation by *S. cretica* was low, suggesting a suitable time for planting corn. A relatively high number of *C. agamemnon* and *O. nubilalis* eggs were laid during the third planting period compared to other periods. Parasitism percentages by *Trichogramma* were high on all planting dates using 30 cards/acre. At season's end, the numbers of *C. agamemnon* and *O. nubilalis* larvae were significantly reduced on the *Trichogramma* released plots. An overall reduction in corn borer larvae on combined treatment resulted in increased yields. EPN and *Trichogramma* together can play a crucial role to control the three corn borers. Trapping and endoparasitic nematophagous fungi isolated from citrus orchards affected nematodes (El-Borai et al., 2009). *Arthrobotrys oligospora, A. dactyloides, A. musiformis* and *Gamsylella gephyropaga,* and endoparasitic *Catenaria* sp. and *Myzocytium* sp. were tested against *S. diaprepesi (Sd), S. glaseri (Sg), S. riobrave, H. zealandica* and *H. indica*. EPN spp. were reduced 56–92% by *G. gephyropaga. Sd/Sg* were unaffected by *Arthrobotrys*, whereas *A. musiformis* reduced all other EPN and *A. oligospora* reduced the numbers of all other species apart from *H. indica.* Both endoparasites reduced the recovery of all EPN by 82% apart from *H. indica*.

11.8.3 Pathogens

Ensheathed (EnJ) and desheathed (DeJ) IJs exposed to the insecticides acephate, dichlorvos, methomyl, oxamyl, or permethrin were compared for normal sinusoidal movement, uncoordinated motion, twitching, convulsion or formation of a pretzel shape, an inactive 'S' posture with fine twitching, or a quiescent straight posture. DeJ displayed these movements at lower concentrations of each insecticide than did EnJ. Insecticide-treated EnJ caused lower mortality to cutworm, *S. litura*, than did EnJ alone but caused greater mortality than insecticides alone. Nictation of DeJ was suppressed at low concentrations, apart from acephate and permethrin. Nictating EnJ or DeJ killed host insects faster than did non-nictating juveniles. Insecticides enhancing nictating behaviour may be used for mixed applications (Ishibashi and Takii, 1993). In an interaction study, combined application of *M. anisopliae* CLO 53 with *H. megidis* and *S. glaseri* increased mortality of third-instar *Hoplia philanthus* in an additive/synergistic way. Larvae exposed to *M. anisopliae* for 3–4 weeks before nematodes caused strong synergism in both species. Reproduction was unaffected. An antagonistic effect on reproduction was caused at higher concentrations of *M. anisopliae* with *H. megidis* (Ansari et al., 2004). *B. thuringiensis* caused 100% mortality of *Simulium ochraceum* Wlk larvae but only for short distances downstream. Increased dosages did not increase the downstream effectiveness in the field. The early-instar larvae were the most susceptible. Releases of *S. feltiae* IJs were either not ingested or were injured during ingestion by the mouthparts of the simuliid larvae and caused no mortality (Gaugler et al. 1983). Combinations of *B. thuringiensis* subsp. *japonensis* Buibui (Btj) and *H. bacteriophora, S. glaseri* or *S. kushidai* caused additive or greater than additive mortalities of scarabaeids, *Cyclocephala hirta* and *C. pasadenae* when exposed to Btj for at least 7 days before the addition of nematodes. This interaction was not observed with *S. kushidai* (Koppenhöfer and Kaya, 1997).

A similar additive effect was produced for control of *S. exigua* in soybean with *S. carpocapsae* and NPV (*SeM* NPV) combination (Gothama et al., 1995). *S. feltiae* progeny contained virus particles in the lumen. Bednarek (1986) reported that the development of *S. feltiae* in gypsy moth, *Lymantria dispar*, was adversely affected whereas Kaya and Burlando

(1989) found successful development and reproduction in 68% of moribund larvae of beet armyworm. Competitive interaction in a single host may not occur. *S. carpocapsae* may infect an *Se*NPV infected larva, which is unsuitable for virus development since the nematode development will kill the larva in 1–2 days and one nematode is capable of killing the larva.

The reproductive success of *S. carpocapsae* or *H. bacteriophora* in larvae of *S. exigua* previously exposed to *B. bassiana* was assessed. *H. bacteriophora* and *B. bassiana* resulted in a higher total mortality of *S. exigua* in soil than when treated with nematode or fungus alone. Treatments with *S. carpocapsae*/*H. bacteriophora* and fungus were better than with either alone. Inundative releases of EPN where *B. bassiana* occurs may result in greater control of soil-borne insect pests than application of either alone (Barbercheck and Kaya, 1991). Gill and Raupp (1994) reported that bagworm, *Thyridopteryx ephemeraeformis* on *Thuja occidentalis* was controlled using formulations of neem, carbaryl, acephate, cyfluthrin, *S. carpocapsae* and *S. feltiae*, and *B. t* var. *kurstaki*. Neem gave a 36–56% reduction. The nematodes, either alone or with oil or antidesiccant, gave 91–100% of control. The synthetic pyrethroid cyfluthrin gave 100% control; carbaryl and acephate gave 83% and 86%, respectively. *Bacillus* and the nematodes provided intermediate levels of control with both *Steinernema* spp.

S. feltiae, *H. heliothidis* and *B. bassiana* possess broad and overlapping host ranges. In general, the period of lethal infection (PLI) for larvae exposed to nematodes and *B. bassiana* simultaneously was shorter than that for larvae exposed to either pathogen alone, or exposed to both sequentially. *B. bassiana* and nematodes rarely coproduced progeny in dually infected hosts. In nematodes alone, or *B. bassiana* and nematodes applied at the same time or 1 day later, nematodes prevented or inhibited the growth of *B. bassiana* in the insect if nematodes were applied within 24h after the application of *B. bassiana*. *B. bassiana* was detrimental to the development of *S. feltiae* and *H. heliothidis* when applied to the insect more than 48h before nematodes. Temperature influenced the development in dually infected larvae. In early sequential treatments, *B. bassiana* more likely developed to the exclusion of nematodes at 15°C, while nematodes were likely to develop in these treatments at 22°C and 30°C. Hence, even though dual infections can result in a decreased PLI compared to singly infected hosts, antagonistic interactions between *B. bassiana* and EPN can adversely affect pathogen development and progeny production (Barbercheck and Kaya, 1990).

Cypermethrin, *B. thuringiensis* subsp. *kurstaki*, *S. feltiae*, NPV, and *B. bassiana* were evaluated against *H. zea*. In the lab, cypermethrin and *B. thuringiensis* were initially significantly different from control despite concentrations. Later, *S. feltiae* showed a significantly higher mortality rate than the control. The virus was highly effective between 2 and 7 days after treatment, killing all larvae. Cypermethrin was more effective in the greenhouse (Carrano-Moreira and All, 1995). Red palm weevil, *Rhyncophorus ferrugineus* (Olivier), a major pest of palms, was controlled with imidacloprid and *S. carpocapsae* in a chitosan formulation in the field. *B. bassiana* was found naturally infecting pupae of *R. ferrugineus*. The potential of *B. bassiana* and *S. carpocapsae* as biological control agents has been confirmed. This should help develop an integrated management programme against this pest (Dembilio and Jacas, 2012). Best root protection grub control was obtained in Christmas tree with chemicals bifenthrin, chlorantraniliprole, thiamethoxam, and time-released imidacloprid tablets and *H. bacteriophora* and *S. carpocapsae* (Liesch and Williamson, 2010). The critical components of agricultural practice for maximizing control by biocontrol agents have been identified within a functioning IPM system.

11.8.4 Manures/Fertilizers

S. carpocapsae reduced cutworm, *A. ipsilon*, damage in soil amended with fresh cow manure, composted manure, and urea. Cutworm damage in nematode-treated plots was greater in plots with fresh manure than in plots without fertilizer. Urea and composted manure did not have a detrimental effect on suppression of cutworm by nematode (Shapiro et al., 1999c). Prolonged (10- to 20-day) exposure in the lab to high inorganic fertilizers inhibited nematode infectivity and reproduction, and short (1-day) exposures increased infectivity. *H. bacteriophora* was more sensitive to adverse effects than *S. feltiae* and *S. anomali*. Organic manures increased the densities of *S. feltiae* in the field whereas NPK fertilizer suppressed the densities. Inorganic fertilizers are likely to be compatible in tank mixes, but may interfere when used as inoculative agents for long-term control. Organic manure may encourage nematode establishment and recycling (Bednarek and Gaugler, 1997). Survival, infectivity, and movement of *S. feltiae* All, *S. bibionis* SN, and *H. heliothidis* NC in poultry manure were studied. The majority (70–100%) died within 18h. Exposure to slurry for 6h killed 95% of *M. domestica*, but nematode exposure for 12h caused <40% mortality. The majority of species remained on the surface and that makes them unlikely candidates for biocontrol in this habitat (Georgis et al., 1987). Lee et al. (2009) investigated the effects of herbal extracts (*Daphne genkwa, Eugenia caryophyllata, Quisqualis indica Zingiber officinale*) and *Pharbitis nil, Xanthium strumarium, Desmodium caudatum* on *S. carpocapsae* and *Heterorhabditis* sp., silkworm (*Bombyx mori*), and ground beetles. *D. genkwa* was highly toxic. All of the IJs of Heterorhabditis (HG) were dead after 3 days by *E. caryophyllata* and *Q. indica*. The mortality of Steinernema carpocapsae (ScP) and HG was <10% by *D. genkwa, D. caudatum, E. caryophyllata, Q. indica* and *Z. officinale* at a concentration of 1000ppm 2 days after treatment while the mortality of HG was 62.8% by *D. genkwa* at the same concentration in an X-plate. However, *E. caryophyllata* at a concentration of 1000ppm had no effect on the survival and pathogenicity of HG. *Q. indica* did not affect silkworm reared on mulberry leaves. Mahmoud (2007) evaluated azadirachtin-based botanical insecticides and *S. feltiae* against peach fruit fly, *Bactrocera zonata*. Response reported were synergistic, additive, antagonistic and without any response. Combined use of azadirachtin, NSK extract and 5% NeemAzal T, with EPN offers an integrated approach to increase efficacy.

11.8.5 Farming Practices

Cropping practices affected populations of *S. carpocapsae* as the population was low in fallow and bare plots, and higher in no-tillage maize plots than in conventional-tillage plots. No IJs were recovered 5 months after application from most of the plots apart from those which had received a conventional tillage. *S. carpocapsae* from soil covered with maize and sorghum debris was lower after 30 days at 25°C. *S. carpocapsae* was capable of moving 3.5cm/day horizontally in bare soil plots and 7.5cm/day in rye mulched plots; mulch enhanced movement (Hsiao and All, 1997). Fresh tissue extracts of marigolds, *Tagetes erecta* and *T. patula*, *S. feltiae*, and a combination of these, were found to be toxic to *Delia radicum*, cabbage maggot larvae (Leger and Riga, 2009). Murugan and Vasugi (2011) reported the combined effect of *Azadirachta indica* and *S. glaseri* against subterranean termite, *Reticulitermes flavipes*. Neem at tested concentrations did not affect the survival of nematodes, whereas it had an impact on the survival of worker termites due to the presence of active neem compounds (Azadirachtin, Salanin). Mortality was 40% at 1% NSKE treatment on the 4th day and increased to 70% at 4% NKSE and to 100% mortality with combined treatment of 4% NSKE + 600 *S. glaseri* on the first day. Nematode and neem extract can be used for control of subterranean termites.

Mustard (*Brassica* and *Sinapis* spp.) green manures tilled into the soil preceding crops act as bio-fumigants that are toxic to plant-parasitic nematodes, providing an alternative to fumigants. However, a trend toward lower rates of EPN infection in fields was found where mustard green manures were applied. Two mustard (*Brassica juncea*) cultivars, differing in glucosinolate levels, disrupted the abilities of *S. carpocapsae, S. feltiae, S. glaseri, S. riobrave, H. bacteriophora, H. marelatus*, and *H megidis*, to infect insect hosts with green manure incorporated into field soil. The negative effects developed slowly in soil. Use of mustard bio-fumigants for the control of plant-parasitic nematodes has the potential to interfere with the biocontrol of insect pests using EPNs (Henderson et al., 2009; Ramirez et al., 2009).

The effect of EPN on non-target beneficials is important in an IPM schedule. Given the complexity of the soil food web, *S. carpocapsae* may likely interact with more specificity than just their intended target, infecting alternative hosts or providing food for native predators. From quantification of the nematode effects on soil arthropod and surface arthropod diversity in fields, more isotomid collembolans, and predatory anystid mites and gnaphosid spiders under nematode applied trees indicate direct predation or indirect trophic effects (Hodson et al., 2012). Significantly fewer *Forficula auricularia* (Dermaptera: Forficulidae) and *Blapstinus discolor* (Coleop: Tenebrionidae) were found under treated trees.

In an IPM schedule, Bhagat et al. (2008) tested the efficiency of some biopesticides and insecticides in the management of *A. ipsilon* on Pioneer Maize (K-85) following recommended agronomic practices. Seed treatment with chlorpyrifos, imidacloprid and insecticidal dust application of chlorpyrifos gave higher yields. Biopesticides, viz., *H. indica, M. anisopliae, B. bassiana, S. carpocapsae, S. carpocapsae + B. bassiana, S. carpocapsae + M. anisopliae* and *H. indica + M. anisoliae* provided less protection at earlier stages of seedling growth; however, they were effective at later stages with higher yield. *H. indica + M. anisopliae* treated plots recorded higher yields compared to other biopesticides. Singh et al. (2008) reported on *Bt* formulations Delfin, Dipel, Halt, Biobit, Biolep, Bioasp, botanical insecticide Neemgold, nematode (*S. feltiae*), green commandos and endosulfan insecticide evaluated against lepidopterous pests of cabbage under field conditions in UP, India. Delfin was very effective in reducing the population of cabbage leaf webber, *Crocidolomia binotalis* (67.6%) and diamondback moth, *P. xylostella* (57.1%). Dipel was equally effective and recorded 67.4% and 56.2% reductions, respectively. Endosulfan was effective against tobacco caterpillar *S. litura*, with a 55.4% reduction without any adverse effects on coccinellid populations. Predators and parasitoids can control the abundance or biomass of herbivores with indirect effects on producer communities and ecosystems, but the interplay of multiple natural enemies may yield unexpected dynamics. The ubiquitous field prevalence and rapid life cycle of *S. feltiae* imply its use of widespread, abundant but small-bodied hosts and indicate the lack of direct competition with nematode and pest in trophic cascade. EPN, fungi, and synthetic insecticides such as Proclaim and Spinosad are commonly used in sustainable agriculture. Cuthbertson et al. (2008) integrated chemical insecticides and *S. carpocapsae* to control sweet potato whitefly, *Bemisia tabaci*. Apart from spiromesifen, other chemicals produced acceptable nematode infectivity. Nematodes in combination with thiacloprid and spiromesifen gave higher mortality compared to nematodes alone. The combination was additive. Kulkarni and Paunikar (2009) discussed the temperature and insecticidal tolerance of EPN and placed them as a most sought after biological control tool for developing IPM against forest insect pests. Generally, microbial control agents such as EPN are applied in a curative manner to achieve pest suppression; prophylactic applications are rare. However, a novel approach to biological control was presented by Shapiro-Ilan et al. (2009b), the ability of *S. carpocapsae* (All and Hybrid) to prophylactically protect peach trees from damage caused by the peach tree borer, *Synanthedon exitiosa*. Nematodes were applied three times (at 1.5–3.0 lakh IJs/tree) in 2005, 2006, and 2007. Following applications in 2005 and 2007, the nematode and chemical treatments caused significant damage suppression; damage ranged from 0% in 2005 to 16% in 2007 when treated with *S. carpocapsae* (Hybrid). Damage ranged from 25% (2005) to 41% (2007) when treated with chlorpyrifos. These results indicate that nematodes applied in a preventative manner during *S. exitiosa*'s oviposition period can reduce insect damage to levels similar to that achieved with recommended chemical insecticide treatments.

11.9 MASS PRODUCTION AND FORMULATION

Nematode material should reach the end user in good condition for successful control. Survival of *H. indica* stored at 15°C (5–15°C) was maximum and mortality highest at 5°C. *H. bacteriophora* survived best at 7.5°C and worst at 25°C. Low pH (6 and 4) reduced bacterial growth and prolonged survival. Ascorbic acid had a possible effect on *H. indica* survival. *H. indica* survival was enhanced for cinnamon and clove extracts. Attapulgite, bentonite clays and sponge did not affect survival and infectivity of stored nematodes; aerated water was superior in increasing survival (Ehlers et al., 2000). *In vitro* cultivation, storage and transport require oxygen supply. In a bioreactor, the oxygen uptake rates (OURs) of the *Steinernema* spp. were below 0.5×10^{-3} mmol O_2 per litre per min in the range of 13–17°C. The OURs of *S. glaseri* and *S. carpocapsae* strains were 0.4×10^{-2} and 0.75×10^{-2} mmol O_2 per litre per min at 21°C, 1.5×10^{-2} and 3.2×10^{-2} mmol O_2 per litre per min at 25°C, and 2.8×10^{-2} and 5.8×10^{-2} mmol O_2 per litre per min at 29°C, respectively. However, the OURs were not significantly affected by agitation speed, which ranged from 50 to 150 rpm. The specific OURs (qo2) of *S. glaseri* NC, Dongrae and Mungyeong strains and *S. carpocapsae* were 0.3×10^{-8}, 0.5×10^{-9}, 0.3×10^{-9} and 0.2×10^{-9} mmol O_2 per cell per min at 25°C, respectively. As nematode size and temperature were increased, the qo2 rates also increased (Kim and Park, 1999). Inoculum size is important for optimizing final yields in *in vitro* solid culture; the highest yield for *H. bacteriophora* was obtained with an inoculum of 10^6 IJs per flask, which was 10-fold the optimal inoculum for *S. carpocapsae*. Extremes of high and low inocula ($1–2 \times 10^7$ IJs per flask) demonstrated differences in reproduction and development between *H. bacteriophora* and *S. carpocapsae*. At 10^7 nematodes per flask, the *H. bacteriophora* population doubled whereas *S. carpocapsae* halved. Inoculation of one *H. bacteriophora* IJ per flask gave a final population of about 25×10^6 nematodes in 6 weeks. However, it was not possible to initiate population development of *S. carpocapsae* in flasks by inoculating 2 IJs per flask (Wang JX et al., 1998). Salma and Shahina (2012) evaluated eight species of EPN, viz., *S. pakistanense, S. asiaticum, S. abbasi, S. siamkayai, S. carpocapsae, S. feltiae, H. indica* and *H. bacteriophora* cultured *in vivo* on three insect species. They found that, in *in vivo* culture at the highest concentration, the production of IJs was 60 to 87.4×10^4 IJs from each larva of *G. mellonella*. *S. pakistanense* produced 86.3, 177, and 38 × 10^4 IJs; *S. asiaticum* 61, 112, and 32×10^4 IJs; *S. siamkayai* produced 87, 181, and 36×10^4 IJs; *S. feltiae* produced 60, 122, and 32×10^4 IJs; *S. carpocapsae* produced 65.5, 123, and 38×10^4 IJs; *S. abbasi* produced 80, 148, and 33×10^4 IJs; *H. bacteriophora* produced 80×10^4, 164×10^4, and 35×10^4 IJs; and *H. indica* produced 85, 155, and 33 × 10^4 IJs from each wax moth larva.

The quality and quantity of lipids in *S. glaseri* in *Popillia japonica* (a natural host), *G. mellonella* (a factitious host), and in solid and liquid media were investigated (Abu Hatab et al., 1998). Yield was four times higher in the *in vivo* compared to *in vitro* cultures. Nematodes produced in *P. japonica* accumulated higher lipids (phospholipids and sterols) compared to those grown using wax moth or *in vitro* solid and liquid media, respectively. C:18 fatty acids were predominant in all methods. *In vivo*-produced nematodes had oleic 18:1 acid as the major component; *in vitro*-produced nematodes had a mixture of oleic 18:1 and linoleic 18:2 acids SO IS host or medium dependent. Plant protein (I), animal protein (II), plant and animal protein media (III) and *in vivo* cultured (IV) were compared for the morphometric, fatty acid content, motility, and penetration rate of IJs of *S. carpocapsae*. Highest relative content of fatty acid, and length of IJs were obtained from medium IV, and lowest FA and length from media II and III. Numbers of nematodes that moved a vertical distance of 5cm in the sand column within 48h and the penetration rates into *Galleria* followed the same trend. Quality of EPN was influenced by the cultural medium component. Animal protein present in media had a strong positive effect on quality (Yang et al., 1997). Plant protein medium had more linoleic and linolenic acid and less palmitic acid, while animal protein medium had more palmitic and oleic acids (Jian et al., 1997). Unsaturation of total lipids increased as recycling or storage temperature decreased, due to an increase in polyunsaturated fatty acids (PUFAs) with the decline in palmitic (16:0) and/or stearic (18:0) acids (Jagdale and Gordon, 1997). *S. riobrave* grew and reproduced over a wide temperature range (15, 20, 25 and 30°C) in *Galleria mellonella*. The lipid content of *S. riobrave* varied in amount and composition. PE and PC were the two major constitutive classes of polar lipids whereas triglycerols were the major constitutive classes of neutral lipids. Lipid content of nematodes grown at 15°C was marginally lower than at other growth temperatures. Nematodes accumulated higher proportions of saturated fatty acids when grown at high temperature (30°C). This ability contributes to thermal tolerance (Abu Hatab and Gaugler, 1997). Yield increased with inoculum size; liquid medium for *S. carpocapsae* and *H. bacteriophora* (H06) per 100g yielded 32 × 10^6 A24 (8 × 10^5in 16 days) and 30 × 10^6 H06 (56 × 10^5in 12 days) (Han, 1996). *S. feltiae* were selectively cultured in wax moth: 13 cycles at 22°C, 4 at 10°C or four at 10°C and one cycle at 22°C. After four cycles at 10°C, the LT_{50} was shorter, establishment higher and the size of IJs longer and wider than those after 13 cycles at 22°C. After 4 cycles at 10°C and one at 22°C, the decrease in LT_{50} observed after 4 cycles at 10°C was lost when tested at 7°C and 10°C and partially lost when tested at 12°C and 15°C. Previous culture temperatures did not affect reproduction (Schirocki and Hague, 1997). *In vitro* substrate modifying Wouts' medium with vegetable oils and animal fats, 16ml bacto nutrient broth, 50g pork fat, 12g agar agar and 1000ml distilled water supported the production of *H. indica* IJs with higher infectivity against cotton bollworm larvae (Gokte-Narkhedkar et al., 2005). Medium with a composition: chicken liver 2%, silkworm pupae 20%, lard 5%, soyabean powder 9%, flour 18%, yeast 1% and water 45% supported high populations of *S. carpocapsae* Agriotos, *S. carpocapsae* BJ and *S. glaseri* NC34 (Pan, 1995). Kikuta et al. (2008) cultured *S. carpocapsae* in insect cells under axenic conditions. Eggs put into the established cell line Sf 9, grew, moulted, developed into adults and produced eggs; the life cycle took 6 days and facilitated subcultures. Living insect cells were food for nematodes. Dembilio et al. (2011) found that a chitosan-based formulation of EPN controlled red palm weevil, *Rhynchophorus ferrugineus* Oliv., in *P. theophrasti* 4-year-old palms after 9-day exposure. Curative applications managed to reduce insect activity and helped palms to recover. The quality of commercial nematode products is critical if EPN are to realize their full potential as biological insecticides. The quality of nematodes produced varies *in vivo* or *in vitro*, in solid or liquid diets. Hence a suitable mass production method should be selected depending on the species of nematode. Expanded use of EPN in biocontrol cannot be expected unless field efficacy is increased. Matching nematode species might bridge the efficacy gap among nematodes, chemicals and the strains against those insects they are best adapted to. Attention should be given to protecting the genetic variability of new isolates, and preventing the loss of alleles through lab adaptation. Prediction models may be developed so that nematodes will be used when and where they are likely to be effective. A distinction should be made between 'lab adapted' and 'field adapted' populations. The competition with chemical pesticides remains fierce due to non-competitive costs compared with chemical pesticides and concerns over inconsistent nematode quality. End users will adopt biocontrol agents that provide adequate efficacy. Advances in application, timing and delivery systems and in particular, in selecting optimal target habitats and pests, have narrowed the efficacy gap between chemical and nematode agents.

11.10 CONCLUSIONS

Entomopathogenic nematodes are currently marketed worldwide for use in inundative biological control, where the applied natural enemy population (rather than its offspring) is expected to reduce insect numbers. Unlike classical biological control, in inundative control, natural enemy establishment is not crucial to achieve pest suppression. They are potential biocontrol agents amenable for mass production, handling and application on a large scale and also for integration with other methods in an IPM schedule for control of insect pests in agriculture, horticulture and forestry and could also be adopted for pests of public health and veterinary importance.

References

Abbas, M.S.T., Saleh, M.M.E., 1998. Comparative pathogenicity of *Steinernema abbasi* and *S. riobravae* to *Spodoptera littoralis* (Lepidoptera: Noctuidae). Int. J. Nematol. 8, 43–45.

Abu Hatab, M.A., Gaugler, R., 1997. Influence of growth temperature on fatty acids and phospholipids of *Steinernema riobravis* infective juveniles. J. Therm. Biol. 22, 237–244.

Abu Hatab, M.A., Gaugler, R., Ehlers, R.U., 1998. Influence of culture method on *Steinernema glaseri* lipids. J. Parasitol. 84, 215–221.

Akalach, M., Wright, D.J., 1995. Control of the larvae of *Conorhynchus mendicus* (Col.: Curculionidae) by *Steinernema carpocapsae* and *Steinernema feltiae* (Nematoda: Steinernematidae) in the Gharb area (Morocco). Entomophaga 40, 321–327.

Ansari, M.A., Tirry, L., Moens, M., 2004. Interaction between *Metarhizium anisopliae* CLO 53 and entomopathogenic nematodes for the control of *Hoplia philanthus*. Biol. Control 31, 172–180.

Ashtari, M., Karimi, J., Rezapanah, M.R., Hassani Kakhki, M., 2011. Biocontrol of leopard moth, *Zeuzera pyrina* L. (Lepi: Cossidae) using entomopathogenic nematodes in Iran. IOBC/WPRS Bull 66, 333–335.

Atwa, A.A., El-Sabah, A.F.B., Gihad, M.M., 2009. The effect of different biopesticides on the cabbage white butterfly, *Pieris rapae* (L.) in cauliflower fields. Alexandria J. Agric. Res. 54, 147–153.

Barbercheck, M.E., 1993. Tritrophic level effects on entomopathogenic nematodes. Environ. Entomol. 22, 1166–1171.

Barbercheck, M.E., Kaya, H.K., 1990. Interactions between *Beauveria bassiana* and the entomogenous nematodes, *Steinernema feltiae* and *Heterorhabditis heliothidis*. J. Invert. Pathol. 55, 225–234.

Barbercheck, M.E., Kaya, H.K., 1991. Competitive interactions between entomopathogenic nematodes and *Beauveria bassiana* (Deuteromycotina: Hyphomycetes) in soil borne larvae of *Spodoptera exigua* (Lepi: Noctuidae). Environ. Entomol. 20, 707–712.

Barbercheck, M.E., Wang, J., Hirsh, I.S., 1995. Host plant effects on entomopathogenic nematodes. J. Invert. Pathol. 66, 169–177.

Barrett, J., 1991. Anhydrobiotic nematodes. Agric. Zool. Rev 4, 161–176.

Baur, M.E., Kaya, H.K., Gaugler, R., Tabashnik, B., 1997. Effects of adjuvants on entomopathogenic nematode persistence and efficacy against *Plutella xylostella*. Biocont. Sci. Technol. 7, 513–525.

Bednarek, A., 1986. Development of the *Steinernema feltiae* entomogenous nematode in the conditions of occurrence in the insect's body cavity of other pathogens. Ann. Warsaw Agric. Univ. SGGW AR 20, 69–74.

Bednarek, A., Gaugler, R., 1997. Compatibility of soil amendments with entomopathogenic nematodes. J. Nematol. 29, 220–227.

Bhagat, R.M., Khajuria, M.K., Uma, S., Monobrullah, M., Kaul, V., 2008. Efficacy of biopesticides and insecticides in controlling maize cutworm in Jammu. J. Biol. Control 22, 99–106.

Bhaskaran, R.K.M., Sivakumar, C.V., Venugopal, M.S., 1994. Biocontrol potential of entomopathogenic nematode in control of red hairy caterpillar, *Amsacta albistriga* of groundnut. Indian J. Agric. Sci. 64, 655–657.

Blum, B., Morelli, R.K., Vinotti, V., Ragni, A., 2009. Control of *Curculio nucum,* the hazelnut borer, by entomopathogenic nematodes. Acta Horticult. 845, 567–570.

Boff, M.I.C., Wiegers, G.L., Smits, P.H., 2000. Effect of storage time and temperature on infectivity, reproduction and development of *Heterorhabditis megidis* in *Galleria mellonella*. Nematology 2, 635–644.

Browne, J., Tunnacliffe, A., Burnell, A.M., 2002. Plant desiccation gene found in a nematode. Nature 4, 16–38.

Cabanillas, H.E., Raulston, J.R., 1996. Evaluation of *Steinernema riobravis, S. carpocapsae* and irrigation timing for the control of corn earworm, *Helicoverpa zea*. J. Nematol. 28, 75–82.

Campbell, J.F., Lewis, E.E., Stock, S.P., Nadler, S., Kaya, H.K., 2003. Evolution of host searching in entomopathogenic nematodes (Nematoda: Steinernematidae). J. Nematol. 35, 142–145.

Carrano-Moreira, A.F., All, J., 1995. Screening of bioinsecticides against the cotton bollworm on cotton. Pesq. Agropec. Bras. 30, 307–312.

Chambers, U., Bruck, D.J., Olsen, J., Walton, V.M., 2010. Control of overwintering filbert worm (Lepidoptera: Tortricidae) larvae with *Steinernema carpocapsae*. J. Econ. Entomol. 103, 416–422.

Choo, H.Y., Kaya, H.K., 1991. Influence of soil texture and presence of roots on host finding by *Heterorhabditis bacteriophora*. J. Invert. Pathol. 58, 279–280.

Choo, H.Y., Kaya, H.K., Kim, J.B., Park, Y.D., 1991. Evaluation of entomopathogenic nematodes, *Steinernema carpocapsae* (Steinernematidae) and *Heterorhabditis bacteriophora* (Heterorhabditidae) against rice stem borer *Chilo suppressalis* (Walker) (Lepidoptera: Pyralidae). Korean J. Appl. Entomol. 30, 50–53.

Cookson, E., Blaxter, M.L., Selkirk, M.E., 1992. Identification of the major soluble cuticular proteins of lymphatic filarial nematode parasites (gp29) as a secretory homolog of glutathione peroxidase. Proc. Natl. Acad. Sci. USA 89, 5837–5841.

Cowles, R.S., Villani, M.G., 1994. Soil interactions with chemical insecticides and nematodes used for control of Japanese beetle (Coleoptera: Scarabaeidae) larvae. J. Econ. Entomol. 87, 1014–1021.

Crowe, L.M., 2002. Lessons from nature. The role of sugars in anhydrobiosis. Comp. Biochem. Physiol. 131, 505–513.

Curran, J., 1993. In: Gaugler, R., Kaya, H.K. (Eds.), Entomopathogenic Nematodes in Biological Control CRC Press, Boca Raton, FL, pp. 365.

Curto, G., Boselli, M., Vergnani, S., Reggiani, A., 2007. Effectiveness of entomopathogenic nematodes in the control of sawfly (*Hoplocampa brevis*) in pear orchards. In: Papierok, B. (Ed.), Bull. OILB/SROP, 30, pp. 13–17.

Curto, G., Vai, N., Dallavalle, E., 2008. Control of *Caliroa varipes* (Klug, 1814) (Hymenoptera, Tenthredinidae) with entomopathogenic nematodes, in an urban park of Bologna suburbs. Redia 91, 173–175.

Cuthbertson, A.G.S., Mathers, J.J., Northing, P., Prickett, A.J., Walters, K.F.A., 2008. The integrated use of chemical insecticides and the entomopathogenic nematode, *Steinernema carpocapsae* (Nematoda: Steinernematidae), for the control of sweet potato whitefly, *Bemisia tabaci* (Hemiptera: Aleyrodidae). Insect Sci. 15, 447–453.

Dembilio, O., Jacas, J.A., 2012. Bio-ecology and integrated management of the red palm weevil, *Rhynchophorus ferrugineus* (Coleop: Curculionidae), in the region of Valencia (Spain). Hellenic Plant Prot. J. 5, 1–12.

Dembilio, O., Karamaouna, F., Kontodimas, D.C., Nomikou, M., Jacas, J.A., 2011. Susceptibility of *Phoenix theophrasti* (Palmae: Coryphoideae) to *Rhynchophorus ferrugineus* (Coleoptera: Curculionidae) and its control using *Steinernema carpocapsae* in a chitosan formulation. Spanish J. Agric. Res. 9, 623–626.

Deol, Y.S., Grewal, S.K., Canas, L., Yelnik, M., Grewal, P.S., 2006. An assessment of entomopathogenic nematode delivery through a commercial potting medium and a garden soil. Int. J. Nematol. 16, 186–193.

Downing, A.S., 1994. Effect of irrigation and spray volume on efficacy of entomopathogenic nematodes against white grubs. J. Econ. Entomol. 87, 643–646.

Downing, A.S., Erickson, C.G., Kraus, M.J., 1991. Field evaluation of entomopathogenic nematodes against citrus root weevils (Coleoptera: Curculionidae) in Florida citrus. Fla. Entomol. 74, 584–586.

Ehlers, R.U., Niemann, I., Hollmer, S., Strauch, O., Jende, D., Shanmugasundaram, M., et al., 2000. Mass production potential of the bacto-helminthic biocontrol complex *Heterorhabditis indica-Photorhabdus luminescens*. Biocontrol Sci. Technol. 10, 607–616.

El-Borai, F.E., Bright, D.B., Graham, J.H., Stuart, R.J., Cubero, J., Duncan, L.W., 2009. Differential susceptibility of entomopathogenic nematodes to nematophagous fungi from Florida citrus orchards. Nematology 11, 231–241.

El-Wakeil, N.E., Hussein, M.A., 2009. Field performance of entomopathogenic nematodes and an egg parasitoid for suppression of corn borers in Egypt. Arch. Phytopathol. Plant Prot. 42, 228–237.

Fetoh, B.E.-S.A., Khaled, A.S., El-Nagar, T.F.K., 2009. Combined effect of entomopathogenic nematodes and biopesticides to control the greasy cut-worm, *Agrotis ipsilon* (Hub.) in the strawberry field. Egypt. Acad. J. Biol. Sci. 2, 227–236.

Figueroa, I.W., 1990. Biocontrol of the banana root borer weevil, *Cosmopolites sordidus* (Germar), with steinernematid nematodes. J. Agric. Univ., Puerto Rico. 74, 15–19.

Ganga Visalakshi, P.N., Krishnamoorthy, A., Hussaini, S.S., 2009. Field efficacy of the entomopathogenic nematode *Steinernema carpocapsae* (Weiser, 1955) against brinjal shoot and fruit borer, *Leucinodes orbonalis* Guenee. Nematol. Medit 37, 133–137.

Ganguly, S., Singh, L.K., 2000. *Steinernema thermophilum* sp.n (Rhabditida: Steinernematidae) from India. Int. J. Nematol. 10, 183–191.

Garcia-del-Pino, F., Morton, A., 2008. Efficacy of *Steinernema feltiae* against the leek moth *Acrolepiopsis assectella* in laboratory and field conditions. BioControl 53, 643–650.

Gaugler, R., Kaplan, B., Alvarado, C., Montoya, J., Ortega, M., 1983. Assessment of *Bacillus thuringiensis* serotype 14 and *Steinernema feltiae* (Nematoda: Steinernematidae) for control of the Simulium vectors of onchocerciasis in Mexico. Entomophaga 28, 309–315.

Gaugler, R., Campbell, J.F., McGuire, T.R., 1989. Selection for host finding in *Steinernema feltiae*. J. Invert. Pathol. 54, 363–372.

Gaugler, R., Bednarek, A., Campbell, J.F., 1992. Ultraviolet inactivation of heterorhabditid and steinernematid nematodes. J. Invert. Pathol. 59, 155–160.

Georgis, R., Gaugler, R., 1991. Predictability in biological control using entomopathogenic nematodes. J. Econ. Entomol. 84, 713–720.

Georgis, R., Mullens, B.A., Meyer, J.A., 1987. Survival and movement of insect parasitic nematodes in poultry manure and their infectivity against *Musca domestica*. J. Nematol. 19, 292–295.

Georgis, R., Koppenhöfer, A.M., Lacey, L.A., Bélair, G., Duncan, L.W., Grewal, P.S., et al., 2006. Successes and failures in the use of parasitic nematodes for pest control. Biol. Control 38, 103–123.

Gill, S.A., Raupp, M.J., 1994. Using entomopathogenic nematodes and conventional and biorational pesticides for controlling bagworm. J. Arboric. 20, 318–322.

Glazer, I., 1992. Survival and efficacy of *Steinernema carpocapsae* in an exposed environment. Biocont. Sci. Technol. 2, 101–107.

Glazer, I., Navon, A., 1990. Activity and persistence of entomoparasitic nematodes tested against *Heliothis armigera*. J. Econ. Entomol. 83, 1795–1800.

Glazer, I., Salame, L., 2000. Osmotic survival of the entomopathogenic nematode *Steinernema carpocapsae*. Biol. Control 18, 251–257.

Glazer, I., Gaugler, R., Segal, D., 1991. Genetics of the nematode *Heterorhabditis bacteriophora* strain HP88: The diversity of beneficial traits. J. Nematol. 23, 324–333.

Glazer, I., Salme, L., Segal, D., 1997. Genetic enhancement of nematode resistance in entomopathogenic nematodes. Biocont. Sci. Technol. 7, 199–512.

Gokte-Narkhedkar, N., Lavhe, N.V., Panchbhai, P.R., Mukewar, P.M., Singh, P., 2005. Development of new substrate for mass multiplication of *Heterorhabditis indica* in the control of cotton bollworm *Helicoverpa armigera*. *Int.* J. Nematol. 15, 191–193.

Gothama, A.A.A., Sikorowski, P.P., Lawrence, G.W., 1995. Interactive effects of *Steinernema carpocapsae* and *Spodoptera exigua* nuclear polyhedrosis virus on *Spodoptera exigua* larvae. J. Invert. Pathol. 66, 270–276.

Grenier, E., Catzeflis, F.M., Abad, P., 1997. Genome sizes of the entomopathogenic nematodes, *Steinernema carpocapsae* and *Heterorhabditis bacteriophora*. Parasitology 114, 495–501.

Grewal, P.S., Georgis, R., 1998. Entomopathogenic nematodes. In: Hall, F.R., Menn, J.J. (Eds.), Biopesticides: Use and Delivery. Humana Press, Totowa, NJ, pp. 271–299.

Grewal, P.S., Selvan, S., Gaugler, R., 1994a. Thermal adaptation of entomopathogenic nematodes: niche breadth for infection, establishment, and reproduction. J. Therm. Biol. 19, 245–253.

Grewal, P.S., Lewis, E.E., Gaugler, R., Campbell, J.F., 1994b. Host finding behaviour as a predictor of foraging strategy in entomopathogenic nematodes. Parasitology 108, 207–213.

Grewal, P.S., Lewis, E.E., Gaugler, R., 1997. Response of infective stage parasites (Nematoda: Steinernematidae) to volatile cues from infected hosts. J. Chem. Ecol. 23, 503–515.

Gupta, S., Kaul, V., Shankar, U., Sharma, D., Ahmad, H., 2009. Field efficacy of Steinernematid and Heterorhabditid nematodes against *Pieris brassicae* (L.) on cauliflower. Ann. Plant Prot. Sci. 17, 181–184.

Hallem, E.A., Dilman, A.R., Hong, A.V., Zhang, Y., Yano, A.M., Stephanie, F.D., et al., 2011. A sensory code for host seeking in parasitic nematodes. Curr. Biol. 8, 377–388.

Han, R.C., 1996. The effects of inoculum size on yield of *Steinernema carpocapsae* and *Heterorhabditis bacteriophora* in liquid culture. Nematologica 42, 546–553.

Hara, A.H., Kaya, H.K., 1982. Effects of selected insecticides and nematicides on the *in vitro* development of the entomogenous nematode, *Neoaplectana carpocapsae*. Exp. Entomol 12, 496–501.

Harris, M.A., Begley, J.W., Warkentin, D.L., 1990. *Liriomyza trifolii* suppression with foliar applications of *Steinernema carpocapsae* and abamectin. J. Econ. Entomol. 83, 2380–2384.

Hashmi, G., Glazer, I., Gaugler, R., 1996. Molecular comparisons of entomopathogenic nematodes using Randomly Amplified Polymorphic DNA (RAPD) markers. Fundam. Appl. Nematol. 19, 399–406.

Hashmi, S., Hashmi, G., Gaugler, R., 1995. Genetic transformation of an entomopathogenic nematode by microinjection. J. Invert. Pathol 66, 293–296.

Henderson, D.R., Riga, E., Ramirez, R.A., Wilson, J., Snyder, W.E., 2009. Mustard biofumigation disrupts biological control by *Steinernema* spp. nematodes in the soil. Biol. Control 48, 316–322.

Hodson, A.K., Siegel, J.P., Lewis, E.E., 2012. Ecological influence of the entomopathogenic nematode, *Steinernema carpocapsae*, on pistachio orchard soil arthropods. Pedobiologia 55, 51–58.

Hoekstra, A.F., Golorina, E.A., Buitink, J., 2001. Mechanism of plant desiccation tolerance. Trends Plant Sci. 6, 431–438.

Hsiao, W.F., All, J.N., 1997. Effect of animal manure on the survival and pathogenicity of the entomopathogenic nematode, *Steinernema carpocapsae*. Chinese J. Entomol. 17, 53–65.

Hussaini, S.S. Sanakaranarayanan, C., 2001. Effect of soil type and depth on pathogenicity of indigenous isolates of entomopathogenic nematodes. National Seminar on Emerging Trends in Pests and Diseases and their Management, Oct 11–13, 2001, pp. 57–58.

Hussaini, S.S., Singh, S.P., Parthasarathy, R., 2000a. Storage effects on activity of native EPN populations. Indian J. Nematol. 30, 225–264.

Hussaini, S.S., Singh, S.P., Parthasarathy, R., Shakeela, V., 2000b. Infectivity of native populations of *Steinernema* spp. and *Heterorhabditis indica* and in sand and sandy loam soil columns against *Agrotis ipsilon* (Hufnagel). Ann. Pl. Prot. Sci. 8, 200–205.

Hussaini, S.S., Ansari, M.A., Ahmad, W., Subbotin, S.A., 2001a. Identification of some Indian populations of *Steinernema* species (Nematoda) by RFLP analysis of ITS region of rDNA. Int. J. Nematol. 11, 73–76.

Hussaini, S.S., Satya, K.J., Hussain, M.A., 2001b. Effect of chemically stimulated entomopathogenic nematodes for control of insect pests, Proceedings of the Second National Symposium IPM in Horticultural Crops: New Molecules, Biopesticides and Environment, Oct 17–19, 2001. Institution of Agricultural Technologists, Bangalore, pp. 204–206.

Hussaini, S.S., Singh, S.P., Shakeela, V., 2001c. Effect of different host plants on infectivity and progeny production of entomopathogenic nematodes, (Steinernematidae, Heterorhabditidae: Rhabditidae) in *Agrotis ipsilon*. Proceedings of the Second National Symposium IPM in Horticultural Crops, New molecules, Biopesticides and Environment, Oct 17–19, 2001. Institution of Agricultural Technologists, Bangalore, pp. 223.

Hussaini, S.S., Singh, S.P., Shakeela, V., 2001d. Compatibility of entomopathogenic nematodes (Steinernematidae, Heterorhabditidae: Rhabiditoidea) with selected pesticides and their influence on some biological traits. Entomon 26, 37–44.

Hussaini, S.S., Singh, S.P., Nagesh, M., 2002. In vitro and field evaluation of some indigenous isolates of *Steinernema* and *Heterorhabditis indica* against shoot and fruit borer, *Leucinodes orbonalis*. Indian J. Nematol. 32, 63–65.

Hussaini, S.S., Nagesh, M., Rajeswari, R., Shahnaz Fathima, M., 2004. Effect of pH on survival, pathogenicity and progeny production of some indigenous isolates of entomopathogenic nematodes. Indian J. Nematol. 34, 169–173.

Hussaini, S.S., Shakeela, V., Dar, M.H., 2005a. Influence of temperature on infectivity of entomopathogenic nematodes to black cutworm, *Agrotis ipsilon* (Hufnagel) larvae J. Biol. Control 19, 51–58.

Hussaini, S.S., Nagesh, M., Rajeshwari, R., Dar, M.H., 2005b. Field evaluation of Entomopathogenic nematodes against white grubs (Coleoptera: Scarabaedidae) on turf grass in Srinagar. Ann. Plant Prot. Sci. 13, 190–193.

Hussaini, S.S., Nagesh, M., Rajeswari, R., Shahnaz Fathima, M., 2005c. Effect of adjuvants on survival and pathogenicity of some indigenous isolates of EPN. Indian J. Plant Prot. 32, 111–114.

Hussaini, S.S., Nagesh, M., Rajeswari, R., Manzoor Hussain Dar, 2005d. Effect of antidesiccants on survival and pathogenicity of some indigenous isolates of EPN against *Plutella xylostella*. Ann. Plant Prot. Sci. 13, 179–186.

Hussaini, S.S., Shakeela, V., Krishnamurthy, D.V., 2008. Effect of Entomopathogenic nematodes on invasion, development and reproduction of root knot nematode, *Meloidogyne incognita* in tomato. Indian J. Plant Prot. 36, 114–120.

Ishibashi, N., Takii, S., 1993. Effect of insecticides on movement, nictation, and infectivity of *Steinernema carpocapsae*. J. Nematol. 25, 204–213.

Jagdale, G.B., Gordon, R., 1997. Effect of temperature on the composition of fatty acid in total lipids and phospholipids of EPN. J. Therm. Biol. 22, 245–251.

Jannson, R.K., Gaugler, R., Mannion, C.M., Lecrone, S.H., 1991. Recent advances in biological control of Cylas formicarius with entomopathogenic nematodes. In: Proceeding Caribbean Meetings on Biological Control, Guadeloupe, 5–7 Nov. 1990, pp. 167–182.

Jarowska, M., 1990. Effect of some insecticides on entomophilic nematodes. Zeszyty Problemmowe Postpow Nauk Rolnizych No. 391, 73–79.

Jaworska, M., Ropek, D., 1994. Influence of host-plant on the susceptibility of *Sitona lineatus* L. (Coleop:

Curculionidae) to *Steinernema carpocapsae* Weiser. J. Invert. Pathol. 64, 96–99.

Jian, H., Yang, H., Zhang, G., 1997. Comparative study on fatty acids of *Steinernema carpocapsae* cultured by different media. In: Chinese Agricultural Sciences: For the compliments to the 40th anniversary of the founding of the Chinese Academy of Agricultural Sciences. China Agricultural Scientech Press, Beijing, pp. 115–121.

Joyce, S.A., Burnell, A.M., Powers, T.O., 1994. Characterization of *Heterorhabditis* isolates by PCR amplification of segments of mtDNA and rDNA genes. J. Nematol. 26, 260–270.

Karunakar, G., Easwaramoorthy, S., David, H., 2002. Compatibility of *Steinernema feltiae, S. glaseri* and *Heterorhabditis indica* with certain fertilizers, herbicide and pesticides. Sugar Tech. 4, 123–130.

Kaya, H.K., 1990. Soil ecology. In: Gaugler, R., Kaya, H.K. (Eds.), Entomopathogenic Nematodes. CRC Press, Boca Raton, FL, pp. 93–115.

Kaya, H.K., Burlando, T.M., 1989. Development of *Steinernema feltiae* (Rhabditida: Steinernematidae) in diseased insect hosts. J. Invert. Pathol. 53, 164–168.

Kaya, H.K., Gaugler, R., 1993. Entomopathogenic nematodes. Annu. Rev. Entomol. 38, 181–206.

Kienzle, J., Heinisch, D., Kiefer, J., Trautmann, M., Volk, F., Zimmer, J., et al., 2010. Three years experience with EPNs for the control of overwintering codling moth larvae in different regions of Germany. Ecofruit. 14th Intl. Conf. Org. Fruit-Growing. Proc. Conf. Hohenheim, Germany, 22–24 Feb. 2010, pp. 163–168.

Kikuta, S., Kiuchi, T., Aoki, F., Nagata, M., 2008. Development of an entomopathogenic nematode, *Steinernema carpocapsae,* in cultured insect cells under axenic conditions. Nematology 10, 845–851.

Kim, D.W., Park, S.H., 1999. Characteristics of the oxygen uptake rate of entomopathogenic nematodes *Steinernema* spp. Korean J. Appl. Entomol. 38, 123–128.

Klein, M.G., Georgis, R., 1992. Persistence of control of Japanese beetle (Coleoptera: Scarabaeidae) larvae with steinernematid and heterorhabditid nematodes. J. Econ. Entomol. 85, 727–730.

Kondo, E., Ishibashi, N., 1985. Effects of soil moisture on the survival and infectivity of the entomogenous nematode, *Steinernema feltiae* (DD-136). Proc. Assoc. Plant Prot. Kyushu 31, 186–190.

Kondo, E., Ishibashi, N., 1989. Non-oral infection of *Steinernema feltiae* (DD-136) to the common cutworm, *Spodoptera litura* (Lepidoptera: Noctuidae). Appl. Entomol. Zool. 24, 85–95.

Koppenhöfer, A.M., Kaya, H.K., 1997. Additive and synergistic interaction between entomopathogenic nematodes and *Bacillus thuringiensis* for scarab grub control. Biol. Control 8, 131–137.

Koppenhöfer, A.M., Rodriguez Saona, C.R., Polavarapu, S., Holdcraft, R.J., 2008. Entomopathogenic nematodes for control of *Phyllophaga georgiana* (Coleoptera: Scarabaeidae) in cranberries. Biocont. Sci. Technol. 18, 21–31.

Kulkarni, N., Paunikar, S., 2009. Temperature and insecticidal tolerance of entomopathogenic nematodes: a useful tool for developing IPM strategy against forest insect pests. *National Forestry Conference,* Nov., 9–11, 2009, Forest Research, Institute, Dehradun (Abs), pp. 177–178.

Kulkarni, N., Paunikar, S., Hussaini, S.S., 2009. Tolerance of EPN, Heterorhabditis indica to some common insecticides useful for developing IPM strategy against forest insect pests. Fifth International Conference on Biopesticides: Stakeholders Perspective, 26th–30th April, 2009 at TERI, New Delhi (Abs), p. 174.

Kung, S.P., Gaugler, R., Kaya, H.K., 1990. Influence of soil pH and oxygen on persistence of *Steinernema* spp. J. Nematol. 22, 440–445.

Laznik, Ž., Toth, T., Lakatos, T., Vidrih, M., Trdan, S., 2009. Efficacy of two strains of *Steinernema feltiae* (Filipjev) (Rhabditida: Steinernematidae) against third-stage larvae of common cockchafer (*Melolontha melolontha* [L.], Coleoptera, Scarabaeidae) under laboratory conditions. Acta Agric. Slovenica 93, 3.

Laznik, Ž., Vidrih, M., Trdan, S., 2012. The effect of different entomopathogens on white grubs (Coleoptera: Scarabaeidae) in an organic hay-producing grassland. Arch. Biol. Sci. 64, 1235–1246.

LeBeck, L.M., Gaugler, R., Kaya, H.K., Hara, A.H., Johnson, M.W., 1993. Host stage suitability of the leafminer *Liriomyza trifolii* (Diptera: Agromyzidae) to the entomopathogenic nematode *Steinernema carpocapsae.* J. Invert. Pathol. 62, 58–63.

Lee, D.W., Choi, H.C., Kim, T.S., Park, J.K., Park, J.C., Yu, H.B., et al., 2009. Effect of some herbal extracts on entomopathogenic nematodes, silkworm and ground beetles. Korean J. Appl. Entomol. 48, 335–345.

Lee, K.S., Lee, D.W., Kim, H.H., Lee, S.M., Choo, H.Y., Shin, H.K., 2005. Pathogenicity of entomopathogenic nematodes to *Popillia quadriguttata* (Coleoptera: Scarabaeidae) adult. Korean J. Appl. Entomol. 44, 145–150.

Leger, C., Riga, E., 2009. Evaluation of marigolds and entomopathogenic nematodes for control of the cabbage maggot, Delia radicum. J. Sustain. Agric. 33, 128–141.

Lewis, E.E., Campbell, J.F., Harrison, R., 1993. Response of cruiser and ambusher entomopathogenic nematodes (Nematoda: Steinernematidae) to host volatile cues. Can. J. Zool. 71, 765–769.

Lewis, E.E., Campbell, J.F., Griffin, C., Kaya, H.K., Peters, A., 2006. Behaviour ecology of entomopathogenic nematodes. Biol. Control 38, 86–99.

Liesch, P.J., Williamson, R.C., 2010. Evaluation of chemical controls and entomopathogenic nematodes for control of *Phyllophaga* white grubs in a Fraser fir production field. J. Econ. Entomol. 103, 1979–1987.

Mahmoud, M.F., 2007. Combining the botanical insecticides NSK extract, NeemAzal T 5%, Neemix 4.5% and the entomopathogenic nematode *Steinernema feltiae* Cross N 33 to control the peach fruit fly, *Bactrocera zonata* (Saunders). Plant Prot. Sci. 43, 19–25.

Mason, J.M., Wright, D.J., 1997. Potential for the control of *Plutella xylostella* larvae with entomopathogenic nematodes. J. Invert. Pathol. 70, 234–242.

Mason, J.M., Matthews, G.A., Wright, D.J., 1998. Screening and selection of adjuvants for the spray application of entomopathogenic nematodes against a foliar pest. Crop. Prot. 17, 463–470.

Menti, H., Wright, D.J., Perry, R.N., 2000. Infectivity of populations of the entomopathogenic nematodes *Steinernema feltiae* and *Heterorhabditis megidis* in relation to temperature, age and lipid content. Nematology 2, 515–521.

Molyneux, A.S., 1986. *Heterorhabditis* spp. and *Steinernema* spp.: Temperature and aspects of behavior and infectivity. Exp. Parasitol. 62, 169–180.

Morton, A., Garcia-del-Pino, F., 2008. Field efficacy of the entomopathogenic nematode, *Steinernema feltiae* against the Mediterranean flat-headed root borer *Capnodis tenebrionis*. J. Appl. Entomol. 132, 632–637.

Mukuka, J., Strauch, O., Ehlers, R.U., 2010. Variability in desiccation tolerance among different strains of the entomopathogenic nematode *Heterorhabditis bacteriophora*. Nematology 12, 711–720.

Murugan, K., Vasugi, C., 2011. Combined effect of *Azadirachta indica* and the entomopathogenic nematode *Steinernema glaseri* against subterranean termite, *Reticulitermes flavipes*. J. Entomol. Acarol. Res. 43, 253–259.

Navaneethan, T., Strauch, O., Besse, S., Bonhomme, A., Ehlers, R.U., 2010. Influence of humidity and a surfactant-polymer-formulation on the control potential of the entomopathogenic nematode *Steinernema feltiae* against diapausing codling moth larvae (*Cydia pomonella* L.). BioControl 55, 777–788.

Nderitu, J., Sila, M., Nyamasyo, G., Kasina, M., 2009. Effectiveness of entomopathogenic nematodes against sweet potato weevil (*Cylas puncticollis*) Boheman (Coleop: Apionidae) under semi-field conditions in Kenya. J. Entomol. 6, 145–154.

Nickle, W.R., Shapiro, M., 1994. Effects of 8 brighteners as solar radiation protectants for *Steinernema carpocapsae*, All strain. J. Nematol. 26, 782–784.

Nickle, W.R., Connick Jr., W.J., Cantelo, W.W., 1994. Effects of pesta-pelletized *Steinernema carpocapsae* (All) on Western corn rootworms and Colorado potato beetles. J. Nematol. 26, 249–250.

Nyasani, J.O., Kimenju, J.W., Olubayo, F.M., Wilson, M.J., 2008. Laboratory and field investigations using indigenous entomopathogenic nematodes for biological control of *Plutella xylostella* in Kenya. Int. J. Pest. Manag. 54, 355–361.

Pan, H.Y., 1995. Improvement of culture medium for entomopathogenic nematodes. J. Jilin. Agric. Univ. 17, 16–19.

Parkman, J.P., Frank, J.H., 1992. Use of sound trap to inoculate *Steinernema scapterisci* into pest mole cricket population. Fla. Entomol. 76, 75–82.

Paunikar, S., Kulkarni, N., Mishra, V.K., Tiple, A.D., Hussaini, S.S., 2010. Susceptibility of teak skeletonizer, *Eutectona machaeralis* Walker to *Steinernema carpocapsae*. Hislopia J. 3, 165–170.

Paunikar, S., Mishra, V., Kulkarni, N., 2011. Infectivity of two native EPN strains, *Steinernema* spp. against waxmoth, *Galleria mellonella*. Third Biopesticide International Conference (BIOCICON 2011), Nov., 28–30, 2011, St. Xavier's College, Palayamkotai (T.N.), India (Abs).

Paunikar, S., Mishra, V., Kulkarni, N., Hussaini, S.S., 2012. Tolerance of EPN, *Heterorhabditis indica* to some biopesticides. Pestology 12, 41–49.

Piggott, S., Mason, J., Matthews, G., Wright, D., 1999. Redesigning spray application technology for entomopathogenic nematodes. Meded. Fac. Landbouw. en Toegepaste Biologische-Wetenschappen, Universiteit Gent 64, 813–820.

Poinar Jr., G.O., 1990. Taxonomy and biology of Steinernematidae and Heterorhabditidae. In: Gaugler, R., Kaya, H.K. (Eds.), Entomopathogenic Nematodes in Biological Control. CRC Press, Boca Raton, FL, pp. 23–61.

Poinar Jr, G.O., Karunakar, G.K., David, H., 1992. *Heterorhabditis indicus* n. sp. (Rhabditida: Nematoda) from India: Separation of *Heterorhabditis* spp. by infective juveniles. Fundam. Appl. Nematol. 15, 467–472.

Portillo-Aguilar, C., Villani, M.G., Tauber, M.J., Tauber, C.A., Nyrop, J.P., 1999. Entomopathogenic nematode (Rhabditida: Heterorhabditidae and Steinernematidae) response to soil texture and bulk density. Environ. Entomol. 28, 1021–1035.

Prabhu, S., Sudheer, M.J., 2008. Evaluation of two native isolates of entomopathogenic nematodes *Steinernema* sp. and *Heterorhabditis indica* from Andhra Pradesh against *Amsacta albistriga* in groundnut. J. Biopest. 1, 140–142.

Priya, P., Subramanian, S., 2008. Compatibility of entomopathogenic nematodes *Heterorhabditis indica* and *Steinernema glaseri* with insecticides. J. Biol. Control 22, 225–230.

Radovi, S., 2010. Effect of selected pesticides on the vitality and virulence of the entomopathogenic nematodes, *Steinernema feltiae* (Nematoda: Steinernematidae). Plant Prot. Sci. 46, 83–88.

Rahimi, F.R., McGuire, T.R., Gaugler, R., 1993. Morphological mutant in the entomopathogenic nematode, *Heterorhabditis bacteriophora*. Heridity 84, 475.

Ramirez II, R.A., Henderson, D.R., Riga, E., Lacey, L.A., Snyder, W.E., 2009. Harmful effects of mustard bio-fumigants on entomopathogenic nematodes. Biol. Control 48, 147–154.

Ratnasinghe, G., Hague, N.G.M., 1997. Efficacy of entomopathogenic nematodes against the diamondback moth, *Plutella xylostella*. *Pak.* J. Nematol. 15, 45–53.

Richter, A.R., Fuxa, J.R., 1990. Effect of *Steinernema feltiae* on *Spodoptera frugiperda* and *Heliothis zea* (Lepi: Noctuidae) in corn. J. Econ. Entomol. 83, 1286–1291.

Rovesti, L., Deseo, K.V., 1990. Compatibility of chemical pesticides with the entomopathogenic nematodes, *Steinernema carpocapsae* Weiser and *S. feltiae* Filipjev (Nematoda: Steinernematidae). Nematologica 36, 237–245.

Saleh, M.M.E., Abbas, M.S.T., 1998. Suitability of certain entomopathogenic nematodes for controlling *Zeuzera pyrina* L. (Lepi: Cossidae) in Egypt. Int. J. Nematol. 8, 126–130.

Saleh, M.M.E., Draz, K.A.A., Mansour, M.A., Hussein, M.A., Zawrah, M.F.M., 2009. Controlling the sugar beet beetle *Cassida vittata* with entomopathogenic nematodes. J. Pest. Sci. 82, 289–294.

Salma, J., Shahina, F., 2012. Mass production of eight Pakistani strains of entomopathogenic nematodes (Steinernematidae and Heterorhabditidae). Pak. J. Nematol. 30, 1–20.

Sankar, M., Sethuraman, V., Palaniyandi, M., Prasad, J.S., 2009. Entomopathogenic nematodes, *Heterorhabditis indica* and its compatibility with other biopesticides on the greater wax moth (*Galleria mellonella* L.). Indian J. Sci. Technol. 2, 57–62.

Schalk, J.M., Bohac, J.R., Dukes, P.D., Martin, W.R., 1993. Potential of non-chemical control strategies for reduction of soil insect damage in sweet potato. J. Am. Soc. Hortic. Sci. 118, 605–608.

Schirocki, A.G., Hague, N.G.M., 1997. The effect of selective culture of *Steinernema feltiae* at low temperature on establishment, pathogenicity, reproduction and size of infective juveniles. Nematologica 43, 481–489.

Schulte, M.J., Martin, K., Buchse, A., Sauerborn, J., 2009. Entomopathogens (*Beauveria bassiana* and *Steinernema carpocapsae*) for biological control of bark-feeding moth, *Indarbela dea* on field-infested litchi trees. Pest Manag. Sci. 65, 105–112.

Sezhian, N., Sivakumar, C.V., Venugopal, M.S., 1996. Alteration of effectiveness of *Steinernema carpocapsae* against *Spodoptera litura* larvae on sunflower by addition of an insect phagostimulant. Indian J. Nematol. 26, 77–81.

Shakeela, V., Hussaini, S.S., 2006. Susceptibility of tobacco cutworm, *Spodoptera litura* Fabricius to some indigenous isolates of entomopathogenic nematodes. J. Ecofriendly Agric. 1, 64–67.

Shakeela, V., Hussaini, S.S., 2009. Influence of soil type on infectivity and persistence of indigenous isolates of entomopathogenic nematodes *H. indica* and *S. carpocapsae*. J. Biol. Control 23, 63–72.

Shapiro, D.I., Cate, J.R., Pena, J., Hunsberger, A., McCoy, C.W., 1999a. Effects of temperature and host age on suppression of *Diaprepes abbreviatus* (Coleop: Curculionidae) by entomopathogenic nematodes. J. Econ. Entomol. 92, 1086–1092.

Shapiro, D.I., Obrycki, J.J., Lewis, L.C., Jackson, J.J., 1999b. Effects of crop residue on the persistence of *Steinernema carpocapsae*. J. Nematol. 31, 517–519.

Shapiro, D.I., Lewis, L.C., Obrycki, J.J., Abbas, M., 1999c. Effects of fertilizers on suppression of black cutworm (*Agrotis ipsilon*) damage with *Steinernema carpocapsae*. J. Nematol. 31, 690–693.

Shapiro-Ilan, D.-I., Campbell, J.F., Lewis, E.E., Elkon, J.M., Kim-Shapiro, D.B., 2009a. Directional movement of steinernematid nematodes in response to electrical current. J. Invert. Pathol. 100, 134–137.

Shapiro-Ilan, D.I., Cottrell, T.E., Mizell III, R.F., Horton, D.L., Davis, J., 2009b. A novel approach to biological control with entomopathogenic nematodes: prophylactic control of the peach tree borer, *Synanthedon exitiosa*. Biol. Control 48, 259–263.

Singh, D.K., Ram, Singh, Dwivedi, R.K., 2008. Evaluation of bio-pesticides against lepidopterous pests of cabbage. Ann. Plant Prot. Sci. 16, 316–319.

Singh, S.P., 1993. Effectiveness of an indigenous nematode against citrus butterfly. J. Insect Sci. 6, 107–108.

Sirjani, F.O., Lewis, E.-E., Kaya, H.K., 2009. Evaluation of entomopathogenic nematodes against the olive fruit fly, *Bactrocera oleae* (Diptera: Tephritidae). Biol. Control 48, 274–280.

Sirjusingh, C., Mauleon, H., Kermarrec, A., 1991. Compatibility and synergism between entomopathogenic nematodes and pesticides for control of *Cosmopolites sordidus* Germar. In: Proceedings of Caribbean Meetings on Biological Control, Guadeloupe, 5–7 Nov. 1990, pp. 183–192.

Sitaramaiah, S., Gunneswara Rao, S., Hussaini, S.S., Venkateswarlu, P., Nageswara Rao, S., 2002. Use of entomopathogenic nematode *Steinernema carpocapsae* against *Spodoptera litura* Fab. in tobacco nursery. In: Tandon, P.L., Ballal, C.R., Jalali, S.K., Rabindra, R.J. (Eds.), Biological Control of Lepidopteran Pests. Precision Phototype Services, Bangalore, pp. 211–213.

Solomon, A., Salomon, R., Paperna, I., Glazer, I., 2000. Desiccation stress of entomopathogenic nematodes induces the accumulation of a novel heat-stable protein. Parasitology 121, 409–416.

Srinivas, P.R., Prasad, J.S., 1991. Record of DD-136 nematode infection on rice leaf folder, *Cnaphalocrosis medinalis*. Indian J. Agric. Sci. 61, 348–349.

Subramanian, S., Rajeswari S., 2002. Foliar persistence of entomopathogenic nematodes. (Singh, R.V., Pankaj., Dhawan, S.C., Gaur, H.S. Eds), Proceedings of the National Symposium on Biodiversity and Management of Nematodes in Cropping Systems for Sustainable Agriculture, Jaipur, India, 11–13 November, 2002, pp. 193–195.

Szalanski, A.L., Taylor, D.B., Mullin, P.G., 2000. Assessing nuclear and mitochondrial DNA sequence variation within *Steinernema* (Rhabditida: Steinernematidae). J. Nematol. 32, 229–233.

Tomalak, M., 1994. Selective breeding of entomopathogenic nematode, *S.feltiae* for improved efficacy in control of mushroom fly, Lycoriella. Biocont. Sci. Technol. 4, 187–198.

Torr, P., Heritage, S., Wilson, M.J., 2004. Vibrations as a novel signal for host location by parasitic nematodes. Int. J. Parasitol. 34, 997–999.

Triggiani, O., Tarasco, E., 2007. Applying entomopathogenic nematodes to *Xanthogaleruca luteola* (Coleop: Chrysomelidae) infested foliage. Redia 90, 29–31.

Tyson, T., Zamora, G.O., Wong, S., Skelton, M., Daly, B., Jones, J.T., et al., 2012. A molecular analysis of desiccation tolerance mechanism in the anhydrobiotic nematode, *Panagrolaimus superus* using expressed sequence tags. BMC Res. notes 5, 68.

Vainio, A., 1994. Effect of pesticides on long-term survival of *Steinernema feltiae* in the field. *Bull. OILB-SROP* 17, 70–76.

Vainio, A., Hokkanen, H., 1990. Side-effects of pesticides on the entomophagous nematode *Steinernema feltiae,* and the entomopathogenic fungi *Metarhizium anisopliae* and *Beauveria bassiana* in the laboratory. Proceedings of the Fifth International Colloquium of Invertebrate Pathology and Microbial Control, Adelaide, Australia, 20–24 August 1990, p. 334.

Wang, J.X., Bedding, R.A., Wang, J.X., 1998. Population dynamics of *Heterorhabditis bacteriophora* and *Steinernema carpocapsae* in *in vitro* monoxenic solid culture. Fund. Appl. Nematol. 21, 165–171.

Wang, X.D., Ishibashi, N., Wang, X.D., 1998. Effects of precedent infection of entomopathogenic nematodes (Steinernematidae) on the subsequent invasion of infective juveniles. Jap. J. Nematol. 28, 8–15.

Wang, Y., Gaugler, R., Wang, Y., 1998. Host and penetration site location by entomopathogenic nematodes against Japanese beetle larvae. J. Invert. Pathol. 72, 313–318.

Wharton, D.A., 2003. The environmental physiology of Antarctic terrestrial nematodes. A review. J. Comp. Physiol. B173, 621–628.

Wharton, D.A., Petrone, L., Duncan, A., McQuillan, A.J., 2008. A surface lipid may control the permeability slump associated with entry into anhydrobiosis in the plant parasitic nematode. *Ditylenchus dipsaci. J. Exp. Biol.* 211, 2901–2908.

Womersley, C.Z., Higa, L.M., Wharton, D.H., 1998. Survival biology. In: Perry, R.N., Wright, D.J. (Eds.), The Physiology and Biochemistry of Free Living and Plant Parasitic nematodes. CAB Int., Wallingford, pp. 271–302.

Xu, J.L., Yang, P., 1992. The application of the codling moth nematode against the litchi stemborer. Acta Phytophylactica Sin. 19, 217–222.

Yang, H.W., Jian, H., Zhang, S.G., Zhang, G.Y., 1997. Quality of the entomopathogenic nematode *Steinernema carpocapsae* produced on different media. Biol. Control 10, 193–198.

Yang, X.F., Yang, H.W., Jian, H., Chen, S.B., Yang, X.F., Yang, H.W., et al., 1999. Effect of emergence time and store period of three *Steinernema* species on their penetrative rate and activity. Chinese J. Biol. Control 15, 62–65.

Zaki, F.N., Awadallah, K.T., Gesraha, M.A., 1997. Competitive interaction between the braconid parasitoid *Meteorus rubens* Nees and the entomogenous nematode, *Steinernema carpocapsae* on larvae of *Agrotis ipsilon* Hufn. (Lep: Noctuidae). J. Appl. Entomol. 121, 151–153.

Zhang, L., Shono, T., Yamanaka, S., Tanabe, H., 1994. Effects of insecticides on the entomopathogenic nematode *Steinernema carpocapsae*. Appl. Entomol. Zool. 29, 539–540.

Zioni, S., Glazer, I., Segal, D., 1992. Phenotype and genetic analysis of a mutant of *Heterorhabditis bacteriophora*. J. Nematol. 24, 359–364.

Züger, M., Bollhalder, F., Andermatt, M., 2005. Control of Codling Moth, *Cydia pomonella* (Lepidoptera: Tortricidae) with nematodes (*Steinernema* spp. and *Heterorhabditis* spp.). 10th European Meeting 'Invertebrate pathogens in biological control: present and future', Bari, Italy, June 2005, pp. 23–29.

CHAPTER

12

Entomopathogenic Viruses and Bacteria for Insect-Pest Control

C.S. Kalha[1], *P.P. Singh*[2], *S.S. Kang*[2], *M.S. Hunjan*[2], *V. Gupta*[1] *and R. Sharma*[1]

[1]**Sher-e-Kashmir University of Agricultural Sciences and Technology of Jammu, Jammu and Kashmir, India,** [2]**Punjab Agricultural University, Ludhiana, India**

12.1 INTRODUCTION

Pest problems are an inevitable part of modern day agriculture. They occur because agro-ecosystems have created less stable natural ecosystems which otherwise govern ecological forces that regulate potential pest species in natural ecosystems. Raising crops in a monoculture thus provides a food resource cycle that allows pest populations to achieve far higher densities than they would in natural environments. A certain cultivation practice can also make the physico-chemical environment more favourable for pest activity, for example through irrigation or the warm conditions found in glasshouses. New cultivars or new crops introduced into a certain area or country may provide food resources for potential pests. Also, the use of broad spectrum insecticides can destroy natural predators that help keep pests under control. In these scenarios, new pest problems arise or existing pests become more serious and cause significant damage to crops, biodiversity and landscape valued at billions of dollars per annum. New strains of plant insect-pests may arise to overcome varietal resistance in crops.

Under the natural scenario, the populations of many arthropods are naturally regulated by entomopathogens such as bacteria and viruses. Entomopathogens have also been used as classical biological control agents of alien insect-pests, and natural pest control by entomopathogens has been enhanced by habitat manipulation. Many farmers and growers are now familiar with the use of predators and parasitoids for biological control of arthropod (insect and mite) pests, but it is also possible to use specific microorganisms that kill arthropods. These include entomopathogenic fungi, nematodes, bacteria and viruses. These are all widespread in the natural environment and cause natural infections in many pest species. Many among these entomopathogens can be mass-produced and formulated for field use to manage pest populations in a manner analogous to chemical pesticides.

D. P. Abrol (Ed): Integrated Pest Management.
DOI: http://dx.doi.org/10.1016/B978-0-12-398529-3.00013-0

12.2 NATURAL OCCURRENCE AND BIODIVERSITY OF ENTOMOPATHOGENIC BACTERIA AND VIRUSES

12.2.1 Entomopathogenic Bacteria

Entomopathogenic bacteria are unicellular prokaryotic organisms having size ranging from less than 1 μm to several μm in length. Bacteria with rigid cell walls are cocci, rod-shaped and spiral while bacteria without cell walls are pleomorphic. More than 100 bacteria have been identified as arthropod pathogens among which, *Bacillus thuringiensis, B. sphaericus, B. cereus* and *B. popilliae* have received most attention as microbial control agents. The majority of bacterial pathogens of insect-pests occur in bacterial families Bacillaceae, Pseudomonadaceae, Enterobacteriaceae, Streptococcaceae, and Micrococcaceae. These families of bacteria usually represent epiphytes or weak pathogens; however, some of them are highly virulent to their respective hosts. Among the entomopathogenic bacteria, much attention has been given to the family Bacillaceae. Some of the bacterial species belonging to the genus *Bacillus* are highly pathogenic to arthropods, such as *Bacillus popilliae*, which causes milky spore disease in scarbaeids, while *B. sphaericus* is highly virulent to mosquitoes. *Bacillus thuringiensis* (*Bt*) is widespread in soil, is a lethal pathogen of a range of orders and is the most widely used entomopathogenic biological control agent. There are at present over 40 *Bt* products available for the control of insect-pests accounting for 1% of the global insecticide market (Evans, 2008). The *Bt* subspecies represents a group of organisms that occur naturally and can be added to an ecosystem to achieve insect control. The commercial *Bt* products may be applied as an insecticide to foliage, soil, water environments and food storage facilities. After application of *Bt* to an ecosystem, the organism may persist as a component of the natural microflora.

Members of this entomopathogenic group of bacteria can be found in most ecological niches. In natural habitats, several *Bt* isolates have no known target, as opposed to early *Bt* isolates, which were known to be pathogenic for insects. This lack of insecticidal activity may be attributed to the loss of ability to produce insecticidal crystalline proteins (ICPs). Or simply, a test insect-pest for the actual target of that isolate is as yet unknown. The current knowledge about the activity of *Bt* populations in the environment is limited, although crop, vegetation and seasonal variations contribute towards numbers and subspecies diversity of *Bt* populations.

12.2.2 Entomopathogenic Viruses

Entomopathogenic viruses are obligate intracellular parasites having either DNA or RNA encapsulated into a protein coat known as capsid to form the virions or nucleocapsids. These viruses have proved to be very effective in managing populations of certain pests such as Lepidoptera and Hymenoptera forest pests in Europe and those introduced into forests in the USA and Canada; also, for controlling the cotton leaf worm, potato tuber worm and greater wax moth larvae. Like entomopathogenic bacteria, they are also very specific to target insects. Diseases caused by entomopathogenic viruses have been known since the 16th century. A disease called *jaundice grasserie*, now identified as a nucleopolyhedrosis virus, was observed in silkworm (*Bombyx mori*) rearing facilities. In 1856, two Italian scientists, Maestri and Cornalia, first described the occlusion bodies (OBs) of silkworms. Steinhaus and his collaborators (1950–1970) tested baculoviruses as biological control agents in the field by applying a nucleopolyhedrovirus (NPV) to control the alfalfa caterpillar (*Colias euwortheme* Boisduval: Lepidoptera). The natural populations of insect viruses belong to many families, some of which occur exclusively in arthropods and/or plants and viruses belonging to these families may vary in the tissue they infect and their ability to cause acute or chronic infections and in the appearance of moribund or dead larvae. In general, viruses are divided into two broad non-taxonomic categories, occluded-viruses and non-occluded-viruses. The first category is the occluded-viruses in which the mature virion particles (virions) are embedded within a protein matrix, forming para-crystalline bodies that are generally referred to as OBs, while the second category is the non-occluded-viruses in which the virions occur freely or occasionally form para-crystalline bodies, characterized by the absence of occlusion body protein interspersed among the virions (Federici, 1999).

Out of a total of 73 known virus families, entomopathogenic viruses have been listed in 13 families as described by Murphy et al. (1995). Among these 13, Baculoviridae family members are the most virulent on different orders of insect-pest including Lepidoptera, Diptera, Hymenoptera, Orthoptera, Isoptera and Neuroptera. Currently, Baculoviridae is divided into two genera: *Nucleopolyhedrovirus* (NPV) and *Granulovirus* (GV) (Francki et al., 1991; Murphy et al., 1995). Virions of NPV and GV are occluded in polyhedral and capsular proteinaceous OBs, respectively. The OBs of GVs are smaller (0.3–0.5 μm in length) than those of the NPVs (0.15–15 μm in diameter) and usually only contain a single enveloped nucleocapsid. The OBs of NPVs contain several hundred virus particles, each of which may contain one (SNPV) or many (MNPV) nucleocapsids. NPVs have limited host ranges, usually being restricted to one host species or genus, with the exception of the NPVs of *Autographa californica* (Speyer), *Anagrapha falcifera* (Kirby) and *Mamestra brassicae* (Linnaeus). GVs are more specific than NPVs as they have been only reported from Lepidoptera (Battu and Arora, 1996; Moscardi, 1999). An example of SNPV is *Trichoplusia ni* SNPV, whereas that of MNPV is *Autographa californica* (AcNPV). Baculoviruses have a large, double-stranded, covalently closed, circular DNA genome of between 88 and 200 kbp. The baculoviruses are characterized by the presence of rod-shaped nucleocapsids, which are further surrounded by a lipoprotein envelope to form virus particles. NPVs produce large particles within the nucleus of an infected cell. The occlusion body is composed of a matrix comprising a 29 kDa protein known as polyhedron. The DNA–protein complex is contained by a rod-shaped nucleocapsid comprising a 39 kDa or 87 kDa capsid protein (King et al., 1994). The size of the virus genome determines the length of the nucleocapsid, which may be 200–400 nm. The width remains constant at about 36 nm. Polyhedra consist largely of a single protein (polyhedrin) of about 30 kDa and formed in the nucleus of infected cells. Virions that have been released from polyhedra are called polyhedra-derived virus in the midgut tissues of susceptible insects, whereas virions that are released from cells without occlusion are called extracellular viruses (ECV) or budded viruses. On the other hand, GVs contain one virion (singly enveloped nucleocapsid) per virus occlusion body or granule. Granulin, the major granule protein, is similar to polyhedrin in function. The baculovirus life cycle involves two distinct forms of virus:

- **i.** *Occlusion/Polyhedra-Derived virus* (ODV/PDV) is present in a protein matrix (polyhedrin or granulin) and is responsible for the primary infection in the midgut epithelial cells of the host.
- **ii.** *Budded virus* (BV) is the non-occluded form released from the infected host cells later during the secondary infection.

12.3 USE OF ENTOMOPATHOGENIC BACTERIA AND VIRUSES AS BIOCONTROL AGENTS

Entomopathogenic bacteria, like other natural enemies, can exert considerable control of target populations (Lacey et al., 2001).

In nature, occurrence of natural epizootics of viral, bacterial and fungal pathogens may be held responsible for the decline in insect-pest populations (Evans, 1986; McCoy et al., 1988). Among the bacterial pathogens, *Bt* is the most studied and exploited one. *Bacillus thuringiensis* (*Bt*) is a spore forming bacterium, with its sporulation generally associated with the synthesis of a proteinaceous protoxin crystal that has insecticidal activities. It has been used for the control of lepidopteran, dipteran and coleopteran insects for over three decades (Sarvjeet, 2000). Ingested crystals of the toxin dissolve within the gut and are cleaved by host proteases to form an active toxin, termed the δ-endotoxin. This binds to receptors in the midgut epithelium to cause the formation of ion pores, leading to gut paralysis. Thus, ingested spores of *Bt* may contribute to bacterial septicaemia. Globally, about 70 *Bt* subspecies are known, which differ in their host preference towards different lepidopteran, dipteran and coleopteran insects. Some strains may also produce exotoxins, which have a wide spectrum of activity including against vertebrates (Lacey and Mulla, 1990).

Many different *Bt* subspecies have been isolated from dead or dying insects especially from the orders Coleoptera, Diptera and Lepidoptera. The carcasses of dead insects often contain large quantities of spores and ICPs. While the dipteran-acting *Bt* subspecies are found in aquatic environments, the coleopteran- and lepidopteran-acting *Bt* subspecies are primarily recovered from soil and phylloplane (Bernhard et al., 1997; Hansen et al., 1998; Itoua-Apoyolo et al., 1996; Kaur and Singh, 2000a,b; Theunis et al., 1998). *Bt* is a ubiquitous soil microbe, however, it is also very frequently recovered from phylloplane (Smith and Couche, 1991). *Bt* is abundant in rich topsoil and rarely subterranean environments. Travers et al. (1987) have given an effective isolation technique to recover a high population of this bacterium from soil.

Many types of *Bt* have been isolated (Brown et al., 1958) which showed variable differences in efficacy against many lepidopteran, dipteran and coleopteran insect species. Their variable activity against different insect species depends upon the type of endotoxins produced by the respective *Bt* isolate. The selection of the new subsp. *B. thuringiensis kurstaki* strain HD-1 (Serotype H3a:3b) that does not produce exotoxins, launched the commercialization of this strain worldwide (Dulmage, 1981). This strain went on to become the most widely used *Bt* insecticide formulated from a bacterium. The *Bt* products are commercially successful and are widely available as liquid concentrates, wettable powders, and ready-to-use dusts and granules. Some products are used to control Indian meal moth larvae in stored grain. Another strain, *Bacillus thuringiensis* var. *aizawai* produces slightly different toxins and is the active ingredient in certain commercially available products such as Certan, Agree and Xentari.

Another group of *Bt* isolates, including those from *Bacillus thuringiensis* var. *san diego* and *Bacillus thuringiensis* var. *tenebrionis*, are toxic to certain beetles. However, their host range is narrow, e.g. *B. thuringiensis* var. *san diego*, sold under the trade names M-Trak, Foil and Novodor, is very active against Colorado potato beetle but is ineffective against corn rootworms and other related species. *Bacillus thuringiensis* var. *israelensis* (*Bti*) is pathogenic to the larvae of certain species of flies and mosquitoes with *Aedes* and *Psorophora* species being the most susceptible.

The first baculovirus to be developed for commercial use was Elcar (Sandoz Inc.), an NPV of *Helicoverpa zea*, primarily developed for use on cotton and registered by the Environmental Protection Agency in USA in 1975 (Ignoffo, 1981). Elcar was active against major *Helicoverpa/Heliothis* species and provided efficient control in sorghum, maize, tomato, chickpea and navy beans (Ignoffo and Couch, 1981; Teakle, 1994). The advent of synthetic pyrethroids in the late 1970s resulted in reduced interest in Elcar and production was stopped in 1982. However, during the last two decades, several GVs and NPVs have been registered in Europe and other parts of the world for use in insect-pest control. In 1996, Biosys introduced GemStar LC, a liquid concentrate formulation of HzNPV for the control of *H. zea* and *H. virescens* (Fabricius) in US cotton. The NPV of soybean caterpillar, *A. gemmatalis*, is the most widely used viral pesticide and is applied annually on approximately 1 million ha of soybean crop in Brazil. The virus is produced directly in the farmers' fields to lower rearing costs (Moscardi, 1999).

12.4 MODE OF ACTION

12.4.1 Entomopathogenic Bacteria

Bt produces a parasporal inclusion, a protein crystal body during sporulation. A large number of related crystal proteins are known and more than one protein type can co-assemble in one crystal. Many distinct crystal protein (*Cry*) genes have been described. The gut epithelium is the primary target tissue for *Bt* delta-endotoxin action. The crystal proteins exert their effect on the host by causing lysis of midgut epithelial cells, which leads to gut paralysis. The insect stops feeding and, if it does not recover, eventually dies. Upon ingestion, the crystals dissolve in the alkaline environment of the midgut and then the protoxin is proteolytically processed to produce the actual toxin. Activation of actual toxin usually involves the removal of a small number of N-terminal amino acid residues along with the cleavage of the C-terminal half (Gill et al., 1992). The activated toxin then binds to specific receptors present on the membranes of the host insect epithelial midgut cells and induces the formation of pores in the membrane of midgut epithelial cells. This is followed by an increase in cell membrane permeability which eventually leads to cell lysis, disruption of gut integrity and finally to the death of the insect from starvation or septicaemia (Adang 1985; Gill et al., 1992; Bauer, 1995).

12.4.2 Entomopathogenic Viruses

The most common route of entry of a virus into an insect host is *per os*. Typically, the initial infection occurs when a susceptible host insect feeds on plants that are contaminated with the occluded form of the virus. When the OBs are ingested by the insect, the protein matrix dissolves in the alkaline environment of the host midgut (pH 8.0), releasing the infective particles (virions or ODV/PDV) into the midgut. Virions (ODV/PDV) enter into the peritrophic membrane either by direct diffusion with microvilli on the brush border midgut columnar epithelial cells or by adsorptive endocytosis. This entry may also be receptor-mediated. In the next step, uncoating of the ODV/PDVs takes place before passing through the nuclear pores. These uncoated ODV/PDVs travel into the nucleus in association with cellular action. DNA of the nucleocapsid is uncoated in the nucleus and the DNA unwinds due to phosphorylation of DNA-binding protein (P6.9). This results in expression and replication of viral DNA through the viral DNA-polymerase enzyme. The newly formed nucleocapsids bud through the nucleus and gain an envelope of the nuclear membrane. This is shed in the cytoplasm and another envelope comprising cytoplasmic membrane and the virus-coded glycoprotein spikes is acquired by budding through the midgut basal membrane. Such forms of the virus are known as Budded viruses (BVs). BVs are released into the haemolymph and undergo rounds of multiplication in the cells of susceptible tissues. The entry into the cells is through cell-mediated endocytosis and GP64/F-protein (Fusion protein). PDVs are produced in the late phase of the infection. Finally, the occlusion body protein (polyhedrin/

granulin) crystallises to form the OBs, which are released into the environment. Viral proteases and chitinases help to disrupt the chitinous exoskeleton, resulting in disintegration and finally the death of the host insect.

Access to the haemocytes allows the BV infection into the haemocytes and other tissues. BV synthesized following the secondary infection of haemocytes and tracheal matrix, initiates further infection of most other tissues of the lepidopteran host. As the infection spreads along the tracheal epithelium from the foci of infection, the virus gains access to various other tissues such as the epidermis and fat body. Appearance of the virus within the fat body and epidermis indicates that *in vivo* spread of the virus is almost complete and that the larva will soon succumb to infection.

The virus replicates within the nuclei of susceptible tissue cells. Tissue susceptibility varies greatly between viruses with some NPVs being capable of infecting almost all tissue types and most GVs being tissue-specific replications (e.g. fat body cell only). The BV initiates infection to other tissues in the haemolymph, e.g. fat bodies, nerve cells, haemocytes. The cells infected in the second round of virus replication in the insect larva also produce BV, but in addition, occlude virus particles within polyhedra in the nucleus. The accumulation of polyhedra within the insect proceeds until the host consists almost entirely of a bag of virus. In the terminal stages of infection, the insect liquefies and thus releases polyhedra, which can infect other insects upon ingestion. A single caterpillar at its death may contain over 10^9 OBs from an initial dose of 1000. The infected larvae exhibit negative geotropism before succumbing to the virus infection, thereby facilitating widespread dissemination. The speed with which death occurs is determined in part by the environmental conditions. Under optimal conditions, the target pest may be killed in 3–7 days, but death may occur in 3–4 weeks when conditions are not ideal (Cunningham, 1995; Flexner and Belnavis, 2000).

12.5 COMMERCIALIZATION AS BIOCONTROL AGENTS

12.5.1 Entomopathogenic Bacteria

During the late 1930s, *Bt* products were first marketed in France (Lambert and Peferoen, 1992). Since then, it has been one of the most consistent and significant biopesticides used. For the commercial delivery of a microbial pesticide, the biocontrol agent must be mass produced and formulated for better shelf life, field delivery and stability. *Bt* products are generally produced using fermentation technology (Bernhard and Utz, 1993). Most commercial products contain ICP and viable *Bt* spores. Large-scale commercial production may lead to partial loss of bioactive components to the environment. This may result in measurable bioactivity of the formulated product especially if the active material is processed through a dryer, due to the exposure of the bioactive components to the high temperatures required for drying. Commercial *Bt* formulations include wettable powders, suspension concentrates, water-dispersible granules, oil miscible suspensions, capsule suspensions and granules (Tomlin, 1997). Quality standards for *Bt* products include limits on the concentration of microbial contaminants and metabolites (Quinlan, 1990). Attempts have been made to improve the performance of *Bt* formulated products for a longer shelf life and effectiveness by modifying stickers and inert spreading materials (Behle et al., 1997; Burges, 1998). Ferrar and Ridway (1995) have reported enhancement of activity of *Bt* by adding feeding stimulants while some workers have used tannic acid to increase its effectiveness (Gibson et al., 1995).

Commercial *Bt* formulations are available as wettable powder, dust, bait and flowable concentrates or granules, suspensions, encapsulations, etc. (Table 12.1) (Brar et al., 2006) and may be applied to foliage, soil or storage facilities. After the application of a *Bt* subspecies to

TABLE 12.1 Commercial Products of *Bt* and Their Usage

Bt subsp. Strain	Trade name	Usage
kurstaki	Able, Bactospeina, Condor, Costar, CRYMAX, Cutlass, Dipel ES, Bactimos L, Futura, Lepinox, Thuricide	Lepidoptera
aizawai	Florbac, Agree, Design, Xentari	Lepidoptera
kurstaki SA-12	Costar	Lepidoptera
kurstaki	Foil, Raven	Lepidoptera/ Coleoptera
kurstaki HD-1	Thuricide, Biobit, Dipel, Foray, Javelin, Vault	Lepidoptera

an ecosystem, the vegetative cells and spores persist as a component of the natural microflora for a long time. The ICPs, however, are rendered biologically inactive within a short time, hours or days. *B.t.* variety *kurstaki* is usually formulated as a stabilized suspension, wettable powder, dust base, dust, bait or flowable concentrate while *B.t.* variety *aizawai* is formulated as a water-dispersible liquid concentrate.

12.5.2 Entomopathogenic Viruses

Nuclear polyhedrosis viruses are being developed for control of lepidopterous larvae. In the USA and Europe, a few baculovirus products are produced commercially for use in field crops (Table 12.2). Companies such as Dupont, Biosys (now Thermo Trilogy), American Cynamid and Agrivirion have active research programmes for development of agricultural-use viral insecticides. For example, Biosys have introduced two baculovirus-based products, Spod-X for control of beet armyworm and GemStar LC for control of tobacco budworm and cotton bollworm. A list of commercially available viral pesticides registered for pest control in different countries is given in Table 12.2.

TABLE 12.2 Commercial Baculovirus Formulations Available for Use in Field Crops

Product Name	Manufacturer	Baculovirus	Pests
Spod-X	Thermo Trilogy	SeNPV	Beet armyworm
GemStar	Thermo Trilogy	HzNPV	*Heliothis/ Helicoverpa*
Elcar	Novartis		
Madmex	Andermatt Biocontrol, Switzerland	CpNPV	Codling moth
Granusal	Behring AG, Werke, Germany		
Caprovirusine	NPP, France		
VPN	Agricola El Sol, Brazil	AgNPV	Velvetbean caterpillar
Gusano	Thermo Trilogy	AcNPV	*Autographa californica*
Spodopterin	NPP, France	SlNPV	*Spodoptera litura*

12.5.2.1 Quality of Baculovirus Preparations

Following the regulation of NPV products such as 'Spod-X', and 'GemStar' for *S. exigua* and *Heliothis* spp., respectively during 1994 in the USA, many other Asian and European countries such as Thailand and Holland allowed the registration of these products (Kolodny-Hirsch and Dimock, 1996). In India too, the interest in commercialization of baculovirus-based insecticides has developed recently and NPV products involving respective baculovirus species from *H. armigera* and *S. litura* are available. However, the widespread use of these products has still not been achieved though the market is huge for *H. armigera* and *S. litura* crop protection products

(HaNPV: 4.26×10^{23} and SlNPV: 1.59×10^{23} viral OBs) to fulfil the needs of at least 10% of the crop area under cotton, chickpea, oilseeds, vegetables, etc. (Sathiah and Jayaraj, 1996).

Sometimes, the quality of commercial NPV preparations is extremely poor and is totally ineffective in killing target pests, especially when field-evaluation reports are evaluated analytically (Grzywacz et al., 1997). Similarly, some NPV products produced in India were ineffective under laboratory conditions when fed via contaminated foliage and contained actual NPV content far below the required quantum, i.e. 6×10^9 OBs making one LE/ml of the product (Battu, 1999). A survey in southern India returned particularly poor results where, in 1996–99, all of the 11 samples examined from a commercial supply of NPV of *H. armigera* had too low levels of viral OBs to be effective (Kennedy et al., 1999). Therefore, quality control of the commercial preparations is extremely important to achieve proper pest control. There could be several reasons for the poor quality but the main drawbacks are related to deficiencies in production techniques and quality control procedures. A problem for producers, customers and regulators is that the standard technique for assessing chemical pesticides through chemical analysis is not appropriate for infective biological agents such as NPV and GV. Even the standard toxicity assessment methodologies applied to rapid action contact pesticides are often inappropriate for the relatively slow acting biopesticides such as NPV and GV, which have to be ingested as viral OBs applied to the pest's food material (Kennedy et al., 1999). The use of LE as a standard measure of NPV activity must be based on actual counts of OBs that can easily be done reliably and efficiently using a haemocytometer on aqueous suspensions through optical microscopy (Battu et al., 1993). Many producers and research workers enthusiastically engaged in entrepreneurships involving propagation of baculoviruses as cottage industries in developing countries are lacking the basic technical training in discriminating between NPV occlusions and artifacts involving cellular debris and the developmental stages of many saprophytic/infectious microorganisms (Battu et al., 1994).

12.6 METHOD OF APPLICATION

12.6.1 Entomopathogenic Bacteria

Bt can be applied using conventional spray equipment. Good spray coverage is absolutely essential as the bacteria must be eaten to be effective against the target insect. Various formulations as listed above have been developed depending on application target and feasibility. Conventional formulations have been substituted by advanced versions such as micro-encapsulations and micro-granules to enhance residual entomotoxicity. Furthermore, for better delivery and efficacy of the product, development of formulations must take into account the biotic (spore concentration and entomotoxicity) and abiotic factors such as UV radiation, temperature, pH, rain, and foliage.

12.6.2 Strategies for Utilization of Entomopathogenic Baculoviruses

There are four basic strategies for using baculoviruses in insect-pest management.

12.6.2.1 Introduction and Establishment

The introduction and establishment of microbials in an environment is intended to result in permanent suppression of the target pest. Most of the successes of viruses in insect control have been by this method. There have been at least 15 successful introductions of viruses, five in crops and 10 in forests. In the 1930s, an NPV was introduced accidentally into Canada along with parasitoids, which were imported from Scandinavia and released for control of pests. Later this NPV was multiplied and applied in selected locations. The NPV was remarkably successful and no control measures have been required against the pest in Canada for the last 50 years. Its success was attributed to relative stability of the forest ecosystem and host populations as well as to efficient horizontal and vertical transmission of the virus. Later, the European pine sawfly, *Neodiprion sertifer* (Geoffroy), and the red-headed pine sawfly, *N. lecontci* (Fitch), were also successfully controlled with one or two introductions of respective NPVs into field populations. An NPV of *Chrysodeixix includens* (Walker) is possibly the best example of a baculovirus implemented as a classical biological control agent in a crop. This NPV was released on 200–250 ha of soybean in the USA and provided control 12–15 years later.

12.6.2.2 Seasonal Colonization

This involves the inoculative release of microbial pathogens to control insect-pests for more than one generation, although subsequent releases are required when the pathogen population declines. It requires efficient replication and transmission of the pathogen in host populations. The most important example is the control of velvetbean caterpillar, *Anticarsia gemmatalis* (Hubner), on soybean by application of AgNPV. The virus is applied on 1 million ha annually in Brazil. The AgNPV occurs naturally in Brazil in *A. gemmatalis* with pathogenesis similar to that of other NPVs. Currently, it is produced directly on the farmer's fields. The procedure involves virus application in soybean infested with *A. gemmatalis* larvae, collection of the dead larvae, and storage in large rooms at −4 to −8°C until processed as a formulation. Cost of the formulated product is about US$0.7/ha and it reaches the farmer at a mean cost of US$1.1–1.5/ha, which is lower than the cost of chemical insecticides (Moscardi, 1999).

12.6.2.3 Environmental Manipulation

This involves changing the host habitat to favour conservation or argumentation of pathogens in a system where they either occur naturally or have been introduced. Modified cultural practices enhance the prevalence of pathogens in insect populations by adding in persistence or assisting their transport from the soil to the insects feeding substrate. These practices include changes in cultivation, grazing, sowing and chemical use to increase natural control of *Wilseana* sp. by NPV in New Zealand pastures. Movement of cattle similarly enhanced NPV transport and natural control of *Spodoptera frugiperda* in Louisiana pastures. Environmental manipulation has also been found useful for enhancing the efficacy of non-occluded virus in the case of rhinoceros beetle on coconut palms. Viral spread and control of the beetle populations are enhanced if some of the dead palms are left standing and the others are piled and overgrown with crops rather than left lying around the plantation.

12.6.2.4 Microbial Insecticides

Most viral pathogens are suitable for use as microbial insecticides. The industry also has maximum interest in this approach, because the multiple applications create the best opportunity for product sales. The NPVs and GVs of lepidopteran caterpillars (Moscardi, 1999) as well as NPVs of several species of sawflies, provide short-term control comparable to that with conventional insecticides.

12.7 FIELD STABILITY AND PERSISTENCE OF ENTOMOPATHOGENIC BACTERIA

12.7.1 Entomopathogenic Bacteria

Field stability and persistence of the augmented population of a bioagent in a particular environment for a required time period is very important for its efficacy against the target pest. *Bt* has been reported to decay in a relatively shorter period after application in the open field by degradation by sunlight and rain (Behle et al., 1997). However, when applied in protective cultivated crops, in greenhouses in particular, the efficacy and persistence increases significantly (Janmaat et al., 2007). Various environmental factors such as plant physiology, rain, pH, ultraviolet radiation and temperature significantly influence efficacy and field stability of *Bt* formulations. UV light is one of the major factors responsible for reduced effectiveness due to inactivation of *Bt* in the environment (Pusztai et al., 1991). The effect of UV radiation also varies under different climatic conditions; for example, the half-life of *Bt* in USA on cotton plants was 30–48h and in Egypt on castor plants, it was 19–40h (Beegle and Yamamoto, 1992; Ragaei, 1990). Rainfall affects the persistence of sprayed formulated product as it leads to wash-off of applied biopesticides from foliage before it starts its action. Optimal *Bt* activity is realized between pH 3 and 11. Shelf life and field persistence of *Bt* formulation are also highly influenced by temperature. It has been observed that temperatures lower than 10°C and higher than 30°C may have deleterious effects on the activity of bacterial pathogens over an extended period of time (Ignoffo, 1992). Unfavourable temperature may cause degradation of the active ingredient in the formulation by heat, or more likely, by reduced insect feeding (Han and Bauce, 2000). In addition to environmental factors, the presence of secondary plant compounds on foliage has a significant impact on persistence of *Bt* on foliage. Specifically, volatiles such as aldehydes, ketones, carboxylic acids and their derivatives present on these leaves have an antibiotic effect on *Bt* spores and sometimes, are the cause of its inactivation (Ferry et al., 2004).

Field stability and persistence of formulated biopesticides can be enhanced by various approaches such as:

- **i.** Encapsulation of biopesticidal materials in a matrix for protection and sustained release of the toxin (Yu and Lee, 1997; Ramstack et al., 1997; Fowler and Feinstein, 1999).
- **ii.** Enhancing foliar retention using corn-starch-based formulations (Tamez-Guerra et al., 2000).
- **iii.** Use of genetically engineered *Pseudomonas fluorescens* strains that produce *Bt* endotoxins. *P. fluorescens* are efficient epiphytes and tolerant to hostile phyllosphere conditions as they possess polysaccharides, proteins, and glycoproteins that help them adhere to crop foliage (Burges, 1998).
- **iv.** Selection of *Bt* strains that are active over a wide range of temperatures and use of encapsulated formulations.

12.7.2 Entomopathogenic Viruses

The persistence, accumulation and denaturation of baculoviruses in the environment are critical factors in determining the successful use of these agents. Entomopathogens are highly susceptible to damage by desiccation, and by exposure to sunlight, or to ultraviolet (UV) radiation (Ignoffo and Batzar 1971; Battu and Ramakrishnan 1989). Formulations of entomopathogens need to be modified to minimize such effects in overall achievements for their better persistence over crop foliage so that pest larvae at various times get an opportunity to ingest their lethal inocula. Angus and Luthy (1971) listed various additives/adjuvants (such as charcoal, India ink, egg-albumin, molasses, and optical brightener) to be used along with formulations of various entomopathogens including baculoviruses.

According to Young and Yearian (1974), the persistence of *Heliothis* NPV was significantly better on tomato (up to 96h) than on soybean and cotton. Furthermore, they observed that persistence was 10 times greater on the calyx and on the inner surface of mature and terminal leaves. The half-life at unprotected sites was 24h, at protected leaf sites 24–48h, and at protected floral sites 86h. Exposure of 0 to 24h could not inactivate the viral potency. It, however, declined drastically with relatively higher subsequent sunshine exposure of 36, 48, 60, 72, 84 and 96h as was evident from respective 96.7, 80.2, 66.5, 55.5, 30.0 and 10.0% observed larval mortality of *H. armigera* in bioassays of residual viral (HaNPV) deposits (Kaushik, 1991). On soybean foliage, Ignoffo et al. (1974) observed the half-life of *Heliothis* NPY to be 2–3 days, while its persistence was detected even after 14 days exposure. Half-life values for the NPY alone and the virus when used with soybean and cotton seed adjuvants were 1.8, 3.5–4.3 and 6.0 days, respectively against *H. zea* on soybean foliage (Smith and Hostetter, 1982). Tuan et al. (1989) reported that weak alkaline dew (pH 8.1) inactivated HaNPV collected from soybean leaves. However, it remained active on the dew from maize, tomato, and asparagus (pH 7.2–7.3).

Heliothis NPV-bait formulations when used on cotton remained active for at least 6 days during hot, dry and sunny weather (McLaughlin et al., 1971). *Heliothis* NPV was known to lose its activity more rapidly on cotton foliage of which some activity was also lost at night. Young and Yearian (1974) reported most rapid inactivation of *Heliothis* NPV on cotton, with little activity remaining after 24h. Dhandapani et al. (1990) reported that addition of crude sugar (15%) to the HaNPV spray fluid increased the persistence of the virus both under natural sunlight and shade. Only low levels of HaNPV remained on sorghum heads at 4 days after application (Young and McNew, 1994). Hugar et al. (1996) also concluded that loss of effectiveness of an NPV of *Mythimna separata* (Walker) on sorghum foliage occurred mainly due to its rapid inactivation by sunlight exposure.

In North Indian conditions, an NPV of *Spilosoma obliqua* Walker lost a total of 33.3% to 50% of its original activity on sunflower foliage within a comparatively short exposure period between 4 and 12h. However, upon exposure to sunlight up to a period of 72h, the virus could still persist with 25% activity. At a maximum exposure period of 4 days, 75% of the activity was lost (Battu and Sidhu, 1992). In the case of groundnut foliage, on the other hand, the same virus, under similar exposure conditions, lost 70% of the original activity within 4 days (Battu and Bakhetia, 1992). The SINPV persisted on sunflower foliage for a period of 6 days with 6.6% of its original activity intact (Kaler, 1996). An evening spraying of NPV of *S. litura* significantly helped to minimize the photo-inactivation of this virus on cotton foliage in addition to allowing its greater ingestion by *S. litura* larvae. The same virus, however, has been reported to persist on banana crop for 1 day in Southern India (Santharam et al., 1978) although it could tolerate sunshine exposure with a severe loss in its virulence up to 8 days (Narayanan et al., 1977).

Certain substrates such as boric acid (Morales et al., 1997), chitinase (Shapiro et al., 1987), extracts of neem tree (Cook et al., 1996), and optical brighteners of the stilbene group (Shapiro, 1995) have enhanced baculovirus activity. Mixtures of baculoviruses with optical brighteners of the stilbene group seem to have excellent potential for use in formulated products because they can enhance viral activity at concentrations as low as 0.01%, reduce time to kill the host, and provide protection against UV solar radiation. These substances have enhanced the activity of NPVs of *A. californica, A. falcifera, A. gemmatalis, H. virescens, H. zea, L. disper*, S. *exigua* and *T. ni* (Shapiro and Argauer, 1997).

Argauer and Shapiro (1997) evaluated eight optical brighteners (Blankophor HRS, P167, BBH, RKH, BSU, DML, LPG and Tinopal LPW) of the stilbene group for their activity as virus enhancers. Five of the eight compounds acted as enhancers and the most active brighteners (BBH, RKH and LPW) reduced LC_{50} of Gypsy moth, *L. dispar* NPV by 800- to 1300-fold. The most effective compounds were those exhibiting the greatest fluorescence. The brightener acts on the insect midgut and has no effect on the virus *per se*. The virus and the optical brightener must be ingested. Within 48h the insects stop feeding, midguts are clear and the gut pH is greatly reduced. The brightener allows the virus to replicate in a non-permissive tissue (columnar cells of the midgut). More importantly, the host spectrum of the baculovirus can be expanded using these compounds. None of the components or derivatives of Tinopal LPW were found to be as active as the parent compound (Shapiro and Argauer, 1997).

12.8 STRAIN IMPROVEMENT

12.8.1 Entomopathogenic Bacteria

A natural occurring *Bt* strain may require genetic improvement before it can be formulated into an effective biopesticide. Any *Bt* strain can be improved to increase toxicity to and the range of target pests and to delay the onset of pest-resistance by including toxins that bind to different sites or have different modes of action. It is generally assumed that total bioactivity of a *Bt* strain is a function of additive and/or synergistic interactions of individual *Cry* proteins present in their proportional amounts, hence strain improvement has been attempted by increasing the copy number and type of *Cry* genes in a strain. Generation of trans-conjugants has been used for *Bt* strain improvement. The high level of expression of a trans-conjugant gene cry3A from *B.t.* subsp. *tenebrionis* was observed in *B.t.* subsp. *kurstaki* HD119 without affecting the native *Cry* gene expression of the latter (Gamel and Piot, 1992). New *Bt* strains with additional *Cry* genes over the native strains have been created through conjugation (Wiwat et al., 1995). In addition to conjugation, another strategy for strain improvement is integration of *Cry* genes into the chromosome of the desired recipient strain. A cry1Aa gene was transferred through phage CP-54 Ber-mediated transduction into *Bt* strains (Lecadet et al., 1992). A cry1C gene from *B.t.* subsp. *aizawai* was transferred into the chromosome of *B.t.* subsp. *kurstaki* HD73 by electroporation (Kalman et al., 1995). These chromosomally integrated *Cry* genes can then be easily transferred through transducing phage to other *Bt* strains to broaden their insecticidal spectrum.

With the ever-growing collection of *Cry* genes and the use of recombinant DNA technology, several *Cry* genes can be introduced and expressed to construct novel *Bt* strains with desired insecticidal activities. Cloned *Cry* genes can be maintained in *Bt* on recombinant plasmids or resident plasmids, or stably integrated into the chromosome by in vivo homologous recombination. To overcome the limitation of persistence in the environment due to hostile phyllosphere conditions, microbial encapsulation was performed by scientists at Mycogen Corporation (USA) by introducing *Cry* genes into a non-pathogenic *Pseudomonas fluorescens* strain which produced crystal proteins. The *P. fluorescens* had better persistence in open fields than their *Bt* counterparts. Similarly, Obukowicz et al. (1986), using transposon Tn5, transferred a *Cry* gene from *B.t.* subsp. *kurstaki* HD-1 to a corn root-colonizing *P. fluorescens* strain to develop pesticidal efficiency similar to that of *B.t.* subsp. *kurstaki* HD-1 against black cutworm (*Agrotis ipsilon*).

Modification for enhanced expression of *Cry* proteins has also been attempted. To counteract the development of pest-resistance, *Cry* proteins can be designed or created through protein engineering to generate *Bt* strains possessing improved insecticidal activity. Expression of *Cry* genes can be further enhanced by altering the regulatory elements in the gene. Furthermore, the yield of a *Cry* protein produced in limited amounts in a naturally occurring strain can be improved through recombinant DNA techniques by manipulation of the controlling elements such as the promoter of its gene, as has been done by Park et al. (1998). They used dual cyt1Aa promoters along with a STAB-SD sequence that stabilized the cry3A transcript-ribosome complex and which resulted in a many-fold increase in the expression of the *cry3Aa* gene.

12.8.2 Entomopathogenic Viruses

Establishment of improved cell lines is the main priority area for identification and development of new strains of baculoviruses. More than 200 cell lines have been established from approximately 70 species of insects. The majority of these cell lines have been described from Lepidoptera, Diptera, Orthoptera, Hemiptera, Coleoptera, and Hymenoptera. Many established cell lines from lepidopteran species have proved to be invaluable tools for the *in vitro* propagation of insect-pathogenic viruses. During the past decade and a half, significant progress has been made in understanding the replication and molecular biology of baculoviruses in cell culture, and these basic studies are providing the basis for understanding the future of virus–host interactions including pathogenicity, host range, virulence and latency (Granados et al., 1987).

More than 14 different multi nucleocapsid NPVs (MNPVs) including that of *A. californica* have been grown in different cell lines. In addition to AcMNPV, NPVs from *Bombyx mori* (Linnaeus), *L. disper*, and *S. frugiperda* grow readily in cell cultures, and are easily plaqued and should be amenable to genetic and molecular biological analysis (Miller, 1987). Until recently, the *H. zea* single nucleocapsid NPV (HzSNPV) was the only SNPV to have been grown in an established cell line. Many insect pathologists believed earlier that SNPVs might be more difficult to grow in cell cultures than MNPVs. However, at least three new SNPVs from *H. armigera* (SNPV) (Zhu and Zhang, 1985), *Orgyia leucostigma* (J.E. Smith) (SNPV) (Sohi et al., 1984) and *T. ni* (SNPV) (Granados et al., 1987) have been propagated *in vitro*.

Granados et al. (1987) established 36 new *T. ni* cell lines from embryonic tissues, 29 such lines supporting replication of *T. ni* SNPV, and it appeared that susceptibility of these lines to this virus was stable. All of the new cell lines were highly susceptible (>95% of cells infected) to AcMNPV infection and several were susceptible to *T. ni* granulosis virus (TnGV). The ability of many of these new cell lines to support the growth of different baculoviruses may be related to the types of tissues used to initiate the cultures. Before 1984, attempts to replicate GVs in primary organ cultures or established cell lines had met with minimal or no success. This was primarily due to the lack of cell viral receptors or missing host enzymes needed for replication. In Germany, Miltenburger et al. (1984) reported the first successful *in vitro* replication of *C. pomonella* GV (CpGV) in primary cell lines from *C. pomonella.* Another development was the successful establishment of several new *T. ni* cell lines, which were susceptible to TnGV (Granados et al., 1987). Even a total of 26 new *T. ni* embryonic cell lines, 15 different cell lines and three sub-lines were susceptible to TnGV as determined by the peroxidase-anti-peroxidase (PAP) assay. This implies that other new cell lines from different insect species could be developed for the growth of new GVs and their subsequent commercial exploitation to produce viral pesticides.

The ultimate goal of research in insect cell culture is the production of viral pathogens in large volume on a commercial scale. A number of satisfactory culture media have been developed for the growth of insect cells. Two methods of large volume cell culturing, i.e. attached cell culture and suspension cell culture from *S. frugiperda* in roller bottles and production of AcMNPV are well known. The principal advantage with these is the economy of space and labour compared to flask cultures (Battu and Arora, 1997; Battu et al., 1993, 1994). The significant achievement in the development of low cost protein-free media is bound to enable the production of viral pesticides (Rabindra and Rajasekaran, 1996). Finally, relatively simple and inexpensive procedures would be required to harvest the viral OBs or the infectious entities in the case of non-occluded baculoviruses.

12.8.2.1 Genetically Modified Baculoviruses for Insect Control

Biotechnology, through recombinant DNA technology, has provided the means of overcoming some of the shortcomings of naturally occurring baculoviruses, while maintaining or enhancing their desirable pest-specific characteristics. Nucleopolyhedro viruses have been genetically altered to enhance the speed with which they kill the target pest. This has been achieved through genetic manipulation of baculoviruses by exchange of genetic material between different baculoviruses or insertion of foreign genes into baculovirus genomes.

Recombinant baculoviruses are constructed in two stages due to the difficulty of manipulating the large genome directly. The foreign gene is incorporated initially into a baculovirus transfer vector. The gene-inserted plasmid is then propagated in the bacterium *Escherichia coli*. Most transfer vectors used are bacterial plasmid University of California (pUC) derivatives, which encode an origin of replication for propagation in *E. coli* and an ampicillin-resistance gene. The pUC fragment is ligated to a small segment of DNA taken from the viral genome. The foreign gene sequence is incorporated into a cloning site downstream of the promoter selected to drive expression. For the second step, the transfer vector is mixed with DNA from the wild-type baculovirus. The engineered DNA is incorporated into the virus via homologous recombination events within the nucleus of cultured insect cells. The baculovirus system allows the precise insertion of foreign DNA without disruption of other genes, unlike genetic engineering in plants, which results in a rather random incorporation of new DNA into the genome.

12.9 ADVANTAGES AND LIMITATIONS OF BACTERIAL AND VIRUS BIOPESTICIDES

12.9.1 Advantages

Similarly to other natural enemies, insect pathogens such as bacteria and viruses can yield considerable control of target populations. The natural populations of these entomopathogens can be augmented after selection, bioassays and improvement in potential strains. Biopesticides are key components of integrated pest management programmes and can reduce the overall insecticidal load on a food, feed or fibre crop significantly. The major advantage of usage of microbes as insecticides is that they are environmentally safe and the formulations of these entomopathogenic bacteria and viruses are easily degraded and do not leave any harmful residues. These can be easily incorporated into organic farming protocols and some of the Baculoviruses can be mass produced by simple cottage industries for use on a limited scale. As they are being adapted through co-evolution to very specific groups of insects, they are highly specific in action and do not cause toxicity or infection in other groups of animals such as birds, animals and humans. Because of their specificity, these biopesticides are regarded as environmentally friendly. In particular, for *Bt*, which works by binding to appropriate receptors on the surface of midgut epithelial cells, any organism that lacks the specific receptors in its gut cannot be affected by it and hence it is safe even to other beneficial arthropods such as pollinators. The effectiveness of these biopesticides can be compared to that of synthetic pesticides and even they can be more effective than chemical pesticides in the long-term.

12.9.2 Limitations

Although these biopesticides have various advantages over the use of chemical pesticides, they are more difficult to mass-produce, as they require specialized substrates for cultivation or even living host insects, hence costing more to produce. Biopesticide is more costly and less readily available than conventional pesticide and also involves more money and time spent obtaining it. Hence, farmers with large cropped areas may find it difficult to consistently use biopesticide. One of the benefits of a biopesticide is its high specificity; however, the greatest strength of a biopesticide is also its greatest weakness. If any pests other than those targeted by the biopesticide invade the crop, they will be immune. This implies that several types of biopesticides may be needed to manage all of the pests. The microbial pesticides are subjected to biotic and abiotic factors of the environment and thus have a finite lifespan where applied and therefore have variable efficacy against the field populations of target pests. Heat, desiccation, or exposure to ultraviolet radiation reduces the effectiveness of several types of microbial insecticides. Consequently, proper timing and application procedures are especially important for some products. And the constant exposure to a toxin creates evolutionary pressure for selection of pests against that toxin. Living organisms evolve and increase their resistance to biological, chemical, physical or any other form of control. If the target population is not exterminated or rendered incapable of reproduction, the surviving population can acquire a tolerance of whatever pressures are brought to bear, resulting in an evolutionary arms race. Moreover, any regular disruption of large insect communities, due to chemical or microbial insecticides, can have long-term deleterious effects on higher trophic levels and ecosystem structure. This may lead to the emergence of secondary pests, for example, as has been reported in the case of cotton where sucking pests have become a serious problem within a few years of adoption of *Bt* cotton. Once the primary pest is brought under control, secondary pests have a chance to emerge due to the lower pesticide applications in *Bt* cotton cultivars. In China, mirids have become a serious problem due to cultivation of *Bt* cotton, while in India, mealy bugs have gained prominence (Zhao et al., 2011).

12.10 BIOSAFETY ISSUES REGARDING USE OF MICROBIAL PESTICIDES

Before commercialization and delivery of biopesticides to agricultural environments, their biosafety, behaviour and impact on ecosystems have to be evaluated. Strict regulations for registration and commercial production of biopesticides are now being followed in many countries and require an in-depth analysis of the environmental impact of a biopesticide. Persistence of *Bt* in the environment is important from both ecological and economical points of view, as reviewed by Otvos and Vanderveen (1993). *Bt* can persist for longer times in soil and water, and although no evidence has been noted, the potential risk of genetic exchange with other related or unrelated bacteria in these ecological niches still remains. There is also a lack of evidence of a direct effect of *Bt* on soil water or plants, but people may always be concerned about the presence of bacteria in water supplies. Non-target organisms such as parasites, predators, and other invertebrates and vertebrates may be exposed to *Bt* either directly by encountering it in the environment (e.g. by eating sprayed leaves and litter) or indirectly, by eating caterpillars which have been infected with *Bt*. Temporary drops in the population of invertebrate parasites and predators that feed on *Bt*-infected insects have been noticed but the drop was primarily due to lack of food supply, rather than *Bt* toxicity. Various workers have indicated no direct effects of commercial formulations on non-target insects (Giroux et al., 1994). Although relatively higher concentrations of *Bt* may be toxic to non-target insects in the laboratory, the rate at which *Bt* formulations are used in the field is lower than for most of those used to initiate high mortality of adult parasites in laboratory experiments. Thus, the probability that adult parasites, pollinators and nectar feeding insects would consume lethal doses of *Bt* while searching for hosts and nectar within a *Bt*-sprayed field might be very low. There has been no documented evidence that *Bt* has any direct or indirect effect on human and animal health. Animals could be exposed to *Bt*-based insecticides through ingesting *Bt* on plants, infected insects, inhaling or dermal contact with *Bt* spray. However, the mode of action of *Bt* indicates that there are no concerns about dermal contact and inhalation in humans and animals. However, concerns always remain with regard to changes in the food chain due to *Bt* applications as they could apply environmental stresses on some non-target species that rely mainly on target insects for a food source.

Before the commercial release of biocontrol agents, several features have to be considered for an adequate assessment of their adverse effects on the ecosystem. For bacterial entomopathogens, their genetic stability and horizontal genetic transfer are the most important factors in addition to their effects on other microbiota and fauna. Development and validation of ecological models predicting the impact of a released biocontrol agent on strain level is also necessary.

12.11 CONCLUSIONS AND FUTURE DIRECTION

Ecofriendly pest and disease management practices are being developed and evaluated globally to reduce the health risks due to higher usage of chemical pesticides in agriculture. In light of this, entomopathogenic bacteria and viruses have wide scope as biocontrol agents in addition to being a very good source for scouting for insecticidal toxin genes. An important benefit of microbial control agents is that they can be used to replace, at least in part, some more hazardous chemical pest control agents. The selective toxicity of these entomopathogenic bacteria and viruses to major insect-pests and their safety to non-target organisms makes them ideal tools for use in integrated pest management (IPM) programmes. These positive trends, however, need to be accompanied by strengthening of research efforts to overcome some of the major limitations in production, use and efficacy of baculoviruses.

The relatively slow speed with which baculoviruses kill their hosts has hampered their effectiveness as well as acceptance by potential users. However, genetic improvement, using traditional methods as well as genetic engineering, may produce strains of baculoviruses with improved pathogenesis and virulence. Recent advances in virus production using insect cell lines offer a way out of this situation. Quality control of commercially produced microbial pesticides is another area requiring urgent attention. It is necessary to maintain the viability and virulence of the pathogens until use. The interaction of these entomopathogens with other methods of pest control should be thoroughly studied to develop stronger IPM strategies.

The future of microbial insecticides appears to be assured as a consequence of pest-resistance and environmental contamination with conventional insecticides. Also, advances in biotechnology should allow the production costs of biopesticides to decrease, in addition to increasing their efficacy. However, there is a pressing need to develop better formulations to enhance their efficacy as biopesticides in the field, although attempts have been made to genetically transform *P. fluorescens* with *Bt* toxin gene to gain better field performance. Many *Bt* subspecies have been registered, while many others have been described but not developed commercially so far. Hence, there is a lot of scope for identification of novel toxin genes from these strains and to further develop them into biopesticides. By using molecular techniques, a desired combination of *Cry* proteins can be pyramided into a particular *Bt* strain with well known safety and production potential, to create a genetically engineered strain to broaden its insecticidal activity and spectrum. *Cry* proteins can also be modified through protein engineering to increase toxicity and the insecticidal spectrum. Use of strong promoters and other regulatory elements can enhance the expression of *Cry* proteins. Further, in asporogenous *Bt* strains, *Cry* proteins are synthesized but viable spores are not made, thus offering an environmental advantage. The discovery of new biopesticidal agents is of utmost importance for tackling the problem of environmental degradation and pest-resistance development.

References

Adang, M.J., Staver, M.J., Rocheleau, T.A., Leighton, J., Barker, R.F., Thompson, D.V., 1985. Characterised full-length and truncated plasmid clones of the crystal protein of *Bacillus thuringiensis* subsp. *kurstaki* HD-73 and their toxicity to *Manduca sexta*. Gene 36, 289–300.

Angus, F.A., Luthy, P., 1971. Formulations of microbial insecticides. In: Burges, H.D., Hussey, N.W. (Eds.), Microbial Control of Insects and Mites. Academic Press, New York, pp. 623–628.

Argauer, R., Shapiro, M., 1997. Fluorescence and relative activities of stilbene optical brightness and enhancers for the Gypsy moth (Lepidoptera: Lymantriidae) baculovirus. J. Econ. Entomol. 90, 416–420.

Battu, G.S., 1999. Viral pathogenesis mediated development of microbial pesticides. In: National Symposium on Emerging Trends in Biotechnological Application for Integrated Pest Management, Chennai, pp. 3–4.

Battu, G.S., Arora, R., 1996. Genetic diversity of baculoviruses: Implications in insect pest management. In: Ananthakrishnan, T.N. (Ed.), Biotechnological Perspectives in Chemical Ecology of Insects. Oxford & IBH Publishing Co. Pvt. Ltd., New Delhi, pp. 179–200.

Battu, G.S., Arora, R., 1997. Insect pest management through microorganisms. In: Dadarwal, K.R. (Ed.), Biotechnological Approaches in Soil Microorganisms for Sustainable Crop Production. Scientific Publishers, Jodhpur, India, pp. 222–246.

Battu, G.S., Bakhetia, D.R.C., 1992. Foliar persistence of the nuclear polyhedrosis virus of *Spilosoma obliqua* (Walker) on groundnut. In: Proceedings of the National Symposium Recent Advances in Integrated Pest management, Ludhiana.

Battu, G.S., Ramakrishnan, N., 1989. Comparative role of various mortality factors in the natural control of *Spilosoma obliqua* (Walker) in Northern India. J. Ent. Res. 13, 38–42.

Battu, G.S., Sidhu R.S., 1992. Persistence of the nuclear polyhedrosis virus of *Spilosoma obliqua* on the sunflower foliage. In: Proceedings of the National Symposium in Recent Advances in Integrated Pest Management, Ludhiana, pp. 98–99.

Battu, G.S., Ramakrishnan, N., Dhaliwal, G.S., 1993. Microbial pesticides in developing countries: Current status and future potential. In: Dhaliwal, G.S., Singh, B. (Eds.), Pesticides: Their Ecological Impact in Developing Countries. Commonwealth Publishers, New Delhi, pp. 270–334.

Battu, G.S., Dhaliwal, G.S., Raheja, A.K., 1994. Biotechnology: Perspectives in insect pest management. In: Dhaliwal, G.S., Arora, R. (Eds.), Trends in Agricultural Insect Pest Management. Commonwealth Publishers, New Delhi, pp. 417–468.

Bauer, L.S., 1995. Resistance: a threat to the insecticidal crystal proteins of *Bacillus thuringiensis*. Fla. Entomol. 78, 414–442.

Beegle, C.C., Yamamoto, T., 1992. Invitation paper: history of *Bacillus thuringiensis* Berliner research and development. Can. Entomol. 124, 587–616.

Behle, R.W., Guire, M.R., Shasha, B.S., 1997. Effect of sunlight and simulated rain on residual activity of *Bacillus thuringiensis* formulations. J. Econ. Entomol. 90, 1560–1566.

Bernhard, K., Utz, R., 1993. Production of *Bacillus thuringiensis* insecticides for experimental and commercial uses. In: Entwistle, P.F., Cory, J.S., Bailey, M.J., Higgs, S. (Eds.), *Bacillus thuringiensis*, an Environmental Biopesticide: Theory and Practice. Wiley and Sons, New York, pp. 255–267.

Bernhard, K., Jarrett, P., Meadows, M., Butt, J., Ellis, D.J., Roberts, G.M., et al., 1997. Natural isolates of *Bacillus thuringiensis*: Worldwide distribution, characterization and activity against insect pests. J. Invertebr. Pathol. 70, 59–68.

Brar, S.K., Verma, M., Tyagi, R.D., Valéro, J.R., 2006. Recent advances in downstream processing and formulations of *Bacillus thuringiensis* based biopesticides. Process Biochem. 41, 323–342.

Brown, E.R., Mady, M.D., Treece, E.L., Smith, C.W., 1958. Differential diagnosis of *Bacillus cereus*, *Bacillus anthrax* and *Bacillus cereus* var. *mycoides*. J. Bacteriol. 75, 499–509.

Burges, H.D., 1998. Formulation of Microbial Biopesticides, Beneficial Microorganisms, Nematodes and Seed

Treatments. Kluwer Academic Publishers, Dordrecht, 412 pp.

Cook, S.P., Webb, R.E., Thorpe, K.W., 1996. Potential enhancement of the gypsy moth (Lepidoptera: Lymantriidae) nuclear polyhedrosis virus with the triterpene azadirachtin. Environ. Entomol. 25, 1209–1214.

Cunningham, J.C., 1995. Baculoviruses as microbial insecticides. In: Reuveni, R. (Ed.), Novel Approaches to Integrated Pest Management. CRC Press, Inc., Boca Raton, FL, pp. 261–292.

Dhandapani, N., Jayaraj, S., Rabindra, R.J., 1990. Influence of sunlight and crop improvement on the efficacy of nuclear polyhedrosis virus against *Heliothis armigera* (Hubner) on groundnut (*Arachis hypogaea* L.). J. Appl. Entomol. 116, 523–526.

Dulmage, H.T., 1981. Insecticidal activities of isolates of *Bacillus thuringiensis* and their potential for pest control. In: Burges, H.D. (Ed.), Microbial Control of Pests and Plant Diseases. Academic Press, London, pp. 191–220.

Evans, H.F., 1986. Ecology and epizootiology of baculoviruses. In: Granados, R.R., Federici, B.A. (Eds.), The Biology of Baculoviruses. Vol. II. Practical Application for Insect Control. CRC Press, Boca Raton, FL, pp. 89–132.

Evans, J., 2008. Biopesticides: from cult to mainstream. Agrow, October 2008, 11–14.

Federici, B.A., 1999. A perspective on pathogens as biological control agents for insect pests. In: Bellows, T.S., Fischer, T.W. (Eds.), Handbook of Biological Control: Principles and Applications of Biological Control. Academic Press, CA, pp. 517–548.

Ferrar, R.R., Ridway, R.L., 1995. Enhancement of activity of *Bacillus thuringiensis* Berliner against four lepidopterous pests by nutrient based phagostimulants. J. Entomol. Sci. 30, 29–42.

Ferry, N., Edwards, M.G., Gatehouse, J.A., Gatehouse, A.M., 2004. Plant-insect interactions: molecular approaches to insect resistance. Curr. Opin. Biotechnol. 15, 155–161.

Flexner, J.L., Belnavis, D.L., 2000. Microbial insecticides. In: Rechcigl, J.E., Rechigl, N.A. (Eds.), Biological and Biotechnological Control of Insect Pests. Lewis Publishers, Boca Raton, FL, pp. 35–61.

Fowler, J.D., Feinstein, B.E., 1999. Insecticidal matrix and process for preparation thereof. US Patent 5,885,603.

Francki, R.I.B., Fauquet, C.M., Knudson, D.L., Brown, F., 1991. Classification and nomenclature of viruses: Fifth Report of the International Committee on Taxonomy of Viruses. Arch. Virol. 2, 117–123.

Gamel, P.H., Piot, J.C., 1992. Characterization and properties of a novel plasmid vector for *Bacillus thuringiensis* displaying compatibility with host plasmids. Gene 120, 17–26.

Gibson, D.M., Greenspan, L.G., Krasnoff, S.B., Ketchum, R.E.B., 1995. Increased efficacy of *Bacillus thuringiensis* subsp. *kurstaki* in combination with tannic acid. J. Econ. Entomol. 88, 270–277.

Gill, S.S., Cowles, E.A., Pietrantonio, P.V., 1992. The mode of action of *Bacillus thuringiensis* δ-endotoxins. Annu. Rev. Entomol. 37, 615–636.

Giroux, S., Coderre, D., Vincent, C., Côté, J.C., 1994. Effects of *Bacillus thuringiensis* var. *san diego* on predation effectiveness, development and mortality of *Coleomegilla maculata lengi* (Col.: Coccinellidae) larvae. Entomophaga 39, 61–69.

Granados, R.R., Dwyer, K.G., Derksen, A.C.G., 1987. Production of viral agents in invertebrate cell cultures. In: Maramorosch, K. (Ed.), Biotechnology in Invertebrate Pathology and Cell Culture. Academic Press, San Diego, USA, pp. 167–181.

Grzywacz, D., McKinley, D., Jones, K.A., Moawad, G., 1997. Microbial contamination in *Spodoptera litura* nuclear polyhedrosis virus produced in insects in Egypt. J. Invertebr. Pathol. 69, 151–156.

Han, E.N., Bauce, E., 2000. Dormancy in the life cycle of the spruce budworm: physiological mechanisms and ecological implications. Rec. Res. Dev. Entomol. 3, 43–54.

Hansen, B.M., Damgaard, P.H., Eilenberg, J., Pederson, J.C., 1998. Characterization of *Bacillus thuringiensis* isolated from leaves and insects. J. Invertebr. Pathol. 71, 106–114.

Hugar, H., Kulkarni, K.A., Lingappa, S., Hugar, P., 1996. Persistence of NPV of *Mythimna separate* (Walker) on sorghum foliage. Karnataka J. Agric. Sci. 9, 51–55.

Ignoffo, C.M., 1981. Living microbial insecticides. In: Norris, J.R., Richmond, M.H. (Eds.), Essays in Applied Microbiology. John Wiley & Sons, New York, pp. 2–31.

Ignoffo, C.M., 1992. Environmental factors affecting persistence of entomo-pathogens. Fla. Entomol. 75, 516–525.

Ignoffo, C.M., Batzer, O.F., 1971. Microencapsulation and ultraviolet protectants to increase sunlight stability of an insect virus. J. Econ. Entomol. 64, 850–853.

Ignoffo, C.M., Couch, T.L., 1981. The nucleopolyhedrosis virus of *Heliothis* species as a microbial insecticide. In: Burges, H.D. (Ed.), Microbial Control of Pests and Plant Diseases 1970–1980. Academic Press, New York, pp. 329–362.

Ignoffo, C.M., Hostetter, D.L., Pinell, R.E., 1974. Stability of *Bacillus thuringiensis* and *Baculovirus heliothis* on soybean foliage. Environ. Entomol. 3, 117–119.

Itoua-Apoyolo, C., Drif, L., Vassal, J.M., De Barjac, H., Bossy, J.P., Leclant, F., et al., 1996. Isolation of multiple species of *Bacillus thuringiensis* from a population of the European Sunflower moth, *Homoeosoma nebulella*. Appl. Environ. Microbiol. 61, 4343–4347.

Janmaat, A.F., Ware, J., Myers, J., 2007. Effects of crop type on *Bacillus thuringiensis* toxicity and residual activity against *Tricopulsiani* in greenhouses. J. Appl. Entomol. 31, 333–337.

Kaler, D., 1996. Utilization of nuclear polyhedrosis virus and *Bacillus thuringiensis* Berliner against *Spodoptera litura* (Fabricius). M.Sc. Thesis, Punjab Agricultural University, Ludhiana.

Kalman, S., Kiehne, K.L., Cooper, N., Reynoso, M.S., Yamamoto, T., 1995. Enhanced production of insecticidal proteins in *Bacillus thuringiensis* strains carrying an additional crystal protein gene in their chromosomes. Appl. Environ. Microbiol. 61, 3063–3068.

Kaur, S., Singh, A., 2000a. Distribution of *Bacillus thuringiensis* isolates in different soil types from North India. Indian J. Ecology. 27, 52–60.

Kaur, S., Singh, A., 2000b. Natural occurrence of *Bacillus thuringiensis* in leguminous phylloplanes in the New Delhi region of India. World J. Microbiol. Biotechnol. 16, 679–682.

Kaushik, H.D., 1991. Studies on nuclear polyhedrosis virus of *Heliothis armigera* (Hubner) on tomato (*Lycopersicon esculentum* Miller), Ph.D. Dissertation, Haryana Agricultural University, Hisar.

Kennedy, J.S., Rabindra, R.J., Sathiah, N., Gryzwacz, D., 1999. The role of standardization and quality control in the successful promotion of NPV insecticides. In: Ignacimuthu, S., Sen, A. (Eds.), Biopesticides and Insect Pest Management. Phoenix Publishing House Pvt. Ltd., New Delhi, pp. 170–174.

King, L.A., Possee, R.D., Hughes, D.S., Atkinson, A.E., Palmer, C.P., Marlow, S.A., et al., 1994. Advances in insect virology. Adv. Insect Physiol. 25, 1–73.

Kolodny-Hirsch, D., Dimock, M., 1996. Commercial development and use of Spod-X, a wild type baculovirus insecticide for beet armyworm. Paper presented at 29th Annual Meeting of Society of Invertebrate Pathology and III Colloquium on Bacillus thuringiensis, Cordoba, Spain.

Lacey, L.A., Mulla, M.S., 1990. Safety of *Bacillus thuringiensis* (H-14) and *Bacillus sphaericus* to non-target organisms in the aquatic environment. In: Laird, M., Lacey, L.A., Davidson, E.W. (Eds.), Safety of Microbial Insecticides. CRC Press, Boca Raton, FL, pp. 169–188.

Lacey, L.A., Frutos, R., Kaya, H.K., Vail, P., 2001. Insect pathogens as biological control agents: Do they have a future? Biol. Control. 21, 230–248.

Lambert, B., Peferoen, M., 1992. Insecticidal promise of *Bacillus thuringiensis*. Facts and mysteries about a successful biopesticide. Bioscience 42, 112–122.

Lecadet, M.M., Chaufaux, J., Ribier, J., Lereclus, D., 1992. Construction of novel *Bacillus thuringiensis* strains with different insecticidal specificities by transduction and by transformation. Appl. Environ. Microbiol. 58, 840–849.

McCoy, C.W., Samson, R.A., Boucias, D.G., 1988. Entomogenous fungi. In: Ignoffo, C.M., Mandava, N.B. (Eds.), Handbook of Natural Pesticides, Vol. V: Microbial Insecticides, Part A: Entomogenous Protozoa and Fungi. CRC Press, Boca Raton, FL, pp. 151–236.

McLaughlin, R.E., Andrews, G.L., Bell, M.R., 1971. Field tests for control of *Heliothis* spp. with a nuclear polyhedrosis virus included in a boll weevil bait. J. Invertebr. Pathol. 18, 304.

Miller, L.K., 1987. Expression of foreign genes in insect cells. In: Maramorosch, K. (Ed.), Biotechnology in Invertebrate Pathology and Cell Culture. Academic Press, San Diego, CA, pp. 295–304.

Miltenburger, H.G., Naser, W.L., Harvey, J.P., 1984. The cellular substrate: a very important requirement for baculovirus *in vitro* replication. Z. Naturforsch. Biosci. 39, 993–1002.

Morales, L., Moscardi, F., Sosa-Gomez, D.R., Paro, F.E., Soldorio, I.L., 1997. Enhanced activity of *Anticarsia gemmatalis* Hub (Lepidoptera: Noctuidae) nuclear polyhedrosis virus by boric acid in the laboratory. Ann. Soc. Entomol. 26, 115–120.

Moscardi, F., 1999. Assessment of the application of baculoviruses for the control of Lepidoptera. Annu. Rev. Entomol. 44, 257–289.

Murphy, F.A., Faquet, C.M., Bishop, D.H.L., Gabrial, S.A., Jarvis, A.W., Martelli, G.P. (Eds.), 1995. Classification and Nomenclature of Viruses: Sixth Report of the International Committee on Taxonomy of Viruses. Springer-Verlag, Berlin.

Narayanan, K., Govindarajan, R., Jayraj, S., 1977. Preliminary observations on the persistence of nuclear polyhedrosis virus of *Spodoptera litura* (F.). Madras Agric. J. 64, 487–488.

Obukowicz, M.G., Perlak, F.J., Kusano-Kretzmer, K., Mayer, E.J., Watrud, L.S., 1986. Integration of the delta endotoxin gene of *Bacillus thuringiensis* into the chromosome of root colonizing strains of pseudomonads using Tn5. Gene 45, 327–331.

Otvos, I.S., Vanderveen, S., 1993. Environmental report and current status of *Bacillus thuringiensis* var. *kurstaki*. Use for control of forest and agricultural insect pests. British Columbia Forestry Canada Rep.

Park, H.W., Ge, B., Bauer, L.S., Federici, B.A., 1998. Optimization of cry3A yields in *Bacillus thuringiensis* by use of sporulation dependent promoters in combination with the STAB-SDS mRNA sequence. Appl. Environ. Microbiol. 64, 3932–3938.

Pusztai, M., Fast, P., Gringorten, L., Kaplan, H., Lessard, T., Carey, P.R., 1991. The mechanism of sunlight mediated inactivation of *Bacillus thuringiensis* crystals. Biochem. J. 273, 43–47.

Quinlan, R.J., 1990. Registration requirements and safety considerations for microbial pest control agents in the European Economic Community. In: Laird, M., Lacey, L.A., Davidson, E.W. (Eds.), Safety of microbial pesticides. CRC Press, Boca Raton, FL, pp. 11–18.

Rabindra, R.J., Rajasekaran, B., 1996. Insect cell cultures: A tool in the basic research, biotechnology and pest control. In: Anathakrishnan, T.N. (Ed.), Biotechnological Perspectives in Chemical Ecology of Insects. Oxford & IBH Publishing Co. Pvt. Ltd., New Delhi, pp. 223–239.

Ragaei, M., 1990. Studies on the effect of *Bacillus thuringiensis* on the greasy cutworm *Agrotis ipsilon* (Rott.). Ph.D. Thesis, University of Cairo.

Ramstack, J.M., Herbert, P.F., Strobel, J., Atkins, T.J., 1997. Preparation of biodegradable microparticles containing a biologically active agent. US Patent 5,650,173.

Santharam, G., Regupathy, A., Easwaramoorthy, S., Jayaraj, S., 1978. Effectiveness of nuclear polyhedrosis virus against field populations of *Spodoptera litura* (F.) on banana *Musca paradisica* L. Indian J. Agric. Sci. 48, 676–678.

Sarvjeet, K., 2000. Molecular approaches towards development of novel *Bacillus thuringiensis* biopesticides. World J. Microbiol. Biotechnol. 16, 781–793.

Sathiah, N., Jayaraj, S., 1996. Technology for mass production of biopesticides. Silver Jubilee Seminar on Employment Opportunities for Biologists, Chennai, India.

Shapiro, M., 1995. Radiation protection and activity enhancement of viruses. In: Hall, F.R., Barray, J.W. (Eds.), Biorational Pest Control Agents: Formulation and Delivery. American Chemical Society, Washington, DC, USA, pp. 153–164.

Shapiro, M., Argauer, R., 1997. Components of the stilbene brightener Tinopal LPW as enhancers for the Gypsy moth (Lepidoptera: Lymantriidae) baculovirus. J. Econ. Entomol. 90, 899–904.

Shapiro, M., Preisler, H.K., Robertson, J.L., 1987. Enhancement of baculovirus activity on gypsy moth (Lepidoptera: Lymantriidae) by chitinase. J. Econ. Entomol. 85, 1120–1124.

Smith, D.B., Hostetter, D.L., 1982. Laboratory and field evaluations of pathogen-adjuvant treatments. J. Econ. Entomol. 75, 472–476.

Smith, R.A., Couche, G.A., 1991. The phylloplane as a source of *Bacillus thuringiensis*. Appl. Environ. Microbiol. 57, 311–315.

Sohi, S.S., Percy, J., Arif, B.M., Gunningham, J.C., 1984. Replication and serial passage of a singly enveloped baculovirus of *Orgyia leucostigma* in homologous cell lines. Intervirology 21, 50–60.

Tamez-Guerra, P., McGuire, M.R., Behle, R.W., Shasha, B.S., Galan-Wong, L.J., 2000. Assessment of microencapsulated formulations for improved residual activity of *Bacillus thuringiensis*. J. Econ. Entomol. 93, 219–225.

Teakle, R.E., 1994. Virus control of *Heliothis* and other key pests: potential and use, and the local scene. In: C.J. Monsour, S. Reid and R.E. Teakle, (Eds.), Proceedings of the First Symposium on Biopesticides: Opportunities for Australian Industry, University of Australia, Brisbane, pp. 51–56.

Theunis, W., Aguda, R.M., Cruz, W.T., Decock, C., Peferoen, M., Lambert, B., et al., 1998. *Bacillus thuringiensis* isolates from the Philippines. Habitat distribution, d-endotoxin diversity and toxicity to rice stem borers (Lepidoptera: Pyralidae). Bull. Entomol. Res. 88, 335–342.

Tomlin, C.D.S. (Ed.), 1997. The Pesticide Manual, 11th ed. British Crop Protection Council, Farnham, Surrey.

Travers, R., Martin, P., Reichelderfer, C., 1987. Selective process for efficient isolation of soil *Bacillus* spp. Appl. Environ. Microbiol. 53, 1263–1266.

Tuan, S.J., Tang, J.C., Hov, R.F., 1989. Factors affecting pathogenicity of NPV preparations to the corn earworm, *Heliothis armigera*. Entomophaga 34, 541–549.

Wiwat, C., Panbangred, W., Mongkolsuk, S., Pantuwatana, S., Bhumiratana, A., 1995. Inhibition of a conjugation-like gene transfer process in *Bacillus thuringiensis* subsp. *Israelensis* by the anti-S-layer protein antibody. Curr. Microbiol. 30, 69–75.

Young, S.Y., McNew, R.W., 1994. Persistence and efficacy of four nuclear polyhedrosis viruses for corn earworm (Lepidoptera: Noctuidae) on heading grain sorghum. J. Entomol. Sci. 29, 370–380.

Young, S.Y., Yearian, W.C., 1974. Persistence of *Heliothis* NPV on foliage of cotton, soybean and tomato. Environ. Entomol. 3, 253–255.

Yu, J.Y., Lee, W.C., 1997. Microencapsulation of pyrrolnitrin from *Pseudomonas cepacia* using gluten and casein. J. Ferment. Bioeng. 84, 444–448.

Zhao, J.H., Ho, P., Azadi, H., 2011. Benefits of Bt cotton counterbalanced by secondary pests? Perceptions of ecological change in China. Environ. Monit. Assess. 173, 985–994.

Zhu, G., Zhang, H., 1985. The multiplication characteristics of *Heliothis armigera* in the established cell lines. Abstr. 3rd Int. Cell Cult. Congr, p. 66.

CHAPTER

13

The Bioherbicide Approach to Weed Control Using Plant Pathogens

Karen L. Bailey

Saskatoon Research Centre, Saskatchewan, Canada

13.1 ECONOMICALLY IMPORTANT WEEDS

WEEDS! Their baneful existence irks every gardener and agriculturalist. A weed is a plant growing where it is not wanted and in competition with a desirable, cultivated plant. In reality, weeds are more than just a nuisance; weeds have severe economic impacts and threaten the global food and natural ecosystems. Weeds compete with crops for moisture, nutrients, sunlight, and space. Weed seeds in harvested crops lower the quality, along with the monetary value. Weed outbreaks are relatively constant, occurring in the same fields year after year due to latent dormancy in the seed and the accumulation of weed seed banks in the soil (Gianessi and Sankula, 2003).

Crops vary in their ability to compete with weeds. Generally, crops such as flax or lentil are poor competitors, whereas corn and soybean are more competitive. Competitiveness is influenced by plant architecture, foliar light interception, and synchrony of weed-crop emergence. Broadleaved weeds are more competitive than grass weeds. For example, common cocklebur (*Xanthium pensylvanicum* Wallr.) reduced soybean yield by 80% at a density of 9 plants/m^2 whereas giant foxtail (*Setaria faberi* Herrm.) reduced yield by only 10% from 6 plants/m^2 (Gianessi and Sankula, 2003). Water-use efficiency is the amount of plant biomass obtained per unit water, and water-use by weeds is a competitive factor because it is equal to or greater than water-use by crops. The water-use efficiency values (mg plant dry weight/ml of water) in soybean ranged from 1.09 to 3.98 and the values for common weeds were 1.47–4.40 for common cocklebur and 8.65 (no range given) for smooth pigweed (*Amaranthus hybridus* L.) (Norris, 1996). Parker (2003) reported that lambsquarter (*Chenopodium album* L.) required 79 gallons of water to produce 1 pound of dry matter compared to corn and wheat which required only 42 and 67 gallons, respectively. If the water used by weeds is not replaced by rainfall or irrigation, then crop losses will occur.

Holm et al. (1977) compiled an inventory of the principal weeds in major crops distributed worldwide and categorized them into two groups: 'the 18 most serious weeds in the

D. P. Abrol (Ed): Integrated Pest Management.
DOI: http://dx.doi.org/10.1016/B978-0-12-398529-3.00014-2

approximate order in which they are troublesome to the world's agriculturalists' and the weeds that are 'troublesome for man in cultivated crops, pastures, and waterways' (Table 13.1). It has been estimated that 227 weed species are responsible for 90% of crop losses in world agriculture (Riches, 2001). The loss in attainable production of rice, wheat, barley, maize, potatoes, soybeans, cotton and coffee due to weeds is about $76.3 billion worldwide (Bhowmik, 1999). Not all weeds are equally distributed throughout the world as some are more regional in nature, whereas others are more associated with specific crops. For example, yield losses in cotton, rice, and maize are estimated to be 18–20% in developing countries, whereas the losses are only 9–10% in the industrialized nations (Terry, 1996). Left uncontrolled, natural infestations of Russian thistle (*Salsola iberica* Sennen & Pau) reduced spring wheat yield up to 50% (Young, 1988). Soybean yields were reduced by 60% from season-long, high density infestations of common ragweed (*Ambrosia artemisiifolia* L.) and smartweed (*Polygonum pensylvanicum* L.) (Coble and Ritter, 1978; Coble et al., 1981). A season-long density of two weeds per 30 cm of corn row reduced yield by 10% from giant foxtail, 11% from lambsquarters (*Chenopodium album* L.), and 22% from common cocklebur (*Xanthium strumarium* L.) (Beckett et al., 1988). Weeds are clearly pests that threaten our global food supply if left unchecked and require some management to preserve adequate yield and quality for a consistent food supply.

13.2 CHANGING SOCIETAL VIEWS TO CONVENTIONAL WEED CONTROL PRACTICES

Aside from cultivation and crop rotation as practices used to control weeds, the use of synthetic herbicides became very common over the past 60 years due to the ease of application, high efficacy and low cost. The rapid adoption of herbicides such as 2,4-D (2,4-dichloro phenoxy acetic acid) and glyphosate have dominated all other control practices used on 80–100% of all major crops (Bhowmik, 1999). In western Canada, 2,4-D was applied to only 40 ha in 1946 and by 1962, this exceeded 10 million ha (Holm and Johnson, 2009). On a global scale, 44% of all pesticides sold are herbicides, but in countries such as Canada and the USA, agriculture is the primary herbicide market accounting for 80% of all pesticides sold (Bailey and Mupondwa, 2006; Bailey et al., 2009; Fishel, 2007). The economic impact of not having 2,4-D to use in the USA is estimated to be $1.6 billion resulting from 37% higher weed control costs, 36% from decreased yield, and 27% higher commodity prices (Bhowmik, 1999).

Aside from the economic issues, there are also environmental issues to consider with the use of herbicides. The use of herbicides with conservation tillage or no-till systems has greatly reduced soil erosion and surface runoff in soils around the world, providing an important environmental benefit (Anderson and Lafond, 2010). Yet herbicide use may be environmentally detrimental when it results in the build-up of herbicide-resistant weed populations which may occur from overuse of a specific herbicide and poor cropping system diversity (Beckie, 2009; Heap, 1997). Some herbicides are persistent in soil, such as the photosynthetic inhibitors of Group 5 and amino acid inhibitors of Group 2 (Holm and Johnson, 2009). Additionally, poor application technique may result in drift contaminating watersheds and other non-target areas (Wolf, 2009).

A third consideration with regard to herbicide use involves the urban–rural interface where municipal boundaries encroach upon the domain of agriculture, making the general public more aware of the practices being used and raising concerns about soil-water quality and residues in the food (Bailey et al., 2010). As a result, the demand for pesticide-free

TABLE 13.1 The World's Worst Weeds

Weed Species	Family	Impact	Weed Species	Family	Impact
GROUP 1: THE 18 MOST SERIOUS WEEDS IN AGRICULTURE					
Cyperus rotundus L.	Sedge	52 crops, 92 countries	*Chenopodium album* L.	Goosefoot	40 crops, 47 countries
Cynodon dactylon L. Pers.	Grass	40 crops, 80 countries	*Digitaria sanguinalis* L. Scop.	Grass	33 crops, 56 countries
Echinochloa crus galli L. Beauv.	Grass	36 crops, 61 countries	*Convolvulus arvensis* L.	Morning glory	32 crops, 44 countries
Echinochloa colonum L. Link	Grass	35 crops, 60 countries	*Avena fatua* L.	Grass	20 crops, 55 countries
Eleusine indica L. Gaertn.	Grass	46 crops, 60 countries	*Amaranthus hybridus* L.	Amaranth	27 crops, 27 countries
Sorghum halepense L. pers.	Grass	30 crops, 53 countries	*Amaranthus spinosus* L.	Amaranth	28 crops, 44 countries
Imperata cylindrica L. Beauv.	Grass	35 crops, 73 countries	*Cyperus esculentus* L.	Sedge	21 crops, 30 countries
Eichhornia crassipes (Mart) Solms.	Pickerel weed	Aquatic Tropics	*Paspalum conjugatum* Berg.	Grass	25 crops, 30 countries
Portulaca oleracea L.	Purslane	45 crops, 85 countries	*Rottboellia exaltata* L.f.	Grass	18 crops, 28 countries
GROUP 2: TROUBLESOME IN CULTIVATED CROPS, PASTURES, AND WATERWAYS					
Ageratum spp.	Aster	36 crops, 46 countries	*Lolium temulentum* L.	Grass	14 crops, 38 countries
Agropyron repens L. Beauv.	Grass	32 crops, 40 countries	*Mikania cordata* Burm. Robins.	Aster	10 crops, 23 countries
Anagallis arvensis L.	Primrose	22 crops, 39 countries	*Mimosa* spp.	Mimosa	22 crops, 38 countries
Argemone mexicana L.	Poppy	15 crops, 30 countries	*Panicum maximum* Jacq.	Grass	20 crops, 42 countries
Axonopus compressus Sw. Beauv	Grass	13 crops, 27 countries	*Panicum repens* L.	Grass	19 crops, 27 countries
Bidens pilosa L.	Aster	31 crops, 40 countries	*Paspalum dilatatum* Poir.	Grass	14 crops, 28 countries
Brachiaria mutica Forsk.	Grass	23 crops, 34 countries	*Pennisetum clandestinum* Hochs.	Grass	14 crops, 36 countries
Capsella bursa-pastoris L. Medic.	Mustard	32 crops, 50 countries	*Pennisetum purpureum* Schum.	Grass	9 crops, 25 countries

(Continued)

TABLE 13.1 (Continued)

Weed Species	Family	Impact	Weed Species	Family	Impact
Cenchrus echinatus L.	Grass	18 crops, 35 countries	*Phragmites australis* Cav. Trin.	Grass	Many crops, worldwide
Ceratophyllum demersum L.	Hornwort	Aquatic	*Pistia stratiotes* L.	Arum	Aquatic areas
Chromolaena odorata L. RM King	Aster	13 crops, 23 countries	*Plantago* spp.	Plantago	Many crops, worldwide
Cirsium arvense L. Scop.	Aster	27 crops, 37 countries	*Polygonum convolvulus* L.	Buckwheat	25 crops, 41 countries
Commelina benghalensis L.	Spiderwort	25 crops, 28 countries	*Rumex* spp.	Buckwheat	16 crops, 37 countries
Cyperus difformis L.	Sedge	5 crops, 46 countries	*Salvinia auriculata* Aublet	Salvinia	Aquatic, 22 countries
Cyperus iria L.	Sedge	17 crops, 22 countries	*Setaria verticillata* L. Beauv.	Grass	18 crops, 38 countries
Dactylotenium aegyptium L. Beauv.	Grass	19 crops, 45 countries	*Setaria viridis* L. Beauv.	Grass	29 crops, 35 countries
Digitaria adscendens Henr.	Grass	22 crop, 19 countries	*Sida acuta* Burm.	Mallow	20 crops, 30 countries
Digitaria scalarum Schweinf. Chiov.	Grass	All crops, East Africa	*Solanum nigrum* L.	Nightshade	37 crops, 61 countries
Eclipta prostrata L. L.	Aster	17 crops, 35 countries	*Sonchus oleraceus* L.	Aster	Row crops, 56 countries
Equisetum arvense L.	Horsetail	Poaceae, World	*Spergula arvensis* L.	Pink	25 crops, 33 countries
Euphorbia hirta L.	Spurge	15 crops, 47 countries	*Sphenoclea zeylanica* Gaertn.	Sphenochlea	1 crop, 17 countries
Galinsoga parviflora Cav.	Aster	Many, 38 countries	*Stellaria media* L. Cyrill.	Pink	20 crops, 50 countries
Galium aparine L.	Madder	19 crops, 31 countries	*Striga lutea* Lour.	Figwort	Many crops, 35 countries
Heliotropium indicum L.	Borage	15 crops, 28 countries	*Tribulus terrestris* L.	Caltrop	21 crops, 37 countries
Lantana camara L.	Verbena	14 crops, 47 countries	*Xanthium strumarium* L.	Aster	8 crops, 39 countries

Adapted from Holm et al. (1977).

food is growing. More than 60% of Canadians interviewed (*N* = 1935) in a survey believed that it was less harmful to eat produce if biological control had been used for pest management instead of synthetic chemical pesticides (McNeil et al., 2010). In some countries, such as in Canada, various municipalities and provinces have banned the use of herbicides for cosmetic purposes in urban areas (Bailey et al., 2010). These changes in the public's attitude on the acceptance of herbicides (and all pesticides) and the introduction of government policies for pesticide reduction, has presented an opportunity for the development of new weed control technologies that have reduced risks and are suitable for organic food production.

13.3 WHAT ARE BIOHERBICIDES?

Broadly defined, bioherbicides are weed control products that are derived from living organisms, including any natural products they produce during their growth, that suppress weed populations (Bailey et al., 2010; Glare et al., 2012; Kiewnick, 2007). The biological origins of most bioherbicides are microbial (bacteria, fungi, virus, nematodes), plant-derived products (corn gluten meal), or minerals (oils). This discussion focuses on microbials as the active ingredient in a bioherbicide.

Bioherbicides are considered to be a type of inundative biological control, meaning that the methods of application and situations for use are very similar to those for conventional herbicides. Ideally, one or two applications of a bioherbicide would provide season-long weed control but the following year, additional applications must be made again. Bioherbicides may be used in natural settings such as pastures, roadsides, and forests, as well as in cultivated situations such as turfgrass, orchards, and row crops. They are applied as granules or sprays using traditional pest control application technology. Hence, they are herbicides with biological origins.

Bioherbicides are considered by regulatory authorities as pest control products that have reduced risk over conventional herbicides (Bailey et al., 2010; Environmental Protection Agency (EPA), 2012a). There are several key features that help to make bioherbicides less risky (Table 13.2). Bioherbicides target specific weeds in specific situations and do not cause harm to crops in which they are applied. They may target a single weed or multiple weeds, but the potential effects on non-target hosts are well understood and are managed through the biology of the bioherbicide and restrictions associated with their application. For example, *Phoma macrostoma* Montagne was evaluated for causing damage and mortality to 94 species including non-target plants and target weeds from 34 plant families (Bailey et al., 2011a). Based on this information, several prospects for weed control in certain agricultural crops, turfgrass and ornamentals, and agro-forestry were identified. Presently, it has received registration in Canada and the US for controlling a number of broadleaved weeds in turfgrass (Pest Management Regulatory Agency (PMRA) 2011b; EPA, 2012b). Additional research is being done to collect data to support an agricultural registration (Bailey et al., 2013).

Bioherbicides are effective but they are not intended to be used as 'stand alone' products as they are most effective when incorporated into integrated weed management programmes considering various cultural and agronomic practices. For example, there was a 10–15% improvement in dandelion control when *P. macrostoma* was applied with commercial fertilizer to lawns compared to unfertilized lawns (Bailey et al., 2013). Efficacy is also strongly influenced by local environmental parameters such as temperature, moisture, and soil type. *Colletotrichum gloeosporioides* (Penzig) Saccardo f.sp. *malvae* controlled round-leaved mallow (*Malva pusilla* Smith) by 90% in 13 trials and 60% in three trials, but exhibited no control in four trials that had less than optimal temperature,

TABLE 13.2 Key Features of a Bioherbicide

Key Features	Description	Comment
Host specificity	May be highly host specific or very broad spectrum, depending on the weed and circumstances; but effects on non-targets and how the bioherbicide could contact non-targets must be understood	Weeds are unlike other pests as there are usually multiple weed species within an area that need control; the removal of one species provides an opportunity for another weed to immediately replace it
Crop tolerance	No effect on crops or other plants growing in the area of application	Part of non-target assessment and specific use patterns; helps to restrict how the bioherbicide may be used and reduce risk to non-targets
Efficacy	Control is greater than 80% weed reduction and suppression shows 60–80% weed reduction	Efficacy may be determined by mortality, biomass, seedling germination, root and shoot growth, physical symptoms
Environmental fate	Low dispersion, persistence limited to the growing season only with no carry over between seasons, and limited reproduction and survival	These traits impose a natural biological containment system reducing the risk of spread, altering the background soil biota, and limiting genetic-based changes
Temperature and moisture spectrum	Should mimic the optimal conditions for weed growth	Mesophyllic microbes should not have growth beyond 37°C to ensure non-infectious to humans and other mammals
Mode of action	Multiple modes	Usually a combination of physical and chemical origins; helps to reduce risk of developing weed resistance
Toxicology	Low toxicity, low re-entry time, no-harvest interval	Required by regulatory authorities; specific tests to assess infectivity, pathogenicity, and toxic or mutagenic effects; usually conducted by approved independent 3rd party contractors to prevent bias

leaf-wetness, and inoculum concentration (Boyetchko et al., 2007).

The environmental fate of registered bioherbicides is usually short, thus resulting in lower exposure and reducing the risk of environmental pollution. Ideally, persistence is limited to one growing season with no carry over to the following year, there is minimal off-site movement or dispersion after application and only asexual reproduction occurs, limiting the genetic diversification of the organism. An example of a bioherbicide with these characteristics is *P. macrostoma* (Zhou et al., 2004, 2005). When these conditions are not met, more extensive testing is required. *Chondrostereum purpureum* (Pers.) Pouzar is a ubiquitous organism throughout North America infecting various tree species and other woody deciduous brush. It is a weak pathogen but has a tetrapolar mating system with multiple alleles at two mating type loci; thus it is capable of producing genetic recombinants. This fungus was being studied as a bioherbicide for controlling woody brush on roadsides and under power lines in remote forested areas (Hintz, 2007). During the regulatory consultation phase before registration, concerns were raised that inundative applications would increase the endemic populations of the fungus and that spore production would affect non-target hosts and pollute waterways. Scientific evidence concluded these fears were unfounded, with the exception of a label restriction to prevent application of the bioherbicide within 50 m of fruit trees and ornamentals which was of little concern since this product was only being used in remote, natural forested areas.

Biopesticides often have multiple modes of action which helps delay the development of weed resistance. The modes of action may be physical or chemical in origin including physical force, enzymatic degradation, and the production of toxins and growth regulators. For example, *Pseudomonas fluorescens* strain BRG 100 is a bacterium that inhibits root growth of annual grass weeds via the production of two cyclic lipodepsipeptides called pseudophomins A and B (Pedras et al., 2003). Interestingly, these compounds also had antifungal activity, but weed root inhibition was greater with pseudophomin A and antifungal activity was greater with pseudophomin B. Researchers studying *Alternaria cirsinoxia* Simmons & Mortensen could not discern whether cell penetration by the fungus was by physical or enzymatic means. However, the presence of the fungus in the plant cells initiated a strong defence reaction in the Canada thistle host (*Cirsium arvense* L. Scop.) which eventually prevented *A. cirsinoxia* from being developed into a bioherbicide (Bailey, 2004). Studies on the mode of action and infection process of *P. macrostoma* showed that this fungus produced phytotoxins that caused root inhibition and photobleaching of foliage (Graupner et al., 2003). Microscopic imaging also showed that the host recognized the presence of the pathogen by restricting pathogen growth to the root outer-epidermal layers in resistant hosts and allowing penetration to the root cortex in susceptible hosts (Bailey et al., 2011b).

To ensure safety, toxicology testing with bioherbicides includes pathogenicity, infectivity, irritation, and toxicity studies using mammals, birds, fish, arthropods and other insects, and non-target plants (Bailey and Mupondwa, 2006). The toxicology evaluation process is based on data gathered from specific tests conducted at arm's length by third party contractors and is followed by a risk assessment evaluation. In Canada and the USA, the scientific testing is based on maximum hazard levels and the risk assessment is independently done by regulatory scientists and not the registrant; in the EU, the assessment is based on interactions between the hazard and exposure, and the registrant provides the risk assessment summary to the regulator (Kiewnick, 2007). So far, all registered bioherbicides in North America have been found to have low toxicity, thus requiring only minimum restrictions with use (i.e. 4 h re-entry interval, no-harvest restrictions, and no residue tolerance limits) (EPA 2012a; Bailey and Falk, 2011).

13.4 FROM START TO FINISH: UNDERSTANDING THE DISCOVERY AND DEVELOPMENT PROCESS

Developing a bioherbicide from discovery to commercialization is a long journey that can take from 10 to 15 years and is filled with many hurdles that can impede progress along the way. It is a challenging field requiring multidisciplinary teamwork to tackle the scientific complexities of biological systems and, the interactions with the environment, and yet survive the rigorous scrutiny of business and market evaluations. There are significant costs associated with the development of a bioherbicide, such as conducting the basic science for product development, completing the toxicology and environmental studies for regulatory assessment, and then setting up the manufacturing process, with the monies all spent before having a product for sale. Therefore it is important to have a clear idea of the stages and steps required to logically and swiftly move through

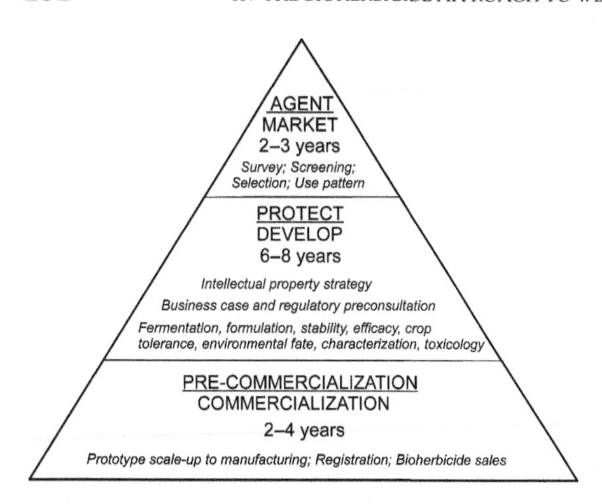

FIGURE 13.1 The discovery and development process of a microorganism culminating in the introduction of a commercial bioherbicide product for sale to the public.

the process. This section will summarize the bioherbicide innovation chain that was initially proposed by Bailey et al. (2009) and then modified with a Stage and Gate model used by industry (Bailey and Falk, 2011).

13.4.1 Discovery and Proof-of-Concept

The first stage in the process involves exploration, ideation, then testing the ideas to prove the concepts, and finally making the selection of the best candidate (Figure 13.1). Sounds simple enough, but is it? The process of deciding whether you begin with random exploration for an agent or have selected a predetermined market before exploration begins may be the first crucial step. It is the age old question of which comes first – the chicken or the egg (Figure 13.1)? For example, BioMal® was an accidental discovery of a fungus causing severe disease on a relatively minor weed (round-leaved mallow) that was not effectively controlled with the available chemical herbicides. So the fungus was developed and registered as a bioherbicide, but when it came time to sell the product, the market conditions had changed and it was no longer a viable business venture (Boyetchko et al., 2007). It has been argued that if the target pest is not a priority market then it does not matter whether the agent works or not because the cost of development is too high for small market niches (Boyetchko, 2005; Charudattan, 2005). Hallett (2005) suggested that some niche systems may work only because society places a high value on control of those weeds: parasitic weeds such as *Striga* spp., allergenic weeds such as ragweed, and narcotic weeds such as poppy. The market conditions are more important than the agent and the market must be large enough to support the recovery costs and return a profit (Bailey et al., 2009; Kiewnick, 2007).

After ideation, agent exploration, screening and selection may begin. Not all traditional plant pathogens will make a good bioherbicide; sometimes weaker pathogens or endophytes may have unique traits that are not apparent upon first examination. Collego™ is an example of a traditional pathogen being developed as a successful bioherbicide (Bowers, 1986; Templeton et al., 1989). The fungus, *Colletotrichum gloeosporioides* (Penz.) f.sp. *aeschynomene*, was a virulent pathogen of northern joint vetch (*Aeschynomene virginica* L.) in rice (*Oryza sativa* L.) causing severe necrosis and dieback. On the other hand, as previously mentioned, *C. purpureum* was a weak pathogen of alder but was made into an effective bioherbicide called Chontrol® (Hintz, 2007). *Phoma macrostoma* was initially isolated from leaf lesions and was assumed to be a foliar pathogen; however, it was not effective as a bioherbicide until it was applied to the soil where it then caused weed mortality (Bailey and Falk, 2011).

At the early stages of screening, Koch's postulates must be fulfilled for any microbial agent selected. Then the pathogen is evaluated for growth and efficacy characteristics under different temperature, moisture, and light regimes; optimal sites for infection; preliminary host range and crop tolerance; and mass production potential (Bailey, 2010). Agents that grow and reproduce above 37°C (i.e. indicating potential human infectivity), produce mycotoxins, or do not grow or sporulate abundantly on artificial media need to be screened out of the programme quickly. If the temperature-moisture optima do not match with the profile for the target weed then this may indicate a poor ecological fit. The idea is to eliminate poor and mediocre agents as quickly as possible and then concentrate on those with the best features.

13.4.2 Technology Development and Transfer

This stage works towards collecting the data required to prepare a regulatory submission, protecting the knowledge and ideas, and testing that the market place will be robust enough to warrant several years of research before commercialization (Figure 13.1). Before releasing any scientific information publicly, care must be taken to protect the knowledge and ideas that were developed. This may be achieved by filing for patents, although some aspects of knowledge are better preserved through the use of trade secrets. Bailey and Mupondwa (2006) provide a detailed explanation on how to protect your intellectual property and the costs associated with protection. If the technology has not been properly protected, then its market value will dramatically decrease. There may be some aspects of a technology that are patent protected, such as the novel use of an agent or a novel metabolite, but other aspects may be kept as trade secrets for the industry partner, such as fermentation and formulation strategies.

Rigorous scientific data must be collected for the genetic and biological characterization of the agent which may require the development of molecular tools for characterization and tracing environmental fate in field release trials (Bailey et al., 2009, 2011b; Zhou et al., 2004, 2005). This stage also requires determining the mode of action and extensive field testing under several environments and years for assessing efficacy and consistency of response to specific weed targets, as well as crop tolerance. This stage works out methods for mass production and integrates fermentation processes with downstream processing and formulation for field application. This is an iterative process whereby various fermentations and formulations must be tested and re-tested before selecting the final one. Toxicology studies are usually initiated nearer to the end of the development phase because they are costly and it must be clear the product has the qualities to succeed. Bailey and Falk (2011) describe a case study on the technology development and assessment of *P. macrostoma* which serves to illustrate the progress and factors considered before its registration.

The business side must evaluate the market size, other competitive products that are or may become available by the time commercialization is reached, and start to estimate costs of production and overall economic feasibility (Bailey, 2010; Bailey et al., 2009). This phase is an economic crunch where money is spent without any foreseeable return. The research, development, and regulatory costs are a huge barrier for smaller companies, with research needs approaching $1–2 million and toxicology testing about $500,000. Although the costs for developing bioherbicides are lower than for conventional chemicals, which may exceed $100 million, the ability to attract financial funding for this period is difficult and this period is often called the 'valley of death' (Bailey and Mupondwa, 2006; Bailey et al., 2009).

13.4.3 Registration and Commercialization

Every country around the world has its own organization and process for registering bioherbicides (Kabaluk et al., 2010). An effective regulatory system combines the following elements: data requirements for human health and safety, value (efficacy), and environmental safety;

clear and predictable procedures for assessing the risk and value; mechanisms for public and industry input; establishing timelines for the process and holding to them; reasonable fees; and enforcement of the legislation and regulations for product use, sale, and distribution. International harmonization to streamline product registration would be the ideal situation but in reality, may be impossible to attain worldwide. However, some harmonization has occurred among countries such as Canada and the USA through the implementation of a joint review process leading to registration in both countries at the same time (Bailey and Mupondwa, 2006). The key data requirements agreed to for joint reviews include information on the origin of the product, derivation and identification, biological properties, manufacturing methods and quality assurance programmes, estimate of potency and product guarantee, unintentional ingredients, storage stability, human health and safety, environmental fate and toxicology, efficacy (reviewed in Canada only), crop tolerance, and value (Bailey, 2010). Expedited timelines are given priority for joint review and the process takes about 2 years if all of the required data are supplied at submission. Registrants are strongly recommended to use the pre-submission regulatory consultation as the outcome of this session will produce a list of data requirements tailored for the specific bioherbicide being put forward.

The home stretch is nearing with the onset of pre-commercial development, the stage to ensure the commercial manufacturing process and downstream processing will deliver a product that works as expected (Bailey and Falk, 2011). Other activities include accumulating enough product from the commercial process to ensure market demands may be met in the region of release, completing package design, organizing the supply chain and delivery dates, and educating the retailers on how to display the new product. Once the product is launched, there is follow-up education for the sales force and consumers as well as monitoring sales and product acceptance. Few bioherbicides make it to this point as it is a long, expensive and challenging process. The next section will highlight those bioherbicides that made it successfully through registration and describe where they are used today.

13.5 BIOHERBICIDES REGISTERED WORLDWIDE

Although the total number of biopesticides registered worldwide is increasing, bioherbicides constitute the smallest fraction of these pest control products (Ash, 2010; Bailey et al., 2010; Glare et al., 2012). In 2001, Charudattan reported that there were eight successfully registered or commercially available bioherbicides in the world (Charudattan, 2001). About 10 years later, Kabaluk et al. (2010) reported on the current worldwide registrations of microbial biopesticides and, surprisingly, the only countries reporting bioherbicides were the Ukraine (one bioherbicide), Canada (three bioherbicides), and the USA (four bioherbicides). These numbers reflect that bioherbicides comprised less than 10% of all biopesticides (i.e. biofungicide, biobactericides, bioinsecticides, and bionematicides) in those countries. Since that report, three additional bioherbicides have been registered in Canada and the USA, but two others had their registrations lapse and are no longer available (Table 13.3). The following section will expand on the historical and current status of the North American bioherbicides presented in Table 13.3.

13.5.1 DeVine®, USA 1981

Phytophthora palmivora Butler strain MVW was first registered as DeVine in 1981 by Abbott Laboratories, IL as a bioherbicide to be applied to citrus crops to control strangler vine (*Morrenia odorata* (Hook. & Arn.) Lindl.). This

TABLE 13.3 Status of Bioherbicides Registered around the World in 2012

Registered Name/Company Name	Microbial Agent	Target Weed	Non-target Crop	Country	Year Registered & Reviewed	Current Status	Reference
Albobacteryn/ Unknown	*Achromobacter album*	Many	Sprouting inhibition	Ukraine	Unknown	Unknown	Kabaluk et al., 2010
DeVine™/ Valent BioSciences Corp	*Phytophthora palmivora*	*Morrenia odorata* Strangler vine	Citrus	USA	1981 2006	Registered; not available	National Pesticide Information Retrieval System http://state.ceris.purdue.edu/; Environmental Protection Agency www.epa.gov/opp00001/biopesticides/ingredients/index.htm
Collego™/ Encore Technologies Lockdown™/ Natural Industries	*Colletotrichum gloeosporioides* f. sp. *aeschynomene*	*Aeschynomene virginica* Northern jointvetch	Rice and soybean	USA	1982 2006	Registered; commercially available	Environmental Protection Agency www.epa.gov/opp00001/biopesticides/ingredients/index.htm
BioMal/ Philom Bios (Novozymes)	*Colletotrichum gloeosporioides* f. sp. *malva*	*Malva pusilla* Round-leaved mallow	Various crops	Canada	1992	Registration lapsed 2006; not available	Boyetchko et al., 2007
Camperico™/ Japan Tobacco	*Xanthomonas campestris*	*Poa annua* Annual bluegrass	Turfgrass	Japan	1997	Unknown	Bellgard, 2008
Woad Warrior™/ Greenville Farms	*Puccina thlaspeos*	*Isatis tinctoria* Dyer's woad	Rangeland; rights of way	USA	2002	Registered; not available	Environmental Protection Agency www.epa.gov/opp00001/biopesticides/ingredients/index.htm
Mycotech Paste™/ Mycoforestis Corp	*Chondrostereum purpureum*	Alders, aspen, hardwoods	Forests; rights of way	Canada	2002	Registration lapsed 2008; not available	Pest Management Regulatory Agency, Label Search www.hc-sc.gc.ca/cps-spc/pest/index-eng.php

(Continued)

TABLE 13.3 (Continued)

Registered Name/Company Name	Microbial Agent	Target Weed	Non-target Crop	Country	Year Registered & Reviewed	Current Status	Reference
Chontrol Paste™/ MycoLogic Inc	*Chondrostereum purpureum*	Alders, aspen, hardwoods	Forests; rights of way	Canada; USA	2004	Registered; commercially available	Bailey, 2010; Hintz, 2007
Smoulder/ Loveland Products Inc	*Alternaria destruens*	*Cuscuta* spp. Dodder	Agriculture; horticulture	USA (MA & WI)	2005	Registered; not available	Environmental Protection Agency www.epa.gov/opp00001/biopesticides/ingredients/index.htm
Sarritor™/ Sarritor Inc	*Sclerotinia minor*	Broadleaved weeds	Turfgrass	Canada	2007 conditional; 2010 full registration	Registered; commercially available	Kabaluk et al., 2010; Watson and Bailey, 2013; Pest Management Regulatory Agency Publication RD2010-08 www.hc-sc.gc.ca/cps-spc/pubs/pest/_decisions/rd2010-08/index-eng.php
Organo-Sol	*Lactobacillus* spp. *Lactococcus* spp.	Broadleaved weeds	Turfgrass	Canada	2010	Registered; commercially available	Kabaluk et al, 2010; Pest Management Regulatory Agency Publication RD2010-10 www.hc-sc.gc.ca/cps-spc/pubs/pest/index-eng.php
Product name not specified/ The Scotts Company	*Phoma macrostoma*	Broadleaved weeds	Turfgrass	Canada USA	2011 conditional; 2012 full registration	Registered; not yet available	Bailey and Falk, 2011; Pest Management Regulatory Agency, Label Search www.hc-sc.gc.ca/cps-spc/pest/index-eng.php
MBI-005 EP/ Marrone Bio Innovations	*Streptomyces* spp.	Broadleaved weeds	Turfgrass	USA	2012	Registered; not yet available	Marrone Bio Innovations Inc (Press release 17 May 2012; Contact rblair@marronebio.com)

product was the outcome of a collaboration that started in 1977 between the company and the Florida Department of Agriculture. DeVine is produced by submerged liquid fermentation and the liquid product is mixed with water and sprayed every other season onto soil under citrus crops after the weed has germinated or while actively growing (EPA, 2006a). The area of use is restricted to five Florida counties where the active ingredient occurs naturally. DeVine was the first bioherbicide developed in the USA and had several features that were difficult for industry to manage. Kenney (1986) discussed these features, such as the time it took to learn how to produce chlamydospores, only achieving a 6 week shelf life, and too much efficacy with weed control lasting as long as 5 years. In 2006, DeVine was still being produced occasionally, but after being registered for such a long time, the EPA required reassessment and reregistration, which was granted to Valent BioSciences Corporation, IL. There is no evidence on the company website that DeVine is commercially available unless they continue to provide it as a goodwill service to the citrus growers in those five Florida counties.

13.5.2 Collego™/LockDown®, USA 1982/2006

Colletotrichum gloeosporioides f.sp. *aeschynomene* strain ATCC 20358 was first registered as Collego in 1982 by Upjohn Company, MI as a bioherbicide to control northern joint vetch in rice (Bowers, 1986; EPA, 1997). That was the outcome of 12 years of collaboration between the company and University of Arkansas (Drs George Templeton and David TeBeest) as well as other researchers from the USDA. The pathogen causes lesions on northern joint vetch that encircle the stem and result in wilt. Even though the weeds do not die, the stem lesions and wilt render the weed non-competitive with the crop. To produce Collego, the fungus is grown in submerged liquid fermentation whereby the spores are harvested and dried as a wettable powder. Upon application, the dried spores are mixed with a rehydrating agent and then water. The product is applied by air and ground spray equipment. Initially, it was easy to obtain large numbers of spores, but these spores did not dry well and had a short shelf life, so considerable research effort was required to develop a commercially viable product (Bowers, 1986). It now has a shelf life of 1 year at room temperature or 3 years under refrigeration. Over the years, Collego has been a low-use, highly specific niche market product, but it has been profitable due to its low production and marketing costs and the fact that there have been few synthetic alternatives (Templeton et al., 1989). By 1997, EPA re-evaluated *C. gloeosporioides* f.sp. *aeschynomene* strain ATCC 20358 under the more recent standards of the day and deemed it was eligible for reregistration (EPA, 1997). In 2006, EPA released the Biopesticide Registration Action Document indicating that the registrant, Agricultural Research Initiatives, Fayetteville, AR, could sell the bioherbicide as LockDown (EPA, 2006b). LockDown has been produced annually since 2008 at a small contract fermentation facility and sold directly to growers (Kelly Cartwright, Agricultural Research Initiatives, personal communication).

13.5.3 BioMal®, Canada 1992

Colletotrichum gloeosporioides f. sp. *malvae* strain ATCC 20767 was registered as the bioherbicide BioMal in 1992 by Philom Bios Inc. to control round-leaved mallow in field crops (Boyetchko et al., 2007). The fungus was discovered in 1982, by researchers with Agriculture and Agri-Food Canada, who determined that it was a good candidate as a bioherbicide because the pathogen was host specific to plants in the family Malvaceae, the fungus was easily grown on artificial medium, and foliar spray applications provided effective weed control by causing severe stem lesions that girdled the stem and wilted the plant under field conditions. By 1985, an agreement was made with Philom Bios, Inc. to collaboratively commercialize the technology and the data submission package was made to PMRA in 1987. The product was available from 1992 to 1994, when production and sales were halted due to commercialization costs and production expenses. These expenses were due to changing market conditions, primarily the introduction of three new synthetic herbicides that were cheaper. Agriculture and Agri-Food Canada researchers sought another industry partner, Encore Technologies, Minnetonka, MN, to pursue reregistration as Mallet WP, but difficulties in manufacturing a consistent product were encountered and commercialization ended. In 2006, PMRA re-evaluated BioMal, deeming it to be safe under the current standards of the day, but with no industry partner the registration lapsed.

13.5.4 Camperico®, Japan 1997

Xanthomonas campestris pv. *poae* is a bacterium isolated from annual bluegrass (*Poa annua* L.) in Japan (Imaizumi et al., 1997). The bacterium enters the host through wounds and migrates to the xylem, which becomes blocked by bacterial exudates resulting in wilting of the host plant. Eighty-nine isolates were evaluated before selecting strain JTP482, which was developed commercially as Camperico by Japan Tobacco Inc., Yokohama (Fujimori, 1999). Since the company did both the discovery and development work, time was of the essence. The researchers spent from 1991 to 1993 collecting diseased annual bluegrass plants from roadsides and turf areas. It then took 3 years to complete the research and achieve commercialization. Although the company wanted to develop a freeze-dried commercial product, they were unable to make it stable in the time frame allocated, so a frozen suspension of cells in fermentation medium was used. When stored at −35°C, Camperico had 1.5 years of stable shelf life and a refrigerated delivery service was designed to ship the product while frozen (at −18°C) to customers. The product was registered in Japan in 1997, but there is no public information on its success or decline. The company website does not currently list Camperico as a product for sale.

13.5.5 Woad Warrior®, USA 2002

Woad Warrior is made from the teliospores of the fungus *Puccinia thlaspeos* C. Shub. 'strain woad' which are applied either as a spray or a powder to control dyer's woad (*Isatis tinctoria* L.) (EPA, 2002). This weed is an invasive species of dry open areas such as those found on farms, rangeland, waste areas and roadsides. It was introduced to the USA from Europe by American colonists, who used it to extract a valuable blue pigment, but the weed spread rapidly in eight semi-arid western states where it remains a problem today. The rust strain was discovered in 1979 in southern Idaho and is highly specific to dyer's woad. The fungus is an obligate pathogen meaning that it can only survive and reproduce on this specific weed host. The teliospores are applied once in the spring (April–May) to first year growth of the weed and the resulting infection interferes with flower and seed formation the following year. The fungus is non-toxic and not infective to mammals, and does not show any adverse effects to birds, fish, insects and non-target plants. The host specificity and ease of application made it very well suited for controlling dyer's woad. Dr Sherman Thomson at Utah State University developed Woad Warrior, which was registered in the USA in 2002 to Greenville Farms, N. Logan, UT. The drawback to this bioherbicide was the method of mass production. Being an obligate pathogen, teliospores had to be harvested from infected plants, which is highly labour intensive. Although Dr Thomson produced the teliospores on his farm, there was insufficient interest from industry to commit to commercialization and marketing the product. Although not available commercially, it is still registered and may be available on a local level with assistance from researchers at the university (Ralph Whitesides, Utah State University, personal communication).

13.5.6 Mycotech™ and Chontrol® Pastes, Canada 2002/2004

The fungal pathogen *Chondrostereum purpureum* prevents stump sprouting and promotes wood decay for vegetation management of woody deciduous trees and bushes (de Jong, 2000; Hintz, 2007). The story of using *C. purpureum* as a bioherbicide is convoluted and unusual. It starts in the Netherlands, where the introduction of American bird cherry or black cherry (*Prunus serotina* Ehrh.) from North America resulted in it becoming an invasive, shrubby weed in the sandy soils of the conifer forests (de Jong, 2000). Cutting the tree did not resolve the problem because the cut stems sprouted more branches. Researchers at the DLO Institute of Agrobiological and Soil Fertility, Wageningen found that when mycelium of *C. purpureum* was placed directly on the cut stem, the cherry tree acquired silver leaf disease, which blocks the vascular system, the host consequently dying (Scheepens and Hoogerbrugge, 1989). Additional research demonstrated that *C. purpureum* could also stop sprouting and caused wood decay leading to the death of many other hardwood species. At this time, the regulatory system for bioherbicides was evolving, and since the fungus was an endemic pathogen in Netherlands forests, Koppert Biological Systems, a Dutch company, started to market a mycelial suspension of *C. purpureum* as BioChon without officially registering the product.

At the same time in Canada, two research groups were independently studying the use of two strains of *C. purpureum* (strain HQ1 from Quebec and strain PFC2139 from British Columbia) for control of weeds in conifer release management and deciduous brush below power lines and other utility rights of way areas (de Jong, 2000). Myco-Forestis Corporation from L'Assumption, QC registered *C. purpureum* HQ1 as Myco-Tech paste for vegetation management anywhere east of the Rocky Mountains in Canada in 2002 and in the USA in 2005. Product was sold for a few years and then the company dissolved in 2007 and the registration lapsed in 2008. There were no reasons found for dissolution of the company, but their paste formulation only had a 3 month shelf life, which is a disadvantage, so production and storage issues may have played a role.

On the other side of Canada, in British Columbia, Dr Ronald Wall from the Pacific Forestry Centre, Victoria, isolated *C. purpureum* strain PFC 2139 from a diseased apple tree. This discovery led to collaboration between the Pacific Forestry Centre, University of Victoria, and the newly formed spin-off company MycoLogic Inc. A unique feature of this strain included very weak pathogenicity, which turned out to be an advantage because this trait conferred security that no harm would be done from field release despite the known broad host range (Hintz, 2007). Extensive environmental fate research demonstrated that releasing this strain throughout North America would have minimal risk of introducing novel virulence alleles outside of its range of origin and that field release of the bioherbicide would not greatly increase the naturally occurring populations. Considerable effort was put in to develop fermentation strategies with long shelf life, which turned out to involve a stirred liquid fermentation first stage followed by a solid state fermentation for the second phase to give adequate titre and longevity (de la Bastide and Hintz, 2007). The product, under the brand name of Chontrol Paste, was submitted under joint review to PMRA and EPA, was approved for registration in 2004, and continues to be sold for vegetation management in North America today.

13.5.7 Smoulder®, USA 2005

Alternaria destruens L. Simmons strain 059 is a fungus that is parasitic to *Cuscuta* species, more commonly known as dodder (EPA, 2005). It was originally isolated as an indigenous pathogen to the USA from swamp dodder in 1986, but controlled several dodder species in crops such as alfalfa, dry bog cranberries, carrots, peppers, tomatoes, eggplant, blueberries, and woody ornamentals. The pathogen infects the weed and suppresses growth either in the spring at weed emergence or late summer when the vines are reaching the top of the cash crop canopy. In order to provide a high level of weed control, the pathogen requires a moist environment with air temperature between 4°C and 35°C. It was determined that the pathogen would not pose any risk to human and other non-target species. *Alternaria destruens* was formulated as both a solid granular (Smoulder G), which should be applied to a moist surface, and a wettable powder (Smoulder WP), whereby a package of water soluble active ingredient is mixed with a package of liquid adjuvant combined with water for late season foliar spraying. The bioherbicide was discovered and developed by Dr Tom Bewick while at the University of Massachusetts Cranberry Experimental Station and product development and registration was undertaken by Loveland Products Inc., Greely CO and Sylvan Bio Inc., Kittanning, PA. The products were approved for registration by the EPA in 2005. Although the products are still registered, neither company lists them as products for sale.

13.5.8 Sarritor®, Canada 2007

Sarritor is made from mycelium of the fungus *Sclerotinia minor* Jagger strain IMI 344141 for post-emergent control of broadleaved weeds in turfgrass (PMRA, 2011a). The fungus has a wide host range for broadleaved plants but is completely nonpathogenic to cereals and grasses. The product is produced by growing the fungus on grain in vented bags using a solid state fermentation system. Once colonized, the grain is dried and the granules may be broadcast to a lawn using a fertilizer-style spreader or a containerized shaker for spot treatments. The product may be used by both home owners and professional lawn care operators. Sarritor requires moderate temperatures and moisture for several days to colonize the weed hosts such as dandelion (*Taraxacum officinale* Weber ex F.H. Wigg.) and plantain (*Plantago major* L.); when the conditions are right, the top growth of the weed turns necrotic and is destroyed. Oxalic acid secreted by the fungus has been shown to be one of the factors responsible for lesion development (Briere et al., 2000). The mycelium of the pathogen does not persist in the soil in the absence of the host and after 4 months, there is no residual activity; this particular strain rarely produces sclerotia, the common overwintering structure of *S. minor*, and thus it cannot persist. Sarritor was not toxic or pathogenic to mammals, birds, fish, honey bees, earthworms, or wild animals. Dr Alan Watson from McGill University, Quebec was the principal scientist who led the team to discover and develop the product. In the mid-1990s, *S. minor* IMI 344141 was one among several fungi being evaluated for biological control of dandelion in a collaborative project involving academia, industry, and government (Stewart-Wade et al., 2002). After 4 years of screening and field evaluation, *S. minor* IMI 344141 was the lead candidate. However, the research priorities of the industry partners changed, and the programme was discontinued. However, a few years later, due to public demand, the Supreme Court of Canada decided that municipalities have the right to regulate whether their community will use pesticides or not. This decision resulted in bans on the use of synthetic herbicides for cosmetic purposes in many urban areas throughout several provinces, opening a market opportunity for

Sarritor as there were no bioherbicides registered for use (which were deemed acceptable under the bans) and few alternative weed control measures. In 2004, a spin-off company (Sarritor Inc. or 4260864 Canada Inc.) was created from the project at McGill University, and together with a consortium of professional lawn care operators, were given the license to exclusively manufacture and sell Sarritor. The product was conditionally registered in Canada in 2007 and after completing a few extra studies, it was given full registration in 2010. Commercial product was available in 2009 and 2010. However, a new bioherbicide made from iron chelated with hydroxyethylenediamine triacetic acid (FeHEDTA) came on the market, which dramatically cut into the professional sales of Sarritor in 2010, so the company is restructuring to focus on domestic markets (Watson and Bailey, 2013).

13.5.9 Organo-Sol®, Canada 2010

Organo-Sol is made from lactic acid bacteria (*Lactobacillus casei* strain LPT-111, *Lactobacillus rhamnous* strain LPT-21, *Lactococcus lactis* ssp. *lactis* strain LL64/CSL, *Lactococcus lactis* ssp. *lactis* strain LL102/CSL, and *Lactococcus lactis* ssp. *cremoris* strain M11/CSL) fermenting dairy products (whey) to produce citric acid and lactic acid (PMRA, 2010). It is grown under a submerged liquid fermentation process, to collect the cells and acidic fermentation by-products, which are then mixed with water and sprayed onto the foliage as either a broadcast or spot treatment. There are no patents related to this technology because this fermentation process is in the public domain, but the use of lactic and citric acids as a bioherbicide must be federally registered under Canadian law. Organo-Sol provides partial suppression of white and red clovers (*Trifolium* spp.), bird's foot-trefoil (*Lotus corniculatus* L.), black medic (*Medicago lupulina* L.), and wood sorrel (*Oxalis acetosella* L.) in established lawns; there may be some yellowing of the turf but recovery is seen in about in 3 weeks. Applications start in May and are repeated every 14 days for up to five applications per season. The product works because the low pH 3.5 allows for penetration into the plant cells whereby it causes necrosis and suppression of plant growth. The most susceptible plants have thin cuticles. It should not be applied to newly seeded grasses as injury will result. Environmental risks are not of concern with Organo-Sol since lactic acid bacteria are widespread in nature, the numbers of bacteria being applied as the bioherbicide are relatively low compared to the natural population, and the citric and lactic acid products are quickly biotransformed in both terrestrial and aquatic environments. Organo-Sol was registered in Canada in 2010 to Lacto-Pro-Tech Inc., St-Hyacinthe, QC and is currently marketed under the trade name Kona™ by AEF Global, Levis, QC (Melanie Greffard, AEF Global, personal communication). Kona is only available to professionals, although in 2014, AEF Global will release a new formulation for domestic use under the trade name of Bioprotec Herbicide™.

13.5.10 Phoma, Canada/USA 2011/2012

Phoma macrostoma strain 94-44B is a fungus that was isolated from Canada thistle and causes shoot and root growth inhibition and severe chlorosis (also called photobleaching) of the foliar parts of many broadleaved plant species (Bailey and Falk, 2011; Watson and Bailey, 2013). The most susceptible plants were in the plant families Asteraceae, Brassicaceae, and Fabaceae whereas the most resistant plant families were Poaceae, Pinaceae, and Linaceae (Bailey et al., 2011b). This fungal strain was targeted for development as a bioherbicide to control broadleaved weeds in turfgrass, agriculture, and agro-forestry. It was grown on grain using solid state fermentation and the infested grain was milled to form granules for broadcast application. During the fermentation process, phytotoxins called macrocidins are produced which have been shown to cause growth inhibition, photobleaching, and mortality (Graupner et al., 2003). *P. macrostoma* strain 94-44B was shown to be not toxic and not infective to mammals, birds, fish, insects, and wild animals. In the soil, it provided weed control for up to 4 months, but after 12 months there was no residual activity and it did not move away from the site of placement (Zhou et al., 2004). This product demonstrated high and consistent efficacy under a broad range of environmental conditions, a long shelf life without requiring stringent temperature control, addressed a broad spectrum of weeds, may be used for both pre-emergent and post-emergent weed control, can be applied in multiple fields of use, and can be used domestically and professionally. The pathogen was discovered by Dr Karen Bailey and Jo-Anne Derby from Agriculture and Agri-Food Canada, Saskatoon and developed with The Scotts Company, USA under a collaborative research agreement. Conditional registration was approved for use on turfgrass in Canada (2011) and full registration in the USA (2012) for use in turfgrass. Pilot-scale manufacturing processes are being developed before product commercialization. Research is ongoing to collect data to support registration for agricultural use.

13.5.11 MBI-005 EP, USA 2012

MBI-005 is made from a natural product compound produced by *Streptomyces acidiscabies* strain RL-110 for control of annual grasses, broadleaf and sedge weeds in turf, ornamentals and crops such as wheat, corn, and rice (EPA, 2012c). The natural compound, thaxtomin A, is a known fast-acting phytotoxin that causes necrosis and prevents cell biosynthesis and division when present at very low levels (i.e. parts per million). It is the first bioherbicide product that is fermented and then heat-treated to kill the bacterial cells before application. The product is not toxic to non-target organisms such as birds, fish, and bees. It may be used commercially in agriculture, nurseries, golf courses, as well as in residential turfgrass. The advantages of MBI-005 are that it has broad spectrum activity as a pre-emergent, killing the weeds as they germinate, as well as selective activity with post-emergent applications in turf and crops. MBI-005, also branded as Opportune®, was discovered and developed by Marrone Bio Innovations Inc., who received notice of registration approval in April 2012, and are preparing to release a commercial product soon (R. Blair, Marrone Bio Innovations, Davis, CA, Press Release 17 May 2012).

13.6 WHAT WILL BE THE ROLE OF BIOHERBICIDES IN THE FUTURE?

The development of science-based technologies for weed control using plant pathogens and other microorganisms has garnered interest and momentum since the 1970s when the principles of biological control were adapted for commercial purposes, as demonstrated by those researchers who developed the first bioherbicides, DeVine and Collego. However, the rate of success has been lower than expected given the number of products commercialized relative to potential biocontrol agents reported. Ash (2010) searched the ISI Web of Science database to find that from 1987 to 2009, there were 509 papers that mentioned 'bioherbicides or mycoherbicides' and that over 335 of the papers contained the phrase 'potential bioherbiocide or mycoherbicide' from which he concluded that there are few attempts at true commercialization of a product. This is probably a fair conclusion given the multidisciplinary and complex nature of the information and techniques required to reach the final goal; the process can appear overwhelming. It is also probably a reflection that many researchers concentrate more on the science side of the technology and advancing careers through publications, forgetting to address important business issues that may make or break the commercialization chain. Lidert (2001) also expressed this view. A common link among all of the bioherbicides that have been registered and sold, even for a short time, is strong industry involvement from early in the development process; the team approach is important.

As our experience with developing bioherbicides increases, we can see that the nature of the challenges encountered back in the 1980s–1990s, which were highly technical (Auld and Morin, 1995), are different from those encountered today (Glare et al., 2012). What has been accomplished to date is continuing collaboration between public researchers and industry, learning how to select better candidates in a shorter period of time, selecting candidates that can provide broader spectrum control, more utilization of the candidate's bioactive compounds, and a greater emphasis on understanding of the markets and costing of processes. What still needs to be accomplished is a better understanding of delivering a product through fermentation, formulation, and application to the target at the pilot scale level. There is a need for more expertise and infrastructure to support novel, cost effective fermentation and formulation. Moving bioherbicides forward in the future also requires continued financial investment from industry R&D as well as public good resources.

Bioherbicides will be a part of our future as demands from society push for less risky pest control products and legislation changes what pest control products are used, and where. It is important to develop products that users want and will adopt. Access to educational resources through industry dealers and university extension programmes, as well as extension demonstrations in highly visible places, will facilitate adoption of bioherbicides and increase additional demand. It is important that users understand that bioherbicides are different from synthetic herbicides. This does not mean bioherbicides are less effective or more difficult to use, but they are derived from living organisms and may require slightly different environmental and storage conditions to make them work optimally. As more bioherbicide products emerge into the marketplace, there will be increased uptake, which will fuel more research, creating a cycle for future development.

References

Anderson, D., Lafond, G.P., 2010. Global perspective of arable soils and major soil associations. Prairie Soils Crops 3, 1–8. <www.prairiesoilsandcrops.ca> (accessed 04.06.12).

Ash, G., 2010. The science, art and business of successful bioherbicides. Biol. Control 52, 230–240.

Auld, B.A., Morin, L., 1995. Constraints in the development of bioherbicides. Weed Technol. 9, 638–652.

Bailey, K.L., 2004. Microbial weed control: An off-beat application of plant pathology. Can. J. Plant Pathol. 26, 239–244.

Bailey, K.L., 2010. Canadian innovations in microbial biopesticides. Can. J. Plant Pathol. 32, 113–121.

Bailey, K.L., Falk, S., 2011. Turning research on microbial bioherbicides into commercial products – A *Phoma* story. Pest Technol. 5 (Special Issue 1), 73–79.

Bailey, K.L., Mupondwa, E.K., 2006. Developing microbial weed control products: Commercial, biological, and technological consideration. In: Singh, H.P., Batish, D.R., Kohli, R.K. (Eds.), Handbook of Sustainable Weed Management. The Haworth Press, New York, pp. 431–474.

Bailey, K.L., Boyetchko, S.M., Peng, G., Hynes, R.K., Taylor, W.G., Pitt, W.M., 2009. Developing weed control technologies with fungi. In: Rai, M. (Ed.), Advances in Fungal Biotechnology. I.K. International Publishing House Pvt. Ltd, New Delhi, pp. 1–44.

Bailey, K.L., Boyetchko, S.M., Längle, T., 2010. Social and economic drivers shaping the future of biological control: A Canadian perspective on the factors affecting the development and use of microbial biopesticides. Biol. Control 52, 221–229.

Bailey, K.L., Pitt, W.M., Falk, S., Derby, J., 2011a. The effects of *Phoma macrostoma* on nontarget plant and target weed species. Biol. Control 58, 379–386.

Bailey, K.L., Pitt, W.M., Leggett, F., Sheedy, C., Derby, J., 2011b. Determining the infection process of *Phoma macrostoma* that leads to bioherbicidal activity on broadleaved weeds. Biol. Control 59, 268–276.

Bailey, K.L., Falk, S., Derby, J., Melzer, M., Boland, G., 2013. The effect of fertilizers on the efficacy of the bioherbicide, *Phoma macrostoma*, to control broadleaved weeds in turfgrass. Biol. Control 65, 147–151.

Beckett, T.H., Stoller, E.W., Wax, L.M., 1988. Interference of four annual weeds in corn (*Zea mays*). Weed Sci. 36, 764–769.

Beckie, H.J., 2009. Herbicide Resistance in Weeds: Influence of Farm Practices. Prairie Soils Crops 2, 17–23. <www.prairiesoilsandcrops.ca> (accessed 04.06.12).

Bellgard, S., 2008. Inundative control using bioherbicides. In: Hayes, L. (Ed.), The Biological Control of Weeds Book Landcare Research. Manaaki Whenua, Lincoln, NZ, p. 4. <http://www.landcareresearch.co.nz/publications/books/biocontrol-of-weeds-book> (accessed 13.08.12).

Bhowmik, P.C., 1999. Herbicides in relation to food security and environment: a global perspective. Indian J. Weed Sci. 31, 111–123.

Bowers, R.C., 1986. Commercialization of Collego™ – An industrialist's view. Weed Sci. 34 (Suppl. 1), 24–25.

Boyetchko, S.M., 2005. Biological herbicides in the future. In: Ivany, J.A. (Ed.), Weed Management in Transition: Topics in Weed Science, Vol. 2. Canadian Weed Science Society, Sainte-Anne-de-Bellevue, QC, pp. 29–47.

Boyetchko, S.M., Bailey, K.L., Hynes, R.K., Peng, G., 2007. Development of the mycoherbicide, BioMal®. In: Vincent, C., Goettel, M.S., Lazarovits, G. (Eds.), Biological Control: A Global Perspective. CAB International, Wallingford, UK, pp. 274–283.

Briere, S.C., Watson, A.K., Hallett, S.G., 2000. Oxalic acid production and mycelial biomass yield of *Sclerotinia minor* for the formulation enhancement of a granular turf bioherbicide. Biocontrol Sci. Technol. 10, 281–289.

Charudattan, R., 2001. Biological control of weeds by means of plant pathogens: significance for integrated weed management in modern agro-ecology. BioControl 46, 229–260.

Charudattan, R., 2005. Ecological, practical, and political inputs into the selection of weed targets: what makes a good biological control agent? Biol. Control 35, 183–196.

Coble, H.D., Ritter, R.L., 1978. Pennsylvania smartweed (*Polygonum pensylvanicum*) interference in soybeans (*Glycine max*). Weed Sci. 26, 556–559.

Coble, H.D., Williams, F.M., Ritter, R.L., 1981. Common ragweed (*Ambrosia artemisiifolia*) interference in soybeans (*Glycine max*). Weed Sci. 29, 339–342.

De Jong, M.D., 2000. The BioChon story: deployment of *Chondrostereum purpureum* to suppress stump sprouting in hardwoods. Mycologist 14, 58–62.

De la Bastide, P.Y., Hintz, W., 2007. Developing the production system for *Chondrostereum purpureum*. In: Vincent, C., Goettel, M.S., Lazarovits, G. (Eds.), Biological Control: A Global Perspective. CAB International, Wallingford, UK, pp. 291–299.

Environmental Protection Agency, 1997. EPA R.E.D. Facts: *Colletotrichum gloeosporioides f. sp. aeschynomene*, EPA-738-F-96-026, April 1997. Biopesticides and Pollution Division (7511C), Office of Pesticide Programs, US EPA, Washington, DC. <http://www.epa.gov/oppbppd1/biopesticides/ingredients/index_cd.htm#c> (accessed 15.08.12).

Environmental Protection Agency, 2002. Biopesticide Registration Action Document: *Puccinia thlaspeos* 'strain woad' (PC Code 006-489). Biopesticides and Pollution Division (7511C), Office of Pesticide Programs, US EPA, Washington, DC, 25pp. <http://www.epa.gov/opp00001/biopesticides/product_lists/new_ai_2002.htm> (accessed 13.08.12).

Environmental Protection Agency, 2005. Biopestcicide Registration Action Document: *Alternaria destruens* strain 059 (PC Code 028301). Biopesticides and Pollution Division (7511C), Office of Pesticide Programs, US EPA, Washington, DC, 30pp. <http://www.epa.gov/opp00001/biopesticides/ingredients/factsheets/factsheet_028301.htm> (accessed 14.08.12).

Environmental Protection Agency, 2006a. EPA R.E.D. Facts: *Phytophthora palmivora* MWV, March 27, 2006. Biopesticides and Pollution Division (7511C), Office of Pesticide Programs, US EPA, Washington, DC. <http://www.epa.gov/oppbppd1/biopesticides/ingredients/index_p-s.htm#p> (accessed 15.08.12).

Environmental Protection Agency, 2006b. *Colletotrichum gloeosporioides* f. sp. *aeschynomene* (226300), Biopesticides Registration Action Document, March 29, 2006. Biopesticides and Pollution Division (7511C), Office of Pesticide Programs, US EPA, Washington, DC. <http://www.epa.gov/oppbppd1/biopesticides/ingredients/index_cd.htm#c> (accessed 15.08.12).

Environmental Protection Agency, 2012a. Pest Wise: Biopesticides Fact Sheet, May 09. Biopesticides and Pollution Division (7511C), Office of Pesticide Programs, US EPA, Washington, DC. <http://www.epa.gov/pesp/htmlpublications/biopesticides_fact_sheet.html> (accessed 06.06.12).

Environmental Protection Agency, 2012b. Pesticide Chemical Search Tool: *Phoma macrostoma*. Biopesticides and Pollution Division (7511C), Office of Pesticide Programs, US EPA, Washington, DC. <http://www.epa.gov/pesticides/chemicalsearch> (accessed 25.05.12).

Environmental Protection Agency, 2012c. Pesticide Product Label System, Details for MBI-005 EP. Biopesticides and Pollution Division (7511C), Office of Pesticide Programs, US EPA, Washington, DC. <http://iaspub.epa.gov/apex/pesticides/f?p=PPLS:8:0::NO::P8_PUID:501804> (accessed 13.08.12).

Fishel, F.M., 2007. Pesticide use trends in the U.S.: Global comparison, Document PI-143. Pesticide Information Office January 2007, Florida Cooperative Extension Service, Institute of Food and Agricultural Sciences, University of Florida, Gainesville, FL.

Fujimori, T., 1999. Bioherbicides: the view from industry. In: Bioherbicides: Fresh Perspectives, Proceedings of a workshop held in conjunction with the 12th Biennial Conference of the Australian Plant Pathology Society, 26th September, 1999, Canberra, AU. ISBN: 0 7347 12685 (Heatherington, S. Auld, B., Eds), pp. 47–51. New South Wales Agriculture, Orange, NSW.

Gianessi, L.P., Sankula, S., 2003. The value of herbicides in US crop production, April 2003. National Center for Food and Agricultural Policy, Washington, <http://www.ncfap.org/pesticideuse.html> (accessed 15.08.12).

Glare, T., Caradus, J., Gelernter, W., Jackson, T., Keyhani, N., Kohl, J., et al., 2012. Have biopesticides come of age? Trends Biotechnol. 30, 250–258.

Graupner, P.R., Carr, A., Clancy, E., Gilbert, J., Bailey, K.L., Derby, J, et al., 2003. The macrocidins: Novel cyclic tetramic acids with herbicidal activity. J. Nat. Prod. 66, 1558–1561.

Hallett, S.G., 2005. Where are the bioherbicides? Weed Sci. 53, 404–415.

Heap, I.M., 1997. The occurrence of herbicide-resistant weeds worldwide. Pestic. Sci. 51, 235–243.

Hintz, W., 2007. Development of *Chondrostereum purpureum* as a mycoherbicide for deciduous brush control. In: Vincent, C., Goettel, M.S., Lazarovits, G. (Eds.), Biological Control: A Global Perspective. CAB International, Wallingford, UK, pp. 284–290.

Holm, F.A., Johnson, E.N., 2009. The history of herbicide use for weed management on the prairies. Prairie Soils Crops 2, 1–11. <www.prairiesoilsandcrops.ca> (accessed 01.06.12).

Holm, L., Plucknett, D.L., Pancho, J.V., Herberger, J.P., 1977. The World's Worst Weeds: Distribution and Biology. The University Press of Hawaii, Honolulu.

Imaizumi, S., Nishino, T., Miyabe, K., Fujimori, T., Yamade, M., 1997. Biological control of annual bluegrass (*Poa annua* L.) with a Japanese isolate of *Xanthomonas campestris* pv. *poae* (JTP-482). Biol. Control 8, 7–14.

Kabaluk, J.T., Svircev, A.M., Goettel, M.S., Woo, S.G., 2010. The use and regulation of microbial pesticides in representative jurisdictions worldwide. IOBC Global <http://www.IOBC_Global.org> (accessed 11.06.12).

Kenney, D.S., 1986. DeVine® – The way it was developed – an industrialist's view. Weed Sci. 34 (Suppl. 1), 15–16.

Kiewnick, S., 2007. Practicalities of developing and registering microbial biological control agents. CAB Reviews: Perspectives in Agriculture, Veterinary Science, Nutrition and Natural Resources 2 (No. 013), 1–11.

Lidert, Z., 2001. Biopesticides: is there a path to commercial success?. In: Vurro, M., Gressel, J., Butt, T., Harman, G.E., Pilgeram, A., St Leger, R.J., Nuss, D.L. (Eds.), Enhancing Biocontrol Agents and Handling Risks. IOS Press, Amsterdam, p. 295.

McNeil, J.N., Cotnoir, P.A., Leroux, T, Laprade, R., Schwartz, J.L., 2010. A Canadian national survey on the public perception of biological control. BioControl 55, 445–454.

Norris, R.F., 1996. Water use efficiency as a method for predicting water use by weeds. Weed Technol. 10, 153–155.

Parker, R., 2003. EM4856 Drought Advisory: Water conservation, weed control go hand in hand. Cooperative Extension Washington State University, Pullman <https://pubs.wsu.edu/ListItems.aspx?Keyword=EM4856> (accessed 29.05.12).

Pedras, M.S., Ismail, N., Quail, J.W., Boyetchhko, S.M., 2003. Structure, chemistry, and biological activity of pseudophomins A and B, new cyclic lipodepsipeptides isolated from the biocontrol bacterium *Pseudomonas fluorescens*. Phytochemistry 62, 1105–1114.

Pest Management Regulatory Agency, 2011a. Evaluation Report ERC2007-02, *Sclerotinia minor* strain IMI 344141. Catalogue Number H113-26/2007-2E-PDF. <http://www.hc-sc.gc.ca/cps-spc/pubs/pest/_decisions/erc2007-02/index-eng.php> (accessed 13.08.12).

Pest Management Regulatory Agency, 2011b. Evaluation Report ERC2011-09, *Phoma macrostoma* strain 94-44B. Catalogue Number H113-26/201109E-PDF. <http://www.hc-sc.gc.ca/cps-spc/pubs/pest/_decisions/erc2011-09/index-eng.php> (accessed 12.04.12).

Pest Management Regulatory Agency, 2010. Registration Decision RD2010-10 *Lactobacillus casei* strain LPT-111, *Lactobacillus rhamnous* strain LPT-21, *Lactococcus lactis* ssp. *lactis* strain LL64/CSL, *Lactococcus lactis* ssp. *lactis* strain LL102/CSL, *and Lactococcus lactis* ssp. *cremoris* strain M11/CSLL. Catalogue number H113-25/2010-10E-PDF. <http://www.hc-sc.gc.ca/cps-spc/pubs/pest/_decisions/rd2010-10/index-eng.php> (accessed 13.08.12).

Riches, C.R., 2001. The World's Worst Weeds, BCPC Symposium Proceedings No. 77. The British Crop Protection Council, Surrey.

Scheepens, P.C. Hoogerbrugge, A., 1989. Control of *Prunus serotina* in forests with the endemic fungus *Chondrostereum purpureum*. In: Delfosse, E.S. (Ed.), Proceedings of the VII International Symposium on Biological Control of Weeds, 6–11 March, 1988, Rome, Italy. Instituto Sperimentale per la Patologia Vegetale, Ministero dell Agricotura e delle Foreste, Rome, Italy, pp. 545–551.

Stewart-Wade, S.M., Green, S., Boland, G.J., Teshler, M.P., Teshler, I.B., Watson, A.K., et al., 2002. *Taraxacum officinale* (Weber), Dandelion (Asteraceae). In: Mason, P.G., Huber, J.T. (Eds.), Biological Control Programmes in Canada, 1981–2000. CAB International, Wallingford, UK, pp. 427–436.

Templeton, G.E., Smith, R.J., TeBeest, D.O., 1989. Perspectives on mycoherbicides two decades after discovery of the Collego® pathogen. In: Delfosse E.S.,

(Ed.), Proceedings of the VII International Symposium on Biological Weed Control, March 6–11, 1988, Rome Italy. Instituto Sperimentale per la Patologia Vegetale, Ministero dell Agricotura e delle Foreste, Rome, Italy, pp. 553–338.

Terry, P.J., 1996. The use of herbicides in the agriculture of developing countries. In: Proceedings of the Second International Weed Control Congress, Copenhagen, Denmark, 25–28 June 1996, Volumes 1–4. Department of Weed Control and Pesticide Ecology, Copenhagen, pp. 601–609.

Watson, A.K., Bailey, K.L., 2013. *Taraxacum officinale* (Weber), Dandelion (Asteraceae). In: Mason, P.G., Gillespie, D.R. (Eds.), Biological Control Programmes in Canada, 2001–2012. CAB International, Wallingford, UK, pp. 383–391.

Wolf, T., 2009. Best management practices for herbicide application technology. Prairie Soils Crops 2, 24–30. <http://www.prairiesoilsandcrops.ca> (accessed 01.06.12).

Young, F.L., 1988. Effect of Russian thistle (*Salsola iberica*) interference on spring wheat (*Triticum aestivum*). Weed Sci. 36, 594–598.

Zhou, L., Bailey, K.L., Derby, J., 2004. Plant colonization and environmental fate of the biocontrol fungus *Phoma macrostoma*. Biol. Control 30, 634–644.

Zhou, L., Bailey, K.L., Chen, C.Y., Keri, M., 2005. Molecular and genetic analysis of geographic variation in isolates of *Phoma macrostoma* used for biological weed control. Mycologia 97, 612–620.

CHAPTER

14

Biological Control of Invasive Insect Pests

Mark G. Wright

University of Hawaii at Manoa, HI, USA

14.1 INTRODUCTION

Species introduced into new environments often become invasive, having deleterious effects on indigenous environments, human activities such as agriculture, and human health, as vectors of diseases. Invasive insect species produce many negative interactions in their new environments, becoming pests of almost all agricultural crops and attacking indigenous plant species. Examples of invasive insect pests of agricultural crops are numerous. In many cases, insect pests have threatened entire agricultural industries, or have placed staple crops of subsistence communities in dire jeopardy. Insects that act as vectors of plant pathogenic viruses are among the most severe pests in agriculture, and a number of these vectors have been instrumental in severely restricting certain crop industries, such as Tomato spotted wilt virus on tomatoes in South Africa and Papaya ringspot virus on papaya in Hawaii. Many invasive insects are direct pests of crops, causing physical damage to plant products, reducing yields, and reducing quality of products (Figure 14.1). Export of agricultural products may also be negatively impacted by insects such as fruit flies (Diptera: Tephritidae), that cause phytosanitary concerns. Natural ecosystems are also impacted by invasive insect species, and Hawaii yields another good example, that of a gall-forming wasp (*Quadrastichus erythrinae*, Eulophidae; Erythrina gall wasp), that attacks an indigenous endemic coral tree, *Erythrina sandwicensis* (Leguminosae). The Erythrina gall wasp, probably from Africa originally, invaded Indian Ocean and Pacific Ocean Islands, and Southeast Asian countries in an unprecedented series of rapid invasions with severe impacts on landscape- and indigenous *Eythrina* trees. There was concern in Hawaii that the indigenous *Erythrina*, considered to be a keystone species in dry lowland forest there, would become extinct because of the depredations of the gall wasps (Figure 14.2).

Invasive insects that become agricultural pests result in estimated economic losses of US$1.3 trillion worldwide annually (Henneberry, 2007). Many insecticide

D. P. Abrol (Ed): Integrated Pest Management.
DOI: http://dx.doi.org/10.1016/B978-0-12-398529-3.00015-4

FIGURE 14.1 Papaya mealybug, *Paracoccus marginatus* (Hemiptera: Pseudococcidae), is a widespread invasive pest of papaya and many indigenous plant species on Pacific islands, but was successfully suppressed by parasitic wasps introduced as biological control agents. *Photo: M. Wright.*

FIGURE 14.2 Erythrina gall wasp, *Quadrastichus erythrinae* (Hymenoptera: Eulophidae), an invasive insect throughout Southeast Asia and Indian Ocean and Pacific Ocean islands, which severely impacted landscape ornamental and endemic *Erythrina* trees, has been effectively suppressed through the introduction of an African parasitic wasp (*Eurytoma erythrinae*, Eurytomidae). *Photo: M. Tremblay.*

applications are used in efforts to manage invasive insect pests. Seeking to improve the sustainability of pest management efforts, and environmental and health issues have driven the adoption of integrated pest management (IPM), including the use of biological control of pest species, in an effort to reduce dependence on insecticides. Other IPM options include the use of physical control of insects, and cultural practices that may be applied to make environments less attractive or susceptible to pests. Physical pest management options include alternatives such as mass trapping of pests, and the use of barriers to exclude pests from crops. While physical insect control measures may be effective, they require sustained upkeep to remain effective. Cultural control options, including the use of tolerant or resistant varieties of crops, mixed cropping systems and planting schedules to take advantage of differential crop susceptibility at different stages and different environmental conditions, can also contribute significantly to the integrated management of insect pests. Cultural management options require considerable planning for effective implementation, and may increase labour costs associated with crop production.

Biological control offers a self-sustaining solution for the suppression of invasive insect pests. Classical biological control (CBC; see below for details), i.e. the introduction of natural enemies of pests from the place of origin of the pest, followed by successful establishment, is a completely sustainable means of achieving pest suppression. Approximately 20% of all biological control projects worldwide are considered to provide complete pest suppression. There are other forms of biological control that offer various levels of sustainability, such as augmentative biological control – the repeated introduction or release of natural enemies into a cropping system – where classical biological control may be unfeasible owing to poor establishment of introduced natural enemies, and other factors such as population density too low to effectively suppress the pest population. Conservation biological control, where indigenous or naturalized natural enemies of pests are conserved within the borders of crop fields, is another alternative approach to implementing biological control in crops. Carefully researched and applied CBC of insect pests is among the most ecologically safe and sustainable practices for the management of invasive insects.

14.2 BIOLOGICAL CONTROL OPTIONS

14.2.1 Classical Biological Control (CBC)

This is probably the best-known form of biological control worldwide. CBC is based on the concept of enemy-release, or enemy free space, experienced by a new invasive species when it is introduced into a new environment. Conversely stated, the newly introduced species experience an environment with minimal biotic resistance, if it has no pre-adapted 'natural enemies' there. Many newly introduced species will not be subject to the mortality factors that suppressed their populations in their place of origin, and can therefore have massive outbreaks in new environments. Upon determination of a new pest establishment, researchers identify the place of origin of the invasive species, and then proceed to seek its natural enemies there, assuming that other organisms that inflict mortality on the natural populations have a suppressive effect on those populations. Exploration efforts are conducted, and mortality factors identified. Natural enemies that are expected to have an effect on the pests' population dynamics are then placed into quarantine in the place of intended introduction, and screened for potential non-target effects, and the presence of pathogens or hyperparasites. Upon determining that the prospective CBC agent is safe for release, it is introduced into the new environment where it will hopefully locate the target pest and become established in the environment, providing sustained suppression of the pest species. The target pest species is highly unlikely to be eradicated by a biological control agent; effective CBC will reduce a pest population to a level below the economic injury level.

14.2.2 Augmentative Biological Control

In cases where CBC is not effective, for example, where establishment of a new natural enemy does not occur, or wild populations of introduced natural enemies are inadequate to effectively suppress a pest species, repeated augmentation of natural enemy populations may offer a useful alternative. Augmentative biological control entails mass rearing of natural enemies in an insect production facility, and releases of large numbers in crop systems at appropriate times to target pests. Augmentative biological control is most effective, and has been most widely adopted, in greenhouse production systems. Different types of augmentative biological control are recognized. Inundative augmentative biological control is used in situations where large numbers of natural enemies are released into a crop system in much the same way a pesticide is applied. The natural enemies are not expected to establish within the environment, but to provide quick short-term suppression of the target pest. Frequent repeated releases may be used. Inoculative augmentative biological control describes an approach where a relatively small number of natural enemies are introduced into a crop system, with the expectation that they will establish for a brief period, perhaps a growing season, and provide suppression of the target pest. A repeat release is made the next season, to establish a temporary natural enemy population in the crop again.

14.2.3 Conservation Biological Control

Cultural practices applied in cropping systems can profoundly influence natural enemies of pests. Manipulation of the crop and field border plants, or pesticide application regimes can be planned to have minimal effects on biological control agents. This is termed conservation biological control, as natural enemies are encouraged to proliferate within, or close to crops, and the applications of insecticides that potentially reduce their populations are planned in such a manner that minimal negative effects occur.

14.2.4 Types of Biological Control Agents

Biological control agents are broadly referred to as 'natural enemies'. Within this broad category, there are many distinct guilds of natural enemies.

14.2.4.1 Predators

Insects that capture and physically devour prey are called predators. Predators are typically larger than their prey, and many are polyphagous – they will prey upon many species that they are physically able to capture and overpower, as well as different developmental stages of prey species. Their impacts are not limited to pest species, and they may prey upon other predators, or even various stages of their own species. Some predatory insects are relatively monophagous, such as *Rodolia cardinalis* (Coleoptera: Coccinellidae; vedalia beetle), which was successfully introduced from Australia to the USA to control cottony cushion scale (*Icerya purchasi*, Hemiptera: Margarodidae).

Larvae and adults of the same species are often predaceous, but not always. Examples of groups of predatory insects that are commonly used in biological control programmes include ladybeetles (Colopetera: Coccinellidae), minute pirate bugs (Hemiptera: Anthocoridae) and hover-fly larvae (Diptera: Syrphidae), among others.

14.2.4.2 Parasitoids

This guild of natural enemies includes parasitic insects that kill their host during their development within, or on, the body of the host. Parasitoids are primarily from the orders Hymenoptera (parasitic wasps), and Diptera (parasitic flies). Endo- and ecto-parasitic lifestyles occur within parasitoid groups. Endoparasitoids develop within the body of the host insect, after the adult female has deposited one or more eggs into the body of the host. They have some diverse and fascinating life histories, such as polyembrony, where a single egg gives rise to many hundreds of parasitic larvae within the host. Endoparasitoid larvae complete their development within the body of their host, feeding on the organs and haemolymph of the host. The host typically dies only once the parasitoid is ready to complete its development. On completion of its metamorphosis, the parasitoid emerges from the host as an adult and proceeds to seek mates, and then new hosts to parasitize. Parasitoids attack specific developmental stages of their hosts. There are species that are exclusively egg parasitoids (such as Trichogrammatidae), egg-larval parasitoids, larval parasitoids, and larval-pupal parasitoids. Egg-larval, or larval-pupal parasitoids are species that initially parasitize the egg or larva, and only complete their metamorphosis in a later development stage, the larva or pupa, respectively. A relatively small number of parasitoid species attack adult insects, for example

Tachinidae flies, which parasitize stink-bugs (Hemiptera: Pentatomidae). Most parasitoids kill their hosts, but some do not, and those might cause adult host sterility, for example. Parasitoids typically impact the behaviour of their hosts, such as causing reduced feeding rate, modified foraging behaviour, or paralysis of the host. Some, such as Braconidae parasitoids of aphids, cause the host to produce a modified body structure; parasitized aphids develop a darkened, hardened body covering, and are known as mummies. Some parasitoids have parthenogenic reproduction, and may be haplo-diploid (diploid females, haploid males; arrhenotokous parthenogenesis), or in some cases, only female progeny are produced (thelytokous parthenogenesis). The latter occurs in species that have symbiotic interactions with bacteria such as *Wolbachia*, which induces unisexual reproduction in its host. Many species of *Trichogramma* (egg parasitoids) have *Wolbachia* associations, and produce exclusively female, or highly female biased populations.

Parasitoids often have relatively complex host seeking behaviour, and because they are intimately associated with the physiology of their hosts, are often finely attuned to specific hosts. Thus parasitoids are frequently oligophagous, or even monophagous, and therefore attractive candidates for potential biological control agents.

14.2.4.3 Insect Pathogens

A diverse range of entomopathogenic microorganisms occurs in nature, and some have been exploited as biological control agents of insects. Entomopathogenic fungi (e.g. *Beauveria bassiana*, *Metarhizium anisopliae*) have been used effectively in the biological control of a spectrum of readily susceptible crop pests such as thrips (Thysanoptera), aphids (Hemiptera) and other soft bodied species that occur in environments conducive to infection. Even migratory brown locusts (*Locustana pardalina*) that occur in dry environments in Africa have been successfully suppressed with *M. anisopliae*. The fungi penetrate the body of the insect host, usually through orifices such as the spiracles, and proceed to colonize the insect with hyphae. Entomopathogenic fungi also produce toxins that result in the death of the host insect. Infected insects typically display modified physiology and behaviour. Locusts infected with *M. anisopliae* have the capacity to induce a fever-like condition as an immune response. Other insects reduce feeding activity and become moribund when infected. Entomopathogenic fungi of the genus *Cordyceps*, some of which infect ants, cause the host to adopt 'zombi-like' behaviour, the latter eventually climbing onto blades of grass or other elevated vegetation parts where they die, and the fungus erupts from the body of the host, resulting in airborne dispersal of the spores, to infect further hosts. A number of entomopathogenic fungi have been developed as commercial formulations, 'bioinsecticides', which can be applied as sprays in augmentative biological control programmes.

Some bacteria, notably varieties of *Bacillus thuringiensis*, are effective agents of insect mortality, and have been developed as commercial bioinsecticides. Entomopathogenic bacteria are typically specific to particular groups of insects (e.g. Lepidoptera, Coleoptera or Diptera). The insect needs to ingest the bacteria while actively feeding, for the bacteria to infect the host. Upon contact with specific conditions in the host's gut, protein crystals produced by the bacteria bind with specific gut receptors, and cause paralysis of the gut. The insect ceases to feed shortly after and dies. The bacteria produce large numbers of spores in the dead insect, which are released back into the environment from the insect cadaver.

A large number of baculoviruses (nucleopolyhedrosis viruses – NPVs, and granuloviruses) infest insects and cause mortality. The insects (typically Lepidoptera larvae), ingest virus particles, and, once in the gut of the insect, these particles move into the epithelia cells of the gut, where replication occurs in the nucleus. The virus causes lysis of the cells, and the host dies shortly after infection. The internal organs of the insect host break down dramatically, and the host becomes a fluid-filled sac, which disperses viral particles upon rupture. Insect pathogenic viruses tend to be highly susceptible to desiccation and UV light exposure, and as a result, their efficacy is often not impressive. Some products using NPVs have been developed for certain applications in insect pest management, most effectively in treating stored product pests, in environments where the virus is protected from exposure to UV radiation.

Entomopathogenic nematodes also play a role in biological control of insects. A number of species in the families Steinernematidae and Heterorhabditidae are used in insect pest management programmes. The nematodes penetrate the body of susceptible insects through various orifices, and once inside the insect host, they release bacteria. The bacteria are pathogenic to the insect, and death of the host results. The nematodes reproduce within the insect's body, and large numbers of juvenile (infective) nematodes are released from the host. A number of commercial formulations of entomopathogenic nematodes are available. These are used in augmentative biological control applications. Entomopathogenic nematodes tend to have quite broad host ranges, and can be used against many pests. They are most effective when targeting soil-dwelling insects, but may be used for foliage-feeding and even wood-boring insects in some cases.

14.3 ECOLOGICAL BASIS FOR BIOLOGICAL CONTROL

Populations of herbivorous insects are regulated by climatic conditions, host-plant quality and abundance (bottom-up regulation), and the effects of biotic mortality factors (predators, parasitoids, entomopathogens) (top-down regulation). Competition between herbivorous insects may seem an obvious source of population regulation, but natural enemies appear to constrain herbivore populations effectively enough that competition is seldom a major factor in undisturbed food webs.

Biotic mortality factors frequently tend to be positively density-dependent – they have a greater effect with increasing host population density, and are desirable options for biological control programmes. Density-dependent effects can have different forms in terms of what impact they have on a population of insects. These different effects are best described by analysing functional and numerical responses of the species interacting. A numerical response in natural enemies refers to an increase in the number of predators or parasitoids present, as a response to an increase in prey density. Numerical increases in natural enemies can be the result of increased reproductive rates, or the aggregation of natural enemies near a burgeoning prey source. Functional response refers to an increase in the number of prey or hosts exploited by a predator or parasitoid, with an increase in prey or host density. These relationships were first explained by Holling (1965), using laboratory experiments that showed that the number of prey exploited during a set period increased with increasing prey density, but then decreased at higher prey densities (Figure 14.3). This response (a plateau in the number of prey eaten) is the result of the predator or parasitoid becoming satiated, or being limited by the handling time required to capture increasing numbers of prey. Up to a certain prey density, and with a Type III response, predators and parasitoids do typically demonstrate (Figure 14.3C) a positive density-dependent effect, but at higher prey densities, they are unable to do so, and the relationship becomes a negative density-dependent one. At high pest densities, predators and parasitoids thus tend to have minimal effect on pests (Figure 14.3E, F), with the rare exception of a Type I response in

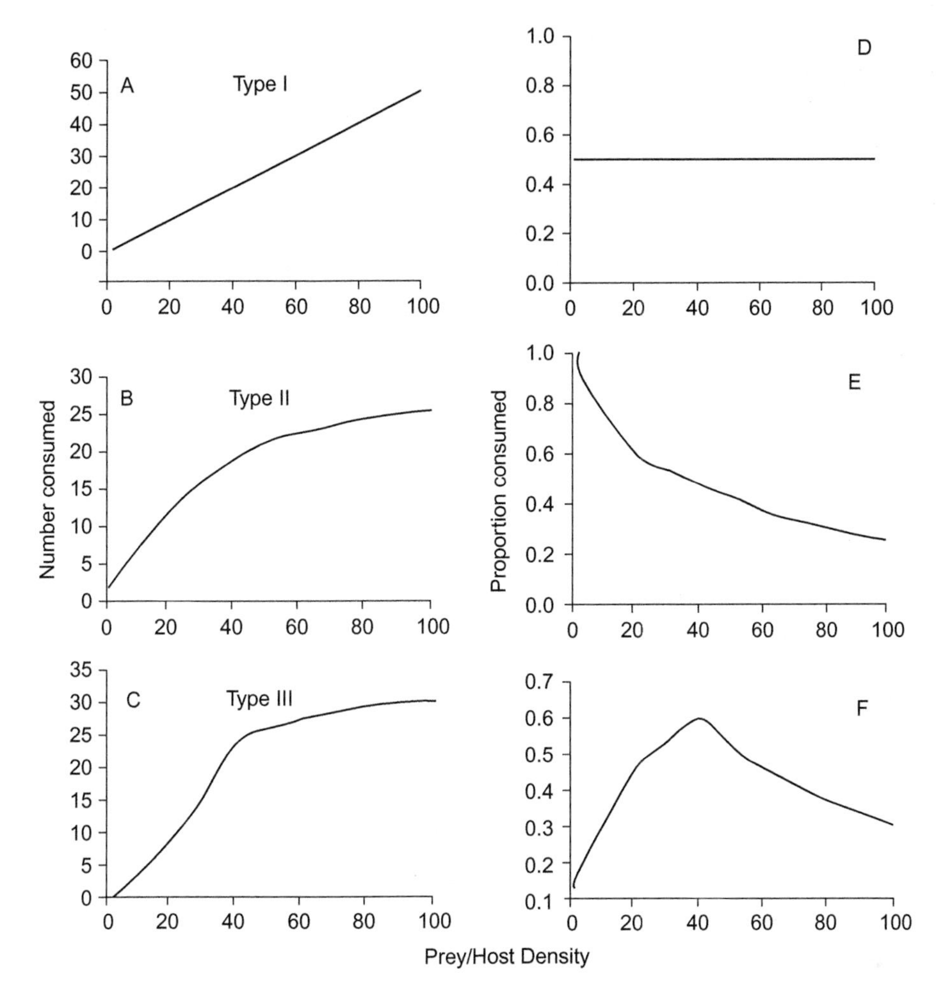

FIGURE 14.3 Functional response curves (A: Type I; B; Type II; C: Type III), showing number of prey (or hosts) consumed or attacked during a given period, and the corresponding proportion of the population consumed (D, E, F), respectively. *Based on Holling (1965).*

a highly efficient natural enemy. An increase in numbers of the predator (numerical response) is required to continue effectively suppressing the prey population once it exceeds a particular density.

The degree of specialization of insect natural enemies also plays a role in determining their effectiveness as biological control agents. Generalist (polyphagous) species, such as many predators, may not demonstrate a numerical response to an increase or decrease in prey density. Specialized (monophagous) species tend to have strong numerical responses. A dramatic decline in host density can result in a reduction in numbers of parasitoids, to the point that local extinction may occur.

Prey or host location ability also plays a role in determining the effectiveness of biological control agents. This is particularly true for augmentative biological control agents, which are released with the intention that they seek out and attack hosts and have an immediate effect. Parasitoids that disperse rapidly and have highly developed host location abilities (e.g. *Trichogramma ostriniae*, Wright et al., 2001) may have good potential for suppressing the target species, and also make such biological control programmes less labour intensive as effort spent on making natural enemy releases is minimized. Some species used in CBC programmes may have limited dispersal capacity, and may require human-mediated dispersal to locate isolated pest populations.

Climate matching is a fundamentally important factor contributing to effective CBC. Attempts to introduce natural enemies into areas with dissimilar climate to the place of origin are unlikely to be very successful. This aspect of selecting biological control agents is not as simple as seeking a locality where the invasive species occurs, finding parts of its distribution with compatible climate to the invaded area, and searching there for natural enemies. It may be important to seek the exact place of origin of the pest species to account for specific population differences in natural enemies associated with the insect across its distribution range. Determining the exact origin of invasive insect species is not a trivial matter, but it has been made easier to achieve with the development of molecular tools and methods for phylogeographic analysis.

There are cases where the exact origin of the invasive insect has not been established, and highly effective biological control has been achieved through the introduction of natural enemies that have no evolutionary association with the target species. This approach is known as 'new-association biological control', and has been suggested to be an approach more likely to yield effective suppression of pest populations. This proposal is based on the theoretical assumption that target species and natural enemies that have an evolutionary history are likely to have developed a relationship where their populations tend to a homeostasis – a balance between mortality and fecundity. Prey and host species evolve behavioural and physiological defences against predators and parasitoids, which allow members of a population to survive the depredations of such natural enemies. Species without evolutionary contact, and recently interacting in a new environment, may have no such mechanisms that result in homeostasis, and the target pest may be reduced to extremely low numbers (Hokkanen and Pimental, 1989).

The ability of any prospective biological control agent to establish in the area of introduction is fundamental to successful CBC. There are factors other than climatic conditions that mediate the likelihood of establishment of introduced species, including number of individuals released, and number of releases made (Hopper and Roush, 1993). Releasing larger numbers of individuals and making multiple releases should increase the likelihood that the biological control agent locates target hosts and establishes breeding populations. It has also been suggested that genetically diverse original stocks of prospective biological control agents should be used to develop the initial batches of insects for release, and that after releases, those genotypes best suited to the new environment will prevail. In some species with pronounced dispersal behaviour, it may be necessary to ensure that adequate numbers of insects are released so that Allee effects (reduced population growth rate at low population density) do not result in extinction of the newly released biological control agent. Some biological control agents (e.g. *Diachasmimorpha tryoni*, Hymenoptera: Braconidae), apparently do not suffer from Allee effects and have established successfully after the release of very few (<20) individuals. Establishment of natural enemies in new environments may also be dependent on the availability of alternative hosts that facilitate overwintering or provide some form of refuge in which parasitoids can persist.

14.4 CONDUCTING A BIOLOGICAL CONTROL PROJECT

Once it has been determined that implementing biological control is a suitable approach for the management of an invasive insect species, biological control researchers set in motion a series of actions that identify potential biological control agents, and determine the environmental safety of those prospective introductions. A series of steps are typically taken, often involving researchers from various countries. The first step is to identify the likely origin of the invasive species of interest. Determination of origin may be done based on taxonomic records and insect collections, and records of previous invasions by the species of interest into other places. DNA sequencing techniques and reconstruction of phylogeographic relationships of invasive insects, compared to insects from putative places of origin, provide accurate methods of determining the origin of a pest population. It is essential to have good taxonomic information available for the target pest, otherwise unsuitable natural enemies may be collected and screened.

The process of exploration for prospective biological control agents can take researchers to distant and interesting places, and can take considerable time to complete. The tasks of exploration entomologists seeking biological control agents are typically not trivial. There are issues of obtaining permits to work in countries that are often not hugely accommodating, and the fieldwork conditions are often logistically challenging. Conducting effective biological control exploration work requires that the researchers have access to suitable laboratory equipment and insect rearing facilities, and also a facility where host plants for the target species can be maintained. If the latter is not a possibility, then samples of natural enemies have to be sent back to the locality of intended introduction, where colonies of the insect can be established in a suitable quarantine facility.

Quarantine is an essential component in the progression of a biological control project. Rearing prospective biological control agents in a quarantine facility ensures that pathogens or hyperparasites possibly associated with imported insects are detected and the likelihood of their release into the environment is limited. Hyperparasitoids (parasitoids that attack insects already parasitized by a species of primary parasite), are among the most disruptive factors that influence biological control effectiveness, and their inadvertent introduction should be avoided fastidiously.

During quarantine, or even in the place of origin, the host range of the natural enemy is investigated. Biological control researchers typically seek monophagous natural enemies, as they pose the least non-target risk, and they are finely attuned to the life cycle and physiology of the target pest. Historically, there has been little emphasis on determining the host range for insect predators and parasitoids intended for release as biological control agents. In the past 40 years however, Hawaii (USA), and countries such as New Zealand and Australia, have implemented strict quarantine requirements on the importation of insect biological control agents, with emphasis on the identification of specialized natural enemies. Host range testing is done using choice- and no-choice trials, where the natural enemy offered either potential non-target species, or combinations of non-target species and the target insect. In places with strict regulations pertaining to host-specificity of biological control agents such as Hawaii, any attack on test species even under no-choice conditions causes rejection of the parasitoid. Species selected for non-target screening usually include species that are phylogenetically close to the target pest, and species that share similar biological characteristics as the target pest. An example of the latter is exposing all gall-forming insects from various orders and families occurring in an area, to a parasitoid being considered for release to suppress a gall-forming wasp (Eulophidae), when there may actually be no indigenous species in that target family with similar biology. Efforts are currently underway to develop improved risk assessment procedures for biological control agents (see below for greater detail). Efforts are often also made during quarantine to estimate the probable effectiveness of a natural enemy on the target pest. If the natural enemy under consideration is unknown to science, taxonomic description and an assessment of the phylogenetic affinities of the natural enemy are made and published in peer-reviewed literature.

Once a species that is appropriately specialized on the target pest is identified, a request is made for permission to release the biological control agent. In the USA, petitions for release have to be made to the United States Department of Agriculture, US Fish and Wildlife Service (where applicable), and the North American Plant Protection Organization (Canada, USA, Mexico). Permission may be required by specific states for release, as is the case in Hawaii. Various countries have different procedures for permitting the release of a new species, and care should be taken to comply with these regulations at all times.

Once permission has been granted for release, adequate numbers of natural enemies need to be produced for release, with careful attention to maintaining high quality of the insects. This includes reducing selection for an 'insectary strain' of the insects, which might have reduced fitness once released. The same concern should be exercised when augmenting established populations of biological control agents (Vorsino et al., 2012a).

Releases of biological control agents should be made using as many individuals as feasible, and by making multiple releases. This strategy should ensure that the likelihood of establishment is maximized (Hopper and Roush, 1993). In some cases, however, very small numbers of insects have been released (e.g. *Diachasmimorpha tryoni* in Hawaii), and in other cases, large repeated releases have been made (e.g. *Trichogramma ostriniae* in the northeastern USA) with no establishment, probably owing to the lack of a suitable overwintering host (Hoffmann et al., 2002). Vorsino et al. (2012b) have cautioned against introducing extreme genetic diversity in biological control agents, and suggest seeking to release strains of natural enemies that are pre-adapted to the place of introduction, determined during exploration and quarantine work, and quantified using genetic data and geographic analyses.

A question still remaining in biological control research is whether multiple or single species should be released to best suppress pest populations. In many cases, the first natural enemy species released in a biological control programme does not provide effective suppression of the target pest, and additional species are introduced. In other cases, however, single species have been highly effective. The challenge in this regard is developing means to predict the likely effect of a new biological control agent on a target pest before release. This work can be conducted in the place of origin of the natural enemy, and in quarantine before release, but it is difficult to provide unequivocal predictions. It is surely preferable to introduce as few species as possible, so this area of investigation deserves continued attention. There are known cases where the introduction of multiple parasitoids targeting a single pest species has resulted in competitive interactions among the parasitoids, resulting in a host shift in the less successful species on the target host. Messing and Wang (2008) showed that competitive interaction with *Fopius arisanus* caused *D. tryoni* (both Braconidae) to expand its host range to include a non-target Tephritidae species, introduced as a biological control agent of *Lantana camara*.

14.5 BENEFITS AND RISKS OF BIOLOGICAL CONTROL

There are some examples of attempted biological control that have been poorly crafted and executed, such as the introduction of mongoose to Hawaii for the control of rats in sugarcane, and other cases (Messing and Wright, 2006). The introduction of mongoose resulted in widespread predation of endemic Hawaiian birds, and no suppression of the target pest. It should be noted though that these early biological control introductions that had negative non-target impacts were made with zero, or limited, oversight from regulatory organizations. After the emergence of concern over non-target impacts of biological control agents, regulation has been improved substantially, and no recent introductions resulting in incidences of non-target impacts are known.

Biological control does pose risks, but certainly no more dire risks than unmitigated invasions by insect pests. Potential risks of biological control agents are primarily ecological in nature, including potential non-target impacts that may occur, and the reality that once introduced, little can be done to remove them from an environment. After introduction into a new environment, biological control agents will very probably undergo some degree of local adaptation, sometimes referred to as 'intelligent pollution' by critics. New developments of biological methods and predictive tools that permit detailed analyses of geographic patterns may offer options for reducing the extent of evolutionary change in introduced agents (Vorsino et al., 2012b). Non-target impacts may either be direct (direct reduction of populations of non-target species), or indirect, involving ecologically complex interactions where shared natural enemies link the dynamics of prey populations to each other (Tack et al., 2011). Although there are a few well-documented valid cases of serious negative non-target impacts of biological control, in many cases there is confusion between actual impact and an agent simply exploiting a non-target species at a low level. There is a need to not only document non-target utilization, but to quantify the extent of impact (Messing and Wright, 2006).

The benefits of biological control of invasive insect pests may be huge. There are many excellent examples of highly successful CBC, resulting in reductions in impacts of pests on agricultural production, natural environments, and reduced dependence on insecticides for pest management. Van Driesche et al. (2008) provide a good overview of the benefits of biological control. One of the most significant benefits of CBC is that a self-sustaining system of pest population suppression is put in place, providing a permanent solution. The benefit-cost ratios of successful biological control programmes are typically large.

In deciding whether to implement a biological control programme, it is necessary to conduct a risk assessment and to be able to weight the benefits of the prospective release to the potential risks. This requires good assessments of the likelihood of success. Making accurate assessments of benefit-risk relationships also requires comprehensive risk analysis. A few options for assessing non-target risk posed by prospective agents have been developed, ranging from a mix of qualitative and quantitative measures, to probabilistic risk modelling that incorporates extensive quantitative ecological data (Wright et al., 2005). The various approaches provide varying degrees of risk quantification, ranging from recommendations of 'release or do not release' to an estimated value for the extent of expected non-target impact. The latter approach includes the use of decision- or precision trees, which estimate the likely outcomes and magnitude of various

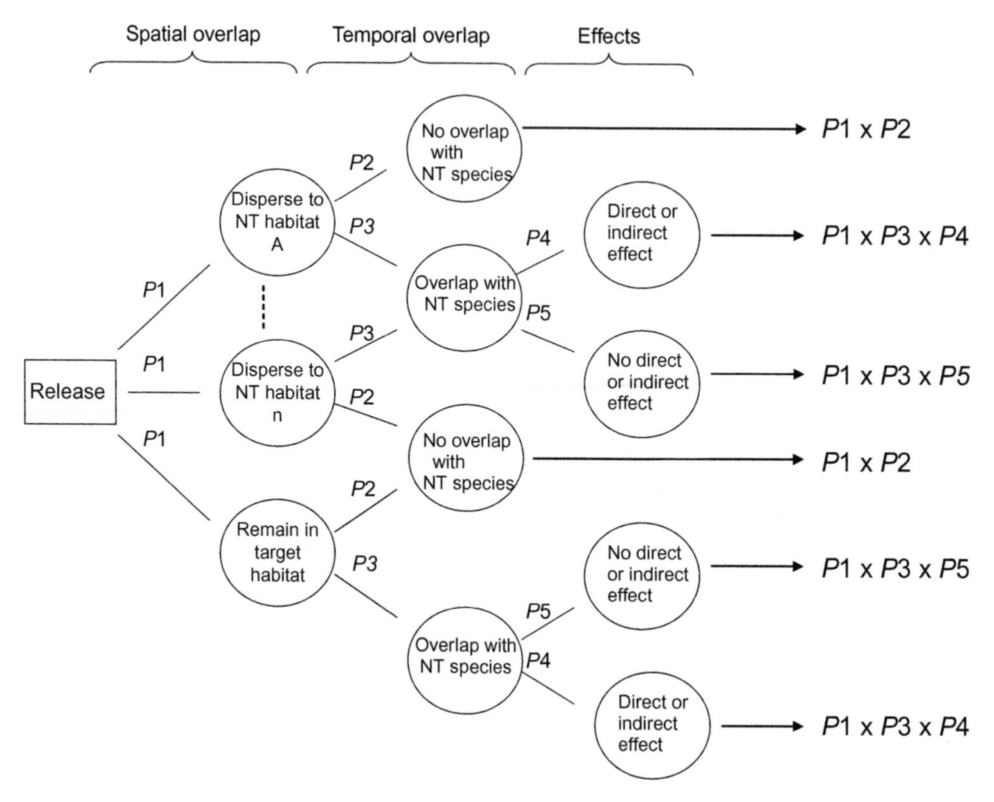

FIGURE 14.4 A conceptual precision tree for estimating non-target (NT) risk posed by prospective biological control agents. Circles contain contingencies (spatial overlap, temporal overlap and direct effects); probability of each contingency occurring is given on connecting lines. The overall probability of impact is estimated by multiplying the *P* values along each branch. *From Kaufman (2008).*

dependent occurrences (behaviours influenced by ecological factors) that lead to attacks on non-target organisms (Figure 14.4).

14.6 INTEGRATING BIOLOGICAL CONTROL OF INVASIVE INSECT SPECIES WITH ALTERNATIVE MANAGEMENT STRATEGIES

For many reasons, biological control may not provide complete suppression of the target pest to a level below the economic injury level, or below a population density that continues to result in environmental impact. In such cases, it is necessary to develop procedures for integrating biological control with other pest suppression techniques. These alternative techniques include the typical IPM options available – cultural, physical and chemical controls. Effort is made to ensure that the application of any of these alternatives, particularly chemical control, does not negatively impact the biological control agents in the system. Some IPM practices, specifically cultural management (viz. modifying the crop environment to make it less attractive to pests), may also render the system more attractive to natural enemies, and offer refugia where beneficial insects may escape applications of insecticides. Many cultural practices actually strive to enhance the ecosystem for natural enemies, through provision of alternative sources of nutrition (pollen, nectar), often to great benefit (Simpson et al., 2011).

Insecticides can be effectively integrated with biological control, if caution is exercised. Newer generation insecticides have become more target-specific than older products, which have been phased out to a large extent. In some cases, careful selection and timing of insecticide sprays can complement the effects of natural enemies without severely compromising the populations of the latter (Hull et al., 1985), through the application of decision systems that include economic and biological information (Musser et al., 2006). A significant component of decision-making includes estimating the effect that natural enemies have on a pest insect population. This can be a complex procedure, as determining the effect of natural enemies on a population is not trivial, there being issues related to dependability of estimates of proportion of insects parasitized, the stage of development of insects sampled and other factors. Hoffmann et al. (1990) developed a method for estimating the in-field parasitism of bollworm eggs, which included a means of accounting for parasitized eggs that could not be diagnosed (owing to the developmental stage of the parasitoid) in the field when estimating parasitism level. This includes a means of adjusting field estimates for undetectable mortality resulting from parasitism before comparing the pest population with the established action threshold (pest level at which remedial action such as insecticide application is taken).

If a decision is made to apply an insecticide, the next important consideration is which compound to choose that will produce the best pest mortality, but the lowest natural enemy mortality. This again is a non-trivial venture, as the range of options for any given pest on a particular crop is probably limited. Older-generation insecticides (e.g. organophosphates, carbamates, pyrethroids) tended to have severe non-target impacts on beneficial insects, and integration with biological control was elusive. Newer generation insecticides, including biological products derived from bacteria (e.g. *Bacillus thuringiensis*), bacterial fermentation products (e.g. Spinosad) and fungi (e.g. *Beauveria bassiana*), offer options that are more readily integrated with beneficial insects such as biological control agents. Many new-generation synthetic insecticides (e.g. neonicotinoids, pyridines, insect growth regulator analogues) have less impact on biological control agents than old-generation compounds, and can be reasonably readily integrated with biological control on crops with suitable registration, and with careful timing of spray applications.

The increasing use of genetically engineered crops (e.g. maize genetically engineered to include *B. thuringiensis* genes, and thus producing cry proteins), has created some concern that there may be unintended non-target effects on natural enemies of pests on the crops. Deleterious impacts on natural enemies in genetically engineered crops seem to be indirect effects however – it is probably the result of reduced prey (pest insect) quality that predators or parasitoids show decreased performance on these crops (Shelton et al., 2009). Long-term studies have shown that generalist predators of aphids increased in abundance in Bt cotton production in China, in large part because of reduced insecticide inputs in those systems (Lu et al., 2012). This offers an optimistic perspective on the integration of biological control, insecticide use, and evolving cultural practices such as the use of genetically engineered crops.

14.7 SUCCESSFUL BIOLOGICAL CONTROL

There are many cases of highly successful biological control of invasive insect pests, and other invasive organisms, such as weeds. Because of space constraints in this chapter, the reader is referred to other literature for comprehensive overviews of biological control programmes. Caltagirone (1981) provides an overview of landmark cases in biological control, dating back to some of the earliest examples. Van Driesche et al. (2008) review some more recent case studies. A good example of the use of biological control to manage an invasive species in a conservation area is that of the introduction of *Rodolia cardinalis* to suppress cottony cushion scale in the Galapagos Islands (Alvarez et al., 2012). Readers interested in studies suggesting negative impacts of biological control are referred to Henneman and Memmott (2001), and Kuris (2003). However, CBC of invasive insect pests has had an overwhelmingly positive impact on the environment and agriculture; for further discussion, see Messing and Wright (2006).

14.8 CONCLUDING REMARKS

Biological control of invasive insect pests requires a comprehensive understanding of natural enemy ecology to facilitate prediction of the effectiveness of prospective biological control agents, and to ensure that non-target impacts are avoided. Research and practice of biological control has made great progress in this regard during the past few decades. There is still space for improvement, particularly in developing predictors of impact on target species. CBC has contributed significantly to reducing the environmental impacts of agricultural pest management, and in conserving indigenous species attacked by invasive insects. CBC is likely to become increasingly important in the future with no evident abatement in the rate of invasion by insect pests worldwide.

References

Alvarez, C.C., Causton, C.E., Hoddle, M.S., Hoddle, C.D., van Driesche, R., Stanek, E.J., 2012. Monitoring the effects of *Rodolia cardinalis* on *Icerya purchasi* populations on the Galapagos Islands. Biocontrol 57, 167–169.

Caltagirone, L.E., 1981. Landmark examples in classical biological control. Annu. Rev. Entomol. 26, 213–232.

Henneberry, T.J., 2007. Insect pest management. In: Pimental, D. (Ed.), Encyclopedia of Pest Management. Taylor and Francis. DOI: 10.1081/E-EPM-120009942.

Henneman, M.L., Memmott, J., 2001. Infiltration of a Hawaiian community by introduced biological control agents. Science 293, 1314–1316.

Hoffmann, M.P., Wilson, L.T., Zalom, F.G., Hilton, R.J., 1990. Parasitism of *Heliothis zea* (Lepidoptera: Noctuidae) eggs: effect on pest management decision rules for processing tomatoes in the Sacramento Valley of California. Environ. Entomol. 19, 753–763.

Hoffmann, M.P., Wright, M.G., Pitcher, S.A., Gardner, J., 2002. Inoculative releases of *Trichogramma ostriniae* (Hymenoptera: Trichogrammatidae) for suppression of *Ostrinia nubilalis* in sweet corn: field biology and population dynamics. Biol. Control 25, 249–258.

Hokkanen, H.M.T., Pimental, D., 1989. New associations in biological control: theory and practice. Can. Entomol. 121, 829–840.

Holling, C.S., 1965. The functional response of predators to prey density and its role in mimicry and population regulation. Mem. Entomol. Soc. Can. 45, 3–60.

Hopper, K.P., Roush, R.T., 1993. Mate finding, dispersal, number released, and the success of biological control introductions. Ecol. Entomol. 18, 321–331.

Hull, L.A., Beers, E.H., Meagher, R.L., 1985. Integration of biological and chemical control tactics for apple pests through selective timing and choice of synthetic pyrethroid insecticides. J. Econ. Entomol. 86, 1355–1358.

Kaufman, L.V., 2008. Non-Target Impacts of Introduced Parasitoids and Validation of Probabilistic Risk Assessment for Biological Control Introductions. Ph.D. Dissertation (Entomology), University of Hawaii at Manoa, Honolulu, HI, USA, pp. 208.

Kuris, A.M., 2003. Did biological control cause extinction of the coconut moth, *Levuana iridescens*, in Fiji? Biol. Inv. 5, 133–141.

Lu, Y., Wu, K., Jiang, Y., Guo, Y, Desneux, N., 2012. Widespread adoption of Bt cotton and insecticide decrease promotes biocontrol services. Nature 487, 362–365.

Messing, R.H., Wang, X.G., 2008. Competitor-free space mediates non-target impact of an introduced biological control agent. Ecol. Entomol. 34, 107–113.

Messing, R.H., Wright, M.G., 2006. Biological control of invasive species: solution or pollution? Front. Ecol. Environ. 4, 132–140.

Musser, F.R., Nyrop, J.P., Shelton, A.M., 2006. Integrating biological and chemical controls in decision making: European corn borer (Lepidoptera: Crambidae) control in sweet corn as an example. J. Econ. Entomol. 99, 1538–1589.

Shelton, A.M., Naranjo, S.E., Romeis, J., Hellmich, R.L., Wolt, J.D., Federici, B.A., et al., 2009. Appropriate analytical methods are necessary to assess non-target effects of insecticidal proteins in GM crops through meta-analysis (response to Andow et al. 2009). Environ. Entomol. 38, 1533–1538.

Simpson, M., Gurr, G.M., Simmons, A.T., Wratten, S.D., James, D.G., Leeson, G., et al., 2011. Attract and reward: combining chemical ecology and habitat manipulation to enhance biological control in field crops. J. Appl. Ecol. 48, 580–590.

Tack, A.J.M., Gripenberg, S, Roslin, T., 2011. Can we predict indirect effects from quantitative food webs? – an experimental approach. J. Anim. Ecol. 80, 108–118.

Van Driesche, R., Hoddle, M., Center, T., 2008. Control of Pests by Natural Enemies: An Introduction to Biological Control. Blackwell Publishing, Singapore.

Vorsino, A.E., Wieczorek, A.M., Wright, M.G., Messing, R.H., 2012a. An analysis of heterosis and outbreeding depression among lab-reared populations of the parasitoid *Diachasmimorpha tryoni* (Cameron) (Hymenoptera: Braconidae); potential implications for augmentative releases. Biol. Control 61, 26–31.

Vorsino, A.E., Wieczorek, A.M., Wright, M.G., Messing, R.H., 2012b. Using evolutionary tools to facilitate the prediction and prevention of host-based differentiation in biological control: a review and perspective. Ann. Appl. Biol. 160, 204–216.

Wright, M.G., Hoffmann, M.P., Chenus, S.A., Gardner, J., 2001. Dispersal behavior of *Trichogramma ostriniae* (Hymenoptera: Trichogrammatidae) in sweet corn fields: implications for augmentative releases against *Ostrinia nubilalis* (Lepidoptera: Crambidae). Biol. Control 22, 29–37.

Wright, M.G., Hoffmann, M.P., Kuhar, T.P., Gardner, J., Pitcher, S.A., 2005. Evaluating risks of biological control introductions: a probabilistic risk-assessment approach. Biol. Control 35, 338–347.

CHAPTER

15

Spiders – The Generalist Super Predators in Agro-Ecosystems

K. Samiayyan

Tamil Nadu Agricultural University, Coimbatore, India

15.1 INTRODUCTION

The order Araneae ranks seventh in global diversity of animals, after the five largest insect orders (Coleoptera, Hymenoptera, Lepidoptera, Diptera, Hemiptera) and Acari among the arachnids (Parker, 1982) in terms of species described or anticipated. Spiders are among the most diverse groups on Earth. Incidentally, earliest records date spiders to some 300 million years ago, 150 million years before flies began buzzing around (Ranjit Lal, 1995). Among these taxa, spiders are exceptional for their complete dependence on predation as a trophic strategy. In contrast, the diversity of insects and mites may result from their diversity in dietary strategies – notably phytophagy and parasitism (Mitter et al., 1988). They are of economic value to man because of their ability to suppress pest abundance in agro-ecosystems. Faced with the need to reduce pesticide usage on the world's crops and optimize natural biological control, a full investigation of the means by which spiders influence pest abundance is long overdue. Also, in recent years, there has been a realization by ecologists that components of agro-ecosystems are tractable to manipulation and that spiders are convenient model organisms. Consequently, there are a growing number of investigations in which spiders in agro-ecosystems are used as tools to gain fundamental insights into the role of generalist predators in community and ecosystem function. There is a little rhyme of English origin, but of uncertain age and derivation which says: 'If you wish to live and thrive, Let a spider run alive'.

15.2 ECOLOGY OF THE SPIDERS – HABIT AND HABITAT

Spiders exist in the most northern islands of the Arctic (Braendegaard, 1946; Jackson, 1930; Leach, 1966), the hottest and most arid of deserts (Cloudsley-Thompson, 1962; Schmoller,

D. P. Abrol (Ed): Integrated Pest Management.
DOI: http://dx.doi.org/10.1016/B978-0-12-398529-3.00016-6

1970), in the depth of caves (Komatsu, 1968), in the intertidal zone of ocean shores (Lamoral, 1968), in bogs and ponds (Judd, 1965), and on high, arid moorlands (Cherrett, 1964), sand dunes (Duffy, 1968; Lowrie, 1948) and flood plains (Berry, 1970; Sudd, 1972). Spiders have even reinvaded the aquatic environment and compete on even terms in the teeming communities in shallow water systems (Vogel, 1970). In all terrestrial environments, spiders occupy virtually every conceivable habitat, including the shelters and artifacts of a host of other animals (Coppel, 1977; Judd, 1965). The ecology of spiders has been well documented by several workers (Dondale, 1961; Dorris, 1970a; Specht and Dondale, 1960; Whitcomb, 1967).

Many factors undoubtedly influence the knitting of spider webs (Barnes, 1929, 1953; Duffy, 1968; Gibson, 1947; Sudd, 1972). Lowrie (1948) and Duffy (1968), both working in dune areas, suggested that the presence of certain plant forms as supporting structures of webs may be more important than microclimate in determining the distribution of some spiders. Many other factors such as temperature (Barnes and Barnes, 1954), humidity (Cherrett, 1964), sunlight (Pointing, 1965), air currents or winds (Cherrett, 1964; Eberhard, 1971), and height above the ground (Dowdy, 1951) have been shown to influence the choice of web site. These factors are undoubtedly closely linked to the physiological needs and tolerances of spider species (Turnbull, 1973). Spiders disperse by walking on the ground, by using silk thread bridges between plants as well as ballooning through the atmosphere from place to place on silken threads.

15.3 IMPORTANT SPIDER TAXONOMIC WORKS

Early taxonomic work focused on faunal and new species descriptions, often collected by the bounty of travellers and explorers. On the global scene, the United States of America leads araneological studies both on systematic and applied research. The notable contributors to the North American spiders are Kaston (1981), Coddington et al. (1990), Levi (1991) and Platnick (1989). Coddington (1990) and Platnick (1977) were instrumental and pioneers in the studies of the systematics of North American spiders using the modern tools of computer and cladistic analysis. Currently, interest is focused on the morphological features of male palp (Sierwald, 1990) and spinneret morphology (Platnick, 1990) for grouping the spiders.

Simon (1875) was the only author who contributed to the taxonomy of spiders in France during the early part of last century and thereafter, work on this line became scanty. Cambridge (1877) made taxonomic studies of the family Lycosidae in London and thereafter the work continued with significant regular publications on the systematics by Comstock (1940) and Roberts (1987) in the UK. Spider studies, particularly on taxonomy, are meagre in Asia. Eskov (1987) and Starabogatov (1985) have made an account of spiders in the former USSR, while Yaginuma (1977) and Barrion and Litsinger (1980) have studied the spiders of Japan and the Philippines, respectively. Raven (1988) was the prime worker on Australian spider systematics and made significant contributions through a series of publications. In New Zealand, Forster (1988) and his co-workers have made a series of publications on spiders representing the Oriental and Australian region. As noted above, it will be a huge task to revise the many genera of spiders from unstudied areas of the world (Coddington and Levi, 1991). The other notable and important workers who contributed to the study of world spider taxonomy include Bristowe (1958), Brignoli (1975), Lehtinen (1967), Levi and Levi (1968), Turnbull (1973) and Preston-Mafham (1991). Selden (1990) contributed to the world fossil spiders and Shear (1986) to the evolution of spider webs and to spider behaviour.

During the eighteenth century, foreign missionaries paved the way for the start of the study of spider taxonomy in the Indian subcontinent. Doleschall (1859) has written an article about the Indian arachnids and this forms the first written note on the Indian Arachnidae. During the period 1888 to 1906, Simon made some studies on the spiders of Andaman, Dhera-Dun, and Kurranchee apart from his general observations on Indian spiders.

About a century ago, Pocock (1900) wrote a volume on Arachnida in the *Fauna of British India* series, in which he stated, 'The spider group contains a vast number of species and is still very imperfectly known, so imperfectly that no satisfactory account of it can at present be given.' In that volume, Pocock dealt with spiders which were common and conspicuous. Proper systematic studies started only during the post-independence period. During this period, work both on systematics and other lines such as ecology, biology and toxicology were started by a few workers. Proper work on taxonomy was started by Tikader (1962). He undertook a monumental work on Indian Araneae and published several books, viz., *The Fauna of India: Spiders: Araneae* Vol. I and II (Tikader 1980, 1982), *Handbook of Indian Spiders* (Tikader 1987) and *Fauna of Khasi and Jaintia Hills* (Tikader 1968). The other very important publications of Tikader were *Spider Fauna of Andaman and Nicobar* (Tikader 1977); *Key to Indian Spiders* (Tikader 1977) and *Spider Fauna of Sikkim* (Tikader 1970).

The important documentation on spider fauna in different agro-ecosystems is summarized in Table 15.1.

15.4 SPIDERS IN COMMUNITIES

Spiders in agro-ecosystems are components of species-rich communities of herbivores, detritivores and natural enemies. The effect of a spider species on a pest population may be enhanced if the spider population increases rapidly in response to a rich supply of nutritious alternative prey (Jeffries and Lawton 1984). However, if the pest species is less preferred than the alternative prey, the net effect of these opposing processes on the level of pest control will be difficult to predict (Bilde and Toft, 1994). Selective predation by spiders in relation to the size of pest taken can alter the mean body size of the pest population, modifying its vulnerability to other size-dependent natural enemies in the community.

15.5 PREY SELECTION BY SPIDERS

Prey selection by the spider has been studied by several ecologists. Savory (1928) stated that 'spiders will eat all kinds of flies, wasps, bees, ants, beetles, earwigs, butterflies, moths and harvestmen, and woodlice and other spiders whenever the opportunity occurs; they show no trace of discrimination'. His views were supported by many records of prey species captured by free living spiders and by feeding tests with caged spiders (Turnbull, 1973). Lycosidae and Pisauridae are generally rapid runners also with good eyesight (Magni et al. 1965). On sighting a potential prey, they orient their body to bring the prey into the centre of vision of two large frontal eyes, charge forward, and seize the prey with the forelegs and chelicerae and subdue it with venom.

15.6 ESTIMATES OF SPIDER SPECIES DIVERSITY

Roughly 34,000 species of spiders had been named by 1988, placed in about 3000 genera and 105 families (Platnick, 1989). A small percentage of those species names will turn out to

TABLE 15.1 Documentation on Spider Fauna in Different Agro-Ecosystems

Country	Agro-ecosystem	References
Australia	Arable land	Thaler and Steiner (1975)
	Meadows	Luhan (1979)
Bangladesh	Rice	Kamal et al. (1990)
Belgium	Cereals	De Clercq (1979); Cottenie and De Clercq (1977)
	Diverse crops	Bosmans and Cottenie (1977)
Bulgaria	Pasture	Delchev and Kajak (1974)
Czechoslovakia	Pasture	Polenec (1968)
	Alfalfa	Miller (cited in Luczak, 1979)
	Sugar-beet	Miller (1974)
Colombia	Rice	Bastidas and Pantoja (1993)
Federal Republic of Germany	Clover	Boness (1958)
	Alfalfa	Boness (1958)
	Cereal	Basedow (1973)
	Vineyard	Kiran (1978)
	Orchards	Kramer (1961)
	Diverse crops	Heydemann (1953)
Finland	Cereals	Huhta and Raatikainen (1974)
France	Alfalfa	Chauvin (1960, 1967)
German Democratic Republic	Alfalfa	Geiler (1963)
	Cereals	Dietrich and Gotze (1974)
	Rape	Beyer (1981)
	Sugar-beet	Beyer (1981)
Great Britain	Pasture	Cherrett (1964)
	Meadows	Duffey (1974)
	Cereals	Fraser (1982); Locket (1978); Vickerman and Sunderland (1975)
	Potato	Dunn (1949)
	Sugar-beet	Thornhill (1983)
	Orchards	Chant (1956)
Hungary	Alfalfa	Balogh and Loksa (1956)
Norway	Strawberry	Taksdal (1973)

(*Continued*)

TABLE 15.1 (Continued)

Country	Agro-ecosystem	References
Poland	Pasture	Delchev and Kajak (1974)
	Meadows	Breymeyer (1967, 1978); Kajak (1960, 1962, 1971, 1978, 1980)
	Alfalfa	Luczak (1975)
	Cereals	Luczak (1979)
	Potato	Czajka and Kania (1976)
	Sugar-beet	Czajka and Goods (1976)
Switzerland	Cereals	Nyffeler and Benz (1981b)
	Rape	Nyffeler and Benz (1979)
Russia	Meadows	Vilbaste (1965)
	Cereals	Ashikbayen (1973)
	Potato	Koval (1976)
Brazil	Sugarcane	Barbosa et al. (1979)
Canada	Over-grazed pasture	Dondale (1977)
	Meadows	Fox and Dondale (1972)
	Orchards	Dondale et al. (1979)
China	Rice	Yang et al. (1990)
Panama	Seed reservation	Breymeyer (1978)
	Pasture	Breymeyer (1978)
	Banana	Harrison (1968)
Philippines	Rice	Barrion and Litsinger (1981a,b)
Peru	Cultivated fields	Aguilar (1965)
	Cotton	Aguilar (1974)
USA	Pasture	Howard and Oliver (1978); Wolcott (1937)
	Meadows	Wolcott (1937)
	Alfalfa	Muniappan and Chada (1970); Wheeler (1973); Yeargan and Cothran (1974)
	Cereals	Horner (1972)
	Grain sorghum	Bailey and Chada (1968)
	Sweet corn	Everly (1938); Blickenstaff and Huggans (1962); Culin and Rust (1980); Pimentel (1961)

(*Continued*)

TABLE 15.1 (Continued)

Country	Agro-ecosystem	References
	Cole	Rogers and Horner (1977)
	Guar	Hensley et al. (1961)
	Sugarcane	Negm and Hensley (1969); Burleigh et al. (1973)
	Rice	Young and Edwards (1990)
	Cotton	Clark and Qlick (1961); Dean et al. (1982); Dorris (1970b); Johnson et al. (1976); Kagan (1943); Leigh and Hunter (1969); McDaniel and Sterling (1982); Pfrimmer (1964); Pieters and Sterling (1974); Shepard and Sterling (1972); Whitcomb and Tadic (1963); Carroll (1980)
	Citrus	Muma (1973, 1975); McCaffrey and Horsburgh (1980)
	Orchards	Specht and Dondale (1960)
Israel	Citrus	Shulov (1938)
	Apple	Mansour et al. (1980a,b,c)
	Orchards	Mansour et al. (1981)
Japan	Cabbage	Kayashima (1960); Suzuki and Okuma (1975)
	Tea	Kaihotsu (1979); Terada et al. (1978)
	Mulberry	Kayashima (1972)
	Citrus	Kaihotsu (1979)
	Orchards	Hukusima (1961)
Korea	Mulberry	Okuma (1973)
Thailand	Orchards	Paik et al. (1973); Dondale (1966); Maclellan (1973)
Egypt	Clover	Negm et al. (1975)
	Cotton	Wiesmann (1955)
Fiji Islands	Coconut	Tothill et al. (1930)
New Guinea	Coffee	Robinson and Robinson (1974)
South Africa	Strawberry	Dippenaar-Schoeman (1976, 1979a,b)
Taiwan	Rice	Chu and Okuma (1970); Kiritani et al. (1972)
Thailand	Rice	Kiritani et al. (1972); Samiayyan (1996)

(*Continued*)

TABLE 15.1 (Continued)

Country	Agro-ecosystem	References
India	Maize	Sharma and Sarup (1979); Singh and Sandhu (1976); Singh et al. (1975); Samiayyan (1996)
	Cotton	Battu and Singh (1975); Muralidharan and Chari (1992)
	Citrus	Sadhana and Kaur (1974); Jandu (1972)
	Grapevines	Sadana and Sandhu (1977); Samal and Misra (1975)
	Coconut	Sathiamma et al. (1987); Dharmaraju (1962); Menon and Pandalai (1958); Mohamed et al. (1982)
	Sugarcane	Singh (1967); Samiayyan (1996)
	Household	Vijayalakshmi (1986)
	Mango	Venkatesan and Rabindra (1992)
	Tobacco	Sitaramaiah et al. (1980)
	Pulses	Samiayyan et al. (2012)
	Sorghum	Hiremath (1989)

be synonyms. Families with over 1000 species described are Salticidae (jumping spiders; ca. 490 genera, 4400 species), Linyphiidae (dwarf or money spiders, sheet web weavers; ca. 400 genera, 3700 species), Araneidae (common or web weavers; ca. 160 genera, 2600 species), Theridiidae (cob web weavers; ca. 50 genera, 2200 species), Lycosidae (wolf spiders; ca. 100 genera, 2200 species), Gnaphosidae (ground spiders; ca. 140 genera, 2200 species), and Thomisidae (crab spiders; ca. 160 genera, 2000 species). Although the aforementioned families are cosmopolitan, the Linyphiids are most diverse in the north temperate regions, whereas the others are most diverse in the tropics or show no particular pattern. Because spiders are not thoroughly studied, estimates of total species diversity are difficult. The fauna of Western Europe (especially England) and Japan are most completely known (Roberts, 1985, 1987; Yaginuma, 1977). The Nearctic fauna is perhaps 80% described (Coddington et al., 1990), New Zealand perhaps 60–70% (Court and Forster, 1988; Forster, 1967; Forster and Blest, 1979; Forster and Wilton, 1973) and Australia perhaps 20% (Raven, 1988). Other areas, especially Latin America, Africa and the Pacific region are much more poorly known. About one-third of all genera (1090 in 83 families) occur in the Neotropics. If the above statistics suggest that 20% of the world fauna have been described, then about 170,000 species of spiders are extant and yet to be discovered. In India, only during the post-independence period was work on spider systematics, ecology and biology started (Tikader, 1962; Samiayyan, 1996).

15.7 FACTORS RESPONSIBLE FOR SPIDERS AS POTENTIAL BIOCONTROL AGENTS

In the pursuit of the identification of selective biocontrol agents, spiders can be considered for the control of insect pests because (i) they are natural entomophagous predators, (ii) they can kill a large number of insects per unit time, (iii) they have high searching ability (especially hunting spiders), and (iv) they predate on a wide variety of insects. They are curious animals and kill a higher number of insects than they actually consume (Greenstone, 1978). Spiders have further suitable adaptations associated with their phylogeny and life style. They have (i) low maintenance energy requirements (Anderson, 1974), (ii) a highly distensible abdomen permitting them to gorge during times of prey abundance (Palanichamy, 1980), (iii) the ability to store large amounts of fat (Collatz and Mommsch, 1975), and (iv) the ability to lower metabolic rate during periods of starvation (Anderson, 1974; Nakamura, 1972) thus surviving for a longer duration. However, the cannibalistic behaviour of spiders imposes problems of mass culture for use in biological control.

Spiders could be mass multiplied if certain precautions are taken, i.e. if the spiderlings are separated and kept individually in specimen tubes and supplied with the first instar caterpillars of the laboratory host, *Corcyra cephalonica* Stainton, until they moult for the third time. At this stage, they can be easily be transported for release on a crop, keeping in view that they do not congregate in one place in the field to safeguard against the expected cannibalism. Spiders have a long life ranging from 1 to 18 years and they can withstand adverse conditions of less food material and can starve for very long periods. After emergence from eggs, spiderlings are also predatory. Spiders also predate on the eggs of insects. None of the spiders is harmful to agriculture and they have a very high rate of feeding. Natural enemies of spiders do not have much detrimental effect on their efficacy as predators. Spiders are placed in the upper trophic levels of the food chain and hence receive a higher accumulation and concentration of pesticides in their fat bodies taken through their insect prey.

15.8 ROLE OF SPIDERS IN AGRO-ECOSYSTEMS

Spider competition potentially occurs in agro-ecosystems because the community structure and physical structure of these engineered ecosystems may promote high densities of a few spider species. How important competition is in moderating the impact of spider populations on pest insect populations will depend on the spider species present in the fields. Wiesmann (1955) attributed an important role to ground dwelling spiders as predators of insect pests with regard to forest ecosystems. Spider microhabitat associations have been found to be frequently linked with patches of abundant prey (Riechert, 1982; Turnbull, 1966), and actual movement from patches of decreasing prey density to those affording higher densities has been reported. In short, generalist predators tend to maintain prey populations at low densities (Kajak, 1978). In another study, the presence of micryphantids (of the family Linyphiidae) in experimental plots resulted in significantly less leaf damage by tobacco cutworm *Spodoptera litura* than was observed in plots from which the spiders had been removed (Yamanaka et al., 1973). Here the primary predatory effect was one of causing the larvae to abandon plants occupied by spiders. Reductions in crop damage through actual spider predation or spider-caused abandonment of plants are also known for greenbug (Horner, 1972), leaf fly (Kayashima, 1961) and leaf hopper-plant hopper (Kiritani and Kakiya, 1975).

In general, spiders are well represented in such systems, though diversities and densities are lower than in natural systems – especially where cultivation is intensive.

Brinjal, okra and tomato with hemipterous insects were also found to harbour three species of lycosids, *Pardosa birmanica* Simon, *Lycosa chaperi* Simon, and *Lycosa himalayensis* Gravely (Anonymous, 1971). Salticids were also reported to prey upon the larvae of *Heliothis* sp. (Whitcomb and Tadic, 1963), and *Choristoneura fumiferana* (Clem.). The larvae of *Hyphantria cunea* (Drury) were preyed upon by clubionids such as *Cheiracanthium inclusum* (Hentz), *Aysha gracilis* (Hentz) and *Anyphaena* sp. and a salticid, *Phidippus putmanii* (Peckman). Raodeo et al. (1973) observed that the social spider *Stegodyphus sarasinorum* was a predator for the lemon butterfly, *Papilio demoleus* Linn. on orange trees in a citrus garden. Singh et al. (1975) revealed that 13 species of spiders belonging to seven families have been recorded preying upon the larvae of *Chilo partellus* in the field. These comprise *Araneus* sp. (Argiopidae), *Cheiracanthium* sp., *Clubiona* sp. (Clubionidae), *Drassodes* sp. (Gnaphosidae), *Heteropoda* sp. (Heteropodidae), *Lycosa* sp. (Lycosidae), *Oxyopes pandae* Tikader, *Oxyopes* sp. (Oxyopidae), and *Marpissa* sp. *Phidippus* sp. (Salticidae), *Araneus* sp., *Clubiona* sp., *Lycosa* sp., *Oxyopes* sp., and *Marpissa* sp. were observed preying upon the nymphs of sugarcane *Pyrilla*, *Pyrilla perpusilla*. Subsequent observations showed that *Araneus* sp. and *Marpissa* sp. along with *Heteropoda* sp., *Phidippus* sp. and *Oxyopes pandae* Tikader predate upon the nymphs of citrus psyllids (Jandu, 1972). Kumar and Monga (1996) revealed that spiders *Zygoballus indica*, *Lyssomanes sikkimensis*, *Myrmarachne bengalensis* and *Lycosa mackenziei* predate on *Idioscopus* sp. and *Drosicha mangiferae* and moderately prefer *Dacus dorsalis* and *Dacus zonatus*. Spiders such as *Oxyopes javanus*, *Argiope aemula*, *Nephila maculata*, *Lycosa psedudoannulata* and *Tetragnatha japanicola* (Joshi et al., 1987) were recorded predating on the larvae of *Cnaphalocrosis medinalis*. The potential of spiders for biological control is only available for the members of Thomisidae, Araneidae, Lycosidae, Oxyopidae, Eresidae, Clubionidae and Hersilidae.

Gertsch (1949) has summarized that spiders are among the dominant predators of any terrestrial community. When the fauna of the soil and plant cover is analyzed, they come to light in vast numbers in such convincing abundance and it is evident that they play a significant part in the life of every habitat. In a pigeon pea ecosystem, 15 spider families including 58 genera/species were recorded from South India and families such as Araneidae, Lycosidae and Salticidae were found to be dominant in occurrence. Maximum numbers of individuals were recorded for *Peucetia viridans* (Sudha et al., 2011). Another study *viz.*, spider diversity in different short duration food legume ecosystems, showed that abundance of spider fauna in black gram, green gram and cow pea was high recording 13 families of spiders including 44 genera/species in both kharif and rabi seasons. Kharif season recorded more species richness, diversity and evenness than rabi (Samiayyan et al., 2012).

In the cotton ecosystem, it was observed by Sivasubramanian et al. (2009) that sucking pests (leaf hoppers, aphids and whiteflies) were consumed by all of the instars of spiders whereas caterpillars (*Helicoverpe armigera* and *Spodoptera litura*) were eaten only by late instars and adult spiders. Among the sucking pests, leaf hoppers were highly preferred by the lynx spider and *H. armigera* was preferred over *S. litura*. In the cotton ecosystem, the hunting spiders preferred prey insects in the order of leaf hoppers>aphids>whiteflies, whereas web spiders preferred prey insects in the order of leaf hoppers>whiteflies>aphids (Vanitha et al., 2009).

15.9 SPIDERS AS PREDATORS OF INSECT EGGS

The phenomenon of egg predation in spiders may be more common than was previously thought. Oophagy has often been overlooked in the past probably due to recognition difficulties in the field. In the majority of these studies, spiders were found preying upon the eggs of lepidopteran pests; however, Richman et al. (1983) provided evidence for spiders feeding on the eggs of Coleoptera. Examples of oophagy of spiders on Insecta are given in Table 15.2.

15.10 PREDATORY POTENTIAL OF SOME SPIDERS

Lycosa pseudoannulata is considered an effective predator of rice hoppers in Taiwan, Japan, and the Philippines (Kenmore et al., 1984). Of the species preyed upon by the spider, 80% were green leafhopper (GLH) and brown planthopper (BPH) (Kiritani et al., 1972; Sasaba et al., 1973). The mean indices of food preference of adult female spiders assessed in the laboratory were 0.49 and 0.60 for BPH and *Nephotettix cincticeps*, respectively (Sasaba et al., 1973). Kiritani et al. (1972) also indicated that the ratio of prey taken by the spider was 5:2 GLH/BPH in Japan. At IRRI (1978), scientists found that *Tetragnatha* sp. and *Argiope* sp. had no apparent effect on prey and *Callitrichia* sp. killed only a few (0.5 nymph/day). In another experiment, *Oxyopes* killed 3 nymphs out of 15 daily for 7 days and *L. pseudoannulata* killed 8 nymphs/day. In the laboratory, Chiu et al. (1974) observed that *Oedothorax insecticeps* Boes. Et. Str., a micryphantid spider, consumed an average of 1.84 and 3.00 adult BPH for second and fourth instar spider nymphs and 3.20 and 2.03 BPH for adult female and adult male spiders, respectively. The average number of BPH preyed on by each adult female was about 1.5 times that preyed on by a male. Dyck et al. (1976) reported that *L. pseudoannulata* killed on average about 15 BPH nymphs per day for 2 weeks under controlled conditions. In India, about 20 species of spiders were found preying on leaf and planthoppers and a single salticid spider consumed on average 18 to 20 adults per day (Samal and Misra, 1975) and similar effect of spiders on plant and leafhoppers were also reported by Samiayyan and Chandrasekaran (1998). A lycosid is known to feed on as many as 20 planthoppers per day. Its voracious appetite makes it a very important natural enemy (Figure 15.1). However, one of the questions often asked about this predator is, 'What will it feed on in the absence of brown planthoppers?' Like other spiders, *Lycosa* and *Oxyopes* do not depend entirely on planthoppers for food. There are many flies in the field, which provide the bulk of the food for spiders (Gavarra and Raros, 1975). Studies in Indonesia have shown the importance of 'neutrals' in supporting a large population of predators in fields. Spiders are found in rice fields before planting, when they survive on these 'neutrals' (Figure 15.2). During the dry season, rice field spiders are known to hide in crevices or in grasses around the field.

15.11 ADDITIONAL FEATURES OF SPIDERS IN PEST CONTROL

15.11.1 Pest Dislodgement

The foraging behaviour of spiders on crop vegetation may disturb pest aggregations and may also cause the disturbed pests to walk or fall off the plants. This can reduce the pest population if the physical conditions on the ground cause rapid mortality (Dill et al., 1990), or if they cannot easily regain the plants, or if they move into danger zones with greater probabilities of attack by natural enemies. It is possible that spiders dislodge pest species belonging to many orders, but the literature emphasizes an effect on Lepidoptera. Dislodgement resulted in death of the pest. Under less extreme

TABLE 15.2 Oophagy by Spiders upon the Eggs of Insects

Spider Taxa	Egg	Habitat	Method	Country	References
Theridiidae *Coleosoma acutiventer* (Keyserling)	DS	Sugarcane	c	USA	Negm and Hensley (1969)
Linyphiidae *Eperigone tridentata* (Emerton)	DS	Sugarcane	c	USA	Negm and Hensley (1969)
Araneidae *Neoscona arabesca* (Walckenaer)	AA	Cotton	c	USA	Gravena and Sterling (1983)
Lycosidae *Lycosa antelucana* (Montgomery)	NO	Soybean	a	USA	McCarty et al. (1980)
Pardosa milvina (Hentz)	DS	Sugarcane	c	USA	Negm and Hensley (1969)
Oxyopidae *Oxyopes salticus* (Hentz)	HV NO	Cotton Soybean	a a	USA	McDaniel and Sterling (1982) McCarty et al. (1980)
Peucetia viridans (Hentz)	AA MS	Cotton Tobacco	a c	USA	Gravena and Sterling (1983) Madden and Chamberlin (1945)
Clubionidae *Cheiracanthium diversum* L. Koch	HS	Cotton	a, c	Australia	Room (1979)
Trachelas deceptus (Banks)	DA	Citrus	c	USA	Richman et al. (1983)
Anyphaenidae *Aysha velox* (Becker)	DA	Citrus	c	USA	Richman et al. (1983)
Thomisidae *Misumenops* sp.	HV	Cotton Cotton	c	USA	McDaniel and Sterling (1982) Lincoln et al. (1967)
Metaphidippus flavipedes (G. and E. Peckham)	CF	Fir	c	USA	Jennings and Houseweart (1978)
Phidippus audax (Hentz)	HZ	Cotton	a	USA	Nuessly (1986)

a = radioisotope studies; c = Direct Observation. Egg taxa: AA, Alabama argillacea (Hiibner); DS, Diatraea saccharalis (Fab.); NO, Noctuidae; HV, Helicoverpa virescens (Fab.); MS, Manduca sexta (L.); HS, Heliothis sp.; HZ, Heliothis zea (Boddie); CF, Choristoneura fumiferana (Clemens); DA, Diaprepes abbreviatus (L.).

FIGURE 15.1 A *lycosid* feeding on a brown planthopper.

FIGURE 15.2 An *oxyopid* feeding on a 'neutral' fly when pest species are scarce.

conditions, the loss of feeding time resulting from dislodgement may be expected to reduce plant damage and also to reduce the rate of increase in the pest population.

15.11.2 Mortality of Insect Pests in Spider Webs

Small pests, such as thrips, midges and aphids, may die by being caught in the webs of large spiders, even when they are ignored by the spider (Nentwig, 1987). Alderweireldt (1994) identified 319 prey items in webs of linyphiid spiders in maize fields in Belgium. Spiders were feeding on only 184 of these prey items. Linyphiidae, Dictynidae, Theridiidae and Agelenidae do not renew their webs daily, and feed infrequently, so these families may contribute to pest control by the action of their webs. First instars of the cereal aphid *Sitobion avenae* did not escape from webs of non-attacking satiated adult female linyphiid spiders, *Lepthyphantes tenuis*. Thus the potential of webs to kill pests, in the absence of spider attack, can be a relevant consideration for biological control.

15.11.3 Wasteful Killing, Partial Consumption and the Wounding of Pests

Under certain circumstances, the spider may kill a pest but subsequently ingests little (partial consumption) or none (variously referred to in the literature as 'superfluous killing' and 'wasteful killing') of the pest's biomass. This is advantageous for pest control because it will result in more pests being killed per unit of spider food demand. These behaviours are usually observed when prey is plentiful (or when a small spider is able to overcome a large prey) and the spider is nearly or completely satiated. There are examples of wasteful killing at high prey density for Clubionidae and Linyphiidae against aphids and for Lycosidae (Samu and Biro, 1993) against flies.

15.12 SPIDER VENOM PEPTIDES AS BIOINSECTICIDES

The chemical complexity of spider venoms is unusual, varying from salts and small organic compounds to large presynaptic neurotoxins (Khan et al., 2006). Based on the mechanism of action and chemical structure, these venoms can be broadly grouped into five classes. These groups include (i) salts and small organic compounds, (ii) linear cytolytic peptides, (iii) disulphide-rich peptide neurotoxins,

(iv) enzymes, and (v) large presynaptic neurotoxins. In contrast with most chemical insecticides, Insecticidal Spider Venom Peptides (ISVPs) are unlikely to be topically active, because, to access their sites of action in the insect nervous system, they would have to penetrate the insect exoskeleton, which comprises an outer lipophilic epicuticle and a heavily sclerotized exocuticle. It is possible that clever peptide analoging, as has been used to confer both oral and topical activity on small insect kinin neuropeptides (Nachman and Pietrantonio, 2010), could be used to engineer topically active ISVPs.

A potential advantage of some ISVPs is that they have novel modes of action compared with extant chemical insecticides; hence, they might be particularly useful for control of arthropod pests that have developed resistance to multiple classes of chemical insecticides. ISVPs can be useful even in situations where they have the same molecular target as an insecticide to which an insect population has developed resistance. This seems counterintuitive, but it is possible because most arachnid toxins act at sites different from those targeted by chemical insecticides. Thus, target-site mutations that confer resistance to chemical insecticides can increase susceptibility to peptide neurotoxins that act on the same target. For example, even though the scorpion toxin AaIT and pyrethroids both target Na_V channels, a pyrethroid-resistant strain of *Heliothis virescens* was more susceptible than non-resistant strains to a recombinant baculovirus expressing AaIT (McCutchen et al., 1997).

15.13 EFFECT OF AGRONOMIC PRACTICES ON SPIDERS

Cultivation regularly destroys vegetation complexity and significantly reduces the local spider community. Many agro-ecosystems are also ephemeral habitats and the existence of spiders depends upon the growing season of annual crops. Srikanth et al. (1997) revealed that the exclusion of selected cultural practices, manual weed control, earthing-up and three detrashing operations significantly ($P < 0.05$) increased the spider population in the latter part of crop growth, particularly the soil associated *Hippasa greenalliae* Blackwall (Lycosidae). Vegetational complexity, particularly involving weeds, has been associated with increased prey and predator (Ali and Reagan, 1985) diversity and abundance on the soil surface and on foliage in sugarcane. The planting and harvesting procedures utilized in agricultural systems are perhaps even more disruptive to spider communities than the use of pesticides. At least once each year, both the habitat and beneficial fauna are destroyed. Aside from the obvious problem with loss of egg sacs and the general suppression of spider numbers, habitat structure is lost, and this is a major determinant of spider community diversity. The extent to which spider control of prey is realized in agro-ecosystems, however, is limited by the disruptive effects of the application of insecticides and the annual harvesting and tilling of the vegetation-ground layer. Both practices represent major sources of mortality to spider populations (Dondale, 1979). Recreational activities, especially walking, have been reported to affect the structure and composition of the spider fauna in dune grasslands to a considerable extent. High trampling intensities strongly reduce the number of spider species. Moderate and low intensities have a more selective effect (Merrett, 1978). Spraying of 5% neem seed kernel extract was less harmful to spiders compared to insecticides on rice against leaf and planthoppers (Samiayyan and Chandrasekaran 1998). Application of bio-fertilizer, *viz.*, azolla @ 2 kg/cent and intercropping of *Sesbania rostrata* in rice (8:1) was found to favour harbouring of a higher number of spiders in the rice ecosystem (Samiayyan and Jayaraj, 2009).

15.14 INTEGRATION OF SPIDERS IN IPM

Natural biological control in irrigated rice at the early crop stages can be mainly attributed to spiders. Orb-weaving spiders are the most abundant spiders assessed across the cropping season, with *Tetragnatha* spp. being the single most common genus in South East Asian countries, apart from the Philippines where *Pardosa pseudoannulata* is the more common species. Heong et al. (1992) found a relative abundance of *P. pseudoannulata* totalling 25–54% of all spiders at five rice sites in the Philippines across the season. In the first 35 days after transplanting, the dominant predators in irrigated rice are the lycosid *P. pseudoannulata* (Bösenberg & Strand) and the linyphiid *Atypena formosana* (Oi) (Sigsgaard et al., 1999: the Philippines; Sahu et al., 1996: Northern Bihar, India). *P. pseudoannulata* is most common among the tillers at the base of the plants. It preys on a wide array of insect pests, including leafhoppers and planthoppers, whorl maggot flies, leaf folders, caseworm and stem borers (Barrion and Litsinger, 1984; Shepard et al., 1987; Rubia et al., 1990). Field densities of both spider species co-vary with hopper densities (Reddy and Heong, 1991). *A. formosana* adults and immatures prefer to live among the rice stem or at the base of rice hills. They have been observed to hunt for nymphs of planthoppers and leafhoppers, Collembola, and small dipterans, such as whorl maggot flies (Barrion and Litsinger, 1984; Shepard et al., 1987; Sigsgaard et al., 1999).

15.14.1 Spiders and their Role in the Irrigated Rice Agro-ecosystem – Detritivores and Organic Material

The population build-up of natural enemies is dependent on the availability of suitable host/prey. The abundant detritivores early in the season may be one key to the success of the current rice agro-ecosystem (Settle et al., 1996). Being polyphagous predators, spiders can prey on alternative prey such as Collembola during fallow periods, thereby maintaining high population levels. (Here the term alternative prey is used to describe all suitable prey other than the target species.) The levels of these alternative prey in turn depend on decaying organic material available in the field. Field and laboratory data from research at the International Rice Research Institute (IRRI) in the Philippines and elsewhere indicate that spiders survive and build up their populations on alternative prey, such as Collembola and dipterans, before the crop is established and in the first weeks after crop establishment (Guo et al., 1995; Settle et al., 1996). Settle et al. (1996) were able to increase the number of detritus feeders, such as collembolans and plankton feeders by adding organic material to the rice field in the treated plots. Most interestingly, the number of spiders increased in the same plots. Plankton feeders in that study included mosquito larvae and chironomid midge larvae, of which many species also feed on detritus (Settle et al., 1996). In a study at IRRI, the addition of rice straw bundles in the rice field after harvest increased the number of *A. formosana* and *P. pseudoannulata* as well as plant and leafhoppers (Shepard et al., 1989). Though the study by Shepard et al. (1989) did not report effects on Collembola density, high Collembola density can be observed in recently cut straw, so probably the beneficial effect was also due to an increase in Collembola. In upland, rice weed residues placed within the rice fields can significantly increase spider densities (Afun et al., 1999). Apart from providing refuges for predators and increasing the density of alternative prey, organic material will also influence plant nutrition, which in turn can influence herbivores feeding on the crop. One can speculate that this in turn could indirectly affect predators.

15.14.2 Bunds and Surrounding Habitats

Between the irrigated rice fields, there are usually bunds, which may be narrow and low and reconstructed often with low and poor vegetation, or which may be wider and higher and with more permanent vegetation. Some bunds are used for growing vegetables or fruits. The bunds surrounding the rice fields provide refugia for predators during fallow periods as well as during farm operations. Bunds may be particularly important as a source of colonization by ground dispersing predators, such as large *P. pseudoannulata* spiderlings and adults, and may be less important for linyphiids, such as *A. formosana*, which colonize the rice field by ballooning. Preliminary results from a study of the directional movement of predators between the rice field and the bund show that *P. pseudoannulata* is an early colonizer of newly established rice, with the highest relative abundance of *P. pseudoannulata* in the bund, stressing the importance of this habitat (Sigsgaard et al., 1999). The same study showed that 3 or 4 weeks after transplanting of rice, the directional movement changed and the early planted field may have become a source of *P. pseudoannulata* to later planted fields. Even within the soil cracks of the fallow rice field, some spiders such as *P. pseudoannulata* are commonly found (Arida and Heong, 1994). The management of bunds can also affect spiders. Grazing of bunds reduced the density of web-building spiders as well as of two hunting spider families, Lycosidae and Oxyopidae, probably due to loss of webbing sites for the web-building spiders and hunting grounds for the hunting spiders (Barrion, 1999). Rice fields are usually intermingled with other crops and habitats such as coconut or banana, and houses, gardens, fallow fields and forests, creating a varied landscape mosaic. Rice is often grown in rotation with vegetables such as onions, or with legumes. Brown planthopper predators, hunting spiders, and the soil fertility imbalance result in luxuriant crop growth conducive to pathogen invasion and reproduction. This is compounded by genetic uniformity of the crop stand, which allows unrestricted spread of the disease from one plant to another, and by continuous year-round cropping that carries over the pathogen to the succeeding seasons.

15.14.3 Studying the Impact of Predators – an Exclusion Cage Experiment in Rice

In exclusion cage experiments, cages were initially cleaned of all arthropods. Pairs of brown planthoppers (one pair per hill) were introduced into the cage. After 24 h, some cages were opened at bottom-most to allow predators in but keeping in the brown planthoppers. One and a half months later, brown planthopper populations had reached very high levels in cages where predators were kept out (closed cages) while populations remained low in cages where predators were present (cages opened). This simple experiment is one of the most effective ways of showing that predators are important in keeping brown planthopper populations low.

Integrated Pest Management (IPM) aims to avoid harming natural crop spiders. For this, IPM attempts to synchronize the timing of spraying of pesticides with the life cycle of the pests, and their natural enemies (predatory spiders and mites) (Bostanian et al., 1984; Volkmar, 1989; Volkmar and Wetzel, 1992). IPM also endeavours to use chemicals that act selectively against pests but not against their enemies. Few studies actually investigate the effects of insecticides other than their direct toxicity (usually LD_{50}) on non-target animals. However, living organisms are finely tuned systems; a chemical does not have to be lethal to threaten the fitness (physical as well as reproductive) of the animal, with unpredictable results on the structure of the biological community (Culin and Yeargan, 1983; Volkmar and Schützel, 1997; Volkmar and

Schier, 2005). Pesticides may affect the predatory and reproductive behaviour of beneficial arthropods short of having direct effects on their survival. Thus, to show that a pesticide is relatively harmless, or indeed has no measurable effect at all, behavioural studies on the effects of sublethal dosages are necessary. Such studies are not often done, presumably because of their costs and methodological difficulties (Vollrath et al., 1990; Volkmar et al., 1998, 2002, 2004).

15.14.4 Effect of Neem Products on Spiders

In fact, neem oil (NO) (3%) and aqueous neem seed kernel extract (NSKE) (5%) were quite safe for spiders, though endosulfan induced 100% mortality of the predators (Fernandez et al., 1992). NSKE, NO or neem cake extract (NCE) (10%) treated rice plots had better recolonization of spider *L. pseudoannulata* than in monocrotophos (0.07%) treated plots after 7 days of treatment (Raguraman, 1987; Raguraman and Rajasekaran, 1996). A similar observation on rice crops was made by Nirmala and Balasubramanian (1999) who studied the effects of insecticides and neem-based formulations on the predatory spiders of rice ecosystems. Lynx spider, *Oxyopes javanus*, was less sensitive to NO (50% EC) than *L. pseudoannulata* (LC_{50} values = 9.73% and 1.18%, respectively) (Kareem et al., 1988; Karim et al., 1992), thereby confirming that NO was the safest pesticide for spiders. In corn fields (Breithaupt et al., 1999) and cabbage fields (Saucke, 1995) in Papua New Guinea, no significant effect was observed against *Oxyopes papuanus* from aqueous NSKEs (2%) or Neem Azal-S treatments. Serra (1992) did not observe adverse effects from NSKE 4% applied on unidentified spiders in tomato fields in the Caribbean. Babu et al. (1998) reported that a combination of seedling root dip in 1% neem oil emulsion for 12h + soil application of neem cake at 500kg/ha + 1% neem oil spray emulsion at weekly intervals gave an effective level of control of green leafhopper (*Nephotettix virescens*) infesting rice (var. Swarna). A combination of neem oil + urea at a ratio of 1:10 when applied three times at the basal, tillering and panicle initiation stages gave a superior level of control of brown planthopper (*Nilaparvata lugens*). The treatments, urea + nimin [neem seed extract] and a seedling root dip with 1% neem oil emulsion + neem cake at 500kg/ha + 1% neem oil spray emulsion at weekly intervals was equally effective against *N. lugens*. All neem products had little effect on predators *L. pseudoannulata* and *Cyrtorhinus lividipennis* (Sontakke, 1993; Babu et al., 1998). NSKE sprays at 5%, 10% and 20% were also substantially safe for spiders and ants in cowpea. Natural biological control in irrigated rice at the early crop stages can mainly be attributed to spiders. Orb-weaving spiders are the most abundant spiders assessed across the cropping season, with *Tetragnatha* spp. being the single most common genus in South East Asian countries, apart from the Philippines where *Pardosa pseudoannulata* is the more common species. Nanda et al. (1996) tested the bioefficacy of neem derivatives against the predatory spiders wolf spider (*L. pseudoannulata*), jumping spider (*Phidippus* sp.), lynx spider (*Oxyopes* sp.), dwarf spider (*Callitrichia formosana*), and orb spider (*Argiope* sp.), and damselflies (*Agriocnemis* sp.) and mirid bug (*C. lividipennis*). It was observed that the neem kernel extract and oil were relatively safer than insecticides to *L. pseudoannulata, Phidippus* sp. and *C. lividipennis* in field conditions (Sithanantham et al., 1997).

15.14.5 Effect of Azolla on Insect Predators

Among cultural practices, wider spacing and low fertilizer levels were found to favour harbouring of more spiders per plant (Vanitha et al., 2009) in the cotton ecosystem. Higher numbers of predators were found in fields with azolla than in those without. Species inhabiting the water surface, which mostly preferred areas with azolla – *Lycosa pseudoannulata* (Boes. Et Str.) (Lycosidae) – were predators of rice hoppers. Spiders inhabiting the rice bund – *Pardosa birmanica* Simon, *Arctosa janetscheki* Buchar, and *Hippasa rimandoi* Barrion (Lycosidae) – moved into fields covered with azolla.

15.14.6 Integration of Host Plant Resistance and Bio Control against BPH

Moderate resistance has also been shown to increase the efficiency of the spider *L. pseudoannulata* feeding on brown planthopper (BPH). This study indicated that combining moderately resistant varieties such as Thriveni (which exert less selection pressure for biotype shifts) with biological and chemical control is a promising means of BPH control. Among the predators, the spider *Lycosa pseudoannulata*, the vellid bug *Microvelia atrolineata*, and the mirid bug *C. lividipennis*, are common in rice (Table 15.3) (Kartohardijono and Heinrichs, 1984).

TABLE 15.3 Population of *N. lugens* as Affected by Varietal Resistance and Predation

Predator	*N. lugens* Mortality (%)			
	Triveni (MR)		TN 1 (S)	
	Without Predators	With Predators	Without Predators	With Predators
Lycosa pseudoannulata (Spider: Lycosidae)	26.9[b,c]	46.6[a]	14.8[d]	36.0[a,b]
Cyrtorhinus lividipennis (Green mirid bug)	8.3[b]	34.2[a]	0.0[c]	10.0[b]

[a], XX
[b], XX
[c], XX
[d], XX.

Integrated management of *N. lugens* shown that varietal resistance and predation are highly compatible. The mortality of *N. lugens* was assessed on the rice varieties PTB 33, TR 64, IR 36, CO 42 and TN 1 with and without the wolf spider during July to August 1988. A significantly high percentage of mortality was recorded on PTB 33 (20.0) followed by IR 64 (15.2) compared to other varieties when non-spiders were introduced. Mortality due to both varietal resistance and predation by spiders was on par in all varieties ranging from 27.6% in susceptible TN 1 to 40.5% in PTB 33. This increased mortality of hoppers due to predation in resistant PTB 33 is simply an additive effect, because when the effect of varietal resistance on *N. lugens* was eliminated using Abbott's formula, the predation rate was not significantly different among varieties (Senguttuvan et al., 1990). Efficiency of spider *L. pseudoannulata* when feeding on BPH on a resistant (ASD7) (Choi, 1979) and a susceptible (TN1) variety showed that the ratio of BPH to spider was 20:1 (Figure 15.3).

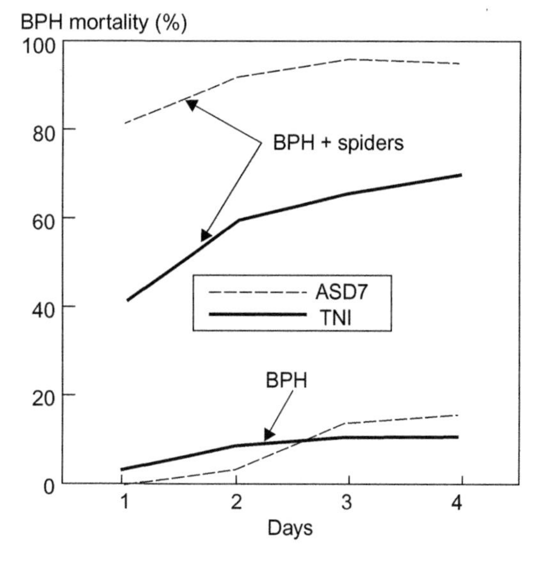

FIGURE 15.3 Efficiency of spider *L. pseudoannulata* when feeding on brown planthopper (BPH) on a resistant (ASD7) and a susceptible (TN1) variety.

15.14.7 Integration of Insecticide with Spiders

The susceptibility of this predator–prey system to an insecticide was examined by computer simulation, assuming different levels of effectiveness of the insecticide in killing predator and prey. The simulation showed that more predators were reduced than prey, even in the case of equal toxicity to both species. The increase in prey species after application of non-selective insecticides was also demonstrated. The effectiveness of the combination of use of a resistant variety together with spiders in controlling hoppers was examined by simulation. The simulation suggested that the presence of a moderate degree of resistance to hoppers is sufficient to control them.

The relative toxicity of insecticides applied topically to BPH, *N. lugens*, to its predator, spider *L. pseudoannulata*, is calculated as: LD_{50} of hopper – LD_{50} of predator. A relative toxicity of 1.0 indicates that the insecticide is relatively more toxic to the predator. The absence of LD_{50} on *Lycosa* up to 5000 ppm perthane indicates its low toxicity to the predator (Figure 15.4).

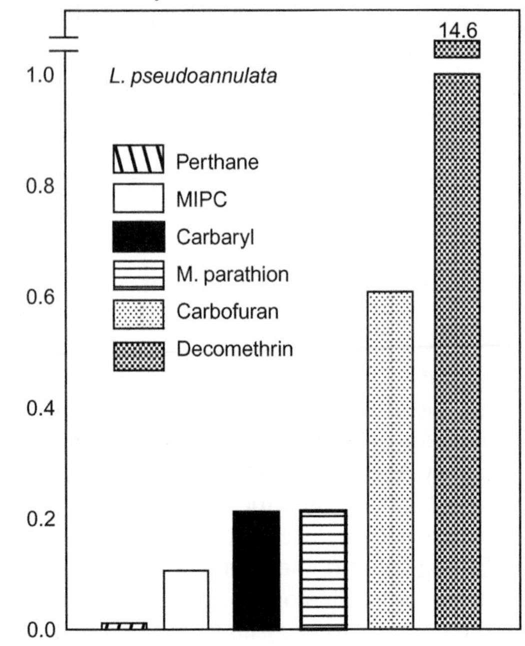

FIGURE 15.4 Relative toxicity of insecticides applied topically to BPH, *N. lugens*, to its predator, spider *L. pseudoannulata*. *Courtesy: Henrichs and Fabellar.*

15.15 STATUS OF ARANEOLOGISTS

Paralleling the loss in taxonomic expertise worldwide, the job situation for systematic araneologists is sufficiently poor that many have left the field and few are entering it. The age structure of systematic araneologists is therefore significantly skewed towards older workers compared to non-systematic araneologists in North America (Coddington et al., 1990). The number of araneologists in non-systematic disciplines has increased much more rapidly and consequently the need for identifications and taxonomic advice has outstripped the ability of systematists to supply it (Riechert et al., 1985). About 24 arachnological societies exist around the world, eight of which publish research journals (Cokendolpher, 1988). The Centre International de Documentation Arachnologique, with about 1000 members, is the major international society for non-acarine arachnid researchers.

15.16 CONCLUSION

Conservation and augmentation of spiders in fields are simple, yet efficient methods of pest control. Their importance as pest control agents is not a recent discovery. On the other hand, farmers from time immemorial have acknowledged their value. Spider species often have complementary niches and so an assemblage of species may be able to attack all growth stages of a pest, thus reducing 'enemy-free' space and improving the prospects for effective biological control. In addition to killing pests, chemical pesticides are taking a heavy toll on useful insects. Ways are being sought to promote the effective use of spiders in biological control but it should be noted that spiders would, for the foreseeable future, be embedded in integrated management systems. Preservation of spiders necessitates abandoning of these pesticides, or spot treatment and rational use of the same. Once pesticides are kept away from the fields, spiders invariably take shelter in the fields, feed on the pests and add to the productivity. Conservation of the diverse spider fauna that is characteristic of agro-ecosystems must be emphasized rather than the life histories, foraging and web-building behaviour of individual spider species.

References

Afun, J.V.K., Johnson, D.E., Russell-Smith, A., 1999. The effects of weed residue management on pests, pest damage, predators and crop yield in upland rice in Cote d'Ivoire. Biol. Agric. Hortic. 17 (1), 47–58.

Aguilar, P.G., 1965. Notas sobre las aranas en el campo cultiva do. Rev. Peruana Ent. 8, 81–83.

Aguilar, P.G., 1974. Aranas del campo cultivado-1: poblacion de araneidos en algodonales de Canete. Huaura of Rimae. (Spiders of crop fields-1: population of aranelias in cotton plantations in Canete Huaura and Rimac). Rev. Peruana Ent. 17, 21–27.

Alderweireldt, M., 1994. Habitat manipulations increasing spider densities in agroecosystems: possibilities for biological control? J. Appl. Entomol. 118, 10–16.

Ali, A.D., Reagan, T.E., 1985. Spider inhabitants of sugarcane ecosystems in Louisiana; an update. Proc. Louisiana Acad. Sci. 48, 18–22.

Anderson, J.F., 1974. Responses to starvation in the spiders *Lycosa lenta* Henz. and *Filistata hibernalis* Henz. Ecology 55, 576–585.

Anonymous, 1971. Annual Report of the Department of Zoology-Entomology for 1970–71. Punjab Agricultural University, Ludhiana, 139 pp.

Arida, G.S., Heong, K.L., 1994. Sampling spiders during the rice fallow period. Int. Rice Res. Notes 19 (1), 20.

Ashikbayen, N.Z., 1973. The life forms of spiders (Araneae) inhabiting wheat fields in the Kustanag region. Entomol. Rev. 52, 335–341.

Babu, G.R., Rao, G.M., Rao, P.A., 1998. Efficacy of neem oil and neem cake for the control of green leafhoppers, brown plant hoppers and their effect on predators of brown plant hoppers. Shashpa 5, 91–94.

Bailey, C.L., Chada, H.L., 1968. Spider populations in grain sorghums. Ann. Ent. Soc. Am. 61, 567–571.

Balogh, J., Loksa, J., 1956. Untersuchungen uber die Zoozonose des Luzernefeldes. Acta Zool. Acad. Sci. Hung. 2, 17–114.

Barbosa, J.T., Riscado, G.M., Filho, M.L., 1979. Population fluctuation of sugarcane froghopper, *Mahanarva posticata* Stal, 1855 and its natural enemies in campus. R. J. in 1977. An. Soc. Entomol. Bras. 8, 39–46.

Barnes, B.M., Barnes, R.D., 1954. The ecology of the spiders of maritime drift lines. Ecology 35, 25–35.

Barnes, R.D., 1929. The ecology of spiders of maritime drift lines. Ecology 35, 25–35.

Barnes, R.D., 1953. The ecological distribution of spiders in non-forested maritime communities at beaufort. North Carolina. Ecol. Monogr 23, 315–337.

Barrion, A., 1999. Ecology of spiders in selected non-rice habitats and irrigated rice fields in two southern Tagalog provinces in the Philippines. Ph.D. Thesis. University of the Philippines at Los Banos.

Barrion, A.T., Litsinger, J.A., 1980. Taxonomy and bionomics of spiders in Philippine rice agro ecosystem: Foundations for future biological control efforts. Paper presented at the 11th Annual Conference of the Pest Control Council of the Philippines, Cebu City, Philippines, 23–26 April 1980, 44 pp.

Barrion, A.T., Litsinger, J.A., 1981a. The spider fauna of Philippine rice agroecosystems I. Dryland. Philipp. Ent. 5, 139–166.

Barrion, A.T., Litsinger, J.A., 1981b. The spider fauna of Phillippine dryland and wetland rice agroecosystems. IRRI. Saturday seminar. April 4. 1981 Los Banos, Philippines.

Barrion, A.T., Litsinger, J.A., 1984. The spider fauna of Philippine rice agro-ecosystems. II. Wetland. Philipp. Entomol. 6, 11–37.

Basedow, T., 1973. Der Einflub Epigaischer Raubathropoden auf die Abundanz phytophager Insekte in der Agarlandshaft. Pedobiologia 13, 410–422.

Bastidas, H., Pantoja, A., 1993. Colombian rice field spiders. Int. Rice. Res. Newsl. 18 (2), 32–33.

Battu, G.S., Singh, B., 1975. A note on spiders predatory on the insect pests of cotton. Sci. Cult. 41, 212–214.

Berry, J.W., 1970. Spiders of the North Carolina piedmont old field communities. J. Elisha Mitchell. Sci. Soc. 86, 97–105.

Beyer, R., 1981. Zur Dynamik der Spinnen- und Weberknechtfauna aufeiner Kulturflache mit wechselendem Pflanzenbestand in Verlaufe von 6 Jahren in Raum heipzig. Paun. Abh. Mus. Tierk. Dresden 8, 119–130.

Bilde, T., Toft, S., 1994. Prey preference and egg production of the carabid beetle *Agonum dorsale*. Entomol. Exp. Appl. 73, 151–156.

Blickenstaff, C.C., Huggans, J.L., 1962. Soybean insects and related arthropods in Missouri. Univ. Missouri Agric. Exp. Stn. Res. Bull. 803.

Boness, M., 1958. Bioconotische untersuchungen uber die Tierwelt von Klee und Luzernefeldern. Z. Morpl. Okol. Tiere 47, 309–373.

Bosmans, R., Cottenie, M.P., 1977. Araignees rares on nouvells pour la faune belge des cultures. Bull. Ann. Soc. R. Belge Entomol. 113, 100.

Bostanian, N.J., Dondale, C.D., Binns, M.R., Pitre, D., 1984. Effects of pesticide use on spiders in Quebec apple orchards. Can. Entomol. 116, 663–675.

Braendegaard, J., 1946. The spiders (Araneae) of east Greenland: a faunistic and zoo-geographical investigation. Medd. Greenland 121, 1–128.

Breithaupt, J., Schmutterer, H., Singh, P.P., 1999. Aqueous neem seed kernel extracts for control of the Asian corn borer, *Ostrinia furnacalis* in Papua New Guinea. In: Juss, V., Singh, R.P., Saxena, R.C. (Eds.), Azadirachta indica. Oxford & IBH Pub. Co. Pvt. Ltd., New Delhi, pp. 191–198.

Breymeyer, A., 1967. Preliminary data for estimating the biological production of wandering spiders. In: Secondary Productivity of Terrestrial Eco System. Petrusewica, K. (Ed.), Warszawa, Krakow. pp. 821–834.

Breymeyer, A., 1978. Analysis of the trophic structure of some grass land ecosystems. Pol. Ecol. Stud. 4, 55–128.

Brignoli, P.M., 1975. Uber die Gruppe der Haplogynae (Araneae). Proc. 6th Int. Arachnol. Congr. (Amsterdam IV. 1974) 1974, 33–38.

Bristowe, W.S., 1958. The World of Spiders. Collins, London, 304 pp.

Burleigh, J.G., Young, J.H., Morrison, R.D., 1973. Strip-croppings effect on beneficial insect and spiders associated with cotton in Oklahoma. Environ. Entomol. 2, 281–285.

Cambridge, O.P., 1877. On some new species of Araneidea, Lycosida family, Podophthalma and Dinopides. Proc. Zool. Soc. Lond., 567.

Carroll, D.P., 1980. Biological notes on the spiders of some citrus groves in central and southern California. Ent. News 91, 147–154.

Chant, D.A., 1956. Predaceous spiders in orchards in south eastern England. J. Hortic. Sci. 31, 35–46.

Chauvin, R., 1960. La fauna du champs cultive et surtout du champ de luzerne. Revue. Zool. Agric. 59, 14–37.

Chauvin, R., 1967. Die welt der Insekten. Kindler, Munchen.

Cherrett, J.M., 1964. The distribution of spiders on the moor house national nature reserve, Westmorland. J. Anim. Ecol. 33, 27–48.

Chiu, S.C., Chu, Y.I., Long, Y.H., 1974. The life history and some bionomic notes on a spider, *Oedothorax insecticeps* Boes, et- St. (Micryphantidae: Araneae). Plant Prot. Bull., Taiwan 16, 153–161.

Choi, S.Y., 1979. Screening methods and sources of varietal resistance. International Rice Research Institute. Brown plant hopper: threat to rice production in Asia. Los Banos, Philippines, pp. 171–186.

Chu, Y.I., Okuma, C., 1970. Preliminary survey on the spider fauna of the paddy fields in Taiwan. Mushi 44, 65–88.

Clark, E.W., Qlick, P.A., 1961. Some predators and scavengers feeding up on pink bollworm moths. J. Econ. Ent. 54, 815–816.

Cloudsley-Thompson, J.L., 1962. Microclimates and their distribution of terrestrial arthropods. Annu. Rev. Entomol. 7, 199–222.

Coddington, J.A., 1990. Cladistics and spider classification: Araneomorph phylogeny and the monophyly of orb weavers (Araneae: Araneomorphae: Orbiculariae). Acta Zool. Fenn. 190, 75–87.

Coddington, J.A., Levi, H.W., 1991. Systematics and evolution of spiders (Araneae). Annu. Rev. Ecol. Syst. 22, 565–592.

Coddington, J.A., Larcher, S.F., Cokendolpher, J.C., 1990. The systematic status of Arachinda, exclusive of Acarina, in North America north of Mexico (Arachinda; Amblypygi, Araneae, Opiliones, Palpigradi, Pseudoscorpiones, Ricinulei, Schizomida, Scorpiones, Solifugae, Uropygi). In: Kosztarab, M., Schaeffer, C.W. (Eds.), Diversity and Dynamics of North American Insect and Arachnid Fauna Va. Polytech. Inst, Blacksburg VA, pp. 5–20.

Cokendolpher, J.C., 1988. Arachnological publications and societies of the world. Am. Arachnol. 38, 10–12.

Collatz, K.G., Mommsch, T., 1975. Physiological conditions and variations of body constituents during the moulting cycle of the spider *Tegnaria atrica* (Ageneidae). Comp. Biochem. Physiol. 52, 465–476.

Comstock, J.H., 1940. The Spider Book. Comstock Publishing Association, New York, 729 pp.

Coppel, H., 1977. Biological Insect Pest Suppression. Springer, Berlin, Heidelberg, New York, 115 pp.

Cottenie, P.I., De Clercq, R., 1977. Study of the Arachnofauna in fields of winter wheat. Parasitica 33, 138–147.

Court, D.J., Forster, R.R., 1988. The spiders of New Zealand. Part VI. Family: Araneidae. Otago Mus. Bull. 6, 68–124.

Culin, J.D., Rust, R.W., 1980. Comparison of the ground surface and foliage dwelling spider communities in a soybean habitat. Environ. Entomol. 9, 577–582.

Culin, J.D., Yeargan, K.V., 1983. The effects of selected insecticides on spiders in alfalfa. J. Kansas Entomol. Soc. 56, 151–158.

Czajka, M., Goods, M., 1976. The spiders (Aranei) of sugarbeet fields in Pawlowice Weilkie near Wroclaw. Pol. Pismo Entomol. 46, 623–629.

Czajka, M., Kania, C., 1976. Spiders (Aranei) jn potato agrocoenosis in Pawlowice Wielkie near Wroclaw in 1971–1974. Pol. Pismo Ent. 46, 623–629.

Dean, D.A., Sterling, W.L., Horner, N.V., 1982. Spiders in eastern Texas cotton fields. J. Arachnol. 10, 251–260.

De Clercq, R., 1979. On the influence of the soil fauna on the aphid population in winter wheat. Int. Symp. 'Integrated Control in Agriculture and Forestry', Vienna, 8–12 Oct. Summaries of main lectures and discussion papers, 79 pp.

Delchev, K., Kajak, A., 1974. Analysis of a sheep pasture ecosystem in the Pieniny Mountains (the Carpathions). XVI. Effect of pasture management on the number and biomass of spiders (Araneae) in two climatic regions (the Pieniny and the Sredna Gara Mountains). Ekol. Pol. 22, 693–710.

Dharmaraju, E., 1962. A check list of parasites, the hyperparasites, predators and pathogens of the coconut leaf eating caterpillar *Nephantis serinopa* Meyrick recorded in Ceylon and in India and their distribution in these countries. Ceylon Cocon. Q. 13, 102–111.

Dietrich, W., Gotze, R., 1974. Okologische Untersuchungen der Spinnenfauna einer Agrobiozonose (unter besonderer Berucksichtigung der Araneae). Unpubl. Dipl-Arb., Sektion Biowissenschaften, Fachbereich Zoologie, Martin- Luther – Univ., Halle.

Dill, L.M., Fraser, A.H.G., 1990. Do Stabilimenta in orb web attracts prey or search spiders. Behav. Ecol. 10, 372–376.

Dippenaar-Schoeman, A.S., 1976. An ecological study of a spider population in strawberries with special reference to the role of Pardosa crassipalpis Purcell (Araneae: Lycosidae) in the control Tetranychus cinnabarinus cjnnabarinus (Boisduval). – M.Sc. thesis, Rand Afrikaans University.

Dippenaar-Schoeman, A.S., 1979a. Spider communities in strawberry beds: seasonal changes in numbers and species composition. Phytophylactica 11, 1–4.

Dippenaar-Schoeman, A.S., 1979b. A simple technique for studying feeding behaviour of spiders on mites. Bull. Brit. Arachnol. Soc. 4, 349.

Doleschall, L., 1859. Tweede Bijdr. Tot be Kenn. D. Arachn. V. D. Ind. Archipel. Acta. Soc. Ind. Neerl. 5 (5), 1–60.

Dondale, C.D., 1961. Life histories of some common spiders from trees and shrubs in Nova Scotia. Can. J. Zool. 38, 777–787.

Dondale, C.D., 1966. The spider fauna (Araneida) of deciduous orchards in the Australian Capital Territory. Austr. J. Zool. 14, 1157–1192.

Dondale, C.D., 1977. Life histories and distribution patterns of hunting spiders (Araneids) in an Ontario meadow. J. Arachnol. 4, 73–93.

Dondale, C.D., Parent, B., Pitre, D., 1979. A 6-year study of spiders (Araneae) in a Quebec apple orchard. Can. Entomol. 111, 377–380.

Dorris, P.R., 1970a. Observations on the impact of certain insecticides on spider populations in a cotton field. Proc. Ark. Acad. Sci. 24, 53–54.

Dorris, P.R., 1970b. Spiders collected from mud–dauber nests in Mississippi. J. Kans. Entomol. Soc. 1, 10–11.

Dowdy, W.W., 1951. Further ecological studies on stratification of the arthropods. Ecology 32, 37–52.

Duffey, E., 1974. Comparative sampling methods for grassland spiders. Bull. Br. Arachnol. Soc. 3, 34–37.

Duffy, E., 1968. Archnological survey of the spider fauna of sand dunes. J. Anim. Ecol. 37, 641–741.

Dunn, J.A., 1949. The parasites and predators of potato aphids. Bull. Entomol. Res. 40, 97–122.

Dyck, V.A., Heinrichs, E.A., Pathak, M.D., Fewer, R., 1976. Insect pest management in rice (IRRI-RPTR-Training Programme). Paper presented at the 7th National Conference of the Pest Control Council of the Philippines. Cagayam de Oro city, May 5–7, p. 19.

Eberhard, W.G., 1971. The ecology of the web of *Uloborous diversus* (Araneae: Uloboridae). Ocecologica 6, 328–342.

Eskov, K.Y., 1987. A new archaeid spider (Chelicerata: Araneae) from the Jurassic of Kazakhstan, with notes on the so-called 'Gondwanan' ranges of recent taxa. Newz J. Geol. Paleont. Abh. 175, 81–106.

Everly, R.T., 1938. Spiders and insects found associated with sweet corn with notes on the food habits of some species. I. Arachnida and Coleoptera. Ohio J. Sci. 38, 134–140.

Fernandez, N.J., Palanginan, E.L., Soon, L.L. and Botrella, D., 1992. Impact of neem on nontarget organisms. Proc. Final Workshop on Botanical Pest Control, IRRI, Los Banos, Philippines, pp. 117–121.

Forster, R.R., 1967. The spiders of New Zealand. Part 1. Otago Mus. Bull. 1, 1–124.

Forster, R.R., 1988. The spiders of New Zealand. Part 6. Family Cyatholipidae. Otago Mus. Bull. 6, 7–34.

Forster, R.R., Blest, A.D., 1979. The spiders of New Zealand. Part 5. Cycloctenidae, Gnaphosidae, Clubionidae, Linyphiidae, Mynogleninae. Otago Mus. Bull. 5, 1–173.

Forster, R.R., Wilton, C.L., 1973. The spiders of New Zealand. Part 4. Agelenidae, Stiphiidae, Amphinectidae, Amaurobiidae, Neolanidae, Ctenidae, Psechridae. Otago Mus. Bull. 4, 1–309.

Fox, C.J.S., Dondale, C.D., 1972. Annotated list of spiders (Araneae) from layfields and their margins in Nova Scotia. Can. Entomol. 104, 1911–1915.

Fraser, A.M., 1982. The role of spiders in checking cereal aphid numbers. Ph.D. thesis. Univ. East Anglia.

Gavarra, M., Raros, R.S., 1975. Studies on the biology of the predatory wolf spider, *Lycosa pseudoannulata* Boesenberg and Strand. Philipp. Entomol 2 (6), 427–444.

Geiler, H., 1963. Die Spinnen – und Weberkneehtfauna mittelddeutscher Felder Die Evertebratenfauna mitteldeutscher Feldkulturenv. Z. ang. Zool. 50, 257–272.

Gertsch, W.J., 1949. American Spiders. D. Van Nostrand Co., Princeton, New Jersey.

Gibson, W.W., 1947. An ecological study of the spiders of a river-terrace forest in western Tennessee. Ohio J. Sci. 47, 38–44.

Gravena, S., Sterling, W.L., 1983. Natural predation on the cotton leafworm (Lepidoptera: Noctuidae). J. Econ. Entomol. 76, 779–784.

Greenstone, M.H., 1978. Numerical response to prey availability of *Pardosa ramulosa* (McCook Araneae: Lycosidae) and its relationship to the role of spiders in the balance of nature. Symp. Zool. Soc. Lond. 42, 183–193.

Guo, Y.J., Wang, N.Y., Jiang, J.W., Chen, J.W., Tang, J., 1995. Ecological significance of neutral insects as a nutrient bridge for predators in irrigated rice arthropod communities. Chin. J. Biol. Control 11, 5–9.

Harrison, J.O., 1968. Some spiders associated with banana plants in Panama. Ann. Entomol. Soc. Am. 61 (4), 878–884 (7).

Hensley, S.D., Mccormick, W.J., Long, W.H., Concienne, J.E., 1961. Field tests with new insecticides for control of sugarcane borers in Lousiana in 1959.1. Econ. Entomol. 54, 1153–1154.

Heydemann, B., 1953. Agarokologische Problematik, dargetan an Untersuchungen uber die Tierwelt der Oberfache der Kulturfelder. Unveroffentl. Diss., Univ. Kiel (Cit Geiler, H., 1963).

Hiremath, I.G., 1989. Survey of sorghum earhead bug and its natural enemies in Karnataka. J. Biol. Cont. 3, 13–16.

Horner, N.V., 1972. *Metaphidippus galathea* as a possible biological control agent. J. Kans. Entomol. Soc. 45, 324–327.

Howard, F.W., Oliver, A.D., 1978. Arthropod populations in permanent pastures treated and untreated with mirex for red imported fire ane control. Environ. Entomol. 7 (6), 901–903 (3).

Huhta, V., Raatikainen, 1974. Spider communities of leys and winter cereal fields in Finland. Ann. Zool. Fenn. 11, 97–104.

Hukusima, S., 1961. Studies on the insect association in crop field. XXI. Notes on spiders in apple orchards. Jap. J. Appl. Ent. Zool. 5, 270–272.

International Rice Research Institute (IRRI), 1978. Control and management of rice pests. Insects Annual Report for 1977. International Rice Research Institute, Los Banos, Philippines, pp. 195–226.

Jackson, C.I., 1930. Results of the Oxford University expedition to Greenland, 1928. Araneae and opiliones collected by Major R.W.G. Hingston: with some notes on Islandic spiders. Ann. Mag. Nat. Hist. 10, 639–656.

Jandu, M.K., 1972. Taxonomic studies on biological observation on the spiders predaceous on citrus pests. M.Sc. (Ag)., Thesis, Punjab Agricultural University, Ludhiana.

Jeffries, M.J., Lawton, J.H., 1984. Enemy free space and structure of ecological communities. J. Ecol. Entomol. 21, 213–217.

Jennings, D.T., Houseweart, M.W., 1978. Spider preys on spruce budworm egg mass. Entomol. News 89, 183–186.

Johnson, A.W., Rosebery, G., Parker, C., 1976. A novel approach to *Striga* and *Orobanche* control using synthetic germination stimulants. Weed Res. 16, 223–227.

Joshi, R.C., Cadappan, E.P., Heinrichs, E.A., 1987. Natural enemies of rice leaf folder *Cnaphalocrosis medinalis* Guenee (Pyralidae: Lepidoptera). A critical review. Agric. Rev. 8 (1), 22–23.

Judd, W.W., 1965. Studies of the Byron Bog in south western Ontario XVIII. Distribution of harvestman and spiders in the bog. Nat. Mus. Can. Nat. Hist. Pap. 28, 1–12.

Kagan, M., 1943. The Araneida found on cotton in central Texas. Ann. Entomol. Soc. Am. 36, 257–258.

Kaihotsu, K., 1979. Spiders and their seasonal fluctuation in a citrus orchard and a tea field in l'4ie Prefecture. Atypus 74, 29–39.

Kajak, A., 1960. Changes in the abundance of spiders in several meadows. Ekol. Pol. A 8, 199–228.

Kajak, A., 1962. Comparison of spider fauna in artificial and natural meadows. Ekol. Pol. A 10, 1–20.

Kajak, A., 1971. Productivity investigation of two types of meadows in the Vistula valley. IX. Production and consumption of field layer spiders. Ekol. Pol. A 19, 197–211.

Kajak, A., 1978. The effect of fertilizers on number and biomass of spiders in a meadow. Symp. Zool. Soc. Lond. 42, 125–129.

Kajak, A., 1980. Do the changes caused in spider communities by the application of fertilizers advance with time? Proceedings International Arachnology Congress Eighth, Vienna, pp. 115–119.

Kamal, N.Q., Odud, A., Begum, A., 1990. The spider fauna in and around the Bangladesh Rice Research Institute farm and their role as predator of rice insect pests. Phillip. Entomol. 8, 771–777.

Kareem, A.A., Saxena, R.C., Malayba, M.T., 1988. Effect of sequential neem (*Azadirachta indica*) treatment on green leafhopper (GLH), rice tungro virus (RTV) infection, predatory mirid and spiders in rice. Int. Rice Res. Newslett. 13, 37.

Karim, R.A.N.M., Chowdhury, M.M.A., Haque, M.N.M., 1992. Current research on neem in rice in Bangladesh, Proc. Final Workshop Botanical Pest Control, IRRI, Los Banos, Philippines, pp. 30–34.

Kartohardjono, A., Heinrichs, E.A., 1984. Population of the brown planthopper, *Nilaparvata lugens* (Stål) (Homoptera: Delphacidae) and its predators on rice varieties with different levels of resistance. Environ. Entomol 13, 359–365.

Kaston, B.J., 1981. Spiders of Connecticut. Bull. Conn. Geol. Nat. Hist. Surv. 70, 1–1020.

Kayashima, I., 1960. Studies on spiders as natural enemies of crop pests. I. Daily activities of spiders in cabbage fields, establishment of spiders in the fields and evaluation of the effectiveness of spiders against crop pests. Sci. Bull. Fac. Agric. Kyushu Univ. 18 (1), 1–24.

Kayashima, I., 1961. Study of the Lynx spider, *Oxyopes sertatus* L. Koch, for biological control of the Crytomerian leaf fly *Contarina inouyei* Mani. Rev. Appl. Entomol. Ser. A 51, 413.

Kayashima, I., 1972. Study on grass spider as a predator to Hyphantnia cunea Drury (Experinent on effectiveness as a predator of Agelena opulenta L. Koch). Acta Arachnol. 24, 60–72.

Kenmore, P.E., Carino, F.O., Perez, C.A., Dyck, V.A., Gutierrez, A.P., 1984. Population regulation of the rice brown planthopper *Nilaparvata lugens* (Stål) within rice fields in the Philippines. J. Plant Prot. Tropics 1, 1–37.

Khan, S.A., Zafar, Y., Briddon, R.W., Malik, K.A., Mukhtar, Z., 2006. Spider venom toxin protects plants from insect attack. Transgenic Res. 15, 349–357.

Kiran, B., 1978. Untersuchung der epigaischen Spinnenfauna (Araneae)m in Rebkulturen. Staats – examensarrbeit in Zoologie. Univ. Freiburg I. Br.

Kiritani, K., Kakiya, N., 1975. An analysis of the predator–prey system in the paddy field. Res. Popul. Ecol. Kyoto Univ. 17, 29–38.

Kiritani, K., Kawahara, S., Sasaba, T., Nakasuji, T., 1972. Quantitative evaluation of predation by spiders on the green leafhopper, *Nephotettix cincticeps* Uhler by sight count method. Res. Popul. Ecol. 13, 187–200.

Komatsu, T., 1968. Cave spiders of Japan II. *Cybeaus dolichocybaes* and *Heterocybeas cybearinae*. Osaka: Arach. Soc. East Asia, 38.

Koval, G., 1976. Biologische Bekampfung des Kartoffelkafers. 4 Zascita Rastenij, Mosakau, 21, 29.

Kramer, P., 1961. Einflub einiger Arthropoden auf Raubmilben (Acari). Z. ang. Zool 48, 257–311.

Kumar, S., Monga, K., 1996. Predaceous spiders of Mango pests and their extent of feeding. Uttar Pradesh J. Zool. 16 (3), 167–168.

Lamoral, B.H., 1968. On the species of the genus Desis Walckenaer, 1837 (Araneae: Amaurobidae) found on the rocky stores of South Africa and South West Africa. Ann. Natal. Mus., 20.

Leach, R.E., 1966. The spiders (Araneida) of Hazen camp 81°49′N, 71°18′W. Quaestiones Entomol 2, 153–212.

Lehtinen, P.T., 1967. Classification of the Cribellate spiders and some allied families with notes on the evolution of the suborder Araneomorpha. Ann. Zool. Fenn. 4, 199–467.

Leigh, T.F., Hunter, R.E., 1969. Predacious spiders in California cotton. Calif. Agr. 23, 4–5.

Levi, H.W., 1991. The neotropical and Mexican species of the orb-weaver genera Araneus, Dubiepeira and Aculepeira (Araneae: Araneidae). Bull. Mus. Comp. Zool. 152 (5), 167–315.

Levi, H.W., Levi, L.R., 1968. A Guide to Spiders and Their Kin. Golden Press, New York. Western Publishing Company Inc., Racine, Wisconsin, 160 pp.

Lincoln, C., Philips, J.R., Whitcomb, W.H., Dowell, G.C., Boyer Jr, W.P., Bell, K.O., 1967. The bollworm tobacco budworm problems in Arkansas and Louisiana. Arkansas Agric. Exp. Stn. Bull. 720, 1–66.

Locket, G.H., 1978. The Leckford survey spiders. Progress report. Jan. 1978.

Lowrie, D.C., 1948. The ecological succession of spiders. Progress report. Jan. 1978.

Luczak, J., 1975. Spider communities of the crop-fields. Pol. Ecol. Stud. 1, 93–110.

Luczak, J., 1979. Spiders in agrocoenoses. Pol. Ecol. Stud. 5, 151–200.

Luhan, U. 1979. Tagesrhythmik und Jahresrhythmik epigaischer Arthropoden (Insbesond, Aranei. Carabidae) einer mesophilon Wiese des Innsbrucker Mittelgebriges Rinn 900 m NN. Osterireich). Univ. off. LA- Hausarbeit in Zoologie. Univ. Innsbruck.

MacLellan, C.R., 1973. Natural enemies of the light brown apple moth, *Epiphyas postvittana*, in the Australian Capital Territory. Can. Entomol. 105, 681–700.

Madden, A.H., Chamberlin, F.S., 1945. Biology of the tobacco hornworm in the southern cigar–tobacco district. USDA Tech. Bull. 896, 1–51.

Magni, F., Papi, F., Savely, H.E., Tongiori, P., 1965. Research on the structure and physiology of the eyes of a lycosid spider. III. Electroretinographic responses to polarized light. Area. Ital. Biol. 103, 146–158.

Mansour, F., Rosen, D., Shulov, A., 1980a. Biology of the spider *Cheriracanthium mildei* (Arachnida: Clubionidae). Entomophaga 25 (3), 237–248.

Mansour, F., Rosen, D., Shulov., A., 1980b. Functional response of the spider *Cheriracanthium mildei* (Arachnida: Clubionidae) to prey density. Entomophaga 25 (3), 313–316.

Mansour, F., Rosen, D., Shulov, A., Plaaut, H.N., 1980c. Evaluation of spiders as biological control agents of *Spodoptera littoralis* larvae on apple in Israel. Acta Ecologica. 1, 225–232.

Mansour, R., Rosen, D., Plaut, H.N., Shulov, A., 1981. Effect of commonly used pesticides on *Cheiracanthium mildei* and other spiders occurring on apple. Phytoparasitica 9, 139–144.

McCaffrey, J.P., Horsburgh, R.L., 1980. The spider fauna of apple trees in central Virginia. Environ. Entomol. 9, 247–252.

McCarty, M.J., Shepard, M., Turnipseed, S.G., 1980. Identification of predacious arthropods in soybeans by using autoradiography. Environ. Entomol. 9, 199–203.

McCutchen, B.F., Hoover, K., Preisler, H.K., Betana, M.D., Herrmann, R., 1997. Interactions of recombinant and wild-type baculoviruses with classical insecticides and pyrethroid resistant tobacco bud worm (Lepidoptera: Noctuidae). J. Econ. Entomol. 90, 1170–1180.

McDaniel, S.G., Sterling, W.S., 1982. Predation of *Heliothis virescens* (F) eggs on cotton in east Texas. Environ. Entomol. 11, 60–66.

Menon, K.P.V., Pandalai, K.M., 1958. The Coconut Palm – A Monograph. Indian Central Coconut Committee, Ernakulam, S. India.

Merrett, P., 1978. Effects of recreation on the spider fauna of dune grasslands Arachnology. Symp. Zool. Soc. Lond. Academic Press, London, 490 pp.

Miller, F. 1974. Pavouci fauna repnych poli v Okoli Chvalkovic a Nakla na Hane- Sb. Praci Prir. Fak. Palack. Univ. Olomouc. 47, Biologica 15, 175–181 (cited in Luczak. J., 1979) – unpublished results (cited in Luczak, J., 1979).

Mitter, C., Farrel, B., Wiegmann, B., 1988. The phylogenetic study of adaptive zone; has phytophagy promoted insect diversification? Am. Nat. 132 (1), 107–128.

Mohammed, U.V.K., Abdurahiman, U.C., Remadevi, O.K., 1982. Coconut caterpillar and its natural enemies, Zoological Monograph No. 2. Dept. of Zoology, University of Calicut, Calicut, Kerla, 162p.

Muma, M.H., 1973. Comparison of ground surface spiders in four central Florida ecosystems. Florida Entomol. 56 (3), 173–196.

Muma, M.H., 1975. Spiders in Florida citrus groves. Florida Entomol. 58 (2), 83–90.

Muniappan, R., Chada, H.L., 1970. Biology of the crab spider *Misumenops celer*. Ann. Entomol. Soc. Am. 63, 1718–1722.

Muralidharan, C.M., Chari, M.S., 1992. Spider fauna activity in cotton field at Anand, Faridabad. Plant Prot. Bull. 44, 26–27.

Nachman, R.J., Pietrantonio, P.V., 2010. Interaction of mimetic analogs of insect kinin neuropeptides with arthropod receptors. Adv. Exp. Med. Biol. 692, 27–48.

Nakamura, K., 1972. The ingestion in wolf spiders II. The expression of degree of hunger and the amount of ingestion in relation to spider's hunger. Popul. Ecol. Kyoto. Univ. 14, 82–96.

Nanda, U.K., Parija, B., Pradhan, N.C., Nanda, B., Dash, D.D., 1996. Bioefficacy of neem derivatives against the insect pest complex of rice. In: Singh, R.P., Chari, M.S., Raheja, A.K., Kraus, W. (Eds.), Neem and Environment 1. Oxford & IBH Pub. Co. Ltd., New Delhi, India, pp. 517–528.

Negm, A.A., Hensley, S.D., 1969. Evaluation of certain biological control agents of the sugarcane borer in Louisiana. J. Econ. Entomol. 62, 1008–1013.

Negm, A.A., Abou-Ghadir, M.F., Salman, A.G.A., 1975. Lur Populationsdynamik und Aktivitätszeit der echten Spinnen (Araneida) auf Kleefeldern in Aegypten. Anz. Schädlingskde. Pflanzenschutz' Unweltschutz 48, 161–163.

Nentwig, W., 1987. The prey of spiders. In: Nentwig, W. (Ed.), Ecophysiology of Spiders. Springer, Berlin, pp. 249–263.

Nirmala, R., Balasubramanian, G., 1999. Residual toxicity effects of insecticides and neem based formulations on the predatory spiders of rice-ecosystem. Neem Newslett. 16, 37–38.

Nuessly, G. S., 1986. Mortality of Heliothis zea eggs affected by predator species, oviposition sites and rain and wind dislodgement. Ph.D. diss., Texas A & M University, College Station, TX.

Nyffeler, M., Benz, G., 1979. Zur Ocologischen Bedeutung der Spinnen der Vegetationsschich von Getreidae and Rapsfeldern bei Zurich (Schweiz). Z. Angew. Entomol. 87, 348–376.

Nyffeler, M., Benz, G., 1981b. Okologische Bedeutung der spinnen als Insektenpradatoren in Wiewen and Getreidenfeldern. Mitt. Dtsch. Ges. allg. angew. Entomol. 3, 33–35.

Okuma, C., Wongsiri, T., 1973. Second report on the spider fauna of the paddy fields in Thailand. Mushi (Fukuoka) 47, 1–17.

Paik, K.Y., Kim, J.S., 1973. Survey on the spider fauna and their seasonal fluctuation in paddy fields of Taegu, Korea. Korean J. Plant Prot. 12, 125–130.

Palanichamy, S., 1980. Ecophysiological studies on a chosen spider Cyrtophora cicatrosa (Araneae: Araneidae). Ph.D. thesis, Madurai Kamaraj University, Madurai, p. 137.

Parker, S.P. (Ed.), 1982. Synopsis and Classification of Living Organisms, vol. 2. McGraw Hill, New York.

Pfrimmer, T.R., 1964. Populations of certain insects and spiders on cotton plants following insecticide applications. J. Econ. Entomol. 57, 640–644.

Pieters, E.P., Sterling, W.L., 1974. A sequential sampling plan for the cotton fleahopper, *Pseudatomoscelis seriatus*. Environ. Entomol. 3, 102–106.

Pimental, D., 1961. The influence of plant spatial patterns on insect populations. Ann. Entomol. Soc. Am. 54, 61–69.

Platnick, N.I., 1977. The hypochiloid spiders: a cladistic analysis, with notes on the Atypoidea (Arachnida: Araneae). Am. Mus. Novitiates 2627, 1–23.

Platnick, N.I., 1989. Advances in Spider Taxonomy, 1981–1987. Manchester Univ. Press, Manchester, UK.

Platnick, N.I., 1990. Spinneret morphology and the phylogeny of ground spiders (Araneae, Gnaphosidae). Am. Mus. Novitiates 2978, 1–42.

Pocock, R.I., 1900. Fauna of British India. Arachnida. Taylor and Francis, London, 1–279.

Pointing, P.J., 1965. Some factors influencing the orientation of the spider *Frontinella communis* (Hentz) in its web (Araneae: Linyphiidae). Can. Entomol. 97, 69–78.

Polenec, A., 1968. Untersuchungen der terrestrischen Arachnidenfauna in slowenischen Kartgebirge. Die Weiden innerhalb der Seslerio autumnalis-Ostryetum-Association bei Divaca. Biol. Vestn. 16, 77–84.

Preston-Mafham, R., 1991. The Book of Spiders and Scorpions. Cresent Books, New York, 284 pp.

Raguraman, S., 1987. Studies on the efficacy of neem products against rice insect pests. M.Sc. thesis, Tamil Nadu Agricultural University, Coimbatore, India.

Raguraman, S., Rajasekaran, B., 1996. Effect of neem products on insect pests of rice and the predatory spider. Madras Agric. J. 83, 510–515.

Ranjit Lal, 1995. In the Spider's Parlour. The Hindu., Sunday, March 5..

Raodeo, A.K., Tikar, D.T., Muqueem, A., 1973. The social spider *Stegodyphus sarasinorum* Karsch. Feeding on the lemon butterfly, *Papilio Demoleus* Linn. J. Bombay Nat. Hist. Soc. 70 (1), 216.

Raven, R.J., 1988. The current status of Australian spider systematics. In: Austin, A.D. Heatner, N.W. (Eds.), Australian Arachnology, No. 5. Aust. Entomol. Soc., Misc. Publ, pp. 1–137.

Reddy, P.S., Heong, K.L., 1991. Co-variation between insects in a rice field and important spider species. Int. Rice Res. Notes 16 (5), 24.

Richman, D.B., Buren, W.F., Whitcomb, W.H., 1983. Predatory arthropods attacking the eggs of *Diaprepes abbreviates* (L.) (Coleoptera: Curculionidae) in Puerto Rico and Florida. J. Ga. Entomol. Soc 18, 335–342.

Riechert, S.E., 1982. Spider interaction strategies: communication Vs coercion. In: Witt, P.N., Rouner, J.S. (Eds.), Spider Communication: Mechanisms and Ecological Significance. Princeton Univ. Press, Princeton, NJ, pp. 281–316.

Riechert, S.E., Uetz, G.W., Abrams, B., 1985. The state of arachnid systematics. Bull. Entomol. Soc. Am. 31, 4–5.

Roberts, M.J., 1985. The Spiders of Great Britain and Ireland, Vols. 1, 3. Colchester, Harley, 256 pp.

Roberts, M.J., 1987. The Spiders of Great Britain and Ireland. Vol. 2. Linyphiidae and Checklist. Martins Essen, Harley, 204 pp.

Robinson, M.H., Robinson, B., 1974. A census of web-building spiders in a coffee plantation at l,Jau, New Guinea, and an assessment of their insecticidal effect. Trop. Ecol. 15, 95–107.

Rogers, C.E., Horner, N.V., 1977. Spiders of guar in Texas and Oklahoma. Environ. Entomol. 6, 523–524.

Room, P.M., 1979. Parasites and predators of Heliothis spp. (Lepidoptera: Noctuidae) in cotton in the Namoi Valley, New South Wales. J. Aust. Entomol. Soc. 18, 223–228.

Rubia, E., Almazan, L., Heong, K., 1990. Predation of yellow stem borer (YSB) by wolf spider. Int. Rice Res. Newsl. 15, 22.

Sadana, G.L., Kaur, M., 1974. Spiders and their prey in citrus orchards. Proceedings Sixty first Indian Science Congress 63.

Sadana, G.L., Sandhu, D., 1977. Feeding intensity of developmental instars of the spider Marpissa ludhianaensis on the fulgorid pest of grapevines. Sci. Cult. 43, 550–551.

Sahu, S., Sing, R., Kumar, P., 1996. Host preference and feeding potential of spiders predaceous on insect pests of rice. J. Entomol. Res. 20, 145–150.

Samal, P., Misra, B.C., 1975. Spiders: the most effective natural enemies of the brown planthopper in rice. Rice Entomol. Newslett. 3, 31.

Samiayyan, K., 1996. Spiders of South India. Ph.D. thesis. TNAU, Coimbatore.

Samiayyan, K., Chandrasekaran, B., 1998a. Influence of botanicals on the spider populations in rice. Madras Agric. J. 85 (7–9), 479–480.

Samiayyan, K., Chandrasekaran, B., 1998b. Prey potential and preference of three rice dwelling spiders (Issued in Feb. 2000). Madras Agric. J. 85 (7–9), 427–429.

Samiayyan, K., Jayaraj, T., 2009. Impact of some intervention practices on spider diversity in rice ecosystem. Madras Agric. J. 96 (7–12), 398–400.

Samiayyan, K., Sudha, V., Radhakrishnan, V., Karuppuchamy, P., Jonathan, E.I., 2012. Spider diversity in different short duration food legume eco systems of Tamil Nadu. In: Second International Symposium of Biopesticides and Ecotoxicological Network (ISBioPEN). September 24–26, 2012, Bangkok, Thailand, pp. 275–279.

Samu, F, Biro, Z., 1993. Functional response, multiple feeding and wasteful killing in a wolf spider (Araneae: Lycosidae). Eur. J. Entomol. 90, 471–476.

Sasaba, T., Kiritani, K., Urabe, T., 1973. A preliminary model to simulate the effect of insecticides on a spider-leafhopper system in the paddy field. Res. Popul. Ecol. 15, 9–22.

Sathiamma, B., Jayapal, S.P., Pillai, G.B., 1987. Observations on spiders (Order:Araneae) predacious on the coconut leaf eating caterpillar *Opisina arenosella* Wlk. (Nephantis serinopa Meyrick) in Kerala: biology of Rhene indicus Tikader (Salticidae) and Cheiracanthium sp. (Clubionidae). Entomon 12, 121–126.

Saucke, H., 1995. Biological and integrated control of diamondback moth Plutella xylostella Lin. and other major pests in brassica crops, SPC/GTZ Biological Control Project: Final Report 1992–1995. Conedobu, Papua New Guinea.

Savory, T.H., 1928. The Biology of Spiders. Sidgwick and Jackson, London, 526 pp.

Schmoller, R., 1970. Terrestrial desert arthropods; fauna and ecology. Biologist 52, 77–98.

Selden, P.A., 1990. Fossil history of the arachnids. Br. Arachnol. Soc. Newslett. 58, 4.

Senguttuvan, T., Gopalan, M., Murali Baskaran, R.K., 1990. Integration of varietal resistance and predation by the wolf spider, *Lycosa pseudoannulata* Boes et. Str. for the management of rice brown plant hopper. Annu. Int. Plant Resist. Insects Newslett. 16, 92.

Serra, A.C, 1992. Untersuchungen zum Einsatz von Niem Samenextrakten in Rahmen integrierter Ansätze zur Bekämpfung von Tomatenschädlingen in der Dominikanischen Republik. Doctoral Thesis, Giessen University, Germany.

Settle, W.H., Ariawan, H., Astuti, E., Cahyana, W., Hakim, A.L., Hindayana, D., et al., 1996. Managing tropical rice pests through conservation of generalist natural enemies and alternative prey. Ecology 77 (7), 1975–1988.

Sharma, V.K., Sarup, P., 1979. Predatory role of spiders in the 'integrated control of the maize stalk borer' Chilo partellus (Swinhoe). J. Entomol. Res. 3, 229–231.

Shear, W.A. (Ed.), 1986. Spiders: Webs, Behaviour and Evolution. Stanford University, Stanford. 544 pp.

Shepard, B.M., Sterling, W.L., 1972. Effects of early season applications of insecticides on beneficial insects and spiders in cotton. Tex. Agric. Exp. Sta. Misc. Publ. 1045.

Shepard, B.M., Barrion, A.T., Litsinger, J.A., 1987. Friends of the Rice Farmer. Helpful Insects, Spiders and Pathogens. International Rice Research Institute, Manila, Philippines.

Shepard, B.M., Rapusas, H.R., Estano, D.B., 1989. Using rice straw bundles to conserve beneficial arthropod communities in the rice field. Int. Rice Res. Notes 14, 30–31.

Shulov, A., 1938. Observations on citrus spiders. Hardilr 11, 206–208.

Sierwald, P., 1990. Morphology and homologous features in the male palpal organ in Pisauridae and other spider families with notes on the taxonomy of Pisauridae. Nemouria 35, 1–59.

Sigsgaard, L., Villareal, S., Gapud, V., Rajotte, E., 1999. Predation rates of *Atypena formosana* (Arachnea: Linyphiidae) on brown planthopper and green leafhopper. Int. Rice Res. Notes 24, 38.

Simon, E., 1875. Les Arachnides de France (Paris) Z, 350 pp.

Simon, E., 1888. Descriptions d'especes et de genres nouveaux de l'Amerique Centrale et des Antiles et observations. Ann. Soc. Entomol. Fr. 7, 203–252.

Simon, E., 1889a. Etude sur les Arachnoides de l'Himalaya recueillis par MM Oldham et wood-Mason. J. Asiat. Soc. Beng 58, 334–344.

Simon, E., 1889b. Etude sur les Arachnoides de l'Himalaya recueillis par MM Oldham et wood-Mason et faisant partie des collections de I'Indian Museum Ire Partie. J. Asiat. Soc. Beng 58 (2), 334–344.

Simon, E., 1892–1895. Histoire Naturella de Araignees 1 (1–4), 1–1084.

Simon, E., 1893a. Etudes arachnologiques, 25e Memoire, XI. Descriptions d'especes at de genres nouvoauz de l'ordre des Araneae. Ann. Soc. Entomol. Fr. 62, 299–330.

Simon, E., 1893b. Arachnids in Voyage de M. E. Simon au Venezuela (December 1887–Avril 1888). 21e Memoire. Ann. Soc. Entomol. Fr. 61, 423–462.

Simon, E., 1895. Histoire Naturelle des Araignes. Tome I, Paris, pp. 761–1084.

Simon, E., 1896. Descriptions d'especes nouvelles de l'ordre des Araneae. Ann. Soc. Entomol. Fr. 65, 465–540.

Simon, E., 1897a. Arachnids recueillis par M. M. Maindron a kurrachee et a Matheran (press Bombay) en 1896. Bull. Mus. Hist. Nat. 7, 289–297.

Simon, E., 1897b. Materiaux pour server a la faune arachnologique de I' Asie Meridionale. V(I). Arachnids recueillis a Dehra-Dun (N. W. Prov.) et dans le Dekka par M. A. Smythies. Mem. Soc. Zool. Fr. 10 (2), 252–262.

Simon, E., 1897c. Arachn. Rec. Par Maindron a Kurrachce a Matheran. Bull. Mus. Hist. Nat. Paris 3 (7), 295.

Simon, E., 1906. Description de quelques Arachnides des bas plateaux de l'Himalaya, Communiques, par le, R. P. Castets (de st. Josephs College a Trichinopoly). Ann. Soc. Entomol. Fr. 75, 306–314.

Singh, G., Sandhu, G.S., 1976. New records of spiders as predators of maize borer and maize jassid. Curr. Sci. 45, 642.

Singh, K., 1967. Heat therapy of sugarcane. Indian Sugar 17, 181.

Sitaramaiah, S., Joshi, B.G., Ramaprasad, G., 1980. New records of spiders and predators of tobacco caterpillar, *Spodoptera litura*. F. Sci. Cult. 46, 29–30.

Sithanantham, S., Otieno, Z.N., Nyarko, K.A., 1997. Preliminary observations on effects of neem sprays on some non target arthropods in cowpea ecosystem. Quatrieme Conf. Int. sur les Ravageurs en Agric., Le Corum, Montpellier, France, pp. 643–648.

Sivasubramanian, P., Vanitha, K., Kavitharaghavan, Z., Banuchitra, R., Samiayyan, K., 2009. Predatory potential of different species of spiders on cotton pests. Karnataka J. Agric. Sci. 22 (Special Issue), 544–547.

Sontakke, B.K., 1993. Field efficacy of insecticide alone and in combination with neem oil against insect pests and their predators in rice. Indian J. Entomol. 55, 260–266.

Specht, H.B., Dondale, C.D., 1960. Spider populations in New Jersey apple orchards. J. Econ. Entomol. 53, 810–814.

Srikanth, J., Easwaramoorthy, S., Kurup, N.K., Santhalakshmi, V., 1997. Spider abundance in sugarcane: Impact of cultural practices, irrigation and post-harvest thrash burning. Biol. Agric. Hortic. 14 (4), 343–356.

Starabogatov, Y.I., 1985. Taxonomic position and the system of the order of spiders (Araneiformes). Proc. Zool. Inst. Acad. Sci. USSR 139, 4–16.

Sudd, J.H., 1972. The distribution of spiders at Spurn Head (E. Yorkshire) in relation to flooding. J. Anim. Ecol. 41, 63–70.

Sudha, V., Samiayyan, K., Radhakrishnan, V., 2011. The diversity of spider fauna in pigeon pea (*Cajanus cajan* L.) Ecosystem. J. Sci. Trans. Environ. Technov. 5 (1), 19–29.

Suzuki, Y., Okuma, C., 1975. Spiders inhabiting the cabbage field. Acta Arachnologica 26 (2), 58–63. ISSN:0001-5202.

Taksdal, S., 1973. Spiders (Araneida) collected in strawberry fields. Norsk. Entomol. Tidsskr 20, 305–307.

Terada, T., Yoshimura, K., Imanishi, M., 1978. Studies on a spider fauna in a tea field (Part III). Comparative studies on spider fauna of a tea field and the adiacent forests. Atypus 72, 25–30.

Thaler, K., Steiner, H.M., 1975. Winter active Spinnen auf einem Acker bei GroBenzersdorf (Niederosterreich). Anz. Schadlingskde., pflanzenschutz, Umweltschutz 48, 184–187.

Thornhill, W.A., 1983. The distribution and probable importance of linyphiid spiders living on the soil surface of sugar-beet fields. Bull. Br. Arachnol. Soc. 6, 127–136.

Tikader, B.K., 1962. Studies on some Indian spiders (Araneae: Archnida). J. Linn. Soc. Lond. Zool. 44, 561–584.

Tikader, B.K., 1968. Studies on spider fauna of Khasi and Jaintia Hills, Assam, India. Part II. J. Assam Sci. Soc. 10, 102–122.

Tikader, B.K., 1970. Spider fauna of Sikkim. Record. Zool. Surv. India 64, 1–83.

Tikader B.K., 1977. Studies on spider fauna of Andaman and Nicobar islands, Indian Ocean. Record. Zool. Surv. India 72, 153–212.

Tikader B.K., 1980. Thomisidae (Crab-spiders). Fauna India (Araneae) 1, 1–247.

Tikader, B.K., 1982. Fauna of India: Arachnid vol. 2: Spiders. Zoological Survey of India.

Tikader, B.K., 1987. Handbook of Indian Spiders. Zoological Survey of India, Calcutta, India.

Tothill, J.D., Taylor, T.H.C., Paine, R.W., 1930. The Coconut Palm Moth in Fiji. Imp. Bur. Ent, London (Spiders p. 153).

Turnbull, A.L., 1966. Seasonal variation in spider abundance in Kuttanad rice agroecosystem, Kerala, India (Araneae). In: Deltshev, C., Stoev, P. (Eds.), European Arachnology 2005. Acta zoologica bulgarica, Suppl. No. 1, 181–190.

Turnbull, A.L., 1973. Ecology of the true spiders (Araneomorphae). Annu. Rev. Entomol. 18, 305–348.

Vanitha, K., Sivasubramanian, P., Kavitharaghavan, Z., Vijayaraghavan, C., Samiayyan, K., 2009. Prey preferences, cross predation and impact of some cultural practices on spiders and their abundance in cotton. J. Agric. Sci. 22 (Spcl. Issue), 548–551.

Venkatesan, S., Rabindra, R.J., 1992. A survey report on the predatory spider population of *mango* leafhopper. South Indian Hortic. 40 (4), 198–201.

Vickerman, G.P., Sunderland, K.D., 1975. Arthropods in cereal crops; nocturnal activity, vertical distribution and aphid predation. J. Appl. Ecol. 12, 755–766.

Vijayalakshmi, K., 1986. Predatory efficiency in giant crab spider -cockroach system - predatory behaviour and biocontrol potential. Proceedings of Oriental Entomology Symposium held at Trivandrum, pp. 132–140.

Vilbaste, A., 1965. Suveaspekti ämblikefaunast kultuurniitudel. (Ueber den Sommeraspekt der Spinnenfauna auf den Kultunriesen.) Eesti NSV Teaduste Akad. Toimetised (Biol.) 14, 329–337.

Vogel, B.R., 1970. Courtship of some wolf spiders. Armadillo Pap. 4, 1–7.

Volkmar, C., 1989. Forschungsbericht 4: Untersuchungenüber die Wirkung von Pflanzenschutzmitteln auf Schadinsekten und Nutzarthropoden. 11 S.

Volkmar, C., Wetzel, T., 1992. Nebenwirkungen von Fungiziden auf räuberische Spinnen in Getreidebeständen. Mitt Biol Bundesanst f. Land- und Forstwirtsch Berlin- Dahlem H 283, 100.

Volkmar, C., Schützel, A. 1997. Spinnengemeinschaften auf einem typischen Ackerbaustandort Mitteldeutschlands und deren Beeinflussung durch unterschiedliche.

Volkmar, C., Schier, A., 2005. Effekte von Maisanbauregime auf epigäische Spinnen. Effects of reduced soil tillage on spider communities. Phytomedizin 35, 17–18.

Volkmar, C., Xylander, E., Wetzel., T., 1998. Zur epigäischen Spinnenfauna im Mitteldeutschen Agrarraum, deren Beeinflussung durch unterschiedliche Pflanzenschutzmaßnahmen und ihre Bedeutung fürden integrierten Pflanzenschutz. Arch. Phytopathol. Pflanzensch. 31, 349–361.

Volkmar, C., Lübke-Al Hussein, M., Jany, D., Hunold, I., Richter, L., Kreuter, T., et al., 2002. Ecological studies on epigeous arthropod populations of transgenic sugar beet at Friemar (Germany). Agric. Ecosyst. Environ. 95, 37–47.

Volkmar, C., Traugott, M., Juen, A., Schorling, M., Freier, B., 2004. Spider communities in *Bt* maize and conventional maize fields. IOBC WPRS Bull 27, 165–170.

Vollrath, F., Fairbrother, W.J., Williams, R.J.P., Tillinghast, E.K., Bernstein, D.T., Gallager, K.S., et al., 1990. Compounds in the droplets of the orb spider's viscid spiral. Nature 345, 526–528.

Wheeler, A.G., 1973. Studies on the arthropod fauna of alfalfa: V. Spiders (Araneida). Can. Entomol. 105 (3), 425–432.

Whitcomb, W.H., 1967. Field studies on the predators of the second instar bollworm, *Heliothis zea* (Boddie) (Lepidoptera: Noctuidae). J. Ga. Entomol. Soc. 2, 113–118.

Whitcomb, W., Tadic, M., 1963. Araneida as predators of the fall webworm. J. Kans. Entomol. Soc. 36, 186–190.

Wiesmann, R., 1955. Unzersuchungen an den predatoren der Baum- wollschadinsekten in Agypten im jabr 1951/52. Acta Trop. 12, 222–239.

Wolcott, G.N., 1937. An animal census of two pastures and a meadow in northern New York. Ecol. Monogr. 7, 1–90. <http://dx.doi.org/10.2307/1943302>.

Yaginuma, T., 1977. A list of Japanese spiders (revised in 1977). Acta Arachnol. 27, 367–406.

Yamanaka, K., Nakasuji, F., Kiritani, K., 1973. Life tables of the tobacco cutworm *Spodoptera litura* and the evaluation of effectiveness of natural enemies. Jpn. J. Appl. Entomol. Zool. 16, 205–214.

Yang, R.C., Wang, N.Y., Mang, K., Chau, Q., Yang, R.R., Chen, S., 1990. Preliminary studies on application of indica photo-thermo genic male sterile line 5460S in hybrid rice breeding. Hybrid Rice 1, 32–34.

Yeargan, K.V., Cothran, W.R., 1974. Population studies of Pardosa ramulosa (McCook) and other common spiders in alfalfa. Environ. Entomol. 3, 989–993.

Young, O.P., Edwards, G.B., 1990. Spiders in United States field crops and their potential effect on crop pests. J. Arachnol. 18, 1–27.

CHAPTER

16

Biotechnological Approaches for Insect Pest Management

V.K. Gupta and Vikas Jindal

Punjab Agricultural University, Punjab, India

16.1 INTRODUCTION

Biotechnology in the context of insect pest management can be defined as the controlled and deliberate manipulation of biological systems to achieve efficient insect pest control. Living organisms have evolved an enormous spectrum of biological capabilities and by choosing appropriate organisms with specific capability, it is possible to obtain meaningful control of such insect pest species. Biotechnology has considerable potential to contribute to sustainable biological elements of integrated pest management (IPM), though biotechnology development to date has been directed at more conventional models for pest control technologies. Until now, most resistance breeding has focused on methods wherein resistance was based on a single gene as these provided high levels of resistance and also the methods were compatible with breeding schemes. However, its gene for gene nature provides possibilities for resistance breakdown through the evolution of resistance breaking pest genotypes. Biotechnology for insect pest management has to some extent been an early by-product of the acquisition of biotechnological know-how, which will have more substantial implications for agriculture than simply improved IPM. Recombinant DNA technology has significantly augmented conventional crop protection by providing dramatic progress in manipulating genes from diverse and exotic sources, and inserting them into microorganisms and crop plants to confer resistance to insect pests and diseases, increased effectiveness of biocontrol agents, and improved understanding of gene action and metabolic pathways. While structural genomics deals with sequence analysis of total genetic information in an organism, efforts in functional genomics are directed at unravelling and understanding the mechanism by which this information is used by an organism. Systematic study of the complete repertoire of metabolites/chemicals of any organism has given birth to a new area of research called 'metabolomics'. Integration of genomics and proteomics with metabolomics will enrich our understanding of the gene-function relationship that can be utilized in achieving crop improvement with a view to higher productivity. The success of any biotechnological intervention for insect pest management also

D. P. Abrol (Ed): Integrated Pest Management.
DOI: http://dx.doi.org/10.1016/B978-0-12-398529-3.00018-X

requires precise identification of target insect pest species and is currently achievable through molecular biology.

16.2 DEVELOPMENTS IN BIOTECHNOLOGY

Crop protection, a major area for improving crop yields through effective management of insect pests and pathogens, has remained the primary target for various developments in biotechnology. These developments could take place due to earlier advancements in genetic engineering and molecular biology, which have resulted in identification, isolation, characterization and modification of resistance genes from diverse biological sources. In this respect, polymerase chain reaction (PCR) providing specific amplification of specific DNA sequences has proved a leading development that has very markedly simplified molecular biology research by substituting many complicated analytical techniques that were earlier considered too indispensable for different purposes. The resulting availability of such fully characterized clean genes in turn led to the development of plant biotechnology making the transgenic expression of such genes possible in crop plants. A number of such genes have already been exploited in different crop plants irrespective of any genetic barrier. However, only a limited number of such genes have afforded desired field resistance to transgenic plants against limited insect pest species. Currently, biotechnology is being applied for precise characterization of insect pest species as well as identification and characterization of novel genes for meaningful pest resistance.

16.3 MOLECULAR MARKERS

The precise characterization of insect pests has become an essential prerequisite for their management. Resulting from diverse types of environmental pressure of current agricultural practices, insect species emerge into morphologically indistinguishable new strains and genetic variants that respond differently to recommended management practices. Molecular biology tools enable the identification of those genomic regions which are equally shared by different but related species and by others that are specific to particular species of the organism under investigation. As these regions serve as specific markers, these are termed molecular markers. In effect, molecular markers help us to determine the existence of specific DNA segments/regions/sequences/genes that are either associated with or control important traits in different pest species. A number of molecular markers have been exploited for molecular analysis of various pest species and their populations. In general, the molecular analysis or screening involves investigating whole or part of the nuclear DNA (nDNA) and/or mitochondrial DNA (mtDNA) (Shaw, 2010). Earlier developed restriction fragment length polymorphism (RFLP) markers have been used extensively for mapping insect resistance genes in different crops (Sun et al., 2006), but, due to involvement of cumbersome procedures, have now been replaced by a number of other markers which utilize PCR to amplify specific regions of interest. Examples of such markers include randomly amplified polymorphic DNAs – RAPDs (Gupta et al., 2006a; Paul et al., 2006; Sethi et al., 2003), sequence characterized amplified regions – SCARs (Gupta et al., 2008, 2010b), amplified fragment length polymorphism – AFLP (Lall et al., 2010), sequence tagged sites – STS (Severson et al., 2002), expressed sequence tags – ESTs, simple sequence repeats – SSRs, single primer amplification reaction – SPAR, simple sequence length polymorphism – SSLP, inter-simple sequence repeats – ISSR (Sandhu et al., 2012), variable number tandem repeats – VNTR (Legendre et al., 2007), cleaved amplified polymorphic sequences – CAPS (Burke et al., 2010), derived cleaved amplified polymorphic sequences – dCAPS, allele specific PCR (AS-PCR, PASA, ASO, ASAP), single-nucleotide amplified polymorphisms – SNAP (Behura, 2006) and single-strand conformational polymorphism – SSCP (Yasukochi, 1998). All such types of molecular markers have already been developed and exploited for rapid and precise identification of both old and newly evolved strains of diverse insect pest species. They have also proved useful in understanding various other important aspects of the insect pest, such as ecology, genetic diversity and host specificity in insects (Dilawari et al., 2002), population structure, tri-trophic interactions, insecticide resistance development (Randhawa et al., 2006; Sharma et al., 2005, 2007; Sherwani et al., 2011) and insect plant relationships (Heckel, 2003). Additional applications of molecular markers to insect science are screening of cotton germplasm for resistance to whitefly transmitted geminvirus (Gupta et al., 2006b; Sabhikhi et al., 2004), host associated genetic variations in whitefly and Trichogramma (Dilawari et al., 2002; Sharma et al., 2008a), genetic relatedness amongst related species (Dilawari et al., 2005; Sharma et al., 2008b), prevalence of endosymbiont alpha-proteobacteria (Rickettsiae) *Wolbachia* in some agriculturally important insects (Sharma et al., 2009), vectoring efficiency of whitefly for geminiviruses (Gupta et al., 2010a) and molecular typing of mealybug *Phenococcus solenopsis* populations (Singh et al., 2012).

16.4 MOLECULAR TAXONOMY

Scientific description and taxonomic characterization of various species are important both for their taxonomic identification as well as cataloguing biological diversity. Though 1.6 million insect species are estimated to exist, the binomial system, based upon morphological descriptions, has been able to describe around 1.0 million insect species during its 250 years of existence. In addition, with the dramatic rise in the species count, the task of species recognition has become more complex as morphological descriptions of species have become much more detailed and taxonomists have become much more specialized. In general, taxonomic identification has proved to be a slow process due to a number of limitations, such as phenotypic plasticity, dependency upon a particular life stage or gender, and the need for a high level of expertise (Hebert et al., 2003). The serious limitations inherent in conventional morphology-based identification systems and the dwindling pool of taxonomists have already signalled the need for a new approach to taxon recognition based upon molecular makeup (Hebert et al., 2003).

'Microgenomic identification systems', permitting species discrimination through the analysis of a small segment of DNA, have already gained broad acceptance and are being used for the identification of various species of viruses, bacteria and protists (Allander et al., 2001). DNA-based identification systems exploiting diversity among specific DNA sequences (molecular signatures or genetic 'barcodes') have also been applied to higher organisms (Trewick, 2000). Although earlier phylogenetic work focused on ribosomal (12S, 16S) DNA (Doyle and Gaut, 2000), the 'mitochondrial cytochrome *c* oxidase I' gene (COX-1 or COI) region has attracted most attention for single gene-based molecular identification of animals including insects (Knowlton and Weigt, 1998). In fact, the evolution of the 'COI gene' has been found to be rapid enough to allow the discrimination of not only closely allied species, but also phylogeographic groups within a single species (Cox and Hebert, 2001). The usefulness of the 'COI' sequence profiles as a single gene phylogenetic signal has been variously established successfully (ca. 100%) in assigning a large number of insect individuals to their respective families, orders, genera and species (Hebert et al., 2004). The effectiveness of the so-called DNA barcodes based on this ~648 bp COI region has also been demonstrated in distinguishing 521 species of Lepidoptera with a resolution of 97.9% (Marshall, 2005; Hajibabaei et al., 2006). Now, COI sequences providing sufficient variability have been globally accepted for barcoding/identification in different genera and species across all of the phyla (Garros et al., 2008).

Based upon the analysis of specimens from diverse families, orders and taxa, Dr Paul Hebert (Biodiversity Institute of Ontario, Canada) first put forward the concept of development of a COI-based species identification system that would provide a reliable, cost-effective and accessible solution to the current problem of species identification, assembly of which would also generate important new insights into the diversification of life and the rules of molecular evolution. Unequivocal acceptance of DNA barcoding as a universal tool for molecular taxonomy signalling a new approach to taxon recognition led to the formation of the 'Consortium for the Barcode of Life' (CBOL) with an ambitious programme of developing DNA barcodes for all of the existing species. CBOL has now become an established international movement of all of the organizations involved in taxonomic research and biodiversity issues. In this direction, the first ever 'The Canadian Barcode of Life Science Symposium' held on 10–11 May 2007 at the University of Guelph witnessed the opening of the Biodiversity Institute of Ontario (BIO) – the home of the 'Canadian Centre for DNA Barcoding' (CCDB) and headquarters of the 'Canadian Barcode of Life Network'.

16.4.1 International Project on Barcode of Life (iBOL)

Based upon the premise that 'Sequence diversity in DNA barcode region (COI) enables both the identification of known species and the discovery of new ones', this project was launched at the '2nd meeting of iBOL Scientific Steering Committee' held on 25 September 2010 at CNN Tower, Toronto, Canada. The iBOL has clear objectives of 'assembling the sequence libraries to identify organisms rapidly and inexpensively'. The specific goals of iBOL are (i) to construct the richly parameterized barcode library needed as the foundation for a DNA-based identification system, (ii) to develop a barcode database for at least 5.0 M specimens representing 500 K species by September 2015, and (iii) to deliver technologies enabling both massive biodiversity screens and point-of-contact identifications. The availability of such an enriched sequence library will enable the creation of a highly effective identification system for commonly encountered species and provide the foundation for a complete barcode reference library for eukaryotes. iBOL has become an assembly of a formidable array of researchers and private sector partners in 26 nations and is based upon the realization that iBOL's research goals can only be achieved through an alliance between nations with high biodiversity and those with the infrastructure for barcode analysis.

16.4.2 Barcode of Life Database System (BOLD)

BOLD is a freely available open access workbench aiding the acquisition, storage, analysis and publication of DNA barcode records. By assembling molecular, morphological and distributional data, it bridges a traditional bioinformatics chasm (Ratnasingh and Hebert, 2007). Class Insecta includes over a million described species with many more millions remaining still undescribed. Contributed to by scientists all over the world, it has been able to assemble 1,334,678 barcode records on 169,989 species of class Insecta. Using these barcodes as reference libraries, the 'Taxonomy browser of BOLD' now provides a 'press key' procedure for identification of unknown species. The current need is to enrich BOLD with 'DNA barcode-based reference libraries' on all of the known and new insect species so that it can serve as a strong scientific platform for molecular identification of different insect species established in the same or new environment(s).

16.4.3 Derived Benefits from DNA Barcode-Based Molecular Taxonomy

DNA barcoding data has already started to prove beneficial to the scientific community. A number of families and genera have faced taxonomical revisions (Hausmann et al., 2009; Liew et al., 2009). Numerous case studies available at BioNET (www.bionet-intl.org) and research publications from different barcoding projects (www.dnabarcodes.org/publications) describe the significance of biosystematics and the correct taxonomy in forensic sciences (Meiklejohn et al., 2011) and in prevention and economic/effective control of a number of insect pests and diseases (Pagès et al., 2009). Additional benefits have come from the correct identification of new and exotic pest species, correct description of new vectors of infectious diseases, establishing authenticity of species, early warning of invasive pest species, and biodiversity benefits from invasive alien species management (dePrins et al., 2009; deWaard et al., 2009). Also included are case studies, which have established how incorrect pest identifications have led to huge losses of life and crop yields and establishment of major pest species. This has also helped in protection of biodiversity by conservation of near extinct species of several life forms and in protecting honeybees and beneficial insects from diseases.

16.5 BIOTECHNOLOGY AND HOST-PLANT RESISTANCE

Plant resistance is the consequence of heritable plant qualities that result in a plant being relatively less damaged than a plant without these qualities. Insect-resistant crop varieties suppress insect pest abundance by elevating their damage tolerance level. The relationship between the insect and plant depends on three kind of resistance, e.g. antibiosis, antixenosis (non-preference), or tolerance. Antibiosis resistance affects the biology of the insect-pest to reduce its abundance and subsequent damage caused by same. It results in increased mortality or reduced longevity and reproduction of the insect. Antixenosis resistance affects the behaviour of an insect-pest and is expressed as non-preference of the insect for a resistant plant. Tolerance is resistance in which a plant is able to withstand or recover from damage caused by insect-pest abundance.

During last 30 years, the major biochemical principles governing such resistance and involved genes have been identified for their directed use through biotechnological approaches, but most emphasis has been given to primary protein products of specific genes. For affording host-plant resistance, genes of primary interest are those whose protein product could be detrimental to the normal growth and development of the target insect-pest/pest group based upon the mechanism of insecticidal action of the gene product. So far, the strategy has focused on the use of open reading frames (ORFs) of target genes as they occur in prokaryotes or cDNAs that are developed *in vitro* against the complex genes of eukaryotes.

The first transgenic plant (tobacco) expressed a cowpea trypsin inhibitor gene (cpti – an insecticidal gene) and produced the inhibitor protein at 1.0% level to provide enhanced protection against the lepidopteran pest *Heliothis virescens* (Hilder et al., 1987). However, subsequently developed transgenic rice (Xu et al., 1996) and potato (Bell et al., 2001) plants with this gene failed to provide sustainable insect protection. In contrast, insecticidal *Cry* family genes from *Bacillus thuringiensis* expressing insecticidal Cry- proteins found more favour for transformation of tobacco and tomato plants (Barton et al., 1987; Fischhoff et al., 1987).

16.5.1 *Bacillus thuringiensis* – Toxins and Biotechnology

Since the discovery of the insecticidal effect of *B. thuringiensis* in the early 20th century, plant protection from insects has involved biological preparations based on specific strains of *B. thuringiensis*. *B. thuringiensis* is a soil bacterium that produces a parasporal crystal made up of Cry- proteins that is toxic to specific group(s) of insects. *B. thuringiensis* occurs as different strains which produce 55 different types of insecticidal Cry- proteins (Cry1 to Cry55) that could be further grouped into three phylogenetically unrelated families (van Frankenhuyzen, 2009). As different Cry- proteins have different target insect specificities, the Cry-group toxins provide a vast range of insecticidal potential against an equally vast range of insect pest species. In general, Cry-toxin (Bt toxin) is produced by *B. thuringiensis* in an inactive form (protoxin ~130 kDa), which is processed into its active form (~70 kDa) by the insect's midgut proteases. The active toxin binds to specific receptors in the midgut linings and causes formation of microscopic pores through which the alkaline midgut content leaks into the haemolymph. This disturbs the normal ionic balance in the insect's body resulting in its ultimate death. This mode of action of Cry-toxins resulted in the development of inbuilt plant protection against insects by producing plants that could produce such toxins (Cry- proteins). By means of genetic engineering and plant biotechnology, the gene-part that encoded the activated Bt toxin could efficiently be transferred from *B. thuringiensis* cells to plants. Tobacco was the first plant in which the *Cry* gene was first found to be expressed. However, such plants produced the active toxin protein to a level that was insufficient to cause effective mortality in the target insect pest. This low level expression of these genes in eukaryotic systems could be attributed to a number of factors, the first of these being the relatively higher A + T content of prokaryotic (*Bacillus*) DNA than in eukaryotic plants. High A + T content is characteristic of splicing and polyadenylation (polyA) sites, mRNA degradation signals, and sites of transcription termination. Secondly, the codon frequencies in genes coding for *Bacillus* toxins differ considerably from those in eukaryotes. In general, plants have an overall preference for G + C content in the 3rd codon position (Perlak et al., 1991) as against A + T in the case of prokaryotic organisms in this degenerate base. Modification of the 3rd codon position from A or T to G or C, in the native *Cry* gene (through chemical synthesis of the gene) resulted in its increased expression in plants. Expression of this synthetic *Cry1Ac* endotoxin gene in transgenic plants showed a 1000-fold increase to result in effective insect control (Riva and Adang, 1996). Similarly, other *Cry* genes have been modified and expressed in tobacco, rice, coffee and other plants to achieve enhanced insect control protein levels (Leroy et al., 2000; Misztal et al., 2004; Shu et al., 2000).

In addition to using synthetic genes with optimized codon usage, efforts have also been made to further improve the expression of heterologous genes through modification of non-coding sequences such as promoters, operators, ribosome binding sites (RBS), 5′ and 3′ UTRs, and polyA terminator sequences. Among these non-coding sequences, promoter sequences are considered most important in deciding the level of expression of a transgene. Cauliflower mosaic virus (CaMV35s) promoter is widely used in transgenic plants and is considered one of the strongest promoters available but with better activity only in dicotyledonous plant cells. On the other hand, rice actin 1 promoter, the maize Emu promoter, and the maize polyubiquitin 1 promoter have proved useful for heterologous expression in monocotyledons.

Synthetic chimeric promoters have also been designed for still better expression in both dicot and monocots (Riva and Adang, 1996). The nucleotide sequences surrounding the N-terminal region of the protein appear particularly sensitive, both to the presence of rare codons and to the identities of the codons immediately adjacent to the initiation codon AUG. There is also some interplay between translation and mRNA stability, which has not been completely understood (Wu et al., 2004). However, reduced translational efficiency may be accompanied by a lower mRNA level because decreased ribosomal protection of the mRNA results in its increased exposure to endo-RNAses. The structure of the 5′ end of the mRNA also exerts a significant effect (Griswold et al., 2003), and strategies using short upstream ORFs for translational coupling of target genes have proved successful in improving the efficiency of expression of some problem genes (Ishida et al., 2002). Molecular understandings on all such various expressional mechanisms will allow meaningful expression of targeted genes for effective crop protection through biotechnological interventions.

Resulting from established procedures in biotechnology, research to transfer insect resistance genes from Bt to crop plants has become a continuous process. Corn, cotton, and potatoes among many commercial crops, have been targeted for Bt insect resistance. So far, most of the insect-resistant transgenic plants have been developed using various Bt δ-endotoxin genes such as *Cry1Ab, Cry1Ac, CryIIa, Cry9c,* and *CryII,* all derived from different strains. Insect-resistant crops have been one of the major successes of applying plant genetic engineering technology to agriculture. Cotton (*Gossypium hirsutum*) resistant to lepidopteran pests and maize (*Zea mays*) resistant to both lepidopteran and coleopteran pests have become widely used in global agriculture and have led to reductions in pesticide usage and lower production costs (Brookes and Barfoot, 2005; Toenniessen et al., 2003). This has led to a reduction in chemical pesticide use of 443 million kg and an additional financial gain for farmers of US $78 billion in the last 15 years (James, 2011). The Cry toxins produced in Bt crops generally target lepidopteran pests, although some also target coleopteran pests (Tabashnik et al., 2008). Commonly referred to as Bt crops (i.e. Bt-corn, Bt-cotton, etc.), they were first released in 1999 in USA and are now grown commercially in many countries. The area under transgenic crops, mainly with genes for insect and herbicide tolerance or in combination, is steadily increasing. More than 30 plant species primarily belonging to soybean, maize, cotton and potato transformed with different *Cry* genes have already been commercialized. Many other transgenic crops such as brinjal, tomato, potato, tobacco, coffee and rice transformed with modified *Cry1Ab, Cry1Ac, Cry1Ie,* etc. for protection against lepidopteran and coleopteran insect pests are in different stages of development and testing. Since the commercialization of biotech crops in 1996, farmers have adopted the technology at such a dramatic rate that year 2012 witnessed the planting of 170.3 million hectares of biotech crops by 17.3 million farmers in 28 countries around the world (James, 2012). The first commercialized Bt crops contained only one Cry toxin. In the second and third generations, scientists have mitigated the risk of possible resistance development through stacking or pyramiding of different genes such as multiple but different *Cry* genes and *Cry* genes combined with other insecticidal proteins, which target different receptors in insect pests but also provide resistance to a wider range of pests (Gatehouse, 2008; Tabashnik et al., 2009). Alternatively, synthetic variants of *Cry* genes have been employed as in the case of MON863 which expresses a synthetic *Cry3Bb1* gene (from *B. thuringiensis* subspecies kumamotoensis) against corn rootworm, which is eight times more effective than the native, non-modified version (Vaughn et al., 2005). Therefore multiple mutations/adaptations need to be made by target pests to develop resistance to this robust new generation of insect-resistant crops.

16.5.2 Non-Bt Genes for Insect Resistance

A parallel search on other possible non-Bt insect-resistant proteins has identified a large number of genes, which can hold potential to interfere with the development and nutritional requirements of different insect-pests. Important gene(s) which have attracted scientific attention for affording similar insect resistance potential in different crop plants are described next.

16.5.3 Vegetative Insecticidal Proteins (VIPs)

VIPs are produced by different bacterial species including *B. cereus* and *B. thuringiensis*. VIP1 and VIP2 are toxic to coleopteran insects and VIP3 is toxic to lepidopteran insects (Zhu et al., 2006). VIP3 protein induces gut paralysis and complete disruption of midgut epithelial cells to ultimately result in larval mortality. However, in contrast to Cry1 type proteins with proven toxicity towards a broader range of lepidopterans, different types of VIP3 proteins (Vip3Aa14, Vip3Ac1, Vip3LB, Vip3AcAa – a chimeric protein) exhibit only low to acute toxicity towards specific lepidopteran pest species (Fang et al., 2007; Sellami et al., 2012) and hence require more critical studies.

16.5.4 Biotin Binding Proteins (Avidin and Streptavidin)

Avidin is a water-soluble tetrameric glycoprotein from chicken egg, which binds strongly to biotin thus making it unavailable for growth and development. Streptavidin is a homologous protein produced by *Streptomyces avidinii*. Avidin was found to be an insecticidal and growth inhibiting dietary protein for five species of Coleoptera (red flour beetle, *Tribolium castaneum*, confused flour beetle, *T. confusum*, sawtoothed grain beetle, *Oryzaphilus surinamensis*, rice weevil, *Sitophilus oryzae*, and lesser grain borer, *Rhyzopertha dominica*) and two species of Lepidoptera (European corn borer, *Ostrinia nubilalis*, and Indianmeal moth, *Plodia interpunctella*) (Morgan et al., 1993). At levels ranging from 10 to 1000 ppm in the diet depending on the species, avidin retarded the growth and caused mortality of all seven species. Avidin and streptavidin are insect growth inhibiting proteins whose genes could potentially be expressed in plants to provide inbuilt plant resistance to insect pests. The toxic effect of expression of these proteins towards insect pests was established by growth retardation leading to effective mortality of larvae of *Spodoptera litura* and *Phthorimaea operculella* (potato tuber moth) by transgenic tobacco plants expressing avidin, streptavidin or their variants (Christeller et al., 2005; Murray et al., 2010). Similarly, transgenic apple expressing either avidin or streptavidin caused decreased growth and increased mortality of the light brown apple moth (*Epiphyas postvittana*) (Markwick et al., 2003).

16.5.5 Chitinases

Chitinase enzymes produced by different microbial, plant and insect species catalyze the hydrolysis of chitin. Chitinase enzymes target chitin in the peritrophic membrane of the midgut causing a reduction in survival and growth (Kabir et al., 2006). Such an attack results in severe and fatal abrasion of the insect's gut lining resulting in disrupted feeding and insect mortality. In this respect, chitinase from an insect source holds better potential as a bioinsecticide than chitinase from other sources as established by mortality of the merchant grain beetle after feeding on a diet containing the Manduca chitinase but not when chitinase was derived from *Serratia* (bacteria), Streptomyces (Actinomycete), or Hordeum (plant) (Wang et al., 1996). By weakening the insect's midgut membrane, chitinases also enhance the toxicity of Bt toxin to cause mortality at sublethal doses of Bt toxin (Ding et al., 1998). Transgenic rapeseed (*Brassica napus*) expressing *M. sexta* chitinase and scorpion insect toxin resulted in increased mortality and reduced growth of *Plutella maculipennis* (Wang et al., 2005).

16.5.6 Proteinase Inhibitors

These proteins interfere with the activity of midgut proteinases that help the insects in drawing their essential nutrients (peptides, amino acids) from feed proteins, thus causing nutritional limitations. Most significant among these are soybean trypsin inhibitor (SBTI), cowpea trypsin inhibitors (CpTI) and polyphenol oxidase. Though these plant defensive proteins have shown promise in addressing specialized insect resistance problems, transgenic plants expressing these plant defensive proteins have generally failed to provide more than partial protection against target insects, attributed to their natural pre-adaptation to such inhibitors.

16.5.7 Bean α-Amylase Inhibitors

The α-amylase inhibitor peptides from some legume seeds impart resistance to coleopteran seed weevils. Expression of α-amylase inhibitor gene from *Phaseolus vulgaris* in seeds of transgenic garden pea (*Pisum sativum*) and other grain legumes caused this inhibitor protein to accumulate up to the 3% level, which could provide significant seed resistance to larvae of bruchid beetles and pea weevil *Bruchus pisorum* (Morton et al., 2000; Shade et al., 1994). This strategy directed toward coleopteran seed herbivores proved a success as the activity retention by inhibitor protein at near neutral gut pH could efficiently interfere with starch digestion to result in nutritional limitations. Despite these results, agricultural deployment of transgenic crops expressing the α-amylase inhibitor gene has not taken place primarily due to safety concerns to mammals (Prescott et al., 2005).

16.5.8 Plant Lectins

Plant lectins represent a class of a heterogeneous group of proteins that specifically bind with sugars to cause sugar limitations (Van Damme et al., 2007, 2008). These glycoproteins constitute direct defence responses in plants against attack by phytophagous insects. Most lectins are multivalent and capable of agglutinating cells and hence are toxic to animal cells. Being extremely stable to proteases, these can exert deleterious effects through their direct effects on gastrointestinal function and growth of the insect (Sadeghi et al., 2006) with minimal effect(s) on non-target organisms. Out of the seven known lectin families (Vandenborre et al., 2011), most insecticidal activity is associated with:

- GNA related lectins: GNA (*Galanthus nivalis* agglutinin), ASAL (*Allium sativum* leaf agglutinin), ASAII (*Allium sativum* bulb agglutinin II), ACA (*Allium cepa* agglutinin) and LOA (*Listera ovata* agglutinin);
- Chitin-binding lectins: WGA (wheat germ agglutinin), UDA (*Urtica dioica* agglutinin) and OSA (*Oryza sativa* agglutinin); and
- Legume lectins: ConA (*Canavalia ensiformis* agglutinin), PHA (*Phaseolus vulgaris* agglutinin), PSA (*Pisum sativum* agglutinin), GS-II (*Griffonia simplicifolia* agglutinin) and BPA (*Bauhinia purpurea* agglutinin).

Many lectins recognize sugar structures that are not present in plants but are common in insect species. This binding specificity of different plant lectins is not directed against simple sugars but against more complex sugar structures such as *O*- and *N*-glycans (Van Damme et al., 2007, 2008). Most of these complex glycans are present as glycoconjugates along the intestinal tract of insects and serve as specific receptors by having a structure complementary to that of the binding site of the lectin. This implies that, in the same insect, glycoconjugates of different nature (e.g. glycoproteins, polysaccharides, glycolipids) but with identical carbohydrate moieties can act as different receptors for the same plant lectin. Thus, any single lectin molecule can also exert different insecticidal effects against different insect orders. Potential exploitation of lectin genes to confer insect resistance in transgenic plants has targeted hemipteran plant pests that are susceptible to lectin toxicity. However, constitutive or phloem-specific expression of the mannose-specific snowdrop lectin (GNA) in transgenic rice plants could provide only a partial resistance (up to 50%) to rice brown planthopper (*Nilaparvata lugens*) and other hemipteran pests resulting from reduced feeding, development, and fertility of the survivors (Foissac et al., 2000). Similarly, the transgenic rice plants expressing ASA-L were shown to decrease transmission of rice tungro virus by its insect vector (Saha et al., 2006). However, besides their direct toxic effect, biotechnology has been able to utilize the capacity of lectins in successfully delivering other insecticidal proteins that normally cannot pass the midgut epithelium into the haemocoel. The classic examples include:

- Inhibition of the feeding and growth of the tomato moth (*L. oleracea*) by a GNA-neuropeptide-allatostatin fusion protein (Fitches et al., 2004),
- Enhanced toxicity of GNA-spider-venom toxin I (SFI1) fusion protein to larvae of the tomato moth, rice brown planthopper (*N. lugens)* and the peach potato aphid (*M. persicae)* (Down et al., 2006; Fitches et al., 2004), and
- Enhanced toxicity of a fusion protein consisting of a GNA-lepidopteran-specific toxin (ButalT) from the South Indian red scorpion (*Mesobuthus tamulus*) to the larvae of the tomato moth (Trung et al., 2006).

However, further progress in research on lectins was limited due to possible consequences to higher animals of ingesting snowdrop lectin, although no adverse effect was seen in rats fed for 90 days on transgenic rice expressing GNA (Poulsen et al., 2007).

16.5.9 Scorpion and Spider Venoms

Venoms of many poisonous insect species such as scorpions and spiders that form a specific class of heterologous proteins, which exert a neurotoxic effect in other insect species, are viewed as potential candidates for developing insect-resistant transgenic plants (Gurevitz et al., 2007; Tedford et al., 2004).

16.5.9.1 Scorpion Venom

Primary interest has been attracted by insect toxin (AaHIT1) purified from the venom of North African scorpion *Androctonus australis* Hector. Transgenic tobacco plants transformed with *AaHIT1* gene could deter aleyrodids from feeding on the leaves (Sultan et al., 2003). When expressed in transgenic cotton, AaHIT1 expression measured at 0.43% of total soluble protein could effectively kill cotton bollworm (*Heliothis armigera*) larvae by 44–98% (Wu et al., 2008). Similar transgenic expression of *AaHIT1* gene in a hybrid poplar clone N-106 (*P. deltoides* x *P. simonii*) expressed significantly improved resistance to first instar larvae of *Lymantria dispar* (Wu et al., 2000). Transgenic plants of *Brassica napus* with an insect-resistant gene combination consisting of insect-specific chitinase (chi) and scorpion toxin gene *BmKiT(Bmk)* from Chinese scorpion *Buthus martensii* Karsch, also showed high-level expression for both chitinase and scorpion toxin proteins and demonstrated high resistance against the diamondback moth, *Plutella maculipennis* (Wang et al., 2005). Zhang et al. (2012), however, has utilized the potential of this gene for prolonging the pupal duration of silkworm *Bombyx mori* Linnaeus, of importance in appropriate management of insect populations in the silk industry.

16.5.9.2 Spider Venoms

With an estimated 100,000 species, spiders are one of the most successful arthropod predators. Their venom causes insect paralysis/lethality through the modulation of ion channels, receptors and enzymes in the insect nervous system (Windley et al., 2012). As a result of having a unique arrangement of disulphide bonds, these peptides are extremely resistant to proteases and hence are orally active in the insect gut and haemolymph. Thus, spider-venom peptides are stand-alone bioinsecticides and transgenes to engineer insect-resistant crops or entomopathogens. These toxins have low ED_{50} values in Orthoptera, Hemiptera, Dictyoptera, Diptera, Coleoptera, Acarina and Lepidoptera with no effect in vertebrates (Bloomquist, 2003; Chong et al., 2007; Khan et al., 2006). ω-Hexatoxin-Hv1a (ω-HXTX-Hv1a), from the venom of Australian funnel-web spiders, holds high affinity and specificity for insect CaV channels (voltage activated channels) and blocks the same (Tedford et al., 2007; Windley et al., 2011). Both topical application of recombinant thioredoxin-ω-HXTX-Hv1a and its transgenic expression result in protection of tobacco plants from *Helicoverpa armigera* and *Spodoptera littoralis* larvae (Khan et al., 2006). The toxin does not have any effect on homologous channels in rat (Tedford et al., 2004). Thus, spider-venom peptides are a rich source of potential bioinsecticides that can combine the desirable attributes of high potency, novel target activity, structural stability and phyletic selectivity. In fact, six different spider toxins from four spider species are under evaluation at different stages for specific toxicity against various insect-pest species belonging to insect orders Blattaria, Diptera, Lepidoptera, and Orthoptera (Windley et al., 2012). These are δ-CNTX-Pn1a and Γ-CNTX-Pn1a from *Phoneutria nigriventer*, κ-HXTX-Hv1c and ω-HXTX-Hv1a from *Hadronyche versuta*, μ-AGTX-Aa1d from *Agelenopsis aperta*, and μ-DGTX-Dc1a from *Diguetia canities.*

16.6 INSECT GROWTH REGULATORS/HORMONES

Insect growth regulators (IGRs), by interfering with normal growth and development, result in the insect's mortality before it reaches adulthood. Because these IGRs work on hormone pathways in insects, they do not have effects on other organisms. Moulting and metamorphosis processes are important biological functions in the growth and development of all insect species, which are precisely regulated by the levels of two major groups of nonpeptidic hormones, juvenile hormones (JHs) and ecdysteroids. The presence of JH during exposure to the ecdysteroids ensures a moult to a like stage, whereas the absence of JH during an ecdysteroid pulse allows metamorphosis. The relative levels of these hormones are precisely regulated by biosynthesis and degradation involving hydrolysis of the methyl ester of JH by soluble esterases and hydration of the epoxide by microsomal epoxide hydrolases. Whereas primary pathways for JH metabolism have been described (DeKort and Granger, 1996), the JH metabolism involves JH esterase (JHE) enzyme. Inhibition of JHE activity reduces the rate of JH degradation to cause thus delayed metamorphosis resulting in giant larvae of *Manduca sexta*. Overexpression of JHE from the embryonic stage of transgenic *B. mori* larvae resulted in larval–pupal metamorphosis only after the third stadium, two stadia earlier than that observed in wild-type insects (Tan et al., 2005). The results indicated that, though JH is not necessary for development during the embryonic and early larval stages of lepidopteran insects, it has a direct fatal influence in later stages of insect development. Though most insect species contain only JH III, JH 0, JH I, and JH II have only been identified in the Lepidoptera (butterflies and moths). The form JHB3 (JH III bisepoxide) appears to be the most important JH in the Diptera, or flies. Establishment of the role of individual JH species in the growth and development of specific insect species through transgenic expression is essential for exploitation of the same for control of these species using biotechnological approaches (Jindra et al., 2012).

16.7 GENETIC ENGINEERING OF BIOLOGICAL CONTROL AGENTS

The list of approved biological control agents that have been widely used in the EPPO regions includes 89 species of insects (74), arachnids (10) and nematodes (5) as established biocontrol agents (OEPP/EPPO, 2002). Besides the above, a wide range of bacteria, fungi and viruses are established pathogens or parasites of various insect-pest species. Many of these have been developed into commercial bioinsecticides and widely exploited on field crops and forests. These agents have shown a lot of promise in terms of host-plant activity, though their efficacy is affected by many biotic and non-biotic factors. There is a great need to elucidate the essence of these factors to improve the overall efficacy of these control agents along with developing novel methods to deliver sufficient inoculum at the target sites. Biotechnology has the potential to manipulate some of the desirable traits to improve the overall field efficacy of these agents as discussed next.

16.7.1 Bacillus thuringiensis

It has now been nearly three decades since the Bt delta-endotoxin gene was first cloned and expressed in *Escherichia coli* demonstrating for the first time the potential of genetic engineering technology for microbial control. Since that time, Bt toxin genes have been cloned and expressed in a wide variety of microorganisms as well as in plants in attempts to improve their delivery and efficacy against insect pests. Progress has been rapid, and in 1991, the United States Environmental Protection Agency's registration of two genetically engineered products (MVP and M-Trak), based on Bt delta-endotoxin expressing *Pseudomonas fluorescens* in which the cells are killed before use, was an important step in the process of commercialization of recombinant or transgenic microbial pesticides. Thus, the microencapsulation of Bt crystals, i.e. 'biopacking' by *P. fluorescens* cell wall, will protect the endotoxin from environmental factors (Dhaliwal et al., 2013). Initially in the late 1990s, Bt strains were genetically improved through conjugation. To improve the delivery of Bt toxins, an organism strain RSI, recently identified as *Bacillus* sp., was found to be an excellent colonizer of cotton phyllosphere. By conjugal transfer, the *Cry1Aa* gene of Bt has been introduced into it. It has been shown that transconjugation colonizes cotton plants for a prolonged period and it protects the plant from attack by *H. armigera* for more than 30 days (Bora et al., 1994).

Recent efforts to improve the insecticidal activity of Bt have been based on the transfer of Bt genes into non-homologous isolates of Bt as a means of combining delta endotoxins to produce either additive or synergistic effects and thus expand the host range. Bt isolate has been genetically transformed with IS232 to deliver the *Cry3A* gene into an isolate producing Cry1A toxin resulting in a strain which displayed insecticidal activity against both Lepidoptera and Coleoptera. A gene *Bel* has been identified in *B. cereus* group genomes which has the potential to increase the insecticidal activity of *B. thuringiensis*-based biopesticides. The combination of *Bel* and *Cry1Ac* increased the mortality rate

2.2-fold (S. Fang et al., 2009). Yue et al. (2005a,b) used integrative and thermosensitive vectors based on Bt transposon *Tn4430* harbouring the *tnpI-tnpA* gene, to integrate *Cry1C* and *Cry3A* genes into the chromosome of *Bt* subsp. *kurstaki* and broadening the insecticidal spectrum of the strain. Very little recent information is available on the successful use of different biotechnological approaches to improve the insecticidal efficacy of *Bacillus* spp. However, with the recent developments in genomics, proteomics and other biotechnological techniques, it would be possible to improve the Bt strains for pest management.

16.7.2 Entomopathogenic Fungi

Entomopathogenic fungi (~750 species) are pathogens of insect pests belonging to different insect orders. Several mycoinsecticides and mycoacaricides have been developed worldwide with most commercial products based on *Beauveria bassiana* (33.9%), *Metarhizium anisopliae* (33.9%), *Isaria fumosorosea* (5.8%), and *B. brongniartii* (4.1%) either registered or undergoing registration (deFaria and Wraight, 2007; Sandhu et al., 2012). Current research has focused on the hyphomycete genera *Metarhizium, Beauveria, Verticillium,* and *Paecilomyces* having a wide host range and high degree of specificity (Kaur et al., 2010; Sandhu et al., 2012). The commercial *B. bassiana*-based mycoinsecticides are relatively stable compared with other biological insect control agents for lepidopteran insect pests (Rana et al., 2008; Thakur et al., 2011). Selection of strains with high virulence requiring continued selection of strain with stable, specific efficacy for a target host has been limited by the high degree of existing genetic diversity (Neves and Hirose, 2005). A useful approach utilizes improvements in wild-type strains by combining the characteristics of different strains and mutants and utilizing optimum growth conditions for optimal conidiospore production utilizing cheap substrates as growth media (Kaur et al., 2010). Genetic improvements could lead to enhanced efficacy of the bioinsecticide and expand its host range. However, essential for the development of a hypervirulent strain is a complete understanding of the remarkable pathology of fungal infections. Molecular biology provides the necessary tools for dissecting the mechanisms of pathogenesis and in the longer term for producing recombinant organisms with new and relevant characteristics. Initial development towards these goals has occurred with *M. anisopliae* and *B. bassiana* (W. Fang et al., 2009; Hegedus et al., 1991). A few genetic transformation systems necessary for the experimental manipulation of virulence genes *in vitro* and *in vivo* have been established (Hasan et al., 2002; Sandhu and Vikrant, 2006). Transformation techniques have been used to isolate specific pathogenic genes, investigate virulence determinants of *M. anisopliae* and *B. bassiana*, and produce a strain with enhanced virulence. Unravelling the molecular mechanisms of fungal pathogenesis in insects will provide the basis for the genetic engineering of entomopathogenic fungi. EST analysis of *B. bassiana* has been studied using cDNA libraries (Cho et al., 2006). EST analyses of two subspecies of *M. anisopliae* revealed distinct patterns of expression of proteases and pathogenicity factors. These expression patterns have led to the ability to examine gene expression during infection of various insect hosts (Freimoser et al., 2005). In this context, more importance has been given to chitinase and proteinase enzymes (Fang et al., 2005) and metabolic pathways governing the production of insect-specific fungal toxins such as beauvericin, bassianolide, isarolides, and beauverolides from *B. bassiana*, and destruxins (DTXs) and cytochalasins from *M. anisopliae* (Lozano-Gutiérrez and España-Luna, 2008). Similarly, expressing a fusion protein with combined protease and chitinase activities increases the virulence of pathogen *B. bassiana* against green peach aphid, *Myzus persicae* (W. Fang et al., 2009).

16.7.3 Baculoviruses

The ability to transform or genetically engineer baculoviruses may provide a means of correcting the perceived defects in the efficiency and marketability of virus pesticides although some of these could also be alleviated with a more sophisticated IPM strategy including early sensing of population dynamics. The recombinant DNA technology has its current applications in inserting foreign genes into insect baculoviruses and achieving their rapid and efficient expression into recipient host systems. A gene *Tox34* from a mite, *Pyemotes tritici*, was cloned and the recombinant baculovirus (vEV-Tox34) produced has the potential to paralyse the insect during infection (Tomalski and Miller, 1991). Similarly, two genetically enhanced isolates of the *Autographa californica* nuclear polyhedrosis virus (AcMNPV) expressing insect-specific neurotoxin genes from the spiders *Diguetia canities* and *Tegenaria agrestis* (designated vAcTalTX-1 and vAcDTX9.2, respectively) have been evaluated for their commercial potential against lepidopteran insects. While vAcTalTX-1 kills faster than vAcDTX9.2, the latter is a faster feeding deterrent, suggesting that it would be more useful in reducing crop damage (Hughes et al., 1997). HaSNPV recombinants HaCXW1, lacking the ecdysteroid UDP-glucosyltransferase (*egt*) gene, and HaCXW2, in which an insect-selective scorpion toxin (AaIT) gene replaced the *egt* gene, were evaluated on *Helicoverpa armigera* in cotton. The recombinant viruses killed cotton bollworm faster compared to wild-type HaSNPV (Sun et al., 2002). Gramkow et al. (2010) constructed two recombinant baculoviruses containing the *ScathL* gene from *Sarcophaga peregrina* (vSynScathL), and the *keratinase* gene from the fungus *Aspergillus fumigatus* (vSyn-Kerat). The time taken to kill using recombinant *Spodoptera frugiperda* larvae is comparatively less by recombinant virus than by wild-type virus. The expression of proteases in infected larvae resulted in destruction of internal tissues late in infection, which could be the reason for the increased viral speed of kill. Similarly, a recombinant AcMNPV was constructed using co-expression of *Bacillus thuringiensis* Cry1-5 crystal protein gene and a Kunitz-type toxin isolated from bumblebee *Bombus ignitus* venom. Recombinant AcMNPV showed an improved insecticidal activity against larvae of *Plutella xylostella* and *Spodoptera exigua* (Choi et al., 2013). Research and development of recombinant baculovirus insecticides are still underway but have not reached the point of successful commercial use. Other polyhedrosis viral genes need to be identified which can be targeted for improving the pathogenicity of viral insecticides. In this regard, genome sequencing of viruses will play an important role. Efforts must be intensified to use the new biotechnological techniques to harvest those baculoviruses with maximum potential.

16.7.4 Entomopathogenic Nematodes

Entomopathogenic nematodes in field conditions are not so effective due to their susceptibility to environment extremes and host finding behaviour. Artificial selection, hybridization, mutagenesis and recombinant DNA technology can be used to improve their efficacy (Dhaliwal et al., 2013). The recent advances in molecular biology techniques have increased efforts to develop genetically engineered EPNs. Hashmi et al. (1995) made the first successful transformed *Heterorhabditis bacteriophora* HP88 by microinjection of plasmid vectors carrying *hsp-16* gene coding for 16-kDa heat shock protein as well as *rol-6* gene coding for roller phenotype.

Recent advances in decoding of the sequence of genomes and progress in looking at many genes, proteins and genetic pathways have led to a great revolution in biological sciences (Jindal et al., 2012a; Koltai, 2009). The first cDNA-sequencing project of the EPNs was reported by Sandhu et al. (2006). Out of 1246

ESTs generated in *H. bacteriophora* strain GPS11, 1072 ESTs were categorized into functional categories and 613 ESTs did not find matches in any of the searched databases, suggesting potentially novel genes in the EPNs. The ESTs were compared to animal-parasitic, human-parasitic, plant-parasitic, and free-living nematodes and 36 ESTs have been identified similar to ESTs of only parasitic nematodes, suggesting that they are involved in parasitic nematode-specific functions (Bai et al., 2007).

Eighty-one unique ESTs were identified in IS-6 strain of *Steinernema feltiae* during desiccation (Gal et al., 2003). Out of these, genes *Sf-ALDH, HSP40, ubq-2* and *glycerol kinase* were upregulated only during 8 to 24h of desiccation, however *Sf-LEA-1* of LEA (late embryogenesis abundant) was highly upregulated even after 24h of desiccation. Hao et al. (2010) analyzed cDNA transcripts expressed in the parasitic phase of *S. carpocapsae* to find genes involved in parasitism. About 32% of total ESTs generated had no homology with any known gene in public databases, which showed the presence of novel genes involved in parasitism of EPN. Recently, the genome of the TTO1 strain of *H. bacteriophora* has been sequenced at the Washington University Genome Sequencing Center in St Louis, MO. In the most comprehensive analysis of 31,485 high quality ESTs of the TTO1 strain, 554 parasitic nematode-specific ESTs were identified (Bai et al., 2009). The completion of the genome sequence of *H. bacteriophora* establishes a solid foundation for the much needed functional genomics studies on the genes that are involved in critical biological processes such as dauer formation, stress resistance, and sex determination. The functions of these genes in *H. bacteriophora* need to be elucidated using RNA interference technology and other means. Ciche and Sternberg (2007) have already indicated the feasibility of postembryonic RNAi in *H. bacteriophora* through soaking, which may be a potent approach for studying gene-function in EPNs. The success of postembryonic RNAi was evidenced by sterility, defective gonads and germ line proliferation in adult animals. Using these genomic resources, it may be possible to generate genetically modified *H. bacteriophora* with improved biological control potential. These approaches may be extended to develop more genetically improved strains of the other EPN species.

16.7.5 Natural Enemies

In spite of the availability of a range of established natural enemies, they often fail to suppress target insect pests due to the effect of detrimental insecticides, and unfavourable environmental conditions promoting asynchronous development to benefit the pest host over the parasitoid. On the other hand, relatively large *Trichogramma* species populations pervade and effectively suppress susceptible and potentially resistant pest species. Such observations are considered integral to the local insecticide resistance management (IRM) strategy for continued, sustainable Bt-transgenic crop production (Davies et al., 2012). The efficacy of bioagents is affected by predisposing ecological factors, particularly temperature and habitat diversity, apart from host-plant interactions. These place demands for research into stress tolerance in natural enemies to abiotic factors such as temperature and drought and the mechanisms governing stress tolerance. Biocontrol agents are very susceptible to pesticides, though in a totally pesticide-free environment and on Bt-transgenic crops, they have been reported to be highly effective (Davies et al., 2012). In India, an endosulfan-tolerant strain of *Trichogramma chilonis* has been developed through natural selection (Jalali et al., 2006). There is considerable scope to improve *Trichogramma*'s biological control potential, via habitat manipulation. *Trichogramma* may prove equally effective in humid regions. However, environmental constraints on *Trichogramma* survival, and those of other natural enemies, require due consideration before their successful application in biological control programmes.

Genetic improvement can be useful when the natural enemy is known to be a potentially effective biological control agent, apart from one limiting factor. Some of the desirable characteristics for transgenic insects include resistance to pathogens and insecticides, adaptation to different environmental conditions, high fecundity and improved host-seeking ability. Biotechnological interventions hold high potential to broaden the host range of natural enemies or enable their production on an artificial diet or non-host insect species that are easy to multiply under laboratory conditions. In addition, there is tremendous scope for developing natural enemies with genes for resistance to pesticides. However, the release of genetically modified insects is likely to face public resistance, being viewed as a potential risk to the environment. This is of particular concern when the same vector transmits several disease causing pathogens, as it might be difficult to develop transgenic individuals incapable of transmitting different pathogens.

16.8 GENETIC CONTROL OF INSECT PESTS THROUGH STERILE INSECT TECHNIQUE

The sterile insect technique (SIT) is a technique in which a large number of sterilized insects is released to reduce mating between fertile wild counterparts. This technique was reported to successfully eradicate the New World screwworm, the tsetse fly, melon fruit fly, Queensland fruit fly, pink bollworm, etc. Still there are chances to improve the efficiency of this technique through development of improved strains for mass rearing and release, molecular markers to identify the released sterile insects in the field, genetic sexing and sterilization (Franz and Robinson, 2011; Morrison et al., 2010). The translation of transgenic SIT methods to agriculturally important insects is now feasible as a result of the development of improved transformation vectors incorporating the *piggyBac* transposable element, and robust transformation markers based on enhanced green fluorescent protein (EGFP) variants. *Mariner, Minos, Hermes* and *piggyback* have been the most widely used transposable elements to date and have allowed transformation of several species of Diptera, Hymenoptera, Lepidoptera and Coleoptera. For monitoring the effectiveness of the SIT programme, discrimination between released sterile and wild insects is critical. The transgenic introduction of a fluorescent transformation marker would help in easy identification of released insects. Fluorescent sperm marking systems were established in mosquito species, viz. *Anopheles gambiae* Giles and *Aedes aegypti* Linnaeus (Schetelig and Wimmer, 2011). During mass rearing, separation of males and females is quite labour intensive, if external morphology and hand sorting are practised. Transgenic sexing systems based on female specific expression of a conditional lethal gene were first developed and tested in *D. melonogaster.* The female lethality was made conditional by using the binary tTA-expression system, which can be suppressed by supplementing food with tetracycline. Similar transgenic sexing systems have been developed for medfly which can kill 99.9% of medfly females. The sterilization, i.e. reproductive sterility, can be achieved by transgene-based embryonic lethality without the need for radiation. A transgenic 100% lethality system for the Tephritid pest species *Ceratitis capitata* (Wiedemann) that causes complete reproductive sterility without the need for radiation has been described. Furthermore, the Tet-off transgenic embryonic sexing system (TESS) for *Anastrepha suspensa* that uses a driver construct having the promoter from the embryo-specific *A. suspensa* serendipity α gene, linked to the Tet-transactivator, was used to drive the expression of a phospho-mutated variant of the pro-apoptotic cell death gene, *Alhid* from *Anastrepha ludens*. This system is highly effective and cost-efficient and eliminates most of the female insects early in embryogenesis (Schetelig and Handler, 2012). Recently, to avoid sterilization by irradiation, the *doublesex* (*dsx*) gene was sequenced and characterized from the pink bollworm (*Pectinophora gossypiella*). The conditional lethal genetic sexing system was developed by sex-alternate splicing in adults of diamondback moth (*Plutella xylostella*) and pink bollworm which increased the SIT of pinkbollworm, as well as broadening SIT-type control to diamondback moth and other Lepidoptera (Jin et al., 2013).

16.9 METABOLIC PATHWAYS AS A SOURCE OF USEFUL GENES AND PRODUCTS

The interaction of insect pest with its susceptible plant host is an intricate balance of complex metabolic pathways inherent both to the insect as well as the host. Any disturbance of these pathways could lead to inability of the insect pest to infest its host plant. Though not completely understood, knowledge of the genes controlling key steps in these pathways could be potential targets for interference through biotechnological interventions. Such genes can be discovered using a variety of approaches, but a routine approach would be generating and sequencing a library of expressed genes. A large number of ESTs are now available in the public databases for several crops, such as *Zea mays*, *Arabidopsis thaliana*, *Oryza sativa*, *Sorghum bicolor*, and *Glycine max*. A comparison of the EST databases from resistant and susceptible plants can reveal the diversity in coding sequences that is expressed due to pest infestation. However, the complexity of the pathways involved makes it difficult to assign specific functions to different differentially expressed genes. Furthermore, such information on specific biochemical pathways in insect species is lacking. To understand the gene functions of a whole organism, functional genomics technology is now focused on high throughput methods using insertion mutant isolation, gene chips or microarrays, and proteomics. These and other high throughput techniques offer powerful new uses for the genes discovered through sequencing (Hunt and Livesey, 2000).

Metabolic engineering, which depends upon metabolite profiling and analysis, holds great potential, in combination with genomics, transcriptomics and proteomics, in facilitating the development of newly emerging technologies such as RNA interference (Singh et al., 2011). RNA interference (RNAi), also known as post-transcriptional gene silencing (PTGS), is a quick, easy, sequence-specific way to establish the biological function of specific genes by *in vivo* 'knocking-down' the expression of such genes (Jindal et al., 2012b). The procedures involve the development of a transcriptomic database for the organism followed by identification of genes and their possible functions in reference to a previously established database on other related insect species. The biological function is then established through specific loss-of-gene function strategy using gene sequence-specific RNAi in living insects. Therefore, RNAi caused by exogenous double-stranded RNA (dsRNA) has emerged as a powerful technique for down-regulating gene expression in insects and various organisms. Injection of dsRNA to suppress expression of the corresponding gene has become a widely used tool in analyzing gene functions. RNAi technology is being exploited in most major crops to derive benefit by down-regulating specific metabolic pathways (Rathore et al., 2012; Regina et al., 2010; Wei et al., 2009).

With successful exploitation in plants, RNAi is being actively extended to specific insect pests. It has been used to explore the post-embryonic functions of heat shock protein HSP90 in regulation of compound eye development in model beetle *Tribolium castaneum* (Knorr and Vilcinskas, 2011). Injection of dsRNAs resulted in a variety of phenotypes among which some were lethal when they interfered with essential developmental processes or metabolism. Upadhyay et al. (2011) performed RNAi-mediated gene silencing of five different genes (actin ortholog, ADP/ATP translocase, alpha-tubulin, ribosomal protein L9-RPL9 and V-ATPase A subunit) in whitefly *B. tabaci* and observed knocking-down of the expression of RPL9 and V-ATPaseA to cause higher insect mortality compared to other genes. Semi-quantitative PCR of the treated insects also established a significant decrease in the level of RPL9 and V-ATPaseA transcripts. Small interfering RNAs (siRNAs) produced during these reactions were also found to be stable in the insect diet for at least 7 days at room temperature. Other similar studies include tissue-specific gene silencing of genes which were expressed in the midgut and salivary glands (Ghanim et al., 2007) and of heat shock protein genes in *Bemisia tabaci* (Lü and Wan, 2011).

The discovery that feeding of dsRNA produces RNAi effects in insects has opened the door for novel approaches in insect biotechnology. RNAi mediated silencing of specific gene(s) in pest insects through plant delivered RNA, offers the possibility to target genes necessary for their development, reproduction, or feeding success. Recent literature provides evidence that the expression of dsRNAs directed against insect genes in transgenic plants causes RNAi effects which can confer protection against insect herbivores (Huang et al., 2006; Zha et al., 2011). Principally, this technology enables engineering of a new generation of pest-resistant GM crops. However, the efficacy of protection and the range of species affected are dependent on the RNAi targets selected.

16.10 CONCLUSIONS

The successful and wide-scale adoption of genetically modified biotech crops worldwide has established the potential of biotechnology in improved crop production. However, the emergence of insect resistance to Bt-cotton occasionally raised concerns on its limited resilience in relation to possible insect resistance development. The future of GM crops however, relies upon the search for new genes, which, by acting differently, could afford similar or supplemented resistance in transgenic plants and has resulted in the identification of a number of genes from diverse sources. Many of these when evaluated have shown significant potential for exploitation in crop protection. Thus, future trends and prospects for biotechnological applications to mediate crop protection against insects include strategies employing stacked genes, modified Bt-toxins, spider/scorpion venom peptides, vegetative insecticidal proteins, lectins, endogenous resistance mechanisms as well as novel approaches. However, while exploiting such strategies, the benefits and risks associated with the adoption of GM insect-resistant crops especially for developing countries and resource-poor small holder farmers need to be kept in mind. Currently, most attention is being focused on vital genes and metabolic pathways that are inherent to the insect pest and host-plant biology. The host–pest relationship being a complex phenomenon, the identification of such genes and the specific role of their function in this complex metabolism has always remained a cumbersome task. Recent developments in biotechnology (RNAi) have provided an efficient means both for identification and functional analysis of new plant genes which are specifically expressed in response to pest attack and others in insect pests that are vital to pest biology. Yellow biotechnology as applied to insects now provides ample opportunities for identification of new genes and analysis of their functions to open a new field for their exploitation in effective insect pest control through transgenic expression in target plants.

References

Allander, T., Emerson, S.U., Engle, R.E., Purcell, R.H., Bukh, J., 2001. A virus discovery method incorporating DNase treatment and its application to the identification of two bovine parvovirus species. Proc. Natl. Acad. Sci. USA 98, 11609–11614.

Bai, X., Adams, B.J., Ciche, T.A., Clifton, S., Gaugler, R., Hogenhout, S.A., et al., 2009. Transcriptomic analysis of the entomopathogenic nematode *Heterorhabditis bacteriophora* TTO1. BMC Genomics 10, 205–218.

Bai, X., Grewal, P.S., Hogenhout, S.A., Adams, B.J., Ciche, T.A., Gaugler, R., et al., 2007. Expressed sequence tag analysis of gene representation in insect parasitic nematode *Heterorhabditis bacteriophora*. J. Parasitol. 93, 1343–1349.

Barton, K.A., Whiteley, H.R., Yang, N.S., 1987. *Bacillus thuringiensis* δ-endotoxin expressed in transgenic *Nicotiana tabacum* provides resistance to lepidopteran insects. Plant Physiol. 85, 1103–1109.

Behura, S.K., 2006. Molecular marker systems in insects: current trends and future avenues. Mol. Ecol. 15, 3087–3113.

Bell, H.A., Fitches, E.C., Marris, G.C., Bell, J., Edwards, J.P., Gatehouse, J.A., et al., 2001. Transgenic GNA expressing potato plants augment the beneficial biocontrol of *Lacanobiaoleracea* (Lepidoptera; Noctuidae) by the parasitoid *Eulophus pennicornis* (Hymenoptera; Eulophidae). Transgenic Res. 10, 35–42.

Bloomquist, J.R., 2003. Mode of action of atracotoxin at central and peripheral synapses of insects. Invertebr. Neurosci. 5, 45–50.

Bora, R.S., Murty, M.G., Shenbagarathai, R., Sekar, V., 1994. Introduction of a lepidopteran-specific insecticidal crystal protein gene of *Bacillus thuringiensis* subsp. *kurstaki* by conjugal transfer into a *Bacillus megaterium* strain that persists in the cotton phyllosphere. Appl. Environ. Microbiol. 60, 214–222.

Brookes, G., Barfoot, P., 2005. GM crops: The global economic and environmental impact – The first nine years 1996–2004. AgBioForum 8, 187–196.

Burke, M.K., Dunham, J.P., Shahrestani, P., Thornton, K.R., Rose, M.R., Long, A.D., 2010. Genome-wide analysis of a long-term evolution experiment with *Drosophila*. Nature 467, 587–590.

Cho, E.M., Liu, L., Farmerie, W., Keyhani, N.O., 2006. EST analysis of cDNA libraries from the entomopathogenic fungus *Beauveria* (Cordyceps) *bassiana*. I. Evidence for stage-specific gene expression in aerial conidia, *in vitro* blastospores and submerged conidia. Microbiology. 152, 2843–2854.

Choi, J.Y., Jung, M.P., Park, H.H., Tao, X.Y., Jin, B.R., Je, Y.H., 2013. Insecticidal activity of recombinant baculovirus co-expressing *Bacillus thuringiensis* crystal protein and Kunitz-type toxin isolated from the venom of bumblebee, *Bombus ignitus*. J. Asia Pac. Entomol. 16, 75–80.

Chong, Y., Hayes, J.L., Sollod, B., Wen, S., Wilson, D.T., Hains, P.G., et al., 2007. The omega-atrcotoxins: selective blockers of insects M-LVA and HVA calcium channels. Biochem. Pharmacol. 74, 623–638.

Christeller, J.T., Malone, L.A., Todd, J.H., Marshall, R.M., Burgess, E.P.J, Philip, B.A., 2005. Distribution and residual activity of two insecticidal proteins, avidin and aprotinin, expressed in transgenic tobacco plants, in the bodies and frass of *Spodoptera litura* larvae following feeding. J. Insect Physiol. 51, 1117–1126.

Ciche, T.A., Sternberg, P.W., 2007. Postembryonic RNAi in *Heterorhabditis bacteriophora*: A nematode insect parasite and host for insect pathogenic symbionts. BMC Dev. Biol. 7, 101.

Cox, A.J., Hebert, P.D.N., 2001. Colonization, extinction and phylogeographic patterning in a freshwater crustacean. Mol. Ecol. 10, 371–386.

Davies, A.P., Carr, C.M., Scholz, B.C.G., Zalucki, M.P., 2012. Using *Trichogramma* Westwood (Hymenoptera: Trichogrammatidae) for insect pest biological control in cotton crops: an Australian perspective. Aust. J. Entomol. 50, 424–440.

DeFaria, M.R., Wraight, S.P., 2007. Mycoinsecticides and mycoacaricides: a comprehensive list with worldwide coverage and international classification of formulation types. Biol. Control 43, 237–256.

DeKort, C.A.D., Granger, N.A., 1996. Regulation of JH titers: the relevance of degradative enzymes and binding proteins. Arch. Insect Biochem. Physiol. 33, 1–26.

DePrins, J., Mozuraitis, R., Lopez-Vaamonde, C., Rougerie, R., 2009. Sex attractant, distribution and DNA barcodes for the Afrotropical leaf-mining moth *Phyllonorycter melanosparta* (Lepidoptera: Gracillariidae). Zootaxa 2281, 53–67.

DeWaard, J.R., Landry, J.F., Schmidt, B.C., Derhousoff, J., McLean, J.A., Humble, L.M., 2009. In the dark in a large urban park: DNA barcodes illuminate cryptic and introduced moth species. Biodivers. Conserv. 18, 3825–3839.

Dhaliwal, G.S., Singh, R., Jindal, V., 2013. A Textbook of Integrated Pest Management. Kalyani Publishers, New Delhi, 617 pp.

Dilawari, V.K., Bons, M.S., Sethi, A., Kumar, P., Gupta, V.K., 2002. RAPD-PCR for studying host specificity of *Trichogramma*. Egg Parasitoid News 14, 9.

Dilawari, V.K., Gupta, V.K., Kumar, P., Sethi, A., 2005. Genetic relatedness in Indian *Trichogramma* spp. J. Insect Sci. 18, 77–84.

Ding, X., Gopalakrishnan, B., Jhonson, L.B., White, F.F., Wang, X., Morgan, T.D., et al., 1998. Insect resistance of transgenic tobacco expressing an insect chitinase gene. Transgenic Res. 7, 77–84.

Down, R.E., Fitches, E.C., Wiles, D.P., Corti, P., Bell, H.A., Gatehouse, J.A., et al., 2006. Insecticidal spider venom toxin fused to snowdrop lectin is toxic to the peach-potato aphid, *Myzus persicae* (Hemiptera: Aphididae) and the rice brown plant hopper, *Nilaparvata lugens* (Hemiptera: Delphacidae). Pest Manag. Sci. 62, 77–85.

Doyle, J.J., Gaut, B.S., 2000. Evolution of genes and taxa: a primer. Plant Mol. Biol. 42, 1–6.

Fang, J., Xu, X.L., Wang, P., Zhao, J.Z., Shelton, A.M., Cheng, J., et al., 2007. Characterization of chimeric *Bacillus thuringiensis* Vip3 toxins. Appl. Environ. Microbiol. 73, 956–961.

Fang, S., Wang, L., Guo, W., Zhang, X., Peng, D., Luo, C., et al., 2009. *Bacillus thuringiensis* bel protein enhances the toxicity of Cry1Ac protein to *Helicoverpa armigera* larvae by degrading insect intestinal mucin. Appl. Environ. Microbiol. 75, 5237–5243.

Fang, W.B., Leng, Y., Xia, K., Jin, J.M., Fan, Y., Feng, J., et al., 2005. Cloning of *Beauveria bassiana* chitinase gene Bb*chit1* and its application to improve fungal strain virulence. Appl. Environ. Microbiol. 71, 363–370.

Fang, W., Feng, J., Fan, Y., Zhang, Y., Bidochka, M.J., St. Leger, R.J., et al., 2009. Expressing a fusion protein with protease and chitinase activities increases the virulence of the insect pathogen *Beauveria bassiana*. J. Invertebr. Pathol. 102, 155–159.

Fischhoff, D.A., Bowdish, K.S., Perlak, F.J., Marrone, P.G., Mccormick, S.M., Niedermeyer, J.G., et al., 1987. Insect tolerant transgenic tomato plants. Biotechnology 5, 807–813.

Fitches, E., Wilkinson, H., Bell, H., Bown, D.P., Gatehouse, J.A., Edwards, J.P., 2004. Cloning, expression and functional characterisation of chitinase from larvae of tomato moth *Lacanobia oleracea*, a demonstration of the insecticidal activity of insect chitinase. Insect Biochem. Mol. Biol. 34, 1037–1050.

Foissac, X., Nguyen, T.L., Christou, P., Gatehouse, A.M.R., Gatehouse, J.A., 2000. Resistance to green leafhopper *Nephotettix virescens* and brown plant hopper *Nilaparvata lugens* in transgenic rice expressing snowdrop lectin *Galanthus nivalis* agglutinin; GNA. J. Insect Physiol. 45, 573–583.

Franz, G., Robinson, A.S., 2011. Molecular technologies to improve the effectiveness of the sterile insect technique. Genetica 139, 1–5.

Freimoser, M., Hu, G., St Leger, R.J., 2005. Variation in gene expression patterns as the insect pathogen *Metarhizium anisopliae* adapts to different host cuticles or nutrient deprivation in vitro. Microbiology 151, 361–371.

Gal, T.Z., Glazer, I., Koltai, H., 2003. Differential gene expression during desiccation stress in the insect-killing nematode *Steinernema feltiae* IS-6. J. Parasitol. 89, 761–766.

Garros, C., Ngugi, N., Githeko, A.E., Tuno, N., Yan, G., 2008. Gut content identification of larvae of the *Anopheles gambiae* complex in western Kenya using a barcoding approach. Mol. Ecol. Resour. 8, 512–518.

Gatehouse, J., 2008. Biotechnological prospects for engineering insect resistant plants. Plant Physiol. 146, 881–887.

Ghanim, M., Kontsedalov, S., Czosnek, H., 2007. Tissue-specific gene silencing by RNA interference in the whitefly *Bemisia tabaci* (Gennadius). Insect Biochem. Mol. 37, 732–738.

Gramkow, A.W., Perecmanis, S., Sousa, R.L.B., Noronha, E.F., Felix, C.R., Nagata, T., et al., 2010. Insecticidal activity of two proteases against *Spodoptera frugiperda* larvae infected with recombinant baculoviruses. Virol. J. 7, 143.

Griswold, K.E., Mahmood, N.A., Iverson, B.L., Georgiou, G., 2003. Effects of codon usage versus putative 5′-mRNA structure on the expression of *Fusarium solani* cutinase in the *Escherichia coli* cytoplasm. Protein Expr. Purif. 27, 134–142.

Gupta, V.K., Dilawari, V.K., Kumar, P., Sethi, A., Bons, M.S., 2006a. Monitoring of Indian whitefly populations with molecular markers for B-biotype. J. Insect Sci. 19, 57–65.

Gupta, V.K., Kumar, P., Sekhon, P.S., Joia, B.S., Dilawari, V.K., 2006b. Molecular screening of cotton germplasm for resistance to whitefly transmitted leaf curl disease through virus specific PCR-amplification. J. Insect Sci. 19, 92–97.

Gupta, V.K., Sharma, P.K., Sharma, R.K., Jindal, J., Dilawari, V.K., 2008. Development of SCAR markers for specific identification of B-biotype of *Bemisia tabaci* (Genn). Ann. Plant Prot. Sci. 16, 274–281.

Gupta, V.K., Sharma, P.K., Singh, S., Jindal, J., Dilawari, V.K., 2010a. Efficiency of *Bemisia tabaci* (Gennadius) populations from different plant-hosts for acquisition and transmission of cotton leaf curl virus. Indian J. Biotechnol. 9, 271–275.

Gupta, V.K., Sharma, R.K., Jindal, V., Dilawari, V.K., 2010b. SCAR markers for identification of host plant specificity in whitefly, *Bemisia tabaci* (Genn.). Indian J. Biotechnol. 9, 360–366.

Gurevitz, M., Karbat, I., Cohen, L., Ilan, N., Kahn, R., Turkov, M., et al., 2007. The insecticidal potential of scorpion β-toxins. Toxicon 49, 473–489.

Hajibabaei, M., Janzen, D.H., Burns, J.M., Hallwachs, W., Hebert, P.D.N., 2006. DNA barcodes distinguish species of tropical Lepidoptera. Proc. Natl. Acad. Sci. USA 103, 968–971.

Hao, Y.J., Motiel, R., Abubucker, S., Mitreva, M., Simoes, N., 2010. Transcripts analysis of the entomopathogenic nematodes *Steinernema carpocapsae* induced *in vitro* with insect haemolymph. Mol. Biochem. Parasitol. 169, 79–86.

Hasan, S., Bhamra, A.K., Sil, K., Rajak, R.C., Sandhu, S.S., 2002. Spore production of *Metarhizium anisopliae* (ENT-12) by solid state fermentation. J. Indian Bot. Soc. 8, 85–88.

Hashmi, S., Hashmi, G., Gaugler, R., 1995. Genetic transformation of an entomopathogenic nematode by microinjection. J. Invertebr. Pathol. 66, 293–296.

Hausmann, A., Hebert, P.D.N., Mitchell, A., Rougerie, R., Sommerer, M., Edwards, T., et al., 2009. Revision of the Australian *Oenochroma vinaria* Guenée (1858) species-complex (Lepidoptera: Geometridae, Oenochrominae): DNA barcoding reveals cryptic diversity and assesses status of type specimen without dissection. Zootaxa 2239, 1–21.

Hebert, P.D.N., Penton, E.H., Burns, J.M., Janzen, D.H., Hallwachs, W., 2004. Ten species in one: DNA barcoding reveals cryptic species in the neotropical skipper butterfly *Astraptes fulgerator*. Proc. Natl. Acad. Sci. USA 101, 14812–14817.

Hebert, P.D.N., Ratnasingham, S., deWaard, J.R., 2003. Barcoding animal life: cytochrome c oxidase subunit 1 divergences among closely related species. Proc. Biol. Sci 270, S96–S99.

Heckel, D.G., 2003. Genomics in pure and applied entomology. Annu. Rev. Entomol. 48, 235–260.

Hegedus, D.D., Pfeifer, T.A., MacPherson, J.M., Khachatourians, G.G., 1991. Cloning and analysis of five mitochondrial tRNA-encoding genes from the fungus *Beauveria bassiana*. Gene 109, 149–154.

Hilder, V.A., Gatehouse, A.M., Sheerman, S.E., Barker, R.F., Boulter, D., 1987. A novel mechanism of insect resistance engineered into tobacco. Nature 300, 160–163.

Huang, G., Allen, R., Davis, E.L., Baum, T.J., Hussey, R.S., 2006. Engineering broad root-knot resistance in transgenic plants by RNAi silencing of a conserved and essential root-knot nematode parasitism gene. Proc. Natl. Acad. Sci. USA 103, 14302–14306.

Hughes, P.R., Wood, H.A., Breen, J.P., Simpson, S.F., Duggan, A.J., Dybas, J.A., 1997. Enhanced bioactivity of recombinant baculoviruses expressing insect-specific spider toxins in lepidopteran crop pests. J. Invertebr. Pathol. 69, 112–118.

Hunt, S.P., Livesey, F.J., 2000. Functional Genomics: A Practical Approach. Oxford University Press, Oxford, UK, 253 pp.

Ishida, M., Muto, A., Ikemura, T., 2002. Overexpression in *Escherichia coli* of the AT-rich trpA and trpB genes from the hyperthermophilic archaeon *Pyrococcus furiosus*. FEMS Microbiol. Lett. 216, 179–183.

Jalali, S.K., Singh, S.P., Venkatesan, T., Murthy, K.S., Lalitha, Y., 2006. Development of endosulfan tolerant strain of an egg parasitoid *Trichogramma chilonis* Ishii (Hymenoptera: Trichogrammatidae). Indian J. Exp. Biol. 44, 584–590.

James, C., 2011. Global Status of Commercialized Biotech/GM Crops. ISAAA Brief No 43. ISAAA, Ithaca, NY, USA. <isaaa.org/resources/publications/briefs/43>.

James, C., 2012. Global Status of Commercialized Biotech/GM Crops. ISAAA Brief No 44. ISAAA, Ithaca, NY, USA. <isaaa.org/resources/publications/briefs/44>.

Jin, L., Walker, A.S., Fu, G., Samuel, T.H., Dafa'alla, T., Miles, A., et al., 2013. Engineered female-specific lethality for control of pest lepidoptera. ACS Synth. Biol. 2, 160–166.

Jindal, V., Bal, H.K., An, R., Grewal, P.S., 2012a. Developments in the genetic improvement of entomopathogenic nematodes. In: Koul, O., Dhaliwal, G.S., Khokhar, S., Singh, R. (Eds.), Biopesticides in Environment and Food Security: Issues and Strategies. Scientific Publishers, Jodhpur, India, pp. 112–141.

Jindal, V., Banta, G., Bharathi, M., Gupta, V.K., 2012b. Potential of RNAi in insect pest management. SBSI Biopestic. Newslett. 16, 3–6.

Jindra, M., Palli, S.R., Riddiford, L.M., 2012. The juvenile hormone signaling pathway in insect development. Annu. Rev. Entomol. 58, 181–204.

Kabir, K.E., Sugimoto, H., Tado, H., Endo, K., Yamanaka, A., Tanaka, S., et al., 2006. Effect of *Bombyx mori* chitinase against Japanese pine sawyer (*Monochamus alternatus*) adults as a biopesticide. Biosci. Biotech. Biochem. 70, 219–229.

Kaur, R., Gupta, V.K., Jindal, V., Blossom, Dilawari, V.K., 2010. Growing *Beauveria bassiana* for cuticle degrading activities and its bioefficacy against greater wax moth *Galleria melonella*. Biopestic. Int. 6, 96–104.

Khan, S.A., Zafar, Y., Briddon, R.W., Malik, K.A., Mukhtar, Z., 2006. Spider venom toxin protects plants from insect attack. Transgenic Res. 15, 349–357.

Knorr, E., Vilcinskas, A., 2011. Post-embryonic functions of HSP90 in *Tribolium castaneum* include the regulation of compound eye development. Dev. Genes Evol. 221, 357–362.

Knowlton, N., Weigt, L.A., 1998. New dates and new rates for divergence across the Isthmus of Panama. Proc. Roy. Soc. Lond. B 265, 2257–2263.

Koltai, H., 2009. Genomics and genetic improvement of entomopathogenic nematodes. In: Stock, S.P., Vandenberg, J., Glazer, I., Boemare, N. (Eds.), Insect Pathogens: Molecular Approaches and Techniques. CAB International, Wallingford, UK, pp. 346–364.

Lall, G.K., Darby, A.C., Nystedt, B., MacLeod, E.T., Bishop, R.P., Welburn, S.C., 2010. Amplified fragment length polymorphism (AFLP) analysis of closely related wild and captive tsetse fly (*Glossina morsitans morsitans*) populations. Parasites Vectors 3, 47.

Legendre, M., Pochet, N., Pak, T., Verstrepen, K.J., 2007. Sequence-based estimation of minisatellite and microsatellite repeat variability. Genome Res. 17, 1787–1796.

Leroy, T., Henry, A-M., Royer, M., Altosaar, I., Frutos, R., Duris, D., et al., 2000. Genetically modified coffee plants expressing the *Bacillus thuringiensis cry1Ac* gene for resistance to leaf miner. Plant Cell Rep. 19, 382–389.

Liew, T-S., Schilthuizen, M., Vermeulen, J.J., 2009. Systematic revision of the genus *Everettia* Godwin-Austen, 1891 (Mollusca: Gastropoda: Dyakiidae) in Sabah, northern Borneo. Zool. J. Linn. Soc. Lond. 157, 515–550.

Lozano-Gutiérrez, J., España-Luna, M.P., 2008. Pathogenicity of *Beauveria bassiana* (Deuteromycotina: Hyphomycetes) against the white grub *Laniifera cyclades* (Lepidoptera: Pyralidae) under field and greenhouse conditions. Fla. Entomol. 91, 664–668.

Lü, Z-C., Wan, F-H., 2011. Using double-stranded RNA to explore the role of heat shock protein genes in heat tolerance in *Bemisia tabaci* (Gennadius). J. Exp. Biol. 214, 764–769.

Markwick, N.P., Docherty, L.C., Phung, M.M., Lester, M.T., Murray, C., Yao, J.L., et al., 2003. Transgenic tobacco and apple plants expressing biotin-binding proteins are resistant to two cosmopolitan insect pests, potato tuber moth and lightbrown apple moth, respectively. Transgenic Res. 12, 671–681.

Marshall, E., 2005. Will DNA bar codes breathe life into classification? Science 307, 1037.

Meiklejohn, K.A., Wallman, J.F., Dowton, M., 2011. DNA-based identification of forensically important Australian Sarcophagidae (Diptera). Int. J. Legal Med. 125, 27–32.

Misztal, L.H., Mostowska, A., Skibinska, M., Bajsa, J., Musial, W.G., Jarmolowski, A., 2004. Expression of modified Cry1Ac gene of *Bacillus thuringiensis* in transgenic tobacco plants. Mol. Biotechnol. 26, 17–26.

Morgan, T.D., Oppert, B., Czapla, T.H., Kramer, K.J., 1993. Avidin and streptavidin as insecticidal and growth inhibiting dietary proteins. Entomol. Exp. Appl. 69, 97–108.

Morrison, N.I., Franz, G., Koukidou, M., Miller, T.A., Saccone, G., Alphey, L.S., et al., 2010. Genetic improvements to the sterile insect technique for agricultural pests. Asia-Pacific J. Mol. Biol. Biotechnol. 18, 275–295.

Morton, R.L., Schroeder, H.E., Bateman, K.S., Chrispeels, M.J., Armstrong, E., Higgins, T.J., 2000. Bean alpha-amylase inhibitor 1 in transgenic peas (*Pisum sativum*) provides complete protection from pea weevil (*Bruchus pisorum*) under field conditions. Proc. Natl. Acad. Sci. USA 97, 3820–3825.

Murray, C., Markwick, N.P., Kaji, R., Poulton, J., Martin, H., Christeller, J.T., 2010. Expression of various biotin-binding proteins in transgenic tobacco confers resistance to potato tuber moth, *Phthorimaea operculella* (Zeller) (fam. Gelechiidae). Transgenic Res. 19, 1041–1051.

Neves, P.M.O.J., Hirose, E., 2005. *Beauveria bassiana* strains selection for biological control of the coffee berry borer, *Hypothenemus hampei* (Ferrari) (Coleoptera: Scolytidae). Neotrop. Entomol. 34, 72–82.

OEPP/EPPO, 2002. List of biological control agents widely used in the EPPO region PM 6/3(2). OEPP/EPPO *Bull.* 32, 447–461. http://dx.doi.org/10.1046/j.1365-2338.2002.00600.x/pdf.

Pagès, N., Muñoz-Muñoz, F., Talavera, S., Sarto, V., Lorca, C., Núñez, J.I., 2009. Identification of cryptic species of Culicoides (Diptera: Ceratopogonidae) in the subgenus Culicoides and development of species-specific PCR assays based on barcode regions. Vet. Parasitol. 165, 298–310.

Paul, M., Gupta, V.K., Dilawari, V.K., Tewari., P.K., 2006. Molecular characterization of *Uscana* spp. (Hymenoptera: Trichogrammatidae) by RAPD analysis of genomic DNA. J. Insect Sci. 19, 4–9.

Perlak, F.J., Fuchs, R.L., Dean, D.A., McPherson, S.L., Fischoff, D.A., 1991. Modification of coding sequence enhances plant expression of insect control protein genes. Proc. Natl. Acad. Sci. USA 88, 3324–3328.

Poulsen, M., Kroghsbo, S., Schroder, M., Wilcks, A., Jacobsen, H., Miller, A., et al., 2007. A 90-day safety study in Wistar rats fed genetically modified rice expressing snowdrop lectin *Galanthus nivalis* (GNA). Food Chem. Toxicol. 45, 350–363.

Prescott, L.M., Harley, J.R., Klein, D.A., 2005. Microbiology. McGraw Hill, New York, 992 pp.

Rana, I.S., Kanojiya, A., Sandhu, S.S., 2008. Mosquito larvicidal potential of fungi isolated from larval mosquito habitats against *Aedes aegypti*. J. Biol. Control 22, 179–183.

Randhawa, P., Gupta, V.K., Joia, B.S., Dilawari, V.K., 2006. Identification of RAPD-PCR markers for ethion resistance in *Bemisia tabaci*. J. Insect Sci. 19, 43–49.

Rathore, K.S., Sundaram, S., Sunilkumar, G., Campbell, L.M., Puckhaber, L., Marcel, S., et al., 2012. Ultralow gossypol cottonseed: generational stability of the seed-specific, RNAi-mediated phenotype and resumption of terpenoid profile following seed germination. Plant Biotech. J. 10, 174–183.

Ratnasingh, S., Hebert, P.D.N., 2007. BOLD: The barcode of life data system. <www.barcodinglife.org>. Mol. Ecol. Notes 7, 355–364.

Regina, A., Kosar-Hashemi, B., Ling, S., Li, Z., Rahman, S., Morell, M., 2010. Control of starch branching in barley defined through differential RNAi suppression of starch branching enzyme IIa and IIb. J. Exp. Bot. 61, 1469–1482.

Riva, G.A., Adang, M.J., 1996. Expression of *Bacillus thuringiensis* δ-endotoxin genes in transgenic plants. Biotechnol. Appl. 13, 5–8.

Sabhikhi, H.S., Sekhon, P.S., Gupta, V.K., Sohu, R.S., 2004. Identification of cotton leaf curl immune genotypes in *Gossypium hirsutum* L. Crop Improv. 31, 66–70.

Sadeghi, A., Van Damme, E.J., Peumans, W.J., Smagghe, G., 2006. Deterrent activity of plant lectins on cowpea weevil *Allosobruchus maculatus* (F.) oviposition. Phytochemistry 18, 2078–2084.

Saha, P., Majumder, P., Dutta, I., Ray, T., Roy, S.C., Das, S., 2006. Transgenic rice expressing *Allium sativum* leaf lectin with enhanced resistance against sap-sucking insect pests. Planta 223, 1329–1343.

Sandhu, S.K., Jagdale, G.B., Hogenhout, S.A., Grewal, P.S., 2006. Comparative analysis of the expressed genome of the infective juvenile entomopathogenic nematode, *Heterorhabditis bacteriophora*. Mol. Biochem. Parasitol. 145, 239–244.

Sandhu, S.S., Vikrant, P., 2006. Evaluation of mosquito larvicidal toxins in the extra cellular metabolites of two fungal genera *Beauveria* and *Trichoderma*. In: Bagyanarayana, G., Bhadraiah, B., Kunwar, I.K. (Eds.), Emerging Trends in Mycology, Plant Pathology and Microbial Biotechnology. B. S. Publications, Hyderabad, India.

Sandhu, S.S., Sharma, A.K., Beniwal, V., Goel, G., Batra, P., Kumar, A., et al., 2012. Myco-biocontrol of insect pests: Factors involved mechanism, and regulation. J. Pathog. http://dx.doi.org/10.1155/2012/126819 2012m. Article ID 126819.

Schetelig, M.F., Handler, A.M., 2012. A transgenic embryonic sexing system for *Anastrepha suspensa* (Diptera: Tephritidae). Insect Biochem. Mol. Biol. 42, 790–795.

Schetelig, M.F., Wimmer, E.A., 2011. Insect transgenesis and the sterile insect technique. In: Vilcinskas, A. (Ed.), Insect Biotechnology. Springer, Dordrecht, NL, pp. 169–194.

Sellami, S., Zghal, T., Cherif, M., Zalila-Kolsi, I., Jaoua, S., Jamoussi, K., 2012. Screening and identification of a *Bacillus thuringiensis* strain S1/4 with large and efficient insecticidal activities. J. Basic Microbiol. http://dx.doi.org/10.1002/jobm.201100653.

Sethi, A., Kumar, P., Jindal, J., Singh, B., Gupta, V.K., Dilawari, V.K., 2003. Monitoring insecticide resistance in whitefly, *Bemisia tabaci* (Genn.) in Indian cotton ecosystem. Resistant Pest Manag. Newslett. 13, 52–55.

Severson, D.W., Meece, J.K., Lovin, D.D., Saha, G., Morlais, I., 2002. Linkage map organization of expressed sequence tags and sequence tagged sites in the mosquito, *Aedes aegypti*. Insect Mol. Biol. 11, 371–378.

Shade, R.E., Schroeder, H.E., Pueyo, J.J., Tabe, L.M., Murdock, L.L., Higgins, T.J.V., et al., 1994. Transgenic pea seeds expressing α-amylase inhibitor of the common bean are resistant to bruchid beetles. Biotechnology 12, 793–796.

Sharma, R.K., Gupta, V.K., Jindal, J., Dilawari, V.K., 2008a. Host associated genetic variations in whitefly *Bemisia tabaci* (Genn). Indian J. Biotechnol. 7, 366–370.

Sharma, R.K., Gupta, V.K., Singh, S., Dilawari., V.K., 2008b. Analysis of the RAPD profiles and ITS-2 sequence of *Trichogramma* species for molecular identification and genetic diversity studies. Pest Manag. Sci. Econ. Zool. 16, 1–8.

Sharma, R.K., Sharma, S., Singh, S., Kumar, R., Gupta, V.K., 2009. Molecular detection of prevalence of an endosymbiont alpha-proteobacteria (Rickettsiae), *Wolbachia* in some agriculturally important insects from Punjab (India). J. Entomol. Res. 33, 189–193.

Sharma, S., Dilawari, V.K., Gupta, V.K., 2005. Inheritance of imidacloprid resistance in cotton whitefly, *Bemisia tabaci* (Hemiptera: Aleyrodidae). J. Insect Sci. 18, 65–71.

Sharma, S., Dilawari., V.K., Gupta, V.K., 2007. Population restructuring in *Bemesia tabaci* in response to imidacloprid selection pressure. J. Insect Sci. 20, 229–235.

Shaw, K.L., 2010. Conflict between nuclear and mitochondrial DNA phylogenies of a recent species radiation: What mtDNA reveals and conceals about modes of speciation in Hawaiian crickets. Proc. Natl. Acad. Sci. USA 99, 16122–16127.

Sherwani, A., Zaki, F.A., Gupta, V.K., Ahmad, M.M., 2011. Molecular analysis of population restructuring in *Panonychus ulmi* (Koch) in response to fanzaquin selection pressure. J. Insect Sci. 24, 23–29.

Shu, Q., Ye, G., Cui, H., Cheng, X., Xiang, Y., Wu, D., et al., 2000. Transgenic rice plants with a synthetic *cry1Ab* gene from *Bacillus thuringiensis* were highly resistant to eight lepidopteran rice pest species. Mol. Breed. 6, 433–439.

Singh, A., Kumar, B., Srivastava, A.K., 2011. Metabolic engineering of crops using RNA interference. Pac. J. Mol. Biol. Biotechnol. 19, 137–148.

Singh, S., Sharma, R., Kumar, R., Gupta, V.K., Dilawari, V.K., 2012. Molecular typing of mealybug *Phenococcus solenopsis* populations from different hosts and locations in Punjab, India. J. Environ. Biol. 33, 539–543.

Sultan, A., Ji, S., Zhu, Y., 2003. Prokaryotic expression and transgenic analysis of a scorpion toxin AaH1T1 gene. Chinese J. Biochem. Mol. Biol. 19, 182–185.

Sun, L.H., Wang, C.M., Su, C.C., Liu, Y.Q., Zhai, H.Q., Wan, J., 2006. Mapping and marker-assisted selection of a brown planthopper resistance gene bph2 in rice (*Oryza sativa* L.). Acta Genet. Sin. 33, 717–723.

Sun, X., Chen, X., Zhang, Z., Wang, H., Bianchi, F.J., Peng, H., et al., 2002. Bollworm responses to release of genetically modified *Helicoverpa armigera* nucleopolyhedroviruses in cotton. J. Invertebr. Pathol. 81, 63–69.

Tabashnik, B.E., Gassmann, A.J., Crowder, D.W., Carrière, Y., 2008. Insect resistance to BT crops: evidence versus theory. Nat. Biotechnol. 26, 199–202.

Tabashnik, B.E., Van Rensburg, J.B.J., Carrière, Y., 2009. Field-evolved insect resistance to BT crops: definition, theory, and data. J. Econ. Entomol. 102, 2011–2025.

Tan, A., Tanaka, H., Tamura, T., Shiotsuki, T., 2005. Precocious metamorphosis in transgenic silkworms overexpressing juvenile hormone esterase. Proc. Natl. Acad. Sci. USA 102, 11751–11756.

Tedford, H.W., Gilles, N., Ménez, A., Doering, C.J., Zamponi, G.W., King, G.F., 2004. Scanning mutagenesis of ω-atracotoxin-Hv1a reveals a spatially restricted epitope that confers selective activity against insect calcium channels. J. Biol. Chem. 279, 44133–44140.

Tedford, H.W., Maggio, F., Reenan, R.A., King, G.F., 2007. A model genetic system for testing the in vivo function of peptide toxins. Peptides 28, 51–56.

Thakur, R., Jain, N., Pathak, R., Sandhu, S.S., 2011. Practices in wound healing studies of plants. J. Evid. Based Complementary Altern. Med. 2011 http://dx.doi.org/10.1155/2011/438056. Article ID 438056.

Toenniessen, G.H., O'Toole, J.C., DeVries, J., 2003. Advances in plant biotechnology and its adoption in developing countries. Curr. Opin. Plant Biotechnol. 6, 191–198.

Tomalski, M.D., Miller, L.K., 1991. Insect paralysis by baculovirus-mediated expression of a mite neurotoxin gene. Nature 352, 82–85.

Trewick, S.A., 2000. Mitochondrial DNA sequences support allozyme evidence for cryptic radiation of New Zealand *Peripatoides* (Onychophora). Mol. Ecol. 9, 269–282.

Trung, N.P., Fitches, E., Gatehouse, J.A., 2006. A fusion protein containing a lepidopteran-specific toxin from the South Indian red scorpion (*Mesobuthus*) and snowdrop lectin shows oral toxicity to target insects. BMC. Biotechnol. 6, 18–29.

Upadhyay, S.K., Chandrashekar, K., Thakur, N., Verma, P.C., Borgio, J.F., Singh, P.K., et al., 2011. RNA interference for the control of whiteflies (*Bemisia tabaci*) by oral route. J. Biosci. 36, 53–161.

Van Damme, E.J.M., Lannoo, N., Peumans, W.J., 2008. Plant lectins. Adv. Bot. Res. 48, 107–209.

Van Damme, E.J.M., Rougé, P., Peumans, W.J., 2007. Carbohydrate-protein interactions: plant lectins In: Kamerling, J.P., Boons, G.J., Lee, Y.C., Suzuki, A., Taniguchi, N., Voragen, A.G.J., (Eds.), Comprehensive Glycoscience – From Chemistry to Systems Biology, vol. 3. Elsevier, Oxford, UK, pp. 563–599.

Van Frankenhuyzen, K., 2009. Insecticidal activity of *Bacillus thuringiensis* crystal proteins. J. Invertebr. Pathol. 101, 1–16.

Vandenborre, G., Smagghe, G., Van Damme, E.J., 2011. Plant lectins as defense proteins against phytophagous insects. Phytochemistry 72, 1538–1550.

Vaughn, T., Cavato, T., Brar, G., Coombe, T., DeGooyer, T., Ford, S., et al., 2005. A method of controlling corn rootworm feeding using a *Bacillus thuringiensis* protein expressed in transgenic maize. Crop Sci. 45, 931–938.

Wang, J.X., Chen, Z.L., Du, J.Z., Sun, Y., Liang, A.H., 2005. Novel insect resistance in *Brassica napus* developed by transformation of chitinase and scorpion toxin genes. Plant Cell Rep. 24, 549–555.

Wang, X., Ding, X., Gopalakrishnan, B., Morgan, T.D., Johnson, L., White, F., et al., 1996. Characterization of a 46kDa insect chitinase from transgenic tobacco. Insect Biochem. Mol. Biol. 26, 1055–1064.

Wei, S., Li, X., Gruber, M.Y., Li, R., Zhou, R., Zebarjadi, A., et al., 2009. RNAi-mediated suppression of DET1 alters the levels of carotenoids and sinapate esters in seeds of *Brassica napus*. J. Agric. Food Chem. 57, 5326–5333.

Windley, M.J., Escoubas, P., Valenzuela, S.M., Nicholson, G.M., 2011. A novel family of insect-selective peptide neurotoxins targeting insect BKCa channels isolated from the venom of the theraphosid spider, *Eucratoscelus constrictus*. Mol. Pharmacol. 80, 1–13.

Windley, M.J., Herzig, V., Dziemborowicz, S.A., Hardy, M.C., King, G.F., Nicholson, G.M., 2012. Spider-venom peptides as bioinsecticides. Toxins 4, 191–227.

Wu, J., Luo, X., Wang, Z., Tian, Y., Liang, A., Sun, Y., 2008. Transgenic cotton expressing synthesized scorpion insect toxin AaHIT gene confers enhanced resistance to cotton bollworm (*Heliothis armigera*) larvae. Biotechnol. Lett. 30, 547–554.

Wu, N.F., Sun, Q., Yao, B., Fan, Y.L, Rao, H.Y., Huang, M.R., et al., 2000. Insect-resistant transgenic poplar expressing AaIT gene. Sheng. Wu. Gong. Cheng. Xue. Bao. 16, 129–133.

Wu, X., Muto, A., Ikemura, T., 2004. Codon optimization reveals critical factors for high level expression of two rare codon genes in *Escherichia coli*: RNA stability and secondary structure but not tRNA abundance. Biochem. Biophys. Res. Commun. 313, 89–96.

Xu, D.P., Xue, Q.Z., McElroy, D., Mawal, Y., Hilder, V.A., Wu, R., 1996. Constitutive expression of a cowpea trypsin inhibitor gene, CpTi, in transgenic rice plants confers resistance to two major rice insect pests. Mol. Breed. 2, 167–173.

Yasukochi, Y., 1998. A dense genetic map of the silkworm, *Bombyx mori*, covering all chromosomes based on 1018 molecular markers. Genetics 150 (4), 1513–1525.

Yue, C., Sun, M., Yu, Z., 2005a. Improved production of insecticidal proteins in *Bacillus thuringiensis* strains carrying an additional *cry1C* gene in its chromosome. Biotechnol. Bioeng. 92, 1–7.

Yue, C., Sun, M., Yu, Z., 2005b. Broadening the insecticidal spectrum of lepidoptera-specific *Bacillus thuringiensis* strains by chromosomal integration of cry3A. Biotechnol. Bioeng. 91, 296–303.

Zha, W., Peng, X., Chen, R., Du, B., He, G., 2011. Knockdown of midgut genes by dsRNA-transgenic plant-mediated RNA interference in the Hemipteran insect *Nilaparvata lugens*. PLOs One 6, e20502.

Zhang, H.K., Cao, G.L., Zhang, X.R., Wang, X.J., Xue, R.Y., Gong, C.L., 2012. Effects of BmKIT 3R gene transfer on pupal development of *Bombyx mori* Linnaeus using a Gal4/UAS binary transgenic system. J. Agric. Food Chem. 60, 3173–3179.

Zhu, C., Ruan, L., Peng, D., Yu, Z., Sun, M., 2006. Vegetative insecticidal protein enhancing the toxicity of *Bacillus thuringiensis* subsp kurstaki against *Spodoptera exigua*. Lett. Appl. Microbiol. 42, 109–114.

CHAPTER

17

Biotechnological and Molecular Approaches in the Management of Non-Insect Pests of Crop Plants

S. Mohankumar, N. Balakrishnan and R. Samiyappan

Tamil Nadu Agricultural University, Coimbatore, India

17.1 INTRODUCTION

Tools and techniques of molecular biology and genetic engineering have provided unprecedented power to manage biotic stresses in an effective way for safe and sustainable agriculture in the 21st century. Biotic stress menaces are the major factors that destabilize crop productivity in agricultural ecosystems. Application of chemical inputs has proven harmful to human health and the environment. Hence there is a need for safer alternatives to manage pests in agricultural ecosystems. Mites and nematodes not only cause direct losses to agricultural produce, but also indirectly play a role as vectors of various plant pathogens. In addition to direct losses, there are extra costs in the form of pesticides applied for pest control. Extensive and indiscriminate usage of chemical pesticides has resulted in environmental degradation, adverse effects on human health and other organisms, eradication of beneficial insects and development of resurgence and resistance to pesticides.

Host-plant resistance is the foundation step for any successful integrated pest management (IPM) method adopted in crop ecosystems. However, in conventional host-plant resistance to pests, progress has been slow, very limited and difficult to achieve. Biotechnology has provided several unique opportunities, *viz.*, access to novel molecules, ability to change the level of gene expression, and the capability to change the expression pattern of genes and develop transgenics with different pesticidal genes. A large number of pesticidal molecules have been reported which are effective against non-insect pests. With the advent of genetic transformation techniques based on recombinant DNA technology, it is now possible to insert genes into the plant genome that confer resistance to specific pests (Bennett, 1994). Biotechnology is not a panacea for solving these non-insect pest problems, but it provides many powerful tools.

D. P. Abrol (Ed): *Integrated Pest Management.*
DOI: http://dx.doi.org/10.1016/B978-0-12-398529-3.00019-1

These tools will be helpful to understand species identification, population variation, development of resistant cultivars, interactions between pest and host plant, chemical ecology, etc. which will lead to better management strategy. This chapter focuses on biotechnological and molecular approaches for the management of agriculturally important mites and nematodes.

17.2 MITES

Chelicerata (horseshoe crabs, scorpions, spiders, ticks, and mites) are the second largest group of arthropods. Extant lineages of chelicerates include Pycnogonida, Xiphosura (horseshoe crabs) and Arachnida (a large group comprising scorpions, spiders and the Acari – ticks and mites). The Acari represent the most diverse chelicerate clade, with over 40,000 described species that exhibit tremendous variations in lifestyle, ranging from parasitic to predatory to plant-feeding. Unlike other arachnids, which are free-living, a large number of acarines have developed intimate associations with other animals/plants. Some free-living mites live in stored food where they often multiply rapidly and attain pest status. Stored-product and house dust mites inhabit human environments, including stored food and animal feed (Thind and Clarke, 2001). The mite contamination includes their allergens (Fernández-Caldas et al., 2008) and transmitted microorganisms (Hubert et al., 2004) of pathogenic importance. Some species of mites cause agricultural damage as they feed on crops, transmit diseases, and cause significant economic losses. Chiefly spider mites, false spider mites, and eriophyid mites feed on vascular tissues of higher plants and through this activity, mites can cause severe losses to field and protected crops. Many of the species of mites are polyphagous in nature and pre-adapted to evolve pesticide resistance. For example, the two-spotted spider mite, *Tetranychus urticae*, is one of the main pests of agricultural crops due to its broad host range. This polyphagous species feeds on more than 1100 plant species, from which about 150 are of great economic value. Thus, it represents a very important pest for field and greenhouse crops, ornamentals, annual and perennial plants all over the world (Migeon and Dorkeld, 2011).

17.2.1 Identification

In pest species, species identification is also important in the development of (biological) pest control strategies. There are only a few taxonomists specialized in morphological identification of mites and their number is decreasing, which adds to the problem of identification. The precise identification of mites is difficult due to the limited number of available morphological characteristics and to the similarity between species. In addition, both sexes of many species are often needed to arrive at precise determinations. For these reasons, molecular methods are increasingly being applied for taxonomic purposes. Molecular methods were also used to determine the origin of invading spider mites (Navajas et al., 1994).

Accumulating large numbers of mite sequences, especially from genes such as mitochondrial cytochrome oxidase I (COI) and the internal transcribed spacer regions (ITS1 and ITS2) of nuclear ribosomal DNA (Navajas et al., 1996a, 1998), can also serve as the scaffold of a molecular method that could simplify spider mite identification. This would facilitate the use of species-specific sequences as DNA 'biological barcodes' for identification purposes. Blasting barcode sequences from an unidentified sample could tell the identifier if there are similar DNA sequences in the database, which were previously obtained from vouchered specimens, therefore leading to simple sample identification (Ben-David et al., 2007).

Recently, Matsuda et al. (2013) identified 10 of the 13 species of *Tetranychus* in the internal transcribed spacer tree. In the cytochrome *c* oxidase subunit I tree, they could identify all 13 known *Tetranychus* species present in Japan. They also indicated that a cryptic species existed in each species. Thus, precise identification is possible when both molecular and morphological tools are used.

17.2.1.1 Molecular Markers

The use of molecular genetic markers, initially protein-based (especially allozymes) in the 1970s, and more recently DNA-based markers, spawned the field of molecular ecology (Avise, 1994). This enabled many hitherto intractable questions concerning such biological parameters to be answered. Often IPM strategies fail because of poor understanding about the fundamental biology of the control agent employed or its target host. This includes basic population structure, movement and migration (dynamics), overwintering, reproduction, sex ratio and host preference. Thus, these markers have provided novel information on founder effects, bottlenecks, migratory ambit, host switches and preference, to name but a few of the applications. Molecular markers allow the fine resolution of ecological interactions to be elucidated.

Among the candidate genomic regions used for analyzing molecular variation, ribosomal DNA (rDNA) sequences have long been recognized as attractive markers for phylogenetic analysis and identification of species. In Eukaryotes, the three genes for the three ribosomal subunits, 28S, 5.8S and 18S are interspersed by internal transcribed spacers 1 (ITS1) and 2 (ITS2) and are arranged in tandemly repeated arrays (Hillis and Dixon, 1991). While the rRNA genes tend to be relatively conserved in evolution, the ITS evolve much faster and can be used to compare closely related taxa (Paskewitz et al., 1993), sometimes down to the intraspecific level (Fritz et al., 1994). Moreover, the presence of highly conserved regions in the coding sequences facilitates the amplification of selected portions of the variable sequence by polymerase chain reaction (PCR).

In acarology, ribosomal sequences represent a valuable source of diagnostic characteristics in species identification for organisms where usable systematic characteristics are scarce: mites are generally very small organisms less than 0.5 mm long. Among the Acari, three groups (Ixodidae, Tetranychidae and Eriophyidae) have been analysed by molecular techniques to clarify some of the problems that their classification poses. A combined analysis of ribosomal ITS2 and mitochondrial sequences was used to construct a new phylogenetic hypothesis for the genus *Tetranychus* (Tetranychidae) (Navajas et al., 1996a, 1997). The ITS2 sequences have also been applied to species identification in the Eriophyidae (Fenton et al., 1995) and the Ixodidae (Wesson et al., 1993) and for the recognition of sibling species in the Tetranychidae (Gotoh et al., 1998).

Phytoseiid mites are important biological control agents and essential components of some pest programmes (McMurtry and Croft, 1997). Some studies have been devoted to defining natural groupings and phenetic relationships among members of the family (Chant and McMurtry, 1994), but no definite classification is yet available. Therefore, the analysis of molecular variations in these mites has important implications for taxonomy, phylogeny, and species diagnostics. Navajas et al. (1999) explored sequence variations of the ITS1 and ITS2 in phytoseiid mites at the specific and intra-specific level that might be analysed in evolutionary studies and for species identification purposes. They emphasized that preliminary work is needed to assess the usefulness of different markers at different taxonomic scales when a new group is analysed. They investigated the level of sequence variation of the nuclear ribosomal spacers ITS1 and 2 and the 5.8S gene in six species of phytoseiid mites. The comparison of ITS1 sequence similarity at the species level might be useful for species identification; however, the value of ITS in taxonomic studies does not extend to the level of the family.

Navajas and Fenton (2000) reviewed the application of molecular markers in mites and ticks and derived new insights into their population structure and taxonomic relationships. Molecular applications have had particular success in facilitating the identification of taxonomically difficult species, understanding population structures and elucidating phylogenetic relationships. Cruickshank (2002) reviewed molecular markers for the phylogenetics of mites and ticks in which he described the properties of the ideal molecular marker and compared these with genes that have been used for phylogenetic studies of mites and ticks. The ITS2 of the nuclear ribosomal gene cluster and the mitochondrial protein-coding gene COI together provide a powerful tool for phylogenetics at low taxonomic levels. The nuclear ribosomal genes 18S and 28S rDNA are equally powerful tools for phylogenetics at the deepest levels within the Acari.

The usefulness of the COI region for delineating tetranychid species has been investigated in several studies (Hinomoto et al., 2001; Hinomoto and Takafuji, 2001; Lee et al., 1999; Navajas et al., 1994, 1996a,b, 1998; Toda et al., 2000; Xie et al., 2006). Ros and Breeuwer (2007) created an extensive COI dataset by analysing the sequences derived from GenBank for the family Tetranychidae, with a wide coverage of the species *T. kanzawai* and *T. urticae* (including *T. cinnabarinus*, which is currently considered synonymous to *T. urticae*). They also elaborated the use of COI for DNA barcoding purposes by considering the intra- and interspecific variation, COI variation in relation to the associated host plant, phylogeographic patterns and the presence of endosymbionts (e.g. *Wolbachia*, *Cardinium*). Carew et al. (2003) studied the species status and population structure of eriophyid mites (bud mite, blister mite and rust mite) in grapevine. Patterns of genetic variation observed using PCR-RFLP of ITS1 confirmed the separate species status of the rust mite, and resolved the species status of bud and blister mites, revealing two closely related but distinct species. Microsatellite markers revealed extensive genetic differentiation between bud mite populations and blister mite populations even at micro-geographical levels, suggesting low movement in these species. The findings indicate that separate control strategies are needed against bud and blister mites.

17.2.1.2 DNA Barcoding

The use of DNA barcodes, short DNA sequences from a standardized region of the genome, has recently been proposed as a tool to facilitate species identification and discovery. In this context, it is important to distinguish between DNA taxonomy and DNA barcoding. DNA taxonomy concerns the circumscription and delineation of species using evolutionary species concepts (Vogler and Monaghan, 2007). DNA barcoding aims at the identification of predefined species and does not address the issue of species delineation *per se* (Monaghan et al., 2005). In DNA barcoding, a short standardized DNA sequence, usually the 5′ end of the mitochondrial cytochrome oxidase subunit I (COI) gene, is used to identify species. DNA barcoding can be used to (i) identify and assign unknown specimens to species that have been previously described and (ii) enhance the discovery of new species using a threshold of sequence divergence (Hebert et al., 2003; Moritz and Cicero, 2004). DNA taxonomy may be based on one or several mitochondrial as well as nuclear DNA regions and can serve as a database for DNA barcoding. In sub-arctic Canada, barcode analysis of 6279 mite specimens (Young et al., 2012) revealed nearly 900 presumptive species of mites with high species turnover between substrates and between forested and non-forested sites. The coupling of DNA barcode results with taxonomic assignments revealed that Trombidiformes compose

49% of the fauna, a larger fraction than expected based on prior studies. This investigation demonstrated the efficacy of DNA barcoding in facilitating biodiversity assessments of hyperdiverse taxa.

Ben-David et al. (2007) concluded that rDNA ITS2 sequence barcodes may serve as an effective tool for the identification of spider mite species and can be applicable as a diagnostic tool for quarantine and other pest management activities and decision-making. Molecular taxonomy of mites using common DNA barcoding markers (sequences of cytochrome oxidase subunit I gene fragment and the D2 region of 28S rDNA) has been well studied by Dabert et al. (2008), Martin et al. (2010) and Skoracka and Dabert (2010). Dabert et al. (2011) studied andropolymorphism (discontinuous morphological variability in males can lead to taxonomic confusion when different male morphs are determined and described as separate species) in *Aclerogamasus* species, by testing variation in a fragment of the mtDNA COI gene and the D2 region of 28S rDNA.

17.2.2 Development of Genes and Molecules for the Effective Management of Mites

17.2.2.1 Bt Cry Proteins

Bacillus thuringiensis (Bt) toxins present the potential for control of pest mites. *Bacillus* spp. spores prolonged the development of tritonymphs of *Dermatophagoides pteronyssinus*, whereas the toxic effect of *B. sphaericus* is higher than that of *Bt* var. *israelensis* (Saleh et al., 1991). Bt maize diets made from kernels of the transgenic maize MON 810, producing Cry1Ab toxin and *Bt* var. *kurstaki*, prolonged the period of development of *Acarus siro*, reduced the average female lifespan, and markedly affected larval and nymphal survival (Dabrowski et al., 2006). Erban et al. (2009) confirmed the toxic effects of *Bt* var. *tenebrionis* with doses higher than $10\,mg\,g^{-1}$ of Bt/diet against stored-product and house dust mites. Although the toxic concentration of Bt with suppressive effects on mites tested was high, Bt application for the protection of human environments is possible (Khetan, 2001). The combined application of Bt with diatomaceous earth (Palyvos et al., 2006), chitinase and soybean trypsin protease inhibitor (Sobotnik et al., 2008), could increase the efficacy of mite control. Genetic engineering and construction of plants producing Cry3A toxin in their products is an alternative to spraying Bt.

17.2.2.2 Protease Inhibitors

The proteolytic digestion in mite species that feed on plants is based mostly on cysteine peptidase activities (Carrillo et al., 2011; Nisbet and Billingsley, 2000). This is consistent with the three-fold proliferation of the cysteine peptidase gene family, mainly of C1A papain and C13 legumain classes, found in the *T. urticae* genome in comparison to other sequenced arthropod species (Grbic et al., 2011). However, serine and aspartic peptidase gene families have also been identified as important peptidases in the spider mite genome, though they are most probably involved in other physiological processes. Peptidase activity is modulated by specific inhibitors that are grouped according to the peptidase type they inhibited (Rawlings et al., 2012). Two of the most abundant plant protease inhibitors are the cystatins (family I25), which inhibit cysteine peptidases C1A and C13, and cereal trypsin/*a*-amylase inhibitors (family I6). Plant protease inhibitors from these two classes have been used as defence proteins against pathogens and pests due to their capability to inhibit heterologous enzymes. Carrillo et al. (2011) have shown that the expression of the barley cystatin HvCPI-6 in maize impaired development and reproductive performance of *T. urticae* by inhibiting their cysteine protease activities. In contrast, experiments developed with tomato plants expressing a glucose oxidase or the soybean Kunitz inhibitor gene enhanced growth of *T. urticae* under greenhouse conditions (Castagnoli et al., 2003).

Pyramiding (stacking) multiple defence genes in one plant has been developed as a method to prevent pest resistance and to improve pest control. The gene pyramiding approach was developed by coexpressing two barley protease inhibitors, the cystatin *Icy6* and the trypsin inhibitor *Itr1* genes, in *Arabidopsis* plants by Agrobacterium-mediated transformation. Single and double transformants were used to assess the effects of spider mite infestation. Double transformed lines showed the lowest damaged leaf area in comparison to single transformants and non-transformed controls (Santamaria et al., 2012).

17.2.2.3 Plant Lectins

One of the most important direct defence responses in plants against the attack by phytophagous insects is the production of insecticidal peptides or proteins. One particular class of entomotoxic proteins present in many plant species is the group of carbohydrate-binding proteins or lectins. A lot of progress was made in the study of a few lectins that are expressed in response to herbivory by phytophagous insects and the insecticidal properties of plant lectins in general. The snowdrop *Galanthus nivalis* agglutinin lectin (GNA) has been shown to be insecticidal against a range of common pests. The crops (potatoes, rice, tobacco, wheat and tomato) incorporated with the genes that encode GNA, showed enhanced resistance against insects. McCafferty et al. (2008) proved that transgenic papaya plants expressing the GNA gene possessed enhanced resistance to carmine spider mites, *T. cinnabarinus*, in a laboratory bioassay.

17.2.2.4 Chitinases

Chitin is an insoluble, structural polysaccharide that is a component in the cell walls of fungi, nematodes, and other organisms including mites (Kramer and Muthukrishnan, 1997). Because of the pivotal role played in insect growth and development by chitin and the chitinolytic enzymes, they are receiving attention for their potential development as biopesticides and microbial biological control agents or as chemical defence proteins in transgenic plants.

McCafferty et al. (2006) observed an increased tolerance to mites under *in vitro* and field conditions in genetically engineered papaya plants expressing the tobacco hornworm (*Manduca sexta*) chitinase protein, through microprojectile bombardment of embryogenic calli. Although the chitinase mode of action is still not well known, it was suggested that it targeted the peritrophic membrane that encloses food in the mid- and hindgut, while the anti-mite activity of lectins was probably mediated by binding to chitin in the peritrophic matrix or by interacting with glycoproteins on the epithelial cells of the mite midgut (McCafferty et al., 2006, 2008).

17.2.3 Silencing of Genes Using RNAi Approach for Developing Pest Resistance Plants

RNA interference (RNAi) represents a powerful method for reverse genetic analysis of gene function. A promising approach to knockout (silence) genes in various plants and animals includes RNA interference (RNAi)-mediated gene silencing (Meister and Tuschl, 2004). Although the RNAi mode of action is conserved, delivery protocols of experimental RNAi vary depending upon the model system. They range from direct delivery to a specific developmental stage (transfection of cell lines or injections into individual embryos) to the introduction of double-stranded RNA (dsRNA) into earlier reproductive stages, thereby allowing gene silencing in the progeny of injected females. Such administration suggests that some species will be sensitive to systemic RNAi effects as has been recorded so far in plants, planarians, *Caenorhabditis elegans*, and insects *Tribolium* and *Oncopeltus* (May and Plasterk, 2005).

Khila and Girbic (2007) developed an RNAi protocol for the two-spotted spider mite *T. urticae* focusing on distal-less (*Dll*), a conserved gene involved in appendage formation in arthropods and revealed that such conserved genes can serve as reliable starting points for the development of functional protocols in non-model organisms. Efforts are under way to control a major devastating mite pest of the honey bee through RNAi. The gene encoding glutathione *S*-transferase, responsible for detoxification of pesticides, was targeted successfully in *Varroa destructor* with dsRNA (Campbell et al., 2010). Soaking mites in a solution of dsRNA in 0.9% NaCl enables high-throughput screening to identify the best targets for control of this pest, even though the exact mechanism of dsRNA uptake is unknown.

Double-stranded RNA was successfully delivered to ticks via injection in various developmental stages and tissues, infection through viral vector, oral ingestion, and incubation via whole body soaking (de la Fuente et al., 2007). Oral RNAi was used to knockdown an anti-complement gene (*isac*) in the blacklegged tick, *Ixodes scapularis*, and it was found that affected nymphs weighed less than those fed a control dsRNA (*lacZ*) and had lower spirochete loads when infected with *Borrelia burgdorferi* (Soares et al., 2005). *Rhipicephalus microplus* is an economically important cattle tick which acts as an ectoparasite and transmits a variety of pathogens, such as *Anaplasma marginale*, and leads to vector-borne infectious diseases. Silencing of a defensin gene, *varisin*, by injection of dsRNA into male ticks reduced their ability to infect calves with *A. marginale* (Kocan et al., 2008).

17.2.4 Molecular Marker-Assisted Selection for Developing Mite-Resistant Plants

Harnessing genetic variability by adopting conventional breeding strategies entails a huge investment of time and resources. To speed up progress in classical breeding programmes, it is important to identify DNA markers related to genomic regions for traits such as resistance to the pests, quality and productivity. Detection and analysis of genetic variation has helped us to understand the molecular basis of various biological phenomena in plants. One of the most significant advances to occur for the development of improved crops is the use of molecular markers to identify and track genes of interest. Polygenic resistance is not associated with a typical mechanism of resistance, but only refers to the number of genes involved in resistance (Lindhout, 2002). In studies by Malik et al. (2003) with wheat curl mite resistance, molecular and cytogenetic analyses located *Cmc4* distally on chromosome 6DS flanked by markers *Xgdm141* (4.1 centimorgans, cM) and *XksuG8* (6.4 cM). Shalini et al. (2007) identified 9 SSR and 4 RAPD markers closely associated with mite resistance in naturally occurring genotypes of coconut. After combined stepwise multiple regression analysis of data, they showed that a combination of five markers could explain 100% association with the mite resistance trait in coconut. The development and validation of a PCR-based marker for gall mite resistance in blackcurrant was done by Brennan et al. (2008). In jute, the SSR markers J-170, HK-64 and HK-89 showed tolerance in more than 70% of the plants individually but in combination it was 100% and they were closely linked to the mite (*Polypagotarsonemus latus* Banks) resistant genes (Ghosh et al., 2010). They also reported that the selection efficiencies of SSR2-004 and RM1358 were 96.8% and 92.7%, respectively, and 99.8% in combination. Recently, Salinas et al. (2013) confirmed the usefulness of Rtu2 Quantitative Trait Loci (QTL) as a valuable target for marker-assisted selection of new spider mite-resistant tomato varieties, but also as a starting point for a better understanding of the molecular genetic functions underlying resistance to *T. urticae*.

17.2.5 Molecular Mechanisms of Pesticide Resistance in Mites

According to the Arthropod Pesticide Resistance Database, two-spotted spider mites have recorded an astonishing 389 cases of resistance, the highest amongst all arthropods (including both insects and mites). This means that spider mites often develop resistance to a pesticide within only 2 to 4 years of its introduction. Identifying a mechanism for the development of pesticide resistance is important for advancing pesticide resistance management for arthropod pests. In spider mites, past genetic and ecological studies have comprehensively suggested that the local concentration of resistance genes (increasing gene frequency in breeding patches) resulting from genetic diversity within habitats based on their biological traits and selection by acaricides, and gene flow from selection sites to surroundings (local and/or regional spread of resistance) are the processes of acaricide-resistance evolution (Osakabe et al., 2009). The two-spotted spider mite, *T. urticae* Koch (Acari: Tetranychidae), is an important agricultural pest with a global distribution. Its phytophagous nature, high reproductive potential and short life cycle facilitate rapid resistance development to many acaricides often after a few applications (Cranham and Helle, 1985; Devine et al., 2001; Keena and Granett, 1990; Stumpf and Nauen, 2001). So far, resistance has been reported in several countries for compounds such as organophosphates (OPs) (Anazawa et al., 2003; Sato et al., 1994), dicofol (Fergusson-Kolmes et al., 1991), organotins (Edge and James, 1986); hexythiazox (Herron and Rophail, 1993), clofentezine (Herron et al., 1993), fenpyroximate (Sato et al., 2004) and abamectin (Beers et al., 1998).

Insensitive AChE causing OP resistance is widespread and has been detected in *T. urticae* strains from Germany (Matsumura and Voss, 1964; Smissaert et al., 1970), Japan (Anazawa et al., 2003) and New Zealand (Ballantyne and Harrison, 1967) and in a few other tetranychid pest species, including *T. cinnabarinus* from Israel (Zahavi and Tahori, 1970) and *T. kanzawai* from Japan (Kuwahara, 1982). Also the insensitivity of AChE to demeton-S-methyl, ethyl paraoxon, chlorpyrifos oxon and carbofuran was identified in a German laboratory strain of *T. urticae* and a field collected strain from Florida (Stumpf et al., 2001). Rates of resistance to structurally diverse pesticides in *T. urticae* are unprecedented, with some field strains resistant to nearly all available compounds (Van Leeuwen et al., 2010). Studies of pesticide resistance in *T. urticae* have focused largely on target-site mutations and on classical detoxifying enzyme systems, such as P450 monooxygenases (P450s), carboxyl/cholinesterases and glutathione-*S*-transferases (Ghadamyari and Sendi, 2009). However, these studies have not been satisfactory for understanding the scope of acaricide resistance in *T. urticae.* Under field conditions, multiresistant strains that are resistant to all commercially available acaricides are often encountered, and strikingly these strains also resist compounds with new modes of action that have never been used in the field (Van Leeuwen et al., 2010). The increasing availability of whole genome sequences and EST databases strongly stimulate mite resistance research. And to obtain new information on target-site genes, cloning and mutagenesis studies will aid in determining the precise nature of the mutations and predicting interactions between mite proteins and acaricides (Van Leeuwen et al., 2012).

17.2.6 Genome Sequencing

The first complete genome of a chelicerate species (*T. urticae*) provides the opportunity for a detailed phylogenomic analysis of arthropods, the most diverse group of animals on Earth. The *T. urticae* genome illustrates the specialized life history of this polyphagous herbivorous pest (Grbic et al., 2011). This mite has the smallest genome of any arthropod determined so far (90 Mbp – 30 times smaller than that of *Ixodes scapularis* (2.3 Gbp)), has only three chromosomes, has a rapid development and is easily reared in the laboratory. The very compact *T. urticae* genome has unique attributes among arthropod genomes with remarkable instances of gene gains and losses. Other chelicerates are much larger (565–7100 Mbp), with the unfinished genome of the tick *Ixodes scapularis* estimated at 2100 Mb (Grbic et al., 2011). Several features of *T. urticae* make this species an excellent candidate to become a genetic model among the Chelicerata. In addition, spider mites are major agricultural pests and are therefore of substantial economic importance and significance for the biotechnology of pest control.

In the future, using genomic resources and advances in next-generation sequencing methods, it is possible to do in-depth analysis in the areas of plant–pest interaction and ecology and biodiversity studies. Using Illumina sequencing, resistant and susceptible strains of the citrus red mite *Panonychus citri* transcriptome were analysed and provided abundant genetic information for further understanding of the molecular mechanisms of acaricide resistance. The above study identified 2701 significantly differentially expressed genes between resistant and susceptible strains. Many potential candidate genes related to special acaricide resistance in *P. citri* have been identified (Liu et al., 2011). Dermauw et al. (2013) constructed an expression microarray that collected genome-wide expression data over a time course ranging from hours to generations after transfer of mites to a new, challenging host. They transferred pesticide susceptible *T. urticae* from bean (a benign host) to tomato (a challenging host) and assessed the mites' differential gene expression over a time course ranging from hours to five generations. They also assessed the constitutive gene expression of multipesticide-resistant *T. urticae* on bean. Marked changes in gene expression were observed in many members of several large gene families previously implicated in the detoxification of harmful compounds, such as the P450 monooxygenase genes (P450s). Among the genes with the most striking expression differences were those belonging to the lipocalin family and the major facilitator superfamily (MFS). Lipocalins are involved in binding small molecules and are thought to mediate interactions with plant compounds. As lipocalins typically bind hydrophobic molecules, they may bind pesticides/allelochemicals in mites, resulting in sequestration of these toxic, generally hydrophobic compounds. The MFS genes encode transporters that might remove toxins or their metabolites from cells. Furthermore, genes encoding intradiol ring-cleavage dioxygenases (ID-RCDs) responded strongly to host transfer (Dermauw et al., 2013). Current functional genomics studies with next-generation sequencing technology in mite genome projects will provide opportunities to move from knowledge and understanding of resistance mechanisms to the practical application in the field, and might deliver novel control tools.

17.3 PLANT-PARASITIC NEMATODES

Members of the phylum Nematoda (round worms) have been in existence for one billion years, and comprise >25,000 described species, many of which are parasites of animals or plants (Blaxter, 2003). As many as 10 million species may have yet to be described. Most nematodes are free-living and feed on bacteria, fungi, and protozoans. Fifteen percent of nematode species described are plant-parasitic. The primary groups of nematodes that cause problems are the root-knot, cyst, burrowing, lesion, foliar, and reniform nematodes. Many plant-parasitic nematodes feed on the roots of plants and cause symptoms such as a reduction in root mass, a distortion of root structure and/or enlargement of the roots. Nematode damage of the plant's root system also provides an opportunity for other plant pathogens to invade the root and thus further weaken the plant. Plant-parasitic nematodes are the primary pathogens of potato, sugar beet, soybean, tomato and other crops (Li et al., 2008), and cause an estimated annual economic loss of $125 billion worldwide (Chitwood, 2003). They exhibit a variety of feeding habits, ranging from migratory browsers to sedentary endoparasites with specialized host associations. Sedentary plant-parasitic nematodes, major agricultural pathogens [cyst nematodes (*Heterodera* and *Globodera* spp.) and root-knot nematodes (*Meloidogyne* spp.)], alone can cause an estimated annual yield loss of $100 billion worldwide (de Almeida Engler et al., 2010). They are difficult to control as they live underground and spend most of their lives in the roots, which can offer protection against chemical nematicides (Li et al., 2008). Migratory endoparasitic nematodes, such as the root-lesion nematodes (*Pratylenchus* spp.) and the burrowing nematodes (*Radopholus* spp.), are also crop pests of widespread economic importance. Improved plant resistance to parasitic nematodes is urgently required to reduce the need for nematicides, which are the most unacceptable class of pesticides widely used in agriculture (Gustafson, 1993; Williamson, 1995), and to overcome the limitations in conventional methods of nematode management. Hence, development of alternative strategies such as biotechnological approaches for the management of the pests is important.

17.3.1 Identification

Accurate identification of nematode species is the foundation of nematological research, quarantine enforcement for regulatory purposes, and nematode management, especially management that does not rely on nematicides. An immediate objective of new techniques in nematology is to increase the ease, accuracy, and speed of species identification. Caswell-Chen et al. (1993) reviewed how biotechnology might influence nematode management avenues of research in applied nematology, including species identification, race and pathotype identification, development of resistant cultivars, nematode–host interactions, nematode population dynamics, establishment of optimal rotations, the ecology of biological control and development of useful biological control agents.

The use of PCR to specifically amplify the internal transcribed spacer regions (ITS1 and ITS2) of ribosomal DNA (rDNA) from individual nematodes provides a potential means to identify nematodes without bias or special training from small samples and potentially from asymptomatic infected plants (Powers et al., 1997). The eukaryotic ITS region is flanked on either side with the 18S and 28S ribosomal DNA subunit. The 18S, 5.8S, and 28S rDNA subunits are coded rRNA regions, which are minimally variable across species because of the selection pressure imposed by their role in translation (Powers et al., 1997). The ITS regions of rDNA, however, are noncoding regions free of strong selection pressure and therefore more variable. The variability of the rDNA ITS regions can be used for comparison of species (Iwahori et al., 1998). Nematologists have used the ITS region to identify species of plant-parasitic, animal-parasitic, and insect-parasitic nematodes (Cherry et al., 1997; Chilton et al., 1995; Fallas et al., 1996; Ibrahim et al., 1994; Joyce et al., 1994; Nasmith et al., 1996; Powers and Harris, 1993; Powers et al., 1997; Stevenson et al., 1995; Szalanski et al., 1997; Thiery and Mugniery, 1996; Vrain et al., 1992; Vrain and McNamara, 1994; Wendt et al., 1995; Williamson, 1992; Zijlstra et al., 1995; Zijlstra et al., 1997).

Universal PCR primer sets located in the flanking, conserved rDNA subunit regions have been used to obtain and identify the ITS1 and ITS2 rDNA regions in the absence of known sequence (Powers et al., 1997). The ability to specifically amplify the ITS region from an individual nematode species suggests that any species, population, or community of nematodes can be analysed both qualitatively and quantitatively using a molecular assay based on the ITS region (Vrain and McNamara, 1994).

A decline in emphasis and specialists within the field of classical taxonomy combined with advances in molecular phylogeny has prompted the development of molecular diagnostic tools for nematodes (Coomans, 2002; Oliveira et al., 2004). Diagnostic protocols need to be objective, sensitive, accurate, reproducible, rapid, user-friendly, and not require the expertise of a trained taxonomist (Oliveira et al., 2004). Molecular diagnostic assays have been developed for different species of plant-parasitic nematodes within the genera *Heterodera, Globodera, Bursaphelenchus, Meloidogyne, Pratylenchus,* and *Xiphinema* (Oliveira et al., 2004; Sheilds et al., 1996; Subbotin et al., 2000; Uehara et al., 1999; Wang et al., 2003; Zijlstra et al., 1995). Although a molecular diagnostic assay for foliar nematodes does not currently exist, *A. fragariae* was used as a control species of Aphelenchidae in reference to phylogenetic analysis and diagnosis of *Bursaphelenchus* nematodes using RFLP and rDNA sequence (Ferris et al., 1993; Iwahori et al., 1998).

Important linkages with collateral data such as digitized images, video clips and specimen voucher web pages are being established on GenBank and NemATOL, the nematode-specific Tree of Life database. The growing DNA taxonomy of nematodes has led to their use in testing specific short sequences of DNA as a barcode for the identification of all nematode species (Powers, 2004). The list of nematodes (around 180) having significance at the quarantine level in Europe are listed at http://www.q-bank.eu/Nematodes/.

A proper molecular toolbox for identifying nematode species should consider as many useful loci as possible, especially when the currently available nuclear loci (18S and 28S) have low resolution at the species level. The amplification across a wide taxonomic range, the ease of sequence alignment and the variability pattern render the I3-M11 partition of COI a good candidate to increase the identification of marine nematode species, provided there is a good reference database. There is a strong indication that nematode DNA barcodes should be thoroughly screened to infer their origin and homology state. Furthermore, digital vouchering of nematode specimens is required before molecular analyses, particularly in those studies that are intended to produce barcodes for new nematode species. Through this approach, a reliable reference database can be built (Derycke et al., 2010).

17.3.2 Development of Genes and Molecules for the Effective Management of Nematodes

17.3.2.1 Bt Cry Proteins Effective against Nematodes

The toxicity of *Bacillus thuringiensis* (*Bt*) towards several groups of soil invertebrates other than pterygota, e.g. Acarin, Nematoda, Collembola, and Annelida, has also been demonstrated (Addison, 1993). *Bt* is a rod-shaped, Gram-positive, spore-forming bacterium that forms parasporal crystals during the stationary phase of growth (Schnepf et al., 1998). The crystal proteins produced by some of *B. thuringiensis* are pore-forming toxins which are lethal against insects, mites and nematodes (de Maagd et al., 2001). Wei et al. (2003) demonstrated that six phylogenetically diverse nematodes were susceptible to *Bt* crystal proteins and thus proved that the phylum Nematoda is a target of *Bt* crystal proteins. Nematicidal activity of natural strains of *B. thuringiensis* has been observed against different life stages of the nematodes by several authors.

Wei et al. (2003) found that Cry5B, Cry6A, Cry14A, and Cry21A were toxic to the tested nematode species (*Caenorhabditis elegans, Pristionchus pacificus, Distolabrellus veechi, Panagrellus redivivus* and *Acrobeloides* sp.) and Cry5A, Cry6B, and Cry12A crystal proteins were not toxic. They also revealed that the inefficacy of Cry5A, Cry6B, and Cry12A crystal proteins may be due to lower dosage of the proteins and that they may be toxic some other nematodes. A study conducted on the characterization of nematode-effective strains for *Cry* genes in 70 Iranian Bt isolates based on PCR analysis using 12 specific primers for *Cry5, Cry6, Cry12, Cry13, Cry14* and *Cry21* genes encoding proteins active against nematodes (*M. incognita*) and two free-living nematodes (*Chiloplacus tenuis* and *Acrobeloides enoplus*) revealed that 31.5% contain a minimum of one nematode-active *Cry* gene. Strains containing the *Cry6, Cry14, Cry21* and *Cry5* genes were 22.8, 14.2, 4.2 and 2.8% of the isolates, respectively. Isolates YD5 and KON4 at 2×10^8 CFU/ml concentrations showed 77% and 81% toxicity, respectively on *M. incognita*. Maximum mortality was recorded for isolates SN1 and KON4 at 2×10^8 CFU/ml concentrations and resulted in 68% and 77% adult deaths of *C. tenuis* and 68% and 72% for *A. enoplus*, respectively (Jouzani et al., 2008).

Mohammed et al. (2008) studied the nematicidal effect of *Bt* toxins against root-knot nematode *M. incognita, in vitro*. The spore/crystal proteins of isolates *Bt*7N, *Bt*Den, *Bt*18, *Bt*K73, *Bt*Soto and *Bt*7 showed the highest nematicidal activities, with the mortality range of 86–100%. Nematicidal activity was found in several families of *B. thuringiensis* crystal proteins, such as Cry5, Cry6, Cry12, Cry13, Cry14, Cry21, and Cry55 (Guo et al., 2008). Cry6A expressed in transgenic roots significantly impaired the ability of *M. incognita* to reproduce (Li et al., 2007b). The nematicidal crystal protein Cry6Aa2, Cry5Ba2, and Cry55Aa1, were isolated from the highly nematicidal *Bt* strain YBT-1518 and the bioassay results showed that these three crystal proteins were highly toxic to second-stage juveniles (J2) of *M. hapla* (Guo et al., 2008), and a combination of Cry6Aa and Cry55Aa showed significant synergistic toxicity against *M. incognita* (Peng et al., 2011). Zhang et al. (2012) showed that 140 kDa *B. thuringiensis* crystal proteins can enter *M. hapla* J2 through the stylet, whereas *Heterodera schachtii* can acquire the crystal proteins below 28 kDa. This has important implications for the design of any transgenic resistance approach against the pests. They also proved the synergistic effect of the combination of different crystal proteins. They observed that Cry6Aa, Cry5Ba and Cry55Aa have different modes of action as they differ in their structures and binding patterns to *M. incognita* proteins.

17.3.2.2 Proteinase Inhibitors Effective against Nematodes

The introduction of pest-specific anti-nutritional factors into crop plants offers a promising approach for the control of a wide range of pests. Protease inhibitors have been used to engineer insect (Boulter et al., 1989; Duan et al., 1996; Hilder et al., 1987; Johnson et al., 1989) and nematode (Hepher and Atkinson, 1992; Urwin et al., 1995, 1997b) resistance. Cowpea trypsin inhibitor expressed in transgenic potato was shown to affect fecundity and the male/female ratio of *Meloidogyne* spp. and *Globodera* spp., respectively (Hepher and Atkinson, 1992). Cysteine proteinase inhibitors (cystatins) represent an attractive option for a safe defence strategy against specific pests because they are the only class of proteinases not expressed in the digestive system of mammals (Atkinson et al., 1995). Rice Oryzacystatin-I (OC-I) or maize cystatins (CC-I) have already been introduced and constitutively expressed in model rice varieties (Hosoyama et al., 1995; Irie et al., 1996), potato (Benchekroun et al., 1995) and tobacco (Masoud et al., 1993), but their effect on nematodes was not evaluated. A variant of Oryzacystatin-I (OCIΔD86) produced by site-directed mutagenesis was shown to mediate nematode resistance when expressed in tomato hairy roots (Urwin et al., 1995) and *Arabidopsis* plants (Urwin et al., 1997a). Despite these efforts, there is no report to date of proteinase inhibitor-mediated transgenic resistance against any nematode in cereals.

Vain et al. (1998) used a genotype-independent transformation system involving particle gun bombardment of immature embryos to genetically engineer rice with a cysteine proteinase inhibitor to impart resistance to nematodes and found that the putative transformants resulted in a 55% reduction in egg production by *M. incognita*. Cystatins are relatively small proteins (~11 kDa). Sedentary endoparasitic nematodes, including root-knot and cyst species, withdraw molecules from their plant-feeding site through a feeding tube that restricts the size of molecules entering the intestine (Berg et al., 2009; Bockenhoff and Grundler, 1994). *H. schachtii* excludes plant-expressed green fluorescent protein (GFP), a 28-kDa protein (Urwin et al., 1997b), whereas *G. rostochiensis* (Goverse et al., 1998) allows entry of GFP.

17.3.2.3 Pathogenesis-Related (PR) Proteins

Two pathways involve the direct production of PR proteins; in one pathway, the production of PR proteins is due to pathogen attack and in the other pathway, PR proteins are produced as a result of wounding, or necrosis-inducing plant pathogens and herbivory by insects, although both pathways can be induced by other mechanisms. The pathogen-induced pathway relies on salicylic acid (SA) that is produced by the plant as a signalling molecule, whereas the wounding pathway relies on jasmonic acid (JA) as the signalling molecule. These compounds and their analogues induce similar responses when they are applied exogenously and, no doubt, there is considerable cross-talk between the pathways (Pieterse et al., 2001). The salicylate- and jasmonate-induced pathways are characterized by the production of a cascade of PR proteins which include antifungals (chitinases, glucanases. and thaumatins), and oxidative enzymes (*viz.*, peroxidases, polyphenol oxidases and lipoxygenases), respectively.

NPR1 (Non-expressor of PR genes-1) has been characterized in great detail in the model system *Arabidopsis* and is shown to be a transcription activator with distinct protein–protein interaction motifs and possesses a bipartite nuclear localization signal (NLS). The *NPR1* gene encodes a protein containing ankyrin repeat domain and a BTB/POZ (broad-complex, tram track, and bric-à-brac/poxvirus, zinc finger) domain, both of which are involved in protein–protein interactions. The induction of systemic acquired resistance (SAR) triggers the monomerization of *NPR1* and its activation. In the activated monomeric form, *NPR1* crosses over to the nucleus and enhances PR gene expression by interacting with the members of TGA family of b-ZIP transcription factors and stimulating their DNA binding activity. This allows the TGA factors to bind to cognate elements in promoters of certain genes and enhance their expression. PR1 is a marker for the activated *NPR1* (Pieterse et al., 2001).

Parkhi et al. (2010) demonstrated that transgenic cotton plants expressing *AtNPR1* exhibited enhanced tolerance to a semi-endoparasitic nematode, *Rotylenchulus reniformis* (reniform nematode) along with resistance to infection caused by some fungal pathogens. Non-expressor of PR genes-1, *NPR1*, has been shown to be a positive regulator of the salicylic acid controlled SAR pathway and modulates the cross-talk between SA and JA signalling. Transgenic tobacco plants expressing *AtNPR1* were studied for their response to infection by the sedentary endoparasitic root-knot nematode, *M. incognita* and enhanced resistance was found (Priya et al., 2011).

17.3.3 Molecular Approaches to Impart Resistance to Nematodes

Basic research in molecular plant nematology is expanding the inventory of knowledge that can be applied to provide crop resistance to parasitic nematodes in an economically and environmentally beneficial manner. McCarter (2008) has provided a detailed discussion on molecular approaches to impart resistance to nematodes. Approaches to transgenic nematode control can be classified as acting (i) on nematode targets (disruption of nematode intestinal function through recombinant plant expression of protease inhibitors or *Bt* toxins), (ii) at the nematode–plant interface (expression of dsRNAs that cause silencing of essential nematode genes and disruption of sensory nervous system function), and (iii) in the plant response (generation of nematicidal metabolites) (Table 17.1). There have been numerous excellent reviews of engineered nematode resistance over the past 10 years (Atkinson et al., 1998a, 1998b, 2003; Burrows and De Waele, 1997; Jung et al., 1998; Lilley et al., 1999a, 1999b; McPherson et al., 1997; Ohl et al., 1997; Stiekema et al., 1997; Thomas and Cottage, 2006; Williamson, 1999; Williamson and Hussey, 1996; Williamson and Kumar, 2006; Zacheo et al., 2007).

17.3.3.1 Molecular Marker-Assisted Selection for Developing Nematode-Resistant Plants

Mapping QTLs is an effective approach for studying plant disease resistance. The first step in map-based cloning is to place molecular markers that lie near a gene of interest and co-segregate with the proposed gene without recombination. It has been shown that QTLs for resistance to race 14 of soybean cyst nematode were found in LGs B1, C1, C2, D1a, D2, E, G, J, and M, in various sources (Wang et al., 2001). The second step is to clone the gene by chromosome walking and sequencing the gene. Determination of QTLs is important for studying epistatic interactions and race specificity. Host-plant resistance is highly effective in controlling crop loss from root-knot nematode (RKN) *M. incognita* infection. Novel sources and enhanced levels of pathogen resistance are desirable for genetic improvement of crop plants. Breeding for optimal resistance must be based on selection of progeny with combinations of genes homozygous for resistance. However, a highly susceptible parent can contribute to nematode resistance via transgressive segregation (Wang et al., 2008). These crosses can derive highly resistant lines, even when both parents have a susceptible phenotype. Such transgressive segregants were used as improved resistance sources in crop breeding (Ulloa et al., 2011).

Since the advent of the first DNA markers, marker-assisted selection (MAS) has been viewed as a promising approach to streamline resistance breeding. Molecular markers are now routinely used in plant cultivar development to assist backcrossing of major genes into elite cultivars and to select alleles with major effects on high-value traits with relatively simple inheritance. Molecular markers have become important tools for enhancing the selection efficiency for various pest resistance traits in precision plant breeding. Traditional breeding efforts are being greatly enhanced through the integration of comparative genomics. In addition to that, by tagging several genes with closely linked molecular markers, MAS strategies facilitate the development of lines with stacked resistance genes, giving the cultivar more durable protection than that afforded by a single resistance gene (William et al., 2007). Also, genes controlling resistance to different races or biotypes of a pest or pathogen, or genes contributing to agronomic or seed quality traits can be pyramided together to maximize the benefit of MAS through simultaneous introgression (Dwivedi et al., 2007). The use of molecular markers is a very reliable method to diagnose the integration of the two genes in the plants. Studies done using MAS to identify nematode resistance are included in Table 17.2.

It is expected that identification of markers associated with nematode resistance and use of MAS will help alleviate some of the difficulties in developing nematode-resistant cultivars.

17.3.3.2 Functional Genomics/RNAi towards Plant-Parasitic Nematode Control

RNA interference (RNAi) represents a major breakthrough in the application of functional genomics for plant-parasitic nematode control. RNAi-induced suppression of numerous genes essential for nematode development, reproduction or parasitism has been demonstrated, highlighting the considerable potential for using this strategy to control damaging pest populations. In an effort to find more suitable and effective gene targets for silencing, researchers are employing functional genomics methodologies, including genome sequencing and transcriptome profiling. Microarrays have been used for studying the interactions between nematodes and plant roots and to measure both plant and nematode transcripts. Furthermore, laser capture microdissection has been applied for the precise dissection of nematode feeding sites (syncytia) to allow the study of gene expression specifically in syncytia. In future, small RNA sequencing techniques will provide more direct information for elucidating small RNA regulatory mechanisms in plants, and specific gene silencing using artificial microRNAs should further improve the potential of targeted gene silencing as a strategy for nematode management (Li et al., 2011).

RNA interference (RNAi) refers to the suppression of gene expression through the use of sequence-specific, homologous RNA molecules. It was first reported in *Caenorhabditis elegans* by Guo and Kemphues (1995). Fire et al. (1998) discovered that the presence of dsRNA, formed from the annealing of sense and antisense strands present in the *in vitro* RNA preparations, is responsible for producing the interfering activity. Later, Bernstein et al. (2001) indicated that Dicer, a ribosome III-like enzyme, was responsible for processing dsRNA to ~21 nucleotide (nt) sequences. The DCR-2/R2D2 complex binds to these small interfering RNA molecules (siRNAs) (Liu et al., 2003), then incorporates siRNAs into a multisubunit complex called the RNA induced silencing complex (RISC), which then dictates the degradation of any RNA molecules sharing sequence complementation (Hammond et al., 2000). This phenomenon of gene silencing occurs in several eukaryotic organisms, including nematodes and higher plants (Hammond et al., 2001). Researchers have been able to add *in vitro*-synthesized RNA directly to cells to obtain a gene knockdown.

To achieve plant-parasitic nematode control using an RNAi strategy, the RNAi mechanism is partially performed *in planta* and partially in the nematodes. siRNAs are generated by Dicer in plants expressing dsRNAs of nematode genes. When nematodes feed on plants, the plant-derived siRNAs or dsRNAs are taken up by nematodes through stylets, and RISC binds siRNAs to induce specific nematode gene degradation. siRNAs are then amplified in nematodes with the help of RNA-dependent RNA polymerase (RDRP) (Chapman and Carrington, 2007; Zamore and Haley, 2005). This siRNA-mediated gene silencing is highly sequence-specific. Even a single base mismatch between a siRNA and its mRNA target prevents gene silencing (Elbashir et al., 2001). The RNAi effect has been documented to spread not only from cell to cell (Fagard and Vaucheret, 2000; Kehr and Buhtz, 2008), but also throughout the plant (Yoo et al., 2004). Limpens et al. (2004) also found that the silencing signal was transported systemically from *Arabidopsis* roots to shoots, although the extent of silencing was limited and greatly variable. The RNAi studies conducted with plant-parasitic nematodes are listed in Table 17.3.

TABLE 17.1 Various Studies on Genetic Engineering and Molecular Approaches for Designing Plants for Management of Plant-Parasitic Nematodes

Approach	Gene/Proteins	Effect	Reference
1. DISRUPTION OF NEMATODE TARGET GENES			
a. Protease inhibitor	Rice oryzacystatin protein (antifeedant/nematicidal proteins)	Inhibits cysteine proteases	Lilley et al. (1999b)
	Cystatins (naturally occurring proteins characterized from edible plants such as rice and maize)	Inhibits protease activity	McPherson et al. (1997)
	Modified version of oryzacystatin, Oc-1ΔD86	Interferes with nematode replication	Urwin et al. (1995)
	Oryzacystatin, Oc-1ΔD86	Expression of Oc-1ΔD86 in *Arabidopsis thaliana* using CaMV35S promoter and infection with the beet cyst nematode *H. schachtii* resulted in adult females that were greatly diminished in size relative to controls. Cryosections of *H. schachtii* recovered from the plants showed diminished cysteine protease activity. Infection of the plants with root-knot nematode *M. incognita* resulted in fewer full size adults	Urwin et al. (1997a)
	Oryzacystatin, Oc-1ΔD86	Transgenic potato plants expressing Oc-1ΔD86 (CaMV35S promoter) potato cyst nematode *G. pallida* in a field trial resulted in a 55–70% decrease in cyst number	Atkinson et al. (2004)
	Serine protease inhibitor	Control of the cereal cyst nematode *H. avenae* in wheat	Vishnudasan et al. (2005)
b. Cry proteins	Cry5B, Cry6A, Cry14A, and Cry21A	Nematicidal against free-living nematodes	Schnepf et al. (2003); Wei et al. (2003)
	Cry5B	Cry5B interacts with the luminal surface of the *C. elegans* intestine via an invertebrate-specific glycolipid, loss of which conveys resistance	Griffitts et al. (2001, 2005)
	Cry6A	Expression of codon-optimized Cry6A in transgenic tomato roots by the CaMV35S promoter reduced egg production by *M. incognita* 56–76%	Li et al. (2007b)
c. Plant protease inhibitors	Found in the latex of papaya and pineapple, disrupt the integrity of the cuticle in parasitic nematodes including *Heligmosomoides polygyrus*	Transgenic expression of such proteases in crop roots could be tested for plant-parasitic nematode control	Stepek et al. (2004)
	Collagenases, chitinases, ribosome inactivating proteins	Collagenases to disrupt the cuticle, chitinases to disrupt egg shells, ribosome inactivating proteins, and patatin, a non-specific lipid acyl hydrolase	Burrows and De Waele (1997); Jung et al. (1998)

d. Disruption of sensory function	Aldicarb-like peptides	Expression of the aldicarb-like peptides as secretory products in transgenic potato resulted in root exudates with acetylcholinesterase-blocking activity, which in greenhouse trials reduced *G. pallida* infection with cyst number declining 36–48% relative to vector controls. Peptide mimics of levamisole also reduced *Globodera* infection in a potato hairy root system	Liu et al. (2005)
	Monoclonal antibody	Monoclonal antibodies were directed against *G. pallida* amphidial secretions; it was found that J2 soaked in these antibodies showed diminished mobility and delayed invasion of potato roots	Fioretti et al. (2002)
e. Nematicidal metabolites	Terpenoids and other flavonoids	Accumulation of a terpenoid aldehyde has been associated with *Meloidogyne* resistance in cotton	Veech and McClure (1977)
		Alfalfa resistant to *Pratylenchus* accumulates the isoflavonoid medicarpin	Baldridge et al. (1998)
		Phytoalexins that may play a role in naturally occurring resistance to plant-parasitic nematodes include coumestrol and psoralidin	Rich et al. (1977)
		Accumulation of flavone-C glycoside O-methyl-apigenin-C-deoxyhexoside-O-hexoside	Soriano et al. (2004a)
		Accumulation of phytoecdysteroid 20-hydroxyecdysone (20E) as a phytoalexin	Soriano et al. (2004b)
		Treatment of spinach with methyl jasmonate increases levels of 20E and decreases plant-parasitic nematode infectivity	Soriano et al. (2004b)
f. Nematicidal and insecticidal compounds from fungi	Macrocyclic lactones	Macrocyclic lactones such as avermectin require a dozen or more synthetic steps to give the nematicidal effect	Yoon et al. (2004)
	Glycosinolates	Glycosinolates from *Brassica* species kill nematodes by breaking down to active isothiocyanates during decomposition in the soil	Chitwood (2002)

(*Continued*)

TABLE 17.1 Continued

Approach	Gene/Proteins	Effect	Reference
2. DISRUPTION AT THE NEMATODE–PLANT INTERFACE			
a. Disruption of nematode pathogenicity factors		Disrupting the function of parasitism gene products may be a way to selectively block nematode infection	
	Nod factor	*M. incognita* J2s appear to produce a secretory factor, *NemF*, similar to a rhizobial Nod factor, which can serve as a signal to the plant at a distance. Mutant plants defective in *Nod* factor reception show diminished galling relative to controls	Bird et al. (2009a); Weerasinghe et al. (2005)
		This effect has been demonstrated in tomato	Ernst et al. (2002); Milligan et al. (1998); Vos et al. (1998)
		This effect has been demonstrated in potato	Paal et al. (2004); Ruben et al. (2006); van der Vossen et al. (2000)
	Cell wall-modifying proteins Endo-1,4-β-glucanase *Gr-Eng-1; Gr-Eng-2; Gt-Eng-1; Gt-Eng-2; Hg-Eng-1; Hg-Eng-2; Hg-Eng-3; Hg-Eng-4; Hg-Eng-5; Hg-Eng-6; Hs-Eng-1; Hs-Eng-2; Mi-Eng-1; Mi-Eng-2 (5A12B); Mi-Eng-3 (8E08B)*	Cell wall degradation in *Globodera* spp., *H. glycines, H. schachtii, M. incognita*	de Meutter et al. (2005); Gao et al. (2004); Goellner et al. (2001); Huang et al. (2003); Rosso et al. (2005); Smant et al. (1998); Vanholme et al. (2004)
	Cellulose-binding domain protein *Mi-Cbp-1* *Hg-Cbp-1* *Hs-Cbp*	Cell wall degradation in *M. incognita* *H. glycines* *H. schachtii*	Ding et al. (1998); Gao et al. (2004); Vanholme et al. (2004)
	Pectate lyase *Gr-Pel-1* *Hg-Pel-1* *Hs-Pel* *Mj-Pel-1* *Mi-Pel-1; Mi-Pel-2 (2B02B)*	Cell wall degradation in *G. rostochiensis* *H. glycines* *H. schachtii* and *M. incognita*	de Boer et al. (2002); Huang et al. (2003); Huang et al. (2006a); Popeijus et al. (2000); Vanholme et al. (2004)
	Polygalacturonase *Mi-Pg-1*	Cell wall degradation in *M. incognita*	Jaubert et al. (2002)
	Chorismate mutase *Mi-Cm-1; Mi-Cm-2; Gp-Cm-1; Hg-Cm-1; Hg-Cm-2; Mj-Cm-1*	Cell wall degradation in *M. incognita, G. pallida* *H. glycines* *M. javanica*	Bekal et al. (2003); Gao et al. (2003); Huang et al. (2006b); Jones et al. (2003); Lambert et al. (1999)
	Chitinase *Hg-Chi-1* *Hs-Chi-1*	Cell wall degradation in *H. glycines, H. schachtii*	Gao et al. (2002); Vanholme et al. (2004)

	Annexin *Hg-Ann-1*	Cell wall degradation in *H. glycines*	Gao et al. (2003)
	Arabinogalactan endo-1,4-β-galactosidase *Hs-Gal-1*	Cell wall degradation in *H. schachtii*	Vanholme et al. (2004)
	Expansin *Gr-Expb-1*	Cell wall degradation in *G. rostochiensis*	Qin et al. (2004); Urszula et al. (2005)
	Mi-1.2 gene in tomato or egg plant	Tomato *Mi*-1.2 gene encodes a leucine-rich repeat protein and confers resistance to three *Meloidogyne* species. *Mi*-1.2 is likely part of a surveillance cascade that detects a specific nematode factor and triggers localized host cell death where giant-cells would normally form near the head of the invading J2 worm	Goggin et al. (2006); Milligan et al. (1998); Vos et al. (1998)
	Hero A (R gene)	Can alter the sex ratio of cyst nematodes with more males being formed, a known nematode stress response	Ernst et al. (2002)
b. Feeding site-specific promoter induces cell death or other site incompatibility	Cauliflower mosaic virus 35S promoter, common in transgenic plant applications, may not be optimal for driving nematode resistance genes, particularly for cyst nematodes	Down-regulation of this promoter in nematode-induced syncytium has been observed for both *H. schachtii* infecting *Arabidopsis thaliana*, and *G. tabacum* infecting tobacco	Bertioli et al. (1999); Goddijn et al. (1993); Urwin et al. (1997b)
	Hs1pro-1	Examples of genes up-regulated in syncytium formed by *H. schachtii* infecting *Arabidopsis*	Thurau et al. (2003)
	AtSUC2 normally expressed in companion cells		Juergensen et al. (2003)
	At17.1 expressed in vascular tissues and root tips		Mazarei et al. (2004)
	FGAM synthase (phosphoribosylformyl-glycinamidine synthase)		Vaghchhipawala et al. (2004)
	ABI3		de Meutter et al. (2005)
	Examples of genes up-regulated in giant cells formed by *M. incognita*	*TobRB7*	Opperman et al. (1994)
		Small heat-shock gene *Hahsp17.7G4* in tobacco	Escobar et al. (2003)
		D-ribulose-5-phosphate-3-epimerase (*RPE)* in *Arabidopsis*	Favery et al. (1998)
	Genes up-regulated in both syncytia and giant-cells	*AtPGM* (co-factor dependent phosphoglycerate mutase enzyme) in *Arabidopsis*	Mazarei et al. (2004)
		Endo-β-1,4-glucanases in tobacco	Goellner et al. (2001)

TABLE 17.2 Studies on Utilization of Molecular Markers in Identifying Resistance to Plant Parasitic Nematodes on Crop Plants

Crop	Nematode	References
Wheat	*Pratylenchus neglectus*	Williams et al. (2002)
Wheat	*P. thornei, P. neglectus*	Zwart et al. (2005, 2006, 2010)
Bread wheat	*P. thornei*	Schmidt et al. (2005)
Barley	*P. neglectus*	Sharma et al. (2011a, 2011b)
Cotton	*M. incognita*	Goodell and Montez (1994)
Cotton	*M. incognita*	Bezawada et al. (2003); Gutiérrez et al. (2010); Hyer et al. (1979); Shen et al. (2010); Ulloa et al. (2010); Wang et al. (2006); Ynturi et al. (2006)
	R. reniformis	Dighe et al. (2009); Gutiérrez et al. (2011); Robinson et al. (2007); Romano et al. (2009); Winter et al. (2007)
Potato	*G. pallida*	Asano et al. (2012); Felcher et al. (2012); Moloney et al. (2010)
Soybean	*H. glycines*	Gao et al. (2004); Kazi et al. (2008); Li et al. (2011); Winter et al. (2007); Wu et al. (2009)

Although transformation of plant-parasitic nematodes represents one approach for deploying RNAi strategy on nematode control, there are no successful reports of engineering these nematodes, partly because of their obligate nature of parasitism. In addition, this approach would have numerous regulatory hurdles for releasing these GMO nematodes into the environment. Li et al. (2011) elaborately discussed different ways (utilizing the capability of nematodes to ingest the preferable size of macromolecules from plants, by soaking nematodes in dsRNA solutions, host-delivered RNAi to silence nematode genes, characterization of target genes and genes associated with mRNA metabolism) to achieve nematode control through the RNAi approach.

The current RNA sequencing methodology produces only siRNAs that correspond to the specific sequence fragments found in the RNAi constructs. The quantity of siRNA species does not increase exponentially because the nematode gene target is not found in the plants (Gheysen and Vanholme, 2007). Moreover, the absence of an endogenous, homologous gene expression in plant cells might induce off-target effects (Bakhetia et al., 2005). To increase the amount of siRNAs in transgenic plants, overexpressing a nematode gene in plants to obtain RNA templates should be a good option. Stacking two or more effective gene RNAi cassettes in one RNAi vector should provide more durable or robust resistance against plant-parasitic nematodes, as engineering an RNAi locus to target multiple genes should make an RNAi-tolerant nematode caused by a random mutation highly unlikely.

MicroRNA (miRNA) is an alternative method for gene silencing. miRNAs are generated by Dicer from short hairpin structure miRNA precursors (pre-miRNA) which are initially derived from longer primary miRNA transcripts (pri-miRNA) (Brodersen and Voinnet, 2009). The mature miRNAs are 20–24 nucleotide long, endogenous, single-stranded small RNA molecules, which are incorporated into RISC to guide mRNA degradation. Traditional RNAi vectors usually contain inverted repeats of more than 100-bp sequences resulting in a population of at least several different species of siRNAs and the potential for 'off-target' gene silencing. These 'off-target'

TABLE 17.3 Various Studies on Genetic Engineering and Molecular Approaches for Designing Plants for the Management of Plant Parasitic Nematodes

Plant Parasitic Nematode	Gene and Method of Delivery of dsRNA	Effect	Reference
Heterodera glycines	hgctl C-type lectin; soaking	41% decrease in no. of established nematodes	Urwin et al. (2002)
H. glycines	hgcp-I cysteine proteinase; soaking	40% decrease in no. of established nematodes	Urwin et al. (2002)
H. glycines	Hg-amp-1, Aminopeptidase; soaking	61% decrease in number of female reproductive	Lilley et al. (2005)
H. glycines	Hg-rps-23, ribosomal protein; soaking	Decrease in J2 viability	Alkharouf et al. (2007)
H. glycines	hg-eng-1, b-1,4-endoglucanase; soaking	Decrease in no. of established nematodes	Bakhetia et al. (2007)
H. glycines	hg-syv46, secreted peptide SYV46; soaking	Decrease in no. of established nematodes	Bakhetia et al. (2007)
H. glycines	Hg-rps-3a, ribosomal protein 3a; in planta	87% reduction in female cysts	Klink et al. (2009)
H. glycines	MSP major sperm protein; in planta	Up to 68% reduction in nematode eggs	Steeves et al. (2006)
H. glycines	Hg-rps-4, ribosomal protein 4; Hg-spk-1, spliceosomal SR protein; in planta	81–88% reduction in female cysts	Klink et al. (2009)
H. glycines	Hg-snb-1, synaptobrevin; in planta	93% reduction in female cysts	Klink et al. (2009)
H. schachtii	4G06, ubiquitin-like; in planta	23–64% reduction in developing females	Sindhu et al. (2009)
H. schachtii	3B05, cellulose-binding protein; in planta	12–47% reduction in developing females	Sindhu et al. (2009)
H. schachtii	8H07, SKP1-like; in planta	>50% reduction in developing females	Sindhu et al. (2009)
H. schachtii	10A06, zinc finger protein; in planta	42% reduction in developing females	Sindhu et al. (2009)
H. glycines	Y25, beta subunit of the COPI complex; in planta	81% reduction in nematode eggs	Li et al. (2010a,b)
H. glycines	Prp-17, pre-mRNA splicing factor; in planta	79% reduction in nematode eggs	Li et al. (2010a)
H. glycines	Cpn-1, unknown protein; in planta	95% reduction in nematode eggs	Li et al. (2010a)
H. glycines	Inverted repeats of 25 different genes of *H. glycines*, in planta	Reduction in nematode eggs	Li et al. (2013)
G. rostochiensis	Gr-ams-1 secreted amphid protein; soaking	Reduced ability to locate and invade roots	Chen et al. (2005)
G. pallida	Flp FMRFamide-like peptides	Inhibition of motility	Kimber et al. (2007)
B. xylophilus	Bx-hsp-1, heat-shock protein 70, cytochrome C; soaking	J2–J3 viability reduced at high temperature	Park et al. (2008)
B. xylophilus	Bx-myo-3 myosin heavy chain, tropomyosin; soaking	J2–J3 viability reduced at high-temperature; abnormal locomotion	Park et al. (2008)

effects in the host plants can become a crucial consideration in RNAi experiments (Gheysen and Vanholme, 2007). Recently, Melito et al. (2010) deployed a miRNA technology to study the role of rhg1 locus LRR-kinase gene for the interaction between soybean and soybean cyst nematode (SCN) and observed that rhg1 locus LRR-kinase had no significant impact on SCN resistance. Taken together, there is little doubt that this technique will play a major role in nematode resistance in the future.

17.3.3.3 Genome Sequencing

Both *M. hapla* and *M. incognita* are highly destructive root-knot nematode species, and their genomes have been sequenced (Bird et al., 2009b). Abad et al. (2008) sequenced the genome of *M. incognita* (total size of assembled sequences is 86 Mb and the estimated size of the genome is 47–51 Mb) and they compared the general features of *M. incognita* with *Brugia malayi* and *C. elegans*. A great improvement in the selection of target genes can be given by the next-generation sequencing (NGS) technologies, with the direct sequence of the cDNA generated from messenger RNA (RNA-seq) (Haas and Zody, 2010). The results of this new method are very useful in the case of non-model organisms for which any genome or transcriptome sequence information is available. Haegeman et al. (2013) investigated the transcriptome of the rice root-knot nematode *M. graminicola* by 454 sequencing of second-stage juveniles as well as mRNA-seq of rice infected tissue. Over 350,000 reads derived from *M. graminicola* preparasitic juveniles were assembled, annotated and checked for homologues in different databases. From infected rice tissue, 1.4% of all reads generated were identified as being derived from the nematode. Using multiple strategies, several putative effector genes were identified: both pioneer genes and genes corresponding to already known effectors. Moreover, the transcriptome can be used to identify possible target genes for RNA interference (RNAi)-based control strategies. Haegeman and co-workers also found that four genes proved to be interesting targets by showing up to 40% higher mortality relative to the control treatment when soaked in gene-specific siRNAs.

17.4 CONCLUSIONS

Plant-parasitic nematodes and mites are major agricultural pests worldwide and are responsible for global agricultural losses. Identification of these minute organisms is crucial both at the quarantine and farm level. Given the current concerns in relation to climate change, food security and the global transport of agricultural commodities, the use of molecular diagnostics for mites and nematodes is highly relevant along with conventional techniques. Molecular methods aid in identification of species, but intra-species variation across their geographic range is very wide. Hence, validation and adaptation of these methods for different geographic regions and different working conditions is a challenging goal. In the future, the routine use of molecular methods that are robust, reliable and inexpensive is foreseeable for the identification of non-insect pests to meet regulatory demands.

Markers linked to QTL and R genes related to resistance have the potential for yielding newer varieties through MAS. RNAi technology, resources of genome sequencing and advances in NGS methods will revolutionize the field of mite and nematode genomics. In relation to non-insect pest management, although there is good scope for biotechnology derived products, there are still questions in developing countries in accepting the products. For development and adoption of biotechnology-based products, high start-up investment, public awareness and acceptance, technology dissemination and proper implementation, and national policies on bio-safety and intellectual property issues are critical and those areas have to be strengthened to obtain real results in the farmer's field. Molecular methods are not a standalone technique for economically viable and environmentally friendly pest management. The judicious integration of traditional and molecular methods will certainly help in developing sustainable pest management practices in the future.

References

Abad, P., Gouzy, J., Aury, J.M., Castagnone-Sereno, P., Danchin, E.G.J., Deleury, E., et al., 2008. Genome sequence of the metazoan plant-parasitic nematode *Meloidogyne incognita*. Nat. Biotechnol. 26 (8), 909–915.

Addison, J.A., 1993. Persistence and non-target effects of *Bacillus thuringiensis* in soil: a review. Can. J. For. Res. 23, 2329–2342.

Alkharouf, N.W., Klink, V.P., Matthews, B.F., 2007. Identification of *Heterodera glycines* (soybean cyst nematode [SCN]) cDNA sequences with high identity to those of *Caenorhabditis elegans* having lethal mutant or RNAi phenotypes. Exp. Parasitol. 115, 247–258.

Anazawa, Y., Tomita, T., Aiki, Y., Kozaki, T., Kono, Y., 2003. Sequence of a cDNA encoding acetylcholinesterase from susceptible and resistant two-spotted spider mite, *Tetranychus urticae*. Insect Biochem. Mol. Biol. 33, 509–514.

Asano, K., Kobayashi, A., Tsuda, S., Nishinaka, M., Tamiya, S., 2012. DNA marker assisted evaluation of potato genotypes for potential resistance to potato cyst nematode pathotypes not yet invading into Japan. Breed. Sci. 62, 142–150.

Atkinson, H.J., Grimwood, S., Johnston, K., Green, J., 2004. Prototype demonstration of transgenic resistance to the nematode *Radopholus similis* conferred on banana by a cystatin. Transgenic Res. 13, 135–142.

Atkinson, H.J., Lilley, C.J., Urwin, P.E., McPherson, M.J., 1998a. Engineering resistance in the potato to potato cyst nematodes. In: Marks, R.J., Brodie, B.B. (Eds.), Potato Cyst Nematodes, Biology, Distribution and Control. CAB International, Wallingford, pp. 209–236.

Atkinson, H.J., Lilley, C.J., Urwin, P.E., McPherson, M.J., 1998b. Engineering resistance to plant-parasitic nematodes. In: Perry, R.N., Wright, D.J. (Eds.), The Physiology and Biochemistry of Free-living and Plant-Parasitic Nematodes. CABI Publishing, Wallingford, pp. 381–405.

Atkinson, H.J., Urwin, P.E., McPherson, M.J., 2003. Engineering plants for nematode resistance. Annu. Rev. Phytopathol. 41, 615–639.

Atkinson, H.J., Urwin, P.E., Hansen, E., McPherson, M.J., 1995. Designs for engineered resistance to root-parasitic nematodes. Trends Biotechnol. 13, 369–374.

Avise, 1994. Molecular Markers. Natural History and Evolution. Chapman and Hall, New York.

Bakhetia, M., Charlton, W.L., Urwin, P.E., McPherson, M.J., Atkinson, H.J., 2005. RNA interference and plant parasitic nematodes. Trends Plant Sci. 10, 362–367.

Bakhetia, M., Urwin, P.E., Atkinson, H.J., 2007. qPCR analysis and RNAi define pharyngeal gland cell-expressed genes of *Heterodera glycines* required for initial interactions with the host. Mol. Plant Microbe Interact. 20, 306–312.

Baldridge, G.D., O'Neill, N.R., Samac, D.A., 1998. Alfalfa (*Medicago sativa* L) resistance to the rootlesion nematode, *Pratylenchus penetrans*: defense-response gene mRNA and isoflavonoid phytoalexin levels in roots. Plant Mol. Biol. 38, 999–1010.

Ballantyne, G.H., Harrison, R.A., 1967. Genetic and biochemical comparisons of organophosphate resistance between strains of spider mites (*Tetranychus* species: Acari). Entomol. Exp. Appl. 10, 231–239.

Beers, E.H., Riedl, H., Dunley, J.E., 1998. Resistance to abamectin and reversion to susceptibility to fenbutatin oxide in spider mite (Acari: Tetranychidae) populations in the Pacific Northwest. J. Econ. Entomol. 91, 352–360.

Bekal, S., Niblack, T.L., Lambert, K., 2003. A chorismate mutase from the soybean cyst nematode Heterodera glycines shows polymorphisms that correlate with virulence. Mol. Plant Microb. Interact. 16, 439–446.

Benchekroun, A., Michaud, D., Nguyen-Quoc, B., Overney, S., Desjardins, Y., Yelle, S., 1995. Synthesis of active oryzacystatin I in transgenic potato plants. Plant Cell Rep. 14, 585–588.

Ben-David, T., Melamed, S., Gerson, U., Morin, S., 2007. ITS2 sequences as barcodes for identifying and analyzing spider mites (Acari: Tetranychidae). Exp. Appl. Acarol. 41, 169–181.

Bennett, J., 1994. DNA-based techniques for control of rice insects and diseases: Transformation, gene tagging and DNA fingerprinting. In: Teng, P.S., Heong, K.L., Moody, K. (Eds.), Rice Pest Science and Management. International Rice Research Institute, Los Banos, Philippines, pp. 147–172.

Berg, R.H., Fester, T., Taylor, C.G., 2009. Development of the root-knot nematode feeding cell. Plant Cell Monogr. 15, 115–152.

Bernstein, E., Caudy, A.A., Hammond, S.M., Hannon, G.J., 2001. Role for a bidentate ribonuclease in the initiation step of RNA interference. Nature 409, 363–366.

Bertioli, D.J., Smoker, M., Burrows, P.R., 1999. Nematode-responsive activity of the cauliflower mosaic virus 35S promoter and its subdomains. Mol. Plant Microbe Interact. 12, 189–196.

Bezawada, C., Saha, S., Jenkins, J.N., Creech, R.G., McCarty, J.C., 2003. SSR marker(s) associated with root-knot nematode resistance gene(s) in cotton. J. Cotton Sci. 7, 179–184.

Bird, D.M., Opperman, C.H., Williamson, V.M., 2009a. Plant infection by root-knot nematode. Plant Cell Monogr. 15, 1–13.

Bird, D.M., Williamson, V.M., Abad, P., McCarter, J., Danchin, E.G., 2009b. The genomes of root-knot nematodes. Annu. Rev. Phytopathol. 47, 333–351.

Blaxter, M.L., 2003. Nematoda: genes, genomes and the evolution of parasitism. Adv. Parasitol. 54, 101–195.

Bockenhoff, A., Grundler, F.M., 1994. Studies on the nutrient uptake by the beet cyst nematode *Heterodera schachtii* by *in situ* microinjection of fluorescent probes into the feeding structure in *Arabidopsis thaliana*. Parasitology 109, 249–254.

Boulter, D., Gatehouse, M.R., Hilder, V., 1989. Use of cowpea trypsin inhibitor (CpTi) to protect plants against insect predation. Biotech. Adv. 7, 489–497.

Brennan, R., Jorgensen, L., Gordon, S., Loades, K., Hackett, C., Russell, J., 2008. The development of a PCR-based marker linked to resistance to the blackcurrant gall mite (*Cecidophyopsis ribis* Acari: Eriophyidae). Theor. Appl. Genet. 118, 205–211.

Brodersen, P., Voinnet, O., 2009. Revisiting the principles of microRNA target recognition and mode of action. Nat. Rev. Mol. Cell Biol. 10 (2), 141–148.

Burrows, P.R., De Waele, D., 1997. Engineering resistance against plant-parasitic nematodes using anti-nematode genes. In: Fenoll, C., Grundler, F.M.W., Ohl, S.A. (Eds.), Cellular and Molecular Aspects of Plant–Nematode Interactions. Kluwer, Dordrecht, pp. 217–236.

Campbell, E.M., Budge, G.E., Bowman, A.S., 2010. Gene-knockdown in the honey bee mite *Varroa destructor* by a non-invasive approach: studies on a glutathione S transferase. Parasit. Vectors 3, 73.

Carew, M.E., Goodisman, M.A.D., Hoffmann, A.A., 2003. Species status and population genetic structure of grapevine eriophyoid mites. Entomol. Exp. Appl. 111, 87–96.

Carrillo, L., Martinez, M., Ramessar, K., Cambra, I., Castanera, P., 2011. Expression of a barley cystatin gene in maize enhances resistance against phytophagous mites by altering their cysteine-proteases. Plant. Cell. Rep. 30, 101–112.

Castagnoli, M., Caccia, R., Liguori, M., Simoni, S., Marinari, S., 2003. Tomato transgenic lines and *Tetranychus urticae*: changes in plant suitability and susceptibility. Exp. Appl. Acarol. 31, 177–189.

Caswell-Chen, E.P., Williamson, V.M., Westerdahl, B.B., 1993. Applied biotechnology in nematology. J. Nematol. 25 (4S), 719–730.

Chant, D.A., McMurtry, J.A., 1994. A review of the subfamilies Phytoseiinae and Typhlodrominae (Acari: Phytoseiidae). Int. J. Acarol. 20, 223–310.

Chapman, E.J., Carrington, J.C., 2007. Specialization and evolution of endogenous small RNA pathways. Nat. Rev. Genet. 8, 884–896.

Chen, Q., Rehman, S., Smant, G., Tones, J.T., 2005. Functional analysis of pathogenicity proteins of the potato cyst nematode *Globodera rostochiensis* using RNAi. Mol. Plant Microbe Interact. 18, 621–625.

Cherry, T., Szalanski, A.L., Todd, T.C., Powers, T.O., 1997. The internal transcribed spacer region of *Belonolaimus* (Nemata: Belonolaimidae). J. Nematol. 29, 23–29.

Chilton, N.B., Gasser, R.B., Beveridge, I., 1995. Differences in a ribosomal DNA sequence of morphologically indistinguishable species within the *Hypodontus macropi* complex (Nematoda: Strongyloidea). Int. J. Parasitol. 25, 647–651.

Chitwood, D.J., 2002. Phytochemical based strategies for nematode control. Annu. Rev. Phytopathol. 40, 221–249.

Chitwood, D.J., 2003. Research on plant-parasitic nematode biology conducted by the United States Department of agriculture – agricultural research service. Pest Manag. Sci. 59, 748–753.

Coomans, A., 2002. Present status and future of nematode systematics. Nematology 4, 573–582.

Cranham, J.E., Helle, W., 1985. Pesticide resistance in Tetranychidae. In: Helle, W. Sabelis, M.W. (Eds.), Spider Mites: Their Biology, Natural Enemies and Control, vol. 1B. Elsevier, Amsterdam, pp. 405–421.

Cruickshank, R.H., 2002. Molecular markers for the phylogenetics of mites and ticks. System. Appl. Acarol. 7, 3–14.

Dabert, J., Ehrnsberger, R., Dabert, M., 2008. *Glaucalges tytonis* sp. (Analgoidea, Xolalgidae) from the barn owl *Tyto alba* (Strigiformes, Tytonidae): compiling morphology with DNA barcode data for taxon descriptions in mites (Acari). Zootaxa 1719, 41–52.

Dabert, M., Bigos, A., Witalinski, W., 2011. DNA barcoding reveals andropolymorphism in *Aclerogamasus* species (Acari: Parasitidae). Zootaxa 3015, 13–20.

Dabrowski, Z.T., Czajkowska, B., Bocinska, B., 2006. First experiments on unintended effects of Bt maize feed on nontarget organisms in Poland. IOBC/WPRS Bull. 29, 39–42.

de Almeida Engler, J., Rodiuc, N., Smertenko, A., Abad, P., 2010. Plant actin cytoskeleton re-modeling by plant parasitic nematodes. Plant Signal. Behav. 5 (3), 213–217.

de Boer, J.M., McDermott, J.P., Wang, X., Maier, T., Qui, F., Hussey, R.S., et al., 2002. The use of DNA microarrays for the developmental expression analysis of cDNA from the oesophageal gland cell region of *Heterodera glycines*. Mol. Plant Pathol. 3, 261–270.

de la Fuente, J., Kocan, K.M., Almazan, C., Blouin, E.F., 2007. RNA interference for the study and genetic manipulation of ticks. Trends Parasitol. 23, 427–433.

de Maagd, R.A., Bravo, A., Crickmore, N., 2001. How *Bacillus thuringiensis* has evolved specific toxins to colonize the insect world. Trends. Genet. 17, 193–199.

de Meutter, J., Robertson, L., Parcy, F., Mena, M., Fenoll, C., Gheysen, G., 2005. Differential activation of *ABI3* and *LEA* genes upon plant-parasitic nematode infection. Mol. Plant Pathol. 6, 321–325.

Dermauw, W., Wybouw, N., Rombauts, S., Menten, B., Vontas, J., Grbic, M., et al., 2013. A link between host plant adaptation and pesticide resistance in the polyphagous spider mite *Tetranychus urticae*. Proc. Natl. Acad. Sci. USA 110 (2), 113–122.

Derycke, S., Vanaverbeke, J., Rigaux, A., Backeljau, T., Moens, T., 2010. Exploring the use of cytochrome oxidase C subunit 1 (COI) for DNA barcoding of free-living marine nematodes. Plos One 5 (10), e13716.

Devine, G.J., Barber, M., Denholm, I., 2001. Incidence and inheritance of resistance to METI-acaricides in European strains of the two-spotted spider mite (*Tetranychus urticae*) (Acari: Tetranychidae). Pest Manag. Sci. 57, 443–448.

Dighe, N., Robinson, A.F., Bell, A., Menz, M., Cantrell, R., Stelly, D., 2009. Linkage mapping of resistance to reniform nematode in cotton (*Gossypium hirsutum* L.) following introgression from *G. longicalyx* (Hutch & Lee). Crop Sci. 49, 1151–1164.

Ding, X., Shields, J., Allen, R., Hussey, R.S., 1998. A secretory cellulose-binding protein from the root-knot nematode (*Meloidogyne incognita*). Mol. Plant Microb. Interact. 11, 952–959.

Ding, L., Spencer, A., Morita, K., Han, M., 2005. The developmental timing regulator *AIN-1* interacts with miRISCs and may target the argonaute protein *ALG-1* to cytoplasmic P bodies in *C. elegans*. Mol. Cell 19, 437–447.

Duan, X., Li, X., Xue, Q., Ado-El-Saad, M., Xu, D., Wu, R., 1996. Transgenic rice plants harboring an introduced potato inhibitor II gene are insect resistant. Biotechnology 14, 494–498.

Dwivedi, S.L., Crouch, J.H., Mackill, D.J., Xu, Y., Blair, M.W., Ragot, M., et al., 2007. The molecularization of public sector crop breeding: progress, problems, and prospects. Adv. Agron. 95, 163–318.

Edge, V.E., James, D.G., 1986. Organotin resistance in *Tetranychus urticae* (Acari: Tetranychidae) in Australia. J. Econ. Entomol. 79, 1477–1483.

Elbashir, S.M., Martinez, J., Patkaniowska, A., Lendeckel, W., Tuschl, T., 2001. Functional anatomy of siRNAs for mediating efficient RNAi in *Drosophila melanogaster* embryo lysate. EMBO J. 20, 6877–6888.

Erban, T., Nesvorna, M., Erbanova, M., Hubert, J., 2009. *Bacillus thuringiensis* var. *tenebrionis* control of synanthropic mites (Acari: Acaridida) under laboratory conditions. Exp. Appl. Acarol. 49, 339–346.

Ernst, K., Kumar, A., Kriseleit, D., Kloos, D.U., Phillips, M.S., Ganal, M.W., 2002. The broad-spectrum potato cyst nematode resistance gene (*Hero*) from tomato is the only member of a large gene family of NBS-LRR genes with an unusual amino acid repeat in the LRR region. Plant J. 31, 127–136.

Escobar, C., Barcala, M., Portillo, M., Almoguera, C., Jordano, J., Fenoll, C., 2003. Induction of the *Hahsp177G4* promoter by root-knot nematodes: involvement of heat-shock elements in promoter activity in giant-cells. Mol. Plant Microbe Interact. 12, 1062–1068.

Fagard, M., Vaucheret, H., 2000. Systemic silencing signal(s). Plant Mol. Biol. 43, 285–293.

Fallas, G.A., Hahn, M.L., Fargette, M., Burrows, P.R., Sarah, J.L., 1996. Molecular and biochemical diversity among isolates of *Radopholus* spp. from different areas of the world. J. Nematol. 28, 422–430.

Favery, B., Lecomte, P., Gil, N., Bechtold, N., Bouchez, D., Dalmasso, A., et al., 1998. *RPE*, a plant gene involved in early developmental steps of nematode feeding cells. EMBO J. 17, 6799–6811.

Felcher, K.J., Coombs, J.J., Massa, A.N., Hansey, C.N., Hamilton, J.P., 2012. Integration of two diploid potato linkage maps with the potato genome sequence. PLoS ONE 7 (4), e36347.

Fenton, B., Malloch, G., Jones, A.T., Amrine Jr, J.W., Gordon, S.C., A'Hara, S., et al., 1995. Species identification of *Cecidophyopsis* mites (Acari: Eriophyidae) from different *Ribes* species and countries using molecular genetics. Mol. Ecol. 4, 383–387.

Fergusson-Kolmes, L.A., Scott, J.G., Dennehy, T.J., 1991. Dicofol resistance in *Tetranychus urticae* (Acari: Tetranychidae): Cross-resistance and pharmacokinetics. J. Econ. Entomol. 84, 41–48.

Fernández-Caldas, E., Puerta, L., Caraballo, L., Lockey, R.F., 2008. Mite allergens. Clin. Allergy Immunol. 21, 161–182.

Ferris, V.R., Ferris, J.M., Faghihi, J., 1993. Variation in spacer ribosomal DNA in some cyst-forming species of plant-parasitic nematodes. Fundam. Appl. Nematol. 16, 177–184.

Fioretti, L., Porter, A., Haydock, P.J., Curtis, R., 2002. Monoclonal antibodies reactive with secreted–excreted products from the amphids and the cuticle surface of *Globodera pallida* affect nematode movement and delay invasion of potato roots. Int. J. Parasitol. 32, 1709–1718.

Fire, A., Xu, S., Montgomery, M.K., Kostas, S.A., Driver, S.E., Mello, C.C., 1998. Potent and specific genetic interference by double-stranded RNA in *Caenorhabditis elegans*. Nature 391, 806–811.

Fritz, G.N., Conn, J., Cockburn, A., Seawright, J., 1994. Sequence analysis of the ribosomal DNA internal transcribed spacer 2 from a population of *Anopheles nuneztovari* (Diptera: Culicidae). Mol. Biol. Evol. 11, 406–416.

Gao, B., Allen, R., Maier, T., McDermott, J.P., Davis, E.L., Baum, T.J., et al., 2002. Characterisation and developmental expression of a chitinase gene in Heterodera glycines. Int. J. Parasitol. 32, 1293–1300.

Gao, B., Allen, R., Maier, T., Davis, E.L., Baum, T.J., Hussey, R.S., 2003. The parasitome of the phytonematode *Heterodera glycines*. Mol. Plant Microb. Interact. 14, 1247–1254.

Gao, B., Allen, R., Davis, E.L., Baum, T.J., Hussey, R.S., 2004. Developmental expression and biochemical properties of a beta-1,4-endoglycanase family in the soybean cyst nematode, *Heterodera glycines*. Mol. Plant Pathol. 5, 93–104.

Ghadamyari, M., Sendi, J.J., 2009. Resistance mechanisms to oxydemeton-methyl in *Tetranychus urticae* Koch (Acari: Tetranychidae). Munis Entomol. Zool. 4 (1), 103–113.

Gheysen, G., Vanholme, B., 2007. RNAi from plants to nematodes. Trends Biotechnol. 25, 89–92.

Ghosh, A., Sharmin, S., Islam, S., Pahloan, S.F., Islam, S., Khan, H., 2010. SSR markers linked to mite (*Polyphagotarsonemus latus* Banks) resistance in Jute (*Corchorus olitorius* L.). Czech. J. Genet. Plant Breed. 46 (2), 64–74.

Goddijn, O.J., Lindsey, K., van der Lee, F.M., Klap, J.C., Sijmons, P.C., 1993. Differential gene expression in nematode-induced feeding structures of transgenic plants harbouring promoter-*gusA* fusion constructs. Plant J. 4, 863–873.

Goellner, M., Wang, X., Davis, E., 2001. Endo-beta-1,4-glucanase expression in compatible plant–nematode interactions. Plant Cell 13, 2241–2255.

Goggin, F.L., Jia, L., Shah, G., Hebert, S., Williamson, V.M., Ullman, D.E., 2006. Heterologous expression of the *Mi-12* gene from tomato confers resistance against nematodes but not aphids in eggplants. Mol. Plant Microbe Interact. 19, 383–388.

Goodell, P.B., Montez, G.H., 1994. Acala cotton tolerance to southern root-knot nematode, Meloidogyne incognita. Proc. of Beltwide Cotton Prod. Res. Conf., Natl. Cotton Council of Am., Memphis, pp. 265–267.

Gotoh, T., Gutierrez, J., Navajas, M., 1998. Molecular comparison of the sibling species *Tetranychus pueraricola* Ehara & Gotoh and *T. urticae* Koch (Acari: Tetranychidae). Entomol. Sci. 1, 55–57.

Goverse, A., Biesheuvel, G.J., Gommers, F.J., Bakker, J., Schots, A., Helder, J., 1998. *In planta* monitoring of the activity of two constitutive promoters, CaMV35S and TR2, in developing feeding cells induced by *Globodera rostochiensis* using green fluorescent protein in combination with confocal laser scanning microscopy. Physiol. Mol. Plant Pathol. 52, 275–284.

Grbic, M., Leeuwen, T.V., Clark, R.M., Rombauts, S., Rouze, P., Grbic, V., et al., 2011. The genome of *Tetranychus urticae* reveals herbivorous pest adaptations. Nature 479, 487–492.

Griffitts, J.S., Haslam, S.M., Yang, T., Garczynski, S.F., Mulloy, B., Morris, H., et al., 2005. Glycolipids as receptors for *Bacillus thuringiensis* crystal toxin. Science 307, 922–925.

Griffitts, J.S., Whitacre, J.L., Stevens, D.E., Aroian, R.V., 2001. Bt toxin resistance from loss of a putative carbohydrate-modifying enzyme. Science 293, 860–863.

Guo, S., Kemphues, K., 1995. *Par-1*, a gene required for establishing polarity in *Caenorhabditis elegans* embryos, encodes a putative Ser/Thr kinase that is asymmetrically distributed. Cell 81, 611–620.

Guo, S., Liu, M., Peng, D., Ji, S., Wang, P., 2008. New strategy for isolating novel nematicidal crystal protein genes from *Bacillus thuringiensis* strain YBT-1518. Appl. Environ. Microbiol. 74, 6997–7001.

Gustafson, D.I., 1993. Pesticides in Drinking Water. Wiley, New York.

Gutiérrez, O.A., Jenkins, J.N., McCarty, J.C., Wubben, M.J., Hayes, R.W., 2010. SSR markers closely associated with genes for resistance to root-knot nematode on chromosomes 11 and 14 of upland cotton. Theor. Appl. Genet. 121, 1323–1337.

Gutiérrez, O.A., Robinson, A.F., Jenkins, J.N., McCarty, J.C., Wubben, M.J., 2011. Identification of QTL regions and SSR markers associated with resistance to reniform nematode in *Gossypium barbadense* L. accession GB713. Theor. Appl. Genet. 122, 271–280.

Haegeman, A., Bauters, L., Kyndt, T., Rahman, M.M., Gheysen, G., 2013. Identification of candidate effector genes in the transcriptome of the rice root knot nematode *Meloidogyne graminicola*. Mol. Plant Pathol. 14, 370–390.

Hammond, S.M., Bernstein, E., Beach, D., Hannon, G.J., 2000. An RNA directed nuclease mediates post-transcriptional gene silencing in *Drosophila* cells. Nature 404, 293–296.

Hammond, S.M., Caudy, A.A., Hannon, G.J., 2001. Post-transcriptional gene silencing by double-stranded RNA. Nat. Rev. Gen. 2, 110–119.

Haas, B.J., Zody, M.C., 2010. Advancing RNA-Seq analysis. Nat. Biotechnol. 28 (5), 421–423.

Hebert, P.D.N., Cywinska, A., Ball, S.L., deWaard, J.R., 2003. Biological identifications through DNA barcodes. Proc. R. Soc. B 270, 313–321.

Hepher, A., Atkinson, H.J., 1992. US patent PCTWO 92/15690.

Herron, G.A., Rophail, J., 1993. Genetics of hexythiazox resistance in two spotted spider mite, *Tetranychus urticae* Koch. Exp. Appl. Acarol. 17, 423–431.

Herron, G.A., Edge, V., Rophail, J., 1993. Clofentezine and hexythiazox resistance in *Tetranychus urticae* Koch in Australia. Exp. Appl. Acarol. 17, 433–440.

Hilder, V.A., Gatehouse, A.M.R., Sheerman, S.E., Barker, R.F., Boulder, D., 1987. A novel mechanism of insect resistance engineered into tobacco. Nature 330, 160–163.

Hillis, D.M., Dixon, M.T., 1991. Ribosomal DNA: molecular evolution and phylogenetic inference. Q. Rev. Biol. 66, 411–453.

Hinomoto, N., Takafuji, A., 2001. Genetic diversity and phylogeny of the Kanzawa spider mite, *Tetranychus kanzawai* in Japan. Exp. Appl. Acarol. 25, 355–370.

Hinomoto, N., Osakabe, M., Gotoh, T., Takafuji, A., 2001. Phylogenetic analysis of green and red forms of the two-spotted spider mite, *Tetranychus urticae* Koch (Acari: Tetranychidae), in Japan, based on mitochondrial cytochrome oxidase subunit I sequences. Appl. Entomol. Zool. 36, 459–464.

Hosoyama, H., Irie, K., Abe, K., Arai, S., 1995. Introduction of a chimeric gene encoding an oryzacystatin-b-glucuronidase fusion protein into rice protoplasts and regeneration of transformed plants. Plant. Cell. Rep. 15, 174–177.

Huang, G., Allen, R., Davis, E.L., Baum, T.J., Hussey, R.S., 2006a. Engineering broad root-knot resistance in transgenic plants by RNAi silencing of a conserved and essential root-knot nematode parasitism gene. Proc. Natl. Acad. Sci. USA 103, 14302–14306.

Huang, G., Dong, R., Allen, R., Davis, E.L., Baum, T.J., Hussey, R.S., 2006b. A root-knot nematode secretory peptide functions as a ligand for a plant transcription factor. Mol. Plant Microbe Interact. 19, 463–470.

Huang, G., Gao, B., Maier, T., Allen, R., Davis, E.L., Baum, T.J., et al., 2003. A profile of putative parasitism genes expressed in the esophageal gland cells of the root-knot nematode *Meloidogyne incognita*. Mol. Plant Microbe Interact. 16, 376–381.

Hubert, J., Stejskal, V., Munzbergova, Z., Kubatova, A., Vanova, M., Zdarkova, E., 2004. Mites and fungi in heavily infested stores in the Czech Republic. J. Econ. Entomol. 97, 2144–2153.

Hyer, A.H., Jorgenson, E.C., Garber, R.H., Smith, S., 1979. Resistance to root-knot nematode in control of root-knot nematode *Fusarium* wilt disease complex in cotton *Gossypium hirsutum*. Crop. Sci. 19, 898–901.

Ibrahim, S.I.L, Perry, R.N., Burrows, P.R., Hooper, D.J., 1994. Differentiation of species and populations of *Aphelenchoides* and of *Ditylenchus angustus* using a fragment of ribosomal DNA. J. Nematol. 26, 412–421.

Irie, K., Hosoyama, H., Takeuchi, T., Iwabuchi, K., Watanabe, H., Abe, M., et al., 1996. Transgenic rice established to express corn cystatin exhibits strong inhibitory activity against insect gut proteinases. Plant Mol. Biol. 30, 149–157.

Iwahori, H., Tsuda, K., Kanzaki, N., Izui, K., Futaki, K., 1998. PCR-RFLP and sequencing analysis of ribosomal DNA of *Bunaphelenchus* nematodes related to pine wilt disease. Fundam. Appl. Nematol. 21, 656–666.

Jaubert, S., Ledger, T.N., Laffaire, J.B., Piotte, C., Abad, P., Rosso, M.N., 2002. Direct identification of stylet secreted proteins from root-knot nematodes by a proteomic approach. Mol. Biochem. Parasitol. 121, 205–211.

Johnson, R., Narvaez, J., An, G., Ryan, C., 1989. Expression of proteinase inhibitors I and II in transgenic tobacco plants: Effects on natural defense against *Manduca sexta* larvae. Proc. Natl. Acad. Sci. USA 86, 9871–9875.

Jones, J.T., Furlanetto, C., Bakker, E., Banks, B., Blok, V., Chen, Q., et al., 2003. Characterization of a chorismate mutase from the potato cyst nematode *Globodera pallida*. Mol. Plant Pathol. 4, 43–50.

Jouzani, Seifinejad, A., Saeedizadeh, A., Nazarian, A., Yousefloo, M., Soheilivand, S., et al., 2008. Molecular detection of nematicidal crystalliferous *Bacillus thuringiensis* strains of Iran and evaluation of their toxicity on free-living and plant-parasitic nematodes. Can. J. Microbiol. 54 (10), 812–822.

Joyce, S.A., Burnell, A.M., Powers, T.O., 1994. Characterization of *Heterorhabditis* isolates by PCR amplification of segments of mtDNA and rDNA genes. J. Nematol. 26, 260–270.

Juergensen, K., Scholz-Starke, J., Sauer, N., Hess, P., van Bel, A.J.E., Grundler, F.M.W., 2003. The companion cell-specific Arabidopsis disaccharide carrier AtSUC2 is expressed in nematode induced syncytia. Plant Physiol. 131, 61–69.

Jung, C., Cai, D., Kleine, M., 1998. Engineering nematode resistance in crop species. Trends Plant Sci. 3, 266–271.

Kazi, S., Shultz, J., Afzal, J., Johnson, J., Njiti, V.N., Lightfoot, D.A., 2008. Separate loci underlie resistance to root infection and leaf scorch during soybean sudden death syndrome. Theor. Appl. Genet. 116 (7), 967–977.

Keena, M.A., Granett, J., 1990. Genetic analysis of propargite resistance in pacific spider mites and two spotted spider mites (Acari: Tetranychidae). J. Econ. Entomol. 83, 655–661.

Kehr, J., Buhtz, A., 2008. Long distance transport and movement of RNA through the phloem. J. Exp. Bot. 59, 85–92.

Khetan, S.K., 2001. Microbial Pest Control. Marcel Dekker Inc, New York, 300pp.

Khila, A., Girbic, M., 2007. Gene silencing in the spider mite *Tetranychus urticae*: dsRNA and siRNA parental silencing of the Distal-less gene. Dev. Genes Evol. 217, 241–251.

Kimber, M.J., McKinney, S., McMaster, S., Day, T.A., Fleming, C.C., Maule, A.G., 2007. *flp* gene disruption in a parasitic nematode reveals motor dysfunction and unusual neuronal sensitivity to RNA interference. FASEB J. 21, 1233–1243.

Klink, V.P., Kim, K.H., Martins, V., MacDonald, M.H., Beard, H.S., Alkharouf, N.W., et al. 2009. A correlation between host-mediated expression of parasite genes as tandem inverted repeats and abrogation of development of female *Heterodera glycines* cyst formation during infection of *Glycine max*. Planta 230, 53–71.

Kocan, K.M., Fuente, J.D.L., Manzano-Roman, R., Naranjo, V., Hynes, W.L., Sonenshine, D.E., 2008. Silencing expression of the *defensin*, *varisin*, in male *Dermacentor variabilis* by RNA interference results in reduced *Anaplasma marginale* infections. Exp. Appl. Acarol. 46, 17–28.

Kramer, K.J., Muthukrishnan, S., 1997. Insect chitinases: molecular biology and potential use as biopesticides. Insect Biochem. Mol. Biol. 27 (11), 887–900.

Kuwahara, M., 1982. Insensitivity of the acetylcholinesterase from the organophosphate resistant kanzawa spider mite, *Tetranychus kanzawai* Kishida (Acarina: Tetranychidae), to organophosphorus and carbamate insecticide. Appl. Entomol. Zool. 17, 486–493.

Lambert, K.N., Allen, K.D., Sussex, I.M., 1999. Cloning and characterization of an esophageal-gland-specific chorismate mutase from the phytoparasitic nematode *Meloidogyne javanica*. Mol. Plant Microbe Interact. 12, 328–336.

Lee, M.L., Suh, S.J., Kwon, Y.J., 1999. Phylogeny and diagnostic markers of six *Tetranychus* species (Acarina: Tetranychidae) in Korea based on the mitochondrial cytochrome oxidase subunit I. J. Asia-Pacific Entomol. 2, 85–92.

Li, J., Todd, T.C., Lee, J., Trick, H.N., 2011. Biotechnological application of functional genomics towards plant-parasitic nematode control. Plant Biotechnol. J. 9, 936–944.

Li, J., Todd, T.C., Oakley, T.R., Lee, J., Trick, H.N., 2010a. Host derived suppression of nematode reproductive and fitness genes decreases fecundity of *Heterodera glycines*. Planta 232 (3), 775–785.

Li, J., Todd, T.C., Trick, H.N., 2010b. Rapid *in planta* evaluation of root expressed transgenes in chimeric soybean plants. Plant. Cell. Rep. 29 (2), 113–123.

Li, J., Todd, T.C., Oakley, T.R., Trick, H.N., 2013. RNA Interference: an alternative approach for soybean cyst nematode resistance. Paper presented at plant and animal genome XXI Conference, January 12–16, 2013 at San Diego, CA, USA, p. 863.

Li, X.Q., Tan, A., Voegtline, M., Bekele, S., Chen, C.S., 2008. Expression of Cry5B protein from *Bacillus thuringiensis* in plant roots confers resistance to root knot nematode. Biol. Control 47, 97–102.

Li, X.Q., Wei, J.Z., Tan, A., Aroian, R.V., 2007. Resistance to root-knot nematode in tomato roots expressing a nematicidal *Bacillus thuringiensis* crystal protein. Plant Biotechnol. 5, 455–464.

Lilley, C.J., Devlin, P., Urwin, P.E., Atkinson, H.J., 1999a. Parasitic nematodes, proteinases and transgenic plants. Parasitol. Today 15, 414–417.

Lilley, C.J., Goodchild, S.A., Atkinson, H.J., Urwin, P.E., 2005. Cloning and characterization of a *Heterodera glycines* aminopeptidase cDNA. Int. J. Parasitol. 35, 1577–1585.

Lilley, C.J., Urwin, P.E., Atkinson, H.J., 1999b. Characterization of plant nematode genes: identifying targets for a transgenic defence. Parasitology 118, S63–S72.

Limpens, E., Ramos, J., Franken, C., Raz, V., Compaan, B., Franssen, H., et al., 2004. RNA interference in *Agrobacterium rhizogenes* transformed roots of Arabidopsis and *Medicago truncatula*. J. Exp. Bot. 55, 983–992.

Lindhout, P., 2002. The perspectives of polygenic resistance in breeding for durable disease resistance. Euphytica 124, 217–226.

Liu, B., Hibbard, J.K., Urwin, P.E., Atkinson, H.J., 2005. The production of synthetic chemodisruptive peptides *in planta* disrupts the establishment of cyst nematodes. Plant Biotech. J. 3, 487–496.

Liu, B., Jiang, G., Zhang, Y., Li, J., Li, X., 2011. Analysis of transcriptome differences between resistant and susceptible strains of the citrus red mite *Panonychus citri* (Acari: Tetranychidae). Plos One 6 (12), e28516.

Liu, Q.H., Rand, T.A., Kalidas, S., Du, F., Kim, H.E., Smith, D.P., et al., 2003. R2D2, a bridge between the initiation and effector steps of the *Drosophila* RNAi pathway. Science 26, 1921–1925.

Malik, R., Brown-Guedira, G.L., Smith, C.M., Harvey, T.L., Gill, B.S., 2003. Genetic mapping of wheat curl mite resistance genes Cmc3 and Cmc4 in common wheat. Crop Sci. 43 (2), 644–650.

Martin, P., Dabert, M., Dabert, J., 2010. Molecular evidence for species separation in the water mite *Hygrobates nigromaculatus* Lebert, (Acari, Hydrachnidia): evolutionary consequences of the loss of larval parasitism. Aquat. Sci. 72, 347–360.

Masoud, S.A., Johnson, L.B., White, F.F., Reeck, G.R., 1993. Expression of a cysteine proteinase inhibitor (oryzacystatin-I) in transgenic tobacco plants. Plant Mol. Biol. 21, 655–663.

Matsuda, T., Fukumoto, C., Hinomoto, N., Gotoh, T., 2013. DNA-based identification of spider mites: molecular evidence for cryptic species of the genus *Tetranychus* (Acari: Tetranychidae). J. Econ. Entomol. 106, 463–472.

Matsumura, T., Voss, G., 1964. Mechanism of malathion and ethyl parathion resistance in the two-spotted spider mite, *Tetranychus urticae*. J. Econ. Entomol. 57, 911–916.

May, R.C., Plasterk, R.H.A., 2005. RNA interference spreading in *C. elegans*. Methods Enzymol. 392, 308–315.

Mazarei, M., Lennon, K.A., Puthoff, D.P., Rodermel, S.R., Baum, T.J., 2004. Homologous soybean and Arabidopsis genes share responsiveness to cyst nematode infection. Mol. Plant Pathol. 5, 409–423.

McCafferty, H.R.K., Moore, P.H., Zhu, Y.J., 2006. Improved *Carica papaya* tolerance to carmine spider mite by the expression of *Manduca sexta* chitinase transgene. Transgenic Res. 15, 337–347.

McCafferty, H.R.K., Moore, P.H., Zhu, Y.J., 2008. Papaya transformed with the *Galanthus nivalis* GNA gene produces a biologically active lectin with spider mite control activity. Plant Sci. 175, 385–393.

McCarter, J.P., 2008. Molecular approaches toward resistance to plant-parasitic nematodes. Plant Cell Monogr. 15, 239–267.

McMurtry, J.A., Croft, B.A., 1997. Life styles of Phytoseiid mites and their role in biological control. Annu. Rev. Entomol. 42, 291–321.

McPherson, M.J., Urwin, P.E., Lilley, C.J., Atkinson, H.J., 1997. Engineering plant nematode resistance by antifeedants. In: Fenoll, C., Grundler, F.M.W., Ohl, S.A. (Eds.), Cellular and Molecular Aspects of Plant–Nematode Interactions. Kluwer, Dordrecht, pp. 237–249.

Meister, G., Tuschl, T., 2004. Mechanisms of gene silencing by doublestranded RNA. Nature 431, 343–349.

Melito, S., Heuberger, A., Cook, D., Diers, B., MacGuidwin, A., Bent, A., 2010. A nematode demographics assay in transgenic roots reveals no significant impacts of the Rhg1 locus LRR-Kinase on soybean cyst nematode resistance. BMC Plant Biol. 10, 104.

Migeon, A., Dorkeld, F., 2011. Spider Mites Web: a comprehensive database for the Tetranychidae. [Accessed November 2012] <http://www.montpellier.inra.fr/CBGP/spmweb>.

Milligan, S.B., Bodeau, J., Yaghoobi, J., Kaloshian, I., Zabel, P., Williamson, V.M., 1998. The root-knot nematode resistance gene *Mi* from tomato is a member of the leucine zipper, nucleotide binding, leucine-rich repeat family of plant genes. Plant Cell 10, 1307–1319.

Mohammed, S.H., Anwer El Saedy, M., Enan, M.R., Ibrahim, N.E., Ghareeb, A., Moustafa, S.A., 2008. Biocontrol efficiency of *Bacillus thuringiensis* toxins against root-knot nematode, *Meloidogyne incognita*. J. Cell Mol. Biol. 7 (1), 57–66.

Moloney, C., Griffin, D., Jones, P.W., Bryan, G.J., McLean, K., Bradshaw, J.E., et al. 2010. Development of diagnostic markers for use in breeding potatoes resistant to Globodera pallida pathotype Pa2/3 using germplasm derived from Solanum tuberosum ssp. Andigena CPC 2802. Theor. Appl. Genet. 120, 679–689.

Monaghan, M.T., Balke, M., Gregory, T.R., Vogler, A.P., 2005. DNA-based species delineation in tropical beetles using mitochondrial and nuclear markers. Phil. Trans. R. Soc. B 360, 1925–1933.

Moritz, C., Cicero, C., 2004. DNA barcoding: promise and pitfalls. Plos Biol. 2, 1529–1531.

Nasmith, C.G., Speranzini, D., Jeng, R., Hubbes, M., 1996. RFLP analysis of PCR amplified ITS and 26S ribosomal RNA genes of selected entomopathogenic nematodes (Steinernematidae, Heterorhabditidae). J. Nematol. 28, 15–25.

Navajas, M., Fenton, B., 2000. The application of molecular markers in the study of diversity in acarology: a review. Exp. Appl. Acarol. 24, 751–774.

Navajas, M., Fournier, D., Lagnel, J., Gutierrez, J., Boursot, P., 1996a. Mitochondrial COI sequences in mites: evidence for variations in base composition. Insect Mol. Biol. 5, 281–285.

Navajas, M., Gutierrez, J., Gotoh, T., 1997. Convergence of molecular and morphological data reveals phylogenetic information in *Tetranychus* species and allows the restoration of the genus *Amphitetranychus* (Acari: Tetranychidae). Bull. Entomol. Res. 87, 283–288.

Navajas, M., Gutierrez, J., Bonato, O., Bolland, H.R., Mapangoudivassa, S., 1994. Intraspecific diversity of the cassava green mite *Mononycellus progresivus* (Acari, Tetranychidae) using comparisons of mitochondrial and nuclear ribosomal DNA-sequences and cross-breeding. Exp. Appl. Acarol. 18, 351–360.

Navajas, M., Gutierrez, J., Lagnel, J., Boursot, P., 1996b. Mitochondrial cytochrome oxidase I in tetranychid mites: a comparison between molecular phylogeny and changes of morphological and life history traits. Bull. Entomol. Res. 86, 407–417.

Navajas, M., Lagnel, J., Fauvel, G., Moraes, G.D., 1999. Sequence variation of ribosomal Internal Transcribed Spacers (ITS) in commercially important Phytoseiidae mites. Exp. Appl. Acarol. 23, 851–859.

Navajas, M., Lagnel, J., Gutierrez, J., Boursot, P., 1998. Species-wide homogeneity of nuclear ribosomal ITS2 sequences in the spider mite *Tetranychus urticae* contrasts with extensive mitochondrial COI polymorphism. Heredity 80, 742–752.

Nisbet, A.J., Billingsley, P.F., 2000. A comparative survey of the hydrolytic enzymes of ectoparasitic and free-living mites. Int. J. Parasitol. 30, 19–27.

Ohl, S.A., van der Lee, F.M., Sijmons, P.C., 1997. Anti-feeding structure approaches to nematode resistance. In: Fenoll, C., Grundler, F.M.W., Ohl, S.A. (Eds.), Cellular and Molecular Aspects of Plant–Nematode Interactions. Kluwer, Dordrecht, pp. 250–261.

Oliveira, C.M.G., Hubschen, J., Brown, D.J.F., Ferraz, L.C.C.B., Wright, F., Neilson, R., 2004. Phylogenetic relationships among *Xiphinema* and *Xiphidorus* nematode species from Brazil inferred from 18S rDNA sequences. J. Nematol. 6, 153–159.

Opperman, C.H., Taylor, C.G., Conkling, M.A., 1994. Root-knot nematode directed expression of a plant root-specific gene. Science 263, 221–223.

Osakabe, M., Uesugi, R., Goka, K., 2009. Evolutionary aspects of acaricide-resistance development in spider mites. Psyche http://dx.doi.org/10.1155/2009/947439. Article ID 947439, 11 pp.

Paal, J., Henselewski, H., Muth, J., Meksem, K., Menendez, C.M., Salamini, F., et al., 2004. Molecular cloning of the potato *Gro1-4* gene conferring resistance to pathotype Ro1 of the root cyst nematode *Globodera rostochiensis*, based on a candidate gene approach. Plant J. 38, 285–297.

Palyvos, N.E., Athanassiou, C.G., Kavallieratos, N.G., 2006. Acaricidal effect of a diatomaceous earth formulation against *Tyrophagus putrescentiae* (Astigmata: Acaridae) and its predator *Cheyletus malaccensis* (Prostigmata: Cheyletidae) in four grain commodities. J. Econ. Entomol. 99, 229–236.

Park, J.E., Lee, K.Y., Lee, S.J., Oh, W.S., Jeong, P.Y., Woo, T., et al., 2008. The efficiency of RNA interference in *Bursaphelenchus xylophilus*. Mol. Cell 26, 81–86.

Parkhi, V., Kumar, V., Campbell, L.M., Bell, A.A., Shah, J., Rathore, K.S., 2010. Resistance against various fungal pathogens and reniform nematode in transgenic cotton plants expressing Arabidopsis *NPR1*. Transgenic Res. 19, 959–975.

Paskewitz, S.M., Wesson, D.M., Collins, F.H., 1993. The internal transcribed spacers of ribosomal DNA in five members of the *Anopheles gambiae* species complex. Insect Mol. Biol. 2, 247–257.

Peng, D., Chai, L., Wang, F., Zhang, F., Ruan, L., 2011. Synergistic activity between *Bacillus thuringiensis* Cry6Aa and Cry55Aa toxins against *Meloidogyne incognita*. Microb. Biotechnol. 4, 794–798.

Pieterse, C.M.J., Ton, J., van Loon, L.C., 2001. Cross-talk between plant defence signaling pathways: boost or burden? Agri. Biotech. Net. 3, 1–18.

Popeijus, H., Overmars, H.A., Jones, J.T., Blok, V.C., Goverse, A., Helder, J., et al., 2000. Degradation of plant cell walls by a nematode. Nature 406, 36–37.

Powers, T.O., 2004. Nematode molecular diagnostics: from bands to barcodes. Annu. Rev. Phytopathol. 42, 367–383.

Powers, T.O., Harris, T.S., 1993. A polymerase chain reaction method for identification of five major *Meloidogyne* species. J. Nematol. 25, 1–6.

Powers, T.O., Todd, T.C., Burnell, A.M., Murray, P.C.B., Fleming, C.C., Szalanski, A.L., et al., 1997. The rDNA internal transcribed spacer region as a taxonomic marker for nematodes. J. Nematol. 29 (4), 441–450.

Priya, D.B., Somasekhar, N., Prasad, J.S., Kirti, P.B., 2011. Transgenic tobacco plants constitutively expressing Arabidopsis NPR1 show enhanced resistance to root-knot nematode, *Meloidogyne incognita*. BMC Res. Notes 4, 231–236.

Qin, L., Kudla, U., Roze, E.H., Goverse, A., Popeijus, H., Nieuwl, J., et al., 2004. Plant degradation: a nematode expansin acting on plants. Nature 427, 30.

Rawlings, N.D., Barrett, A.J., Bateman, A., 2012. MEROPS: the database of proteolytic enzymes, their substrates and inhibitors. Nucleic Acids Res. 40, 343–350.

Rich, J.R., Keen, N.T., Thomason, I.J., 1977. Association of coumestans with the hypersensitivity of lima bean roots to *Pratylenchus scribneri*. Physiol. Plant Pathol. 10, 105–116.

Robinson, A.F., Bell, A.A., Dighe, N.D., Menz, M.A., Nichols, R.L., 2007. Introgression of resistance to nematode *Rotylenchulus reniformis* into upland cotton (*Gossypium hirsutum*) from *Gossypium longicalyx*. Crop Sci. 47, 1865–1877.

Romano, G.B., Sacks, E.J., Stetina, S.R., Robinson, A.F., Fang, D.D., 2009. Identification and genomic location of a reniform nematode (*Rotylenchulus reniformis*) resistance locus (Ren ari) introgressed from *Gossypium aridum* into upland cotton (*G. hirsutum*). Theor. Appl. Genet. 120, 139–150.

Ros, V.I.D., Breeuwer, J.A.J., 2007. Spider mite (Acari: Tetranychidae) mitochondrial COI phylogeny reviewed: host plant relationships, phylogeography, reproductive parasites and barcoding. Exp. Appl. Acarol. 42, 239–262.

Rosso, M.N., Dubrana, M.P., Cimbolini, N., Jaubert, S., Abad, P., 2005. Application of RNA interference to root-knot nematode genes encoding esophageal gland proteins. Mol. Plant Microbe Interact. 18, 615–620.

Ruben, E., Jamai, A., Afzal, J., Njiti, V.N., Triwitayakorn, K., Iqbal, M.J., et al., 2006. Genomic analysis of the *rhg1* locus: candidate genes that underlie soybean resistance to the cyst nematode. Mol. Genet. Genomics 276, 503–516.

Saleh, S., Kelada, N.L., Shaker, N., 1991. Control of European house dust mite Dermatophagoides pteronyssinus (Trouessart) with Bacillus spp. Acarologia 32, 257–260.

Salinas, M., Capel, C., Alba, J.M., Mora, B., Cuartero, J., Fernández-Muñoz, R., et al., 2013. Genetic mapping of two QTL from the wild tomato Solanum pimpinellifolium L. controlling resistance against two-spotted spider mite (Tetranychus urticae Koch). Theor. Appl. Genet. 126, 83–92.

Santamaria, M.E., Cambra, I., Martinez1, M., Pozancos, C., González-Melendi, P., Grbic, V., et al., 2012. Gene pyramiding of peptidase inhibitors enhances plant resistance to the spider mite Tetranychus urticae. Plos One 7 (8), e43011.

Sato, M.E., Miyata, T., da Silva, M., Raga, A., de Souza Filho, M.F., 2004. Selections for fenpyroximate resistance and susceptibility, and inheritance, cross resistance and stability of fenpyroximate resistance in *Tetranychus urticae* Koch (Acari: Tetranychidae). Appl. Entomol. Zool. 39, 293–302.

Sato, M.E., Suplicy Filho, N., de Souza Filho, M.F., Takematsu, A.P., 1994. Resistência do ácaro rajado *Tetranychus urticae* (Acari: Tetranychidae) a diversos acaricidas em morangueiro (Fragaria sp.) nos municípios de Atibaia-SP e Piedade SP. Ecossistema 19, 40–46.

Schmidt, A.L., McIntyre, C.L., Thompson, J.P., Seymour, N.P., Liu, C.J., 2005. Quantitative trait loci for root lesion nematode (*Pratylenchus thornei*) resistance in middle-eastern landraces and their potential for introgression into Australian bread wheat. Aust. J. Agric. Res. 56, 1059–1068.

Schnepf, E., Crickmore, N., Van Rie, J., Lereclus, D., Baum, J., 1998. *Bacillus thuringiensis* and its pesticidal crystal proteins. Microbiol. Mol. Biol. Rev. 62, 775–806.

Schnepf, E.H., Schwab, G.E., Payne, J., Narva, K.E., Foncerrada, L., 2003. Nematicidal Proteins. Mycogen Corporation, USA.

Shalini, K.V., Manjunatha, S., Lebrun, P., Berger, A., Badouin, B., Pirany, N., et al., 2007. Identification of molecular markers associated with mites resistance in coconut (*Cocus nucifera* L.). Genome 50, 35–42.

Sharma, S., Sharma, S., Keil, T., Laubach, E., Jung, C., 2011a. Screening of barley germplasm for resistance to root lesion nematodes. Plant Genet. Resour. 9, 236–239.

Sharma, S., Sharma, S., Kopisch-Obuch, F.J., Keil, T., Laubach, E., Stein, N., et al., 2011b. QTL analysis of root-lesion nematode (Pratylenchus neglectus) resistance in barley. Theor. Appl. Genet. 122, 1321–1330.

Sheilds, R., Fleming, C.C., Stratford, R., 1996. Identification of potato cyst nematodes using the polymerase chain reaction. Fundam. Appl. Nematol. 19, 167–173.

Shen, X., He, Y., Lubbers, E.L., Davis, R.F., Nichols, R.L., 2010. Fine mapping QMi-C11, a major QTL controlling root-knot nematodes resistance in Upland cotton. Theor. Appl. Genet. 121, 1623–1631.

Sindhu, S., Maier, T.R., Mitchum, M.G., Hussey, R.S., Davis, E.L., Baum, T.J., 2009. Effective and specific in planta RNAi in cyst nematodes: Expression interference of four parasitism genes reduces parasitic success. J. Exp. Bot. 60, 315–324.

Skoracka, A., Dabert, M., 2010. The cereal rust mite *Abacarus hystrix* (Acari: Eriophyoidea) is a complex of species: evidence from mitochondrial and nuclear DNA sequences. Bull. Entomol. Res. 100, 263–272.

Smant, G., Stokkermans, J.P., Yan, Y., de Boer, J.M., Baum, T.J., Wang, X., et al., 1998. Endogenous cellulases in animals: isolation of beta-1,4-endoglucanase genes from two species of plant-parasitic cyst nematodes. Proc. Natl. Acad. Sci. USA 95, 4906–4911.

Smissaert, H.R., Voerman, S., Oostenbrugge, L., Renooy, N., 1970. Acetylcholinesterase of organophosphate susceptible and resistant spider mites. J. Agric. Food Chem. 18, 65–75.

Soares, C.A.G., Lima, C.M.R., Dolan, M.C., Piesman, J., Beard, C.B., Zeidner, N.S., 2005. Capillary feeding of specific dsRNA induces silencing of the *isac* gene in nymphal *Ixodes scapularis* ticks. Insect Mol. Biol. 14, 443–452.

Sobotnik, J., Kudlikova-Krizkova, I., Vancova, M., Munzbergova, Z., Hubert, J., 2008. Chitin in the peritrophic membrane of *Acarus siro* (Acari: Acaridae) as a target for novel acaricides. J. Econ. Entomol. 101, 1028–1033.

Soriano, I.R., Asenstorfer, R.E., Schmidt, O., Riley, I.T., 2004a. Inducible flavone in oats (*Avena sativa*) is a novel defense against plant-parasitic nematodes. Phytopathology 94, 1207–1214.

Soriano, I.R., Riley, I.T., Potter, M.J., Bowers, W.S., 2004b. Phytoecdysteroids: a novel defense against plant-parasitic nematodes. J. Chem. Ecol. 30, 1885–1899.

Steeves, R.M., Todd, T.C., Essig, J.S., Trick, H.N., 2006. Transgenic soybeans expressing siRNAs specific to a major sperm protein gene suppress *Heterodera glycines* reproduction. Funct. Plant Biol. 33, 991–999.

Stepek, G., Behnke, J.M., Buttle, D.J., Duce, I.R., 2004. Natural plant cysteine proteinases as anthelmintics? Trends Parasitol. 20, 322–327.

Stevenson, L.A., Chilton, N.B., Gasser, R.B., 1995. Differentiation of *Haemonchus placei* from *H. contortus* (Nematoda: Trichostrongylidae) by the ribosomal DNA second internal transcribed spacer. Int. J. Parasitol. 25, 483–488.

Stiekema, W.J., Bosch, D., Wilmink, A., de Boer, J.M., Schouten, A., Roosien, J., et al., 1997. Towards plantibody-mediated resistance against nematodes. In: Fenoll, C., Grundler, F.M.W., Ohl, S.A. (Eds.), Cellular and Molecular Aspects of Plant–Nematode Interactions. Kluwer, Dordrecht, pp. 262–271.

Stumpf, N., Nauen, R., 2001. Cross-resistance, inheritance, and biochemistry of mitochondrial electron transport inhibitor-acaricide resistance in *Tetranychus urticae* (Acari: Tetranychidae). J. Econ. Entomol. 94, 1577–1583.

Stumpf, N., Zebitz, C.P.W., Kraus, W., Moores, G.D., Nauen, R., 2001. Resistance to organophosphates and biochemical genotyping of acetylcholinesterases in *Tetranychus urticae* (Acari: Tetranychidae). Pestic. Biochem. Physiol. 69, 131–142.

Subbotin, S.A., Halford, P.D., Warry, A., Perry, R.N., 2000. Variation in ribosomal DNA sequences and phylogeny of *Globodera* parasitising solanaceous plants. Nematology 2, 591–604.

Szalanski, A.L., Sui, D.D., Harris, T.S., Powers, T.O., 1997. Identification of cyst nematodes of agronomic and regulatory concern by PCR-RFLP of ITS1. J. Nematol. 29, 255–267.

Thiery, M., Mugniery, D., 1996. Interspecific rDNA restriction fragment length polymorphism in *Globodera* species, parasites of solanaceous plants. Fundam. Appl. Nematol. 19, 471–479.

Thind, B.B., Clarke, P.G., 2001. The occurrence of mites in cereal-based foods destined for human consumption and possible consequences of infestation. Exp. Appl. Acarol. 25, 203–215.

Thomas, C.M., Cottage, A., 2006. Genetic engineering for resistance. In: Perry, R.N., Moens, M. (Eds.), Plant Nematology. CAB International, Wallingford, pp. 255–272.

Thurau, T., Kifle, S., Jung, C., Cai, D., 2003. The promoter of the nematode resistance gene *Hs1pro-1* activates a nematode-responsive and feeding site specific gene expression in sugar beet (*Beta vulgaris* L) and *Arabidopsis thaliana*. Plant Mol. Biol. 52, 643–660.

Toda, S., Osakabe, M., Komazaki, S., 2000. Interspecific diversity of mitochondrial *COI* sequences in Japanese *Panonychus* species (Acari: Tetranychidae). Exp. Appl. Acarol. 24, 82.

Uehara, T., Kushida, A., Momota, Y., 1999. Rapid and sensitive identification of *Pratylenchus* spp. using reverse dot blot hybridisation. Nematology, 549–555.

Ulloa, M., Wang, C., Roberts, P.A., 2010. Gene action analysis by inheritance and quantitative trait loci mapping of resistance to root-knot nematodes in cotton. Plant Breed. 129, 541–550.

Ulloa, M., Wang, C., Hutmacher, R.B., Wright, S.D., Davis, R.M., 2011. Mapping *Fusarium* wilt race1 resistance genes in cotton by inheritance, QTL and sequencing composition. Mol. Genet. Genomics 286, 21–36.

Urszula, K., Qina, L., Milac, A., Kielaka, A., Maissena, C., Overmars, H., et al., 2005. Origin, distribution and 3D-modeling of Gr-EXPB1, an expansin from the potato cyst nematode *Globodera rostochiensis*. FEBS Lett. 579, 2451–2457.

Urwin, P.E., Atkinson, H.J., Waller, D.A., McPherson, M.J., 1995. Engineered oryzacystatin-I expressed in transgenic hairy root confers resistance to *Globodera pallida*. Plant J. 8, 121–131.

Urwin, P.E., Lilley, C.J., Atkinson, H.J., 2002. Ingestion of double stranded RNA by pre-parasitic juvenile cyst nematodes leads to RNA interference. Mol. Plant Microbe Interact. 15, 747–752.

Urwin, P.E., Lilley, C.J., McPherson, M.J., Atkinson, H.J., 1997a. Resistance to both cyst and root-knot nematodes conferred by transgenic *Arabidopsis* expressing a modified plant cystatin. Plant J. 12, 455–461.

Urwin, P.E., Moller, S.G., Lilley, C.J., McPherson, M.J., Atkinson, H.J., 1997b. Continual green-fluorescent protein monitoring of cauliflower mosaic virus 35S promoter activity in nematode-induced feeding cells in *Arabidopsis thaliana*. Mol. Plant Microbe Interact. 10, 394–400.

Vaghchhipawala, Z.E., Schlueter, J.A., Shoemaker, R.C., Mackenzie, S.A., 2004. Soybean *FGAM* synthase promoters direct ectopic nematode feeding site activity. Genome 47, 404–413.

Vain, P., Worland, B., Clarke, M.C., Richard, G., Beavis, M., Liu, H., et al., 1998. Expression of an engineered cysteine proteinase inhibitor (Oryzacystatin-IΔD86) for nematode resistance in transgenic rice plants. Theor. Appl. Genet. 96, 266–271.

van der Vossen, E.A., van der Voort, J.N., Kanyuka, K., Bendahmane, A., Sandbrink, H., Baulcombe, D.C., et al., 2000. Homologues of a single resistance-gene cluster in potato confer resistance to distinct pathogens: a virus and a nematode. Plant J. 23, 567–576.

Van Leeuwen, T., Dermauw, W., Grbic, M., Tirry, L., Feyereisen, R., 2012. Spider mite control and resistance management: Does a genome help? Pest Manag. Sci. 69, 156–159.

Van Leeuwen, T., Vontas, J., Tsagkarakou, A., Dermauw, W., Tirry, L., 2010. Acaricide resistance mechanisms in the two-spotted spider mite *Tetranychus urticae* and other important Acari: A review. Insect Biochem. Mol. Biol. 40 (8), 563–572.

Vanholme, B., De Meutter, J., Tytgat, T., Van Montagu, M., Coomans, A., Gheysen, G., 2004. Secretions of plant-parasitic nematodes: a molecular update. Gene 332, 13–27.

Veech, J.A., McClure, M.A., 1977. Terpenoid aldehydes in cotton roots susceptible and resistant to the root-knot nematode. *Meloidogyne incognita*. J. Nematol. 9, 225–229.

Vishnudasan, D., Tripathi, M.N., Rao, U., Khurana, P., 2005. Assessment of nematode resistance in wheat transgenic plants expressing potato proteinase inhibitor (*PIN2)* gene. Transgenic Res. 14, 665–675.

Vogler, A.P., Monaghan, M.T., 2007. Recent advances in DNA taxonomy. J. Zool. Syst. Evol. Res. 45, 1–10.

Vos, P., Simons, G., Jesse, T., Wijbrandi, J., Heinen, L., Hogers, R., et al., 1998. The tomato *Mi*-1 gene confers resistance to both root-knot nematodes and potato aphids. Nat. Biotechnol. 16, 1365–1369.

Vrain, T.C., McNamara, D.G., 1994. Potential for identification of quarantine nematodes by PCR. EPPO Bull. 24, 453–458.

Vrain, T.C., Wakarchuk, D.A., Levesque, A.C., Hamilton, R.I., 1992. Intraspecific rDNA restriction fragment length polymorphism in the *Xiphinema americanum* group. Fundam. Appl. Nematol. 15, 563–573.

Wang, D., Arelli, P.R., Shoemaker, R.C., Diers, B.W., 2001. Loci underlying resistance to race 3 of soybean cyst nematode in *Glycine soja* plant introduction 468916. Theor. Appl. Genet. 103, 561–566.

Wang, C., Matthews, W.C., Roberts, P.A., 2006. Phenotypic expression of *rkn1*-mediated *Meloidogyne incognita* resistance in *Gossypium hirsutum* populations. J. Nematol. 38, 250–257.

Wang, C., Ulloa, M., Roberts, P.A., 2008. A transgressive segregation factor (*RKN2*) in *Gossypium barbadense* for nematode resistance clusters with gene *rkn1* in *G. hirsutum*. Mol. Gen. Genomics 279, 41–52.

Wang, X., Bosselut, N., Castagnone, C., Voison, R., Abad, P., Esmenjaud, D., 2003. Multiplex polymerase chain reaction identification of single individuals of the Longidorid nematodes *Xiphinema index*, *X. diversicaudatum*, *X. vuittenezi* and *X. italiae* using specific primers from ribosomal genes. Phytopathology 93, 160–166.

Weerasinghe, R.R., Bird, D.M., Allen, N.S., 2005. Root-knot nematodes and bacterial Nod factors elicit common signal transduction events in *Lotus japonicus* root hair cells. Proc. Natl. Acad. Sci. USA 102, 3147–3152.

Wei, J.Z., Hale, K., Carta, L., Platzer, E., Wong, C., Fang, S.C., et al., 2003. *Bacillus thuringiensis* crystal proteins that target nematodes. Proc. Natl. Acad. Sci. USA 100, 2760–2765.

Wendt, K.R., Swart, A., Vrain, T.C., Webster, J.M., 1995. *Ditylenchus africanus* sp. n. from South Africa – a morphological and molecular characterization. Fundam. Appl. Nematol. 18, 241–250.

Wesson, D.M., McLain, D.K., Oliver, J.H., Piesman, J., Collins, F.H., 1993. Investigation of the validity of species status of *Ixodes dammini* (Acari: ixodidae) using rDNA. Proc. Natl. Acad. Sci. USA 90, 10221–10225.

William, H.M., Trethowan, R., Crosby-Galvan, E.M., 2007. Wheat breeding assisted by markers: CIMMYT's experience. Euphytica 157, 307–319.

Williams, K., Taylor, S., Bogacki, P., Pallotta, M., Bariana, H.S., Wallwork, H., 2002. Mapping of the root lesion nematode (*Pratylenchus neglectus*) resistance gene *Rlnn1* in wheat. Theor. Appl. Genet. 104, 874–879.

Williamson, F.A., 1995. Products and opportunities in nematode control. AGROW Reports DS 108, PJB Publications Ltd: Richmond.

Williamson, V.M., 1992. Molecular techniques for nematode species identification. In: Nickle, W.R. (Ed.), Manual of Agricultural Nematology. Marcel Dekker, New York, pp. 107–123.

Williamson, V.M., 1999. Plant nematode resistance genes. Curr. Opin. Plant Biol. 2, 327–331.

Williamson, V.M., Hussey, R.S., 1996. Nematode pathogenesis and resistance in plants. Plant Cell 8, 1735–1745.

Williamson, V.M., Kumar, A., 2006. Nematode resistance in plants: the battle underground. Trends Genet. 22, 396–403.

Winter, S.M., Shelp, B.J., Anderson, T.R., Welacky, T.W., Rajcan, I., 2007. QTL associated with horizontal resistance to soybean cyst nematode in *Glycine soja*. Theor. Appl. Genet. 114, 461–472.

Wu, X., Blake, S., Sleper, D.A., Shannon, J.G., Cregan, P., Nguyen, H.T., 2009. QTL, additive and epistatic effects for SCN resistance in PI 437654. Theor. Appl. Genet. 118 (6), 1093–1105.

Xie, L., Hong, X.-Y., Xue, X-F., 2006. Population genetic structure of the two-spotted spider mite (Acari: Tetranychidae) from China. Ann. Entomol. Soc. Am. 99, 959–965.

Ynturi, P., Jenkins, J.N., McCarty, J.C., Gutierrez, O.A., Saha, S., 2006. Association of root-knot nematode resistance genes with simple sequence repeat markers on two chromosomes in cotton. Crop Sci. 46, 2670–2674.

Yoo, B.C., Kragler, F., Varkonyi-Gasic, E., Haywood, V., Archer-Evans, S., Lee, Y.M., et al., 2004. A systemic small RNA signaling system in plants. Plant Cell 16, 1979–2000.

Yoon, Y., Kim, E., Hwang, Y., Choi, C., 2004. Avermectin: biochemical and molecular basis of its biosynthesis and regulation. Appl. Microbiol. Biotechnol. 63, 626–634.

Young, M.R., Behan-Pelletier, V.M., Hebert, P.D.N., 2012. Revealing the hyperdiverse mite fauna of subarctic Canada through DNA Barcoding. Plos One 7 (11), e48755.

Zacheo, T.B., Melillo, M.T., Castagnone-Sereno, P., 2007. The contribution of biotechnology to root-knot nematode control in tomato plants. Pest Technol. 1 (1), 1–16.

Zahavi, M., Tahori, A.S., 1970. Sensitivity of acetylcholinesterase in spider mites to organophosphorous compounds. Biochem. Pharmacol. 19, 219–225.

Zamore, P.D., Haley, B., 2005. Ribo-gnome: the big world of small RNAs. Science 309, 1519–1524.

Zhang, F., Peng, D., Ye, X., Yu, Z., Hu, Z., Ruan, L., et al., 2012. *In vitro* uptake of 140 kDa *Bacillus thuringiensis* nematicidal crystal proteins by the second stage juvenile of *Meloidogyne hapla*. PLoS One 7 (6), e38534.

Zijlstra, C., Blok, V.C., Phillips, M.S., 1997. Natural resistance: the assessment of variation in virulence in biological and molecular terms. In: Fenoll, C., Grundler, F.M.W., Ohl, S.A. (Eds.), Cellular and Molecular Aspects of Plant–Nematode Interactions. Kluwer, Dordrecht, pp. 153–166.

Zijlstra, C., Lever, A.E.M., Uenk, B.C., VanSilfhout, C.H., 1995. Differences between ITS regions of isolates of root-knot nematodes *Meloidogyne hapla* and *M. chitwoodi*. Phytopathology 85, 1231–1237.

Zwart, R.S., Thompson, J.P., Godwin, I.D., 2005. Identification of quantitative trait loci for resistance to two species of root-lesion nematode (*Pratylenchus thornei* and *P. neglectus*) in wheat. Aust. J. Agric. Res. 56, 345–352.

Zwart, R.S., Thompson, J.P., Milgate, A.W., Bansal, U.K., Williamson, P.M., Raman, H., et al., 2010. QTL mapping of multiple foliar disease and root-lesion nematode resistances in wheat. Mol. Breed. 26, 107–124.

Zwart, R.S., Thompson, J.P., Sheedy, J.G., Nelson, J.C., 2006. Mapping quantitative trait loci for resistance to *Pratylenchus thornei* from synthetic hexaploid wheat in the International Triticeae Mapping Initiative (ITMI) population. Aust. J. Agric. Res. 57, 525–530.

CHAPTER

18

Role of Genetically Modified Insect-Resistant Crops in IPM: Agricultural, Ecological and Evolutionary Implications

S. Mohankumar[1] *and T. Ramasubramanian*[2]

[1]**Tamil Nadu Agricultural University, Coimbatore, India,** [2]**Sugarcane Breeding Institute, Indian Council of Agricultural Research, Coimbatore, India**

18.1 INTRODUCTION

Integrated pest management (IPM) is a concept that arises from the problems posed by the indiscriminate use of insecticides which has led to the problems of resistance, resurgence and residues in the environment. There are an increasing number of reports of resistance of pests to pesticides (Kranthi et al., 2002; Mehrotra, 1989; Saini and Jaglan, 1998). The cumulative number of arthropod species that have developed resistance to pesticides reached 574 in 2012. The red spider mite, *Tetranychus urticae* has developed resistance against as many 93 compounds and ranks first among the resistant arthropod species. This is followed by the diamondback moth *Plutella xylostella* (resistant against 81 compounds) and the green peach aphid *Myzus persicae* (73 compounds). The American bollworm *Helicoverpa armigera* occupies tenth place with resistance against 43 compounds (www.irac-online.org, accessed on 4 January 2013). The number of pest outbreaks has also increased and many innocuous species have attained the status of serious pests (Elzen and Hardee, 2003). Detailed studies have been made in the past on the insecticide-induced resurgence (Chelliah and Heinrichs, 1980; Reissig et al., 1982a,b). The extent of pesticide residues in the environment is another issue of great concern. IPM, the most efficient and cheapest method of pest management, does not upset the balance in nature, delays the development of resistance and minimizes the residue hazards of pesticides (Stern, 1973; Stone and Pedigo, 1972). As a part of sustainable agriculture, IPM must be seen as a component of good agricultural practices (GAP) or integrated crop management (ICM).

IPM has a long history and broad scope including the use of chemical insecticides in combination with improved pest-resistant varieties, employing cultural and biological methods with a focus on managing the pest

D. P. Abrol (Ed): *Integrated Pest Management*.
DOI: http://dx.doi.org/10.1016/B978-0-12-398529-3.00020-8

population in an environmentally benign and most economical manner. Though the concept of IPM was originally proposed as early as 1959 (Stern et al., 1959), it was recognized as a prime requirement for promoting sustainable agriculture and development during the United Nations Conference on Environment and Development (UNCED) held at Rio de Janeiro in 1992. IPM encourages the most compatible and ecologically sound combination of available pest suppression techniques to keep the pest population below the economic threshold level, the level of pest density at which control measures should be initiated to prevent the increasing pest population from reaching economic injury level, i.e. the lowest density that will cause economic damage (Stern, 1973; Stern et al., 1959; Stone and Pedigo, 1972). The term economic injury level was criticized and it was argued that the term action threshold would be more logical (Stylen, 1968). However, the term economic injury level has been retained as it can be more easily understood by farmers. IPM also helps in making plant protection feasible, safe and economical, even for the small and marginal farmers (Campbell et al., 2009; Dhawan et al., 2009; Rapusas et al., 2009; Sharma et al., 2008).

Orr and Ritchie (2004) stressed that the most attractive IPM components for resource poor farmers are host-plant resistance (HPR) and biological pest suppression, the former being better than the latter as it involves no additional expenditure. In general, land races and wild species are potential sources of resistance genes and their utilization as one of the parents in resistance breeding has several limitations including the reduction in economic yield. In some cases, however, it is possible to manipulate the ploidy levels and employ sophisticated techniques such as embryo rescue to transfer resistance genes from more distantly related plant species. Moreover, naturally occurring resistance is often polygenic involving multiple alleles on separate chromosomes and may involve complex genetic mechanisms (Kennedy and Barbour, 1992; Smith, 2005). In addition, polygenic resistance and resistance derived from wild relatives of crops often involve genes having negative, pleiotropic effects or linkages with genes conferring undesirable traits. Breaking these linkages can be difficult and time consuming. In most cases, neither the specific genes coding for resistance nor the underlying chemical or physical mechanisms responsible for resistance are known. Consequently, progeny screening in each generation requires the data on insect populations or damage or bioassay (Smith, 2005). The use of molecular markers tightly linked to resistance genes is helping to improve the selection efficiency, especially for polygenic resistance traits (Smith, 2005; Yencho et al., 2000). Despite several advantages, the success of host-plant resistance as an IPM tool has been constrained by the limited availability of elite cultivars possessing high levels of resistance to key pests. The deployment of recombinant DNA technology to develop insect-resistant crops has paved the way to eliminate this constraint and make HPR a viable component of IPM programmes. It also greatly reduces the time required to develop commercially acceptable insect-resistant cultivars compared to conventional plant breeding procedures. The cultivation of genetically modified crops (GM crops) represents a new 'green revolution', especially for developing countries facing food scarcity, as they could boost the productivity and cope with emerging issues such as climate change and soil exhaustion in addition to helping to conserve natural resources (Conway, 1997; Gliessman, 2000; Ejeta, 2010; Fedoroff et al., 2010).

Since their commercial release in 1996 in USA, and subsequently in other countries, GM crops producing insecticidal proteins of the soil bacterium *Bacillus thuringiensis* (Bt) have helped to manage several insect pests across the globe. The cultivation of Bt crops has substantially reduced the consumption of insecticides on several crops worldwide. In India, after the introduction of Bt-cotton in 2002, the consumption of insecticide for the management of cotton bollworm, *Helicoverpa armigera*, declined rapidly, and currently very few sprays are used in Bt-cotton as against 15–30 sprays before the introduction of Bt-cotton. In 2001 (pre-Bt cotton era), about 9400 metric tonnes of insecticides were used only for the management of cotton bollworm, and this has come down to just 222 metric tonnes in 2011 in India (Kranthi, 2012). The substantial reduction in use of insecticides for bollworm management has resulted in several ecological and environmental benefits. Lu et al. (2012) showed that, in the last 13 years, GM crops delivered significant environmental benefits by reducing insecticide usage by 50% and doubling the level of ladybirds, lacewings and spiders.

Crop- and region-specific IPM programmes have been developed and practised worldwide to suppress pest populations below economic threshold levels (Ahuja et al., 2009; Naved et al., 2008a,b; Sharma et al., 2008). In several instances, lack of elite cultivars that are resistant to key pests creates a vacuum in developing crop-specific IPM programmes and GM crops are certainly ideal candidates to fill the vacuum. The large-scale cultivation of GM crops expressing *Cry* genes is rightly considered as another form of host-plant resistance (Kennedy, 2008), which is a cornerstone in the foundation of IPM (Naranjo et al., 2008). As GM crops become more complex, with multiple traits involved, and as multiple GM crops become established, the capacity to model evolutionary responses within a predictive framework will be critically important. For example, there is potential for both positive and negative ecological and evolutionary feedback between novel crop types, herbivores and weeds, and soil microbial communities (Bais et al., 2006). It is beyond the scope of the present chapter to provide a review of genetic engineering in agriculture and related debates on social, political and ethical issues. However, the relevance of insect-resistant GM crops in IPM with reference to agricultural, ecological and evolutionary implications is discussed in this chapter.

18.2 GM CROPS AS A PART OF HPR IN IPM

Cultivating crop varieties that are less prone to pest attack is an important strategy in IPM. The resistant varieties reduce the cost of cultivation and require less intervention by limiting the damage throughout the crop season barring a few insect biotypes which are capable of surviving and damaging the resistant varieties. With the development of resistance by pests to pesticides and poor adoption of developed and validated IPM technologies, there is renewed interest among the scientific community in breeding resistant cultivars to pests. Host-plant resistance along with natural enemies and cultural practices can play a major role in implementation of crop-specific IPM technologies (Dhaliwal and Singh, 2004). The contribution made by resistant varieties in avoiding yield losses is quite significant and cannot be written off, especially in the case of cereals, oilseeds and pulses (Dhaliwal and Singh, 2004). Use of resistant varieties not only helps us to avoid losses, but also encourages the survival of natural enemies (Sharma and Ortiz, 2002). However, there may be problems if we rely exclusively on plant resistance for insect control as high levels of resistance is in general associated with low yield potential or undesirable quality traits. These inherent problems of conventional plant breeding approaches can easily be eliminated by recombinant DNA technology and there will not be any compromise in yield and quality of the produce obtained from the GM crops. Thus, in a true sense, GM crops are better than the traditional insect-resistant cultivars. Moreover, the vertical resistance in traditional cultivars that is governed by a single major gene can easily be broken down by insects with development of several biotypes. *Mayetiola destructor, Schizaphis graminum, Nilaparvata lugens, Orseolia oryzae* and *Bemisia tabaci* are among the most important pests that have developed an array of biotypes against resistant varieties. The biotypes of *M. destructor*, *S. graminum* and *O. oryzae* are governed by major genes, and the specific pest virulence–host-plant resistance interactions followed a gene-for-gene relationship (Fang et al., 2004). The virulence of *N. lugens* has been reported to be under polygenic control as it is a quantitative character. Thus, vertical resistance in the traditional cultivars is not durable and the development of biotypes cannot be ignored completely when breeding for resistant varieties. On the contrary, the pyramiding of genes from sole or diverse sources in the case of GM crops may offer protection against the target pest for quite a long period and can be compared with the durable horizontal resistance of cultivars developed through conventional plant breeding approaches.

A group of researchers are still sceptical about the widespread cultivation of GM crops and there has been some speculation about the possible negative impacts of GM crops on non-target organisms. Kogan (1998) opined that the excitement about genetic engineering dominates the literature and global management strategies; nevertheless, nothing will have been learnt from past experiences if genetic engineering prevails over all of the other technologies that are also blossoming. It has also been cautioned, 'like in the dark ages, we do not have enough knowledge about the risks that GM crops could pose for wild plant populations' (e.g. through gene flow; Ellstrand, 2003; Andersson and de Vicente, 2010; Dyer et al., 2009), non-target organisms (Dale et al., 2002; Hilbeck and Schmith, 2006), and human health (Schubert, 2002; Finamore et al., 2008; Spiroux de Vendomois et al., 2009).

18.3 CURRENT STATUS OF GM CROPS

Biotechnology holds great promise for improving productivity, profitability and sustainability of farm production systems, including those existing in small and poor farming situations (Brookes and Barfoot, 2006; Cohen, 2005). The recent breakthroughs in biotechnology and genetic engineering such as recombinant DNA technology have opened new vistas for developing resistant cultivars. The introduction of novel genes for pest resistance into crop plants from unrelated plants and microorganisms is an exciting development. The best known example is the incorporation of genes that produce *Bt* endotoxins into crops such as cotton (*Gossypium* sp.), corn (*Zea mays* L.), tobacco (*Nicotiana* sp.), tomato (*Solanum esculentum* L.), potato (*Solanum tuberosum* L.) and rice (*Oryza sativa* L.).

GM crops were first commercialized in 1996 and the acreage under biotech crops increased by an unprecedented 100-fold from 1.7 million hectares in 1996 to over 170 million hectares in 2012. This makes GM crops the fastest adopted crop technology in recent times. Notably, developing countries grew more (52%) of global GM crops in 2012 than industrial countries (48%). Of the 28 countries which planted biotech crops in 2012, 20 were developing and eight were industrial countries; two new countries, Sudan (Bt cotton) and Cuba (Bt maize) planted biotech crops for the first time in 2012. Germany and Sweden did not plant the biotech potato 'Amflora' as it has ceased to be marketed. In 2012, a record 17.3 million farmers grew GM crops and over 15 million (over 90%) were small resource-poor farmers in developing countries. The USA continued to be the lead country with 69.5 million hectares, followed by Brazil with 36.6 million hectares under GM crops. Argentina retained third place with 23.9 million hectares. Canada was fourth at 11.8 million hectares with 8.4 million hectares of canola at a record 97.5% adoption. India was fifth, growing a record 10.8 million hectares of Bt cotton with an adoption rate of 93%. From 1996 to 2011, GM crops saved pesticide usage amounting to 473 million kg active ingredient (a.i.), and increased crop production valued at US$98.2 billion and thus, contributed significantly to food security and environmental safety (James, 2011). In general, use of Bt crops has significantly reduced insecticide use in regions where they have been planted. In cotton, a crop with heavy insecticide usage, Brookes and Barfoot (2006) estimate that the volume of the insecticide active ingredient was reduced by 94.5 million tons in 10 years of Bt crop use. In percentages, the reduction in insecticide use is approximately 50% in Bt cotton compared to non-Bt cotton worldwide (Fitt, 2008). Reductions in insecticide use in corn is not as dramatic, but still estimated at 12.5% less insecticides for the United States in 2005 (Fitt, 2008). Notable benefits for subsistence agriculture in Africa are also reported by Gouse et al. (2006) with the use of these varieties. Other GM crops with the Bt gene have also recorded significant reductions in the use of insecticides, although the area under these crops is much less than that for cotton and corn.

18.4 ROLE OF GM CROPS IN INSECT PEST MANAGEMENT

18.4.1 Case Studies

18.4.1.1 Bt Cotton in India

The role of GM crops in pest management can be very well explained by the spectacular achievement made in cotton production in India during the current decade (2002–2012) after the commercial release of Bt cotton in 2002. Indian cotton farmers were spending more than 43% of the variable costs of cotton production on insecticides during the mid-1990s, around 80% of which was intended for bollworm control and *H. armigera* in particular (ICAC 1998a,b). In 1998–99, at least 14.6% of Indian cotton production was lost to insect (mainly bollworm) damage. The insecticide use on cotton had reached 50% of the total insecticides used in the country and farmers were taking up to 15–30 applications of insecticides (Kranthi, 2012). This indiscriminate use, especially of synthetic pyrethroids, had led to the development of insecticide resistance in *H. armigera*. Cotton cultivation was no longer profitable and was moving towards its darkest era. The Genetic Engineering Approval Committee (GEAC) of India approved the commercial cultivation of three Bt cotton hybrids (MECH-12, MECH-162, and MECH-184) in 2002 when Indian cotton farmers were in desperate need of alternatives to insecticides to sustain cotton production from the ravages from bollworms. The introduction of Bt cotton has revolutionized cotton production and now India is the second largest producer of cotton next to China. Without Bt cotton, it would be impossible to realize this spectacular achievement.

In 2011, Bt cotton was the third most dominant biotech crop worldwide, occupying 17.9 million hectares, equivalent to 11% of the global biotech area. In 2011, Bt cotton was planted in 11 countries, listed in the order of descending hectarage: India, China, Pakistan, Myanmar, Burkina Faso, Brazil, USA, Argentina, Australia, Colombia, and Costa Rica. In 2011–12, the area under Bt cotton reached a record 10.6 million hectares accounting for 87.6% of the area under cotton (12.1 million hectares) in India and 7% of the area under GM crops worldwide (160 million hectares). The area under Bt cotton increased 212-fold in 2011 (10.6 million hectares) from just 50,000 hectares in 2002. Brookes and Barfoot (2012) estimated that India enhanced farm income from Bt cotton by US$9.4 billion in the period 2002 to 2010 and US$2.5 billion in 2010 alone. The country was transformed from a net importer of raw cotton

TABLE 18.1 Impact of Bt Cotton on the Lint Yield of Cotton in Major Bt Cotton Cultivating Countries

Country	Year of Release	GM Cotton Area (lakh hectares)	Total Cotton Area (lakh hectares)	Adoption Rate	Yield Before Release (kg/ha)	Yield in 2011 (kg/ha)	% increase in yield
India	2002	94	111.4	82.45	292	531	82
USA	1996	40	43.3	92.37	602	886	47
China	1997	35	51.5	61.16	890	1326	49
Australia	1996	05	05.9	84.74	1425	1839	29

Adopted from Kranthi, 2012 – Source: International Cotton Advisory Committee, Washington.

TABLE 18.2 Impact of Bt Cotton on Insecticide Usage against Major Pests

	Quantity of insecticides used (metric tonnes)			Monetary value of insecticides used (Rs. in crores)		
Year	Against Bollworms	Against Sucking Pests	Total Insecticides on Cotton	Against Bollworms	Against Sucking Pests	Total Insecticides on Cotton
2001	9410	3312	13,176	747.6	304	1052
2002	4470	2110	6863	415.6	181	597
2003	6599	2909	10,045	680.5	245	925
2004	6454	2735	9367	718.1	314	1032
2005	2923	2688	5914	385.7	263	649
2006	1874	2374	4623	307.4	272	579
2007	1201	3805	5543	287.8	445	733
2008	652	3877	5057	236.7	554	791
2009	500	5816	6726	140.1	694	834
2010	249	7270	7885	122.8	758	880
2011	222	6372	6828	96.3	894	991

Source: Kranthi, 2012.

in 2002 to a net exporter due to increased cotton yield from 308kg per hectare in 2001–02 to 499kg per hectare in 2011–12 (James, 2012). The yield increase was the highest among the four major countries cultivating Bt cotton (Table 18.1). Accordingly, cotton production has increased from 13.6 million bales in 2002–03 to 35.5 million bales in 2011–12. Above all, insecticide usage was significantly reduced from 46% in 2001–02 to 21% of the total insecticide use in 2010 (James, 2011). Before the introduction of Bt cotton, about 9400 metric tonnes of insecticides were used for bollworm control in India and in 2011, only 222 metric tonnes were used for managing the same pest (Table 18.2) (Kranthi, 2012).

18.4.1.2 Bt Maize in USA

A 2010 University of Minnesota study (Hutchinson et al., 2010) on biotech maize, resistant to European corn borer (ECB), reported that 'area-wide suppression dramatically reduced the estimated US$1 billion annual losses caused by the European Corn Borer [ECB].' Importantly, the study reported that Bt maize has even benefited the conventional maize. Widespread planting of Bt maize throughout the Upper Midwest of the USA since 1996 has suppressed the populations of ECB, historically one of maize's primary pests causing losses estimated at approximately US$1 billion per year. Corn borer moths cannot discern between Bt and non-Bt maize, so the pest lays eggs in both Bt and non-Bt maize fields. As soon as the eggs hatch in Bt maize, borer larvae feed and die within 24 to 48h. As a result, corn borer numbers have also declined in neighbouring non-Bt fields by 28–73% in Minnesota, Illinois and Wisconsin. The study also reported similar declines of the pest in Iowa and Nebraska. In the US study, the economic benefit of this area-wide pest suppression was estimated at US$6.9 billion over the 14-year period 1996 to 2009 for the five-state region comprising Minnesota, Illinois, Wisconsin, Iowa and Nebraska. Of the US$6.9 billion, it is noteworthy that non-Bt corn hectares accounted for US$4.3 billion (62%, or almost two-thirds, of the total benefit). The principal benefit of Bt maize is reduced yield losses, resulting from the deployment of Bt maize for which farmers have paid Bt maize technology fees. However, what is noteworthy is that, as a result of area-wide pest suppression, farmers planting non-Bt hectares also experienced yield increases without the cost of Bt technology fees; in fact, non-Bt hectares benefited from more than half (62%) of the total benefits of growing Bt maize in the five states. Importantly, the study noted that 'previous cost-benefit analyses focused directly on Bt maize hectares but this study was the first in the USA to include the value of area-wide pest suppression and the subsequent indirect benefits to farmers planting conventional non-Bt maize.' The study did not consider the benefits for other important Midwestern crops affected by European corn borer, such as sweet corn, potatoes and green beans, which would further increase the economic returns from Bt maize. It is noteworthy that the suppression of the European corn borer was only demonstrable in Minnesota, Illinois and Wisconsin because state entomologists have monitored pest populations for more than 45 years. Pest suppression and related yield benefits may well be occurring to both adopters and non-adopters of Bt maize in other parts of the United States and the rest of the world, but those benefits cannot be documented due to lack of historical benchmark data on pest levels. In summary, this important study confirms that Bt maize delivers more benefits to society than originally realized (James, 2010).

18.4.1.3 Bt Cotton and Bt Rice in China

18.4.1.3.1 BT COTTON

China is the largest producer of cotton in the world with 69% of its 5.0 million hectares under cotton crop occupied by Bt cotton in 2010. In China, Bt cotton has been under cultivation since 1997. In 2010, 6.5 million small farmers in China had already increased their income by approximately US$220 per hectare (equivalent to approximately US$1 billion nationally) with a 10% increase in yield, and a 60% reduction in insecticides, both of which contribute to a more sustainable agriculture and the prosperity of small poor farmers.

18.4.1.3.2 BT RICE

Although China is the largest producer of rice in the world (178 million tonnes of paddy) with 110 million rice households, it needs to increase its rice yield to 7.85 tonnes per hectare by 2030 to feed the estimated 1.6 billion population (Chen et al., 2011). It is estimated that 75% of all rice in China is infested with the rice borer, which is now being targeted by Bt rice. China granted biosafety approval for Bt rice in a landmark decision on 27 November 2009. Biosafety approval was granted to a rice variety Huahui-1 (a restorer line), and a hybrid rice line, Bt Shanyou-63, both of which express Cry1Ab/Cry1Ac and were developed at Huazhong Agricultural University (James, 2009). This approval may have momentous positive implications for GM crops in China and worldwide. Bt rice, if permitted for large-scale commercial release, has the potential to generate benefits of US$4 billion annually from an average yield increase of up to 8%, and an 80% decrease in insecticides (Huang et al., 2005). Though China gave safety approval to Bt rice in 2009, it has not yet begun commercial production. The commercialization of GM rice in China will not be hampered by the international rice trade market and in contrast, it may cause the rest of the world to accept GM crops in the future (Tan et al., 2010).

18.5 ECOLOGICAL IMPLICATIONS

It is a well known fact that virtually no technology can be one hundred percent safe. Although insect-resistant GM crops have the safest bio-safety profiles, the negative impact on non-target organisms and the field control failures due to the evolution of pest resistance reported by several workers cannot be ignored while still expecting to reap the benefits of insect-resistant GM crops in a sustainable manner under IPM.

18.5.1 Impact on Non-Target Organisms

18.5.1.1 Changing Scenario in Pest Status

Though Bt crops do not pose any negative impact directly on non-target minor pests, the near-elimination of one species results in lack of interspecific competition and ultimately results in non-target minor pest infestations becoming more and more severe and their attaining the status of major pests. This is a very common phenomenon observed in most of the Bt cotton growing areas. The recent experience with Bt cotton in China cannot be ignored when thinking of Bt crops. In China, 69% of its 5 million hectares under cotton is occupied by Bt cotton. Bt cotton was commercialized with the aim of managing the bollworm *H. armigera*. Though the bollworm has been under check since the introduction of Bt cotton with a substantial reduction in pesticide use, the population of mirid bugs, which were of little importance in northern China in 1997, has increased 12-fold. Mirids have become a major pest in this region, reducing cotton yields by up to 50% in the absence of pest control measures. Unlike bollworms, mirid bugs are also a major threat to crops such as green beans, cereals, vegetables and various fruits (Lu et al. 2010b), resulting in a greater economic loss. Strictly speaking, this paradigm shift in pest status is not due to Bt cotton per se, but may be attributed to the lack of interspecific competition from bollworms. This problem is gradually increasing in the Indian sub-continent too. The near-elimination of bollworms has resulted in increasing incidences of sucking pests in the Bt cotton growing areas. The quantum of insecticides used to manage the sucking pests reached 6372 metric tonnes in 2011 as against only 2110 metric tonnes in 2002, i.e. during the introduction of Bt cotton in India. As a consequence, the quantum of insecticides used in the cotton ecosystem remains static (6863 metric tonnes in 2002 and 6828 metric tonnes in 2011) though there was a significant reduction in the use of insecticides against bollworms (Table 18.2).

Emerging threats from secondary pests stemming from the cultivation of Bt cotton were not found in Bt maize, although some minor pest species increased in Bt maize in some countries with commercial planting (Hellmich et al., 2008). In Germany, 6-year monitoring of non-target arthropods in Bt maize (expressing the Cry1Ab toxin) did not detect differences in the population densities of aphids, thrips, heteropterans, aphid-specific predators, spiders, and carabids (Schorling and Freier, 2006). The sucking pests, including the cotton aphid, thrips, whitefly, leafhopper and spider mites are the major non-target pests in Bt cotton fields, and are not susceptible to the Bt proteins currently used (Arshad and Suhail, 2010; Manna et al., 2010; Wu and Guo, 2005), and these secondary pest populations have increased and gradually evolved into key pests in the USA, India, China, Australia and other countries (Gouse et al., 2004; Ho and Xue, 2008; Li and Romeis, 2010; Lu et al., 2008; Sharma et al., 2005; Williams, 2006; Wilson et al., 2006; Zhao et al., 2011). Green mirid (*Creontiades dilutus*), green vegetable bug (*Nezara viridula*), leaf hopper (*Austroasca viridigrisea* and *Amrasca terraereginae*), and thrips (*Thrips tabaci, Frankliniella schultzei*, and *Frankliniella occidentalis*) have become more prominent in Australia (Lei et al., 2003; Wilson et al., 2006). The incidence of sucking and other pests such as the mirid bug, mealy bug, thrips, and leaf-eating caterpillar has been increasing gradually since the inception of Bt cotton in 2002 in India (Karihaloo and Kumar, 2009; Nagrare et al., 2009). Field surveys conducted over 10 years in six major cotton-growing provinces (i.e. Henan, Hebei, Jiangsu, Anhui, Shangdong, and Shanxi) in northern China showed that the mirid bugs have progressively increased and acquired pest status in Bt cotton fields in China (Lu et al., 2010a). The spider mites have also been observed to occur at higher levels in Bt cotton during the drought season (Wu and Guo, 2005). Hence, the insecticide increase for the control of these secondary pests is still lower than the reduction in total insecticide use due to Bt cotton adoption (Wang et al., 2009).

18.5.1.2 Impact on Natural Enemies

18.5.1.2.1 IMPACT OF INSECTICIDAL PROTEINS ON THE BIOLOGY OF NATURAL ENEMIES

The negative impacts of Bt crops on natural enemies are generally less than those of broad-spectrum insecticides (Cattaneo et al., 2006; Marvier et al., 2007; Naranjo, 2009; Romeis et al., 2006; Wolfenbarger et al., 2008). Yet, the effects of Bt crops on the population density of specific non-target species can be negative, neutral, or positive (Bourguet et al., 2002; Carrière et al. 2009a; Dively et al., 2004; Naranjo 2005a; Sisterson et al., 2007) (Table 18.3). The impact of insecticidal proteins that are expressed in GM crops on the growth, development, survival and fecundity of natural enemies has been studied either by using the insecticidal proteins directly in the bioassay or by feeding them through GM crop-fed host insect pests. A summary of results obtained is presented in Table 18.3, which indicates that Cry proteins in general do not have any impact on the survival and reproduction of natural enemies. However, notable exceptions are there. For example, Cry1Ab/c had shown a negative impact on the generalist predator *Chrysoperla carnea* (Lawo et al., 2010; Lövei et al., 2009). Another notable example is the negative impact of cotton plants expressing Cry1A + CpTI and Cry1Ac on the ichneumonid parasitoid *Campoletis chlorideae*, a specific solitary parasitoid of *H. armigera* (Liu X.X. et al., 2005a). A few reports of positive impacts of Bt proteins on natural enemies are also available among the literature screened. Maize crop expressing Cry1Ab was shown to exert a positive impact on *Cotesia marginiventris* (Faria et al., 2007). On the other hand, Vojtech et al. (2005) and Ramirez-Romero et al. (2007) have shown the negative impact of Cry1Ab on the same parasitoid in maize. In general, Bt toxins had no effect on the survival or reproduction of predators (Naranjo, 2009).

18.5.1.2.2 IMPACT OF GM CROPS ON THE ABUNDANCE OF NATURAL ENEMIES IN THE FIELD

A large volume of literature pertaining to the impact of Bt cotton or Bt maize on the abundance of natural enemies has revealed that there has been no change in the population densities of natural enemies in Bt crop fields. It is interesting to note that the population densities of natural enemies were sometimes higher in Bt crop fields than what was observed in non-Bt crop fields. A few important field studies on the

TABLE 18.3 Impact of Insecticidal Proteins on Natural Enemies (Predators and Parasitoids)

Predator/Parasitoid	Transgenic Crop and Toxin	Positive/ Negative Impact	Reference
Micromus tasmaniae (Neuroptera: Hemerobiidae)	Potato Cry1Ac9, Cry9Aa2	No effect	Davidson et al., 2006
Stethorus punctillum (Coleoptera: Coccinellidae)	Maize Cry1Ab/Cry3Bb1	No effect	Alvarez-Alfageme et al., 2008; Li and Romeis, 2010
Adalia bipunctata (Coleoptera: Coccinellidae)	Maize Cry1Ab/Cry3Bb1	No effect	Alvarez-Alfageme et al., 2011
Harmonia axyridis (Coleoptera: Coccinellidae)	Potato Cry3A	No effect	Ferry et al., 2007
Propylaea japonica (Coleoptera: Coccinellidae)	Cotton Cry1Ac	Negative	Lövei et al., 2009 Zhang et al., 2006
Poecilus cupreus (Coleoptera: Carabidae)	Maize Cry1Ab	No effect	Alfageme et al., 2009
Nebria brevicollis (Coleoptera: Carabidae)	Potato Cry3A	No effect	Ferry et al., 2007
Atheta coriaria (Coleoptera: Staphylinidae)	Maize Cry1Ab	No effect	García et al., 2010
Chrysoperla carnea (Neuroptera: Chrysopidae)	Cotton Cry1Ab/c	Negative	Lawo et al., 2010; Lövei et al., 2009
Orius insidiosus (Heteroptera: Anthocoridae)	Maize Cry3Bb1	No effect	Duan et al., 2008b
Theridion impressum (Araneae: Theridiidae)	Maize Cry3Bb1	No effect	Meissle and Romeis, 2009
Pirata subpiraticus (Araneae: Lycosidae)	Rice Cry1Ab	Negative	Chen et al., 2009
Campoletis chlorideae (Hymenoptera: Ichneumonidae)	Cotton Cry1A + CpTI, Cry1Ac	Negative	Liu X.X. et al., 2005a
Campoletis sonorensis (Hymenoptera: Ichneumonidae)	Maize Cry1Ab	Negative	Sanders et al., 2007
Diadegma insulare (Hymenoptera: Ichneumonidae)	Broccoli Cry1Ac	No effect	Liu et al., 2011
Microplitis mediator (Hymenoptera: Braconidae)	Cotton Cry1A + CpTI, Cry1Ac	Negative	Liu X.X. et al., 2005b
Cotesia marginiventris (Hymenoptera: Braconidae)	Maize Cry1Ab	Negative	Ramirez-Romero et al., 2007; Vojtech et al., 2005
Cotesia marginiventris (Hymenoptera: Braconidae)	Maize Cry1Ab	Positive	Faria et al., 2007

(*Continued*)

TABLE 18.3 (Continued)

Predator/Parasitoid	Transgenic Crop and Toxin	Positive/ Negative Impact	Reference
Apanteles subandinus (Hymenoptera: Braconidae)	Potato Cry1Ac9, Cry9Aa2	No effect	Davidson et al., 2006
Microplitis mediator (Hymenoptera: Braconidae)	Chinese cabbage Cry1Ac	No effect	Kim et al., 2008
Pteromalus puparum (Hymenoptera: Pteromalidae)	Broccoli Cry1Ac, Cry1C, Cry1Ac + Cry1C	Negative	Chen et al., 2008
Trichogramma ostriniae (Hymenoptera: Trichogrammatidae)	Cry1Ab	No effect	Wang et al., 2007
Harmonia axyridis	Canola Oryzacystatin I	No effect	Ferry et al., 2003
Pterostichus madidus	Canola Oryzacystatin I	Negative	Ferry et al., 2005
Pterostichus melanarius	Canola Oryzacystatin I	No effect	Mulligan et al., 2006
Carabids	Strawberry Cowpea trypsin inhibitor	No effect	Graham et al., 2002
Podisus maculiventris	Potato Cowpea trypsin inhibitor/Barley cystatin HvCPI-1 C68	No effect	Alvarez-Alfageme et al., 2007; Bell et al., 2003
Perillus bioculatus	Oryzacystatin I	No effect	Bouchard et al., 2003a,b
Ctenognathus novaezelandiae	Tobacco Bovine spleen trypsin inhibitor	No effect	Burgess et al., 2008
Diaeretiella rapae	Canola; oryzacystatin I	No effect	Schuler et al., 2001
Eulophus pennicornis	Potato Cowpea trypsin inhibitor	Negative	Bell et al., 2001
Aphidius nigripes/Macrosiphum euphorbiae	Potato Oryzacystatin I	Positive	Ashouri et al., 2001a,b
Aphidius ervi	Oryzacystatin I	No effect	Cowgill et al., 2004
Asaphes vulgaris	Oryzacystatin I	No effect	Cowgill et al., 2004
Osmia bicornis	Canola Oryzacystatin I	No effect	Konrad et al., 2008, 2009
Myzus persicae	Potato Chicken egg white cystatin/oryzacystatin I	No effect	Cowgill et al., 2002
Eupteryx aurata	Potato Chicken egg white cystatin/oryzacystatin I	No effect	Cowgill and Atkinson, 2003
Scheloribates praeincisus	Sugarcane Soybean Bowman-Birk inhibitor/Soybean Kunitz inhibitor	No effect	Simoes et al., 2008

abundance of natural enemies in the Bt versus non-Bt crop fields are discussed here. A detailed investigation in the cotton fields indicated that the population densities of major predator species such as predatory spiders, coccinellids and chrysopids in transgenic Bt cotton fields were not different from those in conventional Bt cotton fields, but were significantly greater than those in conventional cotton fields with pesticides applied (Dhillon and Sharma, 2009; Sharma et al., 2007; Sisterson et al., 2007). In India, it was found that the predatory populations (*Chrysoperla* spp., *Orius* spp., *Coccinella* spp., *Brumus* spp., *Vespa* spp., *Lycosa* spp., and *Araneus* spp.) were similar on Bollgard I, Bollgard II and conventional cotton (Manna et al., 2010). A 3-year field survey showed that the planting of Bt cotton increased the diversity of arthropod communities (Cui et al., 2005; Men et al., 2003). It has been reported that there was no change in the biological control function of natural enemies in Bt cotton fields compared to conventional Bt cotton (Naranjo 2005b; Wolfenbarger et al., 2008; reviewed in Naranjo, 2009). In Germany, a 6-year monitoring of non-target arthropods in Bt maize (expressing the Cry1Ab toxin) showed no differences in the population densities of aphids, thrips, heteropterans, aphid-specific predators, spiders and carabids (Schorling and Freier, 2006). Abundance and species richness of foliage-dwelling spiders (Araneae) were equal or higher in Bt maize fields and adjacent field margins than in non-transgenic maize fields (Ludy and Lang, 2006). Based on a 3-year field investigation, Higgins et al. (2009) reported that there was no impact on the community abundance of non-target arthropods in transgenic Bt maize. Predaceous arthropods were as abundant or more abundant on Bt maize compared to non-Bt maize (Daly and Butin, 2005; Eizaguirre et al., 2006). No effects were found in Cry1Ab and Cry3Bb1 maize on the diversities of macroorganisms and microorganisms in long-term and short-term field studies in the USA (Icoz et al., 2008; Priestley and Brownbridge, 2009; Zeilinger et al., 2010). From a critical analysis of 240 questionnaires from a survey of farmers cultivating MON810 in six European countries in 2009, and through a detailed analysis of more than 30 publications, it has been concluded that there has been a negligible impact from the cultivation of MON810 expressing Cry1Ab on the biodiversity, abundance, or survival of non-target species (Monsanto, 2010). It can be concluded that the biodiversity of non-target arthropods was seemingly more easily affected by the agro-ecological system than the Cry toxin (De la Poza et al., 2005; Farinós et al., 2008). Critical concerns about the ecological risk assessment of transgenic crops still remain, especially for large scale deployment. The study of Lu et al. (2012) confirmed that there was no negative effect of Bt cotton on generalist predators in agricultural landscapes in China. More particularly, they have demonstrated a marked increase in generalist predator population levels and associated biocontrol services linked to decreased insecticide use owing to the widespread adoption of the Bt crop. Lu et al. (2012) provide a comprehensive, long-term and large-scale assessment of the possible ecological and agricultural effects of transgenic crops.

Although a large volume of literature has revealed the neutral or positive impact of Bt crops on the abundance of natural enemies, a few reports showed that the impact was negative. The population densities of parasitic wasps (*Trichogramma confusum*, *Microplitis* spp., *Campoletis chlorideae*, and *Meteorus pulchricornis*) decreased significantly due to poor quality and lower density of *H. armigera* in Bt cotton fields because of the close relationship between parasitoids and their hosts (Wu and Guo, 2005; Xia et al., 2007; Yang et al., 2005). A meta-analysis of data collected from various field studies of Bt-expressing maize and cotton worldwide indicated that the combined abundance of all non-target invertebrates was significantly lower in Bt compared with non-Bt crops, but that abundances were significantly higher in Bt crops if compared with non-Bt crops that had been treated with insecticides (Marvier et al., 2007).

GM crops effectively preserve local populations of various economically important biological control organisms and the only indirect effects on non-target organisms are local reductions in numbers of certain specialist monophagous parasitoids whose hosts are the primary targets of GM crops. Such density-dependent trophic effects will be associated with any effective pest control technology. If employed as part of an IPM philosophy, then GM crops can be very compatible with biological control (Lundgren et al., 2009).

18.5.1.3 Impact of GM Crops on other Non-Target Insects

The impact of GM crops expressing Bt proteins on the growth, development, survival and fecundity of pollinators has been studied either by using the insecticidal proteins directly in the bioassay or by feeding them the pollen of GM crops expressing the insecticidal proteins. The summary of results obtained is presented in Table 18.4, and indicates that the Cry proteins in general do not have any negative impact on the longevity, feeding and learning behaviour, and development of hypopharyngeal glands of honey bees. However, a 7-day oral exposure of cotton pollen expressing Cry1Ac + CpTI toxins disturbed the feeding behaviour of honey bees (Han et al., 2010a) and a lack of long-term exposure did not provide enough evidence to support this result. It has been confirmed that the pollen of Bt corn has adverse effects on the survival of monarch butterflies. However, several researchers reported that the effect of Bt maize on the monarch butterfly was negligible because of limited exposure and low toxicity of Bt maize pollen to monarch larvae (Table 18.4).

18.5.1.4 Impact of GM Crops on Soil Organisms

Bt toxins from GM crops can enter the soil in three different ways: (i) plant pollen deposited in and around Bt crop fields during anthesis, (ii) root exudates and (iii) plant residues after harvest (Heckmann et al., 2006; Li et al., 2007; Vaufleury et al., 2007; Zwahlen et al., 2007). After reaching the soil, Bt toxins can bind to clay particles and humic substances, which renders them resistant to biodegradation, but with retention of larvicidal activity (Clark et al., 2006; Saxena et al., 2010; Viktorov, 2008; Zwahlen et al., 2003). Bt proteins from GM crops breakdown relatively rapidly in the early stage after entering soil, and only a small amount may remain for long and hence, the Bt proteins do not bio-accumulate in the soil (Ahmad et al., 2005; Daudu et al., 2009; Icoz and Stotzky, 2008, Icoz et al., 2008; Li et al., 2007; Margarit et al., 2008; Rauschen et al., 2008; Shan et al., 2008; Zurbrugg et al., 2010). It has been reported by several workers that Cry proteins are not toxic to several soil macroorganisms such as mites, collembola, earthworm, and snails (Ahmad et al., 2005; Bai et al., 2010; Heckmann et al., 2006; Honemann et al., 2008; Honemann and Nentwig, 2009; Liu et al., 2009b; Vaufleury et al., 2007; Vercesi et al., 2006; Zwahlen et al., 2007). However, a laboratory study revealed the negative impact of Bt maize on growth and egg hatchability of snails. The risk involved was, however, not well-established (Kramarz et al., 2009).

Field investigations suggested that crop management practices and/or environmental conditions (e.g. heavy rainfall during the growing season) and pesticide use had the greatest impact on these species' diversity and evenness, rather than the crop itself (Bt or isoline) (Birch et al., 2007; Cortet et al., 2007; Griffiths et al., 2007a). Although Castaldini et al. (2005) identified differences in rhizospheric eubacterial communities, mycorrhizal colonization, soil respiration, bacterial communities,

TABLE 18.4 Impact of GM Crops on Other Non-Target Insects

Nature of Bioassay/Field Survey	Non-target Insect	Impact	Reference
Feeding tests with Bt plant pollen	Honey bees	No effect was observed on the longevity, feeding and learning behaviour, development of hypopharyngeal glands, superoxide dismutase activity, and intestinal bacterial communities of honey bees	Babendreier et al., 2005, 2007; Bailey et al., 2005; Han et al., 2010b; Hofs et al., 2008; Liu B. et al., 2005; Liu et al., 2009a; Rose et al., 2007
Seven-day oral exposure of cotton pollen-expressed Cry1Ac + CpTI toxins	Honey bees	Feeding behaviour of honey bees was found to be disturbed. A lack of long-term exposure did not provide enough evidence to support this result	Han et al., 2010a
Meta-analysis of 25 independent studies (Bt proteins used in GM crops to control lepidopteran and coleopteran pests)	Honey bees	There was no negative impact on the survival of honey bee larvae and adults	Duan et al., 2008a
Field survey	Honey bees	No effects were detected on the abundance, diversity, colony activity and development of honey bees	Hofs et al., 2008; Rose et al., 2007
Dusting of pollen from a Bt corn line onto milkweed leaves	Butterflies	Mortality of monarch larvae	Losy et al., 1999
Field survey	Butterflies	Effect of Bt maize on the monarch butterfly was negligible because of limited exposure and low toxicity of Bt maize pollen to monarch larvae	Hellmich et al., 2008; Sears et al., 2001; Stanley-Horn et al., 2001
Laboratory or glasshouse experiments where the test insects were artificially exposed to high levels of insecticidal Bt proteins	Butterflies	Some adverse effects were observed on the development, body weight, and larval behaviour of butterflies besides mortality	Lang and Vojtech, 2006; Mattila et al., 2005; Perry et al., 2010; Prasifka et al., 2007

and mycorrhizal establishment between Bt176 maize and non-Bt maize, the risk was not well-established. Rui et al. (2005) reported that the fortification of pure Bt toxin into rhizospheric soil did not result in significant changes in the numbers of culturable functional bacteria, apart from nitrogen-fixing bacteria when the concentration of Bt toxin was higher than 500 ng/g. The results indicated that Bt toxin was not the direct factor causing a decrease in the numbers of bacteria in the rhizosphere, and other factors may be involved. Hu et al. (2009) reported that there were no consistent statistically significant differences in the numbers of different groups of functional bacteria between rhizosphere soil of Bt and non-Bt cotton in the same field, and no obvious trends in the numbers of the various group of functional bacteria with the increased duration of Bt cotton cultivation. These studies suggest that multiple-year cultivation of transgenic Bt cotton may not affect the functional bacterial populations in rhizosphere soil.

Balachandar et al. (2008) studied the diversity richness of Pink-pigmented facultative methylotrophs (PPFMs) present in the phyllosphere,

TABLE 18.5 Cases of Field-Evolved Resistance

Pest	Country	GM Crop	Reference
Fall armyworm, *Spodoptera frugiperda*	Puerto Rico	Bt corn producing Cry1F	Matten, 2007; Matten et al., 2008
Cereal stem borer, *Busseola fusca*	South Africa	Bt corn producing Cry1Ab	Kruger et al., 2009
Pink bollworm, *Pectinophora gossypiella*	Western India and China	Bt cotton producing Cry1Ac	Bagla, 2010; Dhurua and Gujar, 2011; Wan et al., 2012; Tabashnik and Carrière, 2010
Cotton bollworm, *Helicoverpa zea*	South-eastern United States	Bt cotton producing Cry1Ac and Cry2Ab	Luttrell et al. 2004; Tabashnik et al., 2008, 2009
Bollworm, *Helicoverpa punctigera*	Australia	Bt cotton producing Cry1Ac and Cry2Ab	Downes et al., 2010
Western corn rootworm, *Diabrotica virgifera virgifera*	Iowa, USA	Bt corn expressing Cry3Bb1	Gassmann et al., 2011

rhizoplane and internal tissues of Bt and non-Bt cotton. They observed no differences between Bt- and non-Bt cotton. Sarkar et al. (2009) concluded from their study that there were no negative effects of Bt cotton on the soil quality indicators (microbial biomass carbon, microbial biomass nitrogen, microbial biomass phosphorus, total organic carbon, microbial quotient, potential N mineralization, nitrification, nitrate reductase, acid and alkaline phosphatase activities) and therefore cultivation of Bt cotton appears to pose no risk to soil ecosystem function. Kapur et al. (2010) assessed the culturable and non-culturable microbial diversities in Bt cotton and non-Bt cotton soils to determine the ecological consequences of application of Bt cotton. Their study indicated that Bt cotton had no effect on the diversity of microbial communities. Diversity of experimental fields was similar during the cropping of both Bt cotton and non-Bt cotton. There was no significant adverse effect of Bt crops on soil microorganisms in the laboratory, microcosm, and under field conditions (Baumgarte and Tebbe, 2005; Devare et al., 2007; Griffiths et al., 2005; Griffiths et al., 2007b; Knox et al., 2008; Oliveira et al., 2008; Miethling-Graff et al., 2010; Shen et al., 2006; Tan et al., 2010).

18.6 EVOLUTION OF RESISTANCE IN INSECTS TO GM CROPS

As farmers increasingly plant insect-resistant GM crops, selection pressure for the development of resistance to Bt toxins in target pests is also increasing. As insects are remarkably successful in adapting to toxins and other control tactics (Onstad et al., 2001; Carrière et al., 2005), the evolution of resistance to Bt toxins by target pests threatens the continued efficacy of Bt crops (Gould, 1998; Matten et al., 2008; Tabashnik, 1994; Tabashnik and Carrière 2009). Widespread planting of transgenic insecticidal crops favours the evolution of resistance (Gassmann et al., 2009). Although Bt cotton has helped to suppress pests, decrease insecticide sprays and promote biological control, evolution of resistance to Bt toxins by pests can diminish these benefits. To date, field-evolved resistance has been documented in populations of five lepidopteran species (Table 18.5) (Carrière et al., 2010). Initially rare genetic mutations that confer resistance to Bt toxins are becoming more common as a growing number of pest populations adapt to Bt crops. Rapid adoption of several crop species with incorporated Bt gene(s) has further simplified the management of insect pests. Yet exclusive use of Bt crops has accelerated the selection of resistant insects (Aldridge, 2008). Certainly, the selection pressure placed on pest populations to evolve resistance is more intense in this kind of crop because the pressure imposed is persistent instead of dependent on manual application.

Numerous insect strains have responded to laboratory selection by evolving greater resistance to Bt toxins (Ferré and van Rie, 2002; Gassmann et al., 2009; Tabashnik, 1994), and this is suggestive of the evolutionary potential of pests to evolve resistance to transgenic Bt crops (Gassmann et al., 2009). More importantly, analysis of field populations has revealed the presence of major resistance alleles for resistance to Bt crops in populations of pink bollworm *Pectinophora gossypiella* (Tabashnik et al., 2005), tobacco budworm *Heliothis virescens* (Gould et al., 1997), the corn earworm *Helicoverpa zea* (Burd et al., 2003), and the old-world bollworm *Helicoverpa armigera* (Gassmann et al., 2009; Wu et al., 2006).

18.6.1 Evolution of Resistance to Bt Cotton in Field Populations of *H. armigera* in China – A Case Study

The development of resistance in cotton bollworm, *H. armigera*, against Cry1Ac in China has provided an early warning of resistance to Bt cotton. Based on larval survival at a diagnostic concentration of Cry1Ac for 15 field populations of cotton bollworm sampled in 2010, the frequency of resistant individuals was 1.3% in northern China versus 0.0% in north western China. The most resistant population from Anyang in Henan province of northern China had 2.6% resistant individuals and a 16-fold increase in the concentration of Cry1Ac was needed to kill 50% of the larvae tested (LC_{50}) relative to a susceptible field population from north western China. The frequency of mutations conferring resistance to Cry1Ac was three times higher in northern China than in north western China. Baseline data showed that before Bt cotton was introduced in China, cotton bollworm susceptibility to Cry1Ac was not lower in northern China than in north western China, which implies that exposure to Bt cotton in northern China selected for resistance to Cry1Ac. In addition, populations in northern China were not resistant to a different Bt toxin, Cry2Ab, which implies that resistance to Cry1Ac in northern China is a specific adaptation to Bt cotton producing Cry1Ac. Moreover, the first generation Bt cotton carried only one Bt gene, i.e. *Cry1Ac*. This evidence may be considered as an early warning because the increased frequency of resistance is sufficiently small not yet to have caused major problems for cotton growers.

It has recently been discovered that, in this early stage of field-evolved resistance, diverse mutations confer resistance to Cry1Ac in cotton bollworm populations from northern China. To understand the genetic basis of resistance, researchers have selected for resistance in cotton bollworm by exposing them to Bt toxins in controlled laboratory experiments. The assumption implicit in this approach is that the genetic basis of resistance in the field will be similar to that found in lab-selected strains. It has been found that the most common mutation in field-selected populations of cotton bollworm from northern China was the same as a mutation that was detected with lab-selection. This recessively inherited mutation disrupts a cadherin protein that binds Cry1Ac in the larval gut of susceptible insects. Because binding to cadherin is required for toxicity, this mutation confers resistance. In addition, other recessive mutations have also been found in the cadherin gene that confers resistance to Cry1Ac. All of these mutations are similar to recessive mutations conferring resistance to Cry1Ac that were characterized previously in several species of lepidopteran pests. There were two dominant mutations, one in the cadherin gene and the other in a different gene that has not yet been identified. The discovery of dominantly inherited resistance is important, because dominant resistance is more difficult to manage than recessive resistance. One practical option for responding to the early warning of bollworm resistance to Cry1Ac in China would be switching to transgenic cotton that produces two Bt toxins, Cry1Ac and Cry2Ab.

18.6.2 High-Dose/Refuge Resistance Management Strategy

Resistance development in target insect pests is a major threat to the sustainable use of Bt crops. European corn borer, *Ostrinia nubilalis* (Hubner), south western corn borer, *Diatraea grandiosella* Dyar (both Lepidoptera: Crambidae), tobacco budworm, *Heliothis virescens* Fabricius (Lepidoptera: Noctuidae), and pink bollworm, *Pectinophora gossypiella* (Saunders) (Lepidoptera: Gelechiidae) – the four major target pests of Bt crops in the USA and Canada, remain susceptible to Bt toxins after 15 years of intensive use of Bt maize and Bt cotton. The success in sustaining susceptibility in these major pests is associated with successful implementation of the 'high-dose/refuge' insecticide resistance management (IRM) strategy (Huang et al. 2011). The 'high-dose/refuge' IRM strategy has three fundamental requirements. First, Bt crop cultivars should express a 'high dose' of Bt protein; second, the initial frequency of resistance alleles should be very low; and third, a refuge needs to be maintained nearby in the environment.

Bt plants expressing a sufficiently high concentration of Bt proteins to ensure ≥95% mortality of the heterozygous individuals carrying one copy of a major resistance allele is the prerequisite for the success of a high-dose/refuge strategy (Andow and Hutchison, 1998; US EPA, 2001). The US EPA Scientific Advisory Panel (SAP) on Bt Plant-Pesticides and Resistance Management (US EPA-SAP, 1998; US EPA, 2001) used empirical data to suggest that a working definition of high dose should be 'a dose 25 times the toxin concentration needed to kill Bt-susceptible larvae.' Based on a literature review, Caprio et al. (2000) suggested 50 times the toxin concentration needed to kill 50% of the Bt susceptible larvae as a better standard. This indirect definition of 'high dose' has been used to evaluate the high-dose status of Bt crops in the USA (US EPA, 2001). Bt cotton varieties that are commercially available in North America are presumed to meet the high-dose requirement of the IRM strategy for these four pests (US EPA, 2001). In addition, the frequencies of Bt resistance allele in field populations of *O. nubilalis, D. grandiosella, H. virescens*, and *P. gossypiella* in North America are very low and meet the rare resistance allele requirement of the 'high-dose/refuge' strategy. The refuge plan in the 'high-dose/refuge' IRM strategy has generally been implemented successfully for all Bt maize varieties in the USA and Canada. This relatively high compliance rate in the USA and Canada has undoubtedly contributed to the long-term success of Bt maize in North America. This has required a significant effort from government agencies, crop growers, extension personnel, biotech industries, scientists, and others who are interested in Bt technology (Huang et al. 2011).

Field resistance (including control failure) to Bt crops has been clearly documented in three situations *viz*., Fall armyworm (*S. frugiperda*) in Puerto Rico, African stem borer (*B. fusca*) in South Africa, and pink bollworm in India. The first case of field resistance to Bt crops is the resistance of *S. frugiperda* to Cry1F maize in Puerto Rico. Cry1F maize became commercially available in Puerto Rico in 2003 to control *S. frugiperda*. By 2006, there were reports of unexpected damage to Cry1F-maize by *S. frugiperda* in three regions on the island, which were identified as field resistance (US EPA 2007b). Laboratory bioassays in 2007 documented that insect populations collected from two of these regions (Santa Isabel and Salinas) were >160-fold more tolerant to Cry1F toxin than a laboratory susceptible strain (US EPA 2007b; Storer et al., 2010). As a consequence, Cry1F maize was withdrawn from commercial use in Puerto Rico (US EPA 2007a; Storer et al., 2010). The second case is the resistance of an African stem borer, *B. fusca* to Cry1Ab maize (e.g. Yield Gard maize) in South Africa. The third case of field resistance is in *P. gossypiella* to Cry1AC cotton in Gujarat, India. Gujarat is the second largest cotton-producing state in India with 2.4 million ha cultivated during 2007–2008 (Karihaloo and Kumar, 2009). During the 2008–2009 cropping season, unexpected survival of *P. gossypiella* in several Bt cotton fields was observed in Gujarat (Monsanto, 2010). Although there are still some disagreements, recent laboratory bioassays have confirmed that the field survival was associated with major resistance to Cry1Ac (Tabashnik and Carrière, 2010; Dhurua and Gujar, 2011). Factors associated with these cases of field resistance include failure to use high-dose Bt cultivars and lack of sufficient refuge (Huang et al. 2011). The rapid development of field resistance in *P. gossypiella* in India compared to the absence of resistance in this pest in the USA deserves attention. In the USA, Bt cotton is high dose, resistance is rare, and the refuge has been implemented. In India, Bt cotton may be a seed blend of high-dose and non-expressing plants, resistance may or may not have been rare, and the refuge has probably not been implemented. Clearly, successful implementation of the 'high-dose/refuge' strategy in the USA has delayed field resistance, and the absence of the 'high-dose/refuge' strategy in India has allowed rapid emergence of field resistance (Huang et al. 2011).

Bt cotton producing these two toxins is grown extensively in the United States and Australia. It would be better to switch to Bt cotton that produces additional toxins unrelated to Cry1Ac, such as Vip3A (a vegetative insecticidal protein from Bt), so that the increased frequency of resistance to Cry1Ac already seen in some populations would not reduce the efficacy of the new Bt cotton. The most widely adopted strategy for delaying pest resistance to Bt crops, the refuge strategy, works best when inheritance of resistance is recessive. Refuges consist of plants that do not have a Bt toxin gene and thus allow survival of insects that are susceptible to the toxin. Retrospective analysis of field data indicates that refuges have been particularly effective in slowing recessive resistance. For example, the refuge strategy delayed pink bollworm resistance to Bt cotton in Arizona for more than a decade. Conversely, with dominant resistance, the progeny of mating between resistant and susceptible insects can survive on Bt crops. Therefore, resistance evolves faster as seen with *Helicoverpa zea* resistance to Bt cotton in the south eastern United States and western corn rootworm resistance to Bt corn in Iowa (Gassmann et al., 2009).

It is widely thought that, as in China, India may not need structured refuges since both countries have small highly fragmented farms and different host crops for cotton bollworms that are cultivated alongside cotton, providing natural refuges for the cotton crop (Bambawale et al., 2010; Qiao et al., 2010). Results of the study conducted by Singla et al. (2012), however, support this hypothesis only for South India. In Central and North India, there is a need of structured refuges. On the other hand, in the USA, the optimal structured refugia found were 16% under the 11-year horizon (Livingston et al. 2004). In India, it was 33%, 8%, and 0% for North, Central, and South regions, respectively, under the 11-year horizon. The reason for the high requirement of structured refuges in the USA and North India might be the prevalence of monocropping patterns in these regions. In Central and South India, however, the cropping pattern was mostly multi-cropped. Results also suggested that planting sprayed refugia might be more profitable than planting unsprayed refugia. Sensitivity analysis revealed that the refuge requirements were sensitive to the initial Bt resistance level, the relative proportion of cotton bollworms in natural refuges, and the proportions of heterozygous and homozygous fitnesses in all three regions. Moreover, static refugia were found more profitable compared to dynamic refugia in the North and Central regions. The refuge requirement also varied with polyphagous/monophagous pest status. As compared to cotton bollworms, critical adoption of refugia is necessary in eggplant since the eggplant shoot and fruitborer are oligophagous. In future, to ensure the adherence of refugia as an essential component in IRM, steps must be taken by both GM crop developers and extension agents to inculcate the knowledge and importance of refugia among farmers.

18.7 CONCLUSIONS

Recombinant DNA technology has enabled the direct insertion of foreign genes derived from any kind of living organism into crop plants, allowing them to express new traits including pest-resistance properties. Though there was a concern about the cost of the technology during initial periods, the growing adoption of GM crops both in the developed and developing world will certainly reduce the cost of their development. The use of this approach has resulted in many insect-resistant varieties, and crops expressing Cry toxins derived from Bt have been planted worldwide. Thus far, the laboratory and field studies conducted have shown that the currently used insect-resistant Bt crops generally do not cause apparent unexpected detrimental effects on non-target organisms or their ecological functions, and in several instances, Bt crops increase the abundance of some beneficial insects and improve the natural control of specific pests due to the reduction in pesticide use. The use of Bt crops, such as Bt maize and Bt cotton, results in significant reductions of insecticide application, and clear benefits for the environment and farmers' health. Consequently, Bt crops can be a useful component of IPM systems to protect the crop from targeted pests. In fact, Bt cotton and Bt maize have revolutionized pest control strategies in a number of countries and have made a breakthrough in conventional IPM practices. Certainly, the sole use of Bt crops cannot solve all of the problems pertaining to pest management and they need to be used with other IPM tactics, including chemical pesticides. For example, to control secondary pests such as mirids and spider mites in Bt cotton, chemical control, especially the use of more specific, safer compounds, remains important together with the use of other IPM tactics such as crop rotation and intercropping.

Increasingly, the use of GM crops will require agronomists, ecologists, farmers and policy makers alike to take more of a systems perspective which considers the broader evolutionary consequences of the traits in question. Indeed, current transgenic crops based on Bt can solve lepidopteran or coleopteran pest problems, but secondary pests such as aphids can arise when competition from the target pest is relieved (Cloutier et al., 2008). Insect-resistant transgenic crops have been advertised as being the sole solution to pest problems, whereas IPM practitioners have found that a single solution is rarely sufficient and, moreover, is not durable. An indirect effect of the use of transgenic crops, i.e. the reduced application of broad-spectrum pesticides, will lead to increased abundance of natural enemies and might reduce the number or severity of secondary pest outbreaks and decrease the risk of resistance development because such enemies will also attack the target herbivores. It has already been demonstrated that currently available insect-resistant transgenic plants are compatible with other IPM approaches, such as biological, chemical and cultural control (Smith, 2005; Romeis et al. 2008b). In the future, stacking genes that confer direct toxicity against herbivores and indirect attraction of carnivores through volatile organic chemicals might further enhance the efficiency of pest control and further reduce the risk of pest resistance (Kos et al., 2009). With the adoption of GM crops for IPM, the potential positive factors, *viz.* control of key pests with high efficacy, no resurgence of target pests, non-target pests having the same or lower pest status, no adverse effects on natural enemies, preservation of ecosystem, low risk of pest resistance, preservation of ecological functions, and socio-economic benefits, are some of the compelling reasons for farmers to adopt GM crops. In general, it is advised to incorporate Bt crops as one tool in IPM that uses the best available combination of pest control tactics, including transgenic and conventionally bred host-plant resistance, biological control, crop rotation, and judicious application of insecticide sprays. IPM can extend the efficacy of Bt crops while promoting sustainable agriculture that limits pest damage, optimizes returns to growers, and preserves environmental quality.

More and more research is needed through integration of GM crops as the basic strategy such as conventional host-plant resistance and combining other technologies for successful management of pests, diseases and weeds in an agro-ecosystem. Since farmers remain the primary decision-makers in IPM programmes at the field level, it is worthwhile for the extension service agents of agro-ecosystems to provide access to the widest possible range of appropriate knowledge about the value of newer technologies and their impact on the ecosystem and as much information as possible on their characteristics, costs and optimal use within IPM strategies. Based on the experiences of farmers and the adoption of practices grounded on scientific principles, such technologies will be a part of IPM in the future. In the IPM-based system, alternation of methods/chemicals with different modes of action is advocated. Similarly, in GM crops, gene stacking is practised and newer molecular targets for combating insects are being researched (Mahon et al., 2012), and many successful examples are foreseeable. For example, RNAi-mediated knockdown of gene function has been reported in different insect orders, such as Lepidoptera, Hemiptera, Coleoptera, Diptera and Hymenoptera (Baum et al. 2007; Price and Gatehouse 2008). Under the condition that the dsRNA introduced is highly specific to certain insects to limit the negative effects on non-target organisms, this is a very promising approach for transgenic pest control. However, the approach has only recently been developed and will need to be further optimized before it can be used on a wide scale (Price and Gatehouse 2008).

Hence, there are many potential reasons to believe that current and future GM crops have the greatest potential in pest management when grown under IPM regimes. Responsible use of insect-resistant GM crops (with single gene/stacked genes) has a significant role in IPM to benefit economic, ecological and evolutionary components of sustainable crop production in the future.

References

Ahmad, A., Wilde, G.E., Zhu, K.Y., 2005. Detectability of coleopteran specific Cry3Bb1 protein in soil and its effect on non-target surface and below-ground arthropods. Environ. Entomol. 34, 385–394.

Ahuja, D.B., Ahuja, U.R., Kalyan, R.K., Sharma, Y.K., Dhandapani, A., Meena, P.C., 2009. Evaluation of different management strategies for *Lipaphis pseudobrassicae* (Davis) on *Brassica juncea*. Int. J. Pest Manag. 55 (1), 11–18.

Aldridge, S., 2008. Insect resistance: from mechanisms to management. Assoc. Br. Sci. Writers, Cited July 2009. Available at: <http://www.absw.org.uk/Briefings/insecticide resistance.htm>.

Alvarez-Alfageme, F., Martínez, M., Pascual-Ruiz, S., Castañera, P., Diaz, I., Ortego, F., 2007. Effects of potato plants expressing a barley cystatin on the predatory bug *Podisus maculiventris* via herbivorous prey feeding on the plant. Transgenic Res. 16, 1–13.

Alvarez-Alfageme, F., Ferry, N., Castanera, P., Ortego, F., Gatehouse, A.M.R., 2008. Prey mediated effects of Bt maize on fitness and digestive physiology of the red spider mite predator *Stethorus punctillum* Weise (Coleoptera: Coccinellidae). Transgenic Res. 17, 943–954.

Alvarez-Alfageme, F., Bigler, F., Romeis, J., 2011. Laboratory toxicity studies demonstrate no adverse effects of Cry1Ab and Cry3Bb1to larvae of *Adalia bipunctata* (Coleoptera: Coccinellidae): The importance of study design. Transgenic Res. 20, 467–479.

Andersson, M.S., de Vicente, C.M., 2010. Gene Flow between Crops and Their Wild Relatives. Johns Hopkins University, Baltimore, 564 pp.

Andow, D.A., Hutchison, W.D., 1998. Bt corn resistance management. In: Mellon, M., Rissler, J. (Eds.), Now or Never: Serious New Plans to Save a Natural Pest Control. Union of Concerned Scientists, Two Brattle Square, Cambridge, MA.

Arshad, M., Suhail, A., 2010. Studying the sucking insect pests community in transgenic Bt cotton. Int. J. Agric. Biol. 12, 764–768.

Ashouri, A., Michaud, D., Cloutier, C., 2001a. Recombinant and classically selected factors of potato plant resistance to the Colorado potato beetle, *Leptinotarsa decemlineata*, variously affect the potato aphid parasitoid *Aphidius nigripes*. Biocontrol 46, 401–418.

Ashouri, A., Michaud, D., Cloutier, C., 2001b. Unexpected effects of different potato resistance factors to the Colorado potato beetle (Coleoptera: Chrysomelidae) on the potato aphid (Homoptera: Aphididae). Environ. Entomol. 30, 524–532.

Babendreier, D., Kalberer, N.M., Romeis, J., Fluri, P., Mulligan, E., Bigler, F., 2005. Influence of Bt transgenic pollen, Bt-toxin and protease inhibitor (SBTI) ingestion on development of the hypopharyngeal glands in honeybees. Apidologie 36, 585–594.

Babendreier, D., Joller, D., Romeis, J., Bigler, F., Widmer, F., 2007. Bacterial community structures in honeybee intestines and their response to two insecticidal proteins. FEMS Microbiol. Ecol. 59, 600–610.

Bagla, P., 2010. Hardy cotton-munching pests are latest blow to GM crops. Science 327 (5972), 1439.

Bai, Y.Y., Yan, R.H., Ye, G.Y., Huang, F.N., Cheng, J.A., 2010. Effects of transgenic rice expressing *Bacillus thuringiensis* Cry1Ab protein on ground-dwelling collembolan community in postharvest seasons. Environ. Entomol. 39, 243–251.

Bailey, J., Scott Dupree, C., Harris, R., Tolman, J., Harris, B., 2005. Contact and oral toxicity to honey bees in Ontario, Canada. Apidologie 36, 623–633.

Bais, H.P., Weir, T.L., Perry, L.G., Gilroy, S., Vivanco, J., 2006. The role of root exudates in rhizosphere interactions with plants and other organisms. Annu. Rev. Plant Biol. 57, 233–266.

Balachandar, D., Raja, P., Nirmala, K., Rithyl, T.R., Sundaram, S.P., 2008. Impact of transgenic Bt-cotton on the diversity of pink-pigmented facultative methylotrophs. World J. Microbiol. Biotechnol. 24, 2087–2095.

Bambawale, O.M., Tanwar, P.K., Sharma, O.P., Bhosle, B.B., Lavekar, R.C., et al., 2010. Impact of refugia and integrated pest management on the performance of transgenic (*Bacillus thuringiensis*) cotton (*Gossypium hirsutum*). Indian J. Agric. Sci. 80, 730–736.

Baum, J.A., Bogaert, T., Clinton, W., Heck, G.R., Feldmann, P., Ilagan, O., et al. 2007. Control of coleopteran insect pests through RNA interference. Nat. Biotechnol. 25, 1322–1326.

Baumgarte, S., Tebbe, C.C., 2005. Field studies on the environmental fate of the Cry1Ab Bt-toxin produced by transgenic maize (MON810) and its effect on bacterial communities in the maize rhizosphere. Mol. Ecol. 14, 2539–2551.

Bell, H.A., Fitches, E.C., Down, R.E., Ford, L., Marris, G.C., Edwards, J.P., et al., 2001. Effect of dietary cowpea trypsin inhibitor (CpTI) on the growth and development of the tomato moth *Lacanobia oleracea* (Lepidoptera: Noctuidae) and on the success of the gregarious ectoparasitoid *Eulophus pennicornis* (Hymenoptera: Eulophidae). Pest Manag. Sci. 57, 57–65.

Bell, H.A., Down, R.E., Fitches, E.C., Edwards, J.P., Gatehouse, A.M.R., 2003. Impact of genetically modified potato expressing plant-derived insect resistance genes on the predatory bug *Podisus maculiventris* (Heteroptera: Pentatomidae). Biocontrol Sci. Technol. 13, 729–741.

Birch, A.N.E., Griffiths, B.S., Caul, S., Thompson, J., Heckmann, L.H., Krogh, P.H., et al., 2007. The role of laboratory, glasshouse and field scale experiments in understanding the interactions between genetically modified crops and soil ecosystems: A review of the ecogen project. Pedobiologia 51, 251–260.

Bouchard, É., Michaud, D., Cloutier, C., 2003a. Molecular interactions between an insect predator and its herbivore prey on transgenic potato expressing a cysteine proteinase inhibitor from rice. Mol. Ecol. 12, 2429–2437.

Bouchard, É., Cloutier, C., Michaud, D., 2003b. Oryzacystatin I expressed in transgenic potato induces digestive compensation in an insect natural predator via its herbivorous prey feeding on the plant. Mol. Ecol. 12, 2439–2446.

Bourguet, D., Chaufaux, J., Micoud, A., Delos, M., Naibo, B., Bombarde, F., et al., 2002. *Ostrinia nubilalis* parasitism and the field abundance of non-target insects in transgenic *Bacillus thuringiensis* corn (*Zea mays*). Environ. Biosaf. Res. 1, 49–60.

Brookes, G., Barfoot, P., 2006. Global impact of biotech crops: Socio-economic and environmental effects in the first ten years of commercial use. AgBio Forum 9, 139–151.

Brookes, G., Barfoot, P., 2012. Forthcoming. GM Crops: Global Socio-economic and Environmental Impacts 1996–2010. PG Economics Ltd, Dorchester, UK.

Burd, A.D., Gould, F., Bradley, J.R., Van Duyn, J.W., Moar, W.J., 2003. Estimated frequency of nonrecessive Bt resistance genes in bollworm, Helicoverpa zea (Boddie) (Lepidoptera: Noctuidae), in eastern North Carolina. J. Econ. Entomol. 96, 137–142.

Burgess, E.P.J., Philip, B.A., Christeller, J.T., Page, N.E.M., Marshall, R.K., Wohlers, M.W., 2008. Tri-trophic effects of transgenic insect-resistant tobacco expressing a protease inhibitor or a biotin-binding protein on adults of the predatory carabid beetle *Ctenognathus novaezelandiae*. J. Insect Physiol. 54, 518–528.

Campbell, P.L., Leslie, G.W., McFarlane, S.A., Berry, S.D., Rhodes, R., Van Antwerpen, R., et al., 2009. An investigation of IPM practices for pest control in sugarcane. Proc. S. Afr. Sugar Technol. Assoc. 82, 618–622.

Caprio, M.A., Sumerford, D.V., Simms, S.R., 2000. Evaluating transgenic plants for suitability in pest and resistance management programs. In: Lacey, L., Kaya, H. (Eds.), Field Manual of Techniques in Invertebrate Pathology. Kluwer Academic Publishers, Boston, pp. 805–828.

Carrière, Y., Ellers-Kirk, C., Kumar, K., Heuberger, S., Whitlow, M., et al., 2005. Long-term evaluation of compliance with refuge requirements for Bt cotton. Pest Manag. Sci. 61, 327–330.

Carrière, Y., Crowder, D.W., Tabashnik, B.E., 2010. Evolutionary ecology of insect adaptation to Bt crops. Evol. Appl. 3, 561–573.

Castaldini, M., Turrini, A., Sbrana, C., Benedetti, A., Marchionni, M., Mocali, S., et al., 2005. Impact of Bt corn on rhizospheric and soil eubacterial communities and on beneficial mycorrhizal symbiosis in experimental microcosms. Appl. Environ. Microb. 71, 6719–6729.

Cattaneo, M.G., Yafuso, C., Schmidt, C., Huang, C., Rahman, M., Olson, C., et al., 2006. Farm-scale evaluation of the impacts of transgenic cotton on biodiversity, pesticide use, and yield. Proc. Natl. Acad. Sci. USA 103, 7571–7576.

Chelliah, S., Heinrichs, E.A., 1980. Factors affecting insecticide-induced resurgence of the brown planthopper, *Nilaparvata lugens*, on rice. Environ. Entomol. 9, 773–777.

Chen, M., Zhao, J.Z., Shelton, A.M., Cao, J., Earle, E.D., 2008. Impact of single-gene and dual-gene Bt broccoli on the herbivore *Pieris rapae* (Lepidoptera: Pieridae) and its pupal endoparasitoid *Pteromalus puparum* (Hymenoptera: Pteromalidae). Transgenic Res. 17, 545–555.

Chen, M., Ye, G.Y., Liu, Z.C., Fang, Q., Hu, C., Peng, Y.F., et al., 2009. Analysis of Cry1Ab toxin bioaccumulation in a food chain of Bt rice, an herbivore and a predator. Exotoxicology 18, 230–238.

Chen, M., Shelton, A., Ye, G., 2011. Insect-resistant genetically modified rice in China: from research to commercialization. Annu. Rev. Entomol. 56, 81–101.

Clark, B.W., Prihoda, K.R., Coats, J.R., 2006. Subacute effects of transgenic Cry1Ab *Bacillus thuringiensis* corn litter on the isopods *Trachelipus rathkii* and *Armadillidium nasatum*. Env. Tox. Chem. 25, 2653–2661.

Cloutier, C., Boudreault, S., Michaud, D., 2008. Impact de pommes de terre resistantes au doryphore sur les arthropodes non vises: une meta-analyse des facteurs possiblement en cause dans lechec dune plante transgeniquet. Cahiers Agric. 17 (4), 388–394.

Cohen, J.I., 2005. Poorer nations turn to publicly developed GM crops. Nat. Biotechnol. 23, 27–33.

Conway, G.R., 1997. The Doubling Green Revolution: Food for All in the 21st Century. Penguin, London.

Cortet, J., Griffiths, B.S., Bohanec, M., Demsar, D., Andersen, M.N., Caul, S., et al., 2007. Evaluation of effects of transgenic Bt maize on microarthropods in a European multi-site experiment. Pedobiologia 51, 207–250.

Cowgill, S.E., Atkinson, H.J., 2003. A sequential approach to risk assessment of transgenic plants expressing protease inhibitors: effects on non target herbivorous insects. Transgenic Res. 12, 439–449.

Cowgill, S.E., Wright, C., Atkinson, H.J., 2002. Transgenic potatoes with enhanced levels of nematode resistance do not have altered susceptibility to non target aphids. Mol. Ecol. 11, 821–827.

Cui, J.J., Xia, J.Y., Luo, J.Y., Wang, C.Y., 2005. Studies on the diversity of insect community in the transgenic CrylAc plus CpTI cotton field. J. Henan Agric. Univ. 39, 151–154.

Dale, P.J., Clarke, B., Fontes, E.M., 2002. Potential for the environmental impact of transgenic crops. Nature Biotechnol. 20, 567–575.

Daly, T., Butin, G.D., 2005. Effect of *Bacillus thuringiensis* transgenic corn for lepidopteran control on non-target arthropods. Environ. Entomol. 34, 1292–1301.

Daudu, C.K., Muchaonyerwa, P., Mnkeni, P.N.S., 2009. Litterbag decomposition of genetically modified maize residues and their constituent *Bacillus thuringiensis* protein (Cry1Ab) under field conditions in the central region of the Eastern Cape, South Africa. Agric. Ecosyst. Environ. 134, 153–158.

Davidson, M.M., Butler, R.C., Wratten, S.D., Conner, A.J., 2006. Impacts of insect-resistant transgenic potatoes on the survival and fecundity of a parasitoid and an insect predator. Biol. Control 37, 224–230.

De la Poza, M., Pons, X., Farinos, G.P., Lopez, C., Ortego, F., Eizaguirre, M., et al., 2005. Impact of farm-scale Bt maize on abundance of predatory arthropods in Spain. Crop Prot. 24, 677–684.

Devare, M., Londono-R, L.M., Thies, J.E., 2007. Neither transgenic Bt maize (MON863) nor tefluthrin insecticide adversely affect soil microbial activity or biomass: A 3-year field analysis. Soil Biol. Biochem. 39, 2038–2047.

Dhaliwal, G.S., Singh, R., 2004. Host Plant Resistance to Insects: Concepts and Application. Panima Publishing Corporation, New Delhi.

Dhawan, A.K., Singh, S., Kumar, S., 2009. Integrated Pest Management (IPM) helps reduce pesticide load in cotton. J. Agric. Sci. Technol. 11, 599–611.

Dhillon, M.K., Sharma, H.C., 2009. Impact of Bt-engineered cotton on target and non-target arthropods, toxin flow through different trophic levels and seed cotton yield. Karnataka J. Agric. Sci. 22, 462–466.

Dhurua, S., Gujar, G.T., 2011. Field-evolved resistance to Bt toxin Cry1Ac in the pink bollworm, *Pectinophora gossypiella* (Saunders) (Lepidoptera: Gelechiidae), from India. Pest Manag. Sci. 67, 898–903.

Dively, G.P., Rose, R., Sears, M.K., Hellmich, R.L., Stanley-Horn, D.E., Calvin, D.D., et al. 2004. Effects on monarch butterfly larvae (Lepidoptera: Danaidae) after continuous exposure to Cry1Ab-expressing corn during anthesis. Env. Entomol. 33, 1116–1125.

Downes, S., Mahon, R.J., Rossiter, L., Kauter, G., Leven, T., et al., 2010. Adaptive management of pest resistance by *Helicoverpa* species (Noctuidae) in Australia to the Cry2Ab Bt toxin in Bollgard II® cotton. Evol. Appl. 3, 574–584.

Duan, J.J., Marvier, M., Huesing, J., Dively, G., Huang, Z.Y., 2008a. A meta-analysis of effects of Bt crops on honey bees (Hymenoptera: Apidae). PLoS ONE 3, 1–6.

Duan, J.J., Teixeira, D., Huesing, J.E., Jiang, C.J., 2008b. Assessing the risk to non-target organisms from Bt corn resistant to corn rootworms (Coleoptera: Chrysomelidae): Tier-I testing with *Orius insidiosus* (Heteroptera: Anthocoridae). Environ. Entomol. 37, 838–844.

Dyer, G.A., Serratos-Hernández, J.A., Perales, H.R., Gepts, P., Piñeyro-Nelson, A., Chávez, A., et al., 2009. Dispersal of transgenes through maize seed systems in Mexico. PLoS ONE 4 (5), 1–9.

Eizaguirre, M., Albajes, R., Lopez, C., Eras, J., Lumbierres, B., Pons, X., 2006. Six years after the commercial introduction of Bt maize in Spain: Field evaluation, impact and future prospects. Transgenic Res. 15, 1–12.

Ejeta, G., 2010. African green revolution needn't be a mirage. Science 327, 831–832.

Ellstrand, N.C., 2003. Dangerous Liaisons? When Cultivated Plants Mate with Their Wild Relatives. Johns Hopkins University, Baltimore.

Elzen, G.W., Hardee, D.D., 2003. United States Department of Agriculture – Agricultural Research Service research on managing insect resistance to insecticides. Pest Manag. Sci. 59, 770–776.

Fang, L., Qiang, F., Feng-Xiang, L., 2004. Biotypes of insect pests and their genetic mechanisms. Acta Entomol. Sin. 47 (5), 670–678.

Faria, C.A., Wäcker, F.L., Pritchar, J., Barrett, D.A., Turlings, T.C.J., 2007. High susceptibility of Bt maize to aphids enhances the performance of parasitoids of lepidopteran pests. PLoS ONE 2, e600.

Farinós, G.P., DelaPoza, M., Hernandez-Crespo, P., Ortego, F., Castanera, P., 2008. Diversity and seasonal phenology of the above-ground arthropods in conventional and transgenic maize crops in Central Spain. Biol. Control 44, 362–371.

Fedoroff, N.V., Battisti, D.S., Beachy, R.N., Cooper, P.J.M., Fischhoff, D.A., Hodges, C.N., et al., 2010. Radically rethinking agriculture for the 21st century. Science 327, 833–834.

Ferré, J., van Rie, J., 2002. Biochemistry and genetics of insect resistance to *Bacillus thuringiensis*. Annu. Rev. Entomol. 47, 501–533.

Ferry, N., Raemaekers, R.J., Majerus, M.E., Jouanin, L., Port, G., Gatehouse, J.A., et al., 2003. Impact of oilseed rape expressing the insecticidal cysteine protease inhibitor oryzacystatin on the beneficial predator *Harmonia axyridis* (multicoloured Asian ladybeetle). Mol. Ecol. 12, 493–504.

Ferry, N., Jouanin, L., Ceci, R., Mulligan, E.A., Emami, K., 2005. Impact of oilseed rape expressing the insecticidal serine protease inhibitor, mustard trypsin inhibitor-2 on the beneficial predator *Pterostichus madidus*. Mol. Ecol. 14, 337–349.

Ferry, N., Mulligan, E.A., Majerus, M.E.N., Gatehouse, A.M.R., 2007. Bi-trophic and tritrophic effects of Bt Cry3A transgenic potato on beneficial, non-target, beetles. Transgenic Res. 16, 795–812.

Finamore, A., Roselli, M., Britti, S., Monastra, G., Ambra, R., Turrini, A., et al., 2008. Intestinal and peripheral immune response to MON810 maize ingestion in weaning and old mice. J. Agric. Food Chem. 56, 11533–11539.

Fitt, G.P., 2008. Have Bt crops led to changes in insecticide use patterns and impacted IPM?. In: Romeis, J., Shelton, A.M., Kennedy, G.G. (Eds.), Progress in Biological Control. Volume 5. Integration of Insect-Resistant Genetically Modified Crops within IPM Programs. Springer Press, New York; Chapter 11.

García, M., Ortego, F., Castanera, P., Farinos, G.P., 2010. Effects of exposure to the toxin Cry1Ab through Bt maize fed-prey on the performance and digestive physiology of the predatory rove beetle *Atheta coriaria*. Biol. Control 55, 225–233.

Gassmann, A.J., Carrière, Y., Tabashnik, B.E., 2009. Fitness costs of insect resistance to *Bacillus thuringiensis*. Annu. Rev. Entomol. 54, 147–163.

Gassmann, A.J., Petzold-Maxwell, J.L., Keweshan, R.S., Dunbar, M.W., 2011. Field-evolved resistance to Bt maize by western corn rootworm. PLoS ONE 6, e22629.

Gliessman, S.R. (Ed.), 2000. Agroecosystem Sustainability – Developing Practical Strategies. CRC Press, Boca Raton, FL, USA.

Gould, F., 1998. Sustainability of transgenic insecticidal cultivars: integrating pest genetics and ecology. Annu. Rev. Entomol. 43, 701–726.

Gould, F., Anderson, A., Jones, A., Sumerford, D., Heckel, D.G., Lopez, J., 1997. Initial frequency of alleles for resistance to Bacillus thuringiensis toxins in field populations of *Heliothis virescens*. Proc. Natl. Acad. Sci. USA 94, 3519–3523.

Gouse, M., Pray, C., Schimmelpfenning, D., 2004. The distribution of benefits from Bt cotton adoption in South Africa. AgBioForum 7, 187–194.

Gouse, M., Pray, C., Schimmelpfennig, D., Kirsten, J., 2006. Three seasons of subsistence insect-resistant maize in South Africa. Have smallholders benefited? AgBioForum 9, 15–22.

Graham, J., Gordon, S.C., Smith, K., McNicol, R.J., McNicol, J.W., 2002. The effect of the cowpea trypsin inhibitor in strawberry on damage by vine weevil under field conditions. J. Hortic. Sci. Biotechnol. 77, 33–40.

Griffiths, B.S., Caul, S., Thompson, J., Birch, A.N.E., Scrimgeour, C., Andersen, M.N., et al., 2005. A comparison of soil microbial community structure, protozoa and nematodes in field plots of conventional and genetically modified maize expressing the *Bacillus thuringiensis* CryIAb toxin. Plant Soil 275, 135–146.

Griffiths, B.S., Caul, S., Thompson, J., Birch, A.N.E., Cortet, J., Andersen, M.N., et al., 2007a. Microbial and microfaunal community structure in cropping systems with genetically modified plants. Pedobiologia 51, 195–206.

Griffiths, B.S., Heckmann, L.H., Caul, S., Thompson, J., Scrimgeour, C., Krogh, P.H., 2007b. Varietal effects of eight paired lines of transgenic Bt maize and near isogenic non-Bt maize on soil microbial and nematode community structure. Plant Biotechnol. J. 5, 60–68.

Han, P., Niu, C.Y., Lei, C.L., Cui, J.J., Desneux, N., 2010a. Quantification of toxins in a Cry1Ac + CpTI cotton cultivar and its potential effects on the honey bee *Apis mellifera* L. Ecotoxicology 19, 1452–1459.

Han, P., Niu, C.Y., Lei, C.L., Cui, J.J., Desneux, N., 2010b. Use of an innovative T-tube maze assay and the proboscis extension response assay to assess sub lethal effects of GM products and pesticides on learning capacity of the honey bee *Apis mellifera* L. Ecotoxicology 19, 1612–1619.

Heckmann, L.H., Griffiths, B.S., Caul, S., Thompsom, J., Pusztai-Carey, M., Moar, W.J., et al., 2006. Consequences for *Protaphorura armata* (Collembola: Onychiuridae) following exposure to genetically modified *Bacillus thuringiensis* (Bt) maize and non-Bt maize. Environ. Pollut. 142, 212–216.

Hellmich, R.L., Albajes, R., Bergvinson, D., Prasifka, J.R., Wang, Z.Y., Weiss, M.J., 2008. The present and future role of insect-resistant genetically modified maize in IPM. In: Romeis, J., Shelton, A.M., Kennedy, G.G. (Eds.), Integration of Insect-Resistant Genetically Modified Crops with IPM Programs. Springer, Berlin, Germany.

Higgins, L.S., Babcock, J., Neese, P., Layton, R.J., Moellenbeck, D.J., Storer, N., 2009. Three-year field monitoring of Cry1F, Event DAS-Ø15Ø7-1, maize hybrids for non-target arthropod effects. Environ. Entomol. 38, 281–292.

Hilbeck, A., Schmith, J.E.U., 2006. Another view on Bt proteins – How specific are they and what else might they do? Biopestic. Int. 2 (1), 1–50.

Ho, P., Xue, D.Y., 2008. Farmers' perceptions and risks of agro-biotechnological innovations in China: Ecological change in Bt cotton? Int. J. Environ. Sustain. Dev. 7, 396–417.

Hofs, J.L., Schoeman, A.S., Pierre, J., 2008. Diversity and abundance of flower-visiting insects in Bt and non-Bt cotton fields of Maputaland (KwaZulu Natal Province, South Africa). Int. J. Trop. Insect Sci. 28, 211–219.

Honemann, L., Nentwig, W., 2009. Are survival and reproduction of *Enchytraeus albidus* (Annelida: Enchytraeidae) at risk by feeding on Bt-maize litter? Eur. J. Soil Biol. 45 351–335.

Honemann, L., Zurbrugg, C., Nentwig, W., 2008. Effects of Bt-corn decomposition on the composition of the soil meso- and macrofauna. Appl. Soil Ecol. 40, 203–209.

Hu, H.Y., Xiao, X.L., Zhang, W.Z., Sun, J.G., Zhang, Q.W., Liu, X.Z., et al. 2009. Effects of repeated cultivation of transgenic Bt cotton on functional bacterial populations in rhizosphere soil. World J. Microbiol. Biotechnol. 25, 357–366.

Huang, J., Hu, R., Scott, R., Pray, C., 2005. Insect-resistant GM rice in farmers' fields: Assessing productivity and health effects in China. Science 308 (5722), 688–690.

Huang, F., Andow, D.A., Buschman, L.L., 2011. Success of the high-dose/refuge resistance management strategy after 15 years of Bt crop use in North America. Entomol. Exp. Appl. 140, 1–16.

Hutchinson, W.D., Burkness, E.C., Mitchell, P.D., Moon, R.D., Leslie, T.W., Fleischer, S.J., et al., 2010. Areawide suppression of European corn borer with Bt maize reaps savings to non-Bt maize growers. Science 330, 222–225.

Icoz, I., Stotzky, G., 2008. Cry3Bb1 protein from *Bacillus thuringiensis* in root exudates and biomass of transgenic corn does not persist in soil. Transgenic Res. 17, 609–620.

Icoz, I., Saxena, D., Andow, D.A., Zwahlen, C., Stotzky, G., 2008. Microbial populations and enzyme activities in soil in situ under transgenic corn expressing Cry proteins from *Bacillus thuringiensis*. J. Environ. Qual. 37, 647–662.

International Cotton Advisory Committee (ICAC), 1998a. Cotton World Statistics. ICAC, Washington, USA.

International Cotton Advisory Committee (ICAC), 1998b. Survey of the Costs of Production of Raw Cotton. ICAC, Washington, USA.

James, C., 2009. ISAAA Brief No.41. Global Status of Commercialized Biotech/GM Crops: 2009. ISAAA, Ithaca, NY.

James, C., 2010. ISAAA Brief No.42. Global Status of Commercialized Biotech/GM Crops: 2010. ISAAA, Ithaca, NY.

James, C., 2011. ISAAA Brief No.43. Global Status of Commercialized Biotech/GM Crops: 2011. ISAAA, Ithaca, NY.

James, C., 2012. ISAAA Brief No.44. Global Status of Commercialized Biotech/GM Crops: 2012. ISAAA, Ithaca, NY.

Kapur, M., Bhatia, R., Pandey, G., Pandey, J., Paul, D., Jain, R.K., 2010. A case study for assessment of microbial community dynamics in genetically modified Bt cotton crop fields. Curr. Microbiol. 61, 118–124.

Karihaloo, J.L., Kumar, P.A., 2009. Bt Cotton in India – A Status Report (2nd edn). Asia-Pacific Consortium on Agricultural Biotechnology (APCoAB), New Delhi, India, pp. 1–56.

Kennedy, G.G., 2008. Integration of insect-resistant genetically modified crops within IPM programs. In: Romeis, J., Shelton, A.M., Kennedy, G.G. (Eds.), Integration of Insect-Resistant Genetically Modified Crops with IPM Programs. Springer, Berlin, Germany, pp. 1–26.

Kennedy, G.G., Barbour, J.D., 1992. Resistance variation in natural and managed systems. In: Fritz, R.S., Simms, E.L. (Eds.), Plant Resistance to Herbivores and Pathogens. The University of Chicago Press, Chicago, pp. 13–41.

Kim, Y.H., Kang, J.S., Kim, J.I., Kwon, M., Lee, S., Cho, H.S., et al., 2008. Effects of Bt transgenic Chinese cabbage on the herbivore *Mamestra brassicae* (Lepidoptera: Noctuidae) and its parasitoid *Microplitis mediator* (Hymenoptera: Braconidae). J. Econ. Entomol. 101 (4), 1134–1139.

Knox, O.G.G., Nehl, D.B., Mor, T., Roberts, G.N., Gupta, V.V.S.R., 2008. Genetically modified cotton has no effect on arbuscular mycorrhizal colonisation of roots. Field Crops Res. 109, 57–60.

Kogan, M., 1998. Integrated pest management: Historical perspectives and contemporary developments. Annu. Rev. Entomol. 43, 243–270.

Kos, M., van Loon, J.J.A., Dicke, M., Vet, L.E.M., 2009. Transgenic plants as vital components of integrated pest management. Trends Biotechnol. 27, 621–627.

Kramarz, P., de Vaufleury, A., Gimbert, F., Cortet, J., Tabone, E., Andersen, M.N., et al., 2009. Effects of Bt-maize material on the life cycle of the land snail *Cantareus asperses*. Appl. Soil Ecol. 42, 236–242.

Kranthi, K.R., 2012. Bt Cotton Questions and Answers. Indian Society for Cotton Improvement, 70 pp.

Kranthi, K.R., Jadhav, D.R., Kranthi, S., Wanjari, R.R., Ali, S.S., Russell, D.A., 2002. Insecticide resistance in five major insect pests of cotton in India. Crop Prot. 21 (6), 449–460.

Kruger, M.J., Van Rensburg, J.B., Van den Berg, J., 2009. Perspective on the development of stem borer resistance to Bt maize and refuge compliance at the Vaalharts irrigation scheme in South Africa. Crop Prot. 28, 684–689.

Lang, A., Vojtech, E., 2006. The effects of pollen consumption of transgenic Bt maize on the common swallowtail, *Papilio machaon* L. (Lepidoptera, Papilionidae). Basic Appl. Ecol. 7, 296–306.

Lawo, N.C., Wackers, F.L., Romeis, J., 2010. Characterizing indirect prey-quality mediated effects of a Bt crop on predatory larvae of the green lacewing, *Chrysoperla carnea*. J. Insect Physiol. 56, 1702–1710.

Lei T., Khan M., Wilson L., 2003. Boll damage by sucking pests: An emerging threat, but what do we know about it? In: Swanepoel A (Ed.), World Cotton Research Conference: Cotton for the New Millennium. Agricultural Research Council–Institute for Industrial Crops, Cape Town, South Africa, pp. 1337–1344.

Li, Y.H., Romeis, J., 2010. Bt maize expressing Cry3Bb1 does not harm the spider mite, *Tetranychus urticae*, or its ladybird beetle predator, *Stethorus punctillum*. Biol. Control 53, 337–344.

Li, Y.H., Wu, K.M., Zhang, Y.J., Yuan, G.H., 2007. Degradation of Cry1Ac protein within transgenic *Bacillus thuringiensis* rice tissues under field and laboratory conditions. Environ. Entomol. 36, 1275–1282.

Liu, B., Xu, C., Yan, F., Gong, R., 2005. The impacts of the pollen of insect-resistant transgenic cotton on honey bees. Biodiv. Conserv. 14, 3487–3496.

Liu, B., Shu, C., Xue, K., Zhou, K.X., Li, X.G., Liu, D.D., et al., 2009a. The oral toxicity of the transgenic Bt + CpTI cotton pollen to honeybees (*Apis mellifera*). Ecotox. Environ. Saf. 72, 1163–1169.

Liu, B., Wang, L., Zeng, Q., Meng, J., Hu, W.J., Li, X.G., et al., 2009b. Assessing effects of transgenic Cry1Ac cotton on the earthworm *Eisenia fetida*. Soil Biol. Biochem. 41, 1841–1846.

Liu, X.X., Sun, C.G., Zhang, Q.W., 2005a. Effects of transgenic Cry1A + CpTI cotton and Cry1Ac toxin on the parasitoid, *Campoketis chlorideae* (Hymenoptera: Ichneumonidae). Insect Sci. 12, 101–107.

Liu, X.X., Zhang, Q.W., Zhao, J.Z., Li, J.C., Xu, B.L., Ma, X.M., 2005b. Effects of Bt transgenic cotton lines on the cotton bollworm parasitoid *Microplitis mediator* in the laboratory. Biol. Control 35, 134–141.

Liu, X.X., Chen, M., Onstad, D., Roush, R., Shelton, A.M., 2011. Effect of Bt broccoli and resistant genotype of *Plutella xylostella* (Lepidoptera: Plutellidae) on development and host acceptance of the parasitoid *Diadegma insulare* (Hymenoptera: Ichneumonidae). Transgenic Res. 20, 887–897.

Livingston, M.J., Carlson, G.A., Fackler, P.L., 2004. Managing resistance evolution in two pests to two toxins with refuge. Am. J. Agric. Econ. 86, 1–13.

Losy, J.E., Rayor, L.S., Cater, M.E., 1999. Transgenic pollen harms monarch larvae. Nat. Biotechnol. 399, 214.

Lövei, G.L., Andow, D.A., Arpaia, S., 2009. Transgenic insecticidal crops and natural enemies: a detailed review of laboratory studies. Environ. Entomol. 38 (2), 293–306.

Lu, Y., Wu, K., Jiang, Y., Guo, Y., Desneux, N., 2012. Widespread adoption of Bt cotton and insecticide decrease promotes biocontrol services. Nature 487 (7407), 362–365.

Lu, Y.H., Qiu, F., Feng, H.Q., Li, H.B., Yang, Z.C., Wyckhuys, K.A.G., et al., 2008. Species composition and seasonal abundance of pestiferous plant bugs (Hemiptera: Miridae) on Bt cotton in China. Crop Prot. 27, 465–472.

Lu, Y.H., Wu, K.M., Jiang, Y.Y., Xia, B., Li, P., Feng, H.Q., et al., 2010a. Mirid bug outbreaks in multiple crops correlated with wide-scale adoption of Bt cotton in China. Science 328, 1151–1154.

Lu, Y.H., Qi, F.J., Zhang, Y.J., 2010b. Integrated Management of Diseases and Insect Pests in Cotton. Golden Shield Press.

Lundgren, J.G., Gassmann, A., Bernal, J., Duan, J.J., Ruberson, J., 2009. Ecological compatibility of GM crops and biological control. Crop. Prot. 28, 1017–1030.

Ludy, C., Lang, A., 2006. A 3-year field-scale monitoring of foliage-dwelling spiders (Araneae) in transgenic Bt maize fields and adjacent field margins. Biol. Control 38, 314–324.

Mahon, R., Downes, S.J., James, B., 2012. Vip3A resistance alleles exist at high levels in Australian targets before release of cotton expressing this toxin. PLoS ONE 7 (6), e39192.

Manna, R.S., Gill, R.S., Dhawan, A.K., Shera, P.S., 2010. Relative abundance and damage by target and non-target insects on Bollgard and Bollgard II cotton cultivars. Crop Prot. 29, 793–801.

Margarit, E., Reggiardo, M.I., Permingeat, H.R., 2008. Bt protein rhizo secreted from transgenic maize does not accumulate in soil. Electron. J. Biotechnol. 11, 1–10.

Marvier, M., McCreedy, C., Regetz, J., Kareiva, P., 2007. A meta-analysis of effects of Bt cotton and maize on non-target invertebrates. Science 316, 1475–1477.

Matten, S., 2007. Review of Dow AgroSciences (and Pioneer HiBreds) Submission (dated July 12, 2007) Regarding Fall Armyworm Resistance to the Cry1F Protein Expressed in TC1507 Herculex I Insect Protection Maize in Puerto Rico. (EPA registrations 68467-2 and 29964-3). Memorandum from S.R. Matten, USEPA/OPP/BPPD to M. Mendelsohn, USEPA/OPP/BPPD, 24 August 2007.

Matten, S.R., Head, G.P., Quemada, H.D., 2008. How governmental regulation can help or hinder the integration of Bt crops within IPM programs. In: Romeis, J., Shelton, A.M., Kennedy, G.G. (Eds.), Integration of Insect Resistant Genetically Modified Crops within IPM Programs. Springer, New York, pp. 27–39.

Mattila, H.R., Sears, M.K., Duan, J.J., 2005. Response of *Danaus plexippus* to pollen of two new Bt corn events via laboratory bioassay. Entomol. Exp. Appl. 116, 31–41.

Mehrotra, K.N., 1989. Pesticide resistance in insect pests: Indian scenario. Pestic. Res. J. 1 (2), 95–103.

Meissle, M., Romeis, J., 2009. The web-building spider *Theridion impressum* (Araneae: Theridiidae) is not adversely affected by Bt maize resistant to corn rootworms. Plant Biotechnol. J. 7, 645–656.

Men, X., Ge, F., Liu, X., Yardim, E.N., 2003. Diversity of arthropod communities in transgenic Bt cotton and non-transgenic cotton agro ecosystems. Environ. Entomol. 32, 270–275.

Miethling-Graff, R., Dockhorn, S., Tebbe, C.C., 2010. Release of the recombinant Cry3Bb1 protein of Bt maize MON88017 into field soil and detection of effects on the diversity of rhizosphere bacteria. Eur. J. Soil Biol. 46, 41–48.

Monsanto, 2010. Annual monitoring report on the cultivation of MON 810 in 2009. Monsanto Europe, Brussels, Belgium. Available at: <http://ec.europa.eu/food/food/biotechnology/docs/annual_monitoring_report_MON810_2009_en.pdf>.

Mulligan, E.A., Ferry, N., Jouanin, L., Walters, K.F.A., Port, G.R., Gatehouse, A.M.R., 2006. Comparing the impact of conventional pesticide and use of a transgenic pest-resistant crop on the beneficial carabid beetle *Pterostichus melanarius*. Pest Manag. Sci. 62, 999–1012.

Nagrare, V.S., Kranthi, S., Biradar, V.K., Zade, N.N., Sangode, V., Kakde, G., et al., 2009. Widespread infestation of the exotic mealybug species, *Phenacoccus solenopsis* (Tinsley) (Hemiptera: Pseudococcidae), on cotton in India. Bull. Entomol. Res. 99, 537–541.

Naranjo, S.E., 2005a. Long-term assessment of the effects of transgenic Bt cotton on the function of the natural enemy community. Environ. Entomol. 34, 1211–1223.

Naranjo, S.E., 2005b. Long-term assessment of the effects of transgenic Bt cotton on the abundance of non target arthropod natural enemies. Environ. Entomol. 34, 1193–1210.

Naranjo S.E., 2009. Impacts of Bt crops on non-target invertebrates and insecticide use patterns. CAB Review: Perspectives in Agriculture, Veterinary Science, Nutrition and Natural Resources 4, 1–11.

Naranjo, S.E., Ruberson, J.R., Sharma, H.C., Wilson, L., Wu, K.M., 2008. The present and future role of insect-resistant genetically modified cotton in IPM. In: Romeis, J., Shelton, A.M., Kennedy, G.G. (Eds.), Integration of Insect-Resistant Genetically Modified Crops With IPM Programs. Springer, Berlin, Germany, pp. 159–194.

Naved, S., Trivedi, T.P., Singh, J., Sardana, H.R., Dhandapani, A., Sohi, A.S., et al., 2008a. Development and promotion of farmers participatory IPM technology in irrigated cotton cropping system – a case study – I. Pestic. Res. J. 20, 38–42.

Naved, S., Trivedi, T.P., Singh, J., Sardana, H.R., Dhandapani, A., 2008b. Development and promotion of IPM package in rainfed Bt cotton and non Bt cotton cropping system – a case study – II. Pestic. Res. J. 20, 43–47.

Oliveira, A.P., Pampulha, M.E., Bennett, J.P., 2008. A two-year field study with transgenic *Bacillus thuringiensis* maize: Effects on soil microorganisms. Sci. Total Environ. 405, 351–357.

Onstad, D.W., Spencer, J.L., Guse, C.A., Levine, E., Isard, S.A., 2001. Modeling evolution of behavioral resistance by an insect to crop rotation. Entomol. Exp. Appl. 100, 195–201.

Orr, A., Ritchie, J.M., 2004. Learning from failure: smallholder farming systems and IPM in Malawi. Agric. Syst. 79, 31–54.

Perry, J.N., Devos, Y., Arpaia, S., Bartsch, D., Gathmann, A., Hails, R.S., et al., 2010. A mathematical model of exposure of non-target lepidoptera to Bt-maize pollen expressing Cry1Ab within Europe. Proc. Biol. Sci. 277, 1417–1425.

Prasifka, P.L., Hellmich, R.L., Prasifka, J.R., Lewis, L.C., 2007. Effects of Cry1Ab-expressing corn anthers on the movement of monarch butterfly larvae. Environ. Entomol. 36, 228–233.

Price, D.R.G., Gatehouse, J.A., 2008. RNAi-mediated crop protection against insects. Trends Biotechnol. 26, 393–400.

Priestley, A.L., Brownbridge, M., 2009. Field trials to evaluate effects of Bt-transgenic silage corn expressing the Cry1Ab insecticidal toxin on non-target soil arthropods in northern New England, USA. Transgenic Res. 18, 425–443.

Qiao, F., Huang, J., Rozelle, S., Wilen, J., 2010. Natural refuge crops, buildup of resistance, and zero-refuge strategy for Bt cotton in China. Sci. China Life Sci. 53, 1227–1238.

Ramirez-Romero, R., Bernal, J.S., Chaufaux, J., Kaiser, L., 2007. Impact assessment of Bt-maize on a moth parasitoid, *Cotesia marginiventris* (Hymenoptera: Braconidae), via host exposure to purified Cry1Ab protein or Bt-plants. Crop Prot. 26, 953–962.

Rapusas, H.R., Santiago, S.E., Ramos, J.M., Roguel, S.M., Miller, S.A., Hammig, M., et al., 2009. Integrated pest management (IPM) technologies: assessment of implementation in rice vegetable cropping systems. Philipp. Entomol. 23, 67–94.

Rauschen, S., Nguyen, H.T., Schuphan, I., Jehle, J.A., Eber, S., 2008. Rapid degradation of the Cry3Bb1 protein from *Diabrotica*-resistant Bt-corn Mon88017 during ensilation and fermentation in biogas production facilities. J. Sci. Food Agric. 88, 1709–1715.

Reissig, W.H., Heinrichs, E.A., Valencia, S.L., 1982a. Insecticide-induced resurgence of the brown planthopper, *Nilaparvata lugens* on rice varieties with different levels of resistance. Environ. Entomol. 11, 165–168.

Reissig, W.H., Heinrichs, E.A., Valencia, S.L., 1982b. Effects of insecticides on Nilaparvata lugens and its predators: spiders, *Microvelia atrolineata* and *Cyrtorhinus lividipennis*. Environ. Entomol. 11, 193–199.

Romeis, J., Meissle, M., Bigler, F., 2006. Transgenic crops expressing *Bacillus thuringiensis* toxins and biological control. Nat. Biotechnol. 24, 63–71.

Romeis, J., Shelton, A.M., Kennedy, G.G., 2008b. Integration of Insect-Resistant Genetically Modified Crops within IPM Programs. Springer, Berlin.

Rose, R., Dively, G.P., Pettis, J., 2007. Effects of Bt corn pollen on honeybees: emphasis on protocol development. Apidologie 38, 368–377.

Rui, Y.K., Yi, G.X., Zhao, J., Wang, B.M., Li, J.H., Zhai, Z.X., et al. 2005. Changes of Bt toxin in the rhizosphere of transgenic Bt cotton and its influence on soil functional bacteria. World J. Microbiol. Biotechnol. 21, 1279–1284.

Saini, R.K., Jaglan, R.S., 1998. Adoption of insect-pest control practices by Haryana farmers in cotton: A survey. J. Cotton Res. Dev. 12 (2), 198.

Sanders, C.J., Pell, J.K., Poppy, G.M., Raybould, A., Garcia-Alonso, M., Schuler, T.H., 2007. Host-plant mediated effects of transgenic maize on the insect parasitoid *Campoletis sonorensis* (Hymenoptera: Ichneumonidae). Biol. Control 40, 362–369.

Sarkar, B., Patra, A.K., Purakayastha, T.J., Megharaj, M., 2009. Assessment of biological and biochemical indicators in soil under transgenic Bt and non-Bt cotton crop in a sub-tropical environment. Environ. Monit. Assess. 156, 595–604.

Saxena, D., Pushalkar, S., Stotzky, G., 2010. Fate and effects in soil of Cry proteins from *Bacillus thuringiensis:* Influence of physico chemical and biological characteristics of soil. Open Toxicol. J. 3, 133–153.

Schorling, M., Freier, B., 2006. Six-year monitoring of non-target arthropods in Bt maize (Cry 1Ab) in the European corn borer (*Ostrinia nubilalis*) infestation area Oderbruch (Germany). J. Verbr. Lebensm. (Suppl, 1), 106–108.

Schubert, D., 2002. A different perspective on GM food. Nat. Biotechnol. 20, 969.

Schuler, T.H., Denholm, I., Jouanin, L., Clark, S.J., Clark, A.J., Poppy, G.M., 2001. Population-scale laboratory studies of the effect of transgenic plants on non-target insects. Mol. Ecol. 10, 1845–1853.

Sears, M.K., Hellmich, R.L., Stanley-Horn, D.E., Oberhauser, K.S., Pleasants, J.M., Mattila, H.R., et al., 2001. Impact of Bt corn pollen on monarch butterfly populations: A risk assessment. Proc. Natl. Acad. Sci. USA 98, 11937–11942.

Shan, G.M., Embrey, S.K., Herman, R.A., McCormick, R., 2008. Cry1F protein not detected in soil after three years of transgenic Bt corn (1507 corn) use. Environ. Entomol. 37, 255–262.

Sharma, H.C., Ortiz, R., 2002. Host plant resistance to insects: an eco-friendly approach for pest management and environment conservation. J. Environ. Biol. 23, 111–135.

Sharma, H.C., Arora, R., Pampapathy, G., 2007. Influence of transgenic cottons with *Bacillus thuringiensis* Cry1Ac gene on the natural enemies of *Helicoverpa armigera.* BioControl 52, 469–489.

Sharma, O.P., Bambawale, O.M., Dhamdapani, A., Tanwar, R.K., Bhosle, B.B., Lavekar, R.C., et al., 2005. Assessing of severity of important diseases of rainfed Bt transgenic cotton in southern Maharashtra. Indian Phytopathol. 58, 483–485.

Sharma, O.P., Garg, D.K., Trivedi, T.P., Satpal, C., Singh, S.P., 2008. Evaluation of pest management strategies in organic and conventional taraori basmati rice. Indian J. Agric. Sci. 78 (10), 862–867.

Shen, R.F., Cai, H., Gong, W.H., 2006. Transgenic Bt cotton has no apparent effect on enzymatic activities or functional diversity of microbial communities in rhizosphere soil. Plant Soil 285, 149–159.

Simoes, R.A., Silva-Filho, M.C., Moura, D.S., Delalibera Jr, I., 2008. Effects of soybean proteinase inhibitors on development of the soil mite *Scheloribates praeincisus* (Acari: Oribatida). Exp. Appl. Acarol. 44, 239–248.

Singla, R., Johnson, P., Misra, S., 2012. Examination of regional-level efficient refuge requirements for Bt cotton in India. AgBioForum 15 (3), 303–314.

Sisterson, M.S., Biggs, R.W., Manhardt, N.M., Carrière, Y., Dennehy, T.J., Tabashnik, B.E., 2007. Effects of transgenic Bt cotton on insecticide use and abundance of two generalist predators. Entomol. Exp. Appl. 124, 305–311.

Smith, C.M., 2005. Plant Resistance to Arthropods: Molecular and Conventional Approaches. Springer Science and Business Media, Dordrecht, the Netherlands.

Spiroux de Vendomois, J., Roullier, F., Cellier, D., Séralini, G.-E., 2009. A comparison of the effects of three GM corn varieties on mammalian health. Int. J. Biol. Sci. 5 (7), 706–726.

Stanley-Horn, D.E., Dively, G.P., Hellmich, R.L., Mattila, H.R., Sears, M.K., Rose, R., et al., 2001. Assessing the impact of Cry1Ab-expressing corn pollen on monarch butterfly larvae in field studies. Proc. Natl. Acad. Sci. USA 98, 11931–11936.

Stern, V.M., 1973. Economic thresholds. Annu. Rev. Entomol. 18, 259–280.

Stern, V.M., Smith, R.F., Bosch van den, R., Hagen, K.S., 1959. The integrated control concept. Hilgardia 29, 81–101.

Stone, J.D., Pedigo, L.P., 1972. Development of economic injury level of the green clover worm on soybean in Iowa. J. Econ. Entomol. 65 (1), 197–201.

Storer, N.P., Babcock, J.M., Schlenz, M., Meade, T., Thompson, G.D., et al., 2010. Discovery and characterization of field resistance to Bt maize: *Spodoptera frugiperda* (Lepidoptera: Noctuidae) in Puerto Rico. J. Econ. Entomol. 103, 1031–1038.

Stylen, E., 1968. Threshold in the economics of insect pest control in agriculture. State Varts Mediterranean 14, 65–79.

Tabashnik, B.E., 1994. Evolution of resistance to *Bacillus thuringiensis*. Annu. Rev. Entomol 39, 47–79.

Tabashnik, B.E., Carrière, Y., 2010. Field-evolved resistance to Bt cotton: bollworm in the U.S. and pink bollworm in India. Southwest. Entomol. 35, 417–424.

Tabashnik, B.E., Biggs, R.W., Higginson, D.M., Henderson, S., Unnithan, D.C., Unnithan, G.C., et al. 2005. Association between resistance to Bt cotton and cadherin genotype in pink bollworm. J. Econ. Entomol. 98, 635–644.

Tabashnik, B.E., Gassmann, A.J., Crowder, D.W., Carrière, Y., 2008. Insect resistance to Bt crops: evidence versus theory. Nat. Biotechnol. 26, 199–202.

Tabashnik, B.E., Van Rensburg, J.B.J., Carrière, Y., 2009. Field-evolved insect resistance to Bt crops: definition, theory, and data. J. Econ. Entomol. 102, 2011–2025.

Tan, F.X., Wang, J.W., Feng, Y.J., Chi, G.L., Kong, H.L., Qiu, H.F., et al., 2010. Bt corn plants and their straw have no apparent impact on soil microbial communities. Plant Soil 329, 349–364.

US EPA (United States Environmental Protection Agency), 2001. Biopesticides Registration Action Document – *Bacillus thuringiensis* Plant-Incorporated Protectants. E: Benefit Assessment. Available at: <http://www.epa.gov/oppbppd1/biopesticides/pips/Bt_brad2/5-benefits.pdf>.

US EPA, 2007a. Review of Dow AgroScience's (and Pioneer Hi- Bred's) submission (dated July 12, 2007) regarding fall armyworm resistance to the Cry1F protein expressed in TC1507 Herculex_ I insect protection maize in Puerto Rico (EPA registrations 68467-2 and 29964-3). DP Barcode: 342635. Decision: 381550. MRID#: 471760-01. US EPA, Washington, DC, USA.

US EPA, 2007b. TC1507 Maize and Fall Armyworm in Puerto Rico, MRID 47176001. USEPA, Washington, DC, USA.

US EPA-SAP (Scientific Advisory Panel), 1998. Report of Subpanel on *Bacillus thuringiensis* (Bt) Plant-Pesticides and Resistance Management. EPA SAP Report. Available at: <http://www.mindfully.org/GE/FIFRA-SAP-Bt.htm>.

Vaufleury, A., Kramarz, P.E., Binet, P., Cortet, J., Caul, S., Andersen, M.N., et al., 2007. Exposure and effects assessments of Bt-maize on non-target organisms (gastropods, microarthropods, mycorrhizal fungi) in microcosms. Pedobiologia 51, 185–194.

Vercesi, M.L., Krogh, P.H., Holmstrup, M., 2006. Can *Bacillus thuringiensis* (Bt) corn residues and Bt-corn plants affect life-history traits in the earthworm *Aporrectodea caliginosa*? Appl. Soil Ecol. 32, 180–187.

Viktorov, A.G., 2008. Influence of Bt-plants on soil biota and pleiotropic effect of δ-endotoxin-encoding genes. Russ. J. Plant Physiol. 55, 738–747.

Vojtech, E., Meissle, M., Poppy, G.M., 2005. Effects of Bt maize on the herbivore *Spodoptera littoralis* (Lepidoptera: Noctuidae) and the parasitoid *Cotesia marginiventris* (Hymenoptera: Braconidae). Transgenic Res. 14, 133–144.

Wan, P., Huang, Y., Wu, H., Huang, M., Cong, S., et al., 2012. Increased frequency of pink bollworm resistance to Bt toxin Cry1Ac in China. PLoS ONE 7 (1), e29975.

Wang, Z.J., Lin, H., Huang, J.K., Hu, R.F., Rozelle, S., Pray, C., 2009. Bt cotton in China: Are secondary insect infestations offsetting the benefits in farmer field? Agric. Sci. China 8, 101–105.

Wang, Z.Y., Wu, Y., He, K.L., Bai, S.X., 2007. Effects of transgenic Bt maize pollen on longevity and fecundity of *Trichogramma ostriniae* in laboratory conditions. Bull. Insectol. 60, 49–55.

Williams M.R., 2006. Cotton insect losses. In: Proceedings of the Beltwide Cotton Conferences, National Cotton Council, Memphis, TN, USA, pp. 1151–1204.

Wilson L., Hickman M., Deutscher S., 2006. Research update on IPM and secondary pests. In: Proceedings of the Thirteenth Australian Cotton Research Conference, Broad Beach, Queensland, Australia, pp. 249–258.

Wolfenbarger, L.L., Naranjo, S.E., Lundgren, J.G., Bitzer, R.J., Watrud, L.S., 2008. Bt crop effects on functional guilds of non-target arthropods: a meta-analysis. PLoS ONE 3, e2118.

Wu, K., Guo, Y., Head, G., 2006. Resistance monitoring of *Helicoverpa armigera* (Lepidoptera: Noctuidae) to Bt insecticidal protein during 2001–2004 in China. J. Econ. Entomol. 99, 893–898.

Wu, K.M., Guo, Y.Y., 2005. The evolution of cotton pest management practices in China. Annu. Rev. Entomol. 50, 31–52.

Xia J.Y., Cui J.J., Luo J.Y., 2007. Resistance of transgenic Bt plus CpTI cotton to *Helicoverpa armigera* Hubner and its effects on other insects in China. World Cotton Research Conference-4 (WCRC-4), Lubbock, Texas, USA, 10–14 September 2007. <http://wcrc.confex.com/wcrc/2007/techprogram/P2207.htm>.

Yang, Y.Z., Yu, Y.S., Ren, L., Shao, Y.D., Qian, K., Zalucki, M.P., 2005. Possible incompatibility between transgenic cottons and parasitoids. Aust. J. Entomol. 44, 442–445.

Yencho, G.C., Cohen, M.B., Byrne, P.F., 2000. Applications of tagging and mapping insect resistant loci in plants. Annu. Rev. Entomol. 45, 393–422.

Zeilinger, A.R., Andow, D.A., Zwahlen, C., Stotzky, G., 2010. Earthworm populations in a northern U.S. corn belt soil are not affected by long-term cultivation of Bt maize expressing Cry1Ab and Cry3Bb1 proteins. Soil Biol. Biochem. 42, 1284–1292.

Zhang, G.F., Wan, F.H., Liu, W.X., Guo, J.Y., 2006. Early instar response to plant-delivered Bt-toxin in a herbivore (*Spodoptera litura*) and a predator (*Propylaea japonica*). Crop Prot. 25, 527–533.

Zhao, J.H., Ho, P., Azadi, H., 2011. Benefits of Bt cotton counter balanced by secondary pests? Perceptions of ecological change in China. Environ. Monit. Assess. 173, 985–994.

Zurbrugg, C., Hönemann, L., Meissle, M., Romeis, J., Nentwig, W., 2010. Decomposition dynamics and structural plant components of genetically modified Bt maize leaves do not differ from leaves of conventional hybrids. Transgenic Res. 19, 257–267.

Zwahlen, C., Hilbeck, A., Howald, R., Nentwig, W., 2003. Effects of transgenic Bt corn litter on the earthworm *Lumbricus terrestris*. Mol. Ecol. 12, 1077–1086.

Zwahlen, C., Hilbeck, A., Nentwig, W., 2007. Field decomposition of transgenic Bt maize residue and the impact on non-target soil invertebrates. Plant Soil 300, 245–257.

CHAPTER

19

Breeding for Disease and Insect-Pest Resistance

S.K. Gupta and M.K. Pandey

S K University of Agricultural Sciences and Technology, Jammu, India

19.1 INTRODUCTION

The cultivation of susceptible varieties has been recognized as the most effective and economical method of reducing crop losses (Stakman and Harrar, 1957). Breeding for resistance is similar to that for other characteristics depending upon the number of genes controlling the trait under consideration. The first step is to search for the resistance source for a particular disease or insect pest in the germplasm within species or in related genera. The next step is to incorporate the resistant genes using various conventional and non-conventional techniques of plant breeding. This breeding for resistance is a lengthy and costly process and therefore, to be cost effective, it should provide durable resistance against disease/insect-pests.

Development of resistant varieties has been receiving more attention recently than in earlier times. Krattiger (1997) reported 15% crop losses due to attack by insect-pests, despite the use of huge amounts of pesticide. In the 19th century, the French wine industry suffered a major setback owing to vine infestation by *Phylloxera* and being forced to resort to the use of a resistant root stock imported from America and which still maintains resistance levels against the aphid-like pest (Painter, 1951). The increase in insect injury that has occurred may be due to the more extensive cultivation of crop varieties over wider areas (Hays et al., 1955). Though many of the established field crop insects are well known, their prevalence is now felt because of the increased acreage of crops, vegetables, fruits, etc. This has led to a greater interest in developing resistant varieties and this is the best method of insect control in crop plants. Such varieties possess the ability to give a higher yield of good quality than susceptible varieties at the same level of insect population. The chapter deals with various aspects of breeding for disease and insect-pest resistance.

D. P. Abrol (Ed): Integrated Pest Management.
DOI: http://dx.doi.org/10.1016/B978-0-12-398529-3.00021-X

19.2 METHODS OF CLASSIFICATION (HAYS ET AL., 1955)

19.2.1 Host Habitation

Under host habitation, the following groups have been recognized by Wardle (1929):

- Insects avoiding only a few plants and having a wide host range including scales and moths, designated as polyphagous.
- Insects mainly living on only one taxonomic unit, e.g. Hessian fly – oligophagy.
- Insects living on some species for part of the year and then on other species (includes seasonal aphids) – oligophagy.
- Some insects avoid all hosts apart from one – monophagy.

19.2.2 Varietal Reaction

Wardle (1929) classified grains and grasses on the basis of reproductive potential of the insect.

i. Normal reproduction – oats and wheat
ii. Reproduction restricted – injury not severe - Kentucky bluegrass, orchard grass
iii. Organisms feed but do not reproduce – rye grass, barley, sorghum, etc.
iv. Organisms do not leave marks – timothy.

19.3 NATURE OF RESISTANCE/ MECHANISM OF RESISTANCE

Painter (1951) proposed three mechanisms of insect resistance:

i. Preference or non-preference
ii. Antibiosis or detrimental effects of the plant on the biology of the insect
iii. Tolerance or the ability of the plant to withstand an insect population.

A resistant variety may employ one or two mechanisms.

TABLE 19.1 Non-Preference Mechanism of Insect Resistance in Important Crops

Host Crop	Insect Pest	Non Preference
Rice	i. Rice stem borer	Lignified stem
	ii. Brown planthopper	Asparagines Red pericarp
Maize	Corn earworm	Hard husk
Brassica	Aphid	Low glucosinolate
Soybean	Potato leaf hopper	Hairiness
Cotton	Jassids	Hairiness of leaves

19.3.1 Preference or Non-Preference

A variety may have some undesirable characteristics for some insects but which are desirable for others (Table 19.1). In wheat crop, a solid stem is not preferred by the stem saw fly whereas a hollow stem is. The characteristics which are associated with non-preference are hairiness, odour, taste, sugar content, smooth leaves, nectar, etc. In *Brassica*, high glucosinolate is preferred by aphids.

19.3.2 Antibiosis

The plant may produce some chemicals which adversely affect the biology of the insect-pest. In some cases, antibiosis may lead to death of the insect.

19.3.3 Tolerance

This mechanism may minimize the damage from the pest. In other words, a tolerant variety will give a higher yield than a susceptible variety despite the insect attack.

Tolerance has been considered to operate in a few *B. juncea* strains (Bakhetia and Sandhu, 1973; Pathak, 1961; Singh et al., 1984). Non preference and antibiosis have been suggested for the observed responses of different cruciferous genotypes in the field (Chander, 1993).

Antibiosis has been considered responsible for effecting longevity, fecundity and nymphal development in different genotypes of *B. nigra*, *B. carinata* and *Eruca sativa* (Chander, 1993; Dilawari, 1985; Dilawari and Dhaliwal, 1988; Kundu and Pant, 1967; Narang, 1982). Non-preference and antibiosis were found to be operative in certain lines of *E. sativa* and *B. juncea* (Abraham, 1991; Abraham and Bhatia, 1992).

19.4 GENETICS OF RESISTANCE

The genetics of resistance to insect-pests is most important for the development of insect-pest resistant varieties. A wide range of genetic systems from monogenic to polygenic control exists in different situations. While studying the inheritance in aphid tolerant cultivars, non-waxy mutant RC 1425 and susceptible Prakash cultivars of *B. juncea*, Yadav et al. (1985) indicated for the first time that the trait for non-waxiness was controlled by a recessive gene. Angadi et al. (1987) studied the inheritance in Indian mustard and suggested the W_1W_1 gene symbol for non-waxiness and w_1w_1 gene for waxiness. Kumar et al. (1990) achieved aphid resistance by a diallel mating system in *B. juncea*. Singh et al. (1991) recorded both additive and non-additive gene effects controlling the inheritance of resistance. A number of genes have been identified as resistant against different insect-pests but resistance broke down within a short timespan.

In rice, a single gene, *Bph-1*, conferred resistance to brown planthopper in a variety IR 26 but this broke down within 2 years. Gallun and Khush (1980) reported the effectiveness of a major gene for resistance to jassids in cotton. Therefore, identification of resistance genes against different insect-pests is of paramount importance for the development of insect-resistant transgenic plants. Briggs and Knowles (1967) mentioned that resistance studies usually require the following:

- A single race of pathogen or insect
- Pure lines of the host to serve as resistant and susceptible parents of crosses
- A uniform and ideal environment of the pathogen or insect, such that resistance and susceptibility are clearly differentiated
- A suitable method of inoculation or infestation so that the disease or insect has an equal opportunity for development on all materials of the study
- A consistent system of classification preferably similar to that used by others conducting similar studies
- Provision to grow self progenies of a cross through F_3 since F_3 families give the best measure of the F_2 genotype
- Repetition of the study with other races or other pure line varieties, or with both.

In the past, a number of genetic studies on disease- and insect-pest-resistant varieties were conducted. After the discovery of the Mendel laws, systematic work on breeding of disease- and insect-pest-resistant varieties started. Disease resistance is known to have been inherited as a monogenic trait following the laws of classical Mendelian genetics. Biffen (1905) demonstrated that the resistance gene for stripe rust in wheat was controlled by a single recessive gene. Then the dominant genes controlling resistance were also found. A few examples have been cited to illustrate the above mechanism.

19.4.1 Genetics of Stem Rust of Wheat

Knott (1964) and Knott and Green (1965) identified 11 genes for resistance which were transferred to variety Marquis by backcrossing. All of the backcross lines of Marquis, each with its own resistance gene, have been used as host differentials to identify races of pathogens.

19.4.2 Hessian Fly (*Phytophaga destructor*) on Wheat

Noble and Suneson (1943) reported from the crosses of resistant variety Dawson and two susceptible varieties Poso and Bigclub, that Dawson had the dominant duplicate gene for resistance. Cartwright and Wiebe (1936) called these genes H_1 and H_2. They also reported a 0% to 2% level of resistance with H_1 and H_2 and 0 to 10% level of resistance with one dominant gene and 83% with no dominant gene. Neither H_1 nor H_2 imparts resistance to Hessian fly in the Corn belt and a third gene H_3 was identified from the variety W38 which imparts resistance to Corn belt type (Caldwell et al., 1946). A fourth recessive gene h_4 was also identified in the variety Java in Californian tests (Suneson and Noble, 1950).

19.5 STRATEGIES AND METHODS OF SCREENING FOR RESISTANCE

The prerequisites to obtain resistance (R) cultivars (cvs) are:

a. Knowledge of the pathogenic variability
b. Development of a screening method able to mimic the conditions met by the plants when exposed to natural sources of inocula in the diverse field environment
c. Availability of usable sources of resistance.

Screening methods for disease resistance (R) should be developed within the framework of a general strategy for resistance. Changes in the frequency of virulent genes among the populations of pathogens inciting disease in above-ground parts of plants are very frequent. The population of such pathogens varies in time and space because of the airborne and seed-borne nature of inocula which facilitates long distance dispersal of their variants. As a result of these situations, breeding for resistance to foliar pathogens is, in general, more difficult than in the case of less mobile pathogens, e.g. soil-borne fungi which are, therefore, more stable.

A screening programme should be initiated with a clear statement of the type of resistance which is sought, i.e. complete resistance (no sporulation of the pathogen) or partial resistance (reduced sporulation of the pathogen) or both, and with at least some knowledge of the pathogenicity and virulence patterns in the pathogen. The durability of resistance can be practically tested only when the R cvs are widely used in space and time. Multilocation cv testing or the challenge of cvs with a large collection of pathogen varieties can help to verify resistance and give timely warning of the possibility of resistance breakdown, but cannot actually be considered as a test for durability of resistance.

Care must be taken in interpreting results of glasshouse or laboratory tests, as the expression of resistance in the field may be considerably modified because of interaction between the microorganism, pathogens and environmental conditions.

For foliar pathogens, the plant material must be adequately challenged with a single race or a pathotype at a realistic inoculum dose to allow disease development, but, at the same time, not obscure minor differences in host response required to identify partial resistance. Use of inocula composed of a mixture of races or naturally infested crop debris of unknown pathotype composition will not be adequate to achieve this objective. To illustrate this, three cvs, each having a single gene for complete resistance to a given race, would be identified only when inoculated singly with each isolate but not if the isolates were use in a mixture. The use of a single race provides the best conditions for the selection of partial resistance in the presence of complete resistance and the selected races should have the broadest possible virulence spectrum to suppress the expression of as many complete R genes as possible. Genotypes with resistance to one virulent race should then be systematically tested in a collection of other isolates.

The identification of cvs with complete resistance is the first step in the development of effective, durable resistance, and genetic analysis of the resistance reaction is essential to reveal similarities and differences in the gene(s) that confer resistance in each genotype. The information is then used to recombine in a single cv several genes known to be effective against a given race and genes effective against all prevalent races in a production region.

19.6 CLASSICAL BREEDING FOR INSECT RESISTANCE

The use of various chemical insecticides has several disadvantages. It increases the cost of cultivation, reduces the population of predators and parasites of insect pests, and leads to environmental pollution and the development of pesticide-resistant biotypes of insects. Thus, the development of resistant varieties provides a cheap method to reduce the input cost thereby increasing the production for human and animal consumption. The presence of gossypol in cotton provides protection against many insects but gossypol is poisonous and makes cottonseed oil unsuitable for humans and animals. The inherent potential of plants to survive in the presence of insect populations has been systematically exploited only during recent times. Painter (1951) has done a lot of work on the mechanism of insect resistance and systematic research on the exploitation of insect resistance in crop plants has started. Considerable similarities exist in the evolution of pathogens and insects in relation to plants, and relatively less effort has been directed to develop pest-resistant compared with disease-resistant varieties. Genetic resistance provides a cheap method which is environmentally friendly. In the biological method, insects are controlled in three ways, i.e. (i) by the use of predators and parasites of insect pests, (ii) by using botanical pesticides such as Neem, Datura, and Ipomea in the form of leaf extracts, and (iii) by using resistant varieties. The biological method is cheap and does not have any adverse effects on the ecosystem, though it is less effective than the chemical method of insect control. Therefore, breeding of insect-resistant varieties requires knowledge about the mechanism of genetic resistance in the host and the virulence of the pest to formulate the appropriate breeding strategies and procedures.

19.7 METHODS OF BREEDING FOR DISEASE AND INSECT-PEST RESISTANCE

Breeding methods depend upon the mode of reproduction of a crop plant and also on a differential reaction to races and biotypes of the host-plant genotypes. Principally, host-plant genotypes are tested in a number of environments across the zones before the selection is made. The breeding methods also vary from crop to crop.

19.7.1 Cross-Pollinated Crops

19.7.1.1 Mass Selection

This is the oldest and the most basic procedure of plant breeding. It consists of three steps:

i. Selection of individual plants for resistance from their source population or the population of plants is inoculated artificially.

ii. The selected plants are massed together and grown in the next generation. Again the population is inoculated and susceptible plants are eliminated before intermating to produce the seed. This procedure is repeated until complete resistance is obtained. Allard (1960) has used this method for the development of resistant varieties against anthracnose and wilt in alfalfa.

19.7.1.2 Recurrent Selection

The main objective is to increase the frequency of resistant genes in the population. Jenkins et al. (1954) used this method for incorporation of resistant genes in *D. turcica* leaf blight disease in corn and reported the effectiveness of three selection cycles. This method has been used by a number of workers to develop resistant varieties against different diseases and insect pests. Penny et al. (1967) found that 2–3 selection cycles produced borer-resistant varieties in corn. Hanson et al. (1972) developed an alfalfa population resistant to four major diseases, namely rust, common leaf spot, bacterial wilt and anthracnose, and resistant to two major insect-pests – spotted alfalfa aphid and potato leaf hopper. The use of this method has led to the release of alfalfa cultivars, namely, Williams, Cherokee, Team and Arc (Hanson et al., 1972). Chahal et al. (1981) used this method to develop resistance against ergot (*Claviceps fungifermis*) and reported an increased proportion of plants with 0–5% ergot severity after 3–4 cycles of selection.

19.7.2 Self-Pollinated Crops

Selection is also one of the important methods used to develop resistant varieties. It is the simplest procedure and involves three main steps:

i. Large numbers of selections are made from the source population.
ii. The progenies are grown from the individual selected plants for different observations. This process is carried out for several years across the environments until the objective is obtained.
iii. The selected genotypes/inbred lines are compared to the existing varieties with respect to the characteristic(s) under consideration.

19.7.2.1 Hybridization

The main objective of hybridization is to transfer the resistant genes from donor parents. The maximum variability is found in F_2 because of regeneration, and homozygosity is achieved after 6–7 generations of selfing.

19.7.2.1.1 PEDIGREE METHOD

This consists of selecting individual F_2 plants for traits such as resistance to disease and insect pests. Selection is made for resistance and the progenies are grown in the next generation. Progenies of the selections are reselected in each succeeding generation until homozygosity is achieved. Artificial epiphytotic conditions are created in the early segregating generation for selection of disease resistance. Pathak (1971) made several selections for resistance to common pests on disease in rice. Blum (1972) screened a number of sorghum lines against smut shoot flies following the pedigree method of breeding.

19.7.2.1.2 BACKCROSS METHOD

If resistance is governed by one or two genes then the backcross method is most appropriate/suitable to transfer the gene from one genetic background to another. When the resistance is dominant, F_1 is backcrossed to the susceptible parent. The progeny of the first backcross generation is tested for resistance. The resistant plants are again backcrossed to recurrent parent (susceptible). After 5–6 generations of backcrossing, plants with almost identical characteristics to the original susceptible variety are obtained in addition to resistance.

When the resistance is recessive, the progeny of each backcross is selfed. At the end of 5–6 backcross generations, progenies are selfed and resistant plants are selected. The newly derived strain is almost identical to the recurrent parent apart from disease resistance.

19.7.3 Mutation Breeding

When the resistant gene is not available in the germplasm, it is advisable to use mutation breeding for altering the susceptible allele(s) into resistant ones without changing the other characteristics of the variety. Murray (1969) recovered 12 mutants with moderate resistance to *Verticillium* in peppermint (*Mentha piperita*). Favret (1971) isolated an induced mutant m 1-0 resistant to *Erysiphe graminis* f. sp. *hordei* in barley. Micke (1983) reported 58 varieties with improved resistance to one or the other disease.

19.7.4 Alien Gene Transfer

This method is used when the resistance gene is not available in the closely related species but is available in wild species. Sears (1972) used the alien gene transfer method through the use of homologous pairing and use of ionizing radiation. Sears (1972) transferred leaf rust resistance from *Agropyron elongatum* into wheat through induced homologous chromosome pairing in the absence of chromosome 5B. Wheat strain TAP had its chromosome 3D substituted by *Agropyron* chromosome carrying the gene for leaf rust resistance. The substitution line was monosomic 5B. It was pollinated by nulli 5B tetra 5D wheat stock. The F_1 plants were nulli 5B-tri 5D-mono 3D-mono 3 Agropyron. In the F_1, 3D-3 Agropyron, which should have paired in the presence of 5B, synapsed in about 30% of the sporocytes. The F_1 was backcrossed to euploid wheat. 77/299 offsprings were resistant to leaf rust, out of which at least 21 were with the 3A-3D transfer and some of the transfer involved less than one arm of *Agropyron* chromosome.

Riley et al. (1968) transferred yellow rust resistance from *Aegilops comosa* to wheat by induction of homologous pairing by suppressing *Aegilops speltoides*. They had a disomic addition line of *A. comosa* chromosome 2M in Chinese spring (2n = 14) with rust resistance. The alien addition line was crossed with a genotype of *A. speltoides* capable of suppressing 5B activity. In the 29 chromosome hybrid, the alien chromosome 2M of *A. comosa* paired with the chromosome of homologous group 2. The F_1 was backcrossed to euploid wheat to recover the chromosome complement of wheat while retaining rust resistance from *A. comosa*. The wheat line so developed was named 'Compair'. Yasumuro et al. (1981) transferred leaf rust resistance from *Agropyron elongatum* by induction of homologous pairing through the ph mutant. Sears (1956) used this technique to transfer rust resistance from *Aegilops umbellulata* to wheat by irradiating a monosomic addition line and released the 'Transfer' cultivar. Sharma and Knott (1966) transferred stem rust resistance from *A. elongatum* to common wheat. Johnson (1966) and Sears (1967) transferred resistance from rye to wheat using this technique. A number of lines, *viz.* T4, Agatha, and Transee, have been developed possessing resistance to various diseases. Rao (1978) transferred genes for stem rust resistance from *A. elongatum* (gene Sr26) and *S. cereale* (Sr27) to susceptible Durum variety Jaya.

19.7.5 Multiline Varieties

These provide long-term control of airborne diseases in self-pollinated crops. By definition, a multiline variety is a population of plants that is agronomically uniform but heterogeneous for genes that condition reaction to a disease organism (Frey and Browning, 1971). This concept was applied by Jensen (1952) to oats and Borlaug and Gibler (1953) and Borlaug (1953) to wheat. The development of multiline varieties is based on two approaches for disease control. One approach is known as the clean crop approach, which was advocated by Borlaug (1959) to control yellow, brown and black rust of wheat. The aim is to keep this crop completely free from disease.

The second approach has been advocated by Frey et al. (1975) and in this approach, none of the lines are completely resistant to all the races of the pathogen. In fact, it has been suggested that, to stabilize the race structure of the pathogen, some susceptibility is essential under the dirty crop approach. Since moderately susceptible lines are also considered, it puts the breeder in an advantageous position as he/she can exercise selection for other morphological traits (Gill et al., 1979).

The first wheat multiline cultivar was Miramar 63, which was released for commercial production under the Colombian Programme (Rockefeller Foundation, 1963, 1964). More than 1200 lines phenotypically similar to the non-recurrent parent but with resistance from 600 non-recurrent parents were developed. Miramar 63 was a mechanical mixture of equal parts of the 10 best lines giving resistance to yellow and black rusts.

Two years after release, two components became susceptible to black rust. Two of these components along with two others were dropped and four new ones from the reserve of 600 lines were added to form another multiline known as Miramar 65. In 1957, Iowa State University initiated the research on multiline in oats against crown rust. Since then, 14 multilines in two maturity groups have been developed and released for cultivation (Frey et al., 1977). Another multiline under the name Tumult was released in the Netherlands (Groenewegen, 1977) based on the resistance sources against yellow rust. In India, the first multiline Sonalika Multiline-1 was developed at Punjab Agricultural University, Ludhiana (Gill and Raupp, 1987). Sonalika Multiline-1 is based on six components having resistance to both yellow and brown rusts. Another Multiline KML 7406 (a mixture of nine components)/or Bithoor was developed at Chandra Shekhar Azad University of Agricultural Sciences and Technology, Kanpur with a background of Kalyan Sona. MLKS-11 is the third multiline developed at IARI, New Delhi.

In additional to conventional methods, the following methods may also be used to develop disease- and insect-pest-resistant plants.

19.7.6 Biotechnological Approaches

Conventional plant breeding has had significant successes in increasing yield and controlling diseases caused by fungi and bacteria. However, the progress in the production of insect-resistant varieties is not encouraging. Weising et al. (1988) reviewed the literature on integration and expression of foreign genes in plants through the Ti plasmid of *Agrobacterium tumifaciens*, Ti plasmid derived plant gene vectors, transformation techniques, structure and expression of genes in their transgenic environment and their impact in crop improvement.

The first widely planted Bt crop cultivars were corn producing Bt toxin Cry1Ab and cotton producing Bt toxin Cry1Ac (Tabashnik et al., 2009). Field-evolved resistance has been documented in some populations of five lepidopteran pests, *viz.* cereal stem borer, *Busseola fusca*, in South Africa to Bt corn producing Cry1Ab (Kruger et al., 2009), armyworm, *Spodoptera frugiperda*, to Bt corn producing Cry1F (Marvier et al., 2008), pink bollworm, *Pectinophora gossypiella*, in western India to Bt cotton producing Cry1Ac (Bagla, 2010), cotton bollworm, *Helicoverpa zea*, in the southeastern United States to Bt cotton producing Cry1Ac and Cry2Ab (Luttrell et al., 2004; Tabashnik et al., 2008a,b, 2009) and bollworm, *Helicoverpa punctigera*, in Australia to Bt cotton producing Cry1Ac and Cry2Ab (Downes et al., 2010).

19.7.6.1 Tissue Culture

Plants regenerated from tissue culture are called somaclones. This technique is used to generate resistance to diseases where the resistance sources are not identified. Moyer and

Collins (1983) have used it to develop a variety of sweet potato using meristem tip culture. Morel and Martin (1952) were the first to demonstrate that virus-free plants can be obtained from virus-infected plants using the technique of meristem culture. In addition, some other systemic pathogens such as mycoplasma, fungi and bacterial diseases can also be eliminated using this technique. Using such techniques, disease-free plants have been obtained in more than 50 species (Hu and Wang, 1983). The meristem culture technique has been extended to a number of species to produce virus-free plants, and is now regularly used to produce virus-free plants in potato, dahlia, strawberry, carnation, chrysanthemum, orchids, etc. Karha and Gamborg (1975) used meristem culture to obtain symptom-free plants from stakes infected with cassava mosaic disease of Indian and Nigerian origin.

Wild species of the cultivated crop plants are useful reservoirs of genetic variability for various economic characteristics such as disease and pest resistance, salt tolerance, high protein and increased biomass. Incorporation of such genetic variability and other desirable characteristics from related species and genera into cultivated varieties involves hybridization between diverse parents. However, in such crosses, several barriers to crossibility are encountered. In such a situation, embryo culture and in-vitro pollination is the most suitable combined technique to avoid the abortion of hybrid embryos. The International Crop Research Institute for Semi-Arid Tropic in India is attempting to incorporate genes for high oil and disease resistance from wild species to cultivated peanut varieties.

19.7.6.2 Gene Pyramiding

Bacterial blight is a major disease of rice caused by *Xanthomonas oryzae* pv. *oryzae* (Xoo). Singh et al. (2001) pyramided three resistant genes *Xa-5, Xa-13* and *Xa-21* into the cultivar 106. The combination of genes provided a wider spectrum of resistance to this pathogen population. Huang et al. (1997) used DNA marker assisted selection to pyramid four BB resistance genes *Xa-4, Xa-5, Xa-13* and *Xa-21*. The pyramided lines showed a wider spectrum of resistance. In transgenics also, gene pyramiding offers effective control because an insect species cannot simultaneously show resistance to a number of toxins produced by different genes deployed in pyramiding since that would require a simultaneous and independent mutational event in a gene encoding the receptor and practically, therefore, the greater the number of genes deployed, the less the possibility of development of resistance. For example, the cotton variety Bollgard 11® has two toxins *viz.* Cry 1Ac controlling tobacco budworm and pink boll worm and Cry 2Ac, controlling corn ear worm (Jackson et al., 2003; Purcell and Perlak, 2004). The second generation dual Bt gene cottons, Bollgard 11® with Cry 1Ac together and Widestrike™ with Cry 1Ac and Cry1 F, express two Bt endotoxins and were introduced to increase the control of *H. zea* which was not satisfactorily controlled by Cry 1A gene alone (Bates et al., 2005; Gahan et al., 2005). A strategy involving a combination of conventional host-plant resistance and noble exotic genes as well as a combination of several genes conferring resistance to a wider array of insect-pests and disease will go a long way in the commercial success of transgenics.

19.8 SYSTEMATIC BREEDING PROGRAMME FOR DEVELOPMENT OF RESISTANT VARIETIES

Inheritance of resistance to insects in plants shows that some resistance traits are controlled by dominant genes and others by recessive genes. Polygenic resistance is relatively more stable but it is difficult to incorporate in high yielding varieties. Screening techniques for each crop are required to evaluate plants for

TABLE 19.2 Resistance Source Identification Criteria Used in Brassica Species against the Mustard Aphid

Technique	Resistance Modality	Remarks
Seedling survival	Non-preference, antibiosis, tolerance	Useful in preliminary screening in screen house
Aphid injury	Resistance, tolerance	Suitable for specific screening on a smaller scale in screen house and also in the field
Aphid population	Non-preference, antibiosis, tolerance	A far more laborious method suited best for screening on a small scale in the field or screen house
Fecundity and development	Antibiosis	Screening of select germplasm preferably in the laboratory or under caged conditions
Grain yield	Tolerance	Best suited for large scale screening in the field or under specially designed cages
Honey-dew secretion	Host-food	Most laborious and time consuming; best suited for laboratory resistance testing of a few selected parameters only

Source: Kalia and Gupta (1997): Recent Advances in Oilseed Brassicas, Kalyani Publishers, New Delhi.

resistance under natural conditions. Screening techniques for rapeseed-mustard crop are mentioned in Table 19.2 as one example. The procedure for evaluating insect resistance in plant genotypes must meet certain criteria as mentioned by Jenkins (1985).

A good supply of eggs, larvae, or adult insects must be available for infesting plants in a breeding nursery. If the insects are reared in the laboratory, they must represent the wild population in vitality, biotype composition, and genetic structure, and they must be nourished so that their behaviour and population capabilities are similar to those in the wild.

A rapid, repeatable rating scale must be developed that is related to the development of the insects, or to the economic damage done by the insect. The strategies to use major gene resistance include: sequential release of varieties with major genes; pyramiding of major genes; rotation of varieties with major genes; development of varietal mixtures and multilines.

Identification of diverse sources of resistance on the basis of reaction to a specific biotype is essential to prolong the life of insect-resistant varieties. The search for new genes conferring resistance has to be continued since resistance genes become ineffective after some time.

19.9 SCREENING TECHNIQUES

Various screening criteria have been employed by different authors for evaluation of cruciferous genotypes to determine the resistance sources. This aspect has been extensively reviewed recently by Bakhetia and Singh (1992) (Table 19.2).

19.9.1 Seedling Survival

This technique was developed by Bakhetia and Bindra (1977) and advocates using the optimum level of aphid population per seedling or plant under the screen house conditions. This technique of resistance screening against the mustard aphid is widely accepted because of its higher efficiency and ease of application at the seedling stage, and thus it

TABLE 19.3 Scoring Methods Used by Different Workers for Aphid Resistance Screening in Rapeseed-Mustard

	Injury Symptoms as Explained in Different Reports		
Injury grade	Pathak (1961), Teotia and Lal (1970)	Rajan (1961)	Bakhetia and Sandhu (1973)
0	Normal	Plants unaffected by aphids	Free from aphid infestation
1	Yellowing in a few leaves	Apparently healthy plants with a few young ones in the flower buds	Normal growth, no yellowing or curling leaves of leaves, a few aphids without injury symptoms, flowering and pod setting almost normal
2	Yellowing and curling of some leaves, medium growth, no stunting	Plant severely infested, inflorescence and pods full of aphids, no visible symptoms of damage	Average growth, flowering and pod setting; curling and yellowing of a few leaves
3	Poor (stunted) growth, and flowers dry up	Plants completely damaged and in different stages of decay	Curling and yellowing of some branches, below average growth, poor flowering with very little pod setting
4	Stunting, plants defoliated and drying up	–	Very poor growth, heavy curling of leaves, stunting of plants; little or no flowering and pod formation
5	–	–	Heavy aphid colonies, severe stunting of plants, curling and yellowing of almost all leaves. No flowering and pod formation. Plants full of aphids

Source: Kalia and Gupta (1997): Recent Advances in Oilseed Brassicas, Kalyani Publishers, New Delhi.

has become an integral part of the mass screening programme.

19.9.2 Aphid Injury

Aphid damage symptoms expressed as injury grades (0–4) were proposed by Pathak (1961) and Teotia and Lal (1970) while Rajan (1961) and Bakhetia and Sandhu (1973) used a different set of injury grades created on the basis of injury symptoms (Table 19.3).

A breeding programme for resistance to insects depends on their behaviour and the nature of the crop involved as well as on the information available at the start. Painter (1951) suggested the following steps for such a programme:

i. Discover possible sources of resistance to insects among varieties and strains locally available.

ii. Search for new germplasms that may carry resistance.
  - **a.** New selections out of older varieties.
  - **b.** Plant introductions out of older varieties.
  - **c.** Varieties of related species of plants, if necessary.

iii. Determination of some of the basic properties of the plants responsible for resistance; at least a separation of the three classes: preference, antibiosis and tolerance.

iv. Hybridization to combine genes for resistance with desirable agronomic characteristics.

v. Study of genetics of resistance to insects where possible.

vi. The testing of resistance in advanced-generation hybrids.

vii. The study of resistance of released varieties in plots and on farms to evaluate as an insect control method.

In cases where selection for resistance to a particular pest is difficult, perhaps because assessment is complex or environmental effects mask the genetic expression of resistance, it may be possible to use indirect selection to achieve the objectives by selecting for early recognizable characteristics which are highly correlated with resistance or pleiotropic effects. Either way, they can provide a convenient method for selecting insect resistance strains.

19.10 SPECIAL CONSIDERATIONS (JENKINS, 1985)

The choice varies with the situation under which the insect is placed, such as monoculture versus a varied crop culture. Another aspect to be considered is the life cycle of the insect. Some insects feed on both crop and weed species. The generation time and population dynamics must be taken into consideration. A further aspect, breeding for insect resistance in the host plant, deals with having an insect supply for infesting the plant breeding nursery. In this context, natural or artificial infestation can be used, but sometimes natural infestation failed to give good results among the plant genotypes. In this situation, it may be possible to enhance the natural infestation or the genotypes may be repeated for 2–3 years to obtain good results. An example in cotton is the nurse crop principle with plant bugs (Laster and Meredith Jr, 1974) whereby garden mustard is inter-planted with cotton to serve as a nurse crop for natural rearing of the population of plant bugs early in the year.

19.11 MORPHOLOGICAL TRAITS VERSUS RESISTANCE

Rai and Sehgal (1975) studied the plant succulence and arrangements of buds in the inflorescence in relation to aphid injury and inferred that tender and thickly packed inflorescence of *B. campestris* varieties were more susceptible than those of *B. juncea* having harder and loose inflorescence. Non-waxiness of *Brassica* plants have been found to impart resistance to aphids (Angadi et al., 1987; Bakhetia et al., 1982; Chatterjee and Sengupta, 1987; Srinivasachar and Malik, 1972; Yadav et al., 1985).

19.12 BIOCHEMICAL CONSTITUENTS VERSUS RESISTANCE

Bakheita et al. (1982) reported a negative relationship of total sugar and sulphur content on *Brassica* species with aphid reproduction and aphid population. Higher concentrations of total sugars and iron are known to impart resistance, while there seems to be an association with nitrogen, zinc, copper and manganese (Narang, 1982). It has also been found that total protein free amino acids, reducing sugars and total sugars were higher in susceptible strains while higher flavonoid content in *B. napus* imparts resistance to aphids (Malik, 1981). Kundu and Pant (1967) reported no influence of total sugars, nitrogen, or free amino acid amides on the growth and development of mustard aphids. Weibull and Melin (1990) reported that total concentrations and individual composition of amino acids had no relationship with the level of aphid resistance (Table 19.4).

19.13 SUMMARY AND CONCLUSIONS

Resistant varieties are very useful in pest management. Varieties reported to be resistant to a particular pest (or pests) should be introduced directly for cultivation in certain areas. Such resistant varieties should be used extensively in a hybridization programme aimed

TABLE 19.4 Biochemical constituents versus aphid resistance in cruciferous crops

Biochemical constituents	Biological effect	Plant species	References
Glucosinolates and singrin	Toxicant/growth inhibition	*Brassica* spp.	Anand (1976) Malik (1981) Rani (1981) Kumar et al. (1996) Chandir (1993)
Total sugars, flavonoids, ascorbic acid, K, S, Fe	Growth inhibition/ toxicant		Rani (1981) Narang (1982) Gill and Bakhetia (1985)
Protein, free amino acid ash, N, P	Stimulants	*Brassica* spp.	Malik (1981, 1988) Gill and Bakhetia (1985)
Free amino acid N, Cu, Zn, Mn	No relation	*Eruca sativa* *Brassica* spp.	Kunder and Pant (1967) Narang (1982) Weibull and Melin (1990)
Gluconapin allyl isothiocynate, volatile isothiocynate	Stimulant, repellent, attractant	*Brassica*. spp. E. *sativa*	Malik (1981) Dilawari (1985) Chandeq (1993)

at incorporating pest resistance into the well adapted varieties of the region. Similarly, species known to carry pest resistance may be used in inter-specific or inter-generic crosses through the genetic and cytogenetic techniques known for chromosomal manipulations. Mutation breeding is also an important tool for developing new varieties with the desired characteristics of pest and disease resistance.

Among the alternative approaches, use of host-plant resistance (HPR) in IPM programmes has been very important over the past two decades, and has resulted in the development and standardization of suitable techniques for germplasm screening in the screen house and fields.

Despite these endless scientific endeavours, it is unfortunate that resistant/tolerant parents have not evolved which could possibly be used in conventional breeding programmes to develop new resistant varieties against the mustard aphid. However, the identification of a resistance imparting gene would solve the problem through more inter-generic transfer of genes from one species to another. The more recent advances in biotechnology have explored the possibility of transferring foreign genes such as NPT II, DHFR (mouse), napin and NOS into *B. napus* and *B. oleracea* plants directly or through *Agrobacterium tumefaciens* mediated gene transfers.

Today, Bt technology can be seen to have provided a powerful tool for tackling some of the important insect-pests, since it is a chemical free and economically viable approach for the control of insect-pests (Gatehouse; 2008; Hilder and Boulter, 1999; De Villiers and Hoisington, 2011; Sanahuja et al., 2011). The first widely used Bt crop cultivars were corn producing Bt toxin Cry1Ac (Tabashnik et al., 2009). There are many other reports about the successful transfer of foreign genes into different crop plants and they showed encouraging results. The available data strongly support the view that insertion of foreign genes does not change the agronomic characteristics of the host plants; hence this technique can become a supplement to conventional crop improvement programmes. Incorporation of the *Bacillus thuringiensis* gene, trypsin inhibitor gene (CpTi) and phaseolin gene into *Brassica* and other crop plants will surely help solve the pest problem because the transgenic plants are reported to have normal morphological traits. In addition, the field of biotechnology will also supplement the conventional breeding programme.

References

Abraham, T., 1991. The biology, significance and control of the maize weevil, *Sitophilus zeamais* Motsch. (*Coleoptera: Curculionidae*) on stored maize. M.Sc. Thesis presented to the School of Graduate Studies of Haramaya University of Agriculture, Ethiopia, 250 pp.

Abraham, V., Bhatia, C.R., 1992. Multiple screening for aphid resistance in Indian mustard, Brassica juncea (Linn) Czern & Coss. Cruciferae Newslett. 16, 1–11.

Allard, R.W., 1960. Principles of Plant Breeding. John Wiley and Sons, Inc., New York.

Anand, I.J., 1976. Oil quality vis-à-vis aphid resistance in Brassicae. Paper presented in the seminar-cum-workshop on oilseed research in India. PKV Nagpur, 5–9 April 1976.

Angadi, S.P., Sinh, J.P., Anand, I.J., 1987. Inheritance of non-waxiness and tolerance to aphid in Indian mustard. J. Oilseed Res. 4, 265–267.

Bagla, P., 2010. Hardy cotton-munching pests are latest blow to GM crops. Science 327, 1439.

Bakhetia, D.R.C., Bindra, O.S., 1977. Screening technique for aphid resistance in Brassica crops. SABRAO J. 9, 91–107.

Bakhetia, D.R.C., Sandhu, R.S., 1973. Differential response of Brassica species/varieties to the aphid, *Lipaphis erysimi* (Kalt.) infestation. J. Res. Panjab Agric. Univ. 10, 272–279.

Bakhetia, D.R.C., Singh, P., 1992. Host plant resistance in oilseeds. In: Dhaliwal, G.S., Dilwari, V.K. (Eds.), Advanced Host Plant Resistance to Insects. Kalyani Publishers, New Delhi, pp. 208–248.

Bakhetia, D.R.C., Sushma R., Ahuja, K.L., 1982. Effect of sulphur nutrition of some Brassica plants on their resistance response to the mustard aphid, L. erysimi (Kalt). Calicut University Research Journal (Special Conference Number), 3–5 May 1982, 36 pp.

Bates, S.L., Zhao, J.-Z., Roush, R.T., Shelton, A.M., 2005. Insect resistance management in GM crops: past, present and future. Nature Biotechnol. 23, 57–62.

Biffen, R.H., 1905. Mendel's law of inheritance and wheat breeding. J. Agric. Sci. 1, 4–48.

Blum, A., 1972. Sorghum breeding for shoot fly resistance in Israel. In: Jotwani, M.G., Young, W.R., (Eds.), Control of Sorghum Shoot Fly. New Delhi: Oxford & IBH Publishing, pp. 180–191.

Borlaug, N.E., 1953. New approach to the breeding of wheat varieties resistant to *Puccinia graminis tritici*. Phytopathology 43, 467.

Borlaug, N.E., 1959. The use of multilineal or composite varieties to control airborne epidemic diseases of self-pollinated crop plants. Proceedings of the First International Wheat Genetics Symposium, University of Manitoba, Winnipeg, Canada, 1958, pp. 12–26.

Borlaug, N.E., Gibler, J.W., 1953. The use of flexible composite wheat varieties to control the constantly changing stem rust pathogen. Abstr. Annu. Meeting Am. Soc. Agron., 81.

Briggs, F.N., Knowles, P.F., 1967. Introduction to Plant Breeding. Reinhold Publishing Corporation, New York.

Caldwell, R.M., Cartwright, W.B., Compton, L.E., 1946. Inheritance of Hessian fly resistance derived from the W38 and Durum P.I. 94587. J. Am. Soc. Agron. 38, 398–409.

Cartwright, W.B., Wiebe, G.A., 1936. Inheritance of resistance to the hessian fly in the wheat crosses Dawson x Poso and Dawson x Big Club. J. Agric. Res. 52, 691–695.

Chahal, S.S., Gill, K.S., Pull, P.S., Singh, N.B., 1981. Effectiveness of recurrent selection for generating ergot resistance in pearl millet. SABRAO J. 13, 184–186.

Chander, H., 1993. Identification and characterization of sources of resistance in some crucifers against the mustard aphid, Lipaphis erysimi (Kaltenbach), Ph.D. Dissertation, Punjab Agricultural University, Ludhiana, 140 pp.

Chatterjee, S.D., Sengupta, K., 1987. Observation on reaction of mustard aphid to white petal and glossy plants of Indian mustard. J. Oilseeds Res. 4, 125–127.

De Villiers, S.M., Hoisington, A.D., 2011. The trends and future of biotechnology crops for insect pest control. Afr. J. Biotechnol. 10, 4677–4681.

Dilawari, V.K., 1985. Role of glucosinolates in host selection, development and multiplication of mustard aphid, L. erysimi (Kalt). Ph.D. Dissertation, Punjab Agric. Univ. Ludhiana, 111 pp.

Dilawari, V.K., Dhaliwal, G.S., 1988. Population build-up of mustard aphid, Lipaphis erysimi (Kaltenbach) on cruciferous cultivars. J. Insect. Sci. 1, 149–153.

Downes, S., Mahon, R.J., Rossiter, L., Kauter, G., Leven, T., Fitt, G., et al., 2010. Adaptive management of pest resistance by *Helicoverpa* species (Noctuidae) in Australia to the Cry2Ab Bt toxin in Bollgard II® cotton. Evol. Appl. 3, 574–584.

Favret, E.A., 1971. Different categories of mutations for disease reaction in the host organism. In: Proceedings of the Symposium, Mutation Breeding for Disease Resistance. IAEA, Vienna, pp. 107–116.

Frey, K.J., Browning, J.A., 1971. Breeding crop plants for disease resistance. In: Proceedings of the Symposium, Mutation Breeding for Disease Resistance. IAEA, Vienna, pp. 45–54.

Frey, K.J., Browning, J.A., Simsons, M.D., 1975. Multiline cultivars in autogamous crop plants. SABRAO J. 7, 113–123.

Frey, K.J., Browning, J.A., Simsons, M.D., 1977. Management systems for host gene to control disease loss. In: The Genetic Base of Epidemics in Agriculture. Ann. NY Acad. Sci. 287, 255–274.

Gahan, L.J., Ma, Y.T., Coble, M.L., Gould, F., Moar, W.J., Heckel, D.G., 2005. Genetic basis of resistance to Cry1Ac and Cry2Aa in Heliothis virescens (Lepidoptera: Noctuidae). J. Econ. Entomol. 43, 701–726.

Gallun, R.L., Khush, G.S., 1980. Genetic factors affecting expression and stability of resistance. In: Maxwell, F.G., Jennings, P.R. (Eds.), Breeding Plant Resistant to Insects. John Wiley, New York, pp. 63–85.

Gatehouse, J., 2008. Biotechnological prospects for engineering insect resistant plants. Plant Physiol. 146, 881–887.

Gill, B.S., Raupp, W.J., 1987. Direct gene transfers from *Aegilops squarrosa* L. to hexaploid wheat. Crop Sci. 27, 445–450.

Gill, K.S., Nanda, G.S., Singh, G., Aujula, S.S., 1979. Multilines in wheat – a review. Indian J. Genet. Plant Breed. 39, 30–37.

Gill, R.S., Bakhetia, D.R.C., 1985. Resistance of some *Brassica napus* and *B. campestris* strains to Lipaphis erysimi (Kalt.). J. Oilseeds Res. 2, 227–239.

Groenewegen, L.J.M., 1977. Multilines as a tool in breeding for reliable yields. Cereal Res. Commun. 5, 125–132.

Hanson, C.H., Busbice, J.H., Hill Jr., R.R., Hunt, O.J., Oakes, A.J., 1972. Directed mass selection for developing multiple pest resistance and conserving germplasm in alfalfa. J. Environ. Qual. 1, 106–111.

Hays, K.H., Immer, R.F., Smith, C.D., 1955. Methods of Plant Breeding. McGraw Hill Company Inc., New York.

Hilder, V.A., Boulter, D., 1999. Genetic engineering of crop plants for insect resistance – a critical review. Crop Prot. 18, 177–191.

Hu, C.Y., Wang, P.J., 1983. Meristem, shoot tip and bud culture. In: Evans, D.A., Sharp, W.R., Ammirato, P.V., Yamada, Y. (Eds.), Handbook of Cell Culture. Vol. 1. Techniques for Propagation and Breeding. MacMillan, New Delhi, pp. 177–227.

Huang, N., Angeles, E.R., Domingo, J., Magpantay, G., Singh, S., Zhang, G., et al., 1997. Pyramiding of bacterial blight resistance genes in rice: marker assisted selection using RFLP and PCR. Theor. Appl. Genet. 95, 313–320.

Jackson, R.E., Bradley, J.R., Van Duyn, W., 2003. Field performance of transgenic cottons expressing one or two Bacillus thuringiensis endotoxins against boll worm, Helicoverpa zea. J. Cotton Sci. 7, 57–61.

Jenkins, J.N., 1985. Breeding for insect resistance. In: Frey, K.J. (Ed.), Plant Breeding 11. Iowa State University Press, Kalyani Publishers, New Delhi.

Jenkins, M.T., Roberts, A.L., Findley Jr., W.R., 1954. Recurrent selection as a method for concentrating genes for resistance to Helminthosporium turcicum leaf blight in corn. Agron. J. 46, 89–94.

Jensen, N.F., 1952. Intervarietal diversifications in oat breeding. Agron. J. 44, 30–34.

Johnson, R., 1966. The substitution of a chromosomes from Agropyron elongatum for chromosomes of hexaploid wheat. Can. J. Genet. Cytol. 8, 279–292.

Kalia, H.R., Gupta, S.K., 1997. Recent Advances in Oilseed Brassicas. Kalyani Publishers, New Delhi, India.

Karha, K.K., Gamborg, O.L., 1975. Elimination of cassiva mosaic disease by meristem culture. Phytopathology 65, 826–828.

Knott, D.R., 1964. The effect on wheat of an Agropyron chromosome carrying rust resistance. Can. J. Genet. Cytol. 6, 500–507.

Knott, D.R., Green, G.J., 1965. The inheritance of stem rust resistance in ten varieties of common wheat. Can. J. Agric. Sci. 36, 174–195.

Krattiger, A.F., 1997. Insect resistance in crops: a case study of *Bacillus thuringiensis* (*Bt*) and its transfer to developing countries. ISAAA Briefs 2, 42.

Kruger, M., van Rensburg, J.B.J., van den Berg, J., 2009. Perspective on the development of stem borer resistance to Bt maize and refuge compliance at the Vaalharts irrigation scheme in South Africa. Crop Prot. 28, 684–689.

Kumar, V., Singh, D., Singh, H., Singh, H., 1990. Combining ability analysis for resistance parameters to mustard aphid (Lipaphis erysimi (Kalt.). Indian J. Agric. Sci. 57, 227–279.

Kundu, G.G., Pant, N.C., 1967. Studies on the Lipaphis erysimi (Kaltenbach) with special reference to insect plant relationship. I. Susceptibility of different varieties of Brassica and Eruca sativa to the mustard aphid infestation. Indian J. Entomol. 29, 241–251.

Laster, M.L., Meredith, Jr., W.R. 1974. Influence of nectariless cotton on insect populations, lint yield, and fiber quality. Info. Sheet 241. Mississippi Agricultural Experiment Station, Mississippi State, MS.

Luttrell, R.G., Ali, I., Allen, K.C., Young, S.Y., Szalanski, A., 2004. Resistance to Bt in Arkansas populations of cotton bollworm. Proceedings of the 2004 Beltwide Cotton Conferences, 5–9 January 2004, National Cotton Council of America, Memphis, pp. 5–9.

Malik, R.S., 1981. Morphological, anatomical and biochemical bases of mustard aphid, Lipaphis erysimi (Kalt.) resistance in cruciferous species. Sver. Utsad. Tidsk. 91, 25–35.

Malik, R.S., 1988. Role of amino acids in relation to aphid (L. erysimi Kalt) resistance in cruciferous species. J. Oilseeds Res. 5 (2), 39–45.

Marvier, M., Carriere, Y., Ellstrand, N., Gepts, P., Kareiva, P., 2008. Harvesting data from genetically engineered crops. Science 320, 452–453.

Micke, A., 1983. Some considerations on the use of induced mutations for improving disease resistance of crop plant. In: Proceedings of the Symposium Induced Mutations for Disease Resistance in Crop Plants. IAEA, Vienna, pp. 3–19.

Morel, G.M., Martin, C., 1952. Guerison de dahlias atteints d'une maladie a virus. C.R. Acad. Sci. (Paris) 235, 1324–1325.

Moyer, J.W., Collins, W.W., 1983. Scarblet sweet potato. Hort. Sci. 18, 111–112.

Murray, M.J., 1969. Successful use of irradiation breeding to obtain Verticillium resistant strains of peppermint (Mentha piperita L.). In: Proceedings of the Symposium on Induced Mutations in Plants. IAEA, Vienna, pp. 345–371.

Narang, D.D., 1982. Studies on the basis of resistance in *Brassica campestris* var. yellow sarson, *B. juncea* and *Eruca sativa* to mustard aphid, Lipaphis erysimi (Kaltenbach). Ph.D. Dissertation, Punjab Agricultural University, Ludhiana, 103 pp.

Noble, W.B., Suneson, C.A., 1943. Differentiation of two genetic factors for resistance to the Hessian fly in Dawson wheat. J. Agric. Res. 67, 27–32.

Painter, R.H., 1951. Insect Resistance in Crop Plants. The University Press of Kansas. Lawrence, 520 pp.

Pathak, M.D., 1961. Preliminary notes on the differential response of yellow and brown sarson and rai to mustard aphid (Lipaphis erysimi Kaltnbach). Indian Oilseeds J. 5, 39–43.

Pathak, M.D., 1971. Resistance to insect-pests in rice varieties. Oryza (Suppl.) 8(2), 135–144.

Penny, L.H., Scott, G.E., Guthrie, W.P., 1967. Recurrent selection of European corn borer resistance in maize. Crop Sci. 7, 407–409.

Purcell, J.P., Perlak, F.J., 2004. Global impact of insect resistant (Bt) cotton. AgBioForum 7, 27–30.

Rai, B., Sehgal, V.K., 1975. Field resistance of Brassica germplasm to mustard aphid, Lipaphis erysimi (Kalt.). Sci. Cult. 41, 444–445.

Rajan, S.S., 1961. Aphid resistance of tetraploid toria. Indian Oilseeds J. 5, 251–255.

Rani, S., 1981. Role of sulphur nutrition in Brassica species for aphid resistance. M.Sc. Thesis, Punjab Agricultural University, Ludhiana, 74 pp.

Rao, M.V.P., 1978. The transfer of alien genes for the stem rust resistance to durum wheat. Proceedings of the fifth International Wheat Genetics Symposium, New Delhi, pp. 338–341.

Riley, R., Chapman, V., Johnson, R., 1968. Introduction of yellow rust resistance of *Aegilops comosa* into wheat by genetically induced homologous recombination. Nature 217, 383–385.

Rockefeller Found Program Agric. Sci. 1963. Annu. Rep. 1962–63: 310.

Rockefeller Found Program Agric. Sci., 1964. Annu. Rep. 1963–64: 285.

Sanahuja, G., Banakar, R., Twyman., R.M., Capell, T., Christou, P., 2011. *Bacillus thuringiensis*: A century of research, development and commercial applications. Plant Biotechnol. J. 9, 283–300.

Sears, E.R., 1956. Transfer of leaf rust resistance from Aegilops umbellulata and wheat Brookhaven. Symp. Brol. Genet. 9, 1–22.

Sears, E.R., 1967. In: Finlay, K.W., Shepherd, K.W. (Eds.), Proceedings of the Third International Wheat Genetics Symposium. Australian Academy of Science, Canberra (1968) 53.

Sears, E.R., 1972. Chromosome engineering in wheat. Stadler Genet. Symp. 4, 23–38.

Sharma, D., Knott, D.R., 1966. The transfer of leaf rust resistance from *Agropyron* to *Triticum* by irradiation. Can. J. Genet. Cytol. 8, 137–143.

Singh, H., Rohilla, H.R., Kalra, V.K., Yadava, T.P., 1984. Response of Brassica varieties sown on different dates to the attack of mustard aphid (*Lipaphis erysimi* (Kalt.)). J. Oilseeds Res. 1, 49–56.

Singh, R.P., Payne, T.S., Rajaram, S., 1991. Characterisation of variability and relationship among components of partial resistance to leaf rust in CIMMYT bread wheats. Theor. Appl. Genet. 82, 674–680.

Singh, S., Sindhu, J.S., Huang, N., Vikal, Y., Li, Z., Brar, D.S., et al., 2001. Pyramiding three bacterial blight resistance genes (Xa-5, Xa-13 and Xa-21) using marker-assisted selection into indica rice cultivar PR106. Theor. Appl. Genet. 102, 1011–1015.

Srinivasachar, D., Malik, R.S., 1972. An induced aphid resistant mutant in turnip rape, *Brassica napus*. Curr. Sci. 40, 311–313.

Stakman, E.C., Harrar, J.G., 1957. Principles of Plant Pathology. Ronald Press Company, New York.

Suneson, C.A., Noble, W.B., 1950. Further differentiation of genetic factors in wheat for resistance to Hessian fly. U.S. Dept. Agric. Tech. Bull., 1004.

Tabashnik, B.E., Gassmann, A.J., Crowder, D.W., Carrière, Y., 2008a. Field-evolved resistance to Bt toxins. Nat. Biotechnol. 26, 1074–1076.

Tabashnik, B.E., Gassmann, A.J., Crowder., D.W., Carrière, Y., 2008b. Insect resistance to Bt crops: Evidence versus theory. Nat. Biotechnol. 26, 199–202.

Tabashnik, B.E., Unnithan, G.C., Masson, L., Crowder, D.W., Li, X., Carrière, Y., 2009. Asymmetrical cross-resistance between *Bacillus thuringiensis* toxins Cry1Ac and Cry2Ab in pink bollworm. Proc. Natl. Acad. Sci. USA 106 (29), 11889–11894.

Teotia, T.P.S., Lal, O.P., 1970. Differential response of different varieties and strains of Brassicae to aphid Lipaphis erysimi (Kalt.). Labdev J. Sci. Technol. 8B, 219–226.

Wardle, R.A., 1929. The Problems of Applied Entomology. Manchester University Press, UK, 587 pp.

Weibull, J., Melin, G., 1990. Free amino acid content of phloem sap from Brassica plants in relation to performance of Lipaphis erysimi. Ann. Biol. 116, 417–423.

Weising, K., Schell, J., Kahl, G., 1988. Foreign genes in plants: transfer, structure, expression and applications. Annu. Rev. Genet. 22, 585–621.

Yadav, A.K., Singh, H., Yadava, T.P., 1985. Inheritance of non-waxy trait in Indian mustard and its reaction to aphids. J. Oilseeds Res. 2, 339–342.

Yasumuro, Y., Morris, R., Sharma, D.C., Schmidt, J.W., 1981. Induced pairing between a wheat (*Triticum aestivum*) and an *Agropyron elongatum* chromosome. Can. J. Genet. Cytol. 23, 49–56.

CHAPTER

20

Integrated Management of Rodent Pests

R.S. Tripathi

Central Arid Zone Research Institute, Jodhpur, India

20.1 INTRODUCTION

Rodents (order: Rodentia) are one of the most successful animals on earth due to their vast breeding potential and adaptability to a variety of living conditions ranging from the snowy heights of 5700m to the extremes of desert. They represent a very diverse group of mammals, including porcupines, squirrels, voles, marmots, rats, gerbils, mice, moles, and rats. Of over 2000 species reported globally, only a limited number of species (~10%) are considered as serious impediments causing significant losses to food production and storage. Many species are also responsible for spreading several dreadful zoonotic diseases to man and his livestock. In many instances, rodents also provide major benefits to the environment, being part of the food web in the ecosystem and also as ecosystem engineers (Dickman, 1999). Rodents also serve as food for many tribal communities. However, problems caused by rodents to food production and public health is very much more to the fore of human awareness the world over. In fact, old Indian scriptures, dating back five millennia, have mentioned rodents as man's enemy. One of the shloka in Atharveda (60.50.1, 3000 BC) reads, "*O Ashwin, kill the burrowing rodents which devastate our food grains. Cut their heads, break their necks, plug their mouth so that they can never destroy our food. Rid mankind of them.*" These rodents probably adapted to commensal life as and when man started a settled life and became an agriculturist. The pest rodents cause 5–10% losses to various production systems such as agriculture, horticulture, forestry and stored food grains (Jain and Tripathi, 2000). In India, rice and wheat are the two main staple grains and suffer a very similar extent of rodent damage. At a moderate 5% level of pre-harvest damage, the losses run to about 7–8 million tonnes annually. In the Sout h East Asian region, the losses to the rice crop alone amount to 10–20% in Indonesia, 2–5% in Malaysia, >10% in Vietnam, 6–7% in Thailand and 5–10% in Bangladesh (Islam et al., 1993; Parshad, 1999; Singleton and Petch, 1994; Singleton et al., 1999a). The tropical and sub-tropical climates are conducive to reproduction and population explosions of rodents (Parshad et al., 1989). Frequently, they maintain high population levels in agricultural and rural situations in the Indian sub-continent where large-scale outbreaks still occur and chronic

D. P. Abrol (Ed): Integrated Pest Management.
DOI: http://dx.doi.org/10.1016/B978-0-12-398529-3.00022-1

annual damage continues unabated (Parshad, 1999; Parshad et al., 1989). Changes in land use patterns, i.e. high value crops replacing traditional crops, diversification of agriculture, deforestation, increased transportation, urbanization, etc. has led to shifts in ecology and behaviour of several rodent pest species which is again an issue of concern for plant protection experts and policy makers. No society or nation can bear such an enormous loss to its food basket, especially when food production in most of the developing nations has remained almost static for quite some time. Therefore, besides several efforts to enhance food production, attention is also focused on protection of food grain from the vagaries of pests, weeds and diseases. Effective management of rodent pests in agro-ecosystems can result in at least a moderate 5% increase in food productivity.

The present day rodent management technologies are mainly based on the use of rodenticides, which may be regarded as a short-term solution to the problem. However, for sustainable management of the pest rodent, other options based on crops/cropping systems, ecology and the behaviour of pest rodents need to be integrated as a package. The concept of ecologically based rodent management has evolved in several countries in recent years (Singleton et al., 1999a). The present chapter discusses the rodent problem, and existing control techniques including integrated rodent pest management approaches in the context of different pest situations in India.

20.2 RODENT DIVERSITY

20.2.1 Characteristics of Rodents

The name rodent is derived from the Latin 'rodere'+dent because all rodents possess a pair of ever growing chisel shaped incisor teeth in each jaw adapted for gnawing. The incisor teeth grow at about 0.4mm/day. Rodents do not have canine teeth, leading to a sizable gap between incisor and cheek teeth called a diastema. The cheek teeth are adapted for vegetal food. The incisors help the rodent to gnaw through any type of hard and soft matter (edible or otherwise). The edible matter is taken in and ground by molars, whereas the non-edible matter is thrown out through diastema. Rodents have less than 22 teeth, with one exception, the silvery mole rat with 28 teeth (Singleton and Dickman, 2001). Their jaw muscles are extremely powerful and modify the skull in various ways, which serves as a good characteristic for classification of the order into families. The hairs are soft or in the form of spines/quills, and the tail is either naked or hairy. The digits of the limbs normally bear claws.

Rodents are generally regarded as small mammals and most of them weigh around 150g. However, the largest rodent in the world, the Capybara (*Hydrochoerus hydrochaeris*), weighs up to 60kg and the smallest, the pygmy jerboa (*Salpingotus* sp.), weighs only a few grams and measures a couple of inches (excluding the tail). The Indian crested porcupine, *Hystrix indica*, is regarded as the largest rodent in India. Rodents may lead an arboreal to subterranean/fossorial life.

20.2.2 Global *vis-à-vis* Indian Diversity

Globally rodents constitute >40% of mammalian species, whereas in India 26% of Indian mammalian species are rodent species. Wilson and Reeder (2005) reported 2277 living rodent species under 481 genera in 33 families and 5 suborders of Rodentia in the world. Among these suborders, Myomorpha comprises the largest number of species (1569) under 326 genera followed by Sciuromorpha (307 species under 61 genera) and Hystricomorpha (290 species under 61 genera). A further two suborders, viz., Castorimorpha and Anomaluromorpha, are comprised of 102 and 9 species under 13 and 4 genera, respectively.

Rodent diversity in India is relatively low. Agarwal (2000) has reported 101 species under 43 genera. Pradhan and Talmale (2011) however included 103 species under 46 genera in three suborders, viz., Sciuromorpha (13 genera and 27 species), Hystricomorpha (2 genera and 3 species), and Myomorpha (31 genera and 73 species). The suborders Sciuromorpha and Hystricomorpha are represented by one family each, i.e. Sciuridae (squirrel family) and Hystricidae (porcupine family), respectively, whereas suborder Myomorpha is comprised of five families, viz., Dipodidae (birch mice), Platacanthomydae (dormice), Spalacidae/Rhizomydae (bamboo rats), Cricetidae (hamsters) and Muridae (voles, rats, mice, gerbils, etc.). The Indian rodent fauna not only include endemic species but also forms that have migrated to India from adjoining areas. Thus, they represent a mixture of Indian, Indochinese, Malayan, Ethiopian and Palaearctic elements.

Not all of the rodent species are pest species. In fact, the majority are non-pests. Analysing the rodent ecosystem relationship, Dickman (1999) highlights the beneficial role played by these rodents in the food web and nutrient cycling, in promoting seed dispersal and providing habitats to other species in forest and grassland ecosystems. Only about 10% of the species may be categorized as pest species and need to be managed. From an analysis of rodent fauna diversity vis-à-vis pestilent species, only four out of 67 species in Australia and five out of 61 species in Western Europe are regarded as pests in agriculture (Leirs, 2003).

20.3 RODENT PEST SPECIES OF INDIA

In India, being a highly diverse country, more than a dozen rodent species are regarded as pests of agriculture (Parshad, 1999; Sridhara and Tripathi, 2005) (Table 20.1). Most of the pest species belong to the family Muridae, followed by three species of Sciuridae and one species in Hystricidae. However, a pest complex of 2–4 rodent species occurs in any particular agro-climatic region/cropping system (Table 20.2). The lesser bandicoot rat, *Bandicota bengalensis*, is the most predominant rodent pest species in India and is well distributed in crop fields and residential areas all over the country, apart from the extremely hot arid regions and islands (Chakraborty, 1992; Jain et al., 1993). During the last few years, the lesser bandicoots have encroached on Indian deserts by establishing their population in the urban locales of Jodhpur (Chaudhary et al., 2005; Idris and Rana, 2001) and in Andaman and Nicobar Islands (Chakraborty and Chakraborty, 2010) and NEH regions (Singh et al., 1995). Other species of national importance in the field are *Millardia meltada*, *Tatera indica* and *Mus booduga*. Two species, viz., *Meriones hurrianae* and *Gerbillus gleadowi*, have limited distribution in the arid areas of western Rajasthan. Similarly, the Himalayan rat, *Rattus nitidus*, Bowers rat, *Rattus bowersi*, and white bellied rat, *Rattus (Niviventer) niviventer* are restricted to the North Eastern Hill (NEH) region only (Singh et al., 1995).

In plantation crops, viz., coconut, cocoa, areca nut, cashew, etc., *Rattus rattus wroughtoni* and *R. r. blanfordi* are major problem species (Bhat and Sujatha, 1986; Bhat, 1992). Among squirrels, *Funambulus pennanti* inhabits fruit orchards in North India (Prakash and Mathur, 1987; Patel et al., 1995) and *F. tristriatus* occurs in plantation crops in South India (Bhat, 1992). The Indian crested porcupine, *Hystrix indica*, is regarded as a pest of tuber crops (Agarwal and Chakraborty, 1992). Another species, *Nesokia indica*, has also been listed as a pest of irrigated crops, fruit orchards and forestry plantations (Jain et al., 1995; Ramesh, 1992; Tripathi and Jain, 1990).

The house rat, *Rattus rattus* and house mouse, *Mus musculus* are the two commensal pest species (Krishnakumari et al., 1992; Sridhara and Tripathi, 2005), however *R. rattus* reported from crop fields in central India

TABLE 20.1 Major Rodent Pest Species in India

Family	Common name	Scientific name	Pest status
Scuiridae	Five Striped/Northern Palm Squirrel	*Funambulus pennanti* (Wroughton, 1905)	Pest of fruit and vegetable crops
	Three Striped/Southern Palm Squirrel	*Funambulus palmarum* (Linn., 1766)	Pest of fruit and plantation crops
	Western Ghat Squirrel	*Funambulus tristriatus* (Water-house, 1873)	Pest of plantation crops
Hystricidae	Indian Crested Porcupine	*Hystrix indica* (Kerr, 1792)	Generally low populations. Damages tuberous crops
Muridae (Sub family: Murinae)	Lesser Bandicoot Rat	*Bandicota bengalensis* (Gray, 1835)	Predominant pest of crops and commensal situations
	Larger Bandicoot Rat	*Bandicota indica* (Bechstein, 1800)	Common pest of crops and damages structures
	Soft Furred Field Rat	*Millardia meltada* (Gray, 1837)	Pest of agricultural crops
	House/Roof/Black Rat	*Rattus rattus* (Linn., 1758)	Most common commensal pest
	Wroughton's Rat	*Rattus rattus wroughtoni* (Hinton, 1919)	Major pest of plantation crops
	Himalayan Rat	*Rattus nitidus* (Hodgson, 1845)	Pest of field crops and storage
	Norway/Sewer Rat	*Rattus norvegicus* (Berkenhout, 1769)	Pest in godowns and stores
	Indian Field Mouse	*Mus booduga* (Gray, 1837)	Pest of agricultural crops
	Brown Spiny Mouse	*Mus platythrix* (Bennet, 1832)	Minor pest of dryland crops
	House Mouse	*Mus musculus* (Linn., 1758)	Most common commensal and agricultural pest
	Short Tailed Mole Rat	*Nesokia indica* (Gray, 1830)	Damages field and fruit crops
	Indian Bush Rat	*Golunda ellioti* (Gray, 1837)	Minor pest of field crops
Muridae (sub family: Gerbillinae)	Indian Gerbil	*Tatera indica* (Hardwickei, 1807)	Major pest of dryland crops
	Indian Desert Gerbil	*Meriones hurrianae* (Jerdon, 1867)	Major pest of arid crops and grasslands
	Hairy Footed Gerbil	*Gerbillus gleadowi* (Murray, 1886)	Occasional pest of arid crops and grasslands

showed atavistic behaviour (Dubey et al., 1992; Khatri et al., 1987; Patel et al., 1992). Other species of economic importance are *Bandicota indica, Rattus norvegicus, Golunda ellioti* and *Mus platythrix* (Parshad, 1999). Fifteen species (Table 20.3) are regarded as endemic, i.e. confined within the political boundaries of India (Mandal and Chakraborty, 1999).

TABLE 20.2 Distribution of Rodent Pests in Different Cropping Systems

Cropping System	Species
DRYLAND/RAINFED CROPS	
Arid zone	*M. hurrianae–G. gleadowi–T. indica*
Semi-arid tracts (North Zone)	*T. indica–M. meltada*
Semi-arid tracts (South Zone)	*T. indica–M. meltada–B. bengalensis*
IRRIGATED CROPS	
Arid and semi-arid tracts (North Zone)	*T. indica–M. meltada–M. hurrianae*
Semi-arid tracts (South Zone)	*B. bengalensis–M. meltada*
Deep water rice	*B. bengalensis–Mus* spp.
PLANTATION CROPS	
Coconut–cocoa–arecanut	*R. r. wroughtoni–F. palmarum–F. tristriatus*
Sugarcane	*B. bengalensis–M. meltada–Mus* spp.
Spices	*F. palmarum–F. tristriatus*
JHUM Crops and Pineapple (NEH Region)	*R. nitidus–B. bengalensis–R. sikkimensis*
Stores/Godowns	*R. nitidus–M. musculus–B. bengalensis*

A brief account of the identification, distribution, habit, habitat, etc. of major rodent pest species (Figure 20.1) is given next.

20.3.1 Squirrels

The Northern Palm or Five Stripped Squirrel, *Funambulus pennanti*, is a medium sized rodent weighing 90 g with a bushy tail. The dorsal side is greyish brown with five distinctly white stripes separated by four off-white bands. It is distributed in India, Pakistan, Nepal and parts of Iran. Within India, its distribution ranges from south of Sikkim to the northern district of Dharwad in Karnataka, from Rajasthan to West Bengal and also in Andaman Island. The squirrel is diurnal and generally lives close to human habitation, orchards, gardens, parks and in areas with a fairly good number of trees. During breeding, females construct nests using twigs, rags, hair, etc. The female reaches sexual maturity by about 6–8 months of age. Although reported to breed throughout the year (Banerjee, 1955, 1957), breeding is generally seasonal from March to September with peaks during March–April and July–September in Rajasthan (Purohit et al., 1966). Gestation lasts for 40–42 days and litter size varies from 1 to 5. The young are weaned after 30 days and reach adult size in 4 months. *F. pennanti* causes severe damage to fruits such as pomegranate, sapota, grapes, guava and jujube. Another species, southern palm squirrel, *F. palmarum*, is the counterpart of *F. pennanti* in southern India. The main difference is morphological as there are three white bands on the dorsal side separated by two off-white bands.

The Western Ghat squirrel, *F. tristriatus*, is the largest species of the genus *Funambulus* weighing around 125 g. The dorsal side has three narrow, white, or pale buff stripes separated by black or brown bands. The central stripe is thinner and shorter than the lateral

TABLE 20.3 Rodent Species Endemic to India

Family	Species and Intra-specific Variation, if any	Distribution
Sciuridae	*Funambulus tristriatus* (three sub sp., viz., *tristriatus, wroughtoni* and *numarius*)	Western Ghats; north to Bombay
	Ratufa indica (four sub sp., viz., *indica, centralis, maxima* and *dealbata*)	Peninsular India
	Biswamoyopterus biswasi	Arunachal Pradesh
Muridae	*Alticola albicaudatus*	Kashmir
	Alticola roylei	North Kumaon and North Himachal Pradesh
	Cremnomys cutchicus (four sub sp., viz. *cutchicus, siva, australis* and *rajput*)	Gujarat, Rajasthan, NW UP, east to Bihar and Orissa, south to Karnataka
	Cremnomys elvira	Known only from Kurumbapatty, Salem (Tamil Nadu)
	Millardia kondana	Sinhgarh Plataeu, Pune (Maharastra)
	Platacanthomys lasiurus	Forested tract of Western Ghats
	Rattus burrus	Islands of Trinkat, Little Nicobar and Great Nicobar
	Rattus palmarum	Nicobar Islands
	Rattus ranjiniae	Trivendrum and Trichur (Kerala)
	Rattus stoicus	South and Little Andamans
	Mus platythrix	Bihar, Peninsular India, Rajasthan, West Bengal
	Mus famulus	Nilgiri, Annamalai and Palani Hills of South India

ones. The belly is whitish with hair bases distinctly grey. It has a limited distribution in western and south western India extending from Mumbai down to Travancore in the south, mostly in the Western Ghats. It breeds year-round with peaks occurring from December to May (summer) with a minimum number of breeding females during June–August, the months of heavy rainfall on the west coast of India. It causes severe losses to cocoa and other plantation crops.

20.3.2 The Indian Crested Porcupine, *Hystrix indica*

The crested porcupine is the largest rodent species in India measuring 680–750 mm in length and weighing 11–18 kg. The neck and upper back are covered with distinct long, stiff, bristle-like hairs called quills (15–30 cm). The body is covered with alternating dark brown and white quills and the tail is covered by short and broad quills. Short, coarse, black hairs thinly cover the ventral surface. The 'rattling quills' on the tail are white in colour, large and open. It is widely distributed in India, Bangladesh, Nepal, Sri Lanka, and Pakistan, extending to Israel, Arabia, and southern and eastern Russia. Porcupines are nocturnal and live in caves, amongst rocks, or in burrows dug by them or by other animals. Burrows can extend up to 18 m in length, and 1.5 m depth, leading to a large square chamber. Porcupines breed throughout the year. Gestation lasts for 109–112 days. Litter size ranges from 1 to 8 (Prakash, 1971). During crop maturity, they lie in thick shrubs near cultivated fields or take

Major rodent pest species of India

Ever-growing incisors: weapon of rodents

High breeding potential: *Tatera indica female* with nine young ones in a litter

Bandicota bengalensis

Millardia meltada

Meriones hurrianae

Mus booduga

FIGURE 20.1 Major rodent pest species of India, rodent damage to different crops and rodent management strategies.

shelter amidst the nearby tall grass, if available. They are known to be chiefly herbivorous, feeding on succulent tubers, bulbs, ripe fruits and the bark of trees. Grain, vegetables of all kinds and roots are also consumed.

20.3.3 The Lesser Bandicoot Rat, *Bandicota bengalensis*

The lesser bandicoot is robust with a round head and a broad muzzle, and weighs

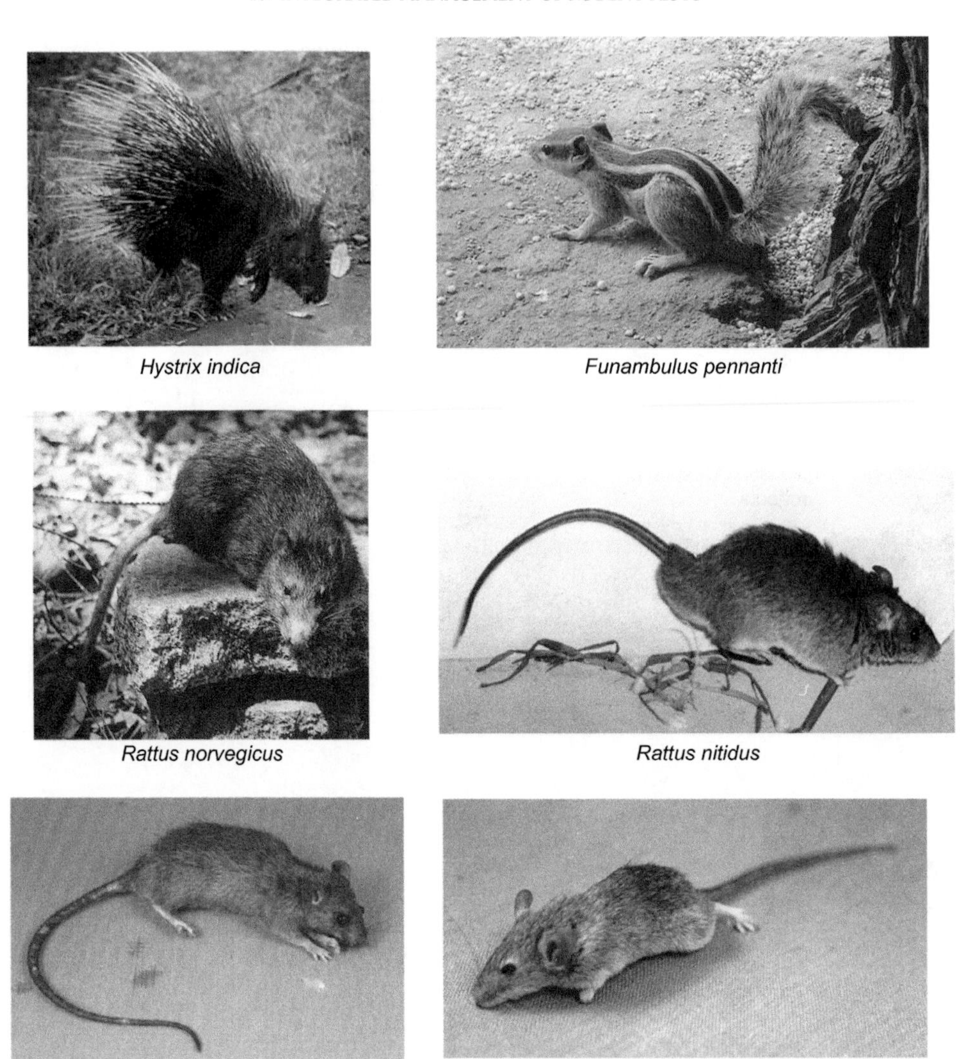

FIGURE 20.1 (*Continued*)

200–350 g. The body is covered with coarse fur which forms black-tipped piles on the dorsal side. The colour on the dorsal side is dark brown but may be blackish, pale brown or reddish. The feet are dark but the digits are paler. The tail is completely dark and occasionally paler below. The belly is grey or light grey and rarely whitish. Apart from extreme arid tracts in western Rajasthan, *B. bengalensis* is widely distributed throughout India, Pakistan, Nepal, Bhutan, Bangladesh, Sri Lanka and South East Asia. The lesser bandicoot rat is a fossorial animal and is well adapted to various habitats and ecological conditions which include cultivated fields, pastures, forests, mountains, inter-tidal mangrove zones, and semi-arid zones and also

FIGURE 20.1 *(Continued)*

as a commensal in towns and cities across India. It also breeds throughout the year, with definite peaks occurring during different seasons across the country. The litter size ranges between 4 and 12. The lesser bandicoot is a serious agricultural pest in India causing extensive damage to food and vegetable crops and coconut nurseries. The burrowing activity of lesser bandicoot rats

Rodent management strategies

Trapping: Buta traps

Smoking the burrows with fumigator

Preparation of fresh poison baits

Burrow baiting

Station baiting (PVC pipes)

Station baiting (coconut husk)

FIGURE 20.1 (*Continued*)

causes damage to roots causing the slow death of trees/plants and irrigation channels.

20.3.4 Larger Bandicoot Rat, *Bandicota indica*

B. indica is a very large rat, head and body normally ranging from 200mm to 366mm. Weight ranges from 500g to 2kg and more. The tail is shorter than the head and body (HB) and is covered with hair throughout its length. The fur is very rough and quite long dorsally. The upper part of the body is dark or blackish brown and the ventrum is grey, drab or dark. In India, its distribution ranges from south of Rajasthan down to the southern tip of India and eastwards too. The species is widely distributed in Bangladesh, Myanmar, Malaysia, Thailand, Vietnam, Sumatra, Java, southern China, Hong Kong, Taiwan and Sri Lanka. The larger bandicoot always lives close to human habitat but never inside the house or inside crop fields. It prefers a habitat that has a lot of garbage for feeding and is close to water bodies. They prefer places close to human dwellings such as compounds, gardens, stables, and poultry and outhouses. The species is nocturnal and fossorial. Burrows are found amidst tall grasses and bushes around marshy areas, often tunnelled through bricks and masonry of poorly constructed houses and huts in villages. Burrows are simple consisting of an unbranched tunnel up to 700cm in length and 6–14cm in diameter. It breeds all year-round but has a seasonal reproduction peak from September to March in India (Chakraborty, 1975). Although there is no estimate of losses, the bandicoots seriously damage crops. In deltaic islands, serious indirect damage is caused to cultivation by their tunnelling of the embankment causing seepage of saline water into rice fields (Chakraborty, 1992).

20.3.5 Indian Gerbil, *Tatera indica*

This is a medium sized rodent weighing about 100–150g. The tail is longer than HB and possesses tufts of hair at the tip (characteristic of all gerbils). The hind feet are longer than the fore feet. The eyes are large and ears round. Generally it is light brown in colour, varying from sandy brown to red on the dorsal side and pure to off-white ventrally. The feet are white in colour. The range of distribution of *T. indica* extends westwards to Iran, Syria, Turkey, Iraq, Arabia, Afghanistan, Pakistan and southwards to Sri Lanka. However, in India, the species is ubiquitous occurring throughout India apart from the hills. The species inhabits open plains, and loose sandy soils of the desert, and is usually found at the edges of cultivation. Burrows are dug near hedges, thickets or under bushes, sometimes also inside the field when conditions are dry. Burrows are easy to notice with beaten paths leading from one opening to another. The burrows are simple with 1–2 branches. Gerbils are nocturnal and their food consists of grain, roots, leaves and grass. *T. indica* breeds throughout the year in arid Rajasthan with maximum littering in the month of August and a minor peak in February (Jain, 1970) with a litter size of 1–9. In Karnataka, peak reproductive activity was seen during October–November with a mean litter size of 1–12. *T. indica* is a serious pest of all crops and is regarded as a reservoir of plague bacteria.

20.3.6 Indian Desert Gerbil, *Meriones hurrianae*

Geographically, the genus *Meriones* occurs from Morocco in North Western Africa to the Indian desert in the east, to Russia through the deserts of China, Mongolia and Manchuria in North China. In India, its species (*M. hurrianae*) is restricted to the northwest desert of Rajasthan, North West Gujarat and Punjab. The adult body weight of *M. hurrianae* is 40–160g, colour is sandy grey to brownish grey dorsally and white to off-white ventrally. The tail is pale with black or dark brown tussles of hair at the tip. Gerbils prefer a sandy habitat, most following the ruderal habitat of arid zones. Their burrows are up to 1m deep and have an emergency exit hole. Burrows can be found not only near cultivation but far away in waste lands, thorny forests, and open desert. Sand dunes, banks, and windblown mounds of sand consolidated beneath desert shrubs and plants are also preferred sites for burrowing. Burrows have cooler temperatures during summer and are warmer in winter. Desert gerbils are diurnal and live in smaller colonies. Although herbivorous, their diet selection is influenced by the availability of food with a definite seasonal pattern. Seeds form the major part of the diet during winter, rhizomes and stems in the summer months, to some extent insects, and in the rainy season, leaves and flowers form the bulk of the food consumed. They breed throughout the year with two peaks in February and July (Prakash, 1981) but a third peak was observed during September–November (Kaul and Ramaswami, 1969) with a litter size of 1–9. It is regarded as a common pest of bajra, wheat, chillies, vegetable crops, grass and other vegetation.

20.3.7 The Hairy Footed Gerbil, *Gerbillus gleadowi*

This is a nocturnal rodent with restricted distribution in western Rajasthan. The dorsum is grayish and the ventrum whitish. They have long tails with tufts of hair at the tip. The feet of *G. gleadowi* are hairy, thus the name hairy footed gerbil. The species breeds during the winter and summer months with a litter size of 2–5. It inhabits sand plains and dunes and attains serious pest status occasionally.

20.3.8 Short Tailed Mole Rat, *Nesokia indica*

This species is heavily built weighing more than 200g. The dorsum is dull to brown with a lighter ventral side. The tail is short and fur is short and rough. The species has a wide distribution, spread over India, Pakistan, Iraq, Iran, Egypt, Russia, Afghanistan, Syria, Chinese Turkistan and Northern Arabia; in India, it is only found in Punjab, Rajasthan, Haryana, Uttar Pradesh, Himachal Pradesh and Delhi. It prefers bunds in cultivated fields along water channels but also occurs in natural vegetation and garden lawns. The species is nocturnal with a bimodal circadian rhythm and is fossorial. It excavates extensive burrows with a heap of soil at the burrow entrance. Although it breeds throughout the year under laboratory conditions, in nature, breeding only occurs during winter with a litter size of 2–5. The species is omnivorous, however it prefers grains, seeds, fruits, tubers, etc. and causes serious damage to crops, fruits and forestry plantations.

20.3.9 The Soft Furred Field Rat, *Millardia meltada*

This species, also referred to as metad, is distributed throughout India, apart from the NEH region. Body colour is light to dark grey dorsally with feet and belly being off-white. The tail is similar to its body colour with dark grey above and pale below. The animals weigh between 40 and 70g. It is one of the most predominant rodent pests in almost all of the states inhabiting crop fields and grasslands, usually choosing the drier patches. It is also reported from ruderal habitats, scrub grassland, gravelly areas and the sandy plains of Rajasthan. It is a nocturnal animal and makes simple burrows which are slightly slanted. In Rajasthan, metads breed throughout the year with peak reproduction occurring in spring and the monsoon season, whereas in south India, its breeding season extends from July to early March with peak reproduction seen during September–November (Govind Raj and Srihari, 1989).

20.3.10 The House Rat, *Rattus rattus*

It is a medium sized rat weighing 150–200g. *R. rattus*, also called the roof rat, black rat, and ship rat, is the most abundant and widely distributed rodent species in India as well as globally. It is characterized by a long tail, slender body and pointed snout. The dorsal fur is mostly blackish in commensal forms and ranges from yellow to brown black with a pale white belly in wild forms. It is nocturnal and colonial and lives in houses, godowns, stores, poultry farms, crop fields, and adjacent to villages. House rats breed throughout the year, reportedly with two peaks of reproduction, viz. March–April and August–September (Krishnakumari et al., 1992) with a litter size of 1–9.

20.3.11 Wroughton's Rat, *R. rattus wroughtoni*

This is a subspecies of *Rattus rattus* with a characteristic white belly. It weighs around 95g and is found abundantly in southern India from the semi-evergreen forests, scrub jungles, teak plantations of Karnataka (Sreenivasan, 1975), throughout the state of Kerala in houses, coconut palms and tree cavities (George et al., 1980) and is a serious pest of coconut in Kerala, Karnataka and Andhra Pradesh and of areca nut and cocoa plantations in Kerala and Karnataka (Bhat et al., 1990; Bhat, 1993; Reddy, 1998). It is an arboreal rodent spending more than 80% of its time in tree tops. It lives in nests constructed in tree holes in forests and either in the interspace of nuts or inside stipules in the spindle portion of coconut. It is a serious pest of plantation crops.

20.3.12 Norway Rat, *Rattus norvegicus*

R. norvegicus is primarily a temperate zone rodent both as a commensal and field pest. Its distribution in India is limited to port cities and larger towns and cities mostly around grain godowns (Roonwal, 1987). In addition, it has also been reported from Meghalaya and Mizoram (Singh et al., 1995). It is nocturnal and commensal in India and weighs 250–350 g. Its body is brownish dorsally and its ventrum is whitish or off-white with a stout tail which is shorter than HB. Although it inhabits sewers elsewhere, in Kolkata and Mumbai, it digs burrows in godowns and is a warehouse pest (Jain et al., 1993).

20.3.13 Himalayan Rat, *Rattus nitidus*

This has been reported from all of the states of NEH region. It is a medium sized rodent weighing 100–175 g. Its dorsum is usually dark brown, occasionally with a darker mid-dorsal patch. Its ventrum is silvery grey or off-white. The tail is wholly dark and naked. It is regarded as a pest under commensal situations, also in fields causing damage to pineapple, rice, maize, etc. in NEH regions.

20.3.14 House Mouse, *Mus musculus*

This is a widely distributed species. The house mouse is a very small rodent weighing 15–20 g. The tail is naked and longer than the HB length. Dorsally, the colour varies from brown to light brown with the belly being whitish or light grey. It is nocturnal in habit nesting in rafters, in crevices in walls, amidst staked undisturbed bags of food grain in godowns, in table drawers, and often lives in fields by digging burrows. The mouse breeds year-round with a litter size of 1–8. Being commensal, it is a nuisance to many products in addition to spilling and spoiling a lot more than eating. In fields, it is known to damage sugarcane, groundnut, etc.

20.3.15 Indian Field Mouse, *Mus booduga*

This tiny mouse is distributed all over India (Rao and Balasubramanyam, 1992). Recently, it has been reported from Andaman and Nicobar Islands (Birah et al., 2012a). The animal weighs between 10 and 15 g. Colour of the dorsum varies from dark brown to lighter sandy brown with under parts whitish to slightly grey. The tail is bicoloured and shorter than HB length. This nocturnal and fossorial mouse makes simple burrows at 50–60 cm depth with characteristic scooped soil near openings. Peak breeding is during September/October and February to June with a litter size of 6–13 (Srivastava, 1968). It occupies second position after *B. bengalensis* in rice, wheat and sugarcane fields in many regions causing serious damage.

20.4 THE RODENT PROBLEM

20.4.1 Why are Rodents a Problem?

Rodents are highly adapted to a variety of habitats, and crop fields and food grain stores provide most conducive environments for field and commensal rodents, respectively. The gnawing habit of rodents owing to their ever-growing incisors, and their vast breeding potential make them one of the most destructive organisms. In addition, rodents possess a great feeding potential as they generally consume 5–10% of their body weight on a daily basis. Agricultural fields serve as a highly productive rodent habitat and crops such as sugarcane, rice, wheat, groundnut and fodder serve as an ideal habitat for rodent pests. Similarly, threshing yards located near crop fields also act as an excellent abode for food and shelter of rodents. For example, the pearl millet crop suffers higher damage by *Meriones hurrianae* when the harvested crop is heaped in the threshing yard than at the crop standing stage (Prakash, 1976).

20.4.2 Arable Crops

A seed, plant, fruit or its produce in any form is always exposed to the vagaries of rodent depredation. It being a herbivore, the choice of food for a rodent is unlimited. No crop is spared by these tiny vertebrates at any stage of their growth. The extent of rodent damage to crops largely depends upon (i) the species involved, (ii) the location and stage of the crop, (iii) the population of the pest, (iv) availability of the crop, and (v) the physical environment. Buckle and Smith (1994) reported that rodents such as *Sigmodon* and *Holochilus* inflict 10–15% damage to rice and sugarcane in South America, whereas *Xerus* and *Praomys* are a major problem (up to 50% damage) in food crops in African countries such as Tanzania and Kenya. In south Asian countries, *Rattus argentiventer* is the major problem species inflicting 2–47% damage to rice. *B. bengalensis* is the predominant rodent in the Indian subcontinent inflicting 20–40% damage to cereals (Pakistan), up to 30% damage to wheat (Bangladesh), and 5–22% damage to rice (India).

The extent of losses (Table 20.4) has been well documented in food crops, sugarcane, pulses and oilseeds (Advani, 1985; Awasthi and Agarwal, 1991; Chopra et al., 1996; Jain and Tripathi, 1992; Kochar and Kaur, 2008; Malhi and Parshad, 1987; Parshad, 1999; Rao, 2003; Srivastava, 1992; Sridhara and Tripathi, 2005; Vyas et al., 2000). Cereals such as wheat and rice suffer pre-harvest rodent damage in the range of 5–15% in different regions. Malhi and Parshad (1987) reported an additional dimension to the rodent problem by recording a loss of 4.3% wheat panicles (equivalent to 1.11 q/ha of wheat) and 4.64% rice panicles (equivalent to 1.72 q/ha of coarse rice) after harvest on the threshing floor.

Sugarcane is highly vulnerable to rodent depredation, experiencing up to 31.0% direct damage in different regions of India. Rodent attack on sugarcane not only decreases the crop yield but also adversely affects sugar recovery to the tune of 25% (Gupta et al., 1968). Similar observations have been reported by Christopher (1987) from Andhra Pradesh with reduced recovery of jaggery (45.6 kg/ha at 20.7% cane damage). The damaged canes develop serious fungal and bacterial infections leading to rotting of affected canes (Parshad, 1897). Being a long duration crop, sugarcane acts as a reservoir habitat for rodents and rodents become a perennial problem not only to sugarcane but to rice/wheat crops grown as a cropping system. Five-year studies from Punjab (2002–07) indicated that rice and wheat crops respectively experienced 0.2–3.0% (mean: 1.13) and 0.5–15.2% (mean: 4.21) rodent damage under monoculture situations, however, when those crops were grown with sugarcane in the surroundings, the damage increased to 5.1% in rice and 10.8% in wheat (Kochar and Kaur, 2008).

Pulses such as mung bean (*Vigna radiata*) and moth bean (*Vigna aconitifolia*) suffered about 3% pod damage in arid regions (Tripathi et al., 2004) and 5–6% in Gujarat (Butani et al., 2006). Similarly, cow pea suffers 4–18% rodent damage. Awasthi and Agarwal (1991) reported a yield loss of 16.5 kg/ha at green pod stage in soybean crop in Madhya Pradesh. Groundnut, an important oilseed crop, also serves as an ideal rodent habitat where rodents registered 4–7% pod damage, besides hoarding 320 g/burrow (Mittal

TABLE 20.4 Rodent Damage and Species Infesting Field Crops in India

Crop	Stage	Damage (%) YL (yield loss kg/ha)	Species	State/Region
Wheat (*Triticum aestivum*)	Seedling to maturity	5.9		Rajasthan
	Pre harvest	18.7–21.3	*Mh, Ti, Rm*	Rajasthan
		6.3–8.2	*Bb, Rm*	Himachal Pradesh
		3.9–5.2	*Bb, Ti*	Punjab
		8.0–10.0	*Bb*	Uttar Pradesh
		3.0–21.0	*Bb, Ti, Mm*	Gujarat
Rice (*Oryza sativa*)	Pre harvest	1.1–17.5	*Bb, Rm*	Punjab
	Pre harvest	98–213 kg/ha	*Bb*	Uttar Pradesh
	Grain formation	9–10	*Bb, Mm, Mb*	Karnataka
	Harvest stage	17.56	*Bb, Mb*	Andhra.Pradesh
	Milky to maturity	4.6–16.8	*Bb, Rn, Mm*	NEH region
Pearl millet (*Pennisetum typhoides*)	Seedling	100 (crop resown)	*Gg*	Rajasthan
	Milky, Grain	Considerable	*Ti, Mh*	Rajasthan
		3.0–12.0	*Bb, Ti, Mm*	Gujarat
Maize (*Zea mays*)	Cobs	9.8	–	Himachal Pradesh
	Cobs	9.1	*Rn, Bb*	Meghalaya
	Seedling	10.7	–	Punjab
	Seedling	50–80	*Bb, Ti, Mm*	Karnataka
	Cob formation	7.0	*Bb, Ti*	Karnataka
	Harvest	12.5	*Ti, Bb, Rr, Mb*	Karnataka
	Cobs	5.0	*Bb, Ti, Mm*	Gujarat
Bengal gram (*Cicer arietinum*)	Pods	2.5	*Mm*	Madhya Pradesh
	Plants and pods	3.0–25.0	*Bb, Ti, Mm*	Gujarat
Soybean (*Glycine max*)	Green pods	27.27	*Bb, Mm, Rr*	Madhya Pradesh
		YL 44.76	*-do-*	Madhya Pradesh
	Pod formation	0.6–3.0	*Mm, Ti*	Karnataka
Groundnut (*Arachis hypogaea*)	Plants and Pods	3.9–19.0	*Ti, Rm,*	Punjab
	Pod setting	4.5	*Bb, Mb, Mm*	Gujarat
	Pod maturity	6.9	*Bb, Ti*	-do-
	Harvesting	7.3	*Mm*	-do-
	Peg formation	30–40	*Bb, Ti*	Karnataka
	Hoarding	2%	*Bb, Mb*	Karnataka

TABLE 20.4 (Continued)

Crop	Stage	Damage (%) YL (yield loss kg/ha)	Species	State/Region
Cotton (*Gossypium* sp.)	Bolls	3.2–23.2	*Ti, Rm*	Gujarat
	Damaged bolls	4.0–6.0	*Bb, Ti, Mm*	Gujarat
Sugarcane (*Saccharum officinarum*)	Partial damage to canes	2.1–21.6	*Bb, Ti, Rm, Mm*	Punjab
	Dried canes	3.2	–	Punjab
	Without lodging	6.8	*Ti, Bb*	Uttar Pradesh
	With lodging	18.9	*Ti, Bb*	Uttar Pradesh

Adapted from Sridhara and Tripathi (2005).
Ti, Tatera indica; Rm, Rattus meltada; Bb, Bandicota bengalensis; Mh, Meriones hurrianae; Mp, Mus platythrix; Rr, Rattus rattus; Ge, Golunda ellioti; Mb, Mus booduga; Mm, Mus musculus; Rn, Rattus nitidus.

and Vyas, 1992; Rao, 2003). During outbreaks (1976 and 1988–89), the groundnut suffered up to 85.42% damage in the Saurashtra region of Gujarat (Shah, 1979; Vyas et al., 2000).

Other field crops such as millet, sorghum and maize suffer extensively due to rodent infestation. A classic example of rodent damage to pearl millet (by *Gerbillus gleadowi*) was reported by Prakash (1976) where gerbils caused complete damage and crops had to be resown in western Rajasthan. Kumar and Mishra (1993) and Singh et al. (1994) registered 9% rodent damage to maize cobs in Himachal Pradesh and Meghalaya, respectively. Similarly, in cotton, rodents devoured about 3.2–23.2% bolls in Gujarat (Rana et al., 1994) and up to 55% plants with rodent damage were noticed by Neelanarayana et al. (1994) from Tamilnadu.

The hoarding behaviour of *B. bengalensis* further intensifies the rodent problem. The earliest report of such a loss was by Wagle (1927) when he recovered 600 rice ear heads from one bandicoot burrow. Deoras (1966) recovered 14 kg of rice, groundnut and millet from 30 bandicoot burrows. Parack and Thomas (1970) stated that bandicoots hoard 54 kg/day, which is five times their food requirements. Fulk (1977) reported hoarding of 100 kg/ha by the lesser bandicoots in Sind province of Pakistan. Rao and Singh (1983) and Sheikher and Malhi (1983) reported that bandicoots hoarded 1.7 kg and 390 g/burrow, respectively. Singh and Saxena (1989) observed about 60% of burrows with hoarded materials. Vegetable crops are equally vulnerable to rodent damage. Cucurbits such as cucumber, musk melon, ridge gourd and sponge gourd suffer up to 10% rodent damage (Butani et al., 2006; Kumar and Pashahan, 1995; Malhi and Parshad, 1992). Among solanaceous vegetables, rodents inflict a fruit damage of 11.1–37.3% (Punjab), 19% (Rajasthan), 2.6–35% (Gujarat) and 5% (Karnataka) to tomato. Similarly, brinjal and potato experience 2–6% rodent damage (Sridhara and Tripathi, 2005). In arid regions, the vegetables experienced 4.4–19% damage by desert rodents (Tripathi et al., 1992). A similar pattern of rodent damage has been reported for other vegetables, such as pea, carrot, cabbage, chillies and beans. Porcupine is also reported to cause serious damage to tuberous crops such as potato (Chandla and Kumar, 1995).

20.4.3 Perennial Crops

Besides the crop losses, rodents inflict severe damage to perennials such as fruit and plantation crops, afforestation plantations, and range/grasslands. In exhaustive reviews on rodent damage, Sridhara and Tripathi (2005) reported that apple, peach, pecan and plum experience 17–40%, 2–7%, 1.6–6.7% and 1–2% rodent damage, respectively in Himachal Pradesh. Similarly, ripe fruits of pomegranate are devoured to the extent of 17–22% by squirrels in Rajasthan (Patel et al., 1995). Date palm is another crop highly vulnerable to squirrel attack, which causes 18–20% fruit damage. Ber (Zuzube), an important fruit crop in arid regions, suffers maximum damage (up to 80%) in nurseries. Among plantation crops, coconut and cocoa are extensively damaged by *R. r. wroughtoni*, *F. palmarum* and *F. tristriatus* in south India. Estimates of rodent damage to nuts account for 21–28.5% in Kerala, 12–15% in Karnataka, 4.5–55% in Lakshdweep and 4.1–5.8% in Andaman and Nicobar Islands (Advani, 1984, 1985; Birah et al., 2012b; Chakravarthy, 1983). Similarly, cocoa pods suffer very high squirrel damage (50–60%) in Tamilnadu and Kerala (Abraham and Remamony, 1999; Bhat et al., 1981). The oil palm cultivation suffered a serious setback due to infestation of *R. rattus* in Andaman and Nicobar Islands as the crop suffered 10–29.5% damage to seedlings and saplings and 57.3% fruit damage (Subiah, 1983).

Rodents extensively damage forestry plantations by debarking and or slicing of tress. The short tailed mole rat *N. indica* was reported to slice the root of arid zone plantations to the tune of 4.4–10% (Tripathi and Jain, 1990). In arid regions, where vast stretches of sandy waste/pasture lands are available, the desert gerbil *M. hurrianae* causes havoc to pasture grasses. Prakash and Mathur (1987) reported a density of 477 gerbils/ha in a pasture, which required 1040 kg/ha feed, whereas the annual productivity of the land was only 1210 kg/ha, thereby leaving almost nothing for desert livestock.

This diurnal rodent was reported to dig up to 14,000 burrows in a plot of 100×100 m (Prakash, 1981) and can loosen 61,500–161,000 kg soil/day/km^2 in cultivated fields in arid regions which can easily be blown away by high velocity winds causing problems of desertification and soil erosion (Sharma and Joshi, 1975).

20.4.4 Rodent Outbreaks

In addition to normal rodent damage of around 5–15% to crops, rodent outbreaks are also reported in certain years when the rodent population explodes due to a variety of ecological factors (Jain and Tripathi, 2000; Rao, 2003). Rodent outbreaks in the NEH regions (Mizoram, Arunanchal Pradesh, Manipur, etc.) leading to famine have been experienced during 1880–83, 1910–12, 1928–29, 1958–59, 1976–77 and 2005–09 (Singh et al., 1995) when rice, maize and other crops suffered heavily to the tune of 75–90%. This outbreak is attributed to the gregarious flowering of certain bamboo species (*Bambusa tulda* and *Melocanna baccifera*) in the region. A similar situation was experienced recently during 2005–09 (Chauhan and Saxena, 1985; Rajendran et al., 2007). The next bamboo flowering related rodent outbreaks are expected somewhere during the period 2023–25. Large-scale devastation of crops has also been reported from other regions. For example, rodent upsurges in Gujarat and Rajasthan during 1901–02, 1913–14, 1970–71, 1975–76 and 1997–98 have been documented by Mittal and Vyas (1992), Prakash and Mathur (1987) and Shah (1979). In South India too, rodent outbreaks, as evident from an upsurge of plague, were recorded in 1826, 1878–79, 1901–02 and 1913–14 (Kinnear, 1919). Rao (2003) has listed 11 such recent events during 1990–2000. Of these, five outbreaks between 1999–2000 occurred in Manipur, Arunanchal Pradesh, Mizoram and Nagaland and other outbreaks took place in Southern India, viz., Cauvery Delta (1994–2001), Godawari Delta (1997–98 and 2001) and Pondicherry (1994). The Saurashtra region of Gujarat also suffered heavy devastation of groundnut, wheat, gram and cotton crop due to a rodent outbreak in 1990–91. Large-scale rodent damage to rice and pulses has been reported from Godawari delta during the last 2–3 years. According to Rao (2003), a sudden upsurge of the rodent population leading to an outbreak may be due to (i) a prolonged dry spell followed by heavy rains, (ii) failure of the monsoon in the preceding years, (iii) flash floods and (iv) gregarious flowering of bamboos.

20.4.5 Post Harvest Losses during Storage

About two-thirds of farm produce is stored by farmers. Rao (2003), quoting the report of a committee constituted by the Government of India, indicated that stored food grain suffered 9.33% loss during storage of which losses due to rodents accounted for 2.5%. House rats, house mice and lesser bandicoots are the major problem species relating to storage. They consume food equivalent to 5–10% of their body weight. There are reports which indicate severe rodent infestations in some premises to the extent of 8 rodents/house (Singh et al., 1990) in villages of Udaipur district (Rahasthan); 10.7 rats/godown of $106\,m^2$ area (Krishnamuthy et al., 1967), near Hapur (Uttar Pradesh), and about 200 bandicoots from godowns of Kolkata (Franz, 1975). In one of the pilot studies, 272,247 rats and mice were killed in 65,433 houses in 80 villages in Gujarat (Chaturvedi et al., 1975). The main reasons for the survival of large populations of rodents in most of the storage premises are inadequate maintenance of buildings coupled with lack of hygiene, poor handling of stored commodities and a serious neglect of rodent proofing. In addition to direct damage, rodents contaminate the stored commodities by their hair, urine and faecal pellets making them unfit for human consumption. The bandicoots and house mice defecate 22 and 13 pellets daily, respectively, and pollute the storage environment (Nimbalkar, 2000).

20.4.6 Public Health

Rodents are known to spread 31 diseases to humans and animals. They are also reported to transmit 11 documented and 12 non-documented hantavirus diseases, however no indigenous hantavirus has been reported from India so far (Biswas and Mittal, 2011). Some human diseases spread by rodents are plague, salmonellosis, rabies, tularaemia, leptospirosis, amoebic dysentery, typhus (scrub and murine), jaundice, trichinosis, lymphocytic choriomeningitis, leishmaniasis, ray fungus, and ringworm. Rodents also transport and host ectoparasites such as mites, lice, fleas, and ticks.

Plague is one such deadly disease, which claimed 12,657,077 human lives from 1896 to 1996 in India (John, 1996; Viliminovic, 1972). The plague epidemic during 1994 in Gujarat resulted in 54 deaths out of 876 cases (WHO, 2000a). Several sylvatic plague foci have been recognized in India in Jammu and Kashmir, Himachal Pradesh, Uttarakahnd, Maharastra, Andhra Pradesh, Karnataka and Tamilnadu states (John, 1996). Rodents are the primary host of plague bacteria (*Yersinia pestis*), which are transmitted to humans by the oriental flea (*Xenopsylla cheopis*), an ectoparasite of rodents. Indian gerbil, *Tatera indica*, is considered to be a reservoir of plague bacteria as gerbils are susceptible to infection but resistant to the disease. Prakash (1991) and Rao (2003) reported that there is a possibility that plague infection present in *T. indica* (the reservoir species) may be transmitted to *B. bengalensis* and *R. rattus* (the susceptible species) through fleas to ruderal habitats. This may lead to an incidence of human plague. Leptospirosis is another zoonotic disease which is spreading to humans in different parts of the country. The

International Leptospirosis Society opined that more than 100,000 severe cases of leptospirosis occur annually worldwide (WHO, 1999). In India, Andaman and Nicobar Islands, Gujarat, Kerala, Orissa, Maharastra and Tamilnadu are the endemic states where this disease is quite prevalent. In addition to rodents, cattle, pigs, dogs and cats are also major hosts of the bacterium, *Leptospira icterohaemorrhagiae*, which is the common sero-group (Rao, 2003). Among rodents, *R. norvegicus, R. rattus* and *B. bengalensis* are associated with this disease (Gangadhar, 1999). The organism lives in the kidney of animals and is excreted through urine. Humans get the infection through abraded skin coming in contact with infected water. People working in rice fields and other waterlogged areas are always vulnerable to the infection. Several cases of leptospirosis have been reported in recent years from Gujarat, Kerala, Orissa and Maharashtra (Rao, 2003; Thakar et al., 2012; WHO, 2000b).

20.4.7 Other Sectors

Many other economic sectors suffer greatly due to rodents. Poultry farms provide a very favourable habitat where rodents maintain large populations. Through burrowing, nibbling and gnawing, feeding, urination, defecation and extensive movements, rodents damage and contaminate poultry farm environments and cause severe economic losses. Rodents are also known to spread several diseases among birds and to poultry keepers (Parshad, 1999). Burrowing and damage to structures increase the maintenance costs of farms. Rodents, especially the commensal ones, viz. *R. rattus, M. musculus* and *B. bengalensis*, are also regarded as a serious menace in many other sectors such as airports, telecommunications, railways, IT, and food and beverage industries. Airports provide very favourable habitats for rodents. Rana and Tripathi (1992a) have reported on the rodent problem in airports in India. Similarly, Witmer and Fantinato (2003) highlighted the problems at airports and suggested measures for their management. The authors reported that, in addition to structural damage, a large population of rodents also attracts raptors, which may cause a direct collision hazard to aircraft moving on the ground (Barras and Seamans, 2002). There are no scientific reports from India about the extent of damage to many of these sectors, but simple nibbling/gnawing of cables by native rodents may cause immense losses. Rana and Tripathi (1992b) have analysed the rodent problem relating to fibre optic telecom cables.

20.5 ECOBIOLOGY OF RODENTS

20.5.1 Population Ecology and Damage Thresholds

Rodents are prolific breeders with a small oestrus (3–7 days) and gestation period (~3 weeks) and a large litter size of up to 20 young ones per female; therefore, they are capable of rapid population growth especially when the conditions are favourable. Such a high fecundity of rodents is countered by several biotic and abiotic factors operating in nature. Rodents sometimes regulate their own population by feeding on their own young-ones (cannibalism) during any type of stress depending upon the carrying capacity of their habitat. In drought years, the rodent population maintains a low profile due to scarcity of natural food; however, the population explodes in a good rainfall year succeeding any drought year. Most of the rodent species exhibit relatively lower population levels during winters and summers (Jain, 1970; Prakash, 1981; Prakash and Kametkar, 1969; Rana and Prakash, 1984; Tripathi, 2005). During monsoon and post monsoon seasons, the rodent pest population attains its peak in arid regions. Based on monthly trapping data

TABLE 20.5 Monthly Population Fluctuations in Desert Rodents (nos per ha)

Rodent Species	Jan	Feb	Mar	Apr	May	Jun	Jul	Aug	Sep	Oct	Nov	Dec
F. pennanti	–	–	–	102	54	61	36	47	28	30	10	6
T. indica	30	39	51	53	63	69	55	74	78	42	40	43
M. hurrianae	–	–	300	–	–	131	–	170	–	–	–	385
M. meltada	–	–	12	–	10	–	12	0	–	10	–	–

After Prakash (1975).

on a few rodent species occurring in the Indian desert, Prakash (1975) has reported that the population of *F. pennanti* gradually decreases from April to October, whereas *T. indica* shows two minor peaks, one during March–April and the other during August–September (Table 20.5). The desert gerbil, *M. hurrianae*, shows a population build-up during winter, which continues until spring, and then their number declines during the summer.

B. bengalensis was reported to show a very rapid population growth rate from 108/ha in November to 446/ha in December during winters in west Bengal (Chakraborty, 1975). This species breed under field situations more than five times in a year (3.24 times in *kharif* and 2.14 in the *rabi* season) with an annual productivity of 48.7 young ones per female during the year 2007–08 in the Godawari delta whereas year-round breeding is observed in urban locales at Jodhpur. *M. meltada* shows peak breeding activity in the monsoon and spring season with an annual productivity of 52.52 young ones/female whereas it was 17.72, 27.03 and 31.0 for *T. indica, R. rattus* and *M. musculus*, respectively (Tripathi, 2005).

Limited studies have been made in India on economic threshold levels for rodent pests. To work out these levels, the rodent pest population needs to be regularly monitored to decide on any management action. This may be done based either on a burrow count or on the actual damage done by pests. Buckle and Rowe (1981) reported that in rice, 15% of hills or 2% of tillers with rodent damage could be treated as the threshold level. Prakash and Mathur (1987) opined that live burrow counts may be a useful index for estimating rodent infestation in crop fields. The Action Plan of the Government of India suggests that a burrow density up to 25/ha can be regarded as low, 25–50/ha as medium and more than 50/ha as severe for deciding management plans. Estimates of damage in wheat fields in Punjab revealed that a loss of Rs. 2.90 could be expected with only 2 bandicoot burrows/ha, and thus rodent control is recommended if there are bandicoot burrows at 2–3 sites per acre of wheat fields (Anonymous, 1990). For damage appraisal, the diagonal transact method, which requires sampling of 25 rice hills and examining healthy and cut tillers, was reported to be a feasible method by Singh and Rao (1982).

20.5.2 Reproduction

Generally, most of the tropical mammals breed during the winter season, whereas the monsoon season is considered to be the most conducive period for peak breeding of rodents in arid regions, due to the plentiful supply of nutritious food, which triggers breeding activity. After winter, again with the advent of spring, a fresh crop of sprouts emerges registering increased breeding activity in certain rodent species. During summer, which is a very tough time for desert fauna, breeding in the majority of desert rodents attains its lowest ebb.

TABLE 20.6 Peak Breeding Season and Litter Size of Rodent Pests

S.No	Rodent Species	Peak Breeding Season	Litter Size
1	*Hystrix indica*	Monsoon and in December (zoo)	1–3
2	*Funambulus pennanti*	March to September March–April and July–September	1–5
3	*Gerbillus gleadowi*	May, June and October to January	2–5 (summer) 5–6 (winter)
4	*Tatera indica*	February, July– August and November	1–9
5	*Meriones hurrianae*	(i) February to April and (ii) July and September to November	2–7
6	*Golunda ellioti gujerati*	March–August	5–10
7	*Rattus rattus*	April–September	1–9
8	*Millardia meltada pallidor*	Monsoon and Spring	3–9
9	*Mus musculus*	All of the year round	1–8
10	*Mus booduga booduga*	September–October and February and June	6–13
11	*Nesokia indica*	January–March and August–October	2–5

Adapted from Tripathi et al. (1992).

Detailed information on various aspects of reproduction is available for only a few species, viz., *B. bengalensis*, *M. hurrianae*, *T. indica*, *Rattus*, *M. meltada*, and *Mus* spp. (Chakraborty, 1992; Ghosh and Taneja, 1968; Jain, 1970; Kaul and Ramaswami, 1969; Prakash, 1969, 1981; Prakash et al., 1971; Prakash and Mathur, 1987; Rana and Prakash, 1984; Rao and Balasubramanyam, 1992; Tripathi et al., 1992). A perusal of data on the peak breeding season of Indian rodents (Table 20.6) indicates a bimodal breeding pattern, and the increase in the rodent population coincides with the cropping patterns. The young ones borne during both the peak breeding periods (monsoon and spring seasons) become mature when the respective *kharif* and *rabi* crops are at flowering/maturity stage which is also considered as a stage highly vulnerable to rodent attack. The phenomenon of superfoetation is also quite common in rats, mice, rabbits, cattle and sheep, however, Rana and Prakash (1979), Jain (1970) and Rana and Prakash (1984) reported repeated conception within 9–12 and 12–15 days of pregnancy by *R. meltada* and *T. indica*, respectively. Cases of pregnancy in lactating mothers have also been reported in *T. indica*, which reveals that gerbils may conceive just after delivery indicating a possibility of post-partum oestrus.

20.5.3 Habitat Use and Behaviour

Some of the behavioural attributes of rodents are presented in Table 20.7. Most of the field rodents are fossorial and nocturnal, which protect them from direct attack by predators and climatic vagaries. The burrows could be very simple (*T. indica*, *M. booduga* and *M. meltada*) or complicated (*B. bengalensis* and *M. hurrianae*). Mole rats (*B. bengalensis* and *N. indica*) keep their burrow openings tightly plugged with soil for safety against natural enemies and flooding. Most of the rodents exhibit neophobic behaviour which persists from 1 to 5 days in different

TABLE 20.7 Behavioural Characteristics of Rodents

S.No	Attributes	Characteristics
1	Sight	Colour blind, but can distinguish between shades. They can discriminate between pattern and size and have good depth perception
2	Taste	Wide food range, however, prefer fresh food but can thrive on garbage and decaying or spoiled food
3	Hearing, smell and touch	Well developed senses; readily distinguish unusual noises and the long whiskers on their muzzle and guard hairs on the body serve as sensitive feelers
4	Balance	Excellent balancing sense enables them to run on pipes, narrow ledges or wires. Long tails act as balancing organ
5	Gnawing	Gnaw to gain entrance to food and to wear down their incisors to keep them in sharpened condition
6	Climbing and swimming	Can climb almost anything they can get their claws to hold. Roof rat is the better climber. Norway rats living in sewers are excellent swimmers
7	Temperament	Bandicoots and Norway rats are much more aggressive than house rats and mouse. Cannibalism is quite common
8	Travel routes	Use fixed pathways, usually moves along the walls, under floors or through thick grass or litter

species. Food and feeding behaviour of rodents indicates their preference for various food items, which helps in selecting an effective carrier for poison bait preparation. Studies on bait preferences have revealed that many of the pest rodent species prefer wheat, rice and pearl millet with slight variations (Rana et al., 1992).

Seasonal movements of rodents indicate that sugarcane provides an excellent abode for rodents as it is a long duration crop with typical phenology and therefore acts as a focal habitat from where the rodents migrate to rice (*kharif*) and wheat (*rabi*) fields after their harvest period (Rao, 1992). Bunds or fields with more weeds are also an ideal place for burrowing rodents. Similarly, denser fields with a higher crop density afford both the cover and energy required for the reproductive activity of field rodents. The mobility of the pest forms a limited social structure based on hierarchy. The members live in small territories called 'home ranges' which depend on the position of food reserves and cover conditions and the presence of other species. The home range of Indian rodents varies from 675 sq.m for *M. musculus* to 1912 sq.m for *T. indica* (Prakash and Mathur, 1987).

20.5.4 Adaptability in Agro-Ecosystems

Rodents are highly adaptable animals and are found from snowy heights of about 5700 m to the extremes of deserts. Field rodents have adapted to various cropping systems/regions. Among the dominant pest species, *B. bengalensis* and *M. booduga* are regarded as mesic, *M. meltada* and *N. indica* as submesic and *M. hurrianae*, *T. indica* and *G. gleadowi* as xeric species and accordingly, they are adapted to different agroclimatic conditions. The xeric rodents in particular show morphological, behavioural and physiological adaptations for overcoming extreme temperatures and food and water scarcity (Goyal and Ghosh, 1992). Generally, the pest status of any species is governed by a combination of cropping pattern of the region vis-à-vis ecobiology of the rodent species. The crop fields are manmade ecosystems with a plant community dominated by one or two crop species. Depending upon the availability of water, climate suitability and soil fertility, different crops are grown in an area, which regularly supply energy-rich food to rodents in a limited area. The availability of food and cover to native rodents is continuous with different crops grown in different seasons. The rodent species that can take advantage of the temporarily abundant food and at the same time somehow overcome the periods when conditions are poor are good candidates for agricultural pest status (Leirs, 2003). In arid regions, rainfed crops experience relatively lower rodent damage than do the irrigated crops, as the area under irrigation is limited and rodents migrate to the irrigated fields in large numbers and cause havoc to such crops.

20.5.5 Survey and Surveillance

Regular surveys and surveillance activities not only provide information about the composition of rodent pest species and their population dynamics, but also help in diagnostics and planning management strategies in any cropping system. It is a most important component of rodent management but is most often overlooked. An accurate population census is not possible due to migratory habits, changing species composition, diversified habitats and different body sizes of native rodents; however, use of signs, tracks, surplus baiting techniques, burrow counts and trappings are some of the methods generally adopted for population estimation. The following two methods are commonly recommended for field level estimations.

20.5.5.1 Live Burrow Count Method

Generally, a live burrow count gives an idea of the number of rodents living in an area. The live burrow or active burrow generally looks fresh, has marks of the rodent, has freshly excavated soil, and the cut parts of various plants, etc. may also be present. Since a rodent may use more than one opening for access, this may give a false picture of the rodent population. Therefore, all of the burrows are plugged in the evening and early in the morning on the next day, the reopened burrows would reveal an almost correct picture of rodent numbers. If the mean number of rodents residing in a burrow is known, the correct population of rodents can be estimated by multiplying this by the number of live burrows. This situation generally arises during the breeding season when more rodents live in a burrow; otherwise, adult rodents occupy a single burrow.

20.5.5.2 Trap Index

There are two methods for evaluation of the rodent population through trapping, i.e. (i) the Removal Method and (ii) the Capture, Mark and Recapture (CMR) Method. Single catch traps are laid in the fields following either trap line/grid methods for at least three nights. In the Removal Method, the rodents trapped are removed and the trap index is computed as follows:

Trap Index (rodents/day/trap) = $M/n \times t$, where n = number of traps used in a trap line, t = number of days and M = total number of rodents trapped.

In the CMR Method, the rodents trapped each night are marked and released, and a second trapping is undertaken 1–2 days after the first trapping would include some marked rodents as well. The population size may be calculated as follows:

$N = M \times n/m$, where N = population size, M = number trapped in the first trapping; n = total number trapped in the second trapping and m = number of rodents re-trapped in the second trapping.

In addition, surveillance activities may also include estimates of damage to crops at different crop growth stages as explained by Buckle (1994), which may be correlated with the number of live burrows per unit area and the trap index.

20.6 RODENT CONTROL TECHNIQUES

20.6.1 Mechanical Control

Mechanical removal of the rodent population from any habitat is mainly done using different types of trap (Figure 20.1). This method is in vogue all over the world, but its success in field rodent control is doubtful. Trapping can no doubt provide information on species composition and density of the pest population, and hence it can be successfully used for survey and monitoring purposes. In indoor habitats, this method is quite effective (Table 20.8), however for fields this method may be used as a follow-up measure after bringing down the rodent pest population by poison baiting. The traps may be of single (Sherman/snap traps) or multiple catch (wonder traps) type. In snap traps, the rodents are killed whereas others traps catch live rodents. Introduction of natural glues or sticky traps have opened new avenues in the field (Tripathi et al., 1994). Rodents use visual and olfactory cues towards traps, therefore camouflaging the traps with bushes, grasses, etc. in fields enhances trapping success considerably (Jain and Tripathi, 2000). Indigenous Tanjor kitty traps in South India and bamboo snap traps in Jhoom fields of the NEH region have proved quite effective (Rao, 2003). Trapping of *B. indica* and *B. bengalensis* in flooded deep water rice in Bangladesh (Islam and Kareem, 1995) and of *B. bengalensis, G. ellioti, M. meltada*, etc. in vegetable crops in Himachal Pradesh, India (Sheikher and Jain, 1992) was found to be an effective method of rodent control.

An eco-friendly mechanical 'burrow fumigator' device has been developed by the All India Network Project on Rodent Control. The device utilizes farm wastes for generating smoke, which is pushed into rodent burrows through an inbuilt blower (Figure 20.1). The rodents die of asphyxiation (Rangareddy et al., 2005). Field efficacy trials in rice fields in Godawari delta, Andhra Pradesh have revealed that this device yields over 90% pest mortality, which is on par with chemical fumigants (aluminium phosphide).

TABLE 20.8 Comparative Efficacy of Different Traps under Indoor Conditions

Habitat	Trap	No. of Traps Used	No. of Rodents Trapped		Trap index
			On 1st day	*In 5 days*	
GLUE TRAP VS. SHERMAN TRAP					
Houses and godowns	Glue trap	30	76	180	118.8
	Sherman trap	30	8	20	13.3
Poultry farms (feed stores)	Glue trap	15	52	80	106.6
	Sherman trap	30	10	25	16.5
GLUE TRAP VS. WONDER TRAP					
Grain godowns	Glue trap	9	5	6	11.6
	Wonder trap	9	4	4	11.0
Poultry feed and egg stores	Glue trap	14	8	25	44.6
	Wonder trap	14	8	25	44.6

Source: Tripathi et al. (1994).

Certain tribal communities in India, viz. Mushars and Nats (Bihar), Yenadis (Andhra Pradesh), *Irulas* and *Kuruwas* (Tamilnadu) and many other tribal communities in NEH states consume rodents. Farmers of Andhra Pradesh and Tamilnadu employ such tribals for physical elimination of field rodents (Jain and Tripathi, 2000; Whitaker, 1987). In many regions, the burrows are flooded with water and escaping rats are killed. In south India, these rat catchers use local traps (*Butta* and *Tanjore* kitty traps) made of palmayra leaves and wooden sticks. Similarly, in the NEH region, several types of indigenous bamboo snap traps are laid along the periphery of *jhoom* fields to trap rodents immigrating from surrounding forests. This practice is often done at crop maturity stage when the rodents' destruction is at its peak; however, this eco-friendly method may prove more effective if performed in a planned schedule.

20.6.2 Habitat Manipulation

This technique is primarily based on ecological concepts wherein the rodents' habitat is manipulated in such a way that this creates stress among native rodent pests. The methods are low cost treatments and involve little modification in crop husbandry practices, such as ploughing, puddling, removal of wild vegetation and refuge of previous crops (Jain and Tripathi, 1992; Pashahan and Sablok, 1987) and reduction in bund size (Sharma and Rao, 1989), etc. It helps in the migration of pest rodents from crop fields. Moreover, the habitat stress created by these practices enhances the chances of rodents falling prey to predators (Green and Taylor, 1975; Hansson, 1975). In sugarcane crop, adoption of techniques such as bunching of standing canes indirectly helps to prevent rodent damage. No truly rodent resistant variety is available globally; sugarcane varieties with a thin cane, soft rind, and low fibre and of the lodging type are more damaged by rodents. Similarly, early maturing varieties of rice are more prone to rodent attack (Parshad, 1999). Synchronous sowing/transplanting in larger areas also reduces rodent damage to a greater extent.

20.6.3 Biological Control

20.6.3.1 Diseases

Many microbial organisms (fungi, bacteria, viruses and protozoans) and macroparasites (helminths and arthropods) are known to cause diseases to a variety of mammalian hosts including rodents, however their potential as biocontrol agents has not been realized thus far. Use of disease agents for control of animals was initially successful through the introduction of myxomatosis in rabbits in Australia, however, after 10 years or so, the rabbits developed genetic resistance to the disease. In the case of rodents, *Salmonella typhimurium* and *S. enteritidis* have been successfully used against them in foreign countries but in India, the pathogens were found to be ineffective against *R. rattus* and *B. bengalensis* yielding 16% and 18% mortality, respectively (Deoras, 1964). Bindra and Mann (1975) also expressed doubts about the success of microbial control because the authors could not record any mortality of *M. (R). meltada* and *B. bengalensis*, whereas in the case of *M. musculus* and *T. indica*, the mortality was over 20% and 40%, respectively. A protozoan parasite, *Trypanosoma evansi*, the causal organism of trypanosomiasis in cattle, has been tried against *R. rattus* and *B. bengalensis* in India. The pathogen recorded two peaks of parasitaemia in both rodent species, with 100% mortality after the second peak (Singla et al., 2003). Due to the presence of pathogens in caudal epididymal fluid and spermatozoa, the authors indicated the possibility of sexual transmission of *T. evansi*. In Australia, Singleton and McCallum (1990) reported that a hepatic nematode, *Cappilaria hepatica*, may prove to be an important agent for controlling mouse plague. In one of the reports during a rodent outbreak in groundnut (1989–90) in Gujarat, Mittal et al. (1991) reported a drastic reduction in the rodent population due to heavy parasitization by a blood sucking lice, *Polyplax spinulosa*. In fact, use of microbes has very little scope for rodent control, particularly due to the inherent possibility of spread of the diseases to humans and livestock. The WHO committee on Zoonosis has also doubted the practical application of microbial rodenticides due to possible public health hazards (Prakash and Mathur, 1987).

20.6.3.2 Predators

Several species of vertebrates, mainly birds and mammals, are listed as natural predators of rodents in the literature (Chaudhary et al., 2001; Deoras, 1964; Neelanarayanan, 1997; Prakash and Mathur, 1987; Tripathi et al., 1992; Whitaker and Dattari, 1986). However, cats in the domestic situation and snakes, owls, mongoose and varanids are the predominant vertebrate predators of rodents (Rao, 2003). Information on the stomach content and faecal matter of predators has indicated that rodents constitute over 75% of the diet of snakes, viz. cobra and Russell viper (Whitaker and Dattari, 1986) and 61% of that of the spotted owlet (Kumar, 1985). Introduction and rehabilitation of barn owls, *Tyto alba*, in oil palm plantations of Malaysia was reported to reduce the rodent damage to oil palm from 19.4% to 1.4% within 2 years (Duckett and Karuppiah, 1991). In the Cauvery Delta of Tamilnadu, India, studies on barn owl as a potential biocontrol agent revealed a predation rate of 1–6 (av. 1.58) rodents/night. *B. bengalensis* (40%) and *M. musculus* (33%) constituted the major prey items of barn owls (Neelanarayanan, 1997). Such predators need to be introduced and rehabilitated for eco-friendly management of rodents. However, in general, the success rate of biological control of rodents through other predators is not very encouraging, because rodents do not constitute the sole diet of most predators and exert relatively lower predatory pressure compared to the faster turnover of rodents. For example, snakes in captivity require only one rodent in 3 days (Whitaker and Dattari, 1986) indicating a very poor potential. Some examples of the introduction of predators, e.g. monitor lizards in Pacific islands to control rodents (Uchida, 1966), Indian mongoose to control *R. norvegicus*, and tame ferret (*Putorius putorius*) for rabbit control in New Zealand (Gibb, 1981), have not yielded a very encouraging success rate.

20.6.4 Botanicals

Some plant products have been found to possess anti-rodent properties. Dixit (1992) has reported antifertility effects in both sexes of desert gerbils, *M. hurrianae*, when they are exposed to various treatments, viz. extracts of flowers of *Malvaviscus conzatii*, fruits of *Sapindus trifoliatus*, leaves of *Cannabis sativa*, seeds of *Abrus precatorius*, roots of *Calotropis gigantea*, seeds of *Argemone mexicana*, etc. Calotropin isolated from roots of *C. gigantia* was a promising interceptive/abortifacient agent in female gerbils, rats and rabbits (Gupta et al., 1990; Sharma et al., 1990). Similarly, plumbagin, isolated from *Plumbago zeylanica*, when given to *M. hurrianae* at 50 mg/kg between days 6 and 9 of pregnancy, resulted in resorption of embryos (Dixit, 1992). Neem leaf powder (5%) showed an antifeedant action on rodents, whereas neem oil repelled the rats to the tune of 18–48%. Neem formulation BBR (3% concentration) recorded a repellency index up to 87% in baits against rodents. Pellet formulation of neem containing eucalyptus oil, developed by CAZRI, Jodhpur revealed complete rejection by *T. indica* even in no-choice tests. Similarly, neem (*Azadirachta indica*) oil emulsion at 5% and 10% in pearl millet baits showed reduced bait intake by Indian gerbil in choice and no choice tests (Tripathi et al., 2006). Jojoba (*Simmondsia chinensis*) seed cake powder (10–20% in baits) registered a repellency index of up to 90% in *T. indica*. The aversion through learning persisted for a week in gerbils (Tripathi et al., 2010). Crude cottonseed oil (5%) also possesses antifertility effects on bandicoots (Singla et al., 2008). Certain plants increase the concentration of certain chemicals such as terpenes and phenolics in reaction to grazing and browsing by animals, including rodents, which leads to natural repellence (Hansson, 1988). Derivatives of alkaloids, cyanogenic glucosides, etc. are likely to disrupt the metabolic processes or even affect the nervous systems of herbivores. These compounds inhibit reproduction and growth in voles, *Microtus montanus* (Berger et al., 1977). Glucosides derived from *Trypterygium wilfordii* have shown antifertility effects on house rats under laboratory conditions. A herbal repellent, 'Indiara M' contacting extracts of several plants (*Ipomea muricata, Acorus calamus, Allium sativum, A. cepa, Brassica* and *Rametha*) when mixed with food (at 6% concentration) resulted in an 80–93% reduction in food intake in *M. hurrianae* and *F. pennanti* showing repellent effects for a limited period (Kashyap and Parveen, 1990). These studies do reveal a good scope for such plant/products in developing a non-toxic approach for rodent management and therefore require in-depth study. Gossypol treated baits have also resulted in reduced sperm motility, live sperm count and sperm concentration.

20.6.5 Pheromones

Rodents are endowed with scent organs and glands whose secretions act as agents of olfactory communication. Sparsely distributed field rodents depend more on olfactory cues than visual or auditory impulses for inter- and intra-specific communications. Specialized integumentary glands constitute one of the major sources of olfaction among mammals. Sebaceous glands are more common than others. Besides skin glands, vagina, urine (possibly through kidney), faeces, accessory glands of male reproductive systems, etc. are other sources of vertebrate pheromones. Studies conducted on scent marking glands in *M. hurrianae, T. indica* and *M. meltada* revealed that the secretions play a great role in maintaining social hierarchy and reproductive manifestations (Idris, 2005; Idris and Tripathi, 2011; Prakash et al., 1998). The exudations of scent marking glands possess phagostimulant properties in desert rodents and therefore can be utilized in formulating attractive poison bait formulations. Similarly, addition of conspecific urine of *M. hurrianae, T. indica* and *M. meltada* significantly enhanced food intake and therefore showed promise in increasing the acceptability and palatability of poison baits. In addition, such exudates have the potential to mask zinc phosphide induced bait/poison shyness in pest rodents. Male urine has been found to possess sex pheromonal properties too as it accelerates puberty in females, and advances vaginal opening by 3 weeks in bandicoots. Body odours released by faeces, urine, saliva, etc. of rodents were found to enhance their trappability as well, when used in traps. Identification and chemical synthesis of the active ingredient of these complex exudates would certainly pave the way for a new era in rodent management.

20.6.6 Chemical Control

20.6.6.1 Chemosterilants

High reproductive potential with a shorter period of sexual maturation of rodents is a major concern in evolving effective and sustainable management technologies and sterilants may be of immense help in this respect. A host of chemicals have been identified globally with an antifertility effect on pest rodents. Some of these are antiestrogen U–11, diphenylindane derivatives, metepa, tepa, tetradifon, furadentin, colchicine, glyzophrol, etc. (Prakash and Mathur, 1987) and others like alpha-chlorohydrin (Ericsson, 1982; Saini and Parshad, 1988). Studies in India by Saini and Parshad (1991) indicated that the chemical at 0.5% in cereal baits is well accepted by

B. bengalensis and in a single day's exposure, mole rats ingested a dose of active ingredient more than its LD_{50} (82 mg/kg). In field trials, it yielded 63–83% mortality of field rodents, and most of the survivors consuming sublethal dosages became permanently sterile (Saini and Parshad, 1993). Alpha-chlorohydrin may therefore prove to be an effective male sterilant at lower dosages. Studies on a germ cell mutagen, ethyl methanesulphonate, have been indicated to affect differentiation, structure and function of spermatozoa of *R. rattus* at dosages of 500 and 625 mg/kg (Kaur and Parshad, 1997). Despite determined efforts worldwide, such compounds have yet to become practically applicable because, in addition to being expensive and time consuming, their delivery to widespread and cryptic pest rodents is highly cumbersome. However, such chemicals do possess hope for the future because of their compatibility with toxic rodenticides and also because they are less controversial from a human/animal rights viewpoint.

20.6.6.2 Rodenticides

Use of rodenticides is the most common, expedient and humane method to control pest rodents and therefore forms the basis of present day rodent management strategy in most parts of the world. They have greater scope in large-scale control operations, since a mixed population of several species are encountered in fields. Rodenticides may be classified as (i) plant origin, or (ii) chemical origin.

The plant origin compounds like red squill and strychnine are historical rodenticides. Red squill, derived from bulbs of *Urginea maritima*, was effective against a wide range of rodents as a scillirocide (a glucoside) and its micronized form was used at 0.05% in baits. In India, it has been evaluated against *R. norvegicus, R. rattus, B. bengalensis* and *T. indica* (Prakash and Mathur, 1987). Similarly, strychnine alkaloid derived from seeds of *Strychnos nux vomica* has also been used as a rodenticide. Its LD_{50} value for *M. hurrianae* and *T. indica* is 5.0 mg/kg. Prakash and Mathur (1987) reported that its acceptability is low due to its bitter taste, fast action and early warning symptoms.

The chemical origin rodenticides are further divided into inorganic and organic compounds. The inorganic rodenticides include arsenic compounds, zinc phosphide, barium carbonate, sodium monofluoroacetate, aluminium phosphide, etc. These are acute poisons and are used as baits, apart from aluminium phosphide, which is used as a burrow fumigant. Owing to their extreme toxicity to non-targets, most of them are not used at present. Among organic rodenticides, RH-787 and alphanaphthyl thio urea (ANTU), Norbormide, etc. are acute in action, whereas other organic rodenticides are mainly chronic poisons comprised of either anticoagulants (warfarin, bromadiolone, etc.) or non-anticoagulants (cholecalciferol, bromethalin, etc.).

20.6.6.2.1 ACUTE RODENTICIDES

Among acute rodent poisons, only zinc phosphide, barium carbonate and aluminium phosphide are registered by the Government of India for common use. Prakash and Mathur (1987) reported that barium carbonate possesses low toxicity with variable efficacy against rodents, is required at higher levels in baits (10–20%) and is therefore not in use presently. Zinc phosphide is the most widely used acute rodenticide in India. It is effective against a wide range of Indian rodent pests. The LD_{50} value ranges between 25 and 40 mg/kg against *T. indica* and *M. hurrianae* (35 mg/kg), *B. bengalensis* and *R. norvegicus* (25 mg/kg), and *R. rattus* (40.1 mg/kg). For *M. musculus*, the values are relatively higher, i.e., 250 mg/kg (Prakash and Mathur, 1987). Zinc phosphide baits are quite stable in air and non-acidic media, but when ingested, the acids present in gastric juices release lethal phosphine gas, which produces necrotic lesions and kidney damage causing death from heart failure. Death may

TABLE 20.9 Persistence of Zinc Phosphide Induced Bait Shyness among Indian Rodents

S.No	Rodent Species	Persistence of Bait Shyness (days)
1	*Funambulus pennanti*	30
2	*Gerbillus gleadowi*	10–15
3	*Tatera indica*	115
4	*Meriones hurrianae*	35
5	*Rattus rattus*	75
6	*Millardia meltada*	135
7	*Mus musculus*	20
8	*Mus booduga*	95
9	*Bandicota bengalensis*	30
10	*Bandicota indica*	105

Source: Jain and Tripathi (2000).

occur within 2h to 2–3 days depending upon the intake. Zinc phosphide is recommended at 2–2.5% concentration in cereal baits. It yields around 60% control success. A major limitation in its frequent use is its high toxicity to non-target species and the development of bait shyness/poison aversion in the target species after sub-lethal consumption (Prakash and Mathur, 1987). Bait shyness may persist for more than 2–3 months in different rodent species (Table 20.9). Bait can be prepared by mixing the poison (2–2.5%) in oil smeared cereals, e.g. wheat, rice, sorghum, and pearl millet. Because of its high toxicity, zinc phosphide is not recommended for commensal situations. For controlling residual rodents (surviving after zinc phosphide treatment), another rodenticide, viz. aluminium phosphide (a fumigant) or a second-generation anticoagulant rodenticide, has been advocated (Rana and Tripathi, 1999).

20.6.6.2.2 CHRONIC RODENTICIDES

(i) Non-anticoagulant rodenticides: Calciferol, cholecalciferol, bromethalin and flupropadine are compounds which have been evaluated for their efficacy on rodents. Calciferol and cholecalciferol are vitamin D_2 and vitamin D_3-based chemicals, respectively and cause hypercalcaemia, leading to death of target rodents. Limited studies on cholecalciferol in India have indicated its good potential (at 0.075% in baits) for causing delayed mortality of test rodents within 4–13 days (Agarwal et al., 1988; Mathur and Jain, 1987). The oral LD_{50} of cholecalciferol is 36mg/kg for adult *R. rattus*. The sex-specific value of LD_{50} (i.e. 30mg/kg for males and 50mg/kg for females) indicated that male rats are more susceptible to cholecalciferol toxicity (Singla et al., 2013). Saini and Parshad (1992) reported 100% mortality of *R. rattus* after 1–2 days, exposure to cholecalciferol (0.075%) bait. The treated rats lose appetite and stop feeding thereby reducing the bait requirements and the risk of secondary poisoning. Bromethalin, another rodenticide, acts on the tissue system and disrupts the process of energy production within the cell. This leads to excessive fluid build-up in the tissues, which affects the nervous impulse leading to paralysis and death of target animals. Mian et al. (1993) from Bangladesh reported that bromethalin is effective against *B. bengalensis*. Parshad (1999) has referred to these chemicals as subacute rodenticides.

(ii) Anticoagulant rodenticides: As the name indicates, this group of rodenticides interfere with the blood coagulation process, leading to internal bleeding and haemorrhage in the target animals. Depending upon their chemistry, these are further divided into indanediones (chlorophacinone, diphacinone, pindone, etc.) and (ii) hydroxycoumarins. The hydroxycoumarin rodenticides are further divided into (i) first generation (coumachlor, coumatetralyl and warfarin) and (ii) second generation (bromadiolone, brodifacoum, difencoum, flocoumafen, difethialone) rodenticides. Based on their efficacy, the first generation anticoagulants are referred to as multi dose and the second

generation anticoagulants are referred to as single dose anticoagulant rodenticides.

Among the first generation anticoagulants, only warfarin, fumarin and coumatetralyl are registered in India. These have shown good potency in the laboratory and field trials (Mathur et al. 1992). Owing to their multi dose requirements (extending to 2–4 weeks) in baits, which causes operational problems, these rodenticides, particularly warfarin and fumarin, have not gained popularity in India and are not available. Coumatetralyl, however, has shown very good efficacy in bait (0.0375%) or as tracking powder (0.75%). *B. bengalensis*, the most predominant rodent pest of India, is highly susceptible to coumatetralyl at 0.0375% in baits (Parshad, 1999). Field applications through burrow or station baiting for 3–5 days yielded 50–70% control of lesser bandicoots (Malhi and Parshad, 1995), however Mathur and Prakash (1984) reported 10–15 days of baiting for similar control success in arid areas. Coumatetralyl has been recommended as cereal mixed baits at 0.0375% for control of commensal as well as field rodents (Rana and Tripathi, 1999).

Among the second generation anticoagulants, several molecules, viz., bromadiolone (Anonymous, 1986), brodifacoum (Jain and Tripathi, 1988), flocoumafen (Jain et al., 1992) and difethialone (Sridhara et al., 2000; Chaudhary et al., 2002; Chaudhary and Tripathi, 2004, 2006) have shown their potency against Indian rodent species (Table 20.10). Compared to first generation anticoagulants, these are more toxic and effective at much lower concentrations in baits (0.0025% for difethialone and 0.005% for others) with a single application. These are also effective against warfarin-resistant rodents (Greaves, 1994) and zinc phosphide induced bait shy rodents (Chaudhary and Tripathi, 2003). Generally, the toxic effect of these chemicals starts from the 3rd day of bait intake and continues for 10–12 days for effective killing of pest rodents. In addition to their small requirements in baits in a single dose and the availability of an effective antidote (vitamin K_1), the second generation anticoagulants have an edge over other rodenticides not only in efficacy but also in safety. Presently, only bromadiolone (0.005%)

TABLE 20.10 Single Dose Toxicity of Grain Baits of Second Generation Anticoagulant Rodenticides

	Percent Mortality and (Mean Days to death) with			
Rodent Species	**Bromadiolone (0.005%)**	**Brodifacoum (0.005%)**	**Flocoumafen (0.005%)**	**Difethialone (0.0025%)**
Bandicota bengalensis	100	100 (6.6)	100	100 (5.8)
Tatera indica	100 (9.5)	90 (6.9)	90 (8.2)	100 (7.9)
Millardia meltada	100 (6.4)	80–100 (6.6–7.1)	100 (5.7)	100
Meriones hurrianae	88	83–100 (4.7–8.0)	100 (7.0)	100 (5.9)
Rattus rattus	100 (7.5)	83–100 (8.0–10.8)	100 (8.6))	100 (5.9)
Mus musculus	91 (8.1)	60–100 (4.6–8.4)	100 (6.8)	–
Mus booduga	83	83 (9.4)	–	–
Funambulus pennanti	–	66 (7.0)	100 (8.1)	100 (6.1)

Source: Tripathi (2009).

is registered by the Government of India for rodent control in fields and commensal situations.

20.7 BAITING TECHNOLOGY

20.7.1 Bait Preparation

'Rodents cannot vomit' is the key for the success of baiting technology. The objective of baiting technology emphasizes that the maximum population of rodents gets access to the baits prepared in the most preferred medium. Normally, two types of options are available: (i) ready to use, and (ii) freshly prepared baits. In India, normally, freshly prepared baits are preferred for field rodent control (Figure 20.1). Based on the behaviour and preferences of rodents and the cropping system, the poison baits are prepared in cereal (pearl millet/wheat/rice/sorghum, etc.) loose baits using edible oils (mustard/groundnut/sesame/coconut oils) as adherent and attractant. The recommended dosages for zinc phosphide, bromadiolone and coumatetralyl, the commercially available rodenticides in India, are 2–2.5%, 0.005% and 0.0375%, respectively in baits. Based on the predominant cereal crops and oilseeds of the region, the cereal and oil are selected and poison powder is mixed accordingly. For zinc phosphide, which is an acute poison, 2–3 days' pre-baiting is recommended for acclimatization of field rodents to overcome neophobia and also to reduce the chances of bait shyness. For anticoagulants, pre-baiting is not required.

20.7.2 Bait Placement

Placement of bait is one of the most important aspects for an effective chemical rodent control strategy. It should, however, be ascertained that the rodent population consists of adults only, which can consume bait material. The bait may be placed either in the burrows or in the bait containers/bait stations (Figure 20.1). Burrow baiting is advisable in field conditions where clear rodent burrows are visible. For this, all of the existing burrow openings should invariably be plugged in the evening and next morning, re-opened/active burrows are treated with pre/poison baits. The treatment of only active burrows saves the poison bait material, labour costs and time. The poison bait is rolled deep inside the active burrows (at 6–15 g/burrow) to avoid any secondary hazards. To assess the control success, the burrows are plugged again after 3–4 days of treatment with zinc phosphide or 15 days after bromadiolone baiting and the reopened/live/active burrows are counted. Application of poison baits can also be done in bait stations. Several types of indigenous bait containers have been used in India for keeping bait. The basic idea of selecting bait containers is that the bait should be easily accessible to the target species only and reduce the hazard to other animals. This will also protect the bait from rain and other weathering. Indigenously, procured items such as mud channels, hollow bamboo pieces, broken pitchers, and coconut shells have been effectively utilized for this purpose.

20.7.3 Burrow Fumigation

Aluminium phosphide, the most common fumigant rodenticide, is available in both tablet and pellet forms, however only the pellet formulation is recommended for rodent burrow fumigation (Rao, 2003). For fumigation, all of the existing burrow openings are plugged with wet mud and aluminium phosphide pellets (at 2 pellets/burrow) are inserted in the active burrows, which should also be plugged with mud to check the escape of lethal gas. All of the nearby burrow openings invariably need to be plugged. The dead rodents are collected the next day and disposed of. The fumigant is more effective in humid zones and irrigated fields with heavy soils. It is never recommended for residential premises/indoor use. Because of its extreme toxicity to non-targets and absence of an antidote, the Government of India has put this chemical under a restriction ban.

20.8 INTEGRATED RODENT PEST MANAGEMENT

Integrated Pest Management (IPM) is a bio-ecological system and is therefore a dynamic one. The concept of IPM has been devised and applied to insects, diseases and to some extent to weeds thus far, however, applying the same principles to rodent pests has several limitations. Rodent management today is mainly a reactive strategy, i.e. it is only effected when the rodent problem is observed. Simply knowing the control methods does not qualify for effective management of rodents, unless the eco-biology and behaviour of the enemy (pest) is well known. Since rodents are highly evolved mammals and have a long association with the wisest mammal, i.e. man, this has made them wiser by learning various tactics to avoid many harmful activities performed by man. Trap avoidance, bait shyness/aversion to acute poisons, and resistance/cross-resistance to anticoagulants are only a few examples of rodents' complex behaviour, which very often questions man's endeavours for their control. Therefore, an IPM approach based on sound eco-biology and ethology of pest species vis-à-vis population reduction technologies having economic viability and sociological acceptance, needs to be evolved.

Although several methods of rodent control have been discussed in the preceding sections, no single method can prove its worth if applied in isolation. Despite the needs of the hour, the real IPM for rodent pests is still not happening, because of several in-built constraints due to the complex biology and behaviour of pest rodents (Buckle and Smith, 1994; Prakash, 1988; Singleton et al., 1999a). Present day rodent control technology mainly depends on the use of rodenticides, which no doubt yields quick results and farmers are also satisfied by seeing the dead rodents or immediate reduction in damage. This success with rodenticidal baiting is short lived and soon the survivors multiply or other rodents immigrate from surrounding areas making the whole exercise of rodent control a futile one. However, rodenticides are going to stay until better options are made available, but based on the present day knowledge on Indian rodents, these can be used judiciously and may be integrated with other methods, e.g. habitat manipulations and trapping so that use of toxic chemicals is at least minimized. Some important considerations are discussed next.

20.8.1 Chemical Control Technology

With the knowledge generated from population dynamics, breeding biology, and the behaviour of pest rodents, one may plan rodenticidal treatments in fields. Jain and Tripathi (2000) opined that, since most of the field rodents show lean breeding periods during the summer (May–June), a community-based anti-rodent campaign may be launched during this period. This provides greater rodent control success because (i) acceptance of poison bait is very high in the absence of green food, (ii) ingestion of poison bait is higher as the majority of the population are at the adult stage (being a lean breeding period), and (iii) farmers are relatively free to undertake the campaign which also reduces the cost of the operation. This treatment takes care of *kharif* (monsoon) crops. A similar operation may be carried out in November–December for *rabi* (winter) crops. Parshad (1999) terms these treatments as prophylactic. During crop growth stages, poison baits are less acceptable due to stiff competition with green food and at this stage, trapping or rat catchers may be employed. Indian farmers, mainly due to poor awareness, avoid any control actions when the crops are not in the fields. Therefore, curative actions have to be taken. It has been observed that, at certain stages of crop growth, rodenticidal bait application yields effective control of rodents, for example, 30–60 days or 42–70 days after transplantation or the panicle initiation stage in rice, the tillering stage in wheat, the flowering and podding stage in groundnut, and during August and October–November in sugarcane (Parshad, 1999). During peak breeding periods, instead of poison baiting, burrow fumigation with chemicals or smoking of burrows with fumigators may be more effective as these eliminate adults, sub adults, juveniles, etc. Burrowing behaviour and the nocturnality/diurnality of a particular species may determine the type of application, such as the timing for burrow baiting and fumigation. For example, identification of live burrows vis-à-vis timing of treatment (late evening for *B. bengalensis* and *T. indica*) and during the morning hours for diurnal rodents such as *M. hurrianae*. This helps rodents to take fresh bait. Similarly, less dosage of fumigant will be required for rodents which make simple burrows (*T. indica*, *M. meltada* and *M. booduga*). Studies on food and feeding behaviour have helped in identifying the bait components. Rodents being seedivorous, cereals are the main bait component of poison baits with oil additives. Even the shyness developed after sub-lethal intake of zinc phosphide can be overcome to some extent by changing the bait components.

In the era of IPM, the present rodenticide-based technologies need immediate refinements, because dependence on toxic chemicals has several limitations in view of environmental, economic and social considerations. What is needed in chemical rodent control tools is relatively safe rodenticides, such as anticoagulants or vitamin D_3-based rodenticides; scheduling of treatments according to the bio-ecology and behaviour of pest rodents vis-à-vis vulnerability of crop stage and suitable bait delivery techniques. Addition of a bitterant, such as denatonium benzoate (10 ppm) in bromadiolone bait makes it safe to non-targets, especially humans without affecting its efficacy. The problem with anticoagulants is the development of resistance in pests. There is an urgent need for a greater number of anticoagulants in India to avoid the dependence on a single chemical, i.e. bromadiolone. As far as bait delivery systems are concerned, paper packet-based baiting, bait stations made of coconut husks, banana sheaths, PVC pipes or bamboo stems, have shown good promise in many parts of the country and therefore may be utilized on a large scale. Among chemical fumigants, a new formulation of aluminium phosphide, which is at the experimental stage, with reduced a.i. (6.0%), was found to have similar potency to that of the existing formulation (56% a.i.). Thus, it holds good promise in reducing the pesticide load on the environment.

20.8.2 Trap Barrier System (TBS)

This is a type of physical control of field rodents by putting fences around crops. Lam (1988) reported that erecting a plastic fence around the main rice crop and keeping multiple catch traps near the holes made in the fences protects the crop from rodents. This system, described as eco-friendly, was named a trap barrier system (TBS). In later years, it was improved by limiting the fencing to around a trap crop sown before the main crop. The trap crop lures the rodents from the surrounding areas and they are trapped in large numbers. The TBS + trap crop provide a halo of protection to the neighbouring rice crops (Singleton et al., 1999b). The system is broadly referred to as ecologically sound rodent management technology when applied on a community basis (community trap barrier system or CTBS) and has proved cost effective in some south Asian countries for rice rat management. In India, results of preliminary trials of erecting fences/barriers of split bamboo around jhoom fields in the NEH region have not been very encouraging. Similarly, studies on TBS in rice in Andhra Pradesh initiated recently have not been successful, but further studies are needed on this aspect under Indian conditions.

20.8.3 Bio-Rodenticides

No such product/formulation is available in India for rodent management, however, a *Salmonella*-based compound, claimed to be specific to rodents, is being used in Cuba, Central America and some countries in SE Asia (IRRI, 2009). Information on its shelf life under humid and tropical field conditions is still sparse. Sporocysts of *Sacrocystis singaporensis*, a protozoan parasite endemic to SE Asia, have proved lethal to rodents (Jackel et al., 1996, 1999). The protozoan is a definitive host in snakes, such as pythons and when the infected rats are fed to snakes, the sporocysts are excreted by snakes through faeces. Field trials by Jackel et al. (1999) in Thailand revealed that bait containing the cysts of these parasites when exposed to rodents (*R. norvegicus, R. tiomanicus* and *B. indica*) yielded 58–92% mortality.

20.8.4 Fertility Control

Manipulation of the fertility of pest rodents may prove to be one of the most effective tools for managing the rodent population in crop fields on a sustainable basis. Reduction in fertility may be achieved either through a non-infectious agent delivered through non-toxic bait or by using infectious bio-control agents in bait. In the first option, some chemicals and plant origin compounds have shown antifertility effects but could not gain sufficient popularity for many obvious reasons. A formulation of plant origin compounds such as glycosides of *Tryptaregium wilfordii* (GTW) is being used in China as a male antifertility agent. Recently, a new concept called 'immunocontraception' is being evolved which envisages the use of the body's immune system to induce immune responses (circulating antibodies or cellular immune effector cells) against reproductive cells or proteins essential for successful gametogenesis, fertilization or implantation, leading to infertility (Chambers et al., 1999). The objective is to develop an antifertility vaccine. Viruses are being tested to deliver the vaccines, which spread naturally through the target pest population. Such methods, being potentially species-specific, may prove eco-friendly and cost effective and thereby revolutionize rodent management technologies in the future.

20.8.5 Community Participation

Due to the immense versatility in terms of adaptability and the wide variation in ecology and ethology of rodents and the extent of losses caused, rodent management is a serious challenge. With changes in agro-climatic conditions and cropping patterns, rodents are showing shifts in their distribution and abundance (Parshad, 1999). Although effective options for their management are available in India, rodents continue to be a serious threat to food and health security. One of the most important constraints is the lack of community participation in rodent management actions. Most farmers show a general neglect towards the problem due to poor awareness, small land holdings, poor education and low economic status. Often, failures of rodent control activities are due to adoption of the wrong methods of bait preparation, and application also discourages the farmers. Social engineering activity on rodent control initiated by the All India Network Project on Rodent Control yielded over 60% success in the adopted villages (Mathur 1992). The Government of India has launched a National Plan on Rodent Pest Management with the major objective of awareness creation, capacity building and rodent control campaigns (Rao, 2010). For creation of trained manpower, a three-phase action plan has been suggested: (i) capacity building through trainers training at apex, middle and field levels for officials and staff extension agencies, (ii) preparation of the community by education/training and field demonstrations, and (iii) management operations in fields through trainee farmers, formation of rodent management squads, and undertaking management operations at regular intervals covering all areas including common property resources.

20.9 CONCLUSIONS

Global climate change, as evident from increasing temperatures and changing rainfall patterns, is likely to accentuate the rodent problem further as these tiny vertebrates would successfully adapt to any such changes. The change in cropping patterns (high value crops replacing traditional crops) and the farming system approach (growing perennials and annuals together) may prove more conducive for rodent life. The introduction of pressurized irrigation systems (sprinklers and drips) provides very favourable microhabitats to native rodents and rodent damage to drip and sprinkler systems is more frequently reported these days. Poly houses/protected agriculture are also threatened by rodent attack. *B. bengalensis*, the predominant and most destructive rodent pest in the country, is spreading its territory and is showing its destructive potential in new areas such as western arid and the NEH regions (Tripathi, 2009). Thus, the challenges raised by pest rodents are manifold. Chemical control methods of India using rodenticides need to be refined in view of their efficacy, safety to non-targets and cost effectiveness. Other methods may be integrated as a package of technology. However, major emphasis is required on the use of non-lethal approaches, such as trap barriers, repellents and biological control with predators, pathogens and antifertility agents. Currently, most of these approaches are at experimental stages. Immunocontraception may prove to be an excellent future tool for the sustainable management of rodent pests in agriculture. The mechanism for effective transfer of rodent management technologies to end users also needs strengthening by exploring various mass media tools for education, training and field demonstrations of the technologies.

References

Abraham, C.C, Remamony, K.S., 1999. Pests that damage cocoa plants in Kerala. Indian Cocoa Arecanut Spices J. 2, 77–81.

Advani, R., 1984. Ecology, biology and control of black rat, *Rattus rattus* in Minicoy Islands. J. Plant. Crops 12, 11–16.

Advani, R., 1985. Rodent pest management in coconut plantation of India and its island. Indian Coconut J. 15, 3–9.

Agarwal, R.K., Chhawara, S.K., Jayraj, K, Sanjeeva, S, Raman, C.P., 1988. Efficacy of cholecalciferol (Vitamin D_3) against house rat, (*Rattus rattus*). Rodent Newslett. 12, 7.

Agarwal, V.C., 2000. Taxonomic studies on Indian Muridae and Hystricidae (Mammalia: Rodentia). *Records of Zoological Survey of India (Occasional paper No 180).* Zoological Survey of India. Kolkata, 177.

Agarwal, V.C., Chakraborty, S., 1992. The Indian Crested Porcupine, *Hystrix indica* (Kerr). In: Prakash, I. Ghosh, P.K. (Eds.), Rodents in Indian Agriculture, Vol. I. Scientific Publishers, Jodhpur, India, pp. 25–30.

Anonymous, 1986. Evaluation of second-generation anticoagulant rodenticides in India. I. Bromadiolone. All India Coordinated Research Project on Rodent Control, Central Arid Zone Research Institute, Jodhpur (Mimeo.)

Anonymous, 1990. Proceedings and recommendation of National Workshop on Planning Strategies for Rodent Pest Management, Central Plant Protection Training Institute, Hyderabad.

Awasthi, A.K., Agarwal, R.K., 1991. Estimation of rodent losses at green pod stage of soybean (*Glycine max*) in Madhya Pradesh. Indian J. Agric. Sci. 61, 159–160.

Banerjee, A., 1955. The family life of a five stripped squirrel, *Funambulus pennanti* Wroughton. J. Bombay Nat. Hist. Soc. 53, 261–265.

Banerjee, A., 1957. Further observations on the family life of the five stripped squirrel, *Funambulus pennanti* Wroughton. J. Bombay Nat. Hist. Soc. 54, 336–343.

Barras, S.C, Seamans, T.W., 2002. Habitat management approaches for reducing wildlife use of airfields. Proc. Vertebrate Pest Conference 20, 309–315.

Berger, P., Sanders, E.H., Gardner, P.H., Negus, N.C., 1977. Phenolic plant compounds functioning as reproduction inhibitors in *Microtus montanus*. Science 195, 575.

Bhat, S.K., 1992. Plantation crops In: Prakash, I. Ghosh, P.K. (Eds.), Rodents in Indian Agriculture, Vol. I. Scientific Publishers, Jodhpur, India, pp. 271–278.

Bhat, S.K., 1993. Rodent and other vertebrate pests management in coconut and cocoa. Tech. Bull. No. 26, Central Plantations Crops Research Institute, Kasaragod (Kerala) India.

Bhat, S.K., Sujatha, A., 1986. A study of the species of small mammals in mixed coconut-cocoa garden in Kerala. Planters Malaysia 62, 428–432.

Bhat, S.K., Nair, C.P.R., Mathew, D.N., 1981. Mammalian pests of cocoa in south India. Trop. Pest Manag. 27, 297–302.

Bhat, S.K., Vidyasagar, P.S.P.V., Sujatha, A., 1990. Records of *Rattus rattus wroughtoni* as a pest of oil palm seedling. Trop. Pest Manag. 36 (75–76).

Bindra, O.S., Mann, G.S., 1975. Ineffectiveness of murine rodent typhus fever bacteria against field rats and field mice at Ludhiana. Indian J. Plant Prot. 3, 98–99.

Birah, A., Tripathi, R.S., Mohan Rao, A.M.K., 2012a. New report of little Indian field mouse, *Mus booduga* (Gray) from Andaman and Nicobar islands. J. Plant. Crops 40 (2), 149–151.

Birah, A., Kumar, A.S., Tripathi, R.S., 2012b. Status of rodent damage to coconut in Andaman and Nicobar Islands. J. Plant. Crops 40 (3), 238–242.

Biswas, S., Mittal, V., 2011. Rodent surveillance and prevention of rodent borne diseases in the context of International Health Regulations – 2005 in international sea ports in India. Rodent Newslett. 35, 2–7.

Buckle, A.P., 1994. Damage assessment and damage surveys. In: Buckle, A.P., Smith, R.H. (Eds.), Rodent Pests and Their Control. CAB International, Wallingford, Oxon, UK, pp. 219–247.

Buckle, A.P., Rowe, F.P., 1981. Rice field rat Project, Malaysia, 1970–80. Center for Overseas Pest Research, London.

Buckle, A.P., Smith, R.H., 1994. Rodent Pests and Their Control. CAB International, Wallingford, Oxon, UK.

Butani, P.G., Vyas, H.J., Kapadia, M.N., 2006. Rodent management in groundnut in Gujarat. In: Sridhara, S. (Ed.), Vertebrate Pests in Agriculture – The Indian Scenario. Scientific Publishers, Jodhpur, India, pp. 193–204.

Chakraborty, R, Chakraborty, S, 2010. Rodents (Mammalia: Rodentia) of Andaman and Nicobar Islands with notes on their conservation and management. Rodent Newslett. 34, 1–8.

Chakraborty, S., 1975. Field observation on biology and ecology of the lesser bandicoot rat, *Bandicota bengalensis* (Gray) in West Bengal. Proceedings of the All India Rodent Seminar, Ahmadabad, pp. 102–112.

Chakraborty, S., 1992. The Indian mole rat or lesser bandicoot rat, *Bandicota bengalensis* Gray In: Prakash, I. Ghosh, P.K. (Eds.), Rodents in Indian Agriculture, Vol. I. Scientific Publishers, Jodhpur, India, pp. 173–192.

Chakravarthy, A.K., 1983. Vertebrate Pest Management in Cardamom and other Crops in Mainad Region. Final Project Report. Indian Council of Agricultural Research, New Delhi.

Chambers, L.K., Malcolm, A.L., Hinds, A., 1999. Biological control of rodents – the case of fertility control using immunocontraception. In: Singleton, G.R., Hinds, L, Leirs, H, Zhibin, Z. (Eds.), Ecologically Based Rodent Management. Australian Center for International Agricultural Research, Canberra, pp. 215–242.

Chandla, V.K., Kumar, S., 1995. Porcupine damage in potato fields in Shillaroo in Himachal Pradesh. Rodent Newslett. 19, 7.

Chaturvedi, G C., Patel, M.J., Patel, M.I., 1975. Rodent control in residential areas. Proceedings of the All India Rodent Seminar, Ahmadabad, pp. 142–147.

Chaudhary, V., Tripathi, R.S., 2003. Bio-efficacy of Difethialone, a second-generation anticoagulant rodenticide for the control of poison shy rodents. Indian J. Agric. Sci. 73, 550–552.

Chaudhary, V., Tripathi, R.S., 2004. Toxicity and acceptability of difethialone baits against *Tatera indica* Hardwicke. Ann. Arid Zone 43 (1), 59–63.

Chaudhary, V., Tripathi, R.S., 2006. Evaluation of single dose efficacy of difethialone – a second generation anticoagulant for the control of rodents inhabiting arid ecosystem. Indian. J. Agric. Sci. 76, 736–739.

Chaudhary, V., Tripathi, R.S., Rana, B.D., 2001. Observation on predation of *Tatera indica* by varanids in arid ecosystem. Rodent Newslett. 25, 5–6.

Chaudhary, V, Tripathi, R.S., Poonia, F.S., 2002. Evaluation of toxicity and acceptability of difethialone treated baits against Indian gerbil, *Meriones hurrianae* Jerdon. Curr. Agric. 26, 31–34.

Chaudhary, V., Tripathi, R.S., Sankhla, A., 2005. Incidence of Bandicota bengalensis in the urban locales of Jodhpur city. Rodent Newslett. 29, 9–11.

Chauhan, N.P.S., Saxena, R.N., 1985. The phenomenon of bamboo flowering and associated increase in rodent population in Mizoram. J. Bombay Nat. Hist. Soc. 82, 644–647.

Chopra, G., Kaur, P., Guraya, S.S., 1996. Rodents: Ecology, Biology and Control. R. Chand & Co., New Delhi.

Christopher, M.J., 1987. Assessment of losses in wheat and sugarcane crops due to rodents in Andhra Pradesh. Ph.D. Thesis, Sri Venkateswara University, Tirupati, p. 101.

Deoras, P.J., 1964. Rats and their control. A review of an account of work done on rats in India. Indian. J. Entomol. 26, 407–408.

Deoras, P.J., 1966. Some observations on the probable damage caused by rats in Bombay. Indian J. Entomol. 35, 415–416.

Dickman, C.R., 1999. Rodent ecosystem relationship: A review. In: Singleton, G.R., Hinds, L, Leirs, H., Zhang, Z. (Eds.), Ecologically Based Rodent Management. Australian Center for International Agricultural Research, Canberra, Australia, pp. 113–133.

Dixit, V.P., 1992. Plant products/non-steroidal compounds affecting fertility in Indian desert gerbil, *Meriones hurrianae*, Jerdon. In: Prakash, I. Ghosh, P.K. (Eds.), Rodents in Indian Agriculture, Vol. I. Scientific Publishers, Jodhpur, India, pp. 595–604.

Dubey, O.P., Awasthi, A.K., Patel, R.K., 1992. Bengal gram: Production losses due to rodents and control strategies. In: Prakash, I., Ghosh, P.K. (Eds.), Rodents in Indian Agriculture. Scientific Publishers, Jodhpur, India, pp. 269–270.

Duckett, J.E., Karuppiah, S., 1991. Progress at PORIM/ Australian Enterprise Berhad Barn Owl Project at Kok Foh Estate up to February 1990. Annual Report of the Negeri Sembilan Planters' Association, pp. 62–70.

Ericsson, R.J., 1982. Alpha-chlorohydrin (Epibloc): a toxicant–sterilant as an alternative in rodent control. Marsh, R.E., (Ed.), Proceedings of the Tenth Vert. Pest Conference, University of California, Davis, USA, pp. 6–9.

Franz, S.C., 1975. The behavioural and ecological milieu of godown bandicoot rats: implications for environmental manipulation. Proceedings of the All India Rodent Seminar, Ahmadabad, pp. 95–101.

Fulk, G.W., 1977. Food hoarding of *Bandicota bengalensis* in a rice field. Mammalia 41, 539–541.

Gangadhar, N.L., 1999. Rodents and Leptospirosis: a global perspective. Proceedings of the First Leptospirosis Conference, Bangalore, pp. 7–14.

George, C.M., Joy, P.J., Abraham, C.C., 1980. On the occurrence of different species of rats in Kerala. Agric. Res. J. Kerala 18, 242–245.

Ghosh, P.K., Taneja, G.C., 1968. Oestrous cycle in the desert rodents, *Tatera indica* and *Meriones hurrianae*. Indian J. Exp. Biol. 6, 54–55.

Gibb, J.A., 1981. What determines the number of small herbivorous mammals. New Zealand J. Ecol. 4, 73.

Govind Raj, G., Srihari, K., 1989. Reproductive biology of the rodent pest, soft furred field rat, Rattus meltada. Proceedings of the International Confefence Biology and Control of Pests of Agriculture and Medical importance, Madurai, India.

Goyal, S.P., Ghosh, P.K., 1992. Adaptations in some Indian rodent species. In: Prakash, I. Ghosh, P.K. (Eds.), Rodents in Indian Agriculture, Vol. I. Scientific Publishers, Jodhpur, India, pp. 357–395.

Greaves, J.H., 1994. Resistance to anticoagulant rodenticides. In: Buckle, A.P, Smith, R.H. (Eds.), Rodent Pests and Their Control. CAB International, Wallingford, Oxon, UK, pp. 197–217.

Green, M.G., Taylor, K.D., 1975. Preliminary experiments on habitat alteration as a means of controlling field rodents in Kenya. Ecol. Bull. Stockholm 19, 175.

Gupta, K.M., Singh, R.A., Misra, S.C., 1968. Economic loss due to rat attack in sugarcane in Uttar Pradesh. Proceedings of the International Symposium on Binomics and Control of Rodents, Kanpur, India, pp. 17–19.

Gupta, R.S., Sharma, N., Dixit, V.P., 1990. Callotropin – A novel compound for fertility control. Ancient Sci. Life 9, 224–230.

Hansson, L., 1975. Effect of habitat manipulation on small rodent populations. Ecol. Bull. Stockholm 19, 163.

Hansson, L., 1988. Natural resistance of plants to pest rodents. In: Prakash, I. (Ed.), Rodent Pest Management. CRC Press, Boca Raton, Florida, pp. 391–398.

IRRI, 2009. Crop Health: Rodent Management Rice Knowledge Bank. International Rice Research Institute, Manila, Philippines, (online version).

Idris, M., 2005. Significance of scent marking gland in desert rodents. In: Tyagi, B.K., Baqri, Q.H. (Eds.), Changing Faunal Ecology in Thar Desert. Scientific Publishers, Jodhpur, India, pp. 267–288.

Idris, M., Rana, B.D., 2001. Occurrence of the Lesser bandicoot rat, *Bandicota bengalensis* in Jodhpur town. Rodent Newslett. 25, 9.

Idris, M., Tripathi, R.S., 2011. Behavioural responses of desert gerbil, *Meriones hurrianae* after removal of scent marking gland. Indian J. Exp. Biol. 49, 555–557.

Islam, Z., Kareem, A.N.M.R., 1995. Rat control by trapping in deep water rice. Int. J. Pest Manag. 41, 229–233.

Islam, Z., Morton, R.G., Jupp, B.P., 1993. The effect of rat damage in deep water rice yields in Bangladesh. Int. J. Pest Manag. 39, 250–254.

Jackel, T., Burgstaller, H., Frank, W., 1996. *Sarcocystis singaporensis*: studies on host specificity, pathogenicity and potential use as biocontrol agent of wild rats. J. Parasitol. 82, 280–287.

Jackel, T., Khoprasert, Y., Endepols., S., Archer-Baumann, C., Suasa-ard, K., Promkerd, P., et al., 1999. Biological control of rodents using *Sarcocystis singaporensis*. Int. J. Parasitol. 29, 1321–1330.

Jain, A.P., 1970. Body weight, sex ratio, age structure and some aspects of reproduction in the Indian gerbil, *Tatera indica indica* Hardwicke in the Rajasthan desert, India. Mammalia 34, 415–432.

Jain, A.P., Tripathi, R.S., 1988. Evaluation of second generation anticoagulant rodenticides in India. II. Brodifacoum All India Coordinated Research Project on Rodent Control. Central Arid Zone Research Institute, Jodhpur (Mimeo), 21 pp.

Jain, A.P., Tripathi, R.S., 1992. Management of rodent pests in oilseed crops. In: Kumar, D., Rai, M. (Eds.), Advances in Oil Seed Research. Scientific Publishers, Jodhpur, India, pp. 360–367.

Jain, A.P., Tripathi, R.S., 2000. An integrated approach to rodent pest management. In: Upadhyaya, R.K., Mukherji, K.G., Dubey, O.P. (Eds.), IPM Systems in Agriculture. Vol. 7. Key Animal Pests. Aditya Books Pvt. Ltd., New Delhi, pp. 275–307.

Jain, A.P., Mathur, M., Tripathi, R.S., 1992. Bio-efficacy of flocoumafen against major desert rodent pests. Indian J. Plant Prot. 20, 81–85.

Jain, A.P., Tripathi, R.S., Rana, B.D., 1993. Rodent Management: The State of Art. Technical Bulletin No 1, All India Coordinated Research Project on Rodent Control, Central Arid Zone Research Institute, Jodhpur, 38 pp.

Jain, A.P., Tripathi, R.S., Patel, N., 1995. Influence of aridity on burrowing and other behavioural traits of *Nesokia indica* Gray. Ann. Arid Zone 34, 49–51.

John, T.J., 1996. Emerging and reemerging bacterial pathogens in India. Indian J. Med. Res. 103, 4–18.

Kashyap, N., Parveen, F., 1990. Trials of herbal powder, Indiara-M as a rodent repellent. Rodent Newslett. 14, 10.

Kaul, D.K., Ramaswami, L.S., 1969. Reproduction in the Indian desert gerbil, *Meriones hurrianae* Jerdon. Acta Zool. 50, 233–248.

Kaur, R., Parshad, V.R., 1997. Ethyl methanesulphonate induces changes in the differentiation, structure and functions of spermatozoa of house rat, *Rattus rattus* L. J. Biosci. 22, 357–365.

Khatri, A.K., Thomas, M., Pachori, R., 1987. Patterns of losses caused by rodents to soybean crop. Indian J. Plant Prot. 15, 103–105.

Kinnear, N.B., 1919. Expected plague of field rats in 1920. J. Bombay Nat. Hist. Soc. 26, 663–666.

Kochar, D.K, Kaur, R., 2008. Rodent damage in wheat and paddy in Punjab. Pestology 32 (9), 44–47.

Krishnakumari, M.K., Urs, Y.L, Mukthabai, K., 1992. The house rat, Rattus rattus Linn. In: Prakash, I. Ghosh, P.K. (Eds.), Rodents in Indian Agriculture, Vol. I. Scientific Publishers, Jodhpur, India, pp. 63–94.

Krishnamuthy, K., Uniyal, V., Singh, J., Pingale, S.V., 1967. Studies on rodents and their control. Part 1. Studies on rat population and losses of food grains. Bull. Grain Technol. 5, 147–153.

Kumar, P., Pashahan, S.C., 1995. Ecology and management of rodents in some vegetable crops in and around Hisar. Rodent Newslett. 19, 5–6.

Kumar, S., Mishra, S.S., 1993. Rat, *Bandicota bengalensis* Gray damage to maize in potato based intercropping system. Rodent Newslett. 17, 4–5.

Kumar, T.S., 1985. A life history of spotted owlet (*Athene brama brama* Temminick) in Andhra Pradesh. Raptor Res. Centre, 241 pp.

Lam, Y.M., 1988. Rice as a trap crop for rice field rats in Malaysia. In: Crabb, A.C., Marsh, R.E. (Eds.), Proc. 13th Vertebrate Pest Conference, University of California, Davis, USA, pp. 123–128.

Leirs, H., 2003. Management of rodents in crops: The pied piper and his orchestra. In: Singleton, G.R, Hinds, L.A., Krebs, C.J., Spratt, D.M. (Eds.), Rats Mice and People: Rodent Biology and Management. Australian Centre for International Agricultural Research, Canberra, Australia, pp. 183–189.

Malhi, C.S., Parshad, V.R., 1992. Rodent damage and control in vegetable crops. Kisan World 19, 29.

Malhi, C.S., Parshad, V.R., 1995. Comparative efficacy of three rodenticides with different baiting methods in wheat and rice crop. Int. Pest Control 37, 55–57.

Malhi, C.S., Parshad, V.R., 1987. Pest damage to harvested wheat and paddy. Indian Farming 37, 29–30.

Mandal, A.K., Chakraborty, S., 1999. Endemic species of rodents in India. Rodent Newslett. 23, 1–2.

Mathur, R.P., 1992. Social engineering activity on rodent control. In: Prakash, I., Ghosh, P.K. (Eds.), Rodents in Indian Agriculture, vol. 1. Scientific Publishers, Jodhpur, India, pp. 605–649.

Mathur, M., Jain, A.P., 1987. Effectiveness of cholecalciferol pellets and wax blocks against *Tatera indica* and *Rattus rattus*. Rodent Newslett. 11, 3.

Mathur, R.P., Prakash, I., 1984. Reduction in the population of Indian desert rodents with anticoagulant rodenticides. Proc. Indian Acad. Sci. 93, 585–589.

Mathur, R.P., Kashyap, N., Mathur, M., 1992. Bioefficacy of anticoagulant rodenticides. In: Prakash, I., Ghosh, P.K. (Eds.), Rodents in Indian Agriculture, vol. 1. Scientific Publishers, Jodhpur, India, pp. 517–583.

Mian, Y., Haque, E., Brooks, J.E.R., Savearie, P.J., 1993. Laboratory evaluation of bromethalin as a rodenticide against *Bandicota bengalensis* Gray. Univ. J. Zool. Rajashahi Univ. 12, 15–19.

Mittal, V.P., Vyas, H.J., 1992. Groundnut: Production losses due to rodents and control strategies. In: Prakash, I. Ghosh, P.K. (Eds.), Rodents in Indian Agriculture, Vol. I. Scientific Publishers, Jodhpur, India, pp. 249–264.

Mittal, V.P., Vyas, H.J., Kotadia, V.S, Virani, V.R., 1991. Rodents: A serious pest of groundnut in Gujarat. Rodent Newslett. 15, 6–7.

Neelanarayana, P., Nagarajan, R., Kanakasabai, R., 1994. Magnitude of rodent depredation in cotton. Rodent Newslett. 18, 3.

Neelanarayanan, P., 1997. Predatory pressure of barn owl (*Tyto alba stertens* Hartert, 1928) on rodent pest – a field evaluation. Ph.D. Thesis, Mayiladuthurai, Bharthidasan University, Thiruchirapalli, 92 pp.

Nimbalkar, S., 2000. Pilot observations in house mouse, *Mus musculus* and lesser bandicoot rat, *Bandicota bengalensis* with special reference to defecation. PG Diploma Dissertation. National Plant Protection Training Institute, Hyderabad, 25 pp.

Parack, D.W., Thomas, J., 1970. The behaviour of the lesser bandicoot rat, *Bandicota bengalensis* (Gray and Hardwicke). J. Bombay Nat. Hist. Soc. 67, 67–68.

Parshad, V.R., 1897. Rodent damage and control in sugarcane. Indian Sugar. 37, 341–345.

Parshad, V.R., 1999. Rodent control in India. Int. Pest Manag. Rev. 4, 97–126.

Parshad, V.R., Kaur, P., Guraya, S.S., 1989. Reproductive cycles of mammals: Rodentia. In: Saidpur, S.K. (Ed.), Reproductive Cycles of Indian Vertebrates. Allied Publishers Limited, New Delhi, pp. 347–408.

Pashahan, S.C., Sablok, V.P., 1987. Effect of removal of wild vegetation on migration of field rodents. HAU J. Res. 17, 12–13.

Patel, N., Jain, A.P., Tripathi, R.S., Praveen, F., Mathur, M., 1995. Losses caused to pomegranate fruits by rodents in arid horticulture. Geobios New Rep. 14, 37–38.

Patel, R.K., Awasthi, A.K., Dubey, O.P., 1992. Rat damage in rice fields under dry field conditions in Madhya Pradesh. Intl. Rice Res. Inst. (IRRI) Newslett. 17, 21.

Pradhan, M.S., Talmale, S.S., 2011. A checklist of Indian Rodent Taxa (Mammalia: Rodentia), Zoological Survey of India (online version), 12 pp.

Prakash, I., 1969. Ecotoxicology and control of Indian desert gerbil, *Meriones hurrianae* (Jerdon) V. Food preference in fields during monsoon. J. Bombay Nat. Hist. Soc. 65, 581–589.

Prakash, I., 1971. Breeding season and litter size of Indian desert rodents. Z. Angew. Zool. 58, 442–454.

Prakash, I., 1975. The population ecology of the rodents of Rajasthan desert, India. In: Prakash, I., Ghosh, P.K. (Eds.), Rodents in Desert Environment. Dr W. Junk BV Publishers, The Hague, pp. 75–116.

Prakash, I., 1976. Rodent Pest Management – Principle and Practices. Central Arid Zone Research Institute Monograph Series No. 4, Jodhpur, 28 pp.

Prakash, I. 1981. Ecology of the Indian Desert Gerbil, *Meriones hurrianae*. Central Arid Zone Research Institute Monograph Series No. 10, Jodhpur, 88 pp.

Prakash, I., 1988. Rodent Pest Management. CRC Press, Boca Raton, Florida, USA, 480 pp.

Prakash, I., 1991. Rodent communities, epidemiological and economic considerations. In: Abrol, I.P., Venkateswarlu, J. (Eds.), Prospects of Indira Gandhi Canal Project. Indian Council of Agricultural Research, New Delhi, pp. 110–115.

Prakash, I., Kametkar, L.R., 1969. Body weight, sex and age factors in population of the northern palm squirrel, *Funambulus pennanti* Wroughton. J. Bombay Nat. Hist. Soc. 66, 99–115.

Prakash, I., Mathur, R.P., 1987. Management of Rodent Pests. Indian Council of Agricultural Research, New Delhi, 133 pp.

Prakash, I., Jain, A.P., Purohit, K.G., 1971. Breeding and postnatal development of the Indian gerbil, *Tatera indica indica* Hardwicke in the Rajasthan desert. *Saug. Mittel* 19, 375–380.

Prakash, I., Idris, M., Kumari, S., 1998. Scent marking behavior of three desert rodents. Proc. Indian Nat. Sci. Acad. B64, 319–334.

Purohit, K.G., Kametkar, L.R., Prakash, I., 1966. Reproductive biology and post natal development in the northern palm squirrel, *Funambulus pennanti* Wroughton. Mammalia 30, 538–546.

Rajendran, T.P., Tripathi, R.S., Dutta, B.C., Bora, D.K., Rao, A.M.K.M., 2007. Rodent pest management in north-east India All India Network Project on Rodent Control (ICAR). Central Arid Zone Research Institute, Jodhpur, 28 pp.

Ramesh, P., 1992. The short tailed mole rat, *Nesokia indica* Gray. In: Prakash, I., Ghosh, P.K. (Eds.), Rodents in Indian Agriculture. Scientific Publishers, Jodhpur, India, pp. 165–172.

Rana, B.D., Prakash, I., 1979. Reproductive biology and population structure of the house shrew, *Suncus murinus sindensis* in western Rajasthan. Saug. Kunde. 44, 333–334.

Rana, B.D., Prakash, I., 1984. Reproduction biology of the soft furred field rat, *Rattus meltada pallidior* (Ryley, 1914) in the Rajasthan desert. J. Bombay Nat. Hist. Soc. 81, 59–70.

Rana, B.D., Tripathi, R.S., 1992a. Rodent problem at Bombay airport. Pestology 16 (4), 22–25.

Rana, B.D., Tripathi, R.S., 1992b. Vulnerability of rodent attack to the fibre optic cables. Pestology 16 (7), 45–47.

Rana, B.D., Tripathi, R.S., 1999. Recent advances in coordinated research on rodent control. All India Coordinated Research Project on Rodent Control (ICAR), Central Arid Zone Research Institute, Jodhpur, India, 48 pp.

Rana, B.D., Jain, A.P., Soni, B.K., 1992. Bait preferences. In: Prakash, I. Ghosh, P.K. (Eds.), Rodents in Indian Agriculture, Vol. I. Scientific Publishers, Jodhpur, India, pp. 419–426.

Rana, B.D., Jain, A.P., Tripathi, R.S., 1994. Fifteen Years of Coordinated Research on Rodent Control. Technical Bulletin No. 3. All India Coordinated Research Project on Rodent Control (ICAR), Central Arid Zone Research Institute, Jodhpur, 141 pp.

Rangareddy, A., Vasanta Bhanu, K., Zaheeruddin, S.M., Raghava Reddy, P., 2005. Burrow fumigator: An eco-friendly device for rodent management. All India Network Project on Rodent Control (ICAR), Central Arid Zone Research Institute, Jodhpur (Mimeo), 6 pp.

Rao, A.M.K., 2010. National Plan on Rodent Pest Management: current initiative of Government of India. Rodent Newslett. 34, 9–12.

Rao, A.M.K.M., 1992. Integrated rodent management In: Prakash, I. Ghosh, P.K. (Eds.), Rodents in Indian Agriculture, Vol. I. Scientific Publishers, Jodhpur, India, pp. 651–668.

Rao, A.M.K.M., 2003. Rodent problem in India and strategies for their management. In: Singleton, G.R., Hinds, L.A., Krebs, C.J., Spratt, D.M. (Eds.), Rats, Mice and People: Rodent Biology and Management Australian Center for International Agricultural Research, Canberra, pp. 203–212.

Rao, A.M.K.M., Balasubramanyam, M., 1992. The Mice, *Mus* spp. In: Prakash, I. Ghosh, P.K. (Eds.), Rodents in Indian Agriculture, Vol. I. Scientific Publishers, Jodhpur, India, pp. 147–164.

Rao, A.M.K.M., Singh, C.E., 1983. Notes on the distribution of rodent damage in Tella Hamsa rice around Hyderabad. Rodent Newslett. 7, 9.

Reddy, R., 1998. Species composition of rodent pests and their management in coconut in Godawari delta. Rodent Newslett. 19, 10.

Roonwal, M.L., 1987. Recent advances in rodentology in India. Records of Zoological Survey of India, Miscellaneous Publication No 105, Zoological Survey of India, Kolkata, 126 pp.

Saini, M.S., Parshad, V.R., 1991. Control of the Indian mole rat with alpha-chlorohydrin: Laboratory studies on bait acceptance and anti fertility effects. Ann. Appl. Biol. 118, 239–247.

Saini, M.S., Parshad, V.R., 1992. Control of Rattus rattus with cholecalciferol: Laboratory acceptance of freshly prepared and ready to use bait formulations. Int. Biodeter. Biodegrad. 30, 87–96.

Saini, M.S., Parshad, V.R., 1993. Field evaluation of alpha-chlorohydrin against Indian mole rat: Studies on toxic and antifertility effects. Ann. Appl. Biol. 122, 153–160.

Saini, M.S., Parshad, V.R., 1988. Laboratory evaluation of a toxicant cum sterilant, alpha-chlorohydrin for the control of Indian mole rat. Ann. Appl. Biol. 113, 307–312.

Shah, M.P., 1979. Population explosion of rodents in groundnut field in Gujarat. Rodent Newslett. 3, 19–20.

Sharma, N., Gupta, R.S., Dixit, V.P., 1990. Anti-implantation/abortifacient activity of plant extracts in female rats. J. Res. Ayur. Siddha 9, 183–193.

Sharma, P.K., Rao, A.M.K.M., 1989. Effect of bund dimension in rodent infestation in rice. Intl. Rice Res. Inst. (IRRI) Newslett. 14, 40.

Sharma, V.N., Joshi, M.C., 1975. Soil excavation by desert gerbils, *Meriones hurrinae* (Jerdon) in Shekhawati region of Rajasthan desert. Ann. Arid Zone 14, 268–273.

Sheikher, C., Jain, S.D., 1992. Efficacy of live trapping as a rodent control technique in vegetable crops. Trop. Pest Manag. 38, 332–335.

Sheikher, C., Malhi, C.S., 1983. Territorial and hoarding behavior in *Bandicota* sp. and *Mus* sp. of Garhwal Himalayas. Proc. Indian Natl. Sci. Acad. B49, 332–335.

Singh, C.D., Rao, A.M.K.M., 1982. Composition of three methods for rodent damage survey in rice fields. Indian J. Plant Prot. 10, 98–99.

Singh, K.N., Singh, Y., Seehra, B.S., Lal, S., 1990. Estimation of rodent population in villages in District Udaipur (India). Bull. Grain Technol. 28, 137–139.

Singh, R., Saxena, Y., 1989. Losses by rodent pests in wheat crop. Rodent Newslett. 13, 5–6.

Singh, Y.P., Kumar, D., Gangwar, S.K., 1994. Status paper on rodents in north-eastern hill region and their management. Rodent Newslett. 18, 3–11.

Singh, Y.P., Gangawar, S.K., Kumar, D., Kushwaha, R.B., Azad Thakur, N.S., Kumar, M., 1995. Rodent Pest and Their Management in North Eastern Hill Region. Monograph No. 37. ICAR Research Complex for NEH Region, Barapani, p. 35.

Singla, N., Parshad, V.R., Singla, L.D., 2003. Potential of *Trypanosama evansi* as a biocide of rodent pests. In: Singleton, G.R., Hinds, L.A., Krebs, C.J., Spratt, D.M. (Eds.), Rats, Mice and People: Rodent Biology and Management. Australian Center for International Agricultural Research, Canberra, pp. 43–46.

Singla, N., Kaur, R., Sehgal, H.S., 2008. Effect of crude cotton seed oil containing gossypol on fertility of male Rattus rattus. National Symposium Recent Advances in Zoological Sciences: Applications to Agriculture, Environment and Human Health. Ludhiana, 107 pp.

Singla, N., Kocher, D.K., Kaur, R., Parshad, V.R., Babbar, B.K, Tripathi, R.S., 2013. Recent Advances in Rodent Research in Punjab. Occasional paper, All India Network Project on Rodent Control, Central Arid Zone Research Institute, Jodhpur, India, p. 56.

Singleton, G.R., Dickman, C., 2001. Rodents. In: Macdonald, D. (Ed.), The New Encyclopedia of Mammals. Andromeda, UK, pp. 578–587.

Singleton, G.R., McCallum, H.F., 1990. The potential of *Capillaria hepatica* to control mouse plagues. Parasitol. Today 6, 190–193.

Singleton, G.R., Petch, D.A., 1994. A review of the biology and management of rodent pests in South East Asia. ACIAR Technical Report No. 30. Australian Center for International Agricultural Research, Canberra, Australia, 65 pp.

Singleton, G.R., Hinds, L.A., Leirs, H., Zhang, Z. 1999a. Ecologically Based Management of Rodent Pests. ACIAR Monograph No. 59, Australian Center for International Agricultural Research, Canberra, Australia, 494 pp.

Singleton, G.R., Sundaramaji, J., Tan, T.Q., Hung, N.Q., 1999b. Physical control of rats in developing countries. In: Singleton, G.R., Hinds, L.A., Leirs, H., Zhang., Z. (Eds.), Ecologically Based Rodent Management. Australian Center for International Agricultural Research, Canberra, Australia, pp. 178–198.

Sreenivasan, M.A., 1975. Trapping of small mammals in relation to vegetation types in Kysanur Forest Disease area, Mysore State, India. J. Bombay Nat. Hist. Soc. 70, 448–492.

Sridhara, S., Tripathi, R.S., 2005. Distribution of Rodents in Indian Agriculture. All India Network Project on Rodent Control (ICAR), Central Arid Zone Research Institute, Jodhpur, 136 pp.

Sridhara, S., Raveendra Babu, T., Ajay, P., 2000. Laboratory and field evaluation of difethialone. AICRP on Rodent Control, UAS, Bangalore (Mimio), 25 pp.

Srivastava, A.S., 1968. Rodent control for increased food production. Rotary Club (West), Kanpur, 152 pp.

Srivastava, D.C., 1992. Sugarcane: Production losses due to rodents and control strategies. In: Prakash, I. Ghosh, P.K. (Eds.), Rodents in Indian Agriculture, Vol. I. Scientific Publishers, Jodhpur, India, pp. 231–248.

Subiah, K.S., 1983. Rodent problem in oilpalm plantations in Hut Bay, Little Andaman and suggested measures. Indian J. Farm Chem. 1, 32–39.

Thakar, G., Panelia, S.K., Tripathi, R.S., Rao, A.M.K.M., 2012. Success story on reduction of leptospirosis incidence with anti rodent campaign in Gujarat. Rodent Newslett. 36, 12–13.

Tripathi, R.S., 2005. Reproduction in desert rodents. In: Tyagi, B.K., Baqri, Q.H. (Eds.), Changing Faunal Ecology in the Thar Desert Scientific Publishers, Jodhpur, pp. 289–304.

Tripathi, R.S., 2009. Rodent management under current scenario of changing land use pattern. In: Symposium Papers: IPM Strategies to Combat Emerging Pests in the Current Scenario of Climate Change. Entomological Society of India, New Delhi, pp. 367–386.

Tripathi, R.S., Jain, A.P., 1990. Rodent damage to desert afforestation plantations. Rodent Newslett. 14, 7–8.

Tripathi, R.S., Jain, A.P., Kashyap, N., Rana, B.D., Prakash, I., 1992. North Western Desert, Special Problem Areas. In: Prakash, I. Ghosh, P.K. (Eds.), Rodents in Indian Agriculture, Vol. I. Scientific Publishers, Jodhpur, India, pp. 357–396.

Tripathi, R.S., Mathur, M., Jain, A.P., Patel, N., 1994. Relative efficacy of glue and other traps for commensal rodent management. Ann. Arid Zone 33, 143–145.

Tripathi, R.S., Chaudhary, V., Idris, M., 2004. Incidence of rodent pests and their management in pulse crops under arid ecosystem. Pestology 28 (6), 67–70.

Tripathi, R.S., Satyaveer, Chaudhary, V., 2006. Evaluation of different formulation of neem for its bio-pesticidal potential against *Tatera indica*. Rodent Newslett. 30, 8–9.

Tripathi, R.S., Chaudhary, V., Singh, S., 2010. Evaluation of bio efficacy of seed powder of Jojoba (*Simmondsia chinensis* [Link] Schneider) against Indian Gerbil *Tatera indica* (Hardwicke). National Conference on Biodiversity of Medicinal and Aromatic Plants: Collection, Characterization and Utilization, Anand, Gujarat, 103 pp.

Uchida, T.A., 1966. Observation on monitor lizard, *Varanus indicus* (Daudin) as a rat control agent on Ifaulk, Western Caronine Islands. Bull WHO 35, 976.

Viliminovic, B., 1972. Plague in South East Asia: a brief historical summary and present geographical distribution. Trans. Roy. Soc. Trop. Med. Hyg. 66, 479–504.

Vyas, H.J., Kotadia, V.S., Butani, P.G., 2000. Glimpses of Research on Rodent Management in Gujarat. All India Coordinated Research Project on Rodent Control, Gujarat Agricultural University, Junagadh, 74 pp.

WHO (World Health Organization), 1999. Leptospirosis worldwide 1999. Weekly Epidemiol. Rec. 74, 237–242.

WHO (World Health Organization), 2000a. Human plague in 1998 and 1999. Weekly Epidemiol. Rec. 75, 338–343.

WHO (World Health Organization), 2000b. Leptospirosis. Weekly Epidemiol. Rec. 75, 217–223.

Wagle, P.V., 1927. The rice rats of lower Sind and their control. J. Bombay Nat. Hist. Soc. 32, 330–338.

Whitaker, R., 1987. Rodent control by Irula tribals. Rodent Newslett. 11, 1–2.

Whitaker, R., Dattari, S., 1986. The role of reptiles in controlling food pests. Hamadryad 11, 23–24.

Wilson, D.E., Reeder, D.M., 2005. Mammalian Species of the World, 3rd Edition. John Hopkins University Press. Baltimore, M.D.

Witmer, G.W., Fantinato, J.W., 2003. Management of rodent populations at airports. Fagerstone, K.A., Witmer, G.W. (Eds.), Proceedings of the Tenth Wildlife Damage Management Conference, 6–9 April 2003, Hot Springs, AR. The Wildlife Damage Management Working Group of The Wildlife Society, Fort Collins, CO, USA, pp. 350–358.

CHAPTER

21

Eco-Friendly Management of Phytophagous Mites

Rachna Gulati

Chaudhary Charan Singh Haryana Agricultural University, Haryana, India

21.1 INTRODUCTION

Mushrooming population, changing climate and demographics, and post harvest losses have eroded many of the gains of surplus food production in many parts of the world. Current estimates also indicate that for the growing population, 70–100% more food will be required by 2050 from the same land area. Simultaneously, development of pesticide resistance (Van Leeuwen et al., 2008), pest resurgence (Gatarayiha et al., 2010b), environmental and human safety concerns, demand for environmentally safe and quality products/organic products have led to alternative methods which are environmentally benign, safe and effective and can reduce the dependency on pesticides. Acaricide resistance in phytophagous mites is a serious phenomenon, especially in spider mites which have a remarkable intrinsic potential for rapid evolution of resistance (Croft and van de Baan, 1988). Spider mites have evolved resistance to more than 80 acaricidals to date, resistance having been reported from more than 60 countries worldwide (Motazedian et al., 2012). These 'non-chemical methods' can replace existing pesticides with improved profitability of production (Erdogan et al., 2012) and without detrimental environmental or human health effects, while maintaining similar levels of pest control with improved yield benefits.

Mite pests are one of the major limiting factors in horticultural and agricultural produce. Plant feeding mites belong to five families, namely Tetranychidae, Tenuipalpidae, Eriophyidae, Tarsonemidae and Tuckerellidae. These plant feeders cause various types of direct damage such as loss of chlorophyll (Figure 21.1), appearance of stippling or bronzing of foliage, stunting of growth, severe defoliation and reduction in yield/marketable produce and indirect damage by acting as vectors of plant diseases, especially members of Tetranychidae and Eriophyidae, causing more loss to growers. The rapid developmental rate,

D. P. Abrol (Ed): *Integrated Pest Management*.
DOI: http://dx.doi.org/10.1016/B978-0-12-398529-3.00023-3

FIGURE 21.1 Chlorotic patches due to spider mite infestation.

FIGURE 21.2 Spider mite webbing on cucumber tendrils.

high reproductive potential, and arrhenotokous parthenogenesis in Tetranychidae allow them to achieve damaging population levels very quickly when growth conditions are good, resulting in an equally rapid decline of host plant quality. The webbings produced by spider mites (Figure 21.2) hamper photosynthetic activity (Bostanian et al., 2003; Dutta et al., 2012), cause biotic stress to its host plant and adversely affect many physiological and biochemical processes (Sivritepe et al., 2009). In addition to yield loss and quality, eriophyid mites are also associated with disease transmission. There is bronzing and rusting symptoms on the lower surface of leaves due to feeding of Tenuipalpid nymphs and adults. Tarsonemid mites usually infest the tender portion of plants and suck the sap from buds, leaves, shoots, flowers and stem sheaths. They cause curling, crinkling and brittleness of foliage but show little leaf symptoms.

Large scale use of chemicals against arthropod pests has adversely affected populations of natural enemies (Ashley, 2003; Gerson and Cohen, 1989), and culminated in environmental pollution, and degradation of soil health and human health (Kumar et al., 2010). Hence, there is an increasing interest in natural pesticides which are derived from plants and microorganisms (Isman, 2006; Isman et al., 2007) because they are generally perceived to be safer than the synthetics. Biological and cultural control being eco-friendly, cost effective and energy efficient could serve as an alternative way to limit chemical treatments which disrupt the biotic balance and lead to secondary pest outbreaks, pest resistance and resurgence.

21.2 CULTURAL CONTROL

Maintenance of field sanitation, balanced use of fertilizers, avoiding monoculture, removal and burning of debris and ratoon crop, adjusting the date of sowing, and pruning technique (removal and destruction of infested material) are some of the important practices which minimize the mite population in various crops. Pruning tea leaves and the infected parts of coconut plants during the peak season significantly reduced the *Brevipalpus phoenicis* (Gope and Das, 1992) and *Aceria guerreronis* (Alenear et al., 1999) population. Similarly, pruning and burning of infested leaves and twigs of horticultural crops effectively reduced *Panonychus ulmi*, and *Aculus schlechtendali* in apple (Koschier, 1997) and *A. litchi* in litchi (Prasad and Singh, 1981). In some cases, a higher predatory population was recorded after pruning (Grafton et al., 1995), which reduces the pest population.

Intercropping with host or non-host crop is also done to reduce the pest population in the target crop. Scanty literature is available regarding various pest mites. Cowpea intercropped with cotton (2:1) reduces the *Tetranychus arabicus* population on cotton (Omar et al., 1993); beans intercropped with maize/onion/garlic showed reduced *T. urticae* damage. Planting of *Azeratus conyzoides* as a ground cover in hilly citrus orchards increased the summer relative humidity and decreased the temperature thereby improving the conditions for natural enemies. These plants provide alternative food and oviposition sites for predatory mites, which suppresses the autumn and spring damage caused by *P. ulmi* (Zhou et al., 1994).

Plant species vary in their degree of acceptance by the mite population. The physico-morphic leaf characteristics may have either a negative or positive influence on the pest population as well as their natural enemies (Krips et al., 1999). Strong webbers such as *Tetranychus* spp. prefer rough surfaces to smooth ones. Leaf hairs could serve as a substrate for attachment of webbing. Likewise, sugarcane varieties with spinous outgrowths on their leaves (Khanna and Ramanathan, 1947) were susceptible to *Oligonychus indicus*. Resistance to mites was attributed to glandular hairs in tomato varieties (Stoner et al., 1968), okra leafed glabrous genotypes in cotton (Wilson, 1992) and fruits with tightly pressed sepals in coconut (Moore and Alexander, 1990). Additionally, plant leaf characteristics such as leaf area, leaf hair density (Misra et al., 1990; Muhammad et al., 2009; Sahayaraj et al., 2003), length of leaf, leaf thickness, and petiole length (Rajaram and Ramamurthy, 2001), play a significant part in the searching capability of mites and their predators (Cedola et al., 2001).

Primary and secondary metabolites found in plants have a role in the defence against herbivores, pests and pathogens. These compounds can function as toxins, deterrents, or digestibility reducers or act as precursors to physical defence systems (Bennett and Wallsgrove, 1994; Balkema-Boomstra et al., 2003). As a result, resistant varieties are developed against mite pests; crop wise resistant varieties are provided in Table 21.1. Regular syringing (using a powerful jet of water from a hose) was also found to be effective in keeping spider mites under control on most ornamental plants. Sterilization of loosened soil with steam at 60°C for 30 min or 80°C for 20 min produced by a soil sterilizer resulted in nearly 100% mortality of bulb mites (Liu et al., 1998) whereas bromide fumigation provided 91% pest control in gladiolus and onion.

21.3 USE OF PREDATORY MITES

Acarine bio-control agents are voracious by nature and possess well-developed searching and dispersal mechanisms (Gulati, 2012). Many are fast runners and can be transported by wind; others locate their prey by kairomones and some utilize their hosts for dispersal. Another important criterion, which increases their chances for inclusion in a successful bio-control programme, is their greater reproductive potential than that of their pests (Gerson and Smiley, 1990). Species of *Blattisocius, Macrocheles, Acarophenax* and *Hemisarcoptes* reproduce more rapidly than their prey. The significance of the sheer size of the acarine predator compared to prey is also considered a

TABLE 21.1 List of Some Important Resistant Varieties/Cultivars in and around the World

Sr. No.	Crop	Mite Pest	Variety/Cultivar
1.	Brinjal	*Polyphagotarsonemus latus*	RHR-58
		Tetranychus telarius	Muthatakeshi, 'Banta'
2.	Chillies	*P. latus*	Jwala
3.	Cotton	*T. urticae*	LAFR 3159, He-BR-S-LA, 3382-6-OSN Pima 5-2, NIAB-999
4.	Maize	*T. urticae*	B-96
5.	Okra	*T. cinnabarinus* *T. telarius*	EC-329390, IC-14-1065
6.	Tomato	*T. cinnabarinus*	Sanjam, *Lycopersicon esculentum, L. peruvianum*
7.	Tea	*Oligonychus coffeae*	MT-18, TR-2027
8.	Coconut	*Eriophyes guerronis*	Kenthali rounded fruits less susceptible
9.	Sugarcane	*O. indicus*	Leaves without hairs

promising characteristic in pest control. Many acarine agents are quite sensitive to prey density and do best when pest populations are strongly clumped whereas others showed enhanced feeding activity with an increased pest population. Predatory mites may lay their eggs in protective locations (e.g. phytoseiids) or where the hatched larvae (e.g. *Pergamasus* sp.) will have easy access to prey.

Phytoseiidae, Ascidae, Anystidae, Bdellidae, Cheyletidae, Cunaxidae, Erythraeidae, Eupalopsellidae, Raphignathidae, Stigmaeidae, Tydeidae, etc. are some of the families associated with plant feeding mites (Table 21.2), Phytoseiidae being the predominant group (Jeppson et al., 1975; McMurtry, 1982). Important predatory mites along with their feeding rate, development period and prey stage consumed were reviewed by Chhillar et al. (2007). Phytoseiid mites have several advantages over other predatory mites because of high fecundity, abundant availability, good searching ability, dispersal rate, adaptability to different ecological niches and a high degree of prey specificity. Generalist predators can feed and develop on various kinds of food: phytophagous mites (such as eriophyids, tarsonemids, tydeids), eggs or immature stages of insects, pollen, nectar and (other) plant exudates, honeydew and fungi (McMurtry and Croft, 1997).

Stigmaied mites are probably next to phytoseiidae as far as predatory efficiency is concerned but these cannot run very fast, hence they are used for controlling slow moving mites and for destruction of mite eggs. Anystid mites are long legged, soft bodied, fast moving mites and start making a whirling movement as soon as touched. These are reared on tetranychids. Cunaxid mites are very strong with thorny mouthparts. They are efficient, fast moving predators which fasten their prey with silken threads secreted by their mouth parts. But their number is limited in nature. So far, no species of high predatory potentiality belonging to Bdellidae and Cheyletidae has been noticed in the field although interactions with phytophagous mites are well documented in the literature.

Studies conducted in vineyards and orchards showed that high densities of many species of phytoseiid mites are observed in the surrounding vegetation (Duso et al., 2004; Toyoshima et al.,

TABLE 21.2 Important Predatory Mites Associated with Plant Feeding Mites

Sr. No.	Predatory Mites	Plant Feeding Mites
1	*Amblyseius finlandicus*	*Brevipalpus phoenicis, Eutetranychus orientalis*
2	*A. alstoniae*	*E. orientalis, Tetranychus macfarlanei, B. phoenicis*
3	*A. (Neoseiulus) fallacis*	*T. urticae, Schizotetranychus andropogoni*
4	*A. andersoni*	*Panonychus ulmi, E. carpini, Colomerus vitis*
5	*A. ovalis*	Tetranychids
6	*A. alstoniae*	*E. orientalis, T. macfarlanei*
7	*A. longispinosus*	*T. cinnabarinus, T. ludeni* eggs
8	*A. tetranychivorus*	*T. ludeni, R. indica*
9	*A. victoriensis*	*T. urticae, Aculus cornutus, C. vitis*
	A. tamatavensis	*Polyphagotarsonemus latus*
10	*Phytoseiulus persimilis*	*T. urticae, T. ludeni, T. fijiensis, T. neocaledonicus, Oligonychus indicus, Raoiella indica*
11	*Typhlodromus pyri*	*T. urticae, Panonychus ulmi, C. vitis*
12	*Agistemus fleschmeri*	*B. obovatus*, tetranychids, *Aceria mangiferae*
13	*A. herbarius*	*O. mangiferus, T. urticae*
14	*A. industani*	*T. ludeni*
15	*A. terminalis*	*Brevipalpus* sp., *Acaphylla theae*
16	*Zetzellia mali*	*Panonychus ulmi, Aculus* sp.
17	*A. indicus*	*T. urticae*
18	*Tencateia* sp.	*Tetranychus* spp.,
19	*Cunaxa setirostris*	*Eutetranychus orientalis, Oligonychus* sp., *O. mangiferus, Schizotetranychus andropogoni*

2008; Tuovinen and Rokx, 1991; Tixier et al. 1998). Ground cover could also be a reservoir for natural enemies by providing refugia and alternative food, as well as places to breed (Gravena et al., 1993). *Amblyseius tamatavensis* and *Proprioseiopsis mexicanus* which feed on *P. latus* in papaya (Collier et al., 2004; Ramos and Rodriguez, 1997) and citrus orchards were abundant in cover crops (Fadamiro et al., 2009) suggesting that ground cover plants may serve as overwintering reservoirs for predacious mites. Likewise, Mailloux et al. (2010) reported wild plant, *Neonotonia wightii*, as the cover crop in citrus acting as a reservoir of phytoseiid mites, sustaining abundant and diverse populations all year round. In a study conducted on a Californian citrus orchard, when *Euseius tularensis* infested cover crops were cut and placed in young citrus trees that had low densities of predacious mites, densities on trees significantly increased.

21.4 FEEDING POTENTIAL

Phytoseiids are considered to be effective predators of phytophagous mites all over the world in many diverse crop ecosystems (Jeppson et al., 1975). Utilization of *Phytoseiulus persimilis* and *Neoseiulus (= Amblyseius) fallacies* as bio-control agents against *T. urticae* on vegetables was reported by several authors (Bravenboer, 1963; Burnett, 1971; Gillespie and Raworth, 2004; Yoldas et al., 1996). *Anystis baccarum* readily preyed upon *Aculus schlechtendali, Panonychus ulmi,* and *Bryobia rubrioculus* in apple orchards. *Typhlodromus pyri* is widespread and abundant on various trees, shrubs, and herbaceous plants (Collyer, 1964) and is common in both intensively managed and wild orchards. Similarly, *P. macropilis* is a widespread predator of both eriophyid and tetranychid mites on several host plants, including unsprayed apple and plum trees. *Zetzellia mali* is also considered a common predator of *A. schlechtendali* and *P. ulmi* on fruit trees (Santos, 1976). *Euseius victoriensis* was successfully used to control the mite pests *P. oleivora, P. latus, Tegolophus australis, Brevipalpus mites,* and *T. urticae* in citrus orchards (Beaulieu and Weeks, 2007). *Colomerus vitis* was successfully checked by phytoseiid mites as indicated by reduced symptoms of erinea (open galls) (Ferragut et al., 2008). Ali et al. (2011) reported that *N. womersleyi* is a highly efficient bio-control agent of *T. macfarlanei* as the maximum daily spider mite consumption was 212.8 eggs, 238.1 larvae, 53.5 protonymphs, and 29.6 deutonymphs. Maximum consumption was recorded in a 5:50 ratio of *A. longispinosus* and *T. urticae* where nymphs fed were 13.50/day and adults fed were 16.54/day (Jeyarani et al., 2012).

21.5 FIELD EFFICACY

Biological control of *T. urticae* by predatory mites has proven effective in greenhouse produced crops (Gough, 1991; Gould and Light, 1971; Kropczynska et al., 1999). Predatory mites thrive in humid environments, whereas pest mites enjoy dry surroundings. To increase humidity, moistening the soil or using a coolmist vaporizer around plants in an enclosed area is found to be beneficial. High humidity is especially beneficial when the first release of predators is made. As a general rule, predatory mites are most effective when released at the first sign of pest mite infestation. Once leaf damage is serious, control is more difficult. If there is heavy infestation, removal of the most affected plant parts and use of knock-down spray (Neem soap) is advocated. Predatory mites should be released 48 h after spraying.

Some phytoseiid species were commercially multiplied in foreign countries and released for spider mite control on vegetables under glasshouse (Hussey and Huffaker, 1976). Phytoseiids have greater efficiency even at low prey densities (Gough, 1991; Hamlen, 1978; Hussey et al., 1965; Kropczynska et al., 1999; McMurtry et al., 1970). When prey/predators were released at a ratio of 50:2 on greenhouse cucumber, *P. persimilis* gave the best control of *T. urticae* after 26 days and subsequently eliminated the prey (Kristova, 1972). French et al. (1976) revealed that when releases were made at a rate of 2–3 predaceous mites/5th plant + 500/10 patches it provided better control of *T. urticae* by *P. persimilis* with a low leaf damage index after 30 days of prey release on tomato. Within one spider mite colony, *P. persimilis* detects its prey by random contact (Jackson and Ford, 1973), but the predator remains in the colony until all prey is eliminated (Sabelis et al., 1984). *P. persimilis* was widely used against *T. urticae* on various vegetable crops grown under greenhouse in France, Spain, Greece, Israel and Italy (Vacante and Nucifora, 1987). Most of the successes for biological control of predatory mites under greenhouses have been reported in the Netherlands and UK (Van Lenteren and Woets, 1988). The combined use of *P. persimilis* and *N. californicus* on short duration potted plants of cucumber could not give better control of *T. cinnabarinus* compared to the single use of *P. persimilis* (Leuprecht, 2000). Mahgoub et al. (2011) reported satisfactory percent reduction of *T. urticae* by *N. californicus* in pepper cultivars.

Mites are shipped as adults and are packed in a wheat bran formulation for easy dispersal. If not released immediately, the culture may be kept for 2–3 days at low temperature. Release rates of *Amblyseius cucumeris* for greenhouse crops are 50–100 predators/cucumber plant, 10–100/pepper plant, 1000/1000 sq.ft bedding and potted plants and 1000/150–200 sq.ft tropical plants. Similarly, for *P. persimilis*, the release rates are 1 predator per plant plus 1–2 per infested leaf of tomato/cucumber, 2000 per 3000 sq.ft for other greenhouse crops, tropical plants, and outdoor gardens, 1000 per 10,000 sq.ft for bedding plants and 5000–20,000 per acre depending on infestation for large agri-businesses. At the first sign of spider mites, *Mesoseiulus longipes* released at one per plant or one per square foot checked the spider mite population. However, later release requires much higher numbers to be effective for spider mites. *Neoseiulus californicus* should be released at 1/sq.ft, 20–40,000/acre, 100,000/ha for spider mites, broad mite, and cyclamen mite. *Euseius victoriensis* was, for example, successfully used in Australia to control the mite pests *P. oleivora, P. latus, Tegolophus australis* (Eriophyidae), *Brevipalpus* spp. (Tenuipalpidae) and *Tetranychus urticae* (Tetranychidae) in citrus orchards (Beaulieu and Weeks, 2007).

21.6 INFLUENCE OF HOST PLANT CHARACTERISTICS ON PREDATORY MITE SURVIVAL

The performance of natural enemies is also influenced by host plant characteristics (Lester et al., 2000). Leaf surface texture and vestiture affect the performance of acarine predators associated with plants. Leaf domatia, minute cavities frequently located at the junctions of primary and secondary veins on the underside of woody dicot leaves, often harbour various predaceous mites (Karban et al., 1995; Pemberton and Turner, 1989). Phytoseiids show higher efficacy on hairy leaves, leaves with more pronounced veins and fruits with rough spurs, which provide them overwintering shelter and better protection from their enemies. Anystids, on the other hand, provide better control on pubescent or glabrous leaves. In addition, many plants provide floral as well as extra floral nectars, pollen and even nutrients from other plant feeders (e.g. homopteran honeydew), natural foods known to be of importance for many generalist predators (McMurtry and Rodriguez, 1987). The quality and quantity of such natural foods differ from plant to plant with no effects on predators. It was demonstrated that a stigmaeid ingesting palm date pollen will produce more than twice as many progeny as one feeding on castor bean pollen.

Plant density and plant architecture influence the distribution of spider mites on a plant species and the ease with which predators can find patches of prey. For example, *P. persimilis* is very efficient on cucumbers that have large leaves and vines that intermingle, but less so on peppers with smaller leaves that do not touch. *P. persimilis* is also less effective on cut rose varieties with fewer leaves because the mites cannot move around as easily on these plants. *N. californicus* is a better choice for control of spider mites on roses, if introduced early. Arranging plants so their leaves are touching may improve biological control on some plant species.

21.7 FACTORS AFFECTING FIELD EFFICACY

Predatory mites have effectively controlled spider mites on chrysanthemum, rose, and other ornamental crops under experimental conditions. However, the need to prevent cosmetic damage on floral or foliage crops may make biological control of mites difficult, especially when pesticides that kill predatory mites are used to suppress other pests and/or diseases. Species selection and release rates vary considerably depending on the plant species and the environmental conditions such as temperature and humidity which influence the growth rate of both predator and prey. *P. persimilis* is an excellent predator of spider mites on low growing plants in humid greenhouses with moderate temperatures. The optimal temperature for developmental time, reproduction and feeding of *P. persimilis* on *T. urticae* was 25–30°C (Bravenboer and Dosse, 1962; Force, 1967). High predator population growth suggests good ability of *P. persimilis* to suppress *T. urticae* populations on commercial vegetable crops at 70 ± 5% R.H. (Mohamed and Omar, 2011). Sucrose, water, kairomones, webbings and plant density enhanced the predaceous efficacy of *P. persimilis* against *T. urticae* (Ashihara et al., 1978; Sabelis and Vander Bean, 1983; Schmidt, 1976; Takafuzi and Chant, 1976). Rasmy and Ellaithy (1988) revealed that a temperature of 35°C gives regular control of *T. urticae* on *P. persimilis* treated cucumber plants as against a control plot where no introduction of predator resulted in a high population of prey. Mohamed and Omar (2011) reported that *P. persimilis* exhibited an increased adult female longevity (24.1 days), good prey suppressing ability, highest daily multiplication rate, R_0/T (2.51), and intrinsic rate of increase, r_m (0.12), when fed on *T. urticae* at a temperature of 25 ± 2°C on commercial vegetable crops at 70 ± 5% R.H.

There are a few crops on which *P. persimilis* cannot be used. For example, *P. persimilis* slips off the stems and leaves of carnations. It does not do well on tomato because the mites become trapped on glandular hairs on the leaf petioles and stems, and are also affected by toxic compounds in the tomato leaf. *P. macropilis* performs better than *P. persimilis* on ornamental plants such as dieffenbachia, dracena, parlor palm and schefflera, under warm, humid conditions. *M. longipes* is frequently used to control spider mites in hot, dry greenhouses on taller plants because it tolerates lower humidity than does *P. persimilis*. *N. californicus* does well on most potted plants in greenhouses with high temperatures and low humidity on cucumber against *T. urticae* (Weintraub and Palevsky, 2008). *G. occidentalis* and *N. californicus* may be better suited for use on semi-permanent greenhouse crops such as rose or gardenia than on short-term vegetable crops. A combination of predators released at regular intervals works best in greenhouses or interior plantscapes with a variety of plants and growing conditions.

The western predatory mite, *Galendromus occidentalis*, is smaller than *P. persimilis* and develops best at cooler temperatures, but it tolerates a wide range of relative humidities (40–80%). It has the capability of regulating spider mite populations at lower densities and for longer periods of time than *P. persimilis*, although it will not control spider mite populations as quickly as *P. persimilis* can. It can also survive for long periods without prey. Several different strains are commercially available, including non-diapausing strains that allow control of spider mites through the winter, when days are short, and strains resistant to a number of organophosphate insecticides. The commercially available *Mesoseiulus longipes* is similar to *P. persimilis* in activity, but can tolerate warmer temperatures – up to 70–90°F – and relative humidity as low as 40%.

21.8 USE OF INSECT PREDATORS

Insect species belonging to the order Coleoptera, Neuroptera, Hemiptera, Thysanoptera and Diptera predate on mite pests (Table 21.3). The feeding potential of beetles was reported by many workers. A single grub of *Scymnus gracilis* and *Oligota* sp. consumes 41–72 eggs and 20 mites/day, respectively whereas adults of both beetles consumed 40–50 mites (adult) of *O. indicus* per day. Hull (1995) reported

TABLE 21.3 Insect Predators Associated with Mites

Insect	Mite	Insect	Mite
COCCINELLIDAE			
Stethorus sp.	*T. urticae, S. cajani*	*S. pauperculus*	*T. macfarlanei, T. papayae, O. indicus, E. orientalis*
S. gilvifrons	*O. indicus*	*S. punctum*	*Tetranychus* sp.
S. keralicus	*Raoiella* sp.	*S. tetranychi*	*R. indica, T. urticae*
S. parcempunctatus	*R. indica*	*S. punctillum*	*T. urticae*
Oligota sp.	*Tetranychus* sp.	*O. oviformes*	*T. urticae*
Scymnus gracilis	*O. indicus, T. urticae*	*Coccinella* sp.	*O. indicus*
C. undecimpunctata	*Tetranychus* sp.	*C. septempunctata*	*P. latens*
Menochilus sp.	*Tetranychus* sp.	*Chilochorus sp*	*Tetranychus* sp.
Brumus suturalis	*O. indicus*	*Micromus timidus*	*T. urticae*
Clanis sorews	*Tetranychus* sp.	Spiders	*Tetranychus* sp.
Adalia sp.	*Tetranychus* sp.	*Jauravia* sp.	*Tetranychus* sp., *R. indica*
CHRYSOPIDS			
Chrysoparla carnea	Citrus red mite	*Chrysopa vulgaris*	European red mite
THRIPS			
Scolothrips sexmaculatus	*T. urticae, P. ulmi*	*S. longicornis*	*T. urticae*
S. indicus	*T. macfarlanei, T. neocaledonicus, O. indicus, E. orientalis*	*Cryptothrips nigripes*	Hibernating mites
Haplothrips faurii	European red mite		
ANTHOCORID BUG			
Anthocoris sp.	*P. ulmi, T. urticae*	Cecidomyid fly, *Arthroconodax occidentalis*	Six spotted mite

that *S. punctum* consumed all stages of mites; adult consumed 75 to 100 mites/day and a larva devoured up to 75 mites/day. The feeding rate of different life stages of *S. punctillum* on the red spider mite were studied by Biswas et al. (2007), who recorded that the fourth instar larva consumed a greater number of eggs, larvae, nymphs and adults of *T. urticae* per day than the first instar larvae.

The genus *Stethorus* (Coleoptera: Coccinellidae), due to its cosmopolitan distribution and its habit of feeding mainly on mite pests of vegetable and fruit crops, is of special interest in acarology. Up to the present time, 31 species of this genus have been reported from different countries. *S. punctillum* was reported from vegetable, ornamental and fruit crops in Canada (Putman and Herne, 1966; Roy

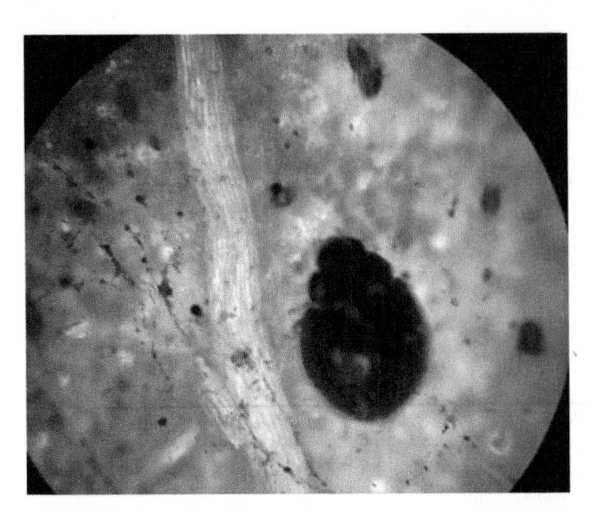

FIGURE 21.3 *Stethorus* feeding on *Tetranychus urticae* stages.

et al., 1999), eastern North America and western America (Gordon, 1985), India (Gulati and Kalra, 2007; Kapur, 1948), Pennsylvania (Wheeler et al., 1973), Spain (Garcia-Mari and Gonzalez-Zamora, 1999), Tashkent (Kapur, 1948), and Uzbekistan (Kapur, 1948) and is associated with tetranychid mites. The developmental stages of the beetle *Stethorus* include eggs, grubs, pupae and adults. Eggs are laid close to primary veins of the leaf (1–10 eggs per leaf), 95% on the under-surface and 5% on the upper surface (Hull, 1995). Grubs emerge after 5 days and pass through four instar stages in about 12 days. The pupation period lasts for about 5 days. Average time from egg deposition to appearance of adult is 23 days. *S. punctillum* was found to be a voracious feeder on *O. ununguis* infesting spruce (Wheeler et al., 1973), *P. ulmi* infesting peach (Putman and Herne, 1966), *T. macdanieli* and *T. urticae* infesting red raspberries (Roy et al., 1999), *T. telarius* and *T. turkestani* infesting cotton (Kapur, 1948), and *T. urticae* infesting strawberries and okra (Figure 21.3) (Garcia-Mari and Gonzalez-Zamora, 1999; Gulati and Kalra, 2007).

21.9 USE OF PREDATORS

In Bio Intensive Pest Management (BIPM), along with botanical insecticides, bioagents are also good alternatives to chemicals. *T. urticae* is mainly controlled by synthetic acaricides (Ako et al., 2006), whereas, in greenhouses, inundative releases of predatory mites such as *Phytoseiulus persimilis* Athias-Henriot have become common practice for combating *T. urticae* outbreaks (Amano and Chant, 1977; Field and Hoy, 1984; Hamlen and Lindquist, 1981). Nonetheless, the control provided by predatory mites was often insufficient and supplementary sprays with selective chemical acaricides have been required (Jacobson et al., 1999).

Stethorus punctillum Weise is a specialized predator of spider mite belonging to the ladybird beetle family (Coccinellidae). Both larvae and adult beetles feed on all stages of spider mites.

Their seasonal abundance was directly correlated with the incidence of mite pests. A high population of beetles during the summer season when the mite population was also high, was reported compared to that in the winter season (Raros and Haramoto, 1974). In Bangalore, the incidence of beetles was noticed throughout the year but they were abundant from January to June (Puttaswamy and Channabasavanna, 1977). In Hisar (Haryana), *S. punctillum* incidence was maximum during August when the mite population was at its peak (Gulati and Kalra, 2007). Geroh et al. (2010) recorded two peaks, first in the fifth week of July (1.0, 1.3, 0.2 beetles/leaf on the ventral surface and 0.3, 0.8, 0.0 on the dorsal surface of the top, middle and bottom strata, respectively), and second, in the first week of August (0.3, 1.2, 0.0 beetles/leaf on the ventral surface and 0.4, 0.7, 0.0 on the dorsal surface of the three strata, respectively) on okra, synchronizing with the pest population.

The feeding potential of various *Stethorus* sp. has been studied by many researchers and they observed that prey was detected by contact (Fleschner, 1950). The grub sucked the inner contents of the chorion of the eggs and discarded the empty shells. The body of mobile stages of mite was first punctured and then their inner contents were sucked. It was observed to be an extra oral digestion in which salivary secretions help in liquefying the body contents of the prey. It was found that 50–100 eggs or 15–17 adults of *P. citri* (Tanaka, 1966) or over 40 females of *T. cinnabarinus* (Dosse, 1967; McMurtry et al., 1970) were needed per day by females of *S. punctillum* to oviposit. Gulati and Kalra (2007) also observed that *S. punctillum* consumed eggs and larvae of *T. urticae* as a whole but sucked the body fluid of nymphs and adults. The grubs and adults consumed 11.2 to 18.2 and 9.0 to 17.4 prey individuals per day, respectively under *in vitro* conditions.

After conducting several *in vitro* studies (Congdon et al., 1993; Gulati and Kalra, 2007; Mathur, 1981), the potential of *Stethorus* spp. was realized. The field efficacy of *Stethorus* spp. in controlling the mite population was reported in the literature by Garcia-Mari and Gonzalez-Zamora (1999), Felland and Hull (1996), Roy et al. (1999) and Gulati and Kalra (2007). Garcia-Mari and Gonzalez-Zamora (1999) recorded that when the population was between 1 and 10 mites per leaflet, the prey/predator ratio should be between 5 and 10 to achieve a decline in the prey population 1–2 weeks later. When the mite population was between 0.1 and 1.0 mites per leaflet, the prey/predator ratio should be between 1 and 5 for an immediate decrease in the prey population. Gulati and Kalra (2007) observed that, under screen house conditions, the ratio of 1:50 predator (adult beetle)/prey (mixed population) resulted in 79.5% control of *T. urticae* 2 days after release on okra leaves.

Due to high feeding, reproductive capacity and synchronization with the pest population, this can rapidly reduce high mite populations to low levels. The predator is highly mobile. Within minutes of release, beetles searched for mites on plants near the release site or flew to neighbouring plants. In combination with *P. persimilis*, *Stethorus* regulated the mite population in the greenhouse at acceptable levels. It was found to be effective for mite control on greenhouse peppers and cucumbers. *Stethorus* sp. at 400–500 beetles per tree reduced the brown mite in avocado. *Clanis sorews*, *Scymnus* sp. and *Brumus suturalis* F. are predaceous on *Oligonychus coffeae*. Other potential predatory coccinellids for mites are *Menochilus sexmaculatus, S. pauperculus, Coccinella septempunctata, Chilochorus nigratus, Brumus suturalis*, etc.

Chrysopids are another group of insects which feed on mites. *Chrysoparla carnea* is reported to consume 1000 to 1500 citrus red mites daily but fails to complete its life cycle on a mite diet. *Chrysopa vulgaris* is known to have better searching ability than *Stethorus* and consumes 30–50 European red mite larvae per hour. The insects belonging to families Coniopterygidae, Anthocoridae and Miridae actively predate upon tetranychids. Among the former, *Conioptyx vicina, Conwentizia hageni* and *C. psociformes* are reported to feed on citrus mite, six spotted mite and European red mite in orchards. *Anthocus muscilus, A. neuromus* and *Orius* sp. are known predators of *P. ulmi, T. urticae* and *P. citri*, respectively. *Blephanidopters angulatus* is also one of the major predators of mites in orchards. Nabidae and Lygaeidae bugs are also reported in the literature as mite feeding insects but their economic importance has not been calculated. Several species of thrips, *Scolothrips sexmaculatus, S. indicus*, and *S. longicornis* are known predators of tetranychids and reduce the pest population rapidly. The larva of *Hyplothrips faurii* consumes approximately 143 eggs of European red mite within 8–10 days of its development. A larva of cecidomyid fly, *Arthroconodaxi occidentalis* consumes 380 mites in 17 days. Flies of other dipteran families such as Dolichopodidae, Empididae and Syrphidae are common feeders on tetranychid mites. Spiders are another important group of mite predators. More than 30 species are known to predate upon mites but cultures of these spiders are neither feasible nor economical. Among these, Araneidae (*Gasteracantha* sp.), Lycosidae (*Pardosa* sp.), Salticidae (*Salticus* sp.), Oxyopidae (*Oxyopes* sp.), Tetragnathidae (*Tetragnatha* sp.) and Clubionidae (*Clubiona* sp.) are reported to be efficient predators of phytophagous mites.

21.10 SMOTHERING AGENTS

Solutions containing petroleum-based horticultural oils, vegetable oils, or agricultural soaps can be applied to control mites. Spider mites and eggs are killed by suffocation when the oil or soap solution smothers them. Petroleum spray oils (PSOs) of various qualities are used to suffocate mite pests, especially *Panonychus ulmi* and *Tetranychus urticae* in horticultural crops (Davidson et al., 1991; Herron et al., 1998). Low phytotoxicity, narrow range, no resistance, and negligible health and environmental concerns have contributed to increase their use.

21.11 USE OF BOTANICALS AND OTHER ECO-FRIENDLY AGENTS

Neem oil (3%), and NSKE (5%) are found to be effective against spider mites. Garlic oil, neem oil and NSKE were also found to be promising against some eriophyid mites, especially coconut mite. Extreme care should be taken with the use of these types of products to limit the chances of phytotoxicity. Petroleum spray oils (PSOs) of various qualities are used to suffocate mite pests, especially *Panonychus ulmi* and *Tetranychus urticae* in horticultural crops (Davidson et al., 1991; Herron et al., 1998). Low phytotoxicity, narrow range, no resistance, and negligible health and environmental concerns have contributed to increase their use. Garlic and neem oil are found to be effective against eriophyid mites whereas medicinal plant extracts (Datura, Withania, Castor, Garlic, Turmeric, Amerbel, etc.) were effective against red spider mites. Extreme care should be taken with the use of these types of products to limit the chances of phytotoxicity.

Gulati et al. (2008) reported the effectiveness of *Withania* extract at 5% and 10% concentrations against *T. urticae* on okra. *Withania* and liquorice extracts tested at three concentrations (2.5, 5 and 7.5% each), had an acaricidal effect against *T. urticae* in a concentration-dependent manner (Geroh, 2007). Neem-based products have no adverse effects in the agro-ecosystem, are economically feasible and have been reported to show adverse effects on the development and reproduction of *T. urticae* (Dimetry et al., 1993; Sundaram and Sloane, 1995) and other mite and insect pests. According to Patil (2005), neem oil (2%) and NSKE (5%) excelled over all other botanical treatments in causing a higher ovicidal and acaricidal action. In other studies as well, neem products significantly reduced the fecundity (Dimetry et al., 1993), and retarded the development of immature stages (Naher et al., 2006) and adults (Sundaram and Sloane, 1995) of *T. urticae.* Similar findings on the adverse effects of neem were reported by Urs (1990). Azadirachtin, the toxic component in neem and responsible for its acaricidal action, was less effective at 10 g/l than *Lolium perenne, Anthemis vulgaris* and *Chenopodium album* in causing mortality in the *T. urticae* population (Yanar et al., 2011).

While studying the management of tetranychid mites in brinjal, Prasanna (2007) reported an 80% reduction in the mite population over untreated control subjected to nimbecidine treatment (0.03%) under field conditions. Other neem-based formulations such as NSKE (5%), neem oil (20 ml/l water) and azadirachtin (0.75 ml/l water) were also found to possess acaricidal activity (88%, 92% and 80%, respectively). On another mite, *Aceria guerreronis*, NSKE (5%) was found to be a significantly superior treatment recording a maximum reduction of 71.46% after one spray, followed by neem oil (67%) and azadirachtin (64.8%) in coconut (Pushpa, 2006). The effectiveness of Nimbecidine, a commercially available formulation of *Azadirachta indica*, against *T. urticae*, was demonstrated on cucumber, causing a 66.8% reduction in the population (Kanika, 2013).

Considering their eco-friendly and non-toxic nature, the botanicals may be recommended for the suppression of mites in perishable goods such as vegetables in alternation with synthetic chemicals. The superiority of the neem product compared to other botanicals may be due to its azadirachtin content, which exhibited high ovicidal, antifeedant and insecticidal growth inhibitory properties resulting in suppression of the mite population (Pushpa, 2006). The significant amounts of phenolics and terpenoids found in botanical extracts constitute the biologically active compounds to be used in pest management (Mehrotra et al., 2011).

21.12 USE OF ANTIMETABOLITES

A number of antimetabolites have been developed within the past 15 years. These include avermectins, pyrroles and milbemycins. Pest mortality results from disruption of metabolism within nerve cells of pest mites. Abamectin, an avermectin, is a mycelial extract of *Streptomyces avermitilis*. Chlorfenapyr is a synthetic pyrrole that has proven extremely effective at suppressing populations of spider mites. Unfortunately, chlorfenapyr exhibits avian toxicity. The milbemycins can be produced by a variety of *Streptomyces* and originally differed from the avermectins only in the C-13 position. Pest mortality results from disruption of metabolism within nerve cells of pest mites. The avermectins are reported to act by blocking neuromuscular transmission. This blockage results in immobilization of the mites. Several groups of investigators have indicated gamma-amino-butyric acid (GABA) release at the target system. GABA is an inhibitory neurotransmitter in vertebrates as well as invertebrates. It has been hypothesized that the avermectins cause a specific prolonged release of GABA.

Avermectin was evaluated against Tydeid mites (*Orthotydeus californicus, O. caudatus*), which are regarded as a quarantine pest on persimmon fruit by regulatory authorities in New Zealand. The most effective treatment is early season applications of mineral oil or avermectin against mites. Other mite species against which avermectin showed its potential are:

Tetranychidae: *Tetranychus urticae, T. pacificus, T. mcdanieli, T. turkestani, Panonychus ulmi, P. citri, Oligonychus pratensis, O. punicae, Eutetranychus hicoriae, Byrobia praetiosa*
Eriophyidae: *Phyllocoptruta oleivora, Eriophyes sheldoni, E. erinea, Epitrimerus pyri, Aculopes lycopersici*
Tenuipalpidae: *Brevipalpus lewisi, B. phoenicis, Dolichotetranychus floridanus*
Tarsonemidae: *Steneotarsonemus bancrofti, S. ananas, S. pallidus, Acarapis woodi.*

Avermectin Bt has shown high activity against several tetranychid and eriophyid mites such as the citrus rust mite, *P. oleivora*, the two spotted spider mite, *T. urticae*, the citrus red mite, *P. citri*, and the strawberry spider mite, *T. turkestani*. In the laboratory, avermectin Bt is toxic to the adult stage of *P. oleivora* at dosage levels less than 1 ppm (McCoy et al., 1982). The compound is also active against *Eutetranychus banksi*.

21.13 USE OF FUNGAL PATHOGENS

Members of the mitosporic (Hyphomycetes) entomopathogens are promising microbial control agents against acari as these fungi invade the host by growing through the external cuticle (Chandler et al., 2000). They can be mass produced using low input technology, easily formulated as myco-pesticides suitable for spraying using conventional chemical spraying equipment (Bateman, 1996), are host specific, less harmful to non-target arthropods and mammals, have no toxic residues, are unlikely to stimulate resistance in target pests and are compatible with chemical insecticides and are, therefore, ideal for integrated pest management (IPM) programme strategies. These are advantages which necessitate increasing the efforts to improve these agents and increase their utility for IPM programmes (Cuperus et al., 2004). Use of fungi as bio-pesticides is considered an attractive strategy in inundation biological control of mites, where the fungal inocula are applied directly to the crop or the target pest and control is achieved exclusively by the released propagules themselves (Eilenberg et al., 2001).

Fungal entomopathogens have been used against a diversity of insect and mite pests (Hajek et al., 2005, 2007). Rabindra et al. (2004) reported 90 genera and 700 species of fungi, representing a large group of entomophorales (*Beauveria* spp., *Aspergillus* spp., *Fusarium* spp., and *Verticillium* spp.) involved with entomopathogenicity. A number of species, e.g. *Hirsutella thompsonii* and *Neozygites floridana*, were specific to Acari whereas other species of entomopathogenic mitosporic fungi killed both Acari and insects (Chandler et al., 2000). The entomopathogenic fungi used for the control of plant feeding mites are *Hirsutella thompsonii* Fisher (Aghajanzadeh et al., 2006; Gardner et al., 1982), *Neozygites floridana* Weiser and Muma (Mietkiewski et al., 1993), *Neozygites adjarica* (Dick and Buschman, 1995), *Metarhizium anisopliae* (Metschnikoff) Sorokin (Chandler et al., 2005), *Beauveria bassiana* (Balsamo) Vuillemin (Alves et al., 2002; Gatarayiha et al., 2010b,c; Geroh, 2011; Irigaray et al., 2003; Tamai et al., 1999) and *Cladosporium cladosporioides*. Among the fungi used as biological control agents (BCAs), *Beauveria bassiana* is a classical entomopathogen which has been extensively used for control of many important pests of various crops around the world (Varela and Morales, 1996). It was reported to constitute 33.9% of all fungal species developed for the control of insects and acarines (Faria and Wraight, 2007). The natural incidence of entomopathogenic fungi on tetranychid mites has been reviewed by many authors (Chandler et al., 2000; van der Geest, 1985).

Initially, *Neozygites* species were observed to cause adult mortality (32–95%) in the citrus red mite, *Panonychus citri*. Subsequently, several other species of spider mites such as *Eutetranychus banksi* and *Tetranychus urticae* were found to be susceptible to *Neozygites* sp. Among mite stages, immature stages of *T. urticae* are more susceptible to *N. floridana* than are adult mites and females. Among adults, females are more susceptible to infection than males. Infection is due to direct penetration of the fungus through the host cuticle by a combination of enzymatic and mechanical processes.

Among Deuteromycetes, *Hirsutella* sp. infections are known from eriophyids. The citrus rust mite, *Phyllocoptruta oleivorus* was found to be most susceptible on grapefruit, going from 5000 mites on a single grapefruit to sudden decimation of its populations (reduced to almost zero) due to *Hirsutella thompsonii*. The spores possess a mucous coat that facilitates adhesion to the host cuticle. The conidia germinate and penetrate all body parts in eriophids, although on spider mites, penetration is usually through the legs. Over 50 species of entomogenous fungi have been reported in the genus *Hirsutella*, but only a few have been reported as pathogens of eriophyids. *H. thompsonii* and *H. necatrix* are two isolates which are effective against spider mites but differ in their pathogenicity. One of the proteins, Hirsutellin A (HtA), inhibits cell growth causing cells to become hypotrophied and disrupting internal organelles and membranes. It caused

100% mortality in *P. oleivora* at a concentration of 100 mg/ml.

H. thompsonii can be easily cultivated on a variety of agar-based and liquid media, in contrast to most Entomophthorales. The fungus was highly pathogenic to *T. cinnabarinus* when it was grown on potato-dextrose agar or on sterile wheat bran, and to *Eutetranychus orientalis* when grown on potato-dextrose agar. *H. kirchneri* was infective towards the eriophyid *P. oleivora, E. orientalis, Panonychus citri, and T. cinnabarinus* (Tetranychidae), and to some degree towards *Hemisarcoptes coccophagus* (Hemisarcoptidae). No infectivity was found towards *Polyphagotarsonemus latus* (Tarsonemidae), *Rhizoglyphus robini, Tyrophagus putrescentiae* (Acaridae) and *Typhlodromus athiasae* (Phytoseiidae). The fungus also has no effect on beneficial insects such as the coccinellids *Coccidophilus citricola* and *Lindorus lophanthae.*

Paecilomyces eriophytis on the blackcurrant mite *Cecidophyopsis ribis*, and *Cephalosporium* sp. on *Phytoptus avellanae* (Eriophyoidea) were also found to be effective. The latter's isolate *C. ribis* also showed a promising result against *P. ulmi, Aceria hippocastani, Cecidophyes galii, A. vaccinii* and *Polyphagotarsonemus latus*.

Infection occurs when hosts come into contact with infectious propagules, usually the conidia (Allee et al., 1990; Fargues and Vey, 1974; Ferron, 1978; Pekrul and Grula, 1979). Conidia adhere to the host's integument (Boucias et al., 1988), germinate and then penetrate the host. Both mechanical pressure and enzymatic lysis of the cuticle are involved in host penetration (Vey and Fargues, 1977). The endoprotease, PR_1, produced by most Deuteromycete fungi, in combination with other enzymes, might play a key role in the penetration process (Charnley and St Leger, 1991; Goettel et al., 1989; St Leger et al., 1987, 1988). Host infection via the mouth parts (Bao and Yendol, 1971; Siebeneicher et al., 1992), the alimentary tract during the consumption of contained food (Broome et al., 1976), the anus after passage through the gut (Allee et al., 1990) and the respiratory system (Clark et al., 1968) have also been reported. After the fungus has entered the body cavity, it invades the haemolymph, producing short filaments or hyphal bodies (blastospores). The invasion was sometimes hampered by defence mechanisms (Charnley, 1992), including phagocytosis and/or encapsulation (Roberts and Humber 1981) which involve the formation of haemocytic aggregates or nodules (Bidochka and Khachatourians, 1987; Dunn, 1986; Hou and Chang, 1985). However, these defences were usually overcome by the fungus, possibly because *B. bassiana* grows more quickly (Figure 21.4) than the pest defences could respond to (Hou and Chang, 1985).

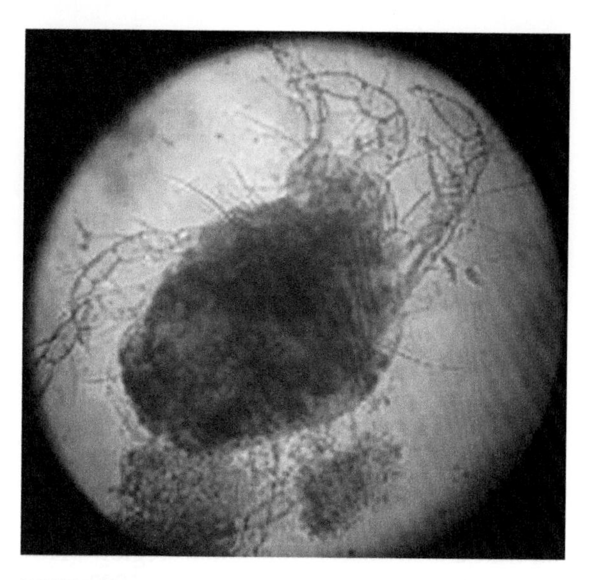

FIGURE 21.4 *Tetranychus urticae* showing mycosis with *Beauveria bassiana* infection.

21.14 VIRULENCE AND HOST SPECIFICITY OF ISOLATES

Virulence and host specificity are two essential elements in the selection of a suitable candidate for microbial control. According to Tigano-Milani et al. (1995), DNA markers provide more detailed genomic information and were not influenced by environmental conditions. Distinctive markers that characterize individual strains are useful to determine the efficacy, survival and spatial temporal distribution in the field. Molecular diagnostics of *B. bassiana* strains have been attempted by studying isozymes (Bridge et al., 1990) and esterase profiles (Varela and Morales, 1996), restriction fragment length polymorphism (RFLP) (Coates et al., 2002), random amplified polymorphic DNA (RAPD) (Cooke et al., 1996; Crowhurst et al., 1995; Geroh, 2011), inter simple sequence repeat (ISSR) markers (Wang et al., 2005), internal-transcribed-spacer ribosomal region (ITS) nucleotide sequences (Muro et al., 2003) and microsatellites (Estrada et al., 2007).

Beauveria bassiana (Balsamo) Vuillemin and *Metarhizium anisopliae* (Metsch.) Sorok. have a wide host range but a difference between host specificity and virulence among strains has been reported (Barreto et al., 2004; Chandler et al., 2000; Dolci et al., 2006; Ferron et al., 1991; Leathers et al., 1993; McCoy et al., 1988; Varela and Morales, 1996; Wang et al., 2005; Wekesa et al., 2005). In the majority of cases, highly virulent strains were grouped in one cluster showing 68–85% similarities (Berretta et al., 1998; Geroh, 2011; Luz et al., 1998). The authors opined that the high similarity could be related to the original host of the strains analysed; however, *B. bassiana* isolates in several instances collected from the same insect species and from the same region were genetically dissimilar (Urtz and Rice, 1997; Berretta et al., 1998). A few reports have revealed no significant correlation between the isolates, host and geographical origin (Kaur and Padmaja, 2008; Muro et al., 2003). In another study, *B. bassiana* strains isolated from aphids (Feng et al., 1990) were found to be highly virulent to *T. cinnabarinus*.

The comparative virulence of *B. bassiana, Metarhizium anisopliae, Aschersonia aleyrodis, Hirsutella* sp., *Isaria farinosa* and *Paecilomyces lilacinus* showed 35–95% mortality against *T. urticae* (Alves et al., 2002; Simova and Draganova, 2003; Tamai et al., 1998, 2002). Wekesa et al. (2005) evaluated 17 isolates of *M. anisopliae* and two isolates of *B. bassiana* against *T. evansi*. All isolates were pathogenic to adult female mites, causing 22.1–82.6% mortalities. Geroh (2011) reported 51.03–65.18% reduction in the *T. urticae* population on okra under field conditions using different strains of *Beauveria bassiana*. Hb-Hyderabad (65.18%) was the most potent in reducing the mite population followed by ITCC-4668 (62.36%), ITCC-5408 (58.55%), ITCC-6063 (54.61%), ITCC-5549 (52.48%) and ITCC-4513 (51.03%).

21.15 LABORATORY BIOASSAYS

Isolates causing more than 70% mortality were subjected to dose–response bioassays. The LC_{50} values ranged between 0.7×10^7 and 2.5×10^7 conidia ml^{-1} and the LT_{50} values of the most active isolates of *B. bassiana* and *M. anisopliae* strains varied between 4.6 and 5.8 days (Wekesa et al. (2005). Concentration-dependent mortality of *T. urticae* was recorded with various concentrations of entomopathogenic fungi by various authors under laboratory conditions. *B. bassiana* (1×10^9 conidia ml^{-1}) and at 10^8 conidia ml^{-1} caused 50% (Tamai et al., 1999) and 77% (Alves et al., 2002) mortality whereas 80–90% and 73% mortality of *T. urticae* was reported with *M. anisopliae* (5×10^7 conidia ml^{-1}) and *Hirsutella* sp. (1.7×10^7 conidia ml^{-1}), respectively (Tamai et al., 2002). Significantly higher mortalities of *T. urticae* eggs, larvae, nymphs and adults (64–94%) were reported with the highest concentration (1×10^{12} conidia ml^{-1}) tested compared to the lowest concentration (1×10^5 conidia ml^{-1}) (Geroh, 2011). Barreto et al. (2004) and Wekesa et al. (2006) also arrived at the same result, i.e. higher efficiency of *B. bassiana* with a higher concentration against cassava green mite, *Mononychellus tanajoa* and tomato spider mite, *T. evansi*, respectively.

The effects of doses of *B. bassiana* and mortality of the mites also differed with the mite stage. Ovicidal action of fungi was more potent against plant feeding mites. Rosas-Acevedo et al. (2003) tested *H. thompsonii* on *T. urticae* egg production and obtained 100% reduction in mite fecundity after 6 days of topical application of the metabolite. They further observed that, depending upon the exudate dose, mites partially recovered within 3 to 6 days post-treatment but produced fewer eggs. Greater than 80% mortality in eggs with *M. anisopliae* and *B. bassiana* against *T. evansi* (Wekesa et al., 2005) and *T. urticae* (Geroh, 2011) were also reported. However, Gatarayiha et al. (2010b,c) observed no infected eggs during the counting of mites even 7 and 14 days after initial spraying. The number of eggs counted on larvae remained the same among the different treatments.

Susceptibility to *B. bassiana* infection decreases with increasing age of motile mites as seen from the lowest values of LC_{50} for larvae, nymphs and adults in different studies. Larvae of *T. urticae* (41.5–94% mortality) (Geroh, 2011) and *T. evansi* (43–48%) (Wekesa et al., 2006) were more susceptible to various concentrations of *B. bassiana* and *M. anisopliae*, compared to nymphs (38.75–92% and 78–82% mortality, respectively). Bugeme et al. (2009) recorded 95.2–99% and 36.5–100% mortality with *B. bassiana* and *M. anisopliae*, respectively against *T. urticae* Koch at 1.0×10^7 conidia ml^{-1} concentration. The main reason for this kind of observation is that the series of moults made in the life cycles of juvenile mites could cause the differential susceptibilities between the two stages. The action of entomopathogenic fungi is contact, i.e. the conidia adheres to its host, germinates and has to penetrate through the cuticle. During moulting, any attached conidia may shed without effecting penetration (Vey and Fargues, 1977).

Lower LC_{50} values in the Direct Spray Bioassay method indicated the higher toxicity of *B. bassiana* and *M. anisopliae* concentrations to spider mites. LC_{50} values ranged from 4.95 to 82.1×10^6 cells ml^{-1} (Tamai et al., 2002), 1.13 to 2.36×10^5 conidia ml^{-1} (Gatarayiha et al., 2011) and 0.3 to 3×10^8 conidia ml^{-1} (Geroh, 2011) against *T. urticae* and 0.7×10^7 to 2.5×10^7 conidia ml^{-1} against *T. evansi* (Wekesa et al., 2005). Irigaray et al. (2003) evaluated the efficacy of Naturalis-L® (a *B. bassiana*-based commercial biopesticide) against the two-spotted spider mite and obtained LC_{50} values in the range of 3184 to 1949 viable conidia ml^{-1}. Lower LT_{50} values reported by the above researchers indicated the rapid infection of the mite with the fungus, which is an important feature for selecting fungal strains as potential biocontrol agents. LT_{50} values of 1.3 and 1.4 days (Tamai et al., 2002), 4.1 to 4.6 days (Bugeme et al., 2009), 108.5 to 122.4 h (Geroh, 2011), 63.8 h (Jeyarani et al., 2011), and 5.5 to 8.6 days (Gatarayiha et al., 2011) were observed with *B. bassiana*, 2.3 to 3.9 days and 110.30 h (Jeyarani et al., 2011) with *Cladosporium cladosporioides* and 2.6 to 8.2 days with *M. anisopliae* (Bugeme et al., 2009). Wekesa et al. (2005) obtained LT_{50} values in the range of 4.6 to 5.8 days against *T. evansi*.

21.16 FIELD APPLICATIONS

With dust formulation of conidia containing 0.5% spores of *B. bassiana* in the field, Dresner (1949) obtained 71% mortality in *T. urticae*. Better control than with chemical pesticide was reported using *B. bassiana* against *T. urticae* infesting chrysanthemum (Alves et al., 1998; Tamai et al., 2002). Four fungal sprays within 14 days reduced mite density from 1.8 to 0.1 mites/leaf. However, the reduction was lower in strawberry (*Fragaria* sp.) than in chrysanthemum, with a mean density of 13 mites/leaflet 21 days after application of 1×10^8 or 5×10^7

conidia ml^{-1}, compared to 43 mites/leaflet in control plots. Spray applications of *B. bassiana* were also reported to reduce the *T. urticae* population in tomato (Gatarayiha et al., 2010c), okra (Geroh, 2011), beans (Kalmath et al., 2007) and eggplants (Gatarayiha et al., 2010b). Chandler et al. (2005) reported up to a 97% reduction in the *T. urticae* population following spraying of Naturalis-L (*B. bassiana* formulation) in glasshouse. Gatarayiha et al. (2010b) reported 56.7, 50.3, 51.7, 25 and 34.6% mortality of *T. urticae* mite with *B. bassiana* 3 weeks after spraying on beans, cucumber, eggplant, maize and tomato, respectively. In the *T. urticae* population, 47–89% mortality was reported with four sprays of 0.7×10^{12} conidia ml^{-1} on okra (Gatarayiha et al., 2011) and 81.6–91% at 10^8 and 10^{10} spores ml^{-1} on cucumber (Kanika, 2013). In *B. bassiana* treated plots, a comparatively higher yield was recorded compared to *T. urticae* infested plots (Geroh, 2011).

21.17 EFFECT OF HOST CROP ON VIRULENCE OF ENTOMOPATHOGENIC FUNGI

The host plant of phytophagous arthropods could influence their susceptibility to entomopathogens (Tanada and Kaya, 1993). The two spotted spider mite is a highly polyphagous arthropod (Rodriguez and Rodriguez, 1987) and has a high capacity to adapt to a range of hosts (Fry, 2004). However, *T. urticae* behaves differently on diverse host plant species (Greco et al., 2006), which might influence its susceptibility to *B. bassiana* infections.

Additionally, plants can also contain chemical compounds inhibitory to entomopathogens which can reduce their infectivity to herbivores (Costa and Gaugler, 1989). Another factor determining the success of foliar applications of microbial control agents is their persistence on plant leaf surfaces (Ignoffo, 1992) and this can also be affected by the crop (Kouassi et al., 2003). When comparing application of Bb on celery and lettuce foliage for the control of tarnished plant bug, *Lygus lineolaris* (Palisot de Beauvois), Kouassi et al. (2003) found better control on lettuce than on celery. This was attributed to longer persistence of *B. bassiana* propagules on the leaves of lettuce than on the leaves of celery. Similarly, in a field experiment with alfalfa and wheatgrass, Inglis et al. (1993) observed that the mortality of grasshopper nymphs due to Bb was correlated with the persistence of the fungal conidia to plant foliage.

Lower control efficacy was found on maize by workers which may be due to low conidial inocula retained on leaves as observed with the number of CFUs recovered immediately after Bb spray applications (Gatarayiha et al., 2010a). Low inoculum levels of a pathogen usually diminish the chances of infection in the host's population (Tanada and Kaya, 1993). Hence, the low number of conidia recovered in maize can probably be attributed to the structural arrangement of leaves and the low density of its canopy, which subsequently may have resulted in a low level of 'collection efficiency' (Mierzejewski et al., 2007). A relatively low efficiency of Bb for mite control on tomato was observed by Chandler et al. (2005). It has been reported that tomato plants possess allelochemicals which can be inhibitory to entomopathogenic fungi (Costa and Gaugler, 1989; Poprawski et al., 2000; Santiago-Álvarez et al., 2006), thus resulting in a low control efficacy of the target pest. The alkaloid tomatine was reported to inhibit the formation of colonies and growth of Bb (Costa and Gaugler, 1989) and *Nomuraea rileyi* (Gallardo et al., 1990) on culture medium. Tomatine also inhibited the germination of Bb on insects when ingested by them (Costa and Gaugler, 1989). Conversely, host plants can also increase the susceptibility of invertebrate herbivores to entomopathogens through dietary stress (Mayer et al., 2002; Tanada and Kaya, 1993). The triterpene cucurbitacin found in cucumber plants and related cucurbit species has been reported to reduce mite survivorship and fitness (Agrawal, 2000), which can influence their susceptibility to Bb infection. The fitness of spider mite can also be influenced by the host acceptance and performance (Agrawal, 2000; Wermelinger et al., 1991).

21.18 EFFECT OF ENVIRONMENTAL FACTORS

Susceptibility of a host (Benz, 1987; Inglis et al., 2001) and persistence of fungal pathogen on plant foliage (Ferron et al., 1991; Ignoffo, 1992; Thompson et al., 2006) are also influenced by environmental factors. It has been reported that the ability of *B. bassiana* propagules to persist in an environment is an important factor in its success as a biological control agent (BCA) and this is most influenced by abiotic factors such as temperature, humidity (Fargues and Luz, 1998) and ultraviolet (UV) light (Inglis et al., 1993). UV light can affect differently the conidia on plant foliage depending on the nature of foliage and the light shielding provided by the plant. Temperature is considered to be one of the environmental factors that influence the virulence of fungal isolates (Ekesi et al., 1999). The importance of temperature in fungal strain selection has been stressed by Fargues et al. (1997) and several other authors (Dimbi et al., 2004; Ekesi et al., 1999). On cucumber crop, *B. bassiana* treatment remained effective between 36°C and 42°C which corroborated the earlier observations of high virulence at higher temperatures (25–35°C) than at lower temperature (20°C) (Bugeme et al., 2009). These studies rejected the earlier hypothesis that *B. bassiana* may not be effective for *T. urticae* control in dry and hot environments (Huffaker et al., 1969). Optimum humidity conditions are met during the night, favouring fungal sporulation and germination. Secondly, low inocula levels of a pathogen usually diminish the chances of infection in the host's population (Tanada and Kaya, 1993). Since epizootic development is density-dependent and high mite density is common on crops, this makes fungi good candidates for spider mite control.

21.19 USE OF ADJUVANTS

B. bassiana is normally applied as conidia. Being hydrophobic, these are difficult to suspend in water during foliar application. Surfactants allow conidia to be suspended uniformly and act as adjuvants. The use of oil as adjuvant in BCA applications is known to protect agents from harmful environmental factors and enhance their activity at the target site (Hong et al., 2005). In foliar sprays for the control of plant pests, *B. bassiana* is usually applied as conidia (Faria and Wraight, 2007). Spider mites feed primarily on the ventral leaf surfaces and produce webs (Morimoto et al., 2006), which protect them from spray applications. It was hypothesized in the study that the use of adjuvants would increase the spread of conidia on the leaf undersides thereby increasing the chance of the target pest to acquire a lethal dose of the infectious propagules. The leaf damage index was substantially reduced from 70% in the unsprayed control to 40% with the application of *Bb* conidia coupled with the adjuvant Break-thru and caused 34–56% mortality of *T. urticae* (Gatarayiha et al., 2010a).

21.20 SAFETY OF BEAUVERIA BASSIANA

Entomopathogenic fungi are promising alternatives to chemical insecticides, and pose no risk to man, domestic animals, wildlife and non-target invertebrates (Strasser et al., 2000; Zimmermann, 2007). Laboratory safety tests of *B. bassiana* revealed that the fungus is essentially non-toxic and non-infectious to vertebrates but caution must be exercised to avoid inhalation of a large quantity of conidia (Copping, 2001; Goettel and Jaronski, 1997). Field studies conducted by other workers (Brinkman and Fuller, 1999; Goettel et al., 1996; Peveling and Weyrich, 1992; Pingel and Lewis, 1996; Reinecke et al., 1990; Tomlin, 1997) also found no evidence of detrimental effects of the fungus and its toxins to humans and non-target organisms. *B. bassiana* fungi secrete a wide range of metabolites, some of which are important medicines or research tools (Vey et al., 2001). The main secondary metabolites produced by *Beauveria* are bassianin, beauvericin, beauveriolides, tenellin and oosporein (Strasser et al., 2000). The amount of oosporein released by *B. brongniartii* into the soil from the formulated product or mycosed insects is relatively small. The concentration of oosporein detected in the soil is usually 2.5 million times lower than that of the pesticides methylbromide and dazomet (Strasser et al., 2000). However, there have been some reports on natural epizootics of *B. bassiana* on non-target organisms, *Neochetina bruchi* (Hustache), a curculionid biocontrol agent of water hyacinth (Chikwenhere and Vestergaard, 2001). Despite a wide host range, evidence to date is that *B. bassiana* can be used with minimal impact on non-target organisms, especially when isolate, selection and spatiotemporal factors are taken into consideration (Vestergaard et al., 2003).

B. bassiana has no detrimental effect on predatory mite, *Neoseiulus cucumeris* (Jacobson et al., 1999) and coccinellid beetle, *Stethorus punctillum* (Geroh, 2007). Ludwig and Oetting (2001) studied the susceptibility of *Phytoseiulus persimilis* Athias-Henriot and *Iphiseius degenerans* (Berlese) to *B. bassiana* (strain JW-1). Under laboratory conditions, the natural enemies were highly susceptible to infection by *B. bassiana* whereas in greenhouse trials, lower infection rates were observed.

Entomopathogenic fungi may contribute to the suppression of spider mite populations in combination with other arthropod natural enemies. The interaction between the fungi and natural enemies may be synergistic/additive, e.g. enhanced transmission and dispersal of spider mite pathogens, or antagonistic (e.g. parasitism/infection, predation and competition) (Ferguson and Stiling, 1996; Roy and Pell, 2000).

21.21 USE OF BACTERIA

Few bacteria have been reported as pathogens of the Acari. *Wolbachia* and *Cardinium* are intracellular bacteria infecting many arthropods and manipulating host reproduction. Ros et al. (2012) found a high level of genetic diversity for *Wolbachia*, with 36 unique strains detected from 64 investigated mite individuals of *Bryobia berlesei, B. kissophila, B. praetiosa, B. rubrioculus, B. sarothamni, Tetranychus urticae*, and *Petrobia harti*. *Wolbachia* cause distorted sex ratios in offspring of spider mites. *Wolbachia* bacteria are transmitted from mother to offspring via the cytoplasm of the egg. When mated to males infected with *Wolbachia* bacteria, uninfected females produce unviable offspring, a phenomenon called cytoplasmic incompatibility (CI).

Field applications of *Bacillus thuringiensis* were successful against *P. citri* (Hall et al., 1971) and *T. pacificus* (Hoy and Ouyang, 1987). Later, Royalty et al. (1990) and Guo et al. (1993) conducted experiments in which two formulations of thuringiensin were tested on *T. urticae*. Results were promising, and indicated that the compound is a potential acaricide. Immature stages are especially susceptible, since these have a high growth rate. Guo et al. (1993) also showed that oviposition started to decline after 2 days and ceased completely after 3–4 days in spider mite at doses of 115 and 211 mg/ml, respectively.

21.22 USE OF VIRUSES

Viruses as biological control agents are known in only a few mites, namely, *P. citri* and *P. ulmi*. In both cases, non-occluded viruses play an important role in the regulation of mite populations in citrus and peach orchards, respectively. Muma (1955) reported the first indication of the occurrence of a virus disease in mites; affected mites showed symptoms of diarrhoea and were stuck to the leaf surface by a smear of black resinous material that was excreted through the anus. A rod-shaped, non-inclusion virus causes the disease in citrus red mite (Reed and Hall, 1972). Field applications were performed by introducing infected immature mites into a peach orchard. A considerable reduction in the population density was obtained. Mites often die with their legs extended downwards when infected with a virus but application of these viruses against mites is feasible.

21.23 COMPATIBILITY OF VARIOUS CONTROL MEASURES

Herbicide application or mowing seems to reduce phytoseiid densities. Reducing mowing frequency could be a good alternative to allow phytoseiid mites to develop in the ground cover. The characterization of dispersal from cover crop to trees is also a key point to determine how cover crop management could directly affect tree biological control (Tixier et al., 2006).

Osborne and Petitt (1985) revealed that applications of insecticidal soap 3 days after release of *P. persimilis* did not adversely affect predator populations and provided enhanced suppression of spider mite populations. A beneficial sublethal effect of imidacloprid on predatory mite, *Amblyseius (Euseius) victoriensis* was reported which showed increased egg production by up to 54% when exposed to the compound. If the effectiveness of lower doses, which are safe to predatory mites, is confirmed, this would open the possibility of using a reduced rate of imidacloprid to counterbalance the stimulatory effect on *T. urticae* by stimulating *G. occidentalis*. Spinosad did not have a toxic effect on *P. persimilis* in a bio-control programme (Pratt and Croft, 2000). Kenneth et al. (2002) revealed that abamectin, hexythiazox, horticultural oil, neem oil, pyridaben, and spinosyn were safest, whereas bifenthrin was toxic to *P. persimilis* at 1, 3, 7, or 14 days after application against *T. urticae*. Cloyd (2007) reported that bifenazate (Floramite), spiromesifen (Judo) and chlorfenapyr (Pylon) were not compatible with *P. persimilis*. The predatory mite, *N. californicus*, is compatible with spinosad (Miles and Dutton, 2003), bifenazate, chlorfenapyr, and spiromesifen (Cloyd et al., 2006). Diflubenzuron (Blumel and Stolz, 1993), spinosad (Holt et al., 2006), pymetrozine, and clofentezine (Cloyd, 2007) are compatible with *P. persimilis*.

Some microbial insecticides are compatible with chemical insecticides and can often be used in combination with them. These are compelling reasons for increasing the efforts to improve these agents and increase their utility for IPM programmes (Cuperus et al., 2004). The efficacies of three microbials (biotrol, dipel and thuricide) and three chemical insecticides (monocrotophos, endosulfan and carbaryl) were compared on four major lepidopterans and their natural enemies. The results showed that the microbials caused mortalities of destructive bollworms and leafroller but allowed the survival of their natural enemies. *B. bassiana* has been found to be compatible with lufenuron 5EC (0.41/ha), thiamethoxan 25WG (100g/ha), methomyl 40SP (1.0kg/ha) and Ethion 50EC (1.51/ha) for insect pest management of major defoliators in soybean. *B. bassiana* (1.25kg/ha) combined with cypermethrin (0.0045%) or chlorpyriphos (0.025%) is as effective and economical as synthetic insecticides alone (cypermethrin 0.009%, endosulfan 0.07% and chlorpyrifos 0.05%) for eco-friendly management of *H. armigera* on cotton.

Combined use of botanicals with microbial pesticides has been reported to increase the efficacy of treatment and delay the development of resistance. Basal application of neem cake in combination with *B. bassiana* (1×10^8 spores/ml) recorded the maximum percent reduction of two spotted spider mite, *T. urticae*, population on okra (Balaji et al., 2007). Similar results were reported by Kumar et al. (2010) who evaluated the efficacy of different organic amendments, viz. farmyard manure, neem cake, and *Pseudomonas fluorescens* in combination with foliar application of plant products, viz. *Ocimum sanctum* (10% leaf extract), neem seed kernel extract (5%), neem oil (3%) and certain entomopathogenic fungal formulations, viz. *Beauveria bassiana, Paecilomyces fumosoroseus*, and *Lecanicillium lecanii* against two spotted spider mite of brinjal. Among the organic amendments, basal application of farmyard manure, *P. fluorescens* and neem cake application in combination with fungal formulations was very effective. Among the botanical pesticides, *O. sanctum* as a foliar application was found to be superior in combination with any of the soil amendments tested. Among the fungal formulations tested as foliar spray, *B. bassiana* was reported to be very promising when combined with any of the soil amendments evaluated. Out of 23 biorational treatments tested against *T. urticae*, a 71.67% reduction in the *T. urticae* population was achieved with neem cake combined with *B. bassiana* (1×10^8 spores/ml) application.

References

Aghajanzadeh, S., Mallik, B., Chandrashekar, S.C., 2006. Bioefficacy of six isolates of *Hirsutella thompsonii* Fisher against citrus rust mite, *Phyllocoptruta oleivora* Ashmead (Acari: Eriophyidae) and two spotted spider mite, *Tetranychus urticae* Koch (Acari: Tetranychidae). Pak. J. Biol. Sci. 9, 871–875.

Agrawal, A.A., 2000. Host-range evolution: adaptation and trade off in fitness of mites on alternative hosts. Ecology 8, 500–508.

Ako, M., Poehling, H.M., Borgemeister, C., Nauen, R., 2006. Effect of imidacloprid on the reproduction of acaricide-resistant and susceptible strains of *Tetranychus urticae* Koch (Acari: Tetranychidae). Pest Manag. Sci. 62, 419–424.

Alenear, J.A., Haji, F.N.P., Moreira, F.R.B., Alenear, J.A., 1999. Coconut mite *Aceria guerreronis* (Keifer) biological aspects, symptoms, damage and control methods. Indian Coconut J. 30, 18.

Ali, M.P., Naif, A.A., Huang, D., 2011. Prey consumption and functional response of a phytoseiid predator, *Neoseiulus womersleyi*, feeding on spider mite, *Tetranychus macfarlanei*. J. Insect Sci. 11, 1–11.

Allee, L.L., Goettel, M.S., Golberg, A., Whitney, H.S., Roberts, D.W., 1990. Infection by *Beauveria bassiana* of *Leptinotarsa decemilineata* larvae as a consequence of faecal contamination of the integument following *per os* inoculation. Mycopathologia 111, 17–24.

Alves, S.B., Tamai, M.A., Lopes, R.B., 1998. Avaliação de *Beauveria bassiana* (Bals.) Vuill. para controle de *Tetranychus urticae* Koch em crisântemo. In: Abstracts 17th Brazil. Congr. Entomol., Rio de Janeiro, 1068 pp.

Alves, S.B., Rossi, L.S., Lopes, R.B., Tamai, M.A., Pereira, R.M., 2002. *Beauveria bassiana* yeast phase on agar medium and its pathogenicity against *Diatraea saccharalis* (Lepidoptera: Crambidae) and *Tetranychus urticae* (Acari: Tetranychidae). J. Invert. Pathol. 81, 70–77.

Amano, H., Chant, D.A., 1977. Life history and reproduction of two species of predacious mites, *Phytoseiulus persimilis* Athias-Henriot and *Amblyseius andersoni* (Chant) (Acarina: Phtoseiidae). Can. J. Zool. 55, 1978–1983.

Ashihara, W., Hamamura, T., Shinkaji, N., 1978. Feeding, reproduction and development of *Phytoseiulus persimilis* Athias-Henriot (Acarina: Phytoseiidae) on various food substances. *Bull. Fruit Tree Res. Stn. Ser.* E 2, 91–98.

Ashley, J.L., 2003. Toxicity of selected acaricides on *Tetranychus urticae* Koch (Tetranychidae: Acari) and *Orius insidiosus* Say (Hemiptera: Anthocoridae) life stages and predation studies with *Orius insidiosus*. M.Sc. Thesis, Faculty of the Virginia Polytechnic Institute and State University.

Balaji, S., Chinniah, C., Kanimozhi, K., Muthiah, C., 2007. Combined effect of organic sources of nutrients and entomopathogenic fungi on the incidence of *Tetranychus urticae* on okra *Abelmoschus esculentus* (L.) Moench. J. Acarol. 17, 38–39.

Balkema-Boomstra, A.G., Zijlstra, S., Verstappen, F.W.A., Inggamer, H., Mercke, P.E., Jongsma, M.A., et al., 2003. Role of cucurbitacin-C in resistance to Spider mite

(*Tetranychus urticae*) in Cucumber (*Cucumis sativus* L.). J. Chem. Ecol. 29 (1), 225–235.

Bao, L.L., Yendol, W.G., 1971. Infection of the eastern subterranean termite, *Reticulitermes flavipes* (Kollar) with the fungus *Beauveria bassiana* (Balsamo) Vuill. Entomophaga 16, 343–352.

Barreto, R.S., Marques, E.J., Gondim Jr., M.G., de Oliveira, J.V., 2004. Selection of *Beauveria bassiana* (Bals.) Vuill. and *Metarhizium anisopliae* (Metsch.) Sorok for the control of the mite, *Mononychellus tanajoa* (Bondar). Sci. Agric. 61, 659–664.

Bateman, R., 1996. Formulation strategies appropriate in ultra-low volume application of biological insecticides. IIBC/WPRS Bull. 19, 29–34.

Beaulieu, F., Weeks, A.R., 2007. Free-living mesostigmatic mites in Australia: their roles in biological control and bioindication. Aust. J. Exp. Agric. 47, 460–478.

Bennett, R.N., Wallsgrove, R.M., 1994. Secondary metabolites in plant defense mechanisms. New Phytol. 127, 617–633.

Benz, S., 1987. Environment. In: Fuxa, J.R., Tanada, Y. (Eds.), Epizootiology of Insect Diseases. Wiley, New York, pp. 177–214.

Berretta, M.F., Lecuona, R.E., Zandomeni, R.O., Grau, O., 1998. Genotypic isolates of the entomopathogenic fungus *Beauveria bassiana* by RAPD with fluorescent labels. J. Invert. Pathol. 71, 145–150.

Bidochka, M.J., Khachatourians, G.G., 1987. Haemocytic defense response of the entomopathogenic fungi in the migratory grasshopper. *Melanoplus sanguinipes*. Entomol. Exp. Appl. 45, 151–156.

Biswas, G.C., Islam, W., Haque, M.M., 2007. Biology and predation of *Sthethorus punctillum* Weise (Coleoptera: Coccinellidae) feeding on *Tetranychus urticae* Koch. J. Biosci. 15, 1–5.

Blumel, S., Stolz, M., 1993. Investigations on the effect of insect growth regulators and inhibitors on the predatory mite *Phytoseiulus persimilis* A.H. with particular emphasis on cyromazine. J. Plant Dis. Prot. 100, 150–154.

Bostanian, N.J., Trudeau, M., Lasnier, J., 2003. Management of the two-spotted spider mite, *Tetranychus urticae* (Acari: Tetranychidae) in eggplant fields. Phytoprotection 84, 1–8.

Boucias, D.G., Pendland, J.C., Latge, J.P., 1988. Nonspecific factors involved in attachment of entomopathogenic deuteromycetes to host insect cuticle. Appl. Environ. Microbiol. 54, 1795–1805.

Bravenboer, L., 1963. Experiments with the predator *Phytoseiulus riegeli* Dosse on glasshouse cucumbers. Mitt. Schweiz. Ent. Ges. 36, 53.

Bravenboer, L., Dosse, G., 1962. *Phytoseiulus riegeli* Dosse als Prädator einiger Schadmilben aus der *Tetranychus urticae* gruppe. Entomol. Exp. Appl. 5, 291–304.

Bridge, P.D., Abraham, Y.J., Corhish, M.C., Prior, C., Moore, D., 1990. The chemotaxonomy of *Beauveria bassiana* (Deuteromycetes: Hyphomycetes) isolates from the coffee borer *Hypothenemus hampei* (Coleoptera: Scolytidae). Mycopathology 111, 85–90.

Brinkman, M.A., Fuller, B.W., 1999. Influence of *Beauveria bassiana* strain GHA on non-target rangeland arthropod populations. Environ. Entomol. 28, 863–867.

Broome, J.R., Sikorowski, P.P., Norment, B.R., 1976. A mechanism of pathogenecity of *Beauveria bassiana* on larvae of the imported ant, *Solenopsis richteri*. J. Invert. Pathol. 28, 87–91.

Bugeme, D.M., Knapp, M., Boga, H.I., Wanjoya, A.K., Maniania, N.K., 2009. Influence of temperature on virulence of fungal isolates of *Metarhizium anisopliae* and *Beauveria bassiana* to the two-spotted spider mite, *Tetranychus urticae*. Mycopathologia 167, 221–227.

Burnett, T., 1971. Prey consumption in acarine predator-prey populations reared in the greenhouse. Can. J. Zool. 49, 903–913.

Cedola, C.V., Sanchez, N.E., Liljesthron, G.G., 2001. Effect of tomato leaf hairiness on functional and numerical response of *Neoseiulus californicus* (Acarina: Phytoseiidae). Exp. Appl. Acarol. 25 (10–11), 819–831.

Chandler, D., Davidson, G., Pell, J.K., Ball, B.V., Shaw, K., Underland, K.D., 2000. Fungal biocontrol of Acari. Biocontrol Sci. Technol. 10, 357–384.

Chandler, D., Davidson, G., Jacobson, R.J., 2005. Laboratory and glasshouse evaluation of entomopathogenic fungi against the two-spotted spider mite, *Tetranychus urticae* (Acari: Tetranychidae), on tomato, *Lycopersicon esculentum*. Biocontrol Sci. Technol. 15, 37–54.

Charnley, A.K., 1992. Mechanisms of fungal pathogenesis in insects with particular reference to locusts. In: Lomer, C.J., Prior, C. (Eds.), Biological Control of Locusts and Grasshoppers. CAB International, Wallingford, UK, pp. 181–190.

Charnley, A.K., St Leger, R.J., 1991. The role of cuticle degrading enzymes in fungal pathogenesis in insects. In: Cole, G.T., Hoch, H.C. (Eds.), The Fungal Spore and Disease Initiation in Plants and Animals. Plenum Press, New York, pp. 267–287.

Chhillar, B.S., Gulati, R., Bhatnagar, P., 2007. Agricultural Acarology. Daya Publiushing House, New Delhi, 355 pp.

Chikwenhere, G.P., Vestergaard, S., 2001. Potential effects of *Beauveria bassiana* (Balsamo) Vuillemin on *Neochetina bruchi* Hustache (Coleoptera: Curculionidae), a biological control agent of water hyacinth. Biol. Control 21, 105–110.

Clark, T.B., Kellen, W.R., Fukuda, T., Lindegren, J.E., 1968. Field and laboratory studies on the pathogenecity of the fungus *Beauveria bassiana* to three genera of mosquitoes. J. Invert. Pathol. 11, 1–7.

Cloyd, R., 2007. Compatibility Conflict: The Use of Pesticides with Biological Control Agents. 2007 AERGC Annual Meeting, 31 July 2007. University of Connecticut, Storrs, CT, 13 pp.

Cloyd, R.A., Galle, C.L., Keith, S.R., 2006. Compatibility of three miticides with the predatory mites *Neoseiulus californicus* McGregor and *Phytoseiulus persimilis* Athias-Henriot (Acari: Phytoseiidae). HortScience 41, 707–710.

Coates, B.S., Hellmich, R.L., Lewis, L.C., 2002. *Beauveria bassiana* haplotype determination based on nuclear rDNA internal transcribed spacer PCR-RFLP. Mycol. Res. 106, 40–50.

Collier, K.F.S., de Lima, J.O.G., Albuquerque, G.S., 2004. Predacious mites in Papaya (Carica papaya L.) orchards: in search of a biological control agent of phytophagous mite pests. Neotrop. Entomol. 33 (6), 799–803.

Collyer, E., 1964. Phytophagous mites and their predators in New Zealand orchards. N. Z. J. Agric. Res. 7, 551–568.

Congdon, B.D., Shanks, C.H., Antonelli, A.L., 1993. Population interaction between *Stethorus punctum picipes* (Coleoptera: Coccinellidae) and *Tetranychus urticae* (Acari: Tetranychidae) in red raspberries at low predator and prey densities. Environ. Entomol. 22, 1302–1307.

Cooke, D.E.L., Kennedy, D.M., Guy, D.C., Russell, J., Unkles, S.E., Duncan, J.M., 1996. Relatedness of group I species of Phyophthora as assessed by randomly amplified polymorphic DNA (RAPDs) and sequences of ribosomal DNA. Mycol. Res. 100, 297–303.

Copping, L.G., 2001. The Biopesticide Manual, third ed. British Crop Protection Council, Farnham, UK.

Costa, S.D., Gaugler, R., 1989. Sensitivity of *Beauveria bassiana* to solanine and tomatine: plant defensive chemicals inhibit an insect pathogen. J. Chem. Ecol. 15, 697–706.

Croft, B.A., van de Baan, H.E., 1988. Ecological and genetic factors influencing evolution of pesticide resistance in tetranychid and phytoseiid mites. Exp. Appl. Acarol. 4, 277–300.

Crowhurst, R.N., King, F.Y., Hawthorne, B.T., Snerson, F.R., Choi-pheng, T., 1995. RAPD characterization of *Fusarium oxysporum* associated with wilt of angsana (*Pterocarpus indicus*) in Singapore. Mycol. Res. 99, 14–18.

Cuperus, G.W., Berberet, R.C., Noyes, R.T., 2004. The essential role of IPM in promoting sustainability of agricultural production systems for future generations. In: Integrated Pest Management: Potential, Constraints and Challenges. CAB International, Wallingford, pp. 265–280.

Davidson, N.A., Dibble, J.E., Flint, M.L., Marer, P.J., Guye, A., 1991. Managing Insects and Mites with Spray Oils. University of California, Oakland.

Dick, G.L., Buschman, L.L., 1995. Seasonal occurrence of a fungal pathogen, *Neozygites adjarica* (Entomophthorales: Neozygitaceae), infecting Banks grass mites, *Oligonychus pratensis*, and two-spotted spider mites, *Tetranychus urticae* (Acari: Tetranychidae), in field corn. J. Kans. Entomol. Soc. 68, 425–436.

Dimbi, S., Maniania, N.K., Lux, S.A., Mueke, J.M., 2004. Effect of constant temperatures on germination, radial growth and virulence of *Metarhizium anisopliae* to three species of African tephritid fruit flies. Biocontrol 49, 83–94.

Dimetry, N.Z., Amer, S.A.A., Reda, A.S., 1993. Biological activity of two neem seed kernel extracts against the two spotted spider mite *Tetranychus urticae* Koch. J. Appl. Entomol. 116, 308–312.

Dolci, P., Guglielmo, F., Secchi, F., Ozino, O.I., 2006. Persistence and efficacy of *Beauveria brongniartii* strains applied as biocontrol agents against *Melolontha melolontha* in the Valley of Aosta (northwest Italy). J. Appl. Microbiol. 100, 1063–1072.

Dosse, G., 1967. Schadmilben des Libanous und ihre Predatoren. Z. Angew. Entomol. 59, 16–48.

Dresner, E., 1949. Culture and use of entomogenous fungi for the control of insects. Contrib. Boyce Thompson Inst. 15, 319–335.

Dunn, P.E., 1986. Biochemical aspects of insects immunology. Annu. Rev. Entomol. 31, 321–339.

Duso, C., Fontana, P., Malagnini, V., 2004. Diversity and abundance of phytoseiids mites (Acari: Phytoseiidae) in vineyards and the surrounding vegetation in northeastern Italy. Acarologia 44, 31–47.

Dutta, N.K., Alam, S.N., Uddin, M.K., Mahmudunnabi, M., Khatun, M.F., 2012. Population abundance of red spider mite in different vegetables along with its spatial distribution and chemical control in brinjal, *Solanum melongena* L. Bangladesh J. Agric. Res. 37 (3), 399–404.

Eilenberg, J., Hajek, A.E., Lomer, C., 2001. Suggestions for unifying the terminology in biological control. Biocontrol 46, 387–400.

Ekesi, S., Maniania, N.K., Ampong-Nyarko, K., 1999. Effect of temperature on germination, radial growth and virulence of *Metarhizium anisopliae* and *Beauveria bassiana* on *Megalurothrips sjostedti*. Biocontrol Sci., Technol. 9, 177–185.

Erdogan, P., Yildirim, A., Sever, B., 2012. Investigations on the effects of five different plant extracts on the two spotted spider mite *Tetranychus urticae* Koch (Arachnida: Tetranychidae). Psyche 10, 1155–1159.

Estrada, M.E., Camacho, M.V., Benito, C., 2007. The molecular diversity of different isolates of *Beauveria bassiana* (Bals.) Vuill. as assessed using microsatellites (ISSRs). Cell. Mol. Biol. Lett. 12 (2), 240–252.

Fadamiro, H.Y., Xiao, Y.F., Nesbitt, M., Childers, C.C., 2009. Diversity and seasonal abundance of predacious mites in Alabama satsuma citrus. Ann. Entomol. Soc. Am. 102 (4), 617–628.

Fargues, J., Luz, C., 1998. Effects of fluctuating moisture and temperature regimes on sporulation of *Beauveria bassiana* on cadavers of *Rhodnius prolixus*. Biocontrol Sci. Technol. 8, 323–334.

Fargues, J., Vey, A., 1974. Modalites dinfection des larves de *Leptinotarsa decemlineata* par *Beauveria bassiana* au cours de la mue. Entomophaga 19, 311–323.

Fargues, J., Ouedraogo, A., Goettel, M.S., Lomer, C.J., 1997. Effects of temperature, humidity and inoculation method on susceptibility of *Schistocerca gregaria* to *Metarhizium flavoviride*. Biocontrol Sci. Technol. 7, 345–357.

Faria, M., Wraight, S.P., 2007. Mycoinsecticides and mycoacaricides: a comprehensive list with worldwide coverage and international classification of formulation types. Biol. Control 43, 237–256.

Felland, C.M., Hull, L.A., 1996. Overwintering of *Stethorus punctum punctum* (Coleoptera: Coccinellidae) in apple orchard ground cover. Environ. Entomol. 25 (5), 972–976.

Feng, J.B., Johnson, J., Kish, L.P., 1990. Survey of entomopathogenic fungi naturally infecting cereal aphids (Homoptera: Aphididae) of irrigated grain crops in southwestern Idaho. Environ. Entomol. 19, 1534–1542.

Ferguson, K.I., Stiling, P., 1996. Non-additive effects of multiple natural enemies on aphid populations. Oecologia 108, 383–402.

Ferragut, F., Gallardo, A., Ocete, R., Lopez, M.A., 2008. Natural predatory enemies of the erineum strain of Colomerus vitis (Pagenstecher) (Acari, Eriophyidae) found on wild grapevine populations from southern Spain (Andalusia). Vitis 47 (1), 51–54.

Ferron, P., 1978. Biological control of insect pests by entomopathogenic fungi. Annu. Rev. Entomol. 23, 409–443.

Ferron, P., Fargues, J., Riba, G., 1991. Fungi as microbial insecticides against pests. In: Arora, D.K., Ajelio, L., Mukerji, K.G. (Eds.), Handbook of Applied Mycology. Dekker, New York, pp. 665–706.

Field, R.P., Hoy, M.A., 1984. Biological control of spider mites on greenhouse roses. Calif. Agric. 38, 29–32.

Fleschner, C.A., 1950. Studies on searching capacity of larvae of three predators of the citrus red mite. Hilgardia 20, 233–265.

Force, D.C., 1967. Effect of temperature on biological control of two-spotted spider mites by *Phytoseiulus persimilis*. J. Econ. Entomol. 60, 1308–1311.

French, N., Parr, W.J., Gould, H.J., Williams, J.J., Simmonds, S.P., 1976. Development of biological methods for the control of *Tetranychus urticae* on tomatoes using *Phytoseiulus persimilis*. Ann. Appl. Biol. 83, 177–189.

Fry, J.D., 2004. Evolutionary adaptation to host plants in a laboratory population of the phytophagous mite *Tetranychus urticae* Koch. Oecologia 8, 559–565.

Gallardo, F., Boethel, D.J., Fuxa, J.R., Richter, A., 1990. Susceptibility of *Heliothis zea* (Boddie) larvae to *Nomureae rileyi* (Farlow) Samson. The effect of alpha-tomatine at the third tritrophic level. J. Chem. Ecol. 16, 1751–1759.

Garcia-Mari, E., Gonzalez-Zamora, J.E., 1999. Biological control of *Tetranychus urticae* (Acari: Tetranychidae) with naturally occurring predators in strawberry plantings in Valencia, Spain. Exp. Appl. Acarol. 23, 487–495.

Gardner, W.A., Oetting, R.D., Storey, G.K., 1982. Susceptibility of the twospotted spider mite, *Tetranychus urticae* Koch, to the fungal pathogen *Hirsutella thompsonii* Fisher. Fla. Entomol. 65, 458–465.

Gatarayiha, M.C., Laing, M.D., Miller, R.M., 2010a. Combining applications of potassium silicate and *Beauveria bassiana* to four crops to control two spotted spider mite, *Tetranychus urticae* Koch. Int. J. Pest Manag. 56 (4), 291–297.

Gatarayiha, M.C., Laing, M.D., Miller, R.M., 2010b. Effects of crop type on persistence and control efficacy of *Beauveria bassiana* against the two spotted spider mite. BioControl 55, 767–776.

Gatarayiha, M.C., Laing, M.D., Miller, R.M., 2010c. Effects of adjuvant and conidial concentration on the efficacy of *Beauveria bassiana* for the control of the two spotted spider mite, *Tetranychus urticae*. Exp. Appl. Acarol. 50, 217–229.

Gatarayiha, M.C., Laing, M.D., Miller, R.M., 2011. Field evaluation of *Beauveria bassiana* efficacy for the control of *Tetranychus urticae* Koch (Acari: Tetranychidae). J. Appl. Entomol. 135 (8), 582–592.

Geroh, M., 2007. Ecology and management of *Tetranychus urticae* Koch on okra, *Abelmoschus esculentus* L., Ph.D. Thesis, CCS HAU, Hisar, 49 pp.

Geroh, M., 2011. Molecular characterization of *Beauveria bassiana* (Balsamo) Vuillemin and its Bioefficacy against *Tetranychus urticae* Koch (Acari: Tetranychidae), Ph.D. Thesis, CCS HAU, Hisar, pp. 51–78.

Geroh, M., Gulati, R., Sharma, S.S., Kaushik, H.D., Aneja, D.R., 2010. Effect of abiotic stresses on population build up of *Tetranychus urticae* Koch and its predator, *Stethorus punctillum* Wiese on okra. Ann. Plant Prot. Sci. 18 (1), 108–113.

Gerson, U., Cohen, E., 1989. Resurgences of spider mites (Acari: Tetranychidae) induced by synthetic pyrethroids. Exp. Appl. Acarol. 6, 29–46.

Gerson, U., Smiley, R.L., 1990. Acarine Biocontrol Agents: An Illustrated Key and Manual. Chapman and Hall, London, 174 pp.

Gillespie, D.R., Raworth., D.A., 2004. Biological control of two-spotted spider mites on greenhouse vegetable crops. In: Heinz, K.M., Van Driesche, R.G., Parrella, M.P. (Eds.), Biocontrol in Protected Culture. Ball Publishers, Batavia, IL, pp. 185–199.

Goettel, M.S., Jaronski, S.T., 1997. Safety and registration of microbial agents for control of grasshoppers and locusts. Mem. Entomol. Soc. Can. 171, 83–99.

Goettel, M.S., St Leger, R.J., Rizzo, N.W., Staples, R.C., Roberts, D.W., 1989. Ultrastructure localization of a cuticle-degrading protease produced by entomopathogenic fungus *Metarhizium anisopliae* during penetration of host cuticle. J. Gen. Microbiol. 135, 2233–2239.

Goettel, M.S., Inglis, G.D., Duke, G.M., Lord, J.C., 1996. Effect of *Beauveria Bassiana* (Mycotech Strain GHA) on Invertebrates in Rangeland and Alfalfa Ecosystems. Report EPA, Washington, DC.

Gope, D., Das, S.C., 1992. Biology and effect of pruning and skiffing on the distribution of the scarlet mite, *Brevipalpus phoenicis* (Geijskes) on tea in North East India. Two and a Bud 39, 23.

Gordon, R.D., 1985. The coccinellidae (Coleoptera) of America North of Mexico. J. New York Entomol. Soc. 93, 1–912.

Gough, N., 1991. Long-term stability in the interaction between *Tetranychus urticae* and *Phytoseiulus persimilis* producing successful integrated control on roses in southeast Queensland. Exp. Appl. Acarol. 12, 83–101.

Gould, H.J., Light, W.I.St.G., 1971. Biological control of *Tetranychus urticae* on stock plants of ornamental ivy. Plant Pathol. 20, 18–20.

Grafton, C.E., Yuling, Q., Quagang, Y.L., 1995. Manipulation of the predaceous mite, *Euseius tularensis* (Acari: Phytoseiidae) with pruning for citrus thrips control. In: Parker, B.L. (Ed.), International Conf. Thysanoptera. Thrips Biology and Management, pp. 251–254.

Gravena, S., Coletti, A., Yamamoto, P.T., 1993. Influence of green cover with *Ageratum conyzoides* and *Eupatorium pauciflorum* on predatory and phytophagous mites in citrus. Bull. IOBC-SROP 16 (7), 104–114.

Greco, N.M., Pereyra, P.C., Guillade, A., 2006. Host-plant acceptance and performance of *Tetranychus urticae* (Acari: Tetranychidae). J. Appl. Entomol. 130, 32–36.

Gulati, R., 2012. Potential of Acarines as bio-control agents in pest management. In: Koul, O., Dhaliwal, G.S., Khokhar, S., Singh, R. (Eds.), Biopesticides in Environment and Food Security Issues and Strategies. Scientific Publishers, Jodhpur, India, pp. 286–313.

Gulati, R., Kalra, V.K., 2007. Predatory potential of *Stethorus punctillum* Weise (Coleoptera: Coccinellidae) against phytophagous mite, *Tetranychus urticae* Koch (Acari: Tetranychidae) on okra. J. Insect Sci. 19, 50–53.

Gulati, R., Sharma, S.K., Dhillon, S., 2008. Field efficacy of botanical extracts against *Tetranychus urticae* Koch (Acarina: Tetranychidae) in okra. J. Med. Aromat. Plant Sci. 30, 105–108.

Guo, Y.-L., Zuo, G.-S., Zhao, J.-H., Wang, N.-Y., Jiang, J.-W., 1993. A laboratory test on the toxicity of thuringiensin to *Tetranychus urticae* (Acari: Tetranychidae) and *Phytoseiulus persimilis* (Acari: Phytoseiidae). Chinese J. Biol. Control 9, 151–155.

Hajek, A.E., McManus, M.L., Delalibera Jr., I., 2005. Catalogue of Introductions of Pathogens and Nematodes for Classical Biological Control of Insects and Mites. USDA, Forest Service FHTET.

Hajek, A.E., McManus, M.L., Delalibera Jr, I., 2007. A review of introductions of pathogens and nematodes for classical biological control of insects and mites. Biol. Control 41, 1–13.

Hall, I.M., Hunter, D.K., Arakawa, K.Y., 1971. The effect of the b-exotoxin fraction of *Bacillus thuringiensis* on the citrus red mite. J. Invertebr. Pathol. 18, 359–362.

Hamlen, R.A., 1978. Biological control of spider mites on greenhouse ornamentals using predaceous mites. Proc. Fla. State Hortic. Soc. 91, 247–249.

Hamlen, R.A., Lindquist, R.K., 1981. Comparison of two *Phytoseiulus* species as predators of two-spotted spider mites on greenhouse ornamentals. Environ. Entomol. 10, 524–527.

Herron, G.A., Beattie, G.A.C., Kallianpur, A., Barchia, I., 1998. A potter spray tower bioassay of two petroleum spray oils against adult female *Panonychus ulmi* (Koch) and *Tetranychus urticae* Koch. (Acari: Tetranychidae). Exp. Appl. Acarol. 22, 553–558.

Holt, K.M., Opit, G., Nechols, J.R., Margolies, D.C., 2006. Testing for non-target effects of spinosad on twospotted spider mites and their predator *Phytoseiulus persimilis* under greenhouse conditions. Exp. Appl. Acarol. 38, 141–149.

Hong, T.D., Edginton, S., Ellis, H.R., de Muro, M.A., Moore, D., 2005. Saturated salt solutions for humidity control and the survival of dry powder and oil formulations of *Beauveria bassiana* conidia. J. Invert. Pathol. 89, 136–143.

Hou, R.F., Chang, J., 1985. Cellular defense response to *Beauveria bassiana* in the silkworm, *Bombyx mori*. Appl. Entomol. Zool. 20, 118–125.

Hoy, M.A., Ouyang, Y.-L., 1987. Toxicity of b-exotoxin of *Bacillus thuringiensis* to *Tetranychus pacificus* and *Metaseiulus occidentalis* (Acari: Tetranychidae and Phytoseiidae). J. Econ. Entomol. 80, 507–511.

Huffaker, C.B., Van de Vries, M., McMurtry, J.A., 1969. The ecology of tetranychid mites and their natural enemies. Annu. Rev. Entomol. 14, 125–144.

Hull, L., 1995. Know Your Friends: *Stethorus punctum*. Midwest Biological Control News Online 2, 12.

Hussey, N.W., Huffaker, C.B., 1976. Spider mites. In: Deluchi, V.L. (Ed.), Studies in Biological Control. Cambridge University Press, Cambridge, pp. 179–236.

Hussey, N.W., Parr, W.J., Gould, H.J., 1965. Observations on the control of *Tetranychus urticae* Koch on cucumbers by the predatory mite *Phytoseiulus riegeli* Dosse. Entomol. Exp. Appl. 8, 271–281.

Ignoffo, C.M., 1992. Environmental factors affecting persistence of entomopathogens. Fla. Entomol. 75, 516–523.

Inglis, G.D., Goettel, M.S., Johnson, D.L., 1993. Persistence of the entomopathogenic fungus, *Beauveria bassiana* on phylloplanes of crested wheatgrass and alfalfa. Biol. Control 3, 258–270.

Inglis, G.D., Goettel, M.S., Butt, T.M., Strasser, H., 2001. Use of hyphomycetous fungi for managing insect pests. In: Butt, T.M., Jachson, C., Magan, N. (Eds.), Fungi as Biocontrol Agents: Progress, Problems and Potential. CABI Publishing, Wallingford, pp. 23–69.

Irigaray, F.J.S.C., Mancebon, M.V., Moreno, P.I., 2003. The entomopathogenic fungus *Beauveria bassiana* and its compatibility with triflumuron: Effects on the two spotted spider mite, *Tetranychus urticae*. Biol. Control 26, 168–173.

Isman, M.B., 2006. Botanical insecticides, deterrents and repellents in modern agriculture and an increasingly regulated world. Annu. Rev. Entomol. 51, 45–66.

Isman, M.B., Machial, C.M., Miresmailli, S., Bainard, L.D., 2007. Essential oilbased pesticides: new insights from old chemistry. In: Ohka-wa, H., Miyagawa, H., Lee, P. (Eds.), Pesticide Chemistry. Wiley, Weinheim, pp. 201–209.

Jackson, G.J., Ford, J.B., 1973. The feeding behaviour of *Phytoseiulus persimilis* Athias-Henriot (Acarina: Phytoseiidae) particularly as affected by certain pesticides. Ann. Appl. Biol. 75, 165–171.

Jacobson, R.J., Croft, P., Fenlon, J., 1999. Response to fenbutatin oxide in populations of *Tetranychus urticae* Koch (Acari: Tetranychidae) in UK protected crops. Crop Prot. 18, 47–52.

Jeppson, L.R., Keifer, H.H., Baker, E.W., 1975. Mites Injurious to Plants. Univ. of Calif. Press, Berkeley, 614 pp.

Jeyarani, S., Banu, J.G., Ramaraju, K., 2011. First record of natural occurrence of *Cladosporium cladosporioides* (fresenius) de vries and *Beauveria bassiana* (Bals.-Criv.) Vuill on two spotted spider mite, *Tetranychus urticae* Koch from India. J. Entomol. 8, 274–279.

Jeyarani, S., Jagannath Singh, R., Ramaraju, K., 2012. Influence of predator density on the efficiency of spider mite predators against two spotted spider mite, *Tetranychus urticae* Koch (Acari: Tetranychidae). Asian J. Biol. Sci. 5, 432–437.

Kalmath, B., Mallik, B., Srinivasa, N., 2007. Occurrence of fungal pathogen *Beauveria bassiana* on Tetranychid mite, *Tetranychus urticae* in Karnataka. Insect Environ. 13, 139–140.

Kanika, 2013. Damage potential of *Tetranychus urticae* Koch and its management in *Cucumis sativus* Linnaeus. Ph.D Thesis, CCS Haryana Agricultural University, Hisar, 116 pp.

Kapur, A.P., 1948. On the old world species of the genus *Stethorus* Weise (Coleoptera: Coccinellidae). Bull. Entomol. Res. 39, 297–320.

Karban, R., English-Loeb, G., Walker, M.A., Thaler, J., 1995. Abundance of phytoseiid mites on *Vitis* species: effects of leaf hairs, domatia, prey abundance and plant phylogeny. Exp. Appl. Acarol. 19, 189–197.

Kaur, G., Padmaja, V., 2008. Evaluation of *Beauveria bassiana* isolates for virulence against *Spodoptera litura* (Fab.) (Lepidoptera: Noctuidae) and their characterization by RAPD-PCR. Afr. J. Microbiol. Res. 2, 299–307.

Kenneth, W.C., Edwin, E.L., Peter, B.S., 2002. Compatibility of acaricide residues with *Phytoseiulus persimilis* and their effects on *Tetranychus urticae*. Hort Science 37, 906–909.

Khanna, K.L., Ramanathan, K.R., 1947. Studies on the association of plant characters and pest incidence. I. Structure of leaf surface and mite attack in sugarcane. Proc. Natl. Inst. Sci. India 13, 327–329.

Koschier, E., 1997. Influence of different pruning measures on the population dynamics of European red mite (*Panonychus ulmi* Koch) and apple rust mite, *Aculus schlechtendali* (Nalepa) on apple trees. Mitteilungen Klosternenburg-Rebe-Und-Wein Obstban-Und-Fruchteverwertugig 47, 44–45.

Kouassi, M., Coderre, D., Todorova, S.I., 2003. Effect of plant type on the persistence of *Beauveria bassiana*. Biocontrol Sci. Technol. 13, 415–427.

Krips, O.E., Kleijn, P.W., Willems, P.E.L., Gols, G.J.Z., Dicke, M., 1999. Leaf hairs influence searching efficiency and predation rate of predatory mite *Phytoseiulus persimilies* (Phytoseiidae: Acarina). Exp. Appl. Acarol. 23 (2), 119–131.

Kristova, E., 1972. The use of *Phytoseiulus persimilis* for the control of spider mites on cucumber in the greenhouses. Rastitelna Zashita 20, 310–334.

Kropczynska, A., Pilko, A., Witul, A., Asshleb, A.M., 1999. Control of two spotted spider mite with *Amblyseius californicus* on croton. IOBC/WPRS (Int. Org. Biol. Integrated Control Noxious Animals Plants, West Palearctic Reg. Section) Bull. 22, 133–136.

Kumar, S.V., Chinniah, C., Muthiah, C., Sadasakthi, A., 2010. Management of two spotted spider mite *Tetranychus urticae* Koch a serious pest of brinjal, by integrating biorational methods of pest control. J. Biopestic. 3 (1), 361–368.

Leathers, T.D., Gupta, S.C., Alexander, N.J., 1993. Mycopesticides: status, challenges and potential. J. Ind. Microbiol. 12, 69–75.

Lester, P.J., Thistlewood, H.M.A., Harmsen, R., 2000. Some effects of pre-release host-plant on the biological control of *Panonychus ulmi* by the predatory mite *Amblyseius fallacies*. Exp. Appl. Acarol. 24 (1), 19–33.

Leuprecht, B., 2000. Occurrence of tomato rust mite in tomatoes under glass. Gemuse Munchen 36, 26–27.

Liu, T.S., Hung, Y.C., Twu, J.S., Hsieh, J.H., 1998. Steam sterilization of soil for the control of bulb mites. Plant Prot. Bull. 40 (241), 249.

Ludwig, S.W., Oetting, R.D., 2001. Susceptibility of natural enemies to infection by *Beauveria bassiana* and impact of insecticides on *Ipheseius degenerans* (Acari: Phytoseiidae). J. Agric. Urban Entomol. 18 (3), 169–178.

Luz, C., Tigano, M.S., Silva, I.G., Cordeiro, C.M., Alhanabi, S.M., 1998. Selection of *Beauveria bassiana* and *Metarhizium anisopliae* isolates to control *Triatoma infestans*. Mem. Inst. Oswaldo Cruz 93 (6), 839–846.

Mahgoub, M.H.A., Abdallah, A.A., El-Saiedy, E.M.A., 2011. Biological control agents against the two-spotted spider mite on four pepper cultivars in greenhouses. Acarines 5 (1), 29–32.

Mailloux, J., Le Bellec, F., Kreiter, S., Tixier, M.S., Dubois, P., 2010. Influence of ground cover management on diversity and density of phytoseiid mites (Acari: Phytoseiidae) in Guadeloupean citrus orchards. Exp. Appl. Acarol. 52, 275–290.

Mathur, L.M.L., 1981. Observations on the predatory efficiency of *Stethorus gilvifrons* on red spider mites. In: Channabasavanna, G.P. (Ed.), Contributions to Acarology in India. Acarol. Soc. India, Bangalore, pp. 188–192.

Mayer, R.T., Inbar, M., McKenzie, C.L., Shatters, R., Borowicz, V., Albrecht, U., et al., 2002. Multitrophic interactions of the silverleaf whitefly, host plants, competing herbivores, and phytopathogens. Arch. Insect Biochem. Physiol. 51, 151–169.

McCoy, C.W., Bullock, R.C., Dybas, R.A., 1982. Avermectin b: a novel miticide active against citrus mites in Florida. Proc. Fla. State Hortic. Soc. 95, 51–56.

McCoy, C.W., Samson, R.A., Boucias, D.G., 1988. Entomogenous fungi. In: Ignoffo, C.M. (Ed.), Handbook of Natural Pesticides: Microbial Insecticides. CRC Press, Florida, pp. 151–236.

McMurtry, J.A., 1982. The use of phytoseiids for biological control: progress and future prospects. In: Hoy, M.A. (Ed.), Recent Advances in Knowledge of the Phytoseiidae. University of California, Division of Agricultural Sciences, Berkeley, CA, pp. 23–28.

McMurtry, J.A., Croft, B.A., 1997. Life-styles of Phytoseiid mites and their roles in biological control. Annu. Rev. Entomol. 42, 291–321.

McMurtry, J.A., Rodriguez, J.G., 1987. Nutritional ecology of phytoseiid mites. In: Slansky, F., Rodriguez, J.G. (Eds.), Nutritional Ecology of Insects, Mites and Spiders. Wiley, Chichester, pp. 609–644.

McMurtry, J.A., Huffaker, C.B., Van de Vire, M., 1970. Ecology of tetranychid mites and their natural enemies: A review. 1. Tetranychid enemies: The biological characters and their impact of spray practices. Hilgardia 40, 331–339.

Mehrotra, V., Mehrotra, S., Kirar, V., Shyam, R., Misra, K., Srivastava, A.K., et al., 2011. Antioxidant and antimicrobial activities of aqueous extract of *Withania somnifera* against methicillin-resistant *Staphylococcus aureus*. J. Microbiol. Biotech. Res. 1 (1), 40–45.

Mierzejewski, K., Reardon, R.C., Thistle, H., Dubois, N.R., 2007. Conventional application equipment: aerial application. In: Lacey, L.A., Kaya, H.K. (Eds.), Field Manual of Techniques in Invertebrate Pathology, second ed. Springer, The Netherlands, pp. 99–126.

Mietkiewski, R., Balazy, S., Van Der Geest, L.P.S., 1993. Observations on a mycosis of spider mites (Acari: Tetranychidae) caused by *Neozygites floridana* in Poland. J. Invertebr. Pathol. 61, 317–319.

Miles, M., Dutton, R., 2003. Testing the effects of spinosad to predatory mites in laboratory, extended laboratory, semi-field and field studies. In: Vogt, H. Heimbach, U. Vinuela, E. (Eds.), Pesticides and Beneficial Organisms, vol. 26. IOBC/WPRS Bull, pp. 9–20.

Misra, K.K., Sarkar, P.K., Das, T.K., Som Choudhary, A.K., 1990. Incidence of *Tetranychus cinnabarinus* (Boisd.) (Acari: Tetranychidae) on some selected accessions of brinjal with special reference to the physical basis of resistance. Indian Agric. 34 (3), 177–185.

Mohamed, O.M.O., Omar, N.A.A., 2011. Life table parameters of the predatory mite, *Phytoseiulus persimilis* Athias-Henriot on four tetranychid prey species (Phytoseiidae–Tetranychidae). Acarines 5, 19–22.

Moore, D., Alexander, L., 1990. Resistance of coconuts in St. Lucia to attack by coconut mites, *Eriophyes guerreronis* Keifer. Trop. Agric. 67, 33–36.

Morimoto, K., Furuichi, H., Yano, S., Osakaba, M.H., 2006. Web-mediated interspecific competition among spider mites. J. Econ. Entomol. 99, 678–684.

Motazedian, N., Ravan, S., Bandani, A.R., 2012. Toxicity and repellency effects of three essential oils against *Tetranychus urticae* Koch (Acari: Tetranychidae). J. Agric. Sci. Tech. 14, 275–284.

Muhammad, H., Afzal, M., Nadeem, S., Nadeem, M.K., 2009. Morphological characters of different cotton cultivars in relation to resistance against Tetranychid mites. Pak. J. Zool. 41 (3), 241–244.

Muma, M.H., 1955. Factors contributing to the natural control of citrus insects and mites in Florida. J. Econ. Entomol. 48, 432–438.

Muro, M.A.D., Mehta, S., David, M., 2003. The use of amplified fragment length polymorphism for molecular analysis of *Beauveria bassiana* isolates from Kenya and other countries, and their correlation with host and geographical origin. FEMS Microbiol. Lett. 229, 249–257.

Naher, N., Islam, T., Haque, M.M., Parween, S., 2006. Effects of native plants and IGRs on the development of *Tetranychus urticae* Koch (Acari: Tetranychidae). Univ. J. Zool. Rajshahi Univ. 25, 19–22.

Omar, H.I.H., Hegab, M.F., Elsorady, A.E.M., 1993. The impact of intercropping cotton and cowpea on pest infestation. Egypt. J. Agric. Res. 71, 709–716.

Osborne, L.S., Petitt, F.L., 1985. Insecticidal soap and the predatory mite, *Phytoseiulus persimilis* (Acari: Phytoseiidae), used in management of the two spotted spider mite (Acari: Tetranychidae) on greenhouse grown foliage plants. J. Econ. Entomol. 78, 687–691.

Patil, R.S., 2005. Investigation on Mite Pests of Solanaceous Vegetable with Special Reference to Brinjal. Ph.D. Thesis, Univ. Agric. Sci., Dharwad, India.

Pekrul, S., Grula, E.A., 1979. Mode of infection of the corn earworm (*Heliothis zea*) by *Beauveria bassiana* as revealed by scanning electromicroscopy. J. Invert. Pathol. 34, 238–247.

Pemberton, R.W., Turner, C.E., 1989. Occurence of predatory and fungivorous mites in leaf domatia. Ann. J. Bot. 76, 105–112.

Peveling, R., Weyrich, J., 1992. Effects of neem oil, *Beauveria bassiana* and dieldrin on non-target tenebrionid beetles in the desert zone of the Republic of Niger. In: Lomer, C.J., Prior, C. (Eds.), Biological Control of Locusts and Grasshoppers. CABI Press, Wallingford, UK, pp. 321–346.

Pingel, R.L., Lewis, L.C., 1996. The fungus *Beauveria bassiana* (Balsamo) Vuillemin in a corn ecosystem: its effect on insect predator *Coleomegilla maculata* De Geer. Biol. Control 6, 137–141.

Poprawski, T.J., Greenberg, S.M., Ciomperlik, M.A., 2000. Effect of host plant on *Beauveria bassiana* and *Paecilomyces fumosoroseus* induced mortality of *Trialeurodes vaporariorum* (Homoptera: Aleyrodidae). Environ. Entomol. 29, 1048–1053.

Prasad, V.G., Singh, R.K., 1981. Prevalence and control of litchi mite, *Aceria litchi* Keifer in Bihar. Indian J. Entomol. 43, 67–75.

Prasanna, K.P., 2007. Seasonal Incidence and Management of Tetranychid Mites in Brinjal. M.Sc. Thesis, University of Agricultural Sciences, Dharwad, India, 35 pp.

Pratt, P.D., Croft, B.A., 2000. Toxicity of pesticides registered for use in landscape nurseries to the acarine biological control agent, *Neoseiulus fallacis*. J. Environ. Hortic. 18, 197–201.

Pushpa V. 2006. Management of Coconut Perianth Mite, *Aceria guerreronis* Keifer. MSc. Thesis, University of Agricultural Sciences, Dharwad, India.

Putman, W.L., Herne, D.H.C., 1966. The role of predators and other biotic agents in regulating the population density of phytophagous mites in Ontario peach orchards. Can. Entomol. 98, 808–820.

Puttaswamy, S.K., Channabasavanna, G.P., 1977. Biology of *Stethorus pauperculus* Weise (Coleoptera: Coccinellidae), a predator of mite. Mysore J. Agric. Sci. 11 (1), 81–89.

Rabindra, R.J., Ballali, C.R., Ramanujan, B., 2004. Biological options for insect pests and nematode management in pulses. In: Singh, M.A., Shiv Kumar, B.B., Dhar, V. (Eds.), Pulses in New Perspective. Indian Society of Pulses Research and Development, Kanpur, India, pp. 400–425.

Rajaram, V., Ramamurthy, R., 2001. Influence of morphological characters on the incidence of chili thrips and mite. Ann. Plant Prot. Sci. 9, 54–57.

Ramos, M., Rodriguez, N., 1997. Acaros fitoseidos Phytoseiidae en el cultivo de la papa: descripcion de una nueva especie. Rev. Prot. Veg. 12 (2), 109–112.

Raros, E.S., Haramoto, F.H., 1974. Biology of *Stethorus siphonulus* Kapur (Coleoptera: Coccinellidae), a predator of spider mites in Hawaii. Proc. Hawaiian Entomol. Soc. 31, 457–465.

Rasmy, A.H., Ellaithy, A.Y.M., 1988. Introduction of *Phytoseiulus persimilis* for the two-spotted spider mite, *T. urticae* control in glasshouses in Egypt (Acarina: Phytoseiidae, Tetranychidae). Entomophaga 33, 435–438.

Reed, D.K., Hall, I.M., 1972. Electron microscopy of a rod-shaped non-inclusion virus infecting the citrus red mite. Panonychus citri. J. Invertebr. Pathol. 20, 272–278.

Reinecke, P., Andersch, W., Stenzel, K., Hartwig, J., 1990. BIO 1020, a new microbial insecticide for use in horticultural crops. In: Proceedings of Brighton Crop Protection Conference 1990: Pests and Diseases, BCPC Publications, Farnham, pp. 49–54.

Roberts, D.W., Humber, R.A., 1981. Entomogenous fungi. In: Cole, G.T., Kenrick, B. (Eds.), Biology of Conidial Fungi, vol. 2. Academic Press, New York, NY, USA, pp. 201–236.

Rodriguez, J.G., Rodriguez, L.D., 1987. Nutritional ecology of phytophagous mites. In: Slansky, F., Rodriguez Jr., J.G. (Eds.), Nutritional Ecology of Insects, Mites, Spiders and Related Invertebrates. Wiley, USA, pp. 177–208.

Ros, V.I., Fleming, V.M., Feil, E.J., Breeuwer, J.A., 2012. Diversity and recombination in Wolbachia and Cardinium from *Bryobia* spider mites. BMC Microbiol. 12 (Suppl. 1), S13.

Rosas-Acevedo, J.L., Boucias, D.G., Lezama, R., Sims, K., Pescador, A., 2003. Exudate from sporulating cultures of *Hirsutella thompsonii* inhibit oviposition by the two-spotted spider mite *Tetranychus urticae*. Exp. Appl. Acarol. 29, 213–225.

Roy, H.E., Pell, J.K., 2000. Interactions between entomopathogenic fungi and other natural enemies: Implications for biological control. Biocontrol Sci. Technol. 10, 737–752.

Roy, M., Brodeur, J., Cloutier, C., 1999. Seasonal abundance of spider mites and their predators in red raspberry in Quebec, Canada. Environ. Entomol. 28 (4), 735–747.

Royalty, R.N., Hall, F.R., Taylor, R.A.J., 1990. Effects of thuringiensin on *Tetranychus urticae* (Acari: Tetranychidae) mortality, fecundity, and feeding. J. Econ. Entomol. 83, 792–798.

Sabelis, M.W., Vander Bean, H.E., 1983. Location of distant spider mite colonies by phytoseiid predators: demonstration of specific Kairomones emitted by *Tetranychus urticae* and *Panonychus ulmi*. Entomol. Exp. Appl. 33, 303–314.

Sabelis, M.W., Vermaat, J.E., Groeneveld, A., 1984. Arrestment responses of the predatory mite *Phytoseiulus persimilis* to steep odour gradients of a kairomone. Physiol. Entomol. 9, 437–446.

Sahayaraj, K., Prabhu, J.N., Martin, P., 2003. Screening of mite resistant okra varieties based on morphology and phytochemistry. Shashpa 10 (2), 147–150.

Santiago-Álvarez, C., Maranhao, E.A., Maranhao, E., Quesada-Moraga, E., 2006. Host plant influences pathogenicity of *Beauveria bassiana* to *Bemisia tabaci* and its sporulation on cadavers. Biocontrol 51, 519–532.

Santos, M.A., 1976. Evaluation of *Zetzellia mali* as a predator of *Panonychus ulmi* and *Aculus schlechtendali*. Environ. Entomol. 5, 187–191.

Schmidt, V.G., 1976. Der einfluss der van den beutetieren hinterlassenen spuren auf such verhalten and sucherfoly von *Phytoseiulus persimilis* (Acarina: Phytoseiidae). Z. Angew. Entomol. 82, 16–18.

Siebeneicher, S.R., Vinson, S.B., Kenerley, C.M., 1992. Infection of the imported fire ant by *Beauveria bassiana* through various routes of exposure. J. Invert. Pathol. 59, 280–285.

Simova, S., Draganova, S., 2003. Virulence of entomopathogenic fungi to *Tetranychus urticae* Koch (Tetranychidae, Acarina). Rasteniev'dni Nauki. 40, 87–90.

Sivritepe, N., Kumral, N.A., Erturk, U., Yerlikaya, C., Kumral, A., 2009. Responses of grapevines to two-spotted spider mite mediated biotic stress. J. Biol. Sci. 9 (4), 311–318.

St Leger, R., Charnely, A.K., Cooper, R.M., 1987. Cuticle-degrading enzymes of entomopathogenic fungi. J. Invert. Pathol. 48, 85–95.

St Leger, R., Durrands, P., Charnely, A.K., Cooper, R.M., 1988. Role of extracellular chymoelastase in the virulence of *Metarhizium anisopliae* for *Manduca sexta*. J. Invert. Pathol. 52, 285–293.

Stoner, A.K., Frank, J.A., Gentile, G.A., 1968. The relationship of glandular hairs on tomatoes to spider mite resistance. Proc. Am. Soc. Hortic. Sci. 93, 532–538.

Strasser, H., Vey, A., Butt, T.M., 2000. Are there any risks in using entomopathogenic fungi for pest control, with particular reference to the bioactive metabolites of *Metarhizium, Tolypocladium* and *Beauveria* species. Biocontrol Sci. Technol. 10, 717–735.

Sundaram, K.M.S., Sloane, L., 1995. Effects of pure and formulated azadirachtin, a neem based biopesticide, on the phytophagous spider mite, *Tetranychus urticae* Koch. J. Environ. Sci. Health 30 (6), 801–814.

Takafuji, A., Chant, D.A., 1976. Comparative studies of two species of predaceous phytoseiid mites (Acarina: Phytoseiidae) with special reference to their responsive studies of two species of predaceous phytoseiid mites (Acarina: Phytoseiidae) with special reference to their responses to the density of their prey. Ibid 17, 255–310.

Tamai, M.A., Alves, S.B., Lopes, R.B., Neves, P.S., 1998. Avaliação de fungos entomopatogênicos para o controle de *Tetranychus urticae* Koch. Abst. 17th Brazil. Congr. Entomol., Rio de Janeiro, 1066 pp.

Tamai, M.A., Alves, S.B., Neves, P.J., 1999. Pathogenicity of *Beauveria bassiana* (Bals.) Vuill. against *Tetranychus urticae* Koch. Sci. Agric. 56, 285–288.

Tamai, M.A., Alves, S.B., Almeida, J.E.M., Faion, M., 2002. Evaluation of entomopathogenic fungi for control of *Tetranychus urticae* Koch (Acari: Tetranychidae). Arq. Inst. Biol. 69, 77–84.

Tanada, Y., Kaya, H.K., 1993. Insect Pathology. Academic Press, London, 276 pp.

Tanaka, M., 1966. Fundamental studies on the utilization of natural enemies in the citrus grove in Japan. I. The bionomics of natural enemies of most serious pests. II. *Stethorus japonicus* H. Kamiya (Coccinellidae), a predator of the citrus red mite, *Panonychus citri* (McGregor). Bull. Hortic. Res. *Stn, Japan, Ser. D.*, 4, 22–24, (in Japanese; English summary).

Thompson, S.R., Brandenburg, R.L., Arends, J.J., 2006. Impact of moisture and UV degradation on *Beauveria bassiana* (Balsamo) Vuillemin conidial viability in turf grass. *Biol. Control* 39, 401–407.

Tigano-Milani, M.S., Gomes, A.C.M.M., Sobral, B.W.S., 1995. Genetic variability among Brazilian isolates of the entomopathogenic fungus, *Metarhizium anisopliae*. J. Invert. Pathol. 65, 206–210.

Tixier, M.S., Kreiter, S., Auger, P., Weber, M., 1998. Colonization of Languedoc vineyards by phytoseiid mites influence of wind and crop environment. Exp. Appl. Acarol. 22, 523–542.

Tixier, M.S., Kreiter, S., Cheval, B., Guichou, S., Auger, P., Bonafos, R., 2006. Immigration of phytoseiid mites from surrounding uncultivated areas into a newly planted vineyard. Exp. Appl. Acarol. 39, 227–242.

Tomlin, C.D.S., 1997. *Beauveria* spp. biocontrol agent. In: Tomlin, C.D.S. (Ed.), A World Compendium – The Pesticide Manual. British Crop Protection Council, Bracknell, pp. 83–86.

Toyoshima, S., Ihara, F., Amano, H., 2008. Diversity and abundance of phytoseiid mites in natural vegetation in the vicinity of apple orchards in Japan. Appl. Entomol. Zool. 43 (3), 443–450.

Tuovinen, T., Rokx, J.A.H., 1991. Phytoseiid mites (Acari: Phytoseiidae) on apple trees and in surrounding vegetation in southern Finland. Densities and species composition. Exp. Appl. Acarol. 12, 35–46.

Urs, D.K.C., 1990. Efficacy of certain plant products in the control of brinjal red spider mite, *Tetranychus urticae*. Fourth National Symposium on Acarology, Calicut, Kerala, 68 pp.

Urtz, B.E., Rice, W.C., 1997. RAPD–PCR characterization of *Beauveria bassiana* isolates from the rice water weevil

Lissorhoptrus oryzophilus. Lett. Appl. Microbiol. 25, 405–409.

Vacante, V., Nucifora, A., 1987. Possibilities and perspectives of biological and integrated control of the two-spotted spider mite in the Mediterranean greenhouse crops. Bull. SROP 10, 170–173.

Van Leeuwen, T., Vanholme, B., Van Pottelberge, S., Van Nieuwenhuyse, P., Nauen, R., Tirry, L., et al., 2008. Mitochondrial heteroplasmy and the evolution of insecticide resistance: non-Mendelian inheritance in action. Proc. Natl. Acad. Sci. USA 105, 5980–5985.

Van Lenteren, J.C., Woets, J., 1988. Biological and integrated pest control in greenhouses. Annu. Rev. Entomol. 33, 239–269.

Van der Geest, L.P.S., 1985. Pathogens of spider mites. In: Helle, W. Sabelis, M.W. (Eds.), Spider Mites, Their Biology, Natural Enemies and Control, vol. 1B. World Crop Pests. Elsevier, Amsterdam, pp. 247–258.

Varela, A., Morales, E., 1996. Characterization of some *Beauveria bassiana* isolates and their virulence towards the coffee berry borer *Hypothenemus hampei*. J. Invert. Pathol. 67, 147–152.

Vestergaard, S., Cherry, A., Keller, S., Gottel, M., 2003. Safety of hyphomycete fungi as microbial control agents. In: Hokkanen, H.M.T., Hajek, A.E. (Eds.), Environmental Impacts of Microbial Insecticides. Kluwer Academic Publishers, The Netherlands, pp. 35–62.

Vey, A., Fargues, J., 1977. Histological and ultrastrutural studies of *Beauveria bassiana* infection in *Leptinotarsa decemlineata* larvae during ecdysis. J. Invert. Pathol. 8, 268–269.

Vey, A., Hoagland, R., Butt, T.M., 2001. Toxic metabolites of fungal biocontrol agents. In: Butt, T.M., Jackson, C., Magan, N. (Eds.), Fungal Biocontrol Agents: Progress, Problems and Potential. CABI, Wallingford, UK, pp. 311–347.

Wang, S., Miao, X., Zhao, W., Huang, B., Fan, M., Li, Z., et al., 2005. Genetic diversity and population structure among strains of the entomopathogenic fungus, *Beauveria bassiana*, as revealed by inter simple sequence repeats (ISSR). Mycol. Res. 109 (12), 1364–1372.

Weintraub, P., Palevsky, E., 2008. Evaluation of the predatory mite, *Neoseiulus californicus*, for spider mite control on greenhouse sweet pepper under hot arid field conditions. Exp. Appl. Acarol. 45, 29–37.

Wekesa, V.W., Maniania, N.K., Knapp, M., Boga, H.I., 2005. Pathogenicity of *Beauveria bassiana* and *Metarhizium anisopliae* to the tobacco spider mite *Tetranychus evansi*. Exp. Appl. Acarol. 36, 41–50.

Wekesa, V.W., Knapp, M., Maniania, N.K., Boga, H.I., 2006. Effects of *Beauveria bassiana* and *Metarhizium anisopliae* on mortality, fecundity and egg fertility of *Tetranychus evansi*. J. Appl. Entomol. 130, 155–159.

Wermelinger, B., Oertli, J.J., Baumgärtner, J., 1991. Environmental factors affecting the life-tables of *Tetranychus urticae* (Acari: Tetranychidae) III. Host-plant nutrition. Exp. Appl. Acarol. 12, 259–274.

Wheeler Jr, A.G., Colburn, R.B., Lehman, R.D., 1973. *Stethorus punctillum* associated with spruce spider mite on ornamentals. Environ. Entomol. 2, 718–720.

Wilson, L.T., 1992. Pest Status and Ecology of Two Spotted Spider Mites on Cotton in Australia and Implications for Management. Ph.D Thesis, Univ. Queensland, Brisbane, Australia, 261 pp.

Yanar, D., Kadio, I., Gokce, A., 2011. Acaricidal effects of different plant parts extracts on two-spotted spider mite (*Tetranychus urticae* Koch). Afr. J. Biotechnol. 10 (55), 11745–11750.

Yoldas, Z., Madadar, N., Gul, A., 1996. Biological control practices against pests in vegetable greenhouses in Izmir (Turkey). Cahiers Options Méditerranéennes 31, 425–434.

Zhou, G.A., Zou, J.J., Huang, S.D., 1994. The effect of *Agaratum conyzoides* interplanting in hilly citrus orchards in Human province on citrus mite and insect populations. Acta Phytophylacica Sin. 21, 57–61.

Zimmermann, G., 2007. Review on safety of the entomopathogenic fungi *Beauveria bassiana* and *Beauveria brongniartii*. Biocontrol Sci. Technol. 17, 553–596.

CHAPTER

22

IPM Extension: A Global Overview

Rajinder Peshin[1], *K.S.U. Jayaratne*[2] *and Rakesh Sharma*[1]

[1]Sher-e-Kashmir University of Agricultural Sciences and Technology of Jammu, Jammu and Kashmir, India, [2]North Carolina State University, NC, USA

22.1 INTRODUCTION

The first modern agricultural extension service came into existence as a result of the outbreak of late blight disease of potato (*Phytophthora infestans*) in Europe in 1845 that persisted until 1851. The effect of potato blight was more severe in Ireland as it was the staple food crop there at that time (Large, 1940). The late blight of potato resulted in deaths, due to starvation, estimated at 1 million and mass migration out of Ireland to the United States and Canada (Nolte, 2002). In 1847, Lord Clarendon, the then British Viceroy, wrote a letter to the president of the Royal Agricultural Improvement Society of Ireland to appoint itinerant lecturers to travel around the most distressed districts to inform and show farmers, in simple terms, how to improve their cultivation and how to grow nutritious root crops other than potatoes. The potato famine also led to the appointment of itinerant agricultural teachers (*Wanderlehrer*) under the auspices of central agricultural societies (Jones and Garforth, 2005). Normally, these itinerant agricultural teachers gave advisory services to the farmers for half of the year travelling around their districts and during the remainder of the year, they taught farmers' sons at agricultural schools (Jones and Garforth, 2005).

Since the outbreak of late blight of potato, pest management (insect pests, diseases, weeds etc.) has transitioned from traditional pest management practices, namely cultural and manual practices (Smith, 1972), to the use of inorganic chemicals such as lead arsenate for insect control in the 1890s to the introduction of synthetic pesticides in the 1940s. The introduction of dichlorodiphenyltrichloroethane (DDT) and benzene hexachloride (BHC) changed the pest management outlook in the scientific community. Pesticides were considered as 'silver bullet' technology for managing pests. Thus, other pest management tactics were relegated to the background. The focus of research during this era shifted from ecological and cultural control to chemicals (Perkins, 1980). In the developed countries, spraying pesticides in a fixed spray schedule was propagated in the 1950s. The extension activities during this period relied on a top-down approach to popularize the use of synthetic pesticides. These activities resulted in popularizing the calendar-based application of pesticides. Farmers reluctant to adopt pesticide intensive

D. P. Abrol (Ed): Integrated Pest Management.
DOI: http://dx.doi.org/10.1016/B978-0-12-398529-3.00026-9

agriculture were termed laggards (Rogers, 1995). Most farmers capitalized on the early boom on pesticides usage and solely relied on them, as the technology was easy to use and propelled by aggressive marketing by the pesticide industry with scientific back-up from agricultural scientists and extensionists. But the desired and functional consequences of a technology go hand in hand with the undesired and dysfunctional consequences (Rogers, 2003). Newsom (1980) called the first 20 years of the synthetic pesticide era as the 'dark ages' of pest control. The dysfunctional consequences of pesticide intensive pest management led to the conceptualization of integrated control and threshold theory for managing pests (Stern et al., 1959). The term 'Integrated Pest Management' (IPM) was used for the first time by Smith and van den Bosch (1967).

22.1.1 Historical Background of IPM Extension in Developed Countries

This section reviews the historical background of IPM extension in the United States (U.S.), Australia, and New Zealand. The US Cooperative Extension provides an example of publicly funded agricultural extension whereas New Zealand acts as an example of privately funded extension. Australia yields an example of a pluralistic extension service. The agricultural extension service in Australia is a combination of private and public extension.

IPM extension in the United States is a success story. The land-grant university system significantly contributed to this success of IPM extension. The Cooperative Extension service organized under the land-grant university system in the US can be considered to be one of the most effective extension systems in the world. It was the result of the Smith-Lever Act of 1914 in that country, which authorized the establishment of Cooperative Extension under the US land-grant universities. The Morrill Act of 1862 led to the establishment of land-grant universities to educate people in agriculture, mechanical arts, and home economics, 'in order to promote the liberal and practical education of the industrial classes in the several pursuits and professions in life' (Morrill Act, 1862, Sec. 4). Initially, one land-grant university was established for each state. The Hatch Act passed in 1887 established collaboration between the United States Department of Agriculture (USDA) and the land-grant universities, and provided federal funds to establish agricultural experimental stations under each land grant-university (Rasmussen, 1989). These agricultural experimental stations were charged with the responsibility to develop agricultural technology useful for farmers. The Second Morrill Act passed in 1890 led to the establishment of 18 land-grant universities for minorities in southern states. The establishment of land-grant universities with experimental stations enabled them to engage in teaching and research. Then, the establishment of Cooperative Extension under the land-grant university system expanded the land-grant university education to the people at the community level. This service mission of land-grant universities immensely contributed to the transfer of agricultural technology from land-grant universities to citizens in every county. Establishment of the land-grant university system with experimental stations and Cooperative Extension can be considered to be the major factor that contributed to agricultural development in the United States.

The US extension system is called the Cooperative Extension Service because of the co-operation among three levels of governments – federal, state, and local – for funding the organization. Since the inception of the organization, this cooperative funding arrangement has contributed to position the Cooperative Extension Service as a strong public organization. Each of these three levels had an influence over the extension system. However, none was dominant over other levels. The Cooperative Extension Service has been flexible and responsive to local needs (Rasmussen, 1989). Generally, the Cooperative Extension Service is organized under the aegis of the college of agriculture of the 1862 state land-grant university. If the state has an 1890 land-grant university, it collaborates with the 1862 land-grant university for delivering the extension service to citizens in the state. Historically, there are two layers of extension personnel in the Cooperative Extension Service. State Extension Specialists comprise one layer of personnel and they are attached to the college of agriculture at the state level. These Extension Specialists are affiliated with their respective academic departments and are responsible for conducting research and transferring knowledge to citizens in the state. Extension Specialists transfer technology to citizens through County Extension Agents. County Extension Agents comprise the second layer of extension personnel of the Cooperative Extension Service. This layer of extension personnel is county-based. Historically, each county has a County Extension Office and a team of Extension Agents to disseminate the knowledge developed at land-grant universities to citizens in the county. Over a period of time, this original staffing arrangement at the county level changed slightly. However, the county-based extension service is still available for citizens in many states in the US. Early extension programmes were mainly focused on improving the quality of life of rural farm families by educating farmers to apply new farming technology, teaching farm women to preserve excess foods, and educating youths about new skills they need to be successful.

The agricultural extension programmes focused on transferring new technology developed at land-grant universities to help farmers address their farming issues and problems. In the early days of Cooperative Extension, agricultural extension programmes were focused on educating farmers on improved crop varieties, improved farming techniques, and improved farming tools. When large tracts of land came under monoculture, insect pests and disease problems became an important issue and the Cooperative Extension Service started to deliver educational programmes related to insect pest and disease control. These early extension programmes helped farmers identify pest and disease problems and recommended chemical pesticides to control pests. During the period 1940–1950, the use of synthetic organic compounds for pest control became common practice in the US. The environmental damage caused by heavy use of synthetic organic compounds such as DDT clearly appeared by the 1950s. Scientists reported development of insect resistance to synthetic organic compounds, destruction of useful organisms, emergence of new pests, and harmful effects on the environment and humans (Allen and Rajotte, 1990; Georghiou, 1986; Task Force Report, 2003). By 1960, environmental damages caused by chemical pest control measures started to be noticed by various environmental groups (Kogan, 1998). Rachel Carson's book 'Silent Spring,' published in 1962, made people aware of this environmental issue. This awareness created a public outcry emphasizing the urgent need for finding alternatives to chemical pest control methods. By this time, many political action groups and organizations had formed to take action on pesticide-related issues (Johnson and Sprenkel, 2008). This public movement inspired scientists to devise alternative methods such as the development of resistant crop varieties for major diseases and identify insect pests and natural enemies of pests. According to Lyons et al., (n.d.), the term integrated control was first introduced by Vernon Stern, Ray Smith, Robert van den Bosch, and Kenneth Hagen, who published an article titled 'The Integrated Control Concept' in 1959. In the US National Academy of Sciences (1969) publication, R. F. Smith used the term 'Integrated Pest Control' (initially in 1962) for the application of different methods for pest control. Later, Smith and van den Bosch (1967) further elaborated this 'Integrated Pest Control' concept with the application of ecological aspects for managing pests. The Cooperative Extension Service started educational programmes related to integrated pest control by the late 1960s in the US. These educational programmes were delivered through County Extension Agents. IPM thereafter started to emerge as a solution to problems caused by chemical pest control (Harris, 2001).

Australia promotes IPM through private and public extension. The history of Australian extension goes back to the early 1900s. State agricultural departments were responsible for providing extension services in the early days. In the late 1800s, some states started agricultural colleges and established agricultural bureaus to help farmers improve their farming activities. For example, the state of South Australia established the first agricultural college and agricultural bureau in the 1880s (Government of South Australia, 2011). Later, the Agricultural Bureau became the state Agriculture Department. The establishment of state departments of agriculture in Australia was due to the economic benefits and social interests associated with agriculture at that time. In 1910, The Commonwealth Government of Australia (currently the Federal Government) proposed state governments to establish producer education programmes through agricultural colleges and experimental farms (Boon, 2009). At the start, state agricultural departments established agricultural experiment farms in different regions. Initially, established agricultural colleges prepared trained people needed to serve in experimental farms. By the 1920s, farm advisory services were started through experimental farms for improving agriculture. Early extension workers were called farm advisors and served as advisors to farmers in all areas of agriculture. Until the mid 1940s, these agricultural research farms were operated with a small number of agricultural researchers. By the end of World War II, state Agriculture Departments were well organized to conduct research in many disciplines and to deliver extension services to farmers. For example, by 1950, the state of South Australia Department of Agriculture had seven divisions including agriculture, horticulture, and animal husbandry (Government of South Australia, 2011). Later, Research and Extension Divisions were formed under the Agriculture Department. Extension Divisions were responsible for the advisory service as well as the regulatory service. Extension used farm visits, training programmes, agricultural publications, and agricultural radio programmes for transferring technology. The Extension Division conducted a coordinated effort by using various means to disseminate agricultural technology including pest control. Research and extension are functions of the Agriculture Department at the state level in Australia. The organization and structure of the Agriculture Department vary across states. Some states use the term 'Department of Primary Industries' for the Department of Agriculture in Australia. Australian agriculture researchers also observed environmental issues such as pesticide resistance created by widespread use of broad-spectrum organochlorine pesticides in the 1960s and started to develop alternatives for pest management. Early IPM research and extension work focused on using biological control methods and selective pesticides as alternatives for broad-spectrum pesticides used by Australian farmers.

In New Zealand, there was a publicly funded agricultural extension service until the late 1980s. It was organized under the Ministry of Agriculture and Fisheries, and evolved from the New Zealand Department of Agriculture. The Department of Agriculture was established in New Zealand in the 1890s. Research and information services were under the Department of Agriculture (Dalton, 2011). An agricultural instruction service had been established in 1923 to provide an extension service to farmers. The public agricultural extension service was charged to provide a technical and farm management advisory service to help farmers. An extension service was provided by the Farm Advisory Division and it was managed by annually set objectives (Hercus, 1991). By 1970, the New Zealand public extension had been well organized. The period between 1970 and 1980 can be considered to be the peak of the publicly funded extension activities in New Zealand (Stantiall and Paine, 2000). Before privatization, the New Zealand public extension service consisted of extension staff organized into eight regions. Farm advisors were the field level extension providers and they delivered information from research to farmers. Agricultural scientists in the research division developed IPM technology and farm advisors delivered IPM technology to farmers. There was a link between farmers and research through extension. Early IPM research and extension focused on reducing the pest management dependency on broad-spectrum pesticides by introducing biological control methods and selective pesticides.

In 1984, the government started to introduce extension privatization process. In this process, the Ministry of Agriculture and Fisheries (MAF) was reorganized into four business centres, namely MAF Technology, MAF Quality Management, MAF Fisheries, and MAF Cooperate Services in 1987 (Hercus, 1991). MAF Technology comprised the previous agriculture research and extension services. Staff reductions took place during this restructuring process. The extension advisory service became a fee-based commercial enterprise after privatization of the agricultural extension service. Development and dissemination of IPM technology became a function of MAF Technology. As a result of privatization, extension priorities and programmes were set based on customer demands. This is considered to be a positive effect of privatization of the extension service. However, there are concerns whether the private extension is reaching small-landholding farmers and providing the necessary advisory service for 'public good'. This commercialization of extension includes a reduction of government services and programmes designed for sustainable environmental practices (Hall et al., 1999) such as IPM.

22.1.2 Historical Background of IPM Extension in Developing Countries

In developing countries, pesticides became a part of the 'Green Revolution' technological paradigm. Small farmers in the Green Revolution lands in Asia adopted pesticides (mainly insecticides) as the sole strategy of pest control. This technology was disseminated through a public extension systems in the developing countries through top-down extension activities. The surprising aspect of this paradigm is that it was propagated as the sole pest control strategy by the extension agencies when the world had taken note of the dysfunctional consequences of pesticide use (Peshin et al. 2009a) brought forward by Rachel Carson in her book 'Silent Spring' in 1962, and by entomologists at The University of California who were developing integrated pest control tactics (Stern et al., 1959).

In rice growing countries of Asia, there were widespread outbreaks of the rice brown planthopper, *Nilaparvata lugens* in the 1970s and 1980s caused by insecticides meant to control it (Kenmore, 2006). This triggered the development of IPM strategies for pest management in rice (Smith, 1972) in developing countries.

The Food and Agriculture Organization (FAO) initiated the Inter-Country Programme (ICP) for the development and application of integrated pest control (IPC) in rice in South and South-East Asia in 1980. From 1977 to 1987, IPM moved from research towards extension. By 1988, the Training and Visit (T&V) extension system in the Philippines, Indonesia, Sri Lanka, Bangladesh, India, Thailand and Malaysia attempted to introduce IPM to rice farmers through their system of 'impact points' or through strategic extension campaigns (Kenmore, 1997). After failing to introduce IPM through the T&V extension system, the introduction of IPM has been fostered by Farmer Field Schools (FFS), which provides 'education with field-based, location-specific research to give farmers the skills, knowledge and confidence to make ecologically sound and cost-effective decisions on crop health'. The FFS training module is based on participatory experiential learning to help farmers develop their analytical skills, critical thinking and creativity, and help them learn to make better decisions (Kenmore, 1997).

The trainer is more of a facilitator rather than an instructor (Roling and van de Fliert, 1994).

Integrated Pest Management Networks (IPMN) (1992–2000) funded by the Swiss Government through the International Rice Research Institute (IRRI) was started in eight countries in the Asian region, namely China, India, Indonesia, Lao People's Democratic Republic, Malaysia, Philippines, Vietnam and Thailand aimed at educating younger generations by modifying the curriculum in schools as well as increasing farmers' IPM participation.

22.2 IPM PROGRAMMES

With the realization of environmental and health issues associated with synthetic organic chemical pesticides, IPM became the buzz word among extension entomologists in many parts of the world in the 1970s. Entomologists, plant pathologists, and weed scientists were engaged in research to devise integrated pest management technologies. Extension systems in some parts of the world worked closely with researchers for the development and dissemination of IPM technology as early as the 1970s. This section describes how the global IPM extension movement has developed over the past four decades in some parts of the world including developed and developing countries.

22.2.1 Developed Countries

This sub-section discusses the IPM extension movement in the USA, Australia, and New Zealand. The United States has been one of the leading countries promoting IPM since the beginning of this movement. The IPM programmes have more than four decades of history in the US (Olson et al., 2003). The United States promotes IPM through research, extension, education, and policies. During the period between 1970 and 1990, IPM programmes were developed and implemented extensively in the United States. Due to the widespread IPM activities during this era, Allen and Rajotte (1990, p.348) termed this period 'the IPM era'. The success of IPM programme in the US is mainly associated with the close working relationship between research and extension services at the federal and state levels. The Cooperative Extension Service and the Agricultural Experiment Station are two major divisions organized under the College of Agriculture at the land-grant university in each state. Since the Research and Cooperative Extension divisions are under the dean of the College of Agriculture, coordination of research and extension functions is at a very good level. This close coordination of research and extension functions is very helpful for achieving the desired results in IPM technology development and dissemination. The US Department of Agriculture (USDA), the US Environmental Protection Agency (EPA), and the National Science Foundation (SNF) were the major federal level institutions responsible for promoting IPM in the US initially. In 1971, the USDA funded a pilot IPM project for tobacco in North Carolina and cotton in Arizona. Later, these IPM projects expanded into other states (Allen and Rajotte, 1990). In 1972, the USDA sponsored an IPM research grant programme. The research conducted under this programme at land-grant universities developed IPM technology for crops such as cotton, soybean, alfalfa, apple, pear, and citrus crops. This research information was utilized for pilot IPM extension programmes initiated by the USDA (Lyons et al., nd). This information was transferred to farmers mainly through Cooperative Extension in each state. The Cooperative Extension Service played a significant role in transferring IPM technology to farmers in the US (Kogan, 1998). Private crop consultants also played an important role in disseminating IPM technology to farmers in the US.

The need for IPM increased with the removal of hazardous pesticides from the US Market (Task Force Report, 2003). Between 1970 and 1980, many states in the US started to introduce regulations to control the use of hazardous pesticides, requiring a licence for pesticide application and advising farmers on pest control methods for minimizing the environmental and health issues associated with chemical pesticides. For example, California introduced an act requiring an environmental impact report before application of any environmentally hazardous pesticide in 1976 (Lyons et al., nd). These regulations made it mandatory for those using pesticides to be responsible in relation to, and to comply with, recommended safety procedures to minimize consumer hazards. Those who violate these regulations were subject to serious fines. Pesticide liabilities may occur with hazardous residues. This policy environment with liability for using chemicals for pest control also contributed to promote IPM among users (Johnson and Sprenkel, 2008). The federal government also initiated various policies to promote IPM technology for pest management. For example, the Clinton Administration established a national goal for implementing IPM on 75% of farmlands in the US by 2000 and started a new IPM initiative to achieve this goal. The USDA coordinated this effort with several federal agencies and allocated about US $170 million (Ratcliffe and Gray, 2004). Four Regional IPM Centers established under this programme contributed to increase research and extension activities regionally. These Centers collaborated with land-grant universities to form a national IPM network for the development and dissemination of IPM technology for all stakeholders. According to an evaluation conducted in 2000, the estimated adoption of IPM technology for managing pests was 86% of the total cotton crop, 86% of vegetables, 78% of soybean, 76% of barley, and 71% of wheat (NASS, 2001).

Some states jointly formed multi-state IPM projects to achieve efficiency of research and extension efforts in the US. For example, in 1974, Alabama, Florida, and Georgia formed a regional IPM project to address pest management issues related to peanuts, cotton, and soybean with funding received from the USDA (Johnson and Sprenkel, 2008). By the 1980s, almost all states had established IPM research and extension programmes with funding support received from the USDA and state governments at their land-grant universities. The IPM research and extension programmes were mainly targeted to increase the application of IPM practices in the field for pest management. Crop scouting became an important part of the IPM programme by the late 1970s. Data gathered from crop scouting programmes were consolidated by Cooperative Extension and used to make IPM decisions by farmers (Harris, 2001). The Consortium for Integrated Pest Management project funded by the USDA in the period 1981–1985 contributed to increase the adoption of crop scouting and economic threshold levels for making pest management decisions (Kogan, 1998).

In the early 1980s, research and extension systems in some states started to use computers for storing, processing, and disseminating information necessary for effective IPM programmes. Researchers developed computer simulation models based on biological, environmental, and agronomical research data. They developed crop growth models, pest population models and weather pattern models. Pest population models were designed to predict pest occurrences with weather changes and predict possible pest outbreaks. This information was useful for estimating expected crop damage and making management decisions. Farmers benefited from these decision-support software programs (Task Force Report, 2003). These computer models were simplified versions of very complicated natural processes. They reflected the existing knowledge about the affecting variables at that time. With the development of new knowledge, these models were updated. By the late 1980s, these computer models had been developed for personal computers (Lyons et al., nd). Initially, these models were used by extension personnel and crop consultants for making integrated pest management decisions for farmers. With the widespread use of personal computers, farmers also started to apply these models as pest management decision support systems. Extension Specialists used computers to transmit useful IPM information from researchers to extension application.

State specialists working in IPM programmes maintain a close communication linkage with County Extension Agents for receiving information about field issues and transmitting new technology to farmers and others involved in US extension. IPM Specialists regularly visit farmers with County Extension Agents. These visits are helpful for transmitting validated technology to farmers very quickly. Some large states divided the state into several areas and hired area IPM specialists to facilitate the development and dissemination of IPM technology more effectively. This staffing arrangement was very helpful to provide the necessary technical support for the field extension staff and farmers in large states. For example, there were seven area IPM specialists serving in California by the period 1980–1981 (Lyons et al., nd). These area IPM specialists were responsible for a relatively small area compared to serving IPM specialists at the state level. Posting IPM specialists at the local level contributed to expedite the IPM technology development, validation, and dissemination process. The area IPM specialists helped State IPM specialists to identify field problems accurately and find solutions quickly. The area IPM specialists had earned an MSc or PhD degree related to pest management disciplines and served as the local IPM resource persons and provided training workshops for County Extension Agents, private crop consultants, and farmers to keep them up to date about IPM technology. Other responsibilities of these area IPM specialists were participating in field experiments initiated by state specialists, conducting local adoptability research, and conducting IPM demonstrations in their multi-county area (Lyons et al., nd). The area IPM specialists collaborated with County Extension Agents and private crop consultants in conducting field demonstrations, collecting data, and evaluating economic threshold levels. Frequent meetings and field demonstrations facilitated the IPM knowledge transfer process from research to practice. This close working arrangement between researchers and extension personnel at the local level contributed to developing locally adoptable IPM technology and convincing local extension officers, private crop consultants, and farmers about the usability of new IPM technology.

In the 1980s, the US Cooperative Extension Service printed publications such as IPM manuals and brochures for pest identification and determining threshold levels. IPM manuals provided information related to biological control guidelines with emphasis on controlling pests using a systems approach. These printed materials were used as complementary information during field days and training workshops. IPM manuals were helpful for Extension Agents, crop consultants, and farmers to make management decisions related to IPM. The use of CD ROMs to disseminate IPM technology was started in the early 1990s. By early 2000, many IPM publications were available in electronic form either saved on a CD ROM or posted on the Internet. Compact discs are used to disseminate IPM technology (Resel and Arnold, 2010). With the growing access to the Internet, the Cooperative Extension Service started to disseminate IPM information through the Internet in early 2000. Almost all Cooperative Extension services have developed their own IPM web pages and posted materials for disseminating information to farmers and other users. With the increased access to the Internet, IPM website usage has increased considerably (Olson et al., 2003). Initially, IPM publications focused only on the information needs of Extension Agents, crop consultants, and farmers. With time, the IPM extension movement in the US expanded its service to other audiences as well. These audiences included residential users, landscapers, students, educators, and park rangers. Currently, users have access to various sources of IPM information through the Internet in the US.

In Australia in the 1960s, emerging environmental issues with heavy use of broad-spectrum pesticides compelled the agriculture industry to consider alternatives for managing pests. Australian agricultural researchers tested biological methods with selective pesticides to control major pests in apple orchards in the 1970s. Almost all organochlorine insecticides were removed from agricultural use in Australia in 1987 (Horne and Page, 2009). The introduction of any biological control agents into Australia is controlled by the Australian Quarantine and Inspection Service. Early release of biological agents for managing pests is managed by government research stations. After testing biological agents for pest management, several private businesses were established in Australia for the production and distribution of biological agents. These companies were responsible for production, quality control, marketing, and advising farmers. Later, these companies formed the Australian Biological Control Association to assure the quality of biological control methods. In the 1980s, agricultural scientists developed pest prediction models and introduced those models for pest management effectively. Most of the IPM strategies were based on biological and cultural control methods and selective pesticides were used if necessary (Horne and Page, 2009). In the 1990s, a community approach to managing a major orchard pest in Victoria was tested by coordinating farmers in a large area by disrupting the mating pattern of the pest and achieved desirable results (Williams and Il'ichev, 2003). This experiment highlights the significance of area-wide coordination of IPM work and a community-based approach to IPM to achieve desirable outcomes. When introducing natural enemies to control pests, entomologists collaborated with farmers to conduct field experiments on their land. The involvement of farmers in IPM research made them aware of beneficial insects and their significance in pest management. This understanding encouraged farmers to avoid routine application of insecticides and monitor levels of beneficial insects before making any strategic application of pesticides. Extension advisors were Bachelor degree or Associate degree holders (Swanson et al., 1990) and provided training programmes to assist growers in understanding the significance of managing biological systems under the concept of applying IPM for pest management. Until the 1990s, the Australian agricultural extension service was an effective, large, publicly funded, production focused, tech-transfer service (Cary, 1998). State Departments of Agriculture were the major IPM extension service providers in Australia. In the 1990s, Australia adopted a pluralistic approach to agricultural extension and the private sector was empowered to deliver IPM programmes. For example, in 1996, IPM Technologies Pty Ltd was created to monitor potato pests and advise farmers how to manage pests with IPM. The farmers, who worked with IPM Technologies Pty Ltd in conducting commercial trials, were easily convinced about the value of IPM and motivated to apply IPM practices (Horne and Page, 2009). Private and public jointly funded brand name groups and Local Best Practice groups were promoted as IPM extension providers in Australia. The 'Best Practice Approach' is based on action research. Since action research leads to understanding the situation while addressing pest management issues within the context of existing issues, it is appropriate for IPM research and extension. This is a participatory approach (Foster et al., 1995) to IPM technology development and adoption. Even after promoting private sector extension, government agencies are still providing extension with some changes. These changes include focusing on human resource development, cost recovery for services, and use of groups for delivering services (Marsh and Pannell, 2000). There are positive factors as well as negative factors associated with private sector IPM extension in Australia. As farmers became more specialized in their operations and needed highly technical customized advice, privatization of extension in Australia opened a tailored advisory option for this growing need. This is a positive factor associated with privatized extension compared to public extension. Public extension provides unbiased, research-based information. However, private businesses are driven by profit and may be biased toward their own products and services. This can be considered to be a negative factor associated with private extension. The Australian extension vision is to strengthen private research and extension for a greater role in meeting farmers' needs. Public institutions are expected to collaborate with private extension to serve farmers efficiently (Marsh and Pannell, 2000).

The farming systems approach was used to develop and disseminate IPM technology. Weather and pest population data were used to predict pest problems and make decisions to use selective pesticides. Extension advisors presented training programmes to educate farmers to understand the importance of managing ecosystems to control pests with biological methods. Printed manuals and newsletters were used to disseminate IPM information. With the widespread use of computers, public and private extension providers used CD-ROMs and the Internet to disseminate IPM information to meet farmers' needs.

Agriculture is an important sector in the New Zealand economy. New Zealand agricultural products are marketed in Europe and other parts of the world. With the growing specialization of New Zealand agriculture, export-market oriented farmers needed a high quality tailored advisory service for their business operations. Export markets demanded stringent quality standards of foods that included pest-free as well as pesticide residue-free food products. If New Zealand farmers wanted to have access to these export markets, they had to follow pest management practices such as IPM to meet market standards. Meeting these specific needs of farmers became a significant challenge to the public extension service with budget limitations. This situation led New Zealand to introduce extension privatization policies in the 1980s.

As in other developed countries, extensive use of broad-spectrum inexpensive insecticides for controlling insect pests was the most common method used by New Zealand farmers since the 1950s until IPM was introduced in the 1960s. As in other parts of the world, the agriculture industry in New Zealand also experienced environmental issues such as the development of pesticide resistance to broad-spectrum pesticides (Suckling, 1996). Detrimental environmental impacts paved the way for the development and implementation of alternative pest management approaches such as IPM practices for managing pests in New Zealand. Attempts had been made to promote IPM in New Zealand since the 1970s, but it was not popular among farmers as a common practice due to their reluctance to give up trusted chemical control methods. IPM did not take off significantly until the 1990s. During this period, consumers and export markets started to demand pesticide-free safe food products (Wearing, 1993). This situation motivated the agriculture industry to pay serious attention to develop and implement the IPM approach for managing pests. As a result of this movement, the horticulture industry started to develop name-brand IPM programmes such as 'KiwiGreen' for kiwi fruits and 'AvoGreen' for Avocados (Suckling et al., 2003, p. 386). Industry funded research and the extension service developed IPM manuals for farmers and consultants. With the development of information technology, email and the Internet were used to deliver apposite IPM data.

22.2.2 Developing Countries

In the developing world, there are two types of agriculture, modern Green Revolution agriculture and second, subsistence agriculture. In India, Green Revolution agriculture is confined to areas with assured irrigation or high rainfall, covering about 103 districts. In the Green Revolution areas, external input usage – fertilizers, high yielding varieties, pesticides – propelled agricultural productivity. Pesticide intensive pest management was the only prescription of researchers and extension specialists for managing pests in rice, cotton, vegetables and other crops. Pest management in the pre-Green Revolution era, in India and other developing countries, was characterized by cultural and manual mechanical practices based on a farmer's lifelong experiences. Experts in this era in most of the developing world (tropical areas) were involved in taxonomy, biology of pests, and the advocacy of cultural practices (Muangirwa, 2002).

22.2.2.1 *India*

There is a great propensity in the developing countries such as India to propagate the technologies developed in the industrialized countries. Such tendencies result in diffusion of technologies which may or may not be suited to the local biophysical and socio-economic conditions of the farmers. Pesticide-intensive pest management was promoted in all type of farming communities, in irrigated/unirrigated areas and in small- and large-landholding farmers. It is worth mentioning here that the average landholding in India is about 1.4 ha and farmers with a holding size of more than 10 ha have been classified as large-landholding farmers. In India, the consumption pattern of pesticides varies from state to state (Peshin et al. 2009a). Pesticide use increased from 5640 tons in the pre-green revolution era to 21,200 tons in 1968–1969 in the green revolution era and reached an all time high of 75,418 tons in 1988–1989. The share of insecticides accounts for 64% of total pesticide use (Peshin et al. 2009a), and increased many-fold in crops such as cotton, rice and vegetables (Peshin, 2009; Peshin et al., 1997; Sharma, 2011).

In India, different organizations, namely the Central Institute for Cotton Research (CICR) Nagpur, the Asian Development Bank (ADB), and the Commonwealth Agricultural Bureau International (CABI) are working for the promotion of IPM (Bambawale et al. 2004). The National Centre for Integrated Pest Management (NCIPM) of the Indian Council of Agricultural Research (ICAR) was established in February 1988 to cater for the plant protection needs of different agro-ecological zones in the country. The first crop centric IPM research and extension work was started in 1974–75 under the Operational Research Project. The FAO Inter-Country programme in India was started in 1993 for rice crops. The Regional Programme on Cotton-IPM was implemented by CABI in 1993 whereas FAO-EU (European Union) IPM programmes for cotton were established in 1999 in six countries including India. The project activities were started in India, China and Vietnam in 2000, and in the remaining countries in 2001 (Ooi, 2003). The National Agricultural Technology Project (NATP) for IPM in 2000, and the Insecticide Resistance Management-based IPM programme were implemented in 10 cotton growing states in India by the Central Institute for Cotton Research (CICR) Nagpur in 2002 (Peshin et al., 2007).

22.2.2.2 *China*

In China, the development of IPM has been a gradual but continual process. Efforts to develop integrated control tactics were initiated in the early 1950s (Ma, 1976). In 1975, the Ministry of Agriculture officially approved IPM as the guiding principle for national plant protection (Li, 1990; Ma, 1976). Since 1983, the Chinese government has funded National IPM Technique Research Projects. This marked the beginning of the era of modern IPM in China (Guo, 1999; Li, 1990). China has been a member of the FAO Inter-Country IPM Programme since 1988. Since then, the FAO Inter-Country rice IPM project, the Asian Development Bank cotton IPM project, and the FAO-EU cotton IPM project have been executed in China. The Farmer-to-Farmer Program took root in China from 1995 to 1999 (CIP-UPWARD, 2003). One of the alternatives adopted by China following the 1992 'Conference of United Nations Environment Programme' was the implementation of IPM and green farming as priority areas in executing sustainable management in China's 21st Century Agenda (Wang, 2000). In 2000, a new IPM Programme focusing on cotton farmers was funded by the EU and managed by FAO. China, with assistance from the United Nations Development Program (UNDP), FAO and other funding countries, has established many farmer field schools (FFS) (Wang, 2000).

22.2.2.3 *Thailand*

In Thailand, the FAO/UNDP funded Programme for Integrated Pest Control in rice was started during 1976–1988 (Sirisingh, 2000). The project began in 1976 on rice, and later on extended to include cotton, sugarcane and vegetables. IPM Implementation in Rice under the Royal Initiative (1998–2000) was carried out by the Department of Agricultural Extension. The activities comprise curriculum development and project planning workshops, the training of trainers, farmer field schools and refresher workshops (Sirisingh, 2000).

22.3 EXTENSION POLICIES IMPACTING DISSEMINATION OF IPM TECHNOLOGIES

For real diffusion of IPM techniques, political support is needed (Larguier, 1997; Nicholls and Altieri, 1997; Williamson, 2003) to allot budgets for applied research on IPM techniques and for subject-matter specialists and field-level extension workers, and for farmers willing to try the techniques so that they can obtain a premium price in the market. Political support is also needed to introduce measures limiting the use of pesticide herbicides or to make them less attractive (taxes, time limitations, licences, etc.).

26.3.1 Developing Countries

The United States has adopted the policies conductive for promoting IPM. These policies include controlling hazardous pesticide manufacturing, handling, and application. These policy measures are in place to ensure the safety of people, environment, and wildlife. The US policies promote the application of IPM for preserving the environment. The USDA Office of Pest Management Policy (OPMP) was established in 1997 to facilitate the implementation of pest management policies in the US. The mandate of this office is to integrate the USDA strategic planning activities related to pest management, coordinate the interagency affairs related to pesticide regulatory process, and supporting agriculture by promoting the development of new pest management approaches such as IPM to meet the needs of the sustainable agricultural system. When pest management policies such as the Food Quality Protection Act of 1996 and the National Road Map for Integrated Pest Management are implemented, OPMP closely works with the Environment Protection Agency (EPA) of the US for interagency coordination. The OPMP communicates across agencies to promote the IPM strategies, communicates with farmers and other stakeholder groups to keep up with current and emerging pest management issues, coordinates and collaborates with regional IPM centers to disseminate IPM information, and provides the EPA and other regulatory agencies with the most accurate data to facilitate regulatory decision making process. The OPMP coordinates the IPM programs among the federal agencies to avoid duplication and achieve efficient programming (OPMP, 2013). In addition to federal policies promoting IPM, states also have adopted various policies to promote IPM as the viable strategy for controlling pests.

According to FAO (2007), Australia and New Zealand don't have a national IPM policy. However, Australia and New Zealand are promoting IPM as a viable pest control measure for sustainable agricultural development. The demand for pesticide residue free foods from the European Union and other countries has compelled Australia and New Zealand to adopt pesticide control policies and procedures for safe food production. This situation has contributed to promote IPM as a practical approach for minimizing the dependency on chemical pest control measures in Australia and New Zealand. In addition to Australian Federal Government, states also have adopted policies conductive for promoting IPM. For example, New South Wales has adopted the *Pesticide Act of 1999* to control the use of pesticides in the state. This act aims to reduce the risks associated with the use of pesticides to human health, the environment, property, industry, and trade (NSWEPA, 2013).

22.3.2 Other Countries

Ineffective agricultural policies are often quoted as an obstacle to the adoption of sustainable agricultural practices. Adoption of IPM practices is closely linked to the prevailing policy on pest management (Munyua, 2003). The conceptualization and implementation of IPM will therefore require identification and elimination of policy constraints (Brady, 1995). If the emphasis in the agricultural policies of the developing countries is on increasing production by the extensive use of chemicals, they will slow down the process for adoption of sustainable agricultural practices (Munyua, 2003). A favourable policy environment must precede successful implementation of IPM practices. Unless governments develop policies that firmly support sustainable agriculture, the response towards implementation of IPM will be slow. It is clear that, without policy reform in pest management and pesticide procurement, it will be difficult for developing countries to overcome constraints that hinder the transition to and implementation of IPM practices (Munyua, 2003). Clear policies such as abolition of subsidies and credit for use of pesticides make it possible to overcome barriers to diffusion of IPM practices. Cuba, Indonesia and the Philippines are examples of countries where IPM has thrived after government declarations to support IPM as the national pest management strategy (Pretty, 2002; Matteson, 1996).

22.3.2.1 India

The Government of India (GOI) after agreeing to the FAO initiated programme for IPM in rice in 1980, re-oriented its plant protection strategy in 1985. The Government took a number of positive steps in promoting IPM which included: development of infrastructure; establishment of Central Integrated Pest Management Centres (CIPMCs); human resource development through a three-tier programme that consisted of season-long training for subject matter specialists, establishment of farmer field schools (FFSs) to train farmers and conducting demonstrations for adoption of field tested IPM technologies; and policy support to promote needs-based pesticides and phase out the use of hazardous pesticides. In the 1990s, India abolished insecticide subsidy and banned pesticides which were highly toxic. It withdrew a US $30 million annual subsidy for insecticides and instead imposed a 10% excise tax on them, which meant US $60 million annual new income for the government, which spends over US $10 million per year on IPM field training (Kenmore, 1997). In 1991, the Central Sector Scheme was started by the Ministry of Agriculture and Cooperation, Government of India for the promotion of IPM in India. In 1994, the Directorate of Plant Protection, Quarantine and Storage, Government of India, the nodal agency for implementing IPM programmes, intensified its efforts and adopted the FFS model for educating farmers through CIPMCs (Peshin et al., 2009a). IPM was envisaged as the main strategy for plant protection in the 'National Agriculture Policy of 2000'. In the XI Plan Period during 2008–2012, an outlay of US $2.8 million was earmarked for conducting 3250 IPM-FFS and state level training programmes out of a total outlay of US $266.7 million for 'Strengthening and Modernizing of Pest Management Approaches' (Peshin et al. 2009a).

22.3.2.2 *China*

The first National Safety Standard for pesticide application was published by the Ministry of Agriculture in 1980. An IPM programme had always been listed among the national key scientific and technological projects in the Sixth Five-Year Plan (1981–85) of the country (Guo, 1997). A regulation banning the use of highly toxic pesticides having high residual effects close to urban areas was established in 1995. In order to enhance the management of the pesticide market as well as protect farmers, the Republic of China issued regulations on pesticide administration in 1997, for guiding the registration, production, management and application of pesticides (Wang, 2000). The Ministry of Agriculture issued Implementation Rules on Regulations on Pesticide Administration in 1999. The two legal texts standardized the behaviour of pesticide production, operation and application in China. Since then, more than 300 national rules on pesticide safety application have been issued for 140 pesticides in 19 crops which strictly limited the application of pesticides harmful to the environment and natural enemies and having easy to leave residues (Zhen-Ying et al., 2003). China also allowed the cultivation of genetically modified crops (only transgenic cotton) that are pest-resistant. However, owing to the debate on risks, the planting area has been limited to only Hubei and Shandong provinces (Wang, 2000).

22.3.2.3 *Indonesia*

In 1986, the Indonesian government issued a comprehensive policy for pest control in which 'Integrated Pest Management' (IPM) was stated as the National Policy that forbids the use of 57 types of wide spectrum pesticides for rice plants. The subsidy on pesticides was gradually reduced to nil at the beginning of 1989. As a follow-up of the policy, the Government has been implementing training and development programmes of National Integrated Pest Management under the responsibility of the National Development Planning Agency (BAPPENAS) (Rustam, 2010). IPM has been a principal government policy in implementing plant conservation activities based on constitution No. 12 of 1992 on the plant cultivation system, Government regulation No. 6 of 1996 on plant conservation, and Declaration of the Agriculture Minister No. 887/Kpts/OT/9/1997 on plant pest organism control. The duty, function and authority of IPM are based on Constitution No. 22 of 1999 on Regional Autonomy, and Regional Regulation No. 25 of 1999 on the implementation of Regional Autonomy (DGHRI, 2003).

22.3.2.4 *Bangladesh*

To promote IPM in the country, several policies have been approved. The National IPM Policy that was approved in 2002 includes: (i) reduced use of pesticides, and use of Integrated Crop Management (ICM) and IPM technologies such as new resistant crop varieties, bio-control agents, etc.; (ii) awareness campaigns for policymakers, farmers, and the public on IPM and integrated nutrients management; (iii) training of farmers and farmer trainers; and (iv) a policy declaration that food safety is a priority over food security. The National Agricultural Policy also includes an IPM component (Gapasin, 2007).

22.3.2.5 *Philippines*

In the Philippines, the national IPM programme, known as *Kasakalikasan* (which means 'Nature is Agriculture's Bounty') was initially led by the national Department of Agriculture, but in 1992, responsibility passed to the municipal and provincial governments. Furthermore, the programmes constantly counter the activities of pesticide dealers, who attempt to influence local politicians (Castillo, 1996).

Thus, it can be concluded that the widespread adoption of IPM relies on several institutional, technical, economic and socio-psychological factors, such as a favourable policy framework, awareness raising with mass media, confidence building, common perception of problems by researchers and farmers, interactive extension, group approach, motivation for both farmers and field advisors, and practical education of advisors.

22.4 EXTENSION SYSTEMS DISSEMINATING IPM

There are various extension systems in different parts of the world disseminating IPM technology to various users. The users of IPM technology include farmers as well as non-farmers. Agricultural extension systems mainly disseminate IPM technology to farmers while other extension systems disseminate IPM technology to the general public. This section reviews what the global extension systems engaged in dissemination of IPM technology are and their strengths and weaknesses for making suggestions to strengthen global IPM extension efforts.

22.4.1 Developed Countries

In the United States, land-grant colleges and the USDA are the two major public institutions responsible for the development and dissemination of IPM technology. These two institutions collaborate with various commodity groups, businesses, and private industries in IPM technology development and the dissemination process. This section discusses how these institutions are organized, supported, and operated to develop and disseminate IPM technology in the US. The IPM extension system in the US has various public institutions and private partners (Figure 22.1). As briefly discussed in section 22.1.1, Cooperative Extension is the major outreach arm of land-grant universities in the US. Each state has a Cooperative Extension service organized under the college of agriculture. When there are two land-grant universities in a state, they collaborate to provide the extension service in the state. There is an extension director at the state level in the college of agriculture. Under the director, there are programme leaders for various extension programmes. The major extension programmes are agriculture, family and consumer sciences, youth development, and community development. Generally, there is a separate programme leader for each of these extension programme areas. However, the arrangement may vary slightly from state to state. The Agriculture Extension Programme Leader is responsible for coordinating agriculture extension programmes in the state. The agriculture programme may include natural resources and horticulture extension programmes as well. The agriculture extension programme leader is responsible for coordinating IPM extension programmes at the state level. There are extension specialists in each of various departments in the college of agriculture.

The major departments responsible for the state IPM programme are entomology, plant pathology, crop science, weed science, animal science, and horticulture. Entomology, plant pathology, and weed science departments comprise the core disciplines of IPM at land-grant colleges (Harris, 2001).

Most of the time, there is an IPM state specialist or coordinator to coordinate the IPM programme at the state level. Normally, extension entomologists, crop scientists, plant pathologists, and weed scientists work closely in state IPM research and extension teams. These specialists are responsible for conducting IPM research related to pest management issues in the state and transferring new IPM technology to potential users and other stakeholders.

The Cooperative Extension Service has an office in each County for providing educational programmes to help citizens meet their learning needs. Most states still use this county-based Cooperative Extension delivery system. However, this original county-based extension structure has been changed in some states. The

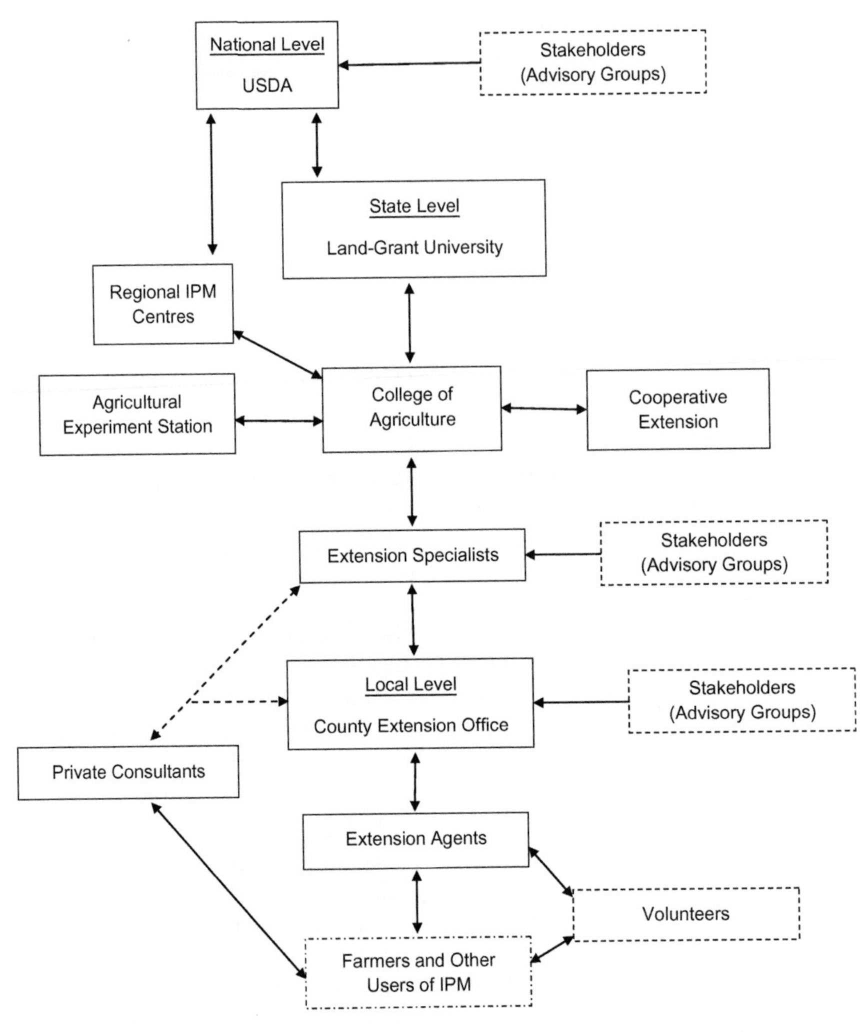

FIGURE 22.1 IPM extension system in the USA.

County Extension Director is in charge of the County Extension office. This individual is responsible for providing educational programming leadership as well as administrative leadership for the county extension staff. The County Extension Director is an experienced individual having an MSc or PhD degree. Typically, the County Extension Office may have a Field Crop Extension Agent, Livestock Extension Agent, Horticulture Extension Agent, Family and Consumer Sciences (FCS) Extension Agent, 4-H and Youth Development Extension Agent, Community Development Extension Agent, and a Forestry Extension Agent. These Extension Agents have a BS or MSc degree related to their assigned responsibility. This typical staffing arrangement varies with the programming needs of the people in that county. For example, an urban county may have a Horticulture Extension Agent, a couple of FCS Agents and

Youth Development Extension Agents due to the needs of the urban setting. Most of the time, Row Crop Extension Agents and Horticulture Extension Agents provide the leading role of conducting the IPM extension programme at the county level. They are responsible for coordinating field days with state Extension Specialists or area Extension Specialists; conducting training programmes for farmers in their county; conducting certification training programmes for pesticide applicators; working with farmers and others to identify pest issues; and implementing the IPM programme at the county level. Livestock Extension Agents are responsible for promoting IPM practices in managing pests related to livestock at the county level. The FCS Extension Agents promote IPM for managing household pests. This includes managing pests using IPM practices in homes, schools, and other public places. 4-H and Youth Development Extension Agents teach IPM concepts to youths through hands-on activities. The above discussed typical arrangement of IPM extension responsibilities at the county level varies with the learning needs of the people in that county. County Extension Offices in most of the states are grouped into Extension districts or regions for administrative purposes. For example, North Carolina has five Extension Districts at present. The District Extension Director provides the overall administrative leadership for the County Extension Directors in the district.

In addition to regular staff, the Cooperative Extension Service uses volunteers to extend the service of the extension to the public. There are different groups of volunteers serving in the Cooperative Extension Service. One group of volunteers are mainly working with the horticulture programme. They are called 'Master Gardeners' and are recruited to extend horticultural education to those who need it. When someone joins the Cooperative Extension Service as a master gardener, that individual has to go through an intensive training programme to become a master gardener. This training provides the necessary knowledge and skills to prepare them as extension volunteers. IPM education is an important part of these training programmes. Master gardeners provide educational programmes to help farmers and others manage their pests using IPM technology. Their service is helpful to reach a large number of people in counties and complement the extension programmes provided by Horticulture Extension Agents. Another group of volunteers serve at the County Extension office as advisory group members. This group of individuals have been recruited from the leaders in the community to represent various client groups of the Cooperative Extension Service. The volunteers serving in County Extension Advisory Groups are mainly expected to advise Extension Agents in needs identification for educational programmes. These Extension Advisory Group Members also assist Extension Agents for marketing programmes to citizens, delivering programmes, and advocating for public support. Advisory groups help identify IPM programme needs at the county level and assist Extension Agents in IPM programme implementation.

At the federal level, IPM research and extension programmes come under the USDA. There are several agencies and institutions organized under the USDA that work collaboratively to support and promote IPM programmes in the US. These institutions include the National Institute of Food and Agriculture (NIFA) or the former Cooperative State Research, Education, and Extension Services (CSREES), Agricultural Research Service (ARS), Natural Resources Conservation Service (NRCS), Animal and Plant Health Inspection Service (APHIS), Forest Service, National Agricultural Statistics Service (NASS), and Economic Research Service (ERS). The National Science Foundation also provides support for promoting IPM. The USDA is responsible for implementing the US government IPM policies through these institutions. It provides funding support through formula funds and various grant programmes for IPM research, education, and extension programmes. In 2000, the USDA established four Regional Integrated Pest Management Centers in Northeast, North Central, Southern, and Western regions as a means to strengthen the IPM programme throughout the US. These centres were established at land-grant universities to facilitate federal and state partnerships for strengthening IPM research, education, and extension efforts. The North Central IPM Center was established at Michigan State University and the University of Illinois. The Northeast IPM Center was established at Pennsylvania State University and Cornell University while the Southern IPM Center was established at the University of Florida, and the Western IPM Center was established at University of California, Davis. The Southern IPM Center was later relocated to North Carolina State University. These regional IPM centres coordinate grant programmes to develop and disseminate IPM technology. Before establishing these regional IPM centres, the National Science Foundation sponsored the establishment of a Center for IPM at North Carolina State University in 1991. These centres collaborate with land-grant universities to function as a national IPM information network to respond to information needs of public and private sector users. The regional IPM centres are responsible for promoting interdisciplinary and multi-organizational IPM research and extension efforts to increase the effectiveness of stakeholder investments; provide IPM information to government agencies and other stakeholders in a timely fashion; respond to emerging issues; and administer the NIFA funded IPM grant programme regionally. These centres collaborate with the National Plant Diagnostic Network, the Animal and Plant Health Inspection Service, ARS, Forest Service, National Plant Health Board, and land-grant universities to address new and emerging pest management issues through research and extension efforts. Regional IPM centres closely work with the land-grant university system for the development and dissemination of IPM technology (USDA, 2009). The USDA organized a Pest Management Stakeholder Forum with the participation of farmers, environmentalists, IPM coordinators, industry representatives, and agency personnel to develop a national road map for IPM in 2002. This road map to IPM was further ratified at the National IPM Symposium in 2003. The National Road Map to IPM was targeted to increase the economic benefits of adopting IPM, reduce health risks, and minimize adverse environmental effects. This programme focused on production agriculture, natural resources and recreational environments, and residential and public areas (Ratcliffe and Gray, 2004).

In Australia, IPM extension is provided by public and private extension services at present. There are various partners at national and state levels. The Department of Agriculture, Fisheries and Forestry (DAFF) is the responsible institution at the federal level. DAFF provides policy guidelines and resources for agricultural research and extension. Until the 1990s, State Departments of Agriculture and Primary Industries were responsible for providing agricultural research, extension, and regulatory services in each state and territory. Australian public extension was a large institution with the focus on technology transfer (Marsh and Pannell, 2000). State Extension Services were divided into regions and district levels for administrative purposes. District offices were staffed with a District Agricultural Officer and Agriculture Extension Officers to deliver extension services locally (Williams, 1968). Public extension was not linked to universities. Field level extension officers were functioning as educators and regulators. This dual function undermined their role as educators. Historically, agriculture research and extension functions were under the state Departments of Agriculture. However, in some states, extension and applied research functions were managed by two separate departments. For example, extension was the function of the Department of Primary Industries in South Australia where research was the function of South Australia Research and Development Corporation (Murray, 1999). This situation had negative effects on technology development and transfer. Government budget shortfalls compelled the Australian public extension service to supplement funding through various cost recovery methods such as fees for services. This public extension service reorganized with the adoption of a policy for promoting a pluralistic extension system in Australia to better serve public needs. With this policy revision, State Departments of Agriculture and Primary Industries became responsible for providing agricultural research and extension services collaboratively with industry, private sector partners, and agricultural colleges at the state level. This pluralistic extension at the state level has various players. With these changes, public extension eliminated a number of extension officer positions and moved from one-on-one extension delivery methods to more group focused methods for delivering service (Murray, 1999). The public extension services used various printed and electronic materials for transferring technical information to farmers. Private extension is gradually taking the responsibility for IPM extension in Australia.

In New Zealand, public extension provided the agricultural extension service until it was reformed in the mid-1980s. The Advisory Services Division of the Ministry of Agriculture and Fisheries was the provider of the agricultural extension service including IPM extension to farmers in New Zealand. This service was provided free of charge until privatization of the extension service. In 1985, New Zealand adopted privatization policies and directed the Advisory Services (Extension) Division to become a commercialized service in 5 years funded by users paying fees. Initially, it was started on a cost recovery and fees for services basis. This was a drastic change for the agricultural extension in New Zealand. During this privatization process, the Advisory Services Division combined with the Agricultural Research Division in 1987 and formed an entity called Ministry of Agriculture and Fisheries (MAF) Technology. The objective of this integration was to establish close research and extension collaboration to achieve cost effectiveness of research investment. During this transition, some of the farm advisors/consultants left MAF Technology. In 1990, the advisory service was reformed into a Management Consultancy Service under the MAF. This consulting service was fully privatized in 1995 and was purchased by New Zealand's largest stock firm. This private consulting service is known as Agriculture New Zealand Ltd. and operated as a commercial business fully owned by a limited liability company (Journeaux and Stephens, 1997). With the privatization, the consulting service increased the number of consultants to meet the need for service. The main business of this consultancy service is providing agricultural advisory services to farmers, technology transfer, rural information gathering, and providing agricultural training courses. Agricultural consultancy service including IPM extension was provided for fees. The government contracted with Agriculture New Zealand Ltd. for some of these services such as rural information gathering.

The Research Division of the MAF with a few other research institutes were reorganized into 10 Crown Research Institutes (CRIs) in 1992. The Public Good Science Fund (PGSF) was established as the centralized government funding for research and administered through the Foundation for Research Science and Technology (FRST). Funding was made available to CRIs through a grant programme administered by FRST. The CRIs received funds from FRST and private industry for conducting research on a commercial basis. Some of these CRIs were responsible for conducting agricultural research including IPM. These centres are promoting collaborative work with the agricultural industry through partnerships (Journeaux and Stephens, 1997).

22.4.2 Developing Countries

22.4.2.1 *India*

There are multiple public extension systems in India (Singh et al., 2010). Indian Council of Agricultural Research (ICAR) institutes, state agricultural universities' extension system and state agricultural departments are all involved in the transfer of technology. The Department of Agriculture and Cooperation under the Union Ministry of Agriculture and the Provincial (state) Departments of Agriculture are primarily responsible for the transfer of technology to farmers. The IPM programmes are implemented by the ICAR institutions, state agricultural universities, the Department of Agriculture and Cooperation Government of India through its Directorate of Plant Protection, Quarantine and Storage (DPPQS), and state departments of agriculture. DPPQS is the nodal agency of the Government of India for disseminating IPM to farmers through its 31 CIPMCs located in 28 states and union territories. In 1993, India adopted the FFS model for dissemination of IPM in relation to rice crops. The ICAR also set up a National Centre for Integrated Pest Management in 1988 to cater for the plant protection needs of different agro-climatic zones.

22.4.2.2 *China*

In China, the Ministry of Agriculture is primarily responsible for the transfer of technology at the national level. At the provincial and county levels, a number of departments are involved in IPM dissemination (Wang, 2000). The National Agro-Tech Extension and Service Center (NATESC), Ministry of Agriculture is the coordinating agency for implementing IPM in rice, cotton, maize, vegetables and fruit crops in China. The international IPM programmes and activities are also coordinated by the NATESC. At the provincial and county levels, a number of departments are involved in IPM implementation.

22.4.2.3 *Bangladesh*

The IPM activities in Bangladesh are conducted by the Department of Agricultural Extension (DAE), Bangladesh Agricultural Research Institute (BARI), Bangladesh Agricultural Research Council (BARC), Bangladesh Agricultural University (BAU), Cotton Development Board (CDB), Bangladesh Water Development Board (BWDB) and Non-governmental organizations (NGOs). The Government-University-NGO model for IPM research and technology transfer is vibrant in Bangladesh (Gapasin, 2007). NGOs are especially active in disseminating IPM technologies and in training farmers in IPM technologies within their development programmes.

22.4.2.4 *Indonesia*

The Indonesian agricultural extension system is large and complex. The Ministry of Agriculture coordinates an array of provincial and district technical units to oversee and implement IPM. The country's lack of strong private and university sectors that can do this puts a huge burden on the Ministry of Agriculture to provide these services.

22.5 EXTENSION STRATEGIES USED TO DISSEMINATE IPM TECHNOLOGIES

Dissemination of IPM technology is quite different from that of technologies such as hybrid corn or an improved rice variety. Hybrid corn itself is a packaged technology. If the farmer has access to information about the economic advantages of hybrid corn, that itself is adequate to transfer this technology. Adopters go through a decision making process before they adopt a new technology. Even with this simple, easy to transfer hybrid corn technology, adopters take considerable time before they make their adoption decision (Rogers, 1983). IPM is not a simple technology. According to the Food and Agriculture Organization, IPM is 'the careful consideration of all available pest control techniques and subsequent integration of appropriate measures that discourage the development of pest populations and keep pesticides and other interventions to levels that are economically justified and reduce or minimize risks to human health and the environment. IPM emphasizes the growth of a healthy crop with the least possible disruption to agro-ecosystems and encourages natural pest control mechanisms' (FAO, 2012, p.1). This description about IPM highlights the fact that it is a combination of technologies used to manage pests, in an economically, socially, and environmentally sustainable manner. Since IPM is a complex technology, it requires a well planned educational effort to facilitate potential users to change their mind to adopt IPM technology (Allen and Rajotte, 1990; Wearing, 1988). For this reason, IPM is considered to be a knowledge-based technology. Farmers' lack of knowledge about the proper application of IPM technology was identified as an important constraint to adoption (Olson et al., 2003). Until farmers fully comprehend the application of IPM technology for managing pests, they will not be able to adopt it. This situation highlights the significance of educating farmers to fully comprehend the concept of IPM for sustainable adoption and effective dissemination. Dissemination of IPM technology requires properly planned extension programmes to educate farmers so that they can fully comprehend the application of technology for managing pests. Properly planned extension education programmes are based on experiential learning concepts and use a variety of effective hands-on learning techniques. According to Dewey, 'the quality of the present experience influences the way in which the principle applies' (Dewey, 1963, p. 37). This experiential learning view asserts that the extent to which farmers are exposed to the IPM experience and internalize that experience determines their future readiness to apply IPM concepts for managing pests. When farmers are exposed to systematic extension education programmes, they will be able to gain hands-on experience and put that experience into practice. According to Kolb (1984), when someone is exposed to a learning experience, he or she goes through an experiential learning cycle that involves gaining the experience, reflection about the experience, internalization of the experience, and acting on that experience. The IPM extension programmes should be able to help farmers go through the experiential learning cycle (Figure 22.2) to change their attitudes, gain knowledge, and develop skills for successful adoption of IPM technology. Extension educators should plan to provide hands-on, minds-in learning experiences enabling farmers to fully understand IPM concepts and apply those practices for managing pests. IPM extension programmes should be based on the concept of experiential learning for achieving desired and lasting results. This section discusses the effective extension education strategies used for the diffusion of IPM technology in different parts of the world and reviews their strengths and weaknesses for making suggestions to improve those methods.

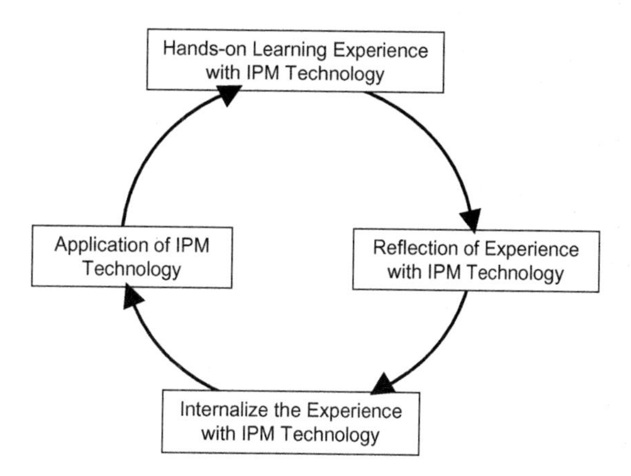

FIGURE 22.2 Experiential learning process of IPM extension programmes.

22.5.1 Extension Strategies Used in Developed Countries

Since the early days, extension in the USA, Australia, and New Zealand used various educational techniques to assure transfer of IPM technology from research to end-users. The most commonly utilized educational methods are one-on-one farm visits, field demonstrations, training workshops, printed materials, telephone calls, mass media, computer programs and the internet.

22.5.1.1 *One-On-One Farm Visits*

This is a common method used since the early days of extension to transfer IPM technology, and it remains the most effective, but the most expensive method. Extension Agents made personal visits to farmers based on a request to provide farm-based advice. In addition to Extension Agents, private agricultural consultants, and volunteers (master gardeners) delivered IPM technology using one-on-one methods. Since the one-on-one extension method provides instantaneous two-way communication, it is a very effective extension strategy. When making one-on-one farm visits, the Extension Agent will be able to clearly understand the farmer's needs and his/her farming situation. As a result, the Extension Agent will be able to customize the advice for the unique needs of the given situation. Also, it provides an opportunity to address additional concerns relating to application of IPM technology. Most of the time, these one-on-one meetings were effective in determining the best methods for farmers when they chose to make IPM decisions. The Extension Agents, crop consultants, and volunteers such as master gardeners in the US have been trained to assess the pest manifestation situation and work with farmers to make the best decision for managing pests under the guidelines of IPM. Extension personnel and farmers were provided various resources to help them make the best decisions for managing pests in the US. These resources included IPM field manuals with coloured pictures, climate data modelling with pest infestation, modelling of crop growth related to major pests, and economic threshold levels for various pests. Since one-on-one extension methods provide an opportunity to address the concerns and issues of the farmer instantly, it facilitates the adoption of technology and minimizes the concerns and risks. Therefore, the one-on-one delivery method can be considered to be one of the most effective methods in delivering IPM technology to new users. However, the major drawback of the one-on-one method is the high personnel and travel costs associated with it. Cooperative Extension uses train-the-trainer programmes to deliver IPM programmes through volunteers such as master gardeners. Training capable volunteers in the community to transfer IPM technology is a viable option to increase the cost effectiveness of the one-on-one method.

22.5.1.2 *Field Demonstrations*

Field demonstrations have been used since the very early days of IPM extension in the USA, Australia, and New Zealand. It is a very effective experiential education method. When Extension Specialists develop new IPM technology, they conduct field demonstrations for Extension Agents, crop consultants, and interested farmers. These field demonstrations are conducted in experiment station fields or collaborating farmers' fields. Conducting field demonstrations in a farmer's field is more appropriate to convince farmers that the demonstrated new IPM technology is feasible in their field conditions. If the field demonstration is conducted in an experimental station field, farmers may tend to think that it is somewhat difficult for them to meet the experimental conditions in their farm situation and this may discourage some farmers from adoption of the new technology. Researchers, Extension Specialists and Extension Agents plan and conduct field demonstrations for disseminating IPM technology in the US. Properly planned field demonstrations are effective means for convincing farmers about IPM technology because field demonstrations provide an opportunity for participants to go through the full experiential learning cycle (Figure 22.2). First the demonstration itself exposes the participants to the technology. Then, the discussion followed by demonstration facilitates participants to reflect on the practical aspects and advantages and disadvantages of applying the technology in their own situation. Handouts distributed at the field demonstration will help participants to further assess the technology and internalize the experience with technology. These handouts refer them to related websites for further information. This information is also helpful for participants when they apply IPM technology to manage pests. Field demonstration is an expensive delivery method, but it is one of the most effective group methods to convince farmers about IPM technology. The use of one field demonstration site to educate many in that area is the best strategy to achieve cost effectiveness of the field demonstrations.

22.5.1.3 *Training Workshops*

Training workshops have been used to educate farmers and others on IPM in the USA, Australia, and New Zealand since the beginning. Cooperative Extension uses IPM training workshops to educate farmers, Extension Agents, agricultural industry people, and volunteers. IPM is a knowledge-based technology. Training workshops are effective in changing participants' attitudes, building their knowledge and skills, and inspiring them to adopt new technology; therefore, training workshops can be considered to be an effective extension strategy for disseminating IPM technology. Training workshops provide an opportunity to use a variety of instructional resources and teaching techniques to facilitate learning and collaboratively address issues and concerns of participants. The use of various instructional tools and discussions enhances learning. This is the major advantage of training workshops. Some of the training workshops are targeted for educating trainers such as Extension Agents, volunteers, and industry personnel. These programmes are called train-the-trainer workshops. These are intensive educational workshops to ensure that participants will have adequate knowledge and skills about IPM. Some of the training programmes are certification programmes and require participants to take tests. Training programmes are extensively used to educate farmers and others how to apply computer simulated modelling programs for making decisions related to IPM. These computer simulation models are based on available research data and provide the best possible options for making management decisions. Participants' input is solicited for the evaluation of training workshops and assessing learning outcomes. Cooperative Extension pays special attention to evaluating training workshops for improvement of training and demonstrating accountability.

22.5.1.4 *Printed Materials*

A variety of printed materials are used in the USA, Australia, and New Zealand to disseminate IPM technology. The most commonly used printed materials include: IPM manuals, brochures, fact sheets, and newsletters. Researchers and Extension Specialists in the US have developed IPM manuals for various crops and pests. These manuals provide detailed information with colour images for identification of pests, diseases, and natural enemies of pests. IPM manuals, brochures, and fact sheets help farmers as well as Extension Agents and agricultural consultants to identify pests, assess pest damages, and make management decisions. IPM newsletters are used to inform farmers about new technology and give timely extension advice for making pest management decisions. The major advantage of printed extension materials is the ability of printed media to reach a large number of people with a minimum cost. Printed materials do not provide two-way communication for addressing any issues or concerns of farmers. This is the major disadvantage of using only printed materials in disseminating IPM technology. Printed material itself is not an effective medium to educate farmers on IPM. However, printed materials can be used to supplement other extension education methods such as demonstrations, one-on-one farm visits, and training workshops for achieving better results.

22.5.1.5 *Telephone and Mass Media*

Farmers in the USA, Australia, and New Zealand use the telephone to obtain advice from the extension and consultants. The Cooperative Extension Service uses the telephone and mass media such as radio for transferring IPM technology. Farmers and other users contact Extension Agents as well as Extension Specialists over the phone to find answers to their pest management issues and receive advice. Since the beginning, the telephone has been the most efficient communication medium for advising farmers and others in the US. The demand for extension advice over the phone increased with the widespread use of cell phones. The invention of the smartphone is further revalorizing the use of phones in transferring IPM technology to stakeholders.

Mass media such as radio broadcasting, farm and business magazines, and television are also utilized for the dissemination of IPM technology. Radio programmes are developed and broadcast with the participation of Extension Specialists and Extension Agents. Timely pest management topics useful for farmers are discussed during these radio broadcasts. Since farmers can listen to the radio while working in the field, they are receptive to IPM related information broadcast through farm radio programmes. Farmers have subscribed to various farm and business magazines. Sometimes, the extension uses these popular farm magazines to transfer technology to farmers. Television is a powerful medium to communicate about IPM technology to many users effectively. However, the use of television in transferring technology is very expensive. The advantage of mass media is the ability to reach a large number of people in a short period of time.

22.5.1.6 *Information Technology*

Extension services in the USA, Australia, and New Zealand use information technology such as the internet and CD ROMs for disseminating IPM technology. The internet has become one of the most cost effective and efficient media for transmitting information to many users. However, there are some farmers who still do not have access to the internet. This is the major limitation of this method in reaching farmers. Cooperative Extension uses the internet to deliver IPM educational programmes; communicate with farmers through emails; and transmit information about pests, natural enemies, and economic threshold levels. In addition to Cooperative Extension, IPM regional centres and eXtension provide IPM information online for users in the US eXtension is a USDA funded extension initiative to bring all the expertise in the US for better serving the needs of people using the internet as the collaborating and information disseminating platform. The eXtension site provides useful information about IPM for Extension Agents as well as farmers and others. With the widespread use of smartphones, the Cooperative Extension Service started to transmit IPM information using phone applications to help users have ready access to information.

22.5.2 Developing Countries

IPM techniques cannot be proposed as a blanket recommendation (Dilts and Hate, 1996). They cannot be invented in Iowa and sold in Argentina. IPM is based on a mix of local knowledge and a modern scientific approach (Dreves, 1996). This requires a location-specific and need-based approach. A wide variety of methods and media are used for delivering IPM technology depending upon the type of information and the stage of implementation (Whalon and Croft,

1982). Different methods used for dissemination of IPM technologies include use of mass media including electronic and print media, field days, extension visits, video movies, plant clinics, IPM clubs, picture songs, IPM websites, farmer field schools, etc. Mass media such as radio broadcasts and television programmes provide general knowledge of IPM practices to farmers. These are useful for creating awareness rather than directly assisting the farmer to use IPM technology (Lumber et al., 1985). Mass media can be very effective for raising awareness at the beginning of a project or programme to attract the attention of the majority of farmers (more specifically smallholders and poorer ones) who otherwise could never be informed about the opportunities offered by IPM (Bhuyan et al., 1995) whereas a more labour-intensive communication strategy based on a group approach (meetings, workshops, demonstrations, etc.), is normally advisable to convince at least the early adopters, who will be followed by the majority in due time. Mass media methods are less intense methods than face-to-face methods, namely farmer field schools. These are inexpensive to produce and transmit but do not provide a great depth of knowledge. These methods have the potential to reach a wide audience.

22.5.2.1 *Radio*

Radios are used to communicate what to do as quickly as possible to all interested farmers (Lagnaoui et al., 2004). They have been used for dissemination of IPM in Vietnam to teach people to avoid insecticide abuse in rice (Bentley et al., 2005).

22.5.2.2 *Picture Songs*

Picture songs is an innovative technique used for dissemination of IPM technologies in Bangladesh. It is a combination of singing, dancing and pictures on a scroll used to disseminate rice IPM technology (Bentley et al., 2005). One NGO in Bangladesh, Shushilan, used picture songs (songs and paintings on a scroll) to reach a large number of rice farmers and teach them about insect pests and natural enemies (Bentley, 2009; Bentley et al., 2005).

22.5.2.3 *Video Movies*

Extension agents show the videos of different IPM practices in communities, and then answer questions from the audience, which allows many people to be trained at once, in a relatively short time. Video movies are used in combination with farmer participatory research and community meetings. A video movie as an extension method is widely popular in Bangladesh (Van Mele et al., 2005).

22.5.2.4 *Mobile Plant Health Clinics*

This is a new extension method being implemented for disseminating IPM technologies in Nicaragua, Bolivia, Uganda and Bangladesh. It is useful for those farmers who cannot get advice about plant health problems as they live in distant areas and cannot bring plant samples. Most of the clinics are 'mobile'. The plant clinics provide a place for personalized consultations between farmers and scientists (Danielson et al., 2006).

22.5.2.5 *Written Material*

Written material including leaflets, folders, bulletins, newsletters, journals, magazines, factsheets and newspapers are used for dissemination of IPM technology throughout the world (Fitt et al., 2009; Padre et al., 2003; Peshin et al. 2009b). In China, newsletters were used to extend pest management recommendations. In India, leaflets and manuals were provided to participating farmers in the Insecticide Resistance Management (IRM)-based IPM programme in cotton in Punjab. This approach has been implemented in 10 cotton growing states since 2002 (Table 22.1).

22.5.2.6 *Reaching a Large Audience*

Street play, a novel method of creating awareness and interest about the judicious

TABLE 22.1 Extension Methods Employed in the IRM-Based IPM Programme in India

Activity
Formation of farmer groups in IRM villages
Meetings starting with sowing of cotton crop
Group meetings every week starting in July up to end of September
Training by master trainer for each of the four windows of the IRM
Strategy – for judicious use of insecticides
Visit to IRM labs set up at PAU research stations
Visit to the selected good IRM farmer's field
Lectures and discussions
Trainers as experts, not the facilitators of the learning process
Training of scouts (local farmers selected as scouts) before the start of the project
Scouts deployed in every village to provide feedback, estimation of pest build-up and organize meetings every week
Visit to farmers' fields for identification of insect-pests and other problems
Information centres established at IRM villages where exhibits and displays kept
Publication and distribution of IRM printed material to farmers
Information centres giving all relevant information on cotton-growing
Street plays organized for creating awareness about IPM

Source: Peshin et al. (2009b).

use of pesticides, has been used extensively in Punjab (Peshin, 2009). Plant health clinics are being used in Nicaragua, Bolivia, Uganda, Bangladesh and some other countries to reach a large audience. They were started in Bolivia in the 1990s (Bentley, 2009). In this, farmers from distant areas bring plant samples and obtain advice from experts (Danielsen et al. 2006). In India, agricultural universities, namely Punjab Agricultural University, Ludhiana and Haryana Agricultural University, Hissar, have permanent plant health clinics housed in the extension directorate to provide personalized consultancy. These clinics remain open for 5 days a week.

22.5.2.7 *Going Public*

Going public is a face-to-face method for a mass audience (Bentley et al., 2005). This is an extension method in which an extensionist goes to a market or another crowded place and delivers a short message and repeats it. It has been used in Bolivia, Bangladesh, Kenya and Uganda (Bentley et al., 2003).

22.5.2.8 *Participatory Learning and Action Research (PLAR)*

Participatory learning and action research (PLAR) developed in West Africa is also based on the FFS philosophy (Bentley, 2009). The PLAR uses weekly meetings with farmers and encourages them to experiment. Groups of farmers work together for several years, and the best findings are adopted by them (Defoer et al., 2004), and disseminated to a wider audience through videos made in the PLAR villages covering the farmer-experimenters (Bentley, 2009).

22.5.2.9 *Agent-Based IPM Diffusion Model (ABM)*

In agent-based models, the actions of each participant are updated at each time step, depending on the selection of IPM practice by the farmer. This model has a preliminary session with farmers as a first contact with them about the theoretical and practical aspects of IPM, then a presentation, and initialization followed by a role playing session and discussion. ABM is thus a powerful tool to advance the application of social psychology theory by stakeholders in rural communities (Smith and Conrey, 2007) and to change individual attitudes (Jacobson et al., 2006), all of which is needed for adoption of IPM.

22.5.2.10 *Farmer Field School (FFS)*

The farmer field school (FFS) is an intensive extension method. Indonesia was the first country to adopt this method in 1989 for implementation of a rice IPM programme (Table 22.2). FFS is a 14-week learning process that starts with the introduction of the IPM FFS programme to farmers and ends with post evaluation testing of farmers (Table 22.3). FFSs are season-long educational courses organized in the field for small (25–30) groups of farmers (Kenmore, 1996). In regular sessions, from planting until harvest, groups of farmers observe and analyse their agro-ecosystem. In FFS, farmers study insect ecology by observing the predatory behaviours of specimens collected from their fields and reared in insect zoos.

Additionally, they collect data that are then analysed in groups, the members of which are encouraged to take informed decisions on plant protection based on the results of the analysis. The process of building farmers' knowledge and confidence in IPM, compared to the conventional top-down delivery of technical recommendations, is more effective in supporting the uptake of ecologically informed farming practices (Bingen, 2004; Mancini et al., 2007). The FFS provide farmers with opportunities to experiment with IPM principles and find locally relevant pest management solutions. The training was first planned for rice farmers, but later on extended to cover crops, namely soybean, corn, potato, cabbage, chilli, shallot, coffee, tea, pepper, cacao, cotton and cashew (Untung, 2006). To date, FFS programmes have been initiated in 78 countries (Braun et al., 2006).

TABLE 22.2 Country-Wide Adoption of Farmer Field School Model in Asia

Country	Year
Indonesia	1989
Vietnam	1992
Philippines	1993
China	1993
India	1994
Bangladesh	1994
Sri Lanka	1995
Cambodia	1996
Laos	1997
Nepal	1998
Thailand	1998

22.6 LESSONS LEARNED FROM GLOBAL IPM EXTENSION EFFORTS

IPM is not a panacea for pest management. However, it is the best approach to manage pests with minimal disturbance to natural processes and the environment. Therefore, the global community is promoting IPM as the appropriate approach to manage pests. This chapter discussed the global IPM extension efforts which have taken place during the last four decades. Despite these efforts, the adoption of IPM technology by farmers and others for managing pests is not yet up to the level it should be at

TABLE 22.3 Weekly Activities of FFS

1ST WEEK
• Registration of farmers • Introduction of IPM/FFS programme to farmers • Field visit
2ND WEEK
• What is IPM? • Objective of IPM • Need and importance of IPM
3RD WEEK
• Ecology and agronomy of crops • Agro-ecosystem analysis • Major pests and disease and their natural enemies • Field visit
4TH WEEK
• Field visit • Identification of insects, pests and their damaging symptoms • Identification of weeds and their management
5TH WEEK
• Field visit • Identification of diseases and their management • Identification of weeds and their management • Identification of rodents and their management
6TH WEEK
• Field visit/pest monitoring • Identification of natural enemies • Pest and defender ratio
7TH WEEK
• Field visit/pest monitoring • ETLs determination • Visual observation on pest and defender
8TH WEEK
• Field visit/pest monitoring • Cultural control practices • AESA by visual/water pan/sweep net method
9TH WEEK
• Mechanical control practices • Installation of traps/bird perches, etc. • Preparation of insect zoo

(Continued)

TABLE 22.3 (Continued)

10TH WEEK
• Biological control • Identification of parasites and predators • Predation activities of natural enemies • Pest monitoring • Conservation of bio-control agents
11TH WEEK
• Chemical control • Ill effect of pesticides • Toxic effect of pesticides on bio-agents • Field visit
12TH WEEK
• Rodent management • Weed management • AESA/P.D.
13TH WEEK
• Comparative study of pests and defenders in IPM and non-IPM fields • Preparation and use of bio-pesticides, i.e. Neem Kernel extract, NPV, Trichoderma, Trichogramma, etc.
14TH WEEK
• Impact of IPM on farmers • Post evaluation testing of farmers • Action plan for following year

(Olson et al., 2003). For example, the World Bank (2005) reported a low level of IPM adoption in developing countries. As discussed earlier in this chapter, IPM is a complex technology. Until users fully comprehend it, they may not be able to adopt it. This situation highlights the need for continued global extension efforts to educate potential users and disseminate IPM technology for sustainable adoption. This section reviews global extension efforts for learning lessons to successfully educate farmers in IPM technology. There are very successful strategies as well as some failed strategies in delivering IPM extension programmes. By determining the successful extension strategies and their characteristics, global extension systems will be able to use those methods. If any failed IPM extension strategies are known, one will be able to avoid those in the future.

Not all IPM extension efforts contribute to positive results. Extension efforts aimed at disseminating IPM information without properly designed experiential education programmes have in some cases failed to convince target audiences to adopt IPM practices. Printed IPM extension materials, newsletters, radio programmes, and television programmes were effective in creating awareness among farmers and other users and providing additional information. However, these delivery methods were not effective in educating potential users and convincing them to adopt IPM practices. This may be because IPM technology is a complex mix of innovations and it requires someone to develop sound knowledge to apply IPM technology for managing pests. Interactive, hands-on experiential learning programmes are needed to educate target audiences and ensure the successful adoption of IPM practices.

22.6.1 Success of IPM Extension Efforts

The relative success of different extension programmes is ultimately judged with respect to knowledge gained and application of this knowledge for improvement of the existing system. This is also the case with IPM programmes and their measure of success (Dent, 1995). The majority of impact studies have concentrated on achievements in insecticide use and yield, under the assumption that good IPM practices lead to reduced spraying frequency and increased production (Baral et al., 2006; Birthal et al., 2000; Peshin and Kalra, 2002). An increase in farmers' knowledge about pest management has been documented by several authors (Praneetvatakul and Waibel, 2006; Peshin, 2009; van Duuren 2003; Reddy and Suryamani, 2005; SEARCA 1999; Sharma, 2011). In a study conducted by Galvan and Kenmore (1991) in Indonesia, IPM trained farmers were able to identify pest problems better and trusted more in their decision making ability. Pesticide use by IPM trained farmers decreased, resulting in a reduction in pesticide expenditure. By adopting IPM practices, the use of and dependence on pesticides has dramatically decreased in areas where training in IPM was imparted to rice farmers. Most studies of the impact of FFS report reduced expenses for inputs, increased yields, and a higher income for farmers (Baral et al., 2006; Birthal et al., 2000; Gajanana et al., 2006; Peshin and Kalra, 2002). However, most of these studies are of pilot programmes, and there is less information on the cost-effectiveness of large-scale IPM programmes (Kelly, 2005). Cotton IPM programmes implemented in India reported reduced insecticide use in the project areas (Sharma et al., 2004; Birthal, 2004; Peshin et al., 2009b), but farmers continue to face problems of being on the 'pesticide treadmill' and insecticide costs as a percentage of the total cost of cultivation continue to increase.

Shamsudin et al. (2010) evaluated the economic benefits of an IPM programme on cabbage crops in Malaysia and the results of the study indicated that there were economic benefits from adoption of the IPM.

22.6.2 Failures of IPM Extension Efforts

The Rice Knowledge Bank (2009) reports two main reasons for IPM programme failures, first, failure on the part of researchers in selecting non-important issues and secondly, inappropriate transfer of IPM technology to potential users. The review of IPM research and extension efforts in the US reveals that the overall US IPM programme has taken measures to prevent these failing factors. These include incorporation of stakeholder inputs for designing IPM technology and use of effective extension delivery approaches to reach potential users.

The major lesson one can learn from the success of IPM programmes in the US is the importance of building collaborative partnerships with all institutions and stakeholders to align IPM policies, research, education, and extension efforts to meet the pest management needs of all stakeholders. This alignment of all efforts creates a conducive environment for successful development and diffusion of IPM technology. Isolated efforts by research or extension without focusing on the needs of stakeholders are futile in disseminating IPM technology to potential users. The available literature (Wearing, 1988) highlights the need to have a close interaction among IPM researchers, extension personnel, the private sector, and farmers for successful development and diffusion of IPM technology. This confirms the concept of an agricultural innovation system given by Hall (2007) in which he has suggested that modern agriculture will sustain itself if there is synergy between research, extension, farmers, traders, private players and other supporting structures, namely banking, transport and marketing agencies.

An extension organization in developing countries faces various challenges in implementation of IPM programmes. Various authors have identified several challenges in implementation of IPM. One reason for the failure of IPM extension efforts in implementation is ineffective agricultural policies. Adoption of IPM practices is closely linked to the prevailing policy on pest management in a given country. The conceptualization and implementation of IPM will therefore, require identification and elimination of policy constraints (Brady, 1995). Sustainable agricultural practices such as IPM cannot thrive where there is a conflict of interests in the government with regard to alternative approaches to agriculture (Munyua, 2003). Clear policies make it possible to overcome barriers to diffusion of IPM practices such as subsidies and credit for use of chemicals. Cuba, Indonesia and the Philippines are examples of countries where IPM has thrived after government declarations to support IPM as the national pest management strategy (Pretty, 2002; Matteson, 1996). In Cuba, the 'alternative model' for agriculture provides an explicit policy direction for replacing pesticides with IPM practices in addition to promoting co-operation among farmers in the communities with regard to sustainable agriculture (Munyua, 2003). In Indonesia, the government's banning of several pesticides in rice and education on biodiversity through farmer field schools have increased the adoption of IPM (Pretty, 2002). Unless governments develop policies that firmly support sustainable agriculture, the response to implementation of IPM will be slow (Munyua, 2003). It is clear that, without policy reform in pest management and pesticide procurement, it will be difficult for developing countries to overcome the constraints that hinder the transition to and implementation of IPM practices (Munyua, 2003).

Secondly, the success of IPM programmes depends on the delivery system, which to a great extent determines the farmers' response (Dreves, 1996; Uhm, 2002). National agricultural extension systems in developing countries generally operate according to the 'technology transfer' (TOT) paradigm, under which agricultural research and development is carried out stepwise by a large, multipurpose hierarchy (Roling and van de Fliert, 1991). IPM requires a facilitation and application process shifting away from the technology-transfer model of adoption of innovations (van Huis and Meerman, 1997; Roling and Wagemakers, 1998; Swanson, 1997). Extension organizations face the challenge to adopt a paradigm shift in information delivery that will create an enabling environment for adoption of sustainable agricultural practices.

Thirdly, working in isolation is unlikely to generate the kind of response needed for adoption of IPM practices. Fragmentation of effort among stakeholders and the lack of institutional and interdisciplinary collaboration are therefore counterproductive to implementation of IPM (Edwards et al., 1991; Funderburk and Higley, 1994; Matteson, 1996; Roling and Wagemakers, 1998; Pretty, 1998; Peshin, 2013).

Fourthly, IPM presents an educational challenge for farmers, and research and extension personnel in developing countries (Edwards et al., 1991; Matteson, 1996). Knausenberger et al. (2001) singled out a lack of IPM training and knowledge as major constraints to the implementation of IPM in Africa within the subsistence and emerging agricultural systems. Other limitations are a lack of specific crop and pest management information as well as access to information on alternative pest control practices. In discussing pest management in Africa, Abate et al. (2000) observed that farmers lacked the biological and ecological information needed for exploratory approaches to pest management. Swanson (1997) noted that Extension Agents in the ranks of subject matter specialists lacked sufficient training and knowledge. When Extension Agents do not have information on IPM practices, they are likely to be sceptical about their role in IPM practices. The IPM extensionists are neither comfortable with the IPM principles nor the farmers' participatory extension methodology (Peshin et al., 2009b). Sharma (2011) reported that an IPM programme implemented in the vegetable crops in India had not resulted in achieving the stated goals because implementation was poor.

Moreover, the fiscal sustainability for large scale IPM extension systems is a cause of concern in developing countries (Feder et al., 1999) and in developed countries (Hanson and Just, 2001). In the developing countries, funding for IPM extension activities in the initial years was mainly from international organizations and donor countries (World Bank, Asian Development Bank, USAID, etc.) for implementing IPM on a pilot basis. Once the effort for scaling-up of IPM extension was initiated, the developing countries lacked both the fiscal means and large numbers of extension entomologists, plant pathologists, and weed scientists to replicate the success achieved on a pilot basis. For example, in India, only 5% of farmers have been covered under IPM-FFS since 1993. The Indonesian IPM programme, cited as a successful example of IPM extension in developing countries, received funding from the World Bank up to 1999 for scaling-up (MOA, 1999) but training quality decreased (Pincus, 2002). In the Philippines, to have one million farmers participate (20% of total farm households) it costs US $47.6 million, indicating that one FFS costs US $47.6 million (SEARCA, 1999). In Indonesia, problems such as funding and other obstacles have affected the quality of IPM-FFS since 1999. This has resulted in the farmers returning to their old methods of routinely spraying pesticides (Pincus, 2002). In China, FFSs have been found to be a very effective approach for achieving higher yields and better pest control in rice, but they are not always successful, and thus improvements in the methodology and organization are still needed (Feder et al., 2004).

22.7 FUTURE STRATEGIES FOR IPM EXTENSION

For an IPM programme to succeed, policy formulation must arise in a new way. Effective policy processes will have to bring together a range of actors and institutions for creative interaction and address multiple realities and unpredictability. What is required is the development of approaches that put participation, negotiation, and mediation at the centre of policy formulation so as to create a much wider common ownership in the programmes. This is a central challenge for sustainability of IPM. For better coordination and to reduce fragmentation of efforts that is observed in promoting IPM practices, it is better to adopt the concept of an agricultural innovation system given by Hall (2007) in which he has suggested that modern agriculture will sustain itself if there is synergy between research, extension, farmers, traders, private players and other supporting structures, namely banking, transport and marketing agencies.

Overcoming fiscal, training and educational constraints in IPM programmes is a major challenge in many developing countries. Higher-quality training is needed in response to the more complex problems there. The participatory non-formal education (NFE) approach is applicable to all agricultural extension subjects because IPM typifies the integrated crop management that is now required to increase agricultural productivity (Byerlee, 1987; Pimbert, 1991; Roling and van de Fliert, 1991). However, policy and institutional change will be necessary before this type of training can be implemented widely and efficiently (Barfield and Swisher, 1994; Roling, 1992). Cost-effective extension methods need to be devised and employed to reach the maximum number of farmers in developing countries. Different types of media need to be exploited in countries such as India to educate farmers about sound agricultural practices. There are certain lobbies in developing and developed countries who are promoters of hard technologies such as pesticides that conflict with supporters of agencies diffusing soft technologies such as IPM and they may resist this change. Thus, there is great need to counter such efforts, neutralizing them with data about the positive environmental and socio-economic impact of IPM techniques.

References

Abate, T., van Huis, A., Ampofo, J.K.O., 2000. Pest management strategies in traditional agriculture: An African perspective. Annu. Rev. Entomol. 45, 631–659.

Allen, W.A., Rajotte, E.G., 1990. The changing role of extension entomology in the IPM era. Annu. Rev. Entomol. 35, 379–397.

Bambawale, O.M., Patil, S.B., Tanwar, R.K., 2004. Cotton IPM at cross roads, Proceedings of International Symposium on Strategies for Sustainable Cotton Production: A Global Vision. UAS, Dharwad, Karnataka, India, pp. 33–36.

Baral, K., Roy, B.C., Rahim, K.M.B., Chatterjee, H., Mondal, P., Mondal, D., et al., 2006. Socio-economic parameters of pesticide use and assessment of impact of an IPM strategy for the control of eggplant fruit and shoot borer in West Bengal, India Technical Bulletin 37. Asian Vegetable Research and Development Centre (AVRDC), Taiwan, 36 pp.

Barfield, C.S., Swisher, M.E., 1994. Integrated pest management: ready for export? Historical context and internationalization of LPM. Food Rev. Int. 10, 215–267.

Bentley, J.W., 2009. At last, a way to involve smallholder farmers in IPM research. Outlooks on Pest Management, August.

Bentley, J.W., Boa, E., Van Mele, P., Almanza, J., Vasquez, D., Eguino, S., 2003. Going public: A new extension method. Int. J. Agric. Sustain. 2, 108–123.

Bentley, J.W., Nuruzzaman, M., Nawaz, Q.W., Haque, M.R., 2005. Picture songs. In: Van Mele, P., Salahuddin, A., Magor, N.P. (Eds.), Innovations in Rural Extension: Case Studies from Bangladesh CABI and IRRI, Wallingford, UK, pp. 115–123.

Bhuyan, R.K., Bordoloi, N., Singha, A.K., 1995. Awareness of farmers towards agricultural programmes, J. Agric. Sci. Soc. North East India 8, 2.

Bingen, J., 2004. Pesticides, politics and pest management towards a political ecology of cotton in Sub-Saharan Africa. In: Moseley, W.G., Logan, B.I. (Eds.), African Environment and Development: Rhetoric, Programs and Realities. Ash gate Publishing Ltd, Aldershot, UK, pp. 111–126.

Birthal, P.S., 2004. Integrated pest management in Indian agriculture: An overview. In: Birthal, P.S., Sharma, O.P (Eds.), Integrated Pest Management in Indian Agriculture. National Centre for Agricultural Economics and Policy Research (NCAP), New Delhi, India, pp. 1–10.

Birthal, P.S., Sharma, O.P., Kumar, S., 2000. Economics of integrated pest management: Evidences and issues. Indian J. Agric. Econ. 55, 644–659.

Boon, K.F., 2009. A critical history of change in agricultural extension and considerations for future policies and programs. Doctoral dissertation. School of Agriculture, Food, and Wine, Adelaide, Australia.

Brady, N.C., 1995. Sustainable agriculture: A research agenda. In: Virmani, S.M., Katyal, J.C., Eswaran, H., Abrol, I.P. (Eds.), Stressed Ecosystems and Sustainable Agriculture Science Publishers, Inc, Lebanon, NH, USA.

Braun, A., Jiggins, J., Roling, N., van den Berg, H., Snijders, P., 2006. A Global Survey and Review of Farmer Field School Experiences. Report Prepared for the International Livestock Research Institute (ILRI), Wageningen, The Netherlands.

Byerlee, D., 1987. Maintaining the momentum in post Green Revolution agriculture: a micro-level perspective from Asia. International Development Paper No. 10, Department of Agricultural Economics, Michigan State University, Lansing, 57 pp.

Cary, J.W., 1998. Issues in public and private technology transfer: The case of Australia and New Zealand. In: Wolf, S.A. (Ed.), Privatization of Information and Agricultural Industrialization. CRC Press, Boca Raton, FL, pp. 185–207.

Castillo, M., 1996. IPM: Institutional constraints and opportunities in the Philippines. In: Waibel, H., Zadoks, J.C., (Eds.), Institutional Constraints to IPM. Paper presented at the XIIIth International Plant Protection Congress (IPPC), The Hague, the Netherlands, 2–7 July 1995. Pesticide Policy Project Publication Series No. 3, March 1996. University of Hannover and GTZ, Hannover, Germany, pp. 55–61.

CIP-UPWARD, 2003. Farmer Field Schools: from IPM to platforms for learning and empowerment. International Potato Center – Users' Perspectives with Agricultural Research and Development, Los Banos, Laguna, Philippines, 87 pp.

Dalton, C., 2011. The death of agricultural research and extension in New Zealand: A personal view. <http://woolshed1.blogspot.com/2011/04/death-agricultural-research-in-new.html> (accessed 14.01.12).

Danielson, S., Boa, E., Bentley, J.W., 2006. Puesto Para Plantas in Nicaragua: A clinic where you can bring your sick plants. Global Plant Clinic. Available at <http://r4d.dfid.gov.uk/PDF/Outputs/Misc_Crop/Book-PPP-Eng.pdf>.

Defoer, T., Wopereis, M.C.S., Idinoba, P., Kadisha, T.K.L., Diack, S., Gaye, M., 2004. Manuel du Facilitateur: Curriculum d'Apprentissage Participatif et Recherche Action (APRA) pour la Gestion Intégrée de la Culture de Riz de Bas-Fonds (GIR) en Afrique Sub-Saharienne. WARDA, CTA, IFDC, CGRAI, Cotonou, Benin.

Dent, D., 1995. Integrated Pest Management. Chapman and Hall, London.

Dewey, J., 1963. Experience and Education. Collier Books, New York.

DGHRI, 2003. Integrated Pest Management (IPM): a new approach to control pest and disease. Annual Report, Department Agricultural Republic of Indonesia. Directorate General Horticulture Republic of Indonesia. Jakarta, 1–6.

Dilts, D., Hate S., 1996. Farmer Field School: Changing Paradigms and Scaling-up. AgREN Network Paper 59b, Overseas Development Institute, London.

Dreves, A.J., 1996. Village-level integrated pest management in developing countries. J. Agric. Entomol. 13 (3), 195–211.

Edwards, C.A., Thurston, H.D., Janke, K., 1991. Integrated pest management for sustainability in developing countries. In: Toward Sustainability: A Plan for Collaborative Research in Agriculture and Natural Resource Management. National Academy Press, Washington, DC, USA, pp.109–133.

FAO, 2007. Plant protection profiles from Asia-Pacific countries. Asia and Pacific Plant Protection Commission and Food and Agriculture Organization of the United Nations. <http://www.fao.org/docrep/010/ag123e/AG123E22.htm> (accessed 29.06.13).

FAO, 2012. Integrated Pest Management. <http://www.fao.org/agriculture/crops/core-themes/theme/pests/ipm/en/> (accessed 20.08.12).

Feder, G., Willett, A., Zipp, W., 1999. Agricultural Extension; Generic Challenges and the Ingredients for Solution. Policy Research Working Paper No. 2129. The World Bank, Washington, DC.

Feder, G., Murgai, R., Quizon, J., 2004. Sending farmers back to school: The impact of farmer field schools in Indonesia. Rev. Agric. Econ. 26, 45–62.

Fitt, G., Wilson, L., Kelly, D., Mensah, R., 2009. Advances with integrated pest management as a component of sustainable agriculture: The case of the Australian cotton industry. In: Peshin, R. Dhawan, A.K. (Eds.), Integrated Pest Management: Dissemination and Impact, 2. Springer Verlag.

Foster, J., Norton, G., Brough, E., 1995. The role of problem specification workshops in extension: An IPM example. J. Ext. 33 (4) <http://www.joe.org/joe/1995august/a1.php> (accessed 20.01.13).

Funderburk, E.F., Higley, L.G., 1994. Management of arthropod pests. In: Hatfield, J.L., Karlen, D.L. (Eds.), Sustainable Agriculture Systems. Lewis Publishers, Boca Raton, pp. 199–228.

Gajanana, T.M., Krishna, P.N., Anupama, H.L., Raghunatha, R., Prasanna Kumar, G.T., 2006. Integrated pest and disease management in tomato: An economic analysis. Agric. Econ. Res. Rev. 19, 269–280.

Galvan, C., Kenmore, P., 1991. IPM made easy. World Education Inc. Report No 29, 8–10.

Gapasin, D.P., 2007. Integrated Pest Management collaborative research support program South Asia (Bangladesh). Site Evaluation Report August, 11–15.

Georghiou, G.P., 1986. The magnitude of the resistance problem. In: Committee on Strategies for the Management of Pesticide Resistant Pest Populations, Pesticide Resistance: Strategies and Tactics for Management. National Academy Press, Washington, DC.

Government of South Australia, 2011. Structure and Development of Extension Services in SA. Department of Agriculture Primary Industries and Resources, Government of South Australia. <http://www.pir.sa.gov.au/aghistory/left_nav/department_of_agriculture_programs/education_and_extension/structure__and__development_of_extension_services_in_sa_dept_of_ag> (accessed 12.01.13).

Guo, F., 1997. The ecological risk analysis of effects of chemicals on environment and its reduction strategies, Proceedings of Integrated Pest Management Theory. Agric. Sci. Tech. Press, Beijing, 149 pp.

Guo, Y.Y., 1999. The measures and results in crop pest disaster reduction in China. In: China Association of Agricultural Sciences, Research Progress in Plant Protection and Plant Nutrition. China Agriculture Press, Beijing, pp. 6–9.

Hall, A.J., 2007. Challenges to Strengthening Agricultural Innovation Systems: Where Do We Go From Here? Working Paper United Nations University, MIERT, Maastricht, <http://www.merit.unu.edu>.

Hall, M.H., Morriss, S.D., Kuiper, D., 1999. Privatization of agricultural extension in New Zealand: Implications for the environment and sustainable agriculture. J. Sustain. Agric. 14 (1), 59–71.

Hanson, J., Just, R., 2001. The potential for transition to paid extension: some guiding economic principles. Paper presented in the AAEA session on compliments and substitutes for traditional extension services: Experiments and experiences. Allied Social Science Association Annual Meetings, January 5–7, New Orleans, LA.

Harris, M.K., 2001. IPM, what has it delivered? A Texas case history emphasizing cotton, sorghum, and pecan. Plant Dis. 85 (2), 112–121. <http://apsjournals.apsnet.org/doi/pdfplus/10.1094/PDIS.2001.85.2.112> (accessed 28.12.12).

Hercus, J.M., 1991. The commercialization of government agricultural extension services in New Zealand. In: Rivera, W.M., Gustafson, D.J. (Eds.), Agricultural Extension: Worldwide Institutional Evolution and Forces for Change. Elsevier, New York, NY, pp. 23–30.

Horne, P., Page, J., 2009. Integrated Pest Management dealing with potato tuber and all other pests in Australian potato crops. In: Kroschel, J., Lacey, L.A. (Eds.), Integrated Pest Management for the Potato Tuber Moth, Phthorimaea operculella (Zeller) – A Potato Pest of Global Importance. Tropical Agriculture 20, Advances in Crop Research 10. Margraf Publishers, Weikersheim, Germany, pp. 111–117.

Jacobson, S.K., McDuff, M.D., Monroe, M.C., 2006. Conservation Education and Outreach Techniques. Oxford University Press, Oxford, UK.

Johnson, F.A., Sprenkel, R.K., 2008. History of Row-crop and Vegetable IPM Extension Programs in Florida. University of Florida. <http://edis.ifas.ufl.edu/images/pdficon_small.gif> (accessed 20.12.12).

Jones, G.E., Garforth, C., 2005. The History and development of agricultural extension. In: Swanson, B.E. (Ed.), Agricultural Extension: A Reference Manual First Indian Edition, Food and Agricultural Organization, Chawla Offset Printers, New Delhi, India.

Journeaux, P., Stephens, P., 1997. The Development of Agricultural Advisory Services in New Zealand. MAF Policy Technical Paper 97/8, Johnson, R.W.M. (Ed.), <http://maxa.maf.govt.nz/mafnet/rural-nz/people-and-their-issues/education/advisory-services/adser.htm> (accessed 20.01.13).

Kelly, L., 2005. The Global Integrated Pest Management Facility: Addressing challenges of globalization: *An independent evaluation of the World Bank's approach to global programs*. Case study. World Bank, Washington, DC.

Kenmore, P., 1996. Integrated pest management in rice. In: Persley, G. (Ed.), Biotechnology and Integrated Pest Management. CAB International, Wallingford, pp. 76–97.

Kenmore, P.E., 1997. A perspective on IPM. LEISA Newslett. 13, 8–9.

Kenmore, P.E., 2006. Delivering on a promise outside the USA, but inside the same globalized world. Presentation at the Fifth National IPM Symposium, 4–6 April, St. Louis, Missouri.

Knausenberger, W.I., Schaefers, G.A., Cochrane, J., Gebrekidan, B. 2001. Africa IPM Link: An Initiative to Facilitate IPM Information Networking in Africa. Retrieved from <http://www.ag.vt.edu/ail/Proceedings/usaid.htm>.

Kogan, M., 1998. Integrated pest management: Historical perspectives and contemporary developments. Annu. Rev. Entomol. 43, 243–270.

Kolb, D.A., 1984. Experiential Learning: Experience as the Source of Learning and Development. Prentice Hall, Englewood Cliffs, NJ.

Lagnaoui, A., Santi, E., Santucci, F., 2004. Strategic communication for integrated pest management. Paper presented at the Annual Conference of the International Association for Impact Assessment in Vancouver, Canada. Agric. Hum. Values, no. 15.

Large, E.C., 1940. The Advance of the Fungi. Dover Publications Inc., New York, USA.

Larguier M., 1997. La protection intégré dans la politique française à l'égard de la protection des cultures. In: Fourth International Conference on Pests in Agriculture, ANPP, Paris.

Li, G.B., 1990. The concept and principle of IPM. In: Li, G.B, Zeng, S.M., Li, Z.Q. (Eds.), Integrated Management of Wheat Pests. China Agricultural Science Tech Press, Beijing, pp. 25–54.

Lumber, M.T., Whalon, M.E., Fear, F., 1985. Diffusion theory and IPM: An example from Michigan apple production. Bull. Entomol. Soc. Am. 31, 40–45.

Lyons, J.M., Flint, M.L., Goodell, P.B., O'Connor-Marer, P., Strand, J. F., (n.d.). A History of the California Statewide IPM program. University of California. <http://www.ipm.ucdavis.edu/IPMPROJECT/HISTORY/ucipm-history.pdf> (accessed 20.12.12).

Ma, S.J., 1976. On the integrated control of agricultural insect pests. Acta Entomol. Sin. 19, 129–141.

Mancini, F., Termorshuizen, A.J., Jiggens, J.L.S., Bruggen, A.H.C., 2007. Increasing the environmental and social sustainability of cotton farming through farmer education in Andhra Pradesh, India. Agric. Syst. 96, 16–25.

Marsh, S.P., Pannell, D.J., 2000. Agricultural extension policy in Australia: The good, the bad and the misguided. Aust. J. Agric. Resour. Econ. 44 (4), 605–627.

Matteson, P.C., 1996. Implementing IPM: Policy and institutional revolution. J. Agric. Entomol. 13 (3), 173–183.

Ministry of Agriculture (MOA), 1999. Laporan Akhir Proyek Pengendalian Hama Terpadu Pusat. Ministry of Agriculture, Jakarta, Indonesia.

Morrill Act, 1862. Morrill Act. 7 United States Congress <http://www.csrees.usda.gov/about/offices/legis/morrill.html> (accessed 12.12.13.).

Muangirwa, C.J., 2002. Pest management in tropical agriculture. In: Pimental, D. (Ed.), Encyclopaedia of Pest Management. Marcel Dekker Inc., New York, pp. 368–372.

Munyua, C.N., 2003. Challenges in Implementing Integrated Pest Management (IPM) Practices: Implications for Agricultural Extension. Proceedings of the Nineteenth Annual Conference, Raleigh, NC, USA.

Murray, M., 1999. A contrast of the Australian and California extension technology transfer process. J. Ext. 37 (2) <http://www.joe.org/joe/1999april/a1.php> (accessed 20.12.13).

NASS (National Agricultural Statistics Service.), 2001. Pest Management Practices: 2000 Summary. USDA, Washington, DC.

National Academy of Sciences, 1969. Principles of Plant and Animal Pest Control, Vol. 3: Insect-Pest Management and Control. National Academy of Sciences, Washington, DC. Network Paper 59b, Overseas Development Institute, London.

Newsom, L.D., 1980. The next rung up the integrated pest management ladder. Bull. Entomol. Soc. Am. 26, 369–374.

Nicholls, C.I., Altieri, M.A., 1997. Conventional agricultural development models and the persistence of the pesticide treadmill in Latin America. Int. J. Sustain. Dev. World Ecol. 4, 2.

Nolte, P., 2002. Historical epidemics: Irish potato famine. In: Pimental, D. (Ed.), Encylopedia of Pest Management. Marcel Dekker, Inc., New York, USA.

NSWEPA, 2013. Pesticide Act 1999. New South Wales. <http://www.epa.nsw.gov.au/pesticides/pesticidesact.htm> (accessed 29.06.13).

Olson, L., Zalom, F., Adkisson, P., 2003. Integrated pest management in the USA. In: Maredia, K.M., Dakouo, D., Mota-Sanchez, D. (Eds.), Integrated Pest Management in the Global Arena. CABI Publishing, Cambridge, MA.

Ooi, PAC, 2003. IPM in cotton: Lessons from IPM. In: Mew, T.W., Brar, D.S., Peng, S., Dewe, D., Hardy, B. (Eds.), Rice Sciences: Innovations and Impacts for Livelihood. IRRI, Los Banos, Philippines, pp. 919–926.

OPMP, 2013. Office of Pest Management policy. United States Department of Agriculture. <http://www.ars.usda.gov/Research/docs.htm?docid=12430> (accessed 29.06.13).

Padre, S., Sudarshana, P., Tripp, R., 2003. Reforming farm journalism: The experience of Adike Pathrike in India. Overseas Development Institute, London, Agricultural Research and Extension Network Paper No. 128.

Perkins, J.H., 1980. The quest for innovation in agricultural entomology. In: Pimental, D., Perkins, J.H. (Eds.), Pest Control Cultural and Environmental Aspects. AAAS Selected Symposium, West View Press, Colorado, pp. 23–80.

Peshin, R., 2009. Evaluation of Insecticide Resistance Management Programme: Theory and Practice. Daya Publishing House, New Delhi, India.

Peshin, R., 2013. Farmers' adoptability of integrated pest management of cotton revealed by a new methodology. Agron. Sustain. Dev. 33, 563–572.

Peshin, R., Kalra, R., 2002. Constraints in the adoption of IPM practices: A case study of Ludhiana district, Punjab. Indian J. Ecol. 29, 49–53.

Peshin, R., Kalra, R., Kaul, V.K., 1997. Impact of training under farmer field school on the adoption of IPM practices in rice crop. Proceedings of Third Agricultural Science Congress. Punjab Agricultural University, Ludhiana, India, 365 pp.

Peshin, R, Dhawan, AK, Vatta, K, Singh, K, 2007. Attributes and socioeconomic dynamics of adopting Bt-cotton. Econ. Polit. Wkly. 42 (52), 73–80.

Peshin, R., Bandral, R.S., Zhang, W., Wilson, L., Dhawan, A.K., 2009a. Integrated pest management: A global overview of history, programs and adoption In: Peshin, R., Dhawan, A.K. (Eds.), Integrated Pest Management: Innovation-Development Process, vol. 1. Springer, Dorchest, The Netherlands.

Peshin, R., Dhawan, A.K., Kranthi, K.R., Singh, K., 2009b. Evaluation of the benefits of an insecticide resistance management programme in Punjab in India. Int. J. Pest Manag. 55, 207–220.

Pincus, J., 2002. State simplification and institution building in bank-financed development project. In: Pincus, J., Winters, J. (Eds.), Reinventing the World Bank. Cornell University Press, Ithaca, New York, pp. 76–100.

Pimbert, M.P., 1991. Designing integrated pest management for sustainable and productive futures. Gatekeeper Series 29. International Institute for Environment and Development, London, 19 pp.

Praneetvatakul, S., Waibel, H., 2006. Farm level and environmental impacts of farmers field schools in Thailand. Working paper No. 7, University of Hannover, Germany.

Pretty, J.N., 1998. Supportive policies and practice for scaling up sustainable agriculture. In: Röling, N., Wagemakers, M.A.E. (Eds.), Facilitating Sustainable Agriculture. Cambridge University Press, UK, pp. 23–45.

Pretty, J., 2002. Agri-Culture: Reconnecting People Land and Nature. Earthscan Publications Ltd., London.

Rasmussen, W.D., 1989. Taking the University to the People: Seventy Five Years of Cooperative Extension. Iowa State University Press, Ames, Iowa.

Ratcliffe, S.T., Gray, M.E., 2004. Will the USDA IPM centers and national IPM roadmap increase IPM accountability? – Responses to the 2001 General Accounting Office report. Am. Entomol. 50 (1), 6–9.

Reddy S.V., Suryamani, M., 2005. Impact of farmer field school approach on acquisition of knowledge and skills by farmers about cotton pests and other crop management practices – Evidence from India. In: Ooi, P.A.C., Praneetvatakul, S., Waibel, H., Walter-Echols, G. (Eds.), The Impact of the FAO EU IPM Programme for Cotton in Asia. Pesticide Policy Project Publication Series n. 9, Hannover University, pp. 61–73.

Resel, E., Arnold, S., 2010. Electronic Integrated Pest Management Program: An educational resources for extension and agricultural producers. J. Ext. 48 (6) <http://www.joe.org/joe/2010december/iw2.php> (accessed 10.01.13).

Rice Knowledge Bank, 2009. Concept 1: Two main reasons for IPM project failure. <http://www.knowledgebank.irri.org/ipm/concept-1-two-main-reasons-for-ipm-project-failure.html> (accessed 15.12.12).

Rogers, E.M., 1983, 1995, 2003. Diffusion of Innovation. The Free Press, New York, USA.

Roling, N., 1992. Facilitating sustainable agriculture: turning policy models upside down. Paper presented at the workshop Beyond Farmer First: Rural People's Knowledge, Agricultural Research and Extension Practice, 27–29 October 1992, Brighton, UK.

Roling, N., Wagemakers, M.A.E., 1998. A new practice: Facilitating sustainable agriculture. In: Roling, N., Wagemakers, M.A.E. (Eds.), Facilitating Sustainable Agriculture. Cambridge University Press, Cambridge, UK, pp. 3–22.

Roling, N., van de Fliert, E., 1991. Beyond the green revolution: capturing efficiencies from integrated agriculture. Indonesia's IPM programme provides a generic case. Informal report, National Programme for Development and Training of Integrated Pest Management in Rice-based Cropping Systems, Jakarta, 72 pp.

Roling, N., van de Fliert, E., 1994. Transforming extension for sustainable agriculture: The case of Integrated Pest Management in rice in Indonesia. Agric. Hum. Values 11, 96–108.

Rustam, R., 2010. Effect of integrated pest management farmer field school (IPMFFS) on farmers' knowledge, farmers groups' ability, process of adoption and diffusion of IPM in Jember district. J. Agr. Ext. Rural Dev. 2, 29–35.

SEARCA, 1999. Integrated pest management training project, World Bank loan 3586-IND: Impact Evaluation Study. Report prepared for the SEAMEO Regional Center for Graduate Study and Research in Agriculture, Laguna, Philippines.

Shamsudin, M.N., Amir, H.M., Radam, A., 2010. Economic benefits of sustainable agricultural production: The case of integrated pest management in cabbage production. Environment Asia 3 (special issue), 168–174.

Sharma, O.P., Bambawale, R.C., Lavekar, R.C., Dhandapani, A., 2004. Integrated pest management in rainfed cotton. In: Birthal, P.S., Sharma, O.P. (Eds.), Integrated Pest Management in Indian Agriculture, Proceedings 11. National Centre for Agricultural Economics and Policy Research and National Centre for Integrated Pest Management, New Delhi, India, pp. 119–128.

Sharma, R., 2011. Evaluation of the Vegetable Integrated Pest Management-Farmer Field School Programme in Jammu Region. Ph.D. dissertation, Sher-E-Kashmir University of Agricultural Sciences and Technology of Jammu, Jammu, India.

Singh, K., Peshin, R., Saini, S.K., 2010. Evaluation of the Agricultural Vocational Training Programmes conducted by the Krishi Vigyan Kendra (Farm Science Centres) in Indian Punjab. J. Agric. Rural Dev. Trop. Subtrop. 111 (2), 65–77.

Sirisingh, S., 2000. Integrated pest management (IPM) and green farming in rural poverty alleviation in Thailand. In: Proceedings of the Regional Workshop on Integrated Pest Management and Green Farming in Rural Poverty Alleviation, Suwon, Republic of Korea, 11–14 October 2000, pp. 98–109.

Smith, E.R., Conrey, F.R., 2007. Agent-based modeling: A new approach for theory building in social psychology. Pers. Soc. Psychol. Rev. 11 (1), 87–104.

Smith, R.F., 1962. Principles of integrated pest control. Proc. N. Cent. Br. Entomol. Soc. Am. 17.

Smith, R.F., 1972. The impact of the green revolution on plant protection in tropical and subtropical areas. Bull. Entomol. Soc. Am. 18 (1), 7–14.

Smith, R.F., van den Bosch, R., 1967. Integrated control. In: Kilgore, W.W., Doutt, R.L. (Eds.), Pest Control: Biological, Physical, and Selected Chemical Methods. Academic Press, New York.

Stantiall, J., Paine, M., 2000. Agricultural Extension in New Zealand: Implications for Australia. Australia Pacific Extension Network. <http://www.regional.org.au/au/apen/2000/4/stantiall.htm> (accessed 14.01.13).

Stern, V.M., Smith, R.F., van den Bosch, R., Hagen, K.S., 1959. The integrated control concept. Hilgardia 29, 81–101.

Suckling, D.M., 1996. The status of insecticide and miticide resistance in New Zealand. In: Bourdôt, W.G., Suckling, D.M. (Eds.), Pesticide Resistance: Prevention and Management. New Zealand Plant Protection Society, Christchurch, pp. 49–58.

Suckling, D.M., McKenna, C., Walker, J.T.S., 2003. Integrated Pest Management in New Zealand horticulture. In: Maredia, K.M., Dakouo, D., Mota-Sanchez, D. (Eds.), Integrated Pest Management in the Global Arena. CABI Publishing, Cambridge, MA, pp. 385–396.

Swanson, B.E., 1997. The changing role of extension in technology transfer. J. Int. Agric. Ext. Educ. 4 (2), 89–92.

Swanson, B.E., Farner, B.J., Bahal, R., 1990. International Directory of Agricultural Extension Organizations. Food and Agriculture Organization of the United Nations, Rome.

Task Force Report, 2003. Integrated Pest Management: Current and Future Strategies. Council for Agricultural Science and Technology, Ames, IA.

USDA, 2009. USDA Regional Integrated Pest Management Centers. <http://www.csrees.usda.gov/nea/pest/in_focus/ipm_if_regional.html> (accessed 12.12.12).

Uhm, K., 2002. Integrated Pest Management. Retrieved from <http://www.agnet.org/libray/abstract/eb470.html>.

Untung, K., 2006. Introduction to integrated pest management. In: Indonesian Penganatar Pengendalian Hama Terpadu, 2nd edn. Gadjah Mada University Press, Yogyakarta.

van Duuren, B. 2003. Report of Consultancy on the Assessment of the Impact of the IPM Program at the Field Level. Integrated pest management farmer training project Cambodia. Royal Government of Denmark, Ministry of Foreign Affairs, Danida and Royal Government of Cambodia, Ministry of Agriculture, Forestry and Fisheries, National IPM Program DAALI, Phnom Penh.

Van Huis, A., Meerman, F., 1997. Can we make IPM work for resource-poor farmers in Sub-Saharan Africa? Int. J. Pest Manag. 43 (4), 313–320.

Van Mele, P., Zakaria, A.K.M., Bentley, J.W., 2005. Watch and learn: Video education for appropriate technology. In: Van Mele, P., Salahuddin, A., Magor, N.P. (Eds.), Innovations in Rural Extension: Case Studies from Bangladesh. CABI and IRRI, Wallingford, UK, pp. 77–88.

Wang Y., 2000. Integrated pest management (IPM) and green farming in rural poverty alleviation in China. In: Proceedings of the Regional Workshop on Integrated Pest Management and Green Farming in Rural Poverty Alleviation, Suwon, Republic of Korea, 11–14 October 2000, pp. 32–40.

Wearing, C.H., 1988. Evaluating the IPM implementation process. Annu. Rev. Entomol. 33, 17–38.

Wearing, C.H., 1993. IPM and the consumer: The challenge and opportunity of the 1990s. In: Cory, S.A., Dall, D.J., Milne, W.M. (Eds.), Pest Control and Sustainable Agriculture. CSIRO, Australia, pp. 21–27.

Whalon, M.E., Croft, B.A., 1982. Apple IPM implementation in North America. Annu. Rev. Entomol. 29, 453–470.

Williams, D.B., 1968. Agricultural Extension: Farm Extension Services in Australia, Britain and the United States of America. Melbourne University Press, Carlton, Victoria.

Williams, D.G., Il'ichev, A.L., 2003. Integrated pest management in Australia. In: Maredia, K.M., Dakouo, D., Mota-Sanchez, D. (Eds.), Integrated Pest Management in the Global Arena. CABI Publishing, Cambridge, MA, pp. 371–384.

Williamson S., 2003. Pesticide provision in liberalized Africa: out of control? AgREN Network.

World Bank, 2005. Sustainable Pest Management: Achievements and Challenges. Report 32714-GBL, Washington, DC <http://www-wds.worldbank.org/external/default/WDSContentServer/WDSP/IB/2005/08/01/000090341_20050801095500/Rendered/PDF/327140GLB.pdf> (accessed 28.12.12).

Zhen-Ying, W., Kang-Lai, H., Jian-Zhou, Z., De-Rong, Z., 2003. Integrated pest management in China. In: Maredia, K.M., Dakoko, D., Mota Sanchez, D. (Eds.), Integrated Pest Management in the Global Arena. CABI Publishing, Cambridge, UK.

CHAPTER

23

Biological Control of Insect Pests in Crops

David Orr and Sriyanka Lahiri
North Carolina State University, NC, USA

23.1 INTRODUCTION

Natural enemies of insects, such as predators and parasitoids, have been recognized and employed for crop pest management for centuries. It was not until early in the 20th century that the term 'biological control' was first used in a pest management context (Smith, 1919). For pest management purposes, the practice of biological control has been defined uniquely for each of the disciplines of entomology and plant pathology. In plant pathology, biological control has been defined as 'the purposeful utilization of introduced or resident living organisms… to suppress the activities and populations of one or more plant pathogens' although the definition has been broadened to include weed pests (Pal and Gardener, 2006). In entomology, biological control is usually distinguished from other biologically based pest-management tactics, especially those involving microbe-derived toxins, behaviour-modifying chemicals, botanically derived chemicals, or plant resistance (Ridgway and Inscoe, 1998). Biological control using insect natural enemies has had many successes in managing weed pests (Julien and Griffiths, 1999), but this review will focus on insect pests. There is a further distinction in entomology, between natural control and biological control. The regulatory actions of beneficial insects on their host or prey populations without human involvement is referred to as natural control and happens regardless of whether humans are aware of it or not. Natural control has been defined as 'the maintenance of a more or less fluctuating population density of an organism within certain definable upper and lower limits over a period of time by the actions of the abiotic and/or biotic environmental factors' (DeBach, 1964). Beneficial insects are only one of many factors, such as weather, food and competition that act to regulate insect populations (Bellows and Fisher, 1999). Natural control should be viewed as an ecological service with significant economic value. For example, in the United States each year, the value of natural control of only native insect pests provided by predatory and parasitic insects was estimated by Losey and Vaughan (2006) at $4.49 billion. This is a strong incentive to enhance natural control by manipulating populations of beneficial insects for biological control of insect pests.

D. P. Abrol (Ed): *Integrated Pest Management*.
DOI: http://dx.doi.org/10.1016/B978-0-12-398529-3.00028-2

Biological control can still be considered a population-level process, involving the use of natural enemy populations to suppress target pest populations (Bellows and Fisher, 1999). There has been debate about expanding the scope and definition of biological control due to technological developments in the methods available for pest management as well as disciplinary differences in terminology (e.g. Eilenberg et al., 2001; Nordlund, 1996). For this review, the definition offered by DeBach (1964) will be used as the 'study, importation, augmentation, and conservation of beneficial organisms (to regulate) population densities of other organisms'.

23.2 APPROACHES TO BIOLOGICAL CONTROL

The three widely accepted general approaches to insect biological control are: importation, augmentation, and conservation. Importation biological control involves introduction of natural enemies into new geographic areas where pests have escaped them through accidental or deliberate introduction. This approach may also be referred to as classical biological control, reflecting its predominance and attention in the historical development of biological control (DeBach, 1964; Bellows and Fisher, 1999). Importation biological control is usually conducted at an institutional level, rather than by growers.

A widely accepted definition of conservation biological control is the 'modification of the environment or existing practices to protect and enhance specific natural enemies or other organisms to reduce the effect of pests' (Eilenberg et al., 2001). Implementation of conservation biocontrol is most commonly achieved through modification of existing pesticide application practices (Ruberson et al., 1998), and so can be thought of as occurring at the farm level. Conservation biological control practices involving environmental modification in agro-ecosystems have been receiving renewed research attention recently and are examined on either a farm or landscape level (e.g. see papers preceded by Jonsson et al., 2008).

Augmentation biological control is usually conducted at a commercial or institutional level, as well as the farm level. It refers to mass production and planned release of either native or non-native biological control agents into crops for management of either native or non-native pests (DeBach, 1964; Bellows and Fisher, 1999). Releases are conducted either inundatively, in which large numbers of enemies are released with the expectation of immediate pest suppression without suppression by the offspring of released individuals, or inoculatively, where the intent is to provide pest suppression over multiple pest generations within a crop cycle or season (van Lenteren, 2003a). The resurgence in interest in urban and residential food gardening and organic production, combined with widespread availability of information and products at retail outlets and on the internet has likely led to greater public awareness of augmentation than other forms of biological control (Cranshaw et al., 1996; Schupp and Sharp, 2012; Warner and Getz, 2008).

23.3 HISTORICAL DEVELOPMENT OF BIOLOGICAL CONTROL IN CROPS

The use of beneficial insects for insect pest management in crops has a long history. This section provides a brief historical description of the development of biological control as a pest management tool in crops.

Ants were the first organisms employed in biological control of pests in agricultural areas, probably because they are easily observed and widespread. A passage in the Talmud suggests the use of ant colonies against one another for control purposes before 200 AD, although there do not appear to be any additional written records of the extent or duration of this practice

(Needham, 1986). Predatory ants appear to have been the first documented biological control agents used in crops, specifically citrus in China and dates in Yemen. Chinese citrus growers reportedly augmented populations of native Asian weaver ants (*Oecophylla smaragdina* F.) by placing their paper nests into trees as early as 300 AD to protect crops from foliage-feeding insects (van Lenteren, 2005). Although these ants were reportedly sold as early as 304 AD, these sales were first observed and reported by western scientists in 1915 and the practice continued until relatively recently (Huang and Yang, 1987). Weaver ants are still recognized for their value in suppressing populations of heteropteran, lepidopteran and coleopteran pests in crops such as cashew, citrus, cocao, coconut, coffee, eucalyptus, litchi, mango and oil palm (van Mele and Cuc, 2000; Way and Khoo, 1992). The use of these ants in current pest management programmes appears to be limited in part due to their aggressive behaviour (Way and Khoo, 1992), and potential interference with pollination (Tsuji et al., 2004). It is also considered an antiquated practice and is more likely to be employed by older farmers (van Mele and Cuc, 2000).

The weaver ant example represents not only the first, but also the longest use of a specific augmentation biological control practice for crop pest management. The concept of augmentation biological control in crops may also have spread long ago as a result of international trade. The practice of bringing colonies of predacious ants (described ambiguously as *Formica animosa* Forskål, 1775) from nearby mountains into date plantations to control herbivorous insects was first observed by a western scientist in 1755, but may have had a much longer history due to trade interactions with China (Needham, 1986).

The *O. smaragdina* example also involved the first use of a conservation biological control practice in a cropping system. From at least 1600 AD onwards, movement of these ants between citrus trees was aided by installing bamboo rods as runways or bridges between trees (DeBach, 1974; Huang and Yang, 1987). Recent conservation biological control practices directed towards *O. smaragdina* involved the use of pesticides to reduce populations of other ant competitors, or interplanting shrubs and ground cover in coconut plantations (van Mele and Cuc, 2000).

Efforts in biological control of crop insect pests in Western countries appear to have begun much later than in Asia with a written proposal by Carl Linnaeus in 1772 suggesting that 'predatory insects should be caught and used for disinfecting crop plants' (Hörstadius, 1974). A variety of written reports in the early 1800s on the value of entomophagous insects in agriculture signalled a more widespread recognition of natural control factors (DeBach, 1974). Erasmus Darwin specifically advocated the use of syrphid flies and coccinellid beetles for aphid management in greenhouses in the early 1800s (Riley, 1931). During the 1800s, the practice of collecting and selling ladybugs for release in European hops, a practice that may have been conducted for centuries, became popular and widespread (Doutt, 1964). Populations of soil-dwelling predators (carabid and staphylinid beetles) were experimentally manipulated in cropping systems by Boisgiraud in 1840 and Antonio Villa in 1844 (Trotter, 1908).

Although the predatory behaviour of insects such as ants was recognized long ago and utilized for pest management, recognition and utilization of the less obvious and more complex phenomenon of parasitism did not occur until much later. The first written reports describing parasitism appeared in the 3rd century and seemed to describe a tachinid fly attacking silkworm larvae in China (Cai et al., 2005). However, the correct interpretation of the life history of this fly (probably *Exorista sorbillans*, the Uzi fly) was not made until 1096, and the first scientific reports of the flies did not appear until 1925 (Cai et al., 2005). It was not until 1669 that hymenopteran parasitism (by braconid wasps) was correctly interpreted in Europe (van Lenteren and Godfray, 2005), and not until 1704 in China (Cai et al., 2005). The much longer time it took for hymenopteran parasitoid life histories to be understood may have been due to their greater complexity, but also lack of economic incentive as there was in the case of silkworms.

Ideas for utilizing parasitoids in crop pest management developed during the 19th century in Europe beginning with a proposal by Hartig in 1827 to collect and store parasitized caterpillars in order to harvest adult wasps for later release to control cabbage butterflies (Sweetman, 1958). This concept was put into practice in 1880 by DeCaux in France, who collected apple buds infested with *Anthonomus pomorum* L. larvae from 800 trees, and held them in gauze covered boxes allowing emerging parasitoids to escape (Sweetman, 1958). Apparently, this successfully controlled weevils for over 10 years.

The concept of deliberately moving natural enemies from one location to another was apparently broadened in scope by C.V. Riley, who distributed parasitoids of the weevil *Conotrachelus nenuphar* (Herbst) around the state of Missouri in 1870 (Doutt, 1964). The first case of intercontinental movement of an arthropod natural enemy appears to have involved the predatory mite *Tyroglyphus phylloxerae*. Populations of the predator were established in France following shipments from the United States in 1873, but it failed to adequately control its target pest, the grape phylloxera, *Daktulosphaira vitifoliae* (Fitch) (Doutt, 1964; Fleschner, 1960). The first intercontinental movement of a parasitoid involved *Cotesia (= Apanteles) glomerata* (L.) that became established in the US following importation from the UK by the US Department of Agriculture in 1883 (Doutt, 1964). This project as well as a variety of other international movements of natural enemies for pest control in the late 1800s did not achieve complete economic control of their intended targets (Fleschner, 1960).

It is widely acknowledged that the first case of complete and sustained economic control of an insect pest by another insect was with the cottony cushion scale, *Icerya purchasi* Maskell, in California during the late 1800s (DeBach, 1974; van den Bosch et al., 1982; Doutt, 1964; Fleschner, 1960). The scale was apparently introduced into California from Australia in 1869 and within 17 years, was a major threat to the viability of the southern California citrus industry (DeBach, 1974). Within 2 years of the first release in 1887 of the vedalia beetle, *Rodolia cardinalis*, *I. purchasi* populations throughout California were under complete control. Along with the beetle, a parasitic fly *Cryptochaetum iceryae* was also established and became the major factor controlling scale populations in coastal areas of the state. This classic example of importation biological control is highlighted in many books (e.g. DeBach, 1964, 1974; Van Driesche and Bellows, 1996), and set the stage for future biological control programmes. Excitement from the success of the *I. purchasi* project kept early biological control efforts focused on importation (DeBach, 1964). Although a number of international shipments of natural enemies for biological control occurred in the late 1800s, none achieved the complete economic control seen with the *I. purchasi* example (Fleschner, 1960).

It was not until the early 1920s that the first sustained, large-scale successful augmentation project began targeting citrophilus mealybug, *Pseudococcus calceolariae*, a citrus pest in southern California with releases of the lady beetle *Cryptolaemus montrouzieri* (Luck and Forster, 2003). This beetle is still reared and sold for pest management (van Lenteren, 2003b). The first successful inoculative natural enemy releases utilized the parasitoid *Encarsia formosa* for management of greenhouse whitefly, *Trialeurodes vaporariorum*, in the UK, which eventually led to development of the commercial augmentation industry in Europe in the 1960s (Pilkington et al., 2010).

The concept of conserving natural enemies by modifying pesticide use practices was developed through the 1950s and 1960s (DeBach, 1964; Newsom and Brazzel, 1968; Stern et al., 1959). Early conservation research projects also evaluated artificial nesting structures, supplemental feeding, provision of alternative hosts, control of antagonistic ants, and modification of agricultural practices (van den Bosch and Telford, 1964). However, the only practical approach to conservation biological control remained pesticide use modification (Ehler, 1998).

23.4 IMPORTATION BIOLOGICAL CONTROL

Hall and Ehler (1979) noted that of 2300 predators and parasitoids introduced into 600 unique situations worldwide from 1890 to 1960, approximately 34% became established. From the same dataset, Hall et al. (1980) noted that releases led to complete pest suppression in 16% of cases, and an additional 42% provided partial pest suppression.

Greathead and Greathead (1992) examined a later dataset of 4769 insect records from the BIOCAT database (see Cock (2010) for more information) and noted that organisms from 30% of releases became established, and 11% exerted substantial control over the target pest. Of the importation projects that demonstrate some degree of suppression over the intended target pest, the percentage that represents complete success has apparently increased since the 1930s (Hokkanen, 1985). However, the overall establishment and success rates do not appear to have changed, with the percentage of agents that establish lying between 20% and 55%, and the percentage of complete successes between 5% and 15% (Gurr and Wratten, 1999; Hall and Ehler, 1979; Hall et al., 1980). These data indicate that the rates of establishment and success have not changed markedly over the last 100 years (Gurr and Wratten, 1999; Hall et al., 1980).

Various methods have been proposed for increasing the success rates of importation biological control. These include numerical ranking of target organisms (e.g. Barbosa and Segarra-Carmona, 1993), or identifying either qualitative or quantitative characteristics of natural enemies that are desirable (e.g. Coppell and Mertens, 1976; Hoelmer and Kirk, 2005). Consideration of the methods in which enemies are deployed has centred around considerations of the type of target pest, the target habitat, introducing parasitoids versus predators, specialists versus generalists, and single versus multiple species introductions (DeBach, 1964; van Driesche and Bellows, 1996).

Several reports have examined empirical evidence to determine quantitatively if any trends or traits exist with beneficial organisms that can be used to more accurately predict success when employed in importation biological control. A meta-analysis by Stiling and Cornelissen (2005) of data gleaned from biological control publications over 10 years from two journals indicated that target pests were most likely to be lepidopteran pests, and that released natural enemies were most likely to be parasitoids. Biological control programmes were more successful at reducing pest abundance when predators were employed, when two or more natural enemies were introduced, and when generalists rather than specialists were utilized. Based on a quantitative analysis of 87 well documented biological control agents in the United States, Kimberling (2004) presented [nonhierarchical] dichotomous tree diagrams that visually describe the life history characteristics that most accurately predict success in controlling target pests. These suggest that monophagous endoparasitoids with female biased sex ratios and more than one generation per year in relation to hosts have the greatest likelihood of success. Kimberling's analysis also suggests that the likelihood of success is diminished if the target is a lepidopteran, and/or is located in a forest system, and/or if native natural enemies are present.

Importation biological control has been described as 'engineering an invasion' (Grandgirard et al., 2008). The equating of biological control agents with invasive species has in some cases led to more restrictive viewpoints on the use of importation biological control

(Sheppard and Raghu, 2005). This combined with language in the Nagoya Protocol document relating to Access and Benefit Sharing (http://www.cbd.int/abs/text/default.shtml) of genetic resources under the UN Convention on Biological Diversity may present additional administrative challenges and barriers to importation biological control (Cock et al., 2010). Given these potential restrictions, the onus is on biological control researchers to ensure not only the safety, but also efficacy and economic value, of releasing exotic natural enemies for pest management.

Cost-benefit analyses are a systematic approach to estimating net gain, and are the most common approach to assessing the economic benefits of importation biological control (Headley, 1985; Tisdell, 1990). Economic analyses of importation biological control projects are uncommon, and focus on projects that have been very successful (Dean et al., 1979; Ervin et al., 1983; Jetter et al., 1997; Kipkoech et al., 2006; Tisdell, 1990; Voegele, 1989; Zeddies et al., 2001). Some of the highest benefit-to-cost ratios of any approach to pest management have been noted for these projects, resulting in savings in the billions of dollars (Tisdell, 1990). An estimated cost-benefit ratio for biological control of the cassava mealybug *Phenacoccus manihoti* in sub-Saharan Africa ranged from 200:1 to 740:1, depending on the market price used for cassava (Zeddies et al., 2001). The cost-benefit ratio estimated for the ash whitefly biocontrol programme in California was between 270:1 and 344:1 (Jetter et al., 1997). Marsden et al. (1980) estimated the expected cost-benefit ratio for three importation biocontrol projects from 1960 to 2000 by CSIRO in Australia as 9.4:1, which compared favourably to the average 2.5:1 ratio noted for all other pest management programmes conducted during the same time period. Although it is tempting to draw generalities from these numbers, it may be of value to keep in mind the low success rate of importation releases. It would be interesting to know the average cost-benefit for importation biocontrol projects that target specific pests, including the successes and failures, when compared with other pest management approaches. The self-sustaining nature of successful importation biocontrol and the immense potential benefits continue to make this pest management approach desirable.

Importation biological control has historically targeted exotic, invasive pest species utilizing exotic natural enemies (DeBach, 1964). While this approach has great potential for pest management in agricultural and conservation applications (Hoddle, 2004), it also has the potential to cause ecological or economic harm (Louda and Stiling, 2004). Concerns include the potential for predation or parasitism of non-target species, competition with native species, community and ecosystem effects, and unexpected effects such as loss of species dependent on the target pest species (Simberloff and Stiling, 1996). These concerns have been thoroughly reviewed (e.g. Bigler et al., 2006; Follett and Duan, 2000; van Lenteren et al. 2006) and their significance and practical impacts debated (Frank, 1998; Simberloff and Stiling, 1996, 1998). Conclusions about the safety and value of importation biocontrol vary depending on individual perspective.

A long-standing example of non-target impacts involves the parasitoid *Cotesia glomerata*, the first intercontinentally transported biological control agent. The parasitoid did not completely control its intended target, the imported cabbageworm *Pieris rapae*, and had a significant impact on a native butterfly *Pieris napioleracea* Harris in the northeastern United States, extirpating it from parts of its native range (Van Driesche, 2008). This parasitoid has continued to have more recent non-target impacts on native butterflies in other areas such as the Canary Islands (e.g. Lozan et al., 2008). A Chinese strain of *Cotesia rubecula* was successfully introduced into the northeast US in 1988 in an effort to improve parasitism and biological control of *P. rapae* (Van Driesche and Nunn, 2002). As a result of intense competition for early instars of *P. rapae*,

C. rubecula displaced populations of *C. glomerata* (Van Driesche, 2008). In an interesting twist, the non-target impacts of *C. glomerata* on the native *P. napi* were actually ameliorated by this displacement since *C. rubecula* has little effect on *P. napi* populations (van Driesche, 2008).

There is only one example of an intentionally imported natural enemy that became a pest. The multicoloured Asian lady beetle, *Harmonia axyridis* was intentionally introduced for biological control in North America several times from the early 1900s until the 1980s (Koch and Galvan, 2008). It provided management of the target pests as well as some others (Koch, 2003). However, as populations expanded, *H. axyridis* has become not only a threat to native biodiversity and possibly ecological services through intra-guild and inter-guild predation, but also a noxious household pest, and minor agricultural pest (Roy and Wajnberg, 2008; Koch and Galvan, 2008).

An example that demonstrates unusual consideration of non-target impacts involved testing vertebrates that may be affected by alkaloids produced by lady beetles. Lincango et al. (2011) demonstrated that native finches in the Galapagos would not be impacted by ingestion of *R. cardinalis* before release of this lady beetle for biological control of the scale insect *I. purchasi* in the islands.

Biological control researchers and practitioners are aware that non-target impacts should be an important consideration in decisions with regard to importation biological control although in practice, the risk potential is not nearly as carefully evaluated as are the benefits (Sheppard and Raghu, 2005). Lynch and Thomas (2000) compiled a database that showed there were relatively few cases where data on non-target effects were collected for biocontrol programmes, although the cases where non-target effects did occur were mostly from very early importation efforts and were mostly relatively minor.

Less obvious non-target impacts resulting from competition, intra-guild predation or interactions with human activities can compromise the integrity and/or functioning of ecosystems and should be included in considerations of impacts before natural enemy releases (Parry, 2009). Of course, when considering the significance of non-target effects of introducing biological control agents, it is important to compare them to ecological and economic effects that would occur in their absence (Cock et al., 2010).

23.5 AUGMENTATION BIOLOGICAL CONTROL

Augmentation biocontrol is estimated to be used on over 17 million hectares worldwide, almost 60% of which is in Russia, and only 1% in Europe and North America combined (van Lenteren, 2000a). Insectary facilities that rear the natural enemies for augmentation are either non-commercial centralized production units that are government or grower-industry owned, or commercial for-profit enterprises (Cock et al., 2010). Non-commercial augmentation involving state, coop and farmer operated insectaries focus their rearing efforts on egg parasitoid wasps in the genus *Trichogramma* primarily for inundative release against lepidopteran pests of agriculture and forestry (van Lenteren, 2003a; van Lenteren and Bueno, 2003; Smith, 1996). Approximately 20 species of *Trichogramma* are reared and regularly released for pest management in at least 22 crops and tree species on an estimated 32 million ha annually (Li, 1994). The relative ease and low cost of rearing *Trichogramma* wasps combined with the fact they kill their insect host in the egg stage, preventing feeding injury, probably explains their widespread use (Wajnberg and Hassan, 1994). Implementation of non-commercial augmentation may be expanding in less developed countries worldwide (van Lenteren and Bueno, 2003). In the somewhat unique case of Cuba, augmentation is very widely used primarily because trade embargos have prevented other pest management tactics from being utilized (Dent, 2005).

Approximately 170 different species are offered for commercial augmentation worldwide, and of the arthropods offered, about half are predators and half parasitoids (Cock et al., 2010; Hunter, 1997). Despite the number of species available as commercial products, only four pest groups (whiteflies, thrips, spider mites and aphids) account for 84% of expenditures (van Lenteren, 2003b). Approximately 75% of commercial sales are in Europe, primarily for pest management in greenhouses, but there are examples of augmentation used in other cropping systems (Cock et al., 2010; van Lenteren, 2003b, 2006; van Lenteren and Bueno, 2003; Luck and Forster, 2003; Shipp et al., 2007). Although commercial sales and hectarage treated represent only a small fraction of total augmentation activity, they have drawn most of the attention of the scientific community. As a result, the future viability of augmentation as a pest management practice is perhaps unduly influenced by the commercial natural enemy industry, and the cost, availability and quality of natural enemies it produces (Bolckmans, 2003; Warner and Getz, 2008).

The scientific foundation, ecological limitations, efficacy and cost effectiveness of augmentation has been called into question and openly debated, especially in comparison with insecticide applications (Collier and van Steenwyk, 2004, 2006; van Lenteren, 2006). The need for repeated applications of augmented natural enemies both within and across crop cycles, similarly to insecticides, is the reason the augmentation industry developed but it may also be its Achilles heel. Parrella et al. (1992) suggested that perceived similarities between augmentation and insecticide applications have diminished interest by the scientific community in this discipline. The large-scale, statistically valid, detailed studies that are required to rigorously evaluate natural enemy augmentation effectively present tremendous logistical difficulties (Luck et al., 1988). Perhaps in part because of this, many studies done to support augmentation have not had adequate experimental designs and have not included insecticides as a treatment comparison (Collier and van Steenwyk, 2004). The strong partnership between government agencies and industry and the commercial augmentation industry in Europe is likely an important factor in the current strength of the industry (van Lenteren, 2000b, 2003b). However, if strong advocates in the scientific community are lost, this situation may change.

Predictability of pest management results has been a concern for some time with all types of augmentation. A lack of supporting data for augmentative releases in the past has prevented development of recommendations for release rates and application methodologies that provide predictable results (Parrella et al., 1992). Another contributing factor to unpredictability and unsatisfactory pest management with augmentation biological control has been the variable or poor quality of mass-reared and released natural enemies and resulting incorrect releases rates (Hoy et al., 1991). Even though serious efforts to improve quality control of natural enemies began decades ago, this remains a serious issue (Bolckmans, 2003). Significant effort has been made to rectify these quality and predictability issues (see articles in van Lenteren, 2003c). However, augmentation may not see more widespread adoption until there is consistent availability of high quality products and sufficient end-user knowledge and support systematology (Bolckmans, 2003).

Worldwide, the commercial augmentation industry is comparatively small, with a market value between €150 and €200 million, or less than 1% of pesticide sales (Cock et al., 2010) and roughly 2% of insecticide sales (Kiely et al., 2004). Over 75% of commercial augmentation sales take place in Europe and North America (van Lenteren and Bueno, 2003). European sales are approximately three times those of the commercial augmentation industry in North America which is also comparatively small, with annual sales in the range of $25–30 million, or roughly 10% of the market for biologically based products, and less than 1% of the insecticide market (van Lenteren, 2003b; Warner and Getz, 2008). Although the industry in Europe remains strong (Cock et al., 2010; van Lenteren, 2003b), the North American industry appears to be facing some significant challenges (Warner and Getz, 2008). These include difficulty obtaining investment capital, static or declining market and product prices, recruitment of cooperating researchers and problems with cross-border shipments (Warner and Getz, 2008). Attitudes in this market may differ from Europe, and consumers in the larger potential markets may be less accepting of greenhouse biocontrol strategies (Wawrzynski et al., 2001). Factors that are often cited in the scientific literature as promoters of augmentation such as the introduction of narrow-spectrum insecticides, pesticide resistance and the expansion of the organic industry apparently offer no benefit to the North American industry (Warner and Getz, 2008). Survey responses suggest that urban and suburban residents in the United States are willing to pay more for biological control products versus chemical insecticides for outdoor pest management (Jetter and Paine, 2004). Indeed, over half of the North American augmentation industries' sales are for outdoor crops (Warner and Getz, 2008). The controlled environments of greenhouses are amenable to augmentation because of greater predictability of pest and natural enemy populations (van Lenteren, 2000b), whereas demand for augmentation products for outdoor crops can be difficult to predict, leading to product shortages and price volatility (Warner and Getz, 2008).

Increased management requirements and expenses mean that commercial augmentation has typically been most successfully adopted by higher value cropping systems such as greenhouses, orchards, nurseries, fruits and vegetables (Hale and Elliott, 2003). However, for small-scale growers in these systems, costs may be a challenge. For equivalent pest management outcomes, *Aphidius colemani* augmentation costs almost five times as much as imidacloprid applications for management of *Aphis gossypii* on greenhouse-grown chrysanthemums, primarily due to shipping costs required for multiple parasitoid releases (Vásquez et al., 2006). One possible way to reduce these costs is the use of banker plants. Banker plants are intended to sustain released natural enemy populations in the absence of target pests or other food resources in a cropping system (Frank, 2010). The use of banker plants in greenhouses has focused on aphid management, using the non-pest aphid *Rhopalosiphum padi* on cereal plants hosting the parasitoid *A. colemani* (Van Driesche et al., 2008). Although significant research is required to provide predictable results with this method for more pests, there is significant potential for expanded use of this approach in the future (Frank, 2010).

Commercial augmentation has been successfully employed in some lower cost crop systems. Sales of *Trichogramma brassicae* wasps make up approximately 10% of the pesticide market for European corn borer, *Ostrinia nubilalis*, management in Western Europe, and these wasps have been shown to be as efficacious as insecticides in the management of this pest (Orr and Suh, 1999). However, this may be a unique situation with a predictable pest population and a natural enemy that can be stored for long periods following production before pre-planned releases on company-determined dates (Orr and Suh, 1999).

Because augmentative releases of natural enemies result in relatively short-term increases in enemy numbers and activity (Lynch and Thomas, 2000; Kuske et al., 2003; van Lenteren et al., 2006), augmentation biological control usually does not face the same scrutiny over potential non-target effects as importation biological control. The exceptions occur mainly where exotic species are involved. For example, a company in New Zealand faced criminal prosecution for failing to follow national laws regulating exotic species by the import, production and sale of a predatory bug, mirid *Macrolophus pygmaeus* in that country (New Zealand Press Association, 2010). Increasing risk-aversion by decision makers and the public (Sheppard and Raghu, 2005), combined with increased regulation and administrative barriers appear to have changed the types of organisms used by the augmentation industry (Cock et al., 2010). Over the last 50 years, the industry offerings in Europe have shifted from 58% exotic species to the current 24% (Cock et al., 2010). The European augmentation industry is focused on products for use in greenhouses (van Lenteren, 2003b) which may reduce the likelihood for non-target impacts. However, in North America, half of industry sales and the driver of industry expansion has been sales to individual homeowners for outdoor releases (Warner and Getz, 2008).

Sales and shipment of native species have also drawn concern. In North America, shipments of the native lady beetle *Hippodamia convergens* often contain significant numbers of beetles that are parasitized or infected with pathogens as a result of beetles being collected from their natural overwintering sites rather than being reared in an insectary (Bjørnson, 2008). This practice may result in increased parasitism or transmission of pathogens to local lady beetle populations where releases are made (Bjørnson, 2008; Saito and Bjørnson, 2008). This is probably an isolated example, since most augmented natural enemies are reared under highly controlled conditions, usually with strict quality control standards in place.

Another concern expressed over shipment of local populations of native species with widespread distributions, such as *H. convergens*, is the potential for 'genetic pollution' (Dubois, 2008). While genetic mixing will not necessarily lead to decreased fitness, the concern is that future phylogenetic studies will not be able to accurately characterize original, natural populations (Dubois, 2008).

Augmentative releases of natural enemies have resulted in clearly documented economic and ecological damage in only one case. The multicoloured Asian lady beetle, *H. axyridis*, was distributed and established in western Europe at least partly through commercial sales (Brown et al., 2008). Although *H. axyridis* can contribute to pest management in some systems, its pest status as a household invader, its impacts on fruit and wine production and its impacts on non-target arthropods are considered to outweigh its benefits in both western Europe and North America (Brown et al., 2008; Koch and Galvan, 2008).

23.6 CONSERVATION BIOLOGICAL CONTROL

Although a variety of crop production practices can influence natural enemy populations (Van Driesche and Bellows, 1996), by far the greatest impacts are from pesticides and the most commonly used method of achieving conservation of natural enemies is through modification of their use (Ruberson et al., 1998). The future potential of habitat management to affect biological control is also significant (Jonsson et al., 2008). This section will focus on these two approaches to conservation biological control.

The goal of conservation biological control is to enhance the activity of resident natural enemy populations, regardless of whether they are native or exotic (Gurr and Wratten, 1999). In the absence of human intervention, natural enemies provide a valuable ecological service. By one estimate, the native predators and parasitoids attacking native pests in the United States avert $4.5 billion in annual crop losses (Losey and Vaughan, 2006). It seems sensible to try to take advantage of this service, but it has been challenging to implement widespread practices, especially those that involve modification of the environment (Ehler, 1998). Even though the most common practice for achieving conservation biocontrol is modification of existing pesticide application practices (Ruberson et al., 1998), this can still be challenging to implement. The justifiable desire for risk reduction by farmers is a significant driver behind the resistance to pesticide use changes (Fernandez-Cornejo et al., 1998). Simply adopting IPM practices can have important effects on the services provided by natural enemy populations. The abundance, diversity, and more importantly the efficacy of predators and parasitoids of Brassica crop pests were significantly greater on farms in the Lockyer valley of Australia that practised IPM compared with those that did not (Furlong et al., 2004).

Potential modifications to make pesticide applications less harmful to natural enemies include treating only when economic thresholds dictate, use of active ingredients and formulations that are selectively less toxic to natural enemies, use of the lowest effective rates of pesticides, and temporal and spatial separation of natural enemies and pesticides (Ruberson et al., 1998). However, if broad-spectrum pesticides are used to treat pest populations that exceed threshold levels, then the use of thresholds alone will not necessarily lead to natural enemy conservation (Ruberson et al., 1998). Studies have used natural enemy numbers to revise economic thresholds to more accurately determine the need for or timing of insecticide treatments within a pest generation (Ostlie and Pedigo, 1987), and predict the need for treatment of a future pest generation (Van Driesche et al., 1994). Formal revised economic thresholds incorporating natural enemy numbers have been developed to incorporate *Trichogramma* parasitism for fruitworm management in tomatoes (Hoffman et al., 1991) and coccinellid numbers for cotton aphid management in cotton (Obrycki et al., 2009). In some cases, informal revised thresholds have been used successfully by crop consultants for decades such as with cotton aphid management in North Carolina cotton production (Orr and Suh, 1999).

Fungicides generally have little or no direct effects on parasitoids and predators (Carmo et al., 2010), so modification of their use does not seem to be needed. Although some herbicides can have direct lethal and sublethal effects on natural enemies (e.g. Carmo et al., 2010), their main effect seems to be indirect, through simplified plant communities and unfavourable changes in microclimate within crop fields (e.g. Taylor et al., 2006). Because of clear reductions in yield, quality, and harvestability of many crops from increased in-field weed populations (Oerke, 2006), practical applications of herbicide use modification may be difficult to achieve. Insecticide effects on natural enemies are well known (Desneux et al., 2007), and ameliorating these effects has long been considered an important component of integrated pest management programmes (Stern et al., 1959; DeBach, 1964; Newsom and Brazzel, 1968). Encouraging behavioural changes in farmers to meet goals for IPM or conservation biological control can be a difficult balancing act between the demands of crop production and the desire for more ecologically and perhaps more economically sound pest management.

Careful use of physiologically selective pesticides can ameliorate pesticide impacts on natural enemies (Van Driesche and Bellows, 1996), and may be the decision-making tool most readily available to growers to reduce pesticide impacts on natural enemies (Ruberson et al., 1998). Various databases and ranking systems have been constructed that have the potential to compare deleterious effects of pesticides (Reus and Leendertse, 2000; van der Werf, 1996), but they have not been in a form that could be used by pest management professionals. To make this information more available and user-friendly, Hoque et al. (2002) developed a beneficial disruption index that was found to be an effective measure of insecticide impacts on beneficial insects in cotton (Mansfield et al., 2006).

Reduced risk and organic pesticides have fewer impacts on natural enemies than traditional synthetic pesticides, but are still toxic and deleterious and their use does not guarantee conservation of beneficials (Bahlai et al., 2010;

Sarvary et al., 2007). Natural enemy activity did not improve in individual crop fields when reduced risk insecticides were used for apple pest management, even when natural habitat was interspersed with cropland (Sarvary et al., 2007). Benefits that might be had from using reduced-risk insecticides may be eliminated if they are sold or used in combination with broad-spectrum insecticides (Ohnesorg et al., 2009).

A significant limiting factor in the life cycle of many natural enemies, especially parasitoids, is the availability of external food resources to sustain host searching and egg development (Heimpel and Jervis, 2005; Vinson, 1998; Wäckers et al., 2008). However, direct evidence of a connection between provision of carbohydrate resources (both floral and honeydew) in the field and increased parasitism and pest management has been difficult to obtain (e.g. Lee and Heimpel, 2008; Wäckers et al., 2008). Direct application of resources to crops in the form of food sprays also have not produced any practical methods for pest management (Wade et al., 2008). A study by Blaauw and Isaacs (2012) noted that natural enemy numbers and activity increased and soybean aphid numbers decreased as planted wildflower plot size increased. Interestingly, a similar positive relationship was seen in wildlife that utilize wildflower plantings (Riddle et al., 2008).

Rather than individual resources on a farm, it appears the composition of landscape may be a more important determinant of natural enemy populations and the ecological services they provide (Woltz et al., 2012). For example, Thies et al. (2003) demonstrated a linear relationship between the percentage of non-crop habitat in a landscape and parasitism of the rape pollen beetle, with 20% being a crucial point for retaining parasitoid services. However, while provision of non-crop habitat for enemies in some cases may increase their activity, it may also have the confounding effect of increasing pest populations (Thies et al., 2005). In a landscape with many competing owners and interests, it may be difficult to manage large areas with a single management objective.

To improve implementation of conservation biocontrol, Fiedler et al. (2008) suggest combining multiple ecological service goals by looking for synergies in various activities such as biodiversity conservation, ecological restoration, human cultural values, tourism, biological control and other ecosystem services. Another approach that has seen some success has been to offer Payment for Ecosystem Services (PES), along with technical assistance to help growers meet broader objectives for conservation of specific services, which can include pest management (Garbach et al., 2012).

The composition of natural enemy communities also appears to be important, and may have important implications for habitat manipulation. Increasing the abundance and diversity of natural enemies does not appear to consistently improve pest suppression, and, in some cases, may have the opposite effect as a result of intraguild predation (Straub et al., 2008). There is also evidence that species evenness in natural enemy communities may be more important than species richness (Straub and Snyder, 2006). Although the mechanisms are not completely clear, it appears that organic farming systems may enhance species evenness and thereby improve pest management (Crowder et al., 2010).

Since conservation biological control focuses on enhancing resident natural enemy populations, it does not face the same scrutiny or have the same potential for non-target impacts that importation or augmentation do (van Lenteren et al., 2006). Still, concerns have been expressed about the potential spillover of generalist predator populations from agricultural systems into natural habitats and negatively impacting native arthropods (Rand and Louda, 2006), and enhancing these populations may create future conflict. Unlike with importation and augmentation biological control, economic analyses of conservation biocontrol efforts are very rare and uniquely difficult to conduct (Cullen et al., 2008), possibly reflecting the lower implementation levels of this pest management approach.

23.7 CONCLUSIONS

Biological control has an extremely long association with agriculture, has played an integral role in crop integrated pest management programmes throughout the world, and seems poised to continue to do so. The increasing numbers of exotic, invasive pests resulting from international trade and travel that reach new geographic areas without their regulating populations of natural enemies present new opportunities for importation biological control to provide long-term cost-effective pest management. Continued concerns over potential non-target impacts and administrative requirements of international agreements such as the Convention on Biological Diversity present challenges, but the demand for biological control services will undoubtedly prompt many new projects. At the same time as the need to increase global food production has intensified (Godfray et al., 2010), there is a more widespread understanding of potential environmental degradation associated with many modern agricultural practices (IAASTD, 2009). There is a growing realization in the environmental community that conservation of biological diversity and ecosystem function requires conservation projects to focus on developed lands, including agro-ecosystems, rather than just undeveloped lands to meet their goals (Brussaard et al., 2010). These trends combined with a developing appreciation for the value of ecological services may present opportunities for implementing conservation biological control on a larger scale. The increasing public demand for 'local' and organic foods may also be creating opportunities for increased implementation of biological control. As it has throughout the history of agriculture, biological control should continue to play an important role in integrated pest management of insect pests in cropping systems worldwide.

References

Bahlai, C.A., Xue, Y., McCreary, C.M., Schaafsma, A.W., Hallett, R.H., 2010. Choosing organic pesticides over synthetic pesticides may not effectively mitigate environmental risk in soybeans. PLoS ONE 5 (6), e11250.

Barbosa, P., Segarra-Carmona, A., 1993. Criteria for the selection of pest arthropod species as candidates for biological control. In: Van Driesche, R.G., Bellows, T.S. (Eds.), Steps in Classical Biological Control. Thomas Say Publications in Entomology, Entomological Society of America, Lanham, MD, pp. 5–23.

Bellows, T.S., Fisher, T.W., 1999. Handbook of Biological Control. Academic Press, San Diego, CA.

Bigler, F., Babendreier, D., Kuhlmann, U. (Eds.), 2006. Environmental Impact of Invertebrates for Biological Control of Arthropods: Methods and Risk Assessment. CABI Publishing, Wallingford.

Bjørnson, S., 2008. Natural enemies of the convergent lady beetle, *Hippodamia convergens* Guérin-Méneville: their inadvertent importation and potential significance for augmentative biological control. Biol. Control. 44, 305–311.

Blaauw, B.R., Isaacs, R., 2012. Larger wildflower plantings increase natural enemy density, diversity, and biological control of sentinel prey, without increasing herbivore density. Ecol. Entomol. 37, 386–394.

Bolckmans, K.J.F., 2003. State of affairs and future directions of product quality assurance in Europe. In: van Lenteren, J.C. (Ed.), Quality Control and Production of Biological Control Agents: Theory and Testing Procedures. CABI, Wallingford, UK, pp. 215–224.

Brown, P.M.J., Adriaens, T., Bathon, H., Cuppen, J., Goldarazena, A., Hägg, T., et al., 2008. *Harmonia axyridis* in Europe: spread and distribution of a non-native coccinellid. BioControl 53, 5–21.

Brussaard, L., Caron, P., Campbell, B., Lipper, L., Mainka, S., Rabbinge, R., et al., 2010. Reconciling biodiversity conservation and food security: scientific challenges for a new agriculture. Curr. Opin. Environ. Sustain. 2, 34–42.

Cai, W., Yan, Y., Li, L., 2005. The earliest records of insect parasitoids in China. Biol. Control 32, 8–11.

Carmo, E.L., Bueno, A.F., Bueno, R.C.O.F., 2010. Pesticide selectivity for the insect egg parasitoid *Telenomus remus*. BioControl 55, 455–464.

Cock, M.J.W., 2010. General news: BIOCAT to be updated and made open access. Biocontrol News Inform. 31 (3), 17N–24N.

Cock, M.J.W., van Lenteren, J.C., Brodeur, J., Barratt, B.I.P., Bigler, F., Bolckmans, K., et al., 2010. Do new Access and Benefit Sharing procedures under the Convention on Biological Diversity threaten the future of biological control? BioControl 55, 199–218.

Collier, T., van Steenwyk, R., 2004. A critical evaluation of augmentative biological control. Biol. Control 31, 245–256.

Collier, T., van Steenwyk, R., 2006. How to make a convincing case for augmentative biological control. Biol. Control 39, 119–120.

Coppell, H.C., Mertens, J.W., 1976. Biological Insect Pest Suppression. Springer Inc., Berlin.

Cranshaw, W., Sclar, D.C., Cooper, D., 1996. A review of 1994 pricing and marketing by suppliers of organisms for biological control of arthropods in the United States. Biol. Control 6, 291–296.

Crowder, D.W., Northfield, T.D., Strand, M.R., Snyder, W.E., 2010. Organic agriculture promotes evenness and natural pest control. Nature 466, 109–112.

Cullen, R., Warner, K.D., Jonsson, M., Wratten, S.D., 2008. Economics and adoption of conservation biological control. Biol. Control 45, 272–280.

DeBach, P. (Ed.), 1964. Biological Control of Insect Pests and Weeds. Chapman and Hall, London.

DeBach, P., 1974. Biological Control by Natural Enemies. Cambridge University Press, London.

Dean, H.A., Schuster, M.F., Boling, J.C., Riherd, P.T., 1979. Complete biological control of *Antonina graminis* in Texas with *Neodusmetia sangwani* (a classic example). Bull. Entomol. Soc. Am. 25, 262–267.

Dent, D., 2005. Overview of agrobiologicals and alternatives to synthetic pesticides. In: Pretty, J. (Ed.), The Pesticide Detox: Towards a More Sustainable Agriculture. Earthscan Publications Ltd, London, pp. 70–82.

Desneux, N., Decourtye, A., Delpuech, J.M., 2007. The sublethal effects of pesticides on beneficial arthropods. Annu. Rev. Entomol. 52, 81–106.

Doutt, R.L., 1964. The historical development of biological control. In: DeBach, P. (Ed.), Biological Control of Insect Pests and Weeds. Chapman and Hall, London, pp. 21–42.

Dubois, A., 2008. The notion of biotic pollution: faunistic, floristic, genetic and cultural pollutions. Bull. Soc. Zool. France 133, 357–382.

Ehler, L.E., 1998. Conservation biological control: past, present, and future. In: Barbosa, P. (Ed.), Conservation Biological Control. Academic Press, New York, pp. 1–8.

Eilenberg, J., Hajek, A., Lomer, C., 2001. Suggestions for unifying the terminology in biological control. BioControl 46, 387–400.

Ervin, R.T., Moffitt, L.J., Meyerdirk, D.E., 1983. Comstock mealybug (Homoptera: Pseudococcidae): cost analysis of a biological control program in California. J. Econ. Entomol. 76, 605–609.

Fernandez-Cornejo, J., Jans, S., Smith, M., 1998. Issues in the economics of pesticide use in agriculture: A review of the empirical evidence. Appl. Econ. Perspect. Pol. 20, 462–488.

Fiedler, A.K., Landis, D.A., Wratten, S.D., 2008. Maximizing ecosystem services from conservation biological control: The role of habitat management. Biol. Control 45, 254–271.

Fleschner, C.A., 1960. Parasites and predators for pest control. In: Reitz, L.P. (Ed.), Biological and Chemical Control of Plant and Insect Pests. Am. Assoc. Adv. Science, Washington, pp. 183–199.

Follett, P.A., Duan, J.J. (Eds.), 2000. Nontarget Effects of Biological Control. Kluwer Academic Publishers, Norwell.

Frank, J.H., 1998. How risky is biological control? Comment. Ecology 79, 1829–1834.

Frank, S.D., 2010. Biological control of arthropod pests using banker plant systems: Past progress and future directions. Biol. Control 52, 8–16.

Furlong, M.J., Shi, Z.H., Liu, Y.Q., Guo, S.J., Lu, Y.B., Liu, S.S., et al., 2004. Experimental analysis of the influence of pest management practice on the efficacy of an endemic arthropod natural enemy complex of the diamondback moth. J. Econ. Entomol. 97, 1814–1827.

Garbach, K., Lubella, M., DeClerck, F.A.J., 2012. Payment for Ecosystem Services: The roles of positive incentives and information sharing in stimulating adoption of silvopastoral conservation practices. Agric. Ecosyst. Environ. 156, 27–36.

Godfray, H.C.J., Beddington, J.R., Crute, I.R., Haddad, L., Lawrence, D., Muir, J.F., et al., 2010. Food security: The challenge of feeding 9 billion people. Science 327, 812–818.

Grandgirard, J., Hoddle, M.S., Petit, J.N., Roderick, G.K., Davies, N., 2008. Engineering an invasion: classical biological control of the glassy-winged sharpshooter, *Homalodisca vitripennis*, by the egg parasitoid *Gonatocerus ashmeadi* in Tahiti and Moorea, French Polynesia. Biol. Invasions 10, 135–148.

Greathead, D.J., Greathead, A.H., 1992. Biological control of insect pests by insect parasitoids and predators: the BIOCAT database. Biocontrol News Inform. 13, 61–68.

Gurr, G., Wratten, S., 1999. Integrated biological control: a proposal for enhancing success in biological control. Int. J. Pest Manag. 45 (2), 81–84.

Hale, M.A., Elliott, D., 2003. Successes and challenges in augmentative biological control in outdoor agricultural applications: A producer's perspective. In: Van Driesche, R.G., (Ed.), Proceedings of the First International Symposium on Biological Control of Arthropods, The Forest Health Technology Enterprise Team (USDA Forest Service), Morgantown, pp. 185–188.

Hall, R.W., Ehler, L.E., 1979. Rate of establishment of natural enemies in classical biological control. Bull. Entomol. Soc. Am. 25, 280–282.

Hall, R.W., Ehler, L.E., Bisabri-Ershadi, B., 1980. Rate of success in classical biological control of arthropods. Bull. Entomol. Soc. Am. 26, 111–114.

Headley, J.C., 1985. Cost benefit analysis: defining research needs. In: Hoy, M.A., Herzog, D.C. (Eds.), Biological Control in Agricultural IPM Systems. Academic Press, New York, pp. 53–63.

Heimpel, G.E., Jervis, M.A., 2005. Does floral nectar improve biological control by parasitoids? In: Wäckers, F., van Rijn, P., Bruin, J. (Eds.), Plant-Provided Food and Plant–Carnivore Mutualism. Cambridge University Press, Cambridge, pp. 267–304.

Hoddle, M., 2004. Restoring balance: using exotic species to control invasive exotic species. Conserv. Biol. 18, 38–49.

Hoelmer, K.A., Kirk, A.A., 2005. Selecting arthropod biological control agents against arthropod pests: Can the science be improved to decrease the risk of releasing ineffective agents? Biol. Control 34, 255–264.

Hoffman, M.P., Wilson, L.T., Zalom, F.G., Hilton, R.J., 1991. Dynamic sequential sampling plan for *Helicoverpa zea* (Lepidoptera: Noctuidae) eggs in processing tomatoes: parasitism and temporal patterns. Environ. Entomol. 20, 1005–1012.

Hokkanen, H.M.T., 1985. Success in classical biological control. CRC Crit. Rev. Plant Sci. 3, 35–72.

Hoque, Z., Dillon, M., Farquharson, B., 2002. Three seasons of IPM in an area wide management group – a comparative analysis of field level profitability. In: Swallow, D., (Ed.), Proceedings of the Eleventh Australian Cotton Conference, ACGRA, Brisbane, pp. 749–755.

Hörstadius, S., 1974. Linnaeus, animals and man. Biol. J. Linn. Soc. 6, 269–275.

Hoy, M.A., Nowierski, R.M., Johnson, M.W., Flexner, J.L., 1991. Issues and ethics in commercial releases of arthropod natural enemies. Am. Entomol. 37, 74–75.

Huang, H.T., Yang, P., 1987. The ancient cultured citrus ant. BioScience 37, 665–671.

Hunter, C.D., 1997. Suppliers of Beneficial Organisms in North America. Publication PM 97-01. California Environmental Protection Agency, Department of Pesticide Regulation, Sacramento, CA.

IAASTD, 2009. The International Assessment on Agricultural Knowledge, Science and Technology for Development (IAASTD). <http://www.unep.org/dewa/Assessments/Ecosystems/IAASTD/tabid/105853/Default.aspx/index.cfm> (accessed 29.05.13.).

Jetter, K., Paine, T.D., 2004. Consumer preferences and willingness to pay for biological control in the urban landscape. Biol. Control 30, 312–322.

Jetter, K., Klonsky, K., Pickett, C.H., 1997. A cost/benefit analysis of the ash whitefly biological control program in California. J. Arboricult. 23, 65–72.

Jonsson, M., Wratten, S.D., Landis, D.A., Gurr, G.M., 2008. Recent advances in conservation biological control of arthropods by arthropods. Biol. Control 45, 172–175.

Julien, M.H., Griffiths, M.W. (Eds.), 1999. Biological Control of Weeds: A World Catalogue of Agents and Their Target Weeds, Fourth edn. Oxford University Press, USA.

Kiely, T., Donaldson, D., Grube, A., 2004. Pesticides Industry Sales and Usage: 2000 and 2001 Market Estimates. U.S. Environmental Protection Agency, Washington, DC.

Kimberling, D.N., 2004. Lessons from history: predicting successes and risks of intentional introductions for arthropod biological control. Biol. Invasions 6, 301–318.

Kipkoech, A.K., Schulthess, F., Yabann, W.K., Maritim, H.K., Mithofer, D., 2006. Biological control of cereal stem borers in Kenya: a cost benefit approach. Ann. Soc. Entomol. France 42, 519–528.

Koch, R.L., 2003. The multicolored Asian lady beetle, *Harmonia axyridis*: A review of its biology, uses in biological control, and non-target impacts. J. Insect Sci. 3, 32.

Koch, R.L., Galvan, T.L., 2008. Bad side of a good beetle: the North American experience with *Harmonia axyridis*. BioControl 53, 23–35.

Kuske, S., Widmer, F., Edwards, P.J., Turlings, T.C.J., Babendreier, D., Bigler, F., 2003. Dispersal and persistence of mass released *Trichogramma brassicae* (Hymenoptera: Trichogrammatidae) in non-target habitats. Biol. Control 27, 181–193.

Lee, J.C., Heimpel, G.E., 2008. Floral resources impact longevity and oviposition rate of a parasitoid in the field. J. Anim. Ecol. 77, 565–572.

Li, L.-Y., 1994. Worldwide use of *Trichogramma* for biological control on different crops: a survey. In: Wajnberg, E., Hassan, S.A. (Eds.), Biological Control With Egg Parasitoids. CABI, Wallingford, UK.

Lincango, M.P., Causton, C.E., Calderón Alvarez, C., Jiménez-Uzcátegui, G., 2011. Evaluating the safety of *Rodolia cardinalis* to two species of Galapagos finch; *Camarhynchus parvulus* and *Geospiza fuliginosa*. Biol. Control 56, 145–149.

Losey, J.E., Vaughan, M., 2006. The economic value of ecological services provided by insects. BioScience 56, 311–323.

Louda, S.M., Stiling, P., 2004. The double-edged sword of biological control in conservation and restoration. Conserv. Biol. 18, 50–53.

Lozan, A.I., Monaghan, M.T., Spitzer, K., Jaros, J., Zurovcova, M., Broz, V., 2008. DNA-based confirmation that the parasitic wasp *Cotesia glomerata* (Braconidae, Hymenoptera) is a new threat to endemic butterflies of the Canary Islands. Conserv. Genet. 9 (6), 1431–1437.

Luck, R.F., Forster, L.D., 2003. Quality of augmentative biological control agents: a historical perspective and lessons learned from evaluating *Trichogramma*. In: van Lenteren, J.C. (Ed.), Quality Control and Production of Biological Control Agents: Theory and Testing Procedures. CABI, Wallingford, UK, pp. 231–246.

Luck, R.F., Shepard, B.M., Penmore, P.E., 1988. Experimental methods for evaluating arthropod natural enemies. Annu. Rev. Entomol. 33, 367–391.

Lynch, L.D., Thomas, M.B., 2000. Nontarget effects in the biocontrol of insects with insects, nematodes and microbial agents: the evidence. Biocontrol News Inform. 21 (4), 117–130.

Mansfield, S., Dillon, M.L., Whitehouse, M.E.A., 2006. Are arthropod communities in cotton really disrupted? An assessment of insecticide regimes and evaluation of the beneficial disruption index. Agric. Ecosyst. Environ 113, 326–335.

Marsden, J.S., Martin, G.E., Parham, D.J., Risdell Smith, T.J., Johnston, B.G., 1980. Returns on Australian Agricultural Research. CSIRO, Canberra.

Needham, J., 1986. Science and Civilisation in China. vol. 6: Biology and Biological Technology, Part 1: Botany. Cambridge University Press, Cambridge.

New Zealand Press Association, 2010. Scientist illegally imported insect. New Zealand Herald, Auckland, New Zealand, August 10, 2010.

Newsom, L.D., Brazzel, J.R., 1968. Pests and their control. In: Elliot, F.C., Hoover, M., Porter, W.K. (Eds.), Advances in Production and Utilization of Quality Cotton: Principles and Practices. Iowa State University Press, Ames, pp. 367–405.

Nordlund, D.A., 1996. Biological control, integrated pest management and conceptual models. Biocontrol News Inform. 17, 35–44.

Obrycki, J.J., Harwood, J.D., Kring, T.J., O'Neil, R.J., 2009. Aphidophagy by Coccinellidae: Application of biological control in agroecosystems. Biol. Control 51, 244–254.

Oerke, E.-C., 2006. Crop losses to pests. J. Agric. Sci. 144, 31–43.

Ohnesorg, W.J., Johnson, K.D., O'Neal, M.E., 2009. Impact of reduced-risk insecticides on soybean aphid and associated natural enemies. J. Econ. Entomol. 102, 1816–1826.

Orr, D.B., Suh, C.P.-C., 1999. Parasitoids and predators. In: Rechcigl, J., Rechcigl, N. (Eds.), Biological and Biotechnological Control of Insect Pests. Ann Arbor Press, Ann Arbor, pp. 3–34.

Ostlie, K.R., Pedigo, L.P., 1987. Incorporating pest survivorship into economic thresholds. Bull. Entomol. Soc. Am. 33, 98–102.

Pal, K.K., Gardener, B., 2006. Biological control of plant pathogens. Plant Health Instructor dx.doi.org/10.1094/PHI-A-2006-1117-02.

Parrella, M.P., Heinz, K.M., Nunney, L., 1992. Biological control through augmentative release of natural enemies: a strategy whose time has come. Am. Entomol. 38, 172–179.

Parry, D., 2009. Beyond Pandora's Box: quantitatively evaluating non-target effects of parasitoids in classical biological control. Biol Invasions 11, 47–58.

Pilkington, L.J., Messelink, G., van Lenteren, J.C., Le Mottee, K., 2010. Protected biological control biological pest management in the greenhouse industry. Biol. Control 52, 216–220.

Rand, T.A., Louda, S.M., 2006. Predator spillover across the agricultural-natural interface: implications for insects in fragmented landscapes. Conserv. Biol. 20 (6), 1720–1729.

Reus, J.A.W.A., Leendertse, P.C., 2000. The environmental yardstick for pesticides: a practical indicator used in the Netherlands. Crop Prot. 19, 637–641.

Riddle, J.D., Moorman, C.E., Pollock, K.H., 2008. The importance of habitat shape and landscape context to northern bobwhite populations. J. Wildl. Manag. 72, 1376–1382.

Ridgway, R.L., Inscoe, M.N., 1998. Mass-reared natural enemies for pest control: trends and challenges. In: Ridgway, R.L., Hoffman, J.H., Inscoe, M.N., Glenister, C.S. (Eds.), Mass-reared Natural Enemies: Application, Regulation, and Needs. Entomological Society of America, Lanham, MD.

Riley, W.A., 1931. Erasmus Darwin and the biological control of insects. Science 73, 475–476.

Roy, H., Wajnberg, E., 2008. From biological control to invasion: the ladybird *Harmonia axyridis* as a model species. BioControl 53, 1–4.

Ruberson, J.R., Nemoto, H., Hirose, Y., 1998. Pesticides and conservation of natural enemies in pest management. In: Barbosa, P. (Ed.), Conservation Biological Control. Academic Press, San Diego, CA, pp. 207–220.

Saito, T., Bjørnson, S., 2008. Effects of a microsporidium from the convergent lady beetle, *Hippodamia convergens* Guérin-Méneville (Coleoptera: Coccinellidae), on three non-target coccinellids. J. Invertebr. Pathol. 99, 294–301.

Sarvary, M.A., Nyrop, J., Reissig, H., Gifford, K.M., 2007. Potential for conservation biological control of the oblique banded leafroller (OBLR) *Choristoneura rosaceana* (Harris) in orchard systems managed with reduced-risk insecticides. Biol. Control 40, 37–47.

Schupp, J.L., Sharp, J.S., 2012. Exploring the social bases of home gardening. Agric. Hum. Values 29, 93–105.

Sheppard, A.W., Raghu, S., 2005. Working at the interface of art and science: How best to select an agent for classical biological control? Biol. Control 34, 233–235.

Shipp, L., Elliott, D., Gillespie, D., Brodeur, J., 2007. From chemical to biological control in Canadian greenhouse crops. In: Vincent, C., Goettel, M.S., Lazarovits, G. (Eds.), Biological Control: A Global Perspective. CABI, Wallingford, UK, pp. 118–127.

Simberloff, D., Stiling, P., 1996. How risky is biological control? Comment. Ecology 77, 1965–1974.

Simberloff, D., Stiling, P., 1998. How risky is biological control? Reply. Ecology 79, 1834–1836.

Smith, H.S., 1919. On some phases of insect control by the biological method. J. Econ. Entomol. 12, 288–292.

Smith, S.M., 1996. Biological control with *Trichogramma*: advances, successes, and potential of their use. Annu. Rev. Entomol. 41, 375–406.

Stern, V.M., Smith, R.F., van den Bosch, R., Hagen, K.S., 1959. The integrated control concept. Hilgardia 29, 81–101.

Stiling, P., Cornelissen, T., 2005. What makes a successful biocontrol agent? A meta-analysis of biological control agent performance. Biol. Control 34, 236–246.

Straub, C.S., Snyder, W.E., 2006. Species identity dominates the relationship between predator biodiversity and herbivore suppression. Ecology 87, 277–282.

Straub, C.S., Finke, D.L., Snyder, W.E., 2008. Are the conservation of natural enemy biodiversity and biological control compatible goals? Biol. Control 45, 225–237.

Sweetman, H.L., 1958. The Principles of Biological Control. WMC Brown Co, Dubuque, USA, 560 pp.

Taylor, R.L., Maxwell, B.D., Boik, R.J., 2006. Indirect effects of herbicides on bird food resources and beneficial arthropods. Agric. Ecosyst. Environ. 116, 157–164.

Thies, C., Steffan-Dewenter, I., Tscharntke, T., 2003. Effects of landscape context on herbivory and parasitism at different spatial scales. Oikos 101, 18–25.

Thies, C., Roschewitz, I., Tscharntke, T., 2005. The landscape context of cereal aphid–parasitoid interactions. Proc. R. Soc. B 272 (no. 1559), 203–210.

Tisdell, C.A., 1990. Economic impact of biological control of weeds and insects. In: Mackauer, M., Ehler, L.E., Roland, J. (Eds.), Critical Issues in Biological Control. Intercept, Andover, pp. 301–316.

Trotter, A., 1908. Due precursori nell' applicazione degli insetti carnivori a difesa delle plante coltivate. Redia 5, 126–132.

Tsuji, K., Hasyim, A., Nakamura, K., 2004. Asian weaver ants, *Oecophylla smaragdina*, and their repelling of pollinators. Ecol. Res. 19, 669–673.

van den Bosch, R., Messenger, P.S., Gutierrez, A.P., 1982. An Introduction to Biological Control. Plenum Press, New York, USA.

van den Bosch, R., Telford, A.D., 1964. Environmental modification and biological control. In: DeBach, P. (Ed.), Biological Control of Insect Pests and Weeds. Chapman and Hall, London, pp. 459–488.

van der Werf, H.M.G., 1996. Assessing the impact of pesticides on the environment. Agric. Ecosyst. Environ. 60, 81–96.

Van Driesche, R.G., 2008. Biological control of *Pieris rapae* in New England: host suppression and displacement of *Cotesia glomerata* by *Cotesia rubecula* (Hymenoptera: Braconidae). Fla. Entomol. 91, 22–25.

Van Driesche, R.G., Bellows Jr., T.S., 1996. Biological Control. Chapman and Hall, New York.

Van Driesche, R.G., Nunn, C., 2002. Establishment of a Chinese strain of *Cotesia rubecula* (Hymenoptera: Braconidae) in New England. Fla. Entomol. 85, 386–388.

Van Driesche, R.G., Elkinton, J.S., Bellows Jr, T.S., 1994. Potential use of life tables to evaluate the impact of parasitism on population growth of the apple blotch leafminer (Lepidoptera: Gracillaridae). In: Maier, C. (Ed.), Integrated Management of Tentiform Leafminers, Phyllonorycter (Lepidoptera: Gracillaridae) spp., in North American Apple Orchards. Thomas Say Publications in Entomology, Entomological Society of America, Lanham, MD.

Van Driesche, R.G., Lyon, S., Sanderson, J.P., Bennett, K.C., Stanek III, E.J., Zhang, R., 2008. Greenhouse trials of *Aphidius colemani* (Hymenoptera: Braconidae) banker plants for control of aphids (Hemiptera: Aphidae) in greenhouse spring floral crops. Fla. Entomol. 91, 583–591.

van Lenteren, J.C., 2000a. Measures of success in biological control of arthropods by augmentation of natural enemies. In: Gurr, G., Wratten, S. (Eds.), Measures of Success in Biological Control. Kluwer Academic Publishers, Dordrecht, pp. 77–103.

van Lenteren, J.C., 2000b. A greenhouse without pesticides: fact or fantasy? Crop Prot. 19, 375–384.

van Lenteren, J.C., 2003a. Need for quality control of mass-produced biological control agents. In: van Lenteren, J.C. (Ed.), Quality Control and Production of Biological Control Agents: Theory and Testing Procedures. CABI, Wallingford, UK, pp. 1–18.

van Lenteren, J.C., 2003b. Commercial availability of biological control agents. In: van Lenteren, J.C. (Ed.), Quality Control and Production of Biological Control Agents: Theory and Testing Procedures. CABI, Wallingford, UK, pp. 167–179.

van Lenteren, J.C. (Ed.), 2003c. Quality Control and Production of Biological Control Agents: Theory and Testing Procedures. CABI, Wallingford, UK.

van Lenteren, J.C., 2005. Early entomology and the discovery of insect parasitoids. Biol. Control 32, 2–7.

van Lenteren, J.C., 2006. How not to evaluate augmentative biological control. Biol. Control 39, 115–118.

van Lenteren, J.C., Bueno, V.H.P., 2003. Augmentative biological control of arthropods in Latin America. BioControl 48, 123–139.

van Lenteren, J.C., Godfray, H.C.J., 2005. European science in the Enlightenment and the discovery of the insect parasitoid life cycle in The Netherlands and Great Britain. Biol. Control 32, 2–7.

van Lenteren, J.C., Bale, J.S., Bigler, F., Hokkanen, H.M.T., Loomans, A.J.M., 2006. Assessing risks of releasing exotic biological control agents of arthropod pests. Ann. Rev. Entomol. 51, 609–634.

van Mele, P., Cuc, N.T.T., 2000. Evolution and status of *Oecophylla smaragdina* (Fabricius) as a pest control agent in citrus in the Mekong Delta, Vietnam. Int. J. Pest Manag. 46, 295–301.

Vásquez, G.M., Orr, D.B., Baker, J.R., 2006. Efficacy assessment of *Aphidius colemani* (Hymenoptera: Braconidae) for suppression of *Aphis gossypii* (Homoptera: Aphididae) in greenhouse-grown chrysanthemum. J. Econ. Entomol. 99, 1104–1111.

Vinson, S.B., 1998. The general host selection behavior of parasitoid hymenoptera and a comparison of initial strategies utilized by larvophagous and oophagous species. Biol. Control 11, 79–96.

Voegele, J.M., 1989. Biological control of *Brontispa longissima* in western Samoa: an ecological and economic evaluation. Agric. Ecosyst. Environ. 27, 315–329.

Wäckers, F.L., van Rijn, P.C.J., Heimpel, G.E., 2008. Honeydew as a food source for natural enemies: Making the best of a bad meal? Biol. Control 45, 176–184.

Wade, M.R., Zalucki, M.P., Wratten, S.D., Robinson, K.A., 2008. Conservation biological control of arthropods using artificial food sprays: Current status and future challenges. Biol. Control 45, 185–199.

Warner, K.D., Getz, C., 2008. A socio-economic analysis of the North American commercial natural enemy industry and implications for augmentative biological control. Biol. Control 45, 1–10.

Wawrzynski, R.P., Ascerno, M.E., McDonough, M.J., 2001. A survey of biological control users in Midwest greenhouse operations. Am. Entomol. 47, 228–234.

Wajnberg, E., Hassan, S.A. (Eds.), 1994. Biological Control with Egg Parasitoids. CABI, Wallingford, UK.

Way, M.J., Khoo, K.C., 1992. Role of ants in pest management. Annu. Rev. Entomol. 37, 479–503.

Woltz, J.M., Isaacs, R., Landis, D.A., 2012. Landscape structure and habitat management differentially influence insect natural enemies in an agricultural landscape. Agric. Ecosyst. Environ. 152, 40–49.

Zeddies, J., Schaab, R.P., Neuenschwander, P., Herren, H.R., 2001. Economics of biological control of cassava mealybug in Africa. Agric. Econ. 24, 209–219.

Index

Note: Page numbers followed by "*f*" and "*t*" refers to figures and tables.

C

F

G

H

I

X

Y

Z

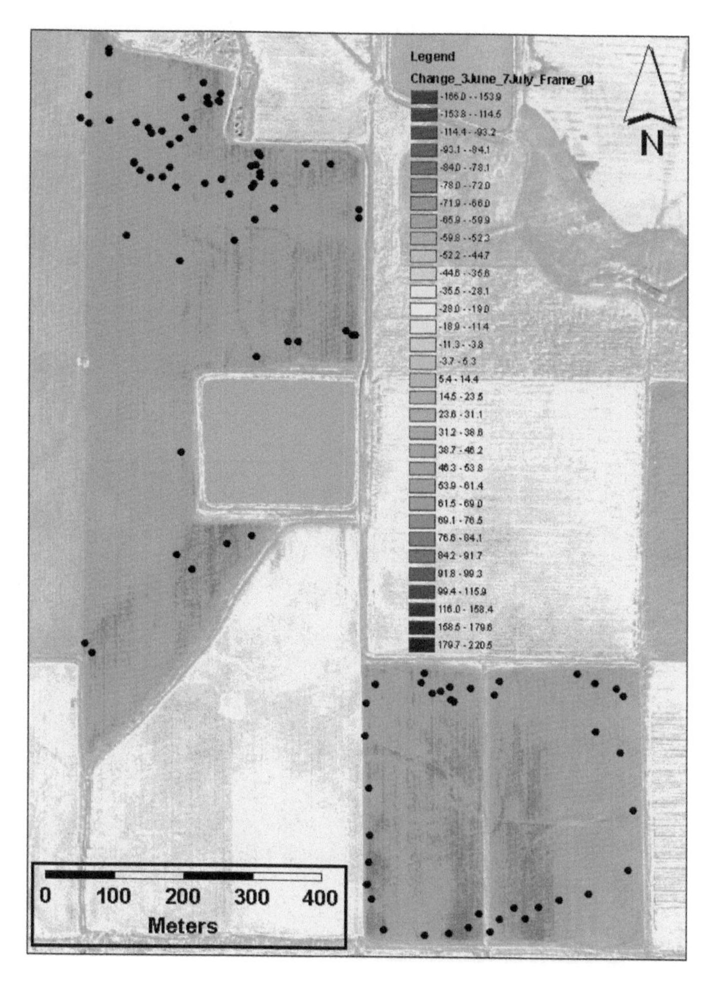

FIGURE 3.9 Example of scouting sites selection by a field scout for the entire 2006 production season for three cotton fields in Noxubee County, MS.

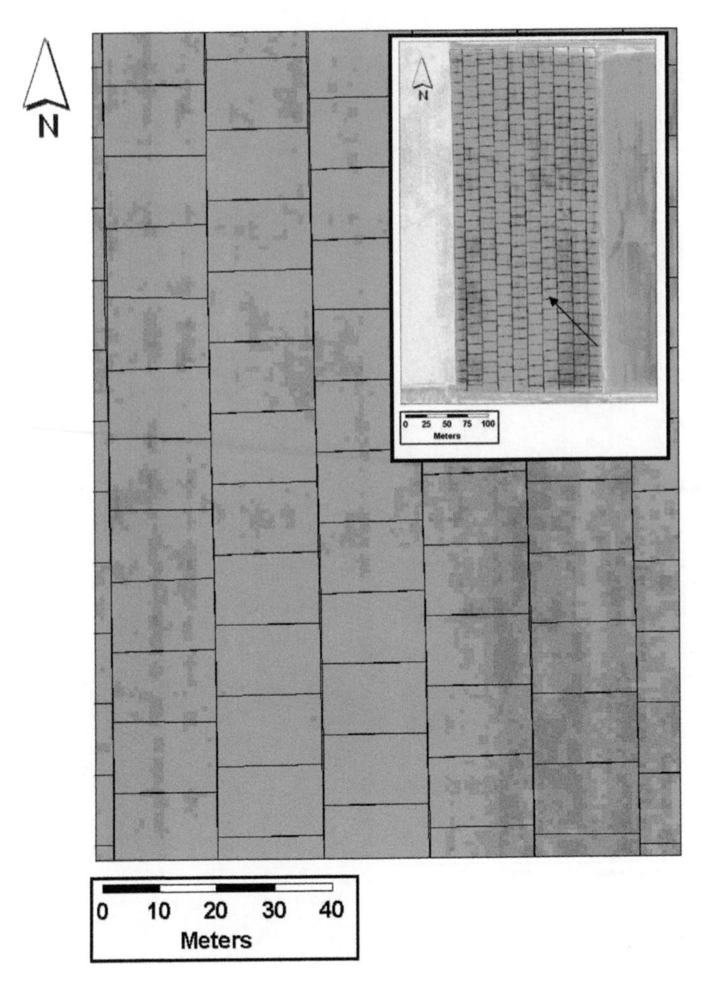

FIGURE 3.10 Example spray grid showing the polygons within sprayer paths that are the basis of assignments for different pesticide rates to build a spatial prescription. Each polygon of the grid is equivalent in area to a stacked belt transect sample of 315 units (or 219.456 m^2).

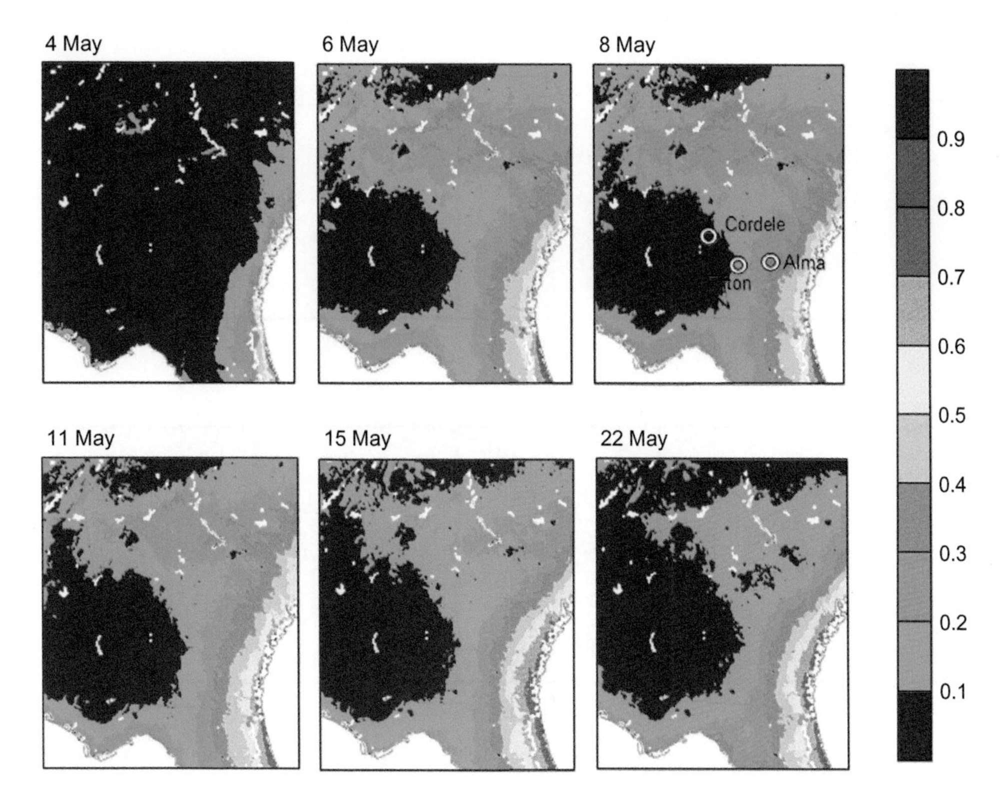

FIGURE 4.1 Spatiotemporal distribution of the probability of occurrence of early leaf spot of peanut (*Arachis hypogaea L.*), a disease caused by *Cercospora arachidicola* S. Hori, during the period from 4 May to 22 May 2007 for Georgia, USA (Olatinwo et al., 2012).

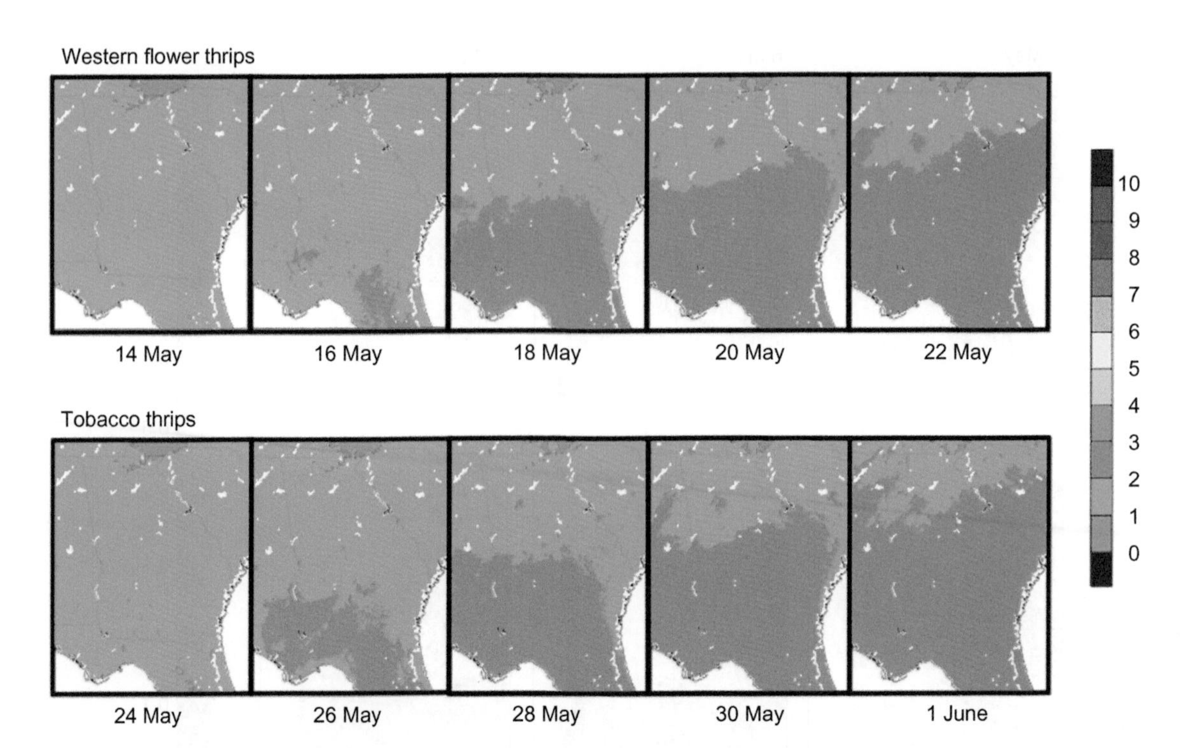

FIGURE 4.2 Spatial and temporal distribution of potential thrips accumulated generations in the southeastern United States using predictions from the Weather Research and Forecasting (WRF) model and the base temperature requirement for tobacco thrips (*Frankliniella fusca*) and western flower thrips (*Frankliniella occidentalis*) for 2007 (Olatinwo et al., 2011).

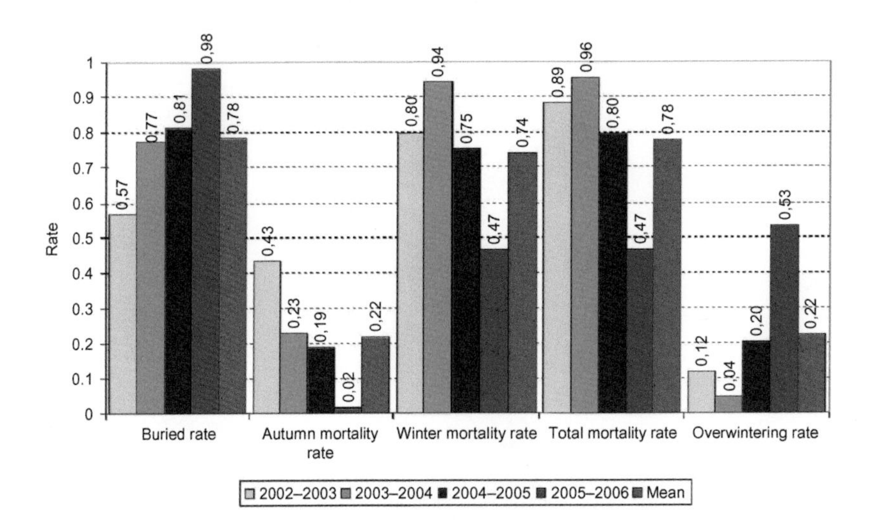

FIGURE 5.2 Buried rate, autumn, winter and total mortality rate and overwintering rate of CPB adult population in trials 2002–2003 to 2005–2006.

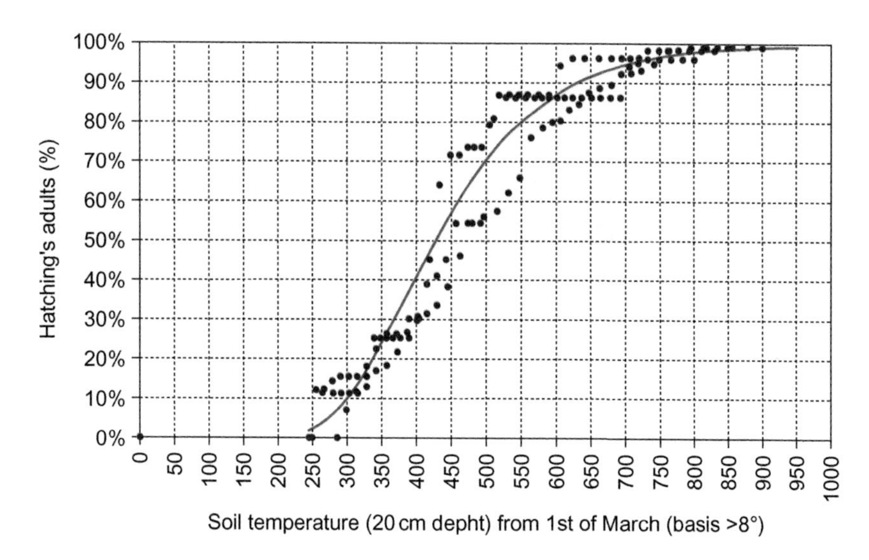

FIGURE 5.3 Scatter plot of the percentage of CPB adults hatching from the soil, expressed as a percentage of the overwintering population, depending on the sum of temperature from 1 March (8°C base level) and the SIMLEP1-Start interpolating Richard function.

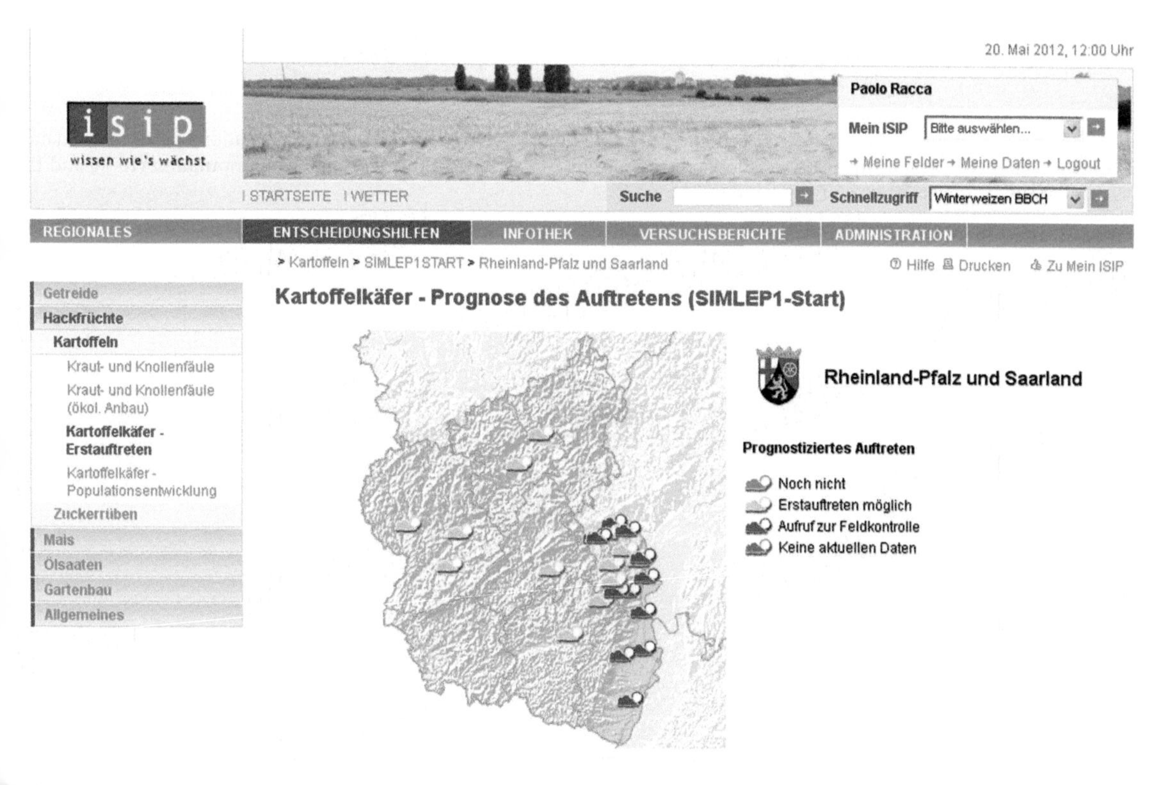

FIGURE 5.6 Output of the SIMLEP1-Start model from the ISIP website for the Rhineland-palatinate region. Clouds indicate weather stations. Cloud colours indicate: green, CPB overwintering population still in diapause; yellow, beginning of hatching of the overwintering population (10% CPB overwintering population have emerged from the soil); red, beginning of the egg laying in potato fields (>91% CPB overwintering population have emerged from the soil).

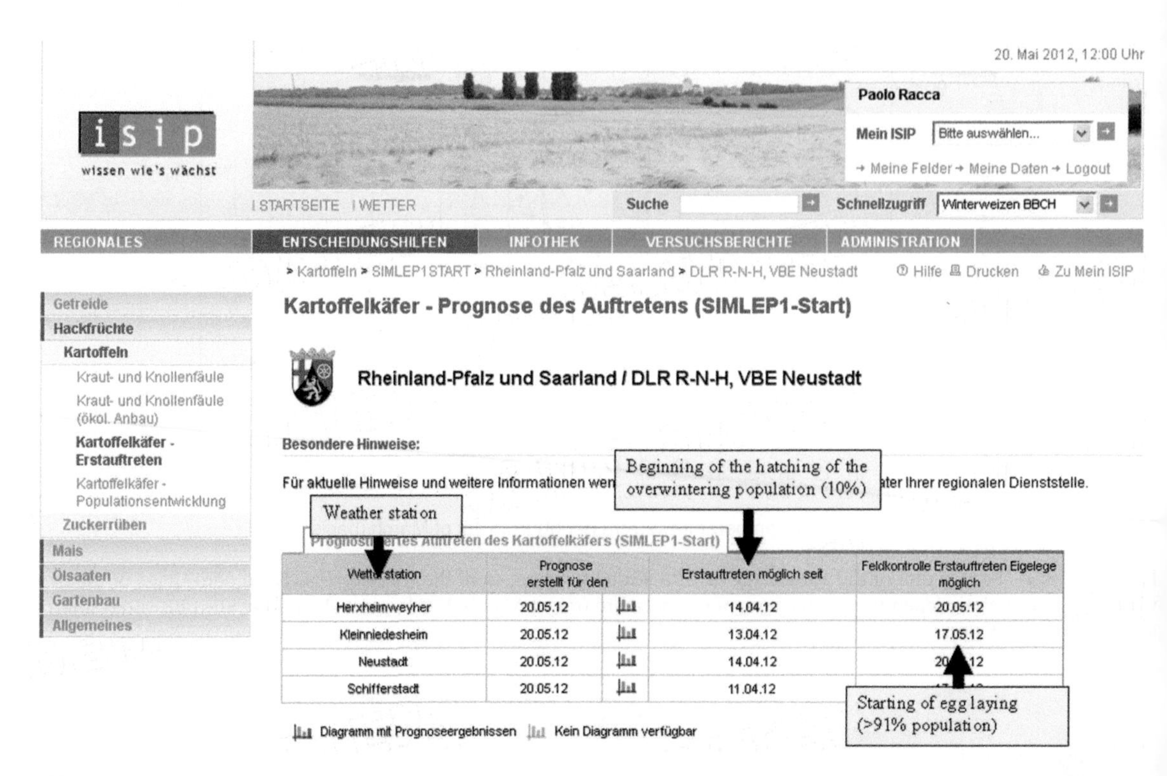

FIGURE 5.7 Output of the SIMLEP1-Start model from the ISIP website for two weather stations in the Rhineland-palatinate region. The first results column shows the beginning of hatching of the overwintering population (10%), and the second column indicates the start of egg laying.

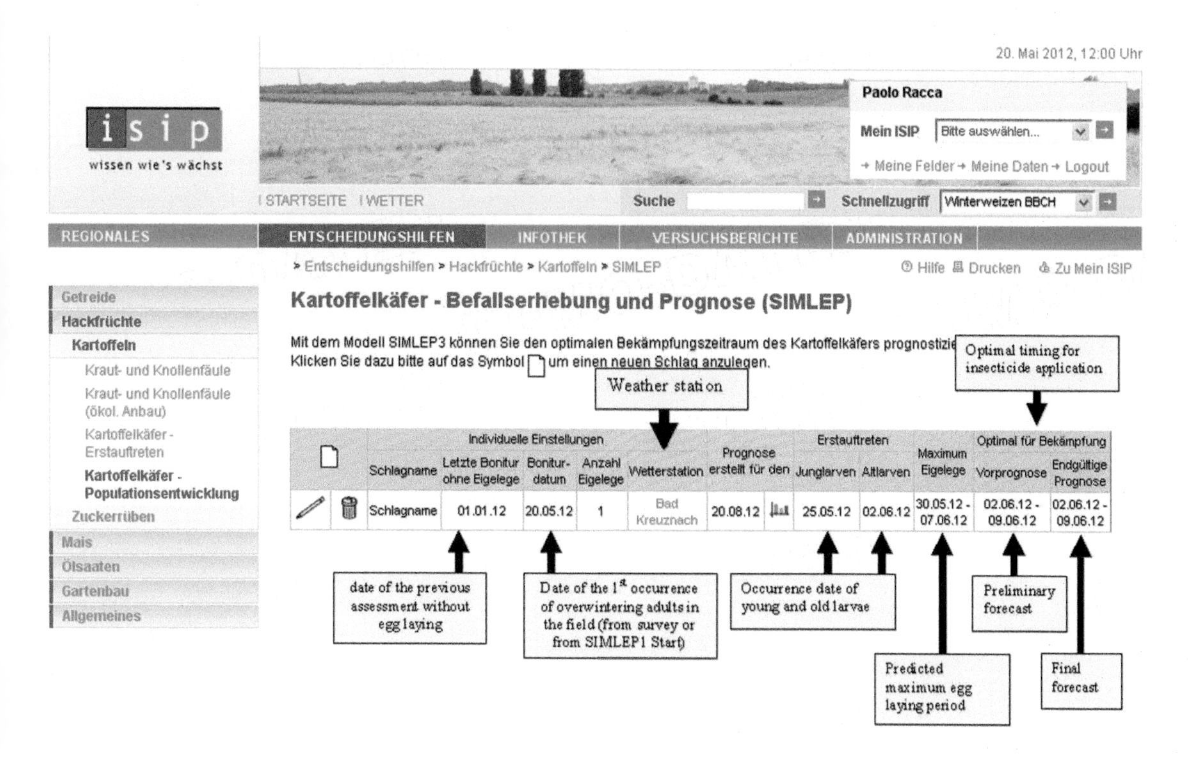

FIGURE 5.8 SIMLEP3 results from the ISIP website.

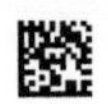

Printed and bound by CPI Group (UK) Ltd, Croydon, CR0 4YY
08/05/2025
01864980-0001